GW01372843

Introduction to

Environmental Science

Introduction to

Environmental Science

Earth and Man

Malcolm Cresser
Lesley Batty
Alistair Boxall
Craig Adams

PEARSON

Harlow, England • London • New York • Boston • San Francisco • Toronto • Sydney • Auckland • Singapore • Hong Kong
Tokyo • Seoul • Taipei • New Delhi • Cape Town • São Paulo • Mexico City • Madrid • Amsterdam • Munich • Paris • Milan

Pearson Education Limited
Edinburgh Gate
Harlow
Essex CM20 2JE
England

and Associated Companies throughout the world

Visit us on the World Wide Web at:
www.pearson.com/uk

First published 2013

ISBN 978-0-13-178932-6

British Library Cataloguing-in-Publication Data
A catalogue record for this book is available from the British Library

Library of Congress Cataloging-in-Publication Data
A catalog record for this book is available from the Library of Congress

ARP impression 98

Typeset in 10/12pt ITC Giovanni by 35
Printed and bound in Great Britain by Ashford Colour Press Ltd

Brief contents

Contents

Lecturer Resources

ON THE WEBSITE

For password-protected online resources tailored to support the use of this textbook in teaching, please visit **www.pearsoned.co.uk/cresser**

Case studies

Guided tour

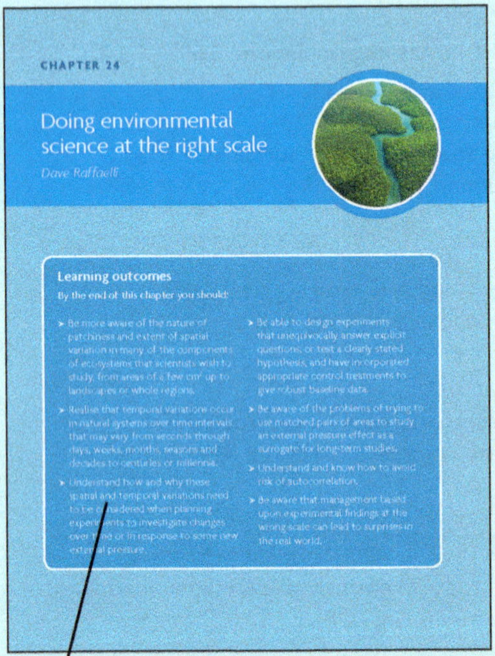
CHAPTER 24

Doing environmental science at the right scale

Dave Raffaelli

Learning outcomes

By the end of this chapter you should:

- Be more aware of the nature of patchiness and extent of spatial variation in many of the components of ecosystems that scientists wish to study, from areas of a few cm² up to landscapes or whole regions.
- Realise that temporal variations occur in natural systems over time intervals that may vary from seconds through days, weeks, months, seasons and decades to centuries or millennia.
- Understand how and why these spatial and temporal variations need to be considered when planning experiments to investigate changes over time or in response to some new external pressure.
- Be able to design experiments that unequivocally answer explicit questions, or test a clearly stated hypothesis, and have incorporated appropriate control treatments to give robust baseline data.
- Be aware of the problems of trying to use matched pairs of areas to study an external pressure effect as a surrogate for long-term studies.
- Understand and know how to avoid risk of autocorrelation.
- Be aware that management based upon experimental findings at the wrong scale can lead to surprises in the real world.

Learning outcomes introduce topics covered and help you to focus on what you *should* have learnt by the end of the chapter, and check back that you *have* learnt it.

The book is richly illustrated with **full colour photos** and **diagrams** not only to illustrate the text but also help you retain ideas and form mental pictures.

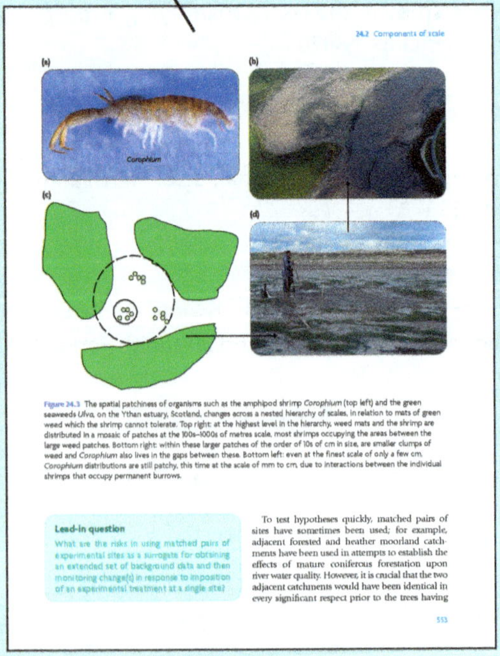
24.2 Components of scale

Figure 24.3 The spatial patchiness of organisms such as the amphipod shrimp *Corophium* (top left) and the green seaweeds *Ulva*, on the Ythan estuary, Scotland, changes across a nested hierarchy of scales, in relation to mats of green weed which the shrimp cannot tolerate. Top right: at the highest level in the hierarchy, weed mats and the shrimp are distributed in a mosaic of patches at the 100s–1000s of metres scale, most shrimps occupying the areas between the large weed patches. Bottom right: within these larger patches of the order of 10s of cm in size, are smaller clumps of weed and *Corophium* also lives in the gaps between these. Bottom left: even at the finest scale of only a few cm, *Corophium* distributions are still patchy, this time at the scale of mm to cm, due to interactions between the individual shrimps that occupy permanent burrows.

Lead-in question

What are the risks in using matched pairs of experimental sites as a surrogate for obtaining an extended set of background data and then monitoring change(s) in response to imposition of an experimental treatment at a single site?

To test hypotheses quickly, matched pairs of sites have sometimes been used; for example, adjacent forested and heather moorland catchments have been used in attempts to establish the effects of mature coniferous forestation upon river water quality. However, it is crucial that the two adjacent catchments would have been identical in every significant respect prior to the trees having

553

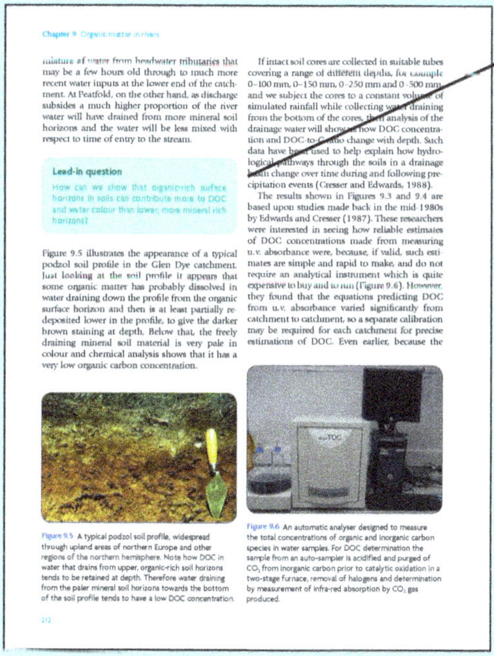

mixture of water from headwater tributaries that may be a few hours old through to much more recent water inputs at the lower end of the catchment. At Peatfold, on the other hand, as discharge subsides a much higher proportion of the river water will have drained from more mineral soil horizons and the water will be less mixed with respect to time of entry to the stream.

Lead-in question

How can we show that organic-rich surface horizons in soils can contribute more to DOC and water colour than lower, more mineral rich horizons?

Figure 9.5 illustrates the appearance of a typical podzol soil profile in the Glen Dye catchment. Just looking at the soil profile it appears that some organic matter has probably dissolved in water draining down the profile from the organic surface horizon and then is at least partially re-deposited lower in the profile, to give the darker brown staining at depth. Below that, the freely draining mineral soil material is very pale in colour and chemical analysis shows that it has a very low organic carbon concentration.

If intact soil cores are collected in suitable tubes covering a range of different depths, for example 0–100 mm, 0–150 mm, 0–250 mm and 0–500 mm and we subject the cores to a constant volume of simulated rainfall while collecting water draining from the bottom of the cores, then analysis of the drainage water will show us how DOC concentration and DOC-to-C ratio change with depth. Such data have been used to help explain how hydrological pathways through the soils in a drainage basin change over time during and following precipitation events (Cresser and Edwards, 1988).

The results shown in Figures 9.3 and 9.4 are based upon studies made back in the mid-1980s by Edwards and Cresser (1987). These researchers were interested in seeing how reliable estimates of DOC concentrations made from measuring u.v. absorbance were, because, if valid, such estimates are simple and rapid to make, and do not require an analytical instrument which is quite expensive to buy and to run (Figure 9.6). However, they found that the equations predicting DOC from u.v. absorbance varied significantly from catchment to catchment, so a separate calibration may be required for each catchment for precise estimations of DOC. Even earlier, because the

Figure 9.5 A typical podzol soil profile, widespread through upland areas of northern Europe and other regions of the northern hemisphere. Note how DOC in water that drains from upper, organic-rich soil horizons tends to be retained at depth. Therefore water draining from the paler mineral soil horizons towards the bottom of the soil profile tends to have a low DOC concentration.

Figure 9.6 An automatic analyser designed to measure the total concentrations of organic and inorganic carbon species in water samples. For DOC determination the sample from an auto-sampler is acidified and purged of CO_2 from inorganic carbon prior to catalytic oxidation in a two-stage furnace, removal of halogens and determination by measurement of infra-red absorption by CO_2 gas produced.

Lead-in questions invite you to think about your own preconceptions as you tackle a section of text, challenge them when necessary, and build on useful intuition.

Policy implications boxes at the end of each chapter summarise the authors' views on the policy implications of the subject matter, enabling you to develop a feel for how policy makers think and act, and ultimately form your own opinion.

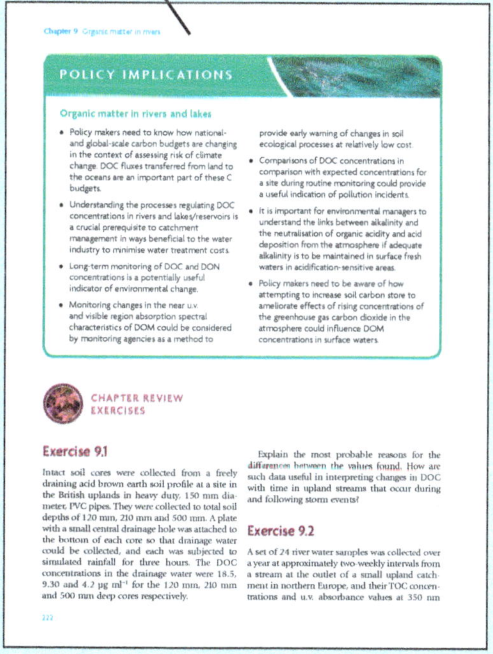

POLICY IMPLICATIONS

Organic matter in rivers and lakes

- Policy makers need to know how national- and global-scale carbon budgets are changing in the context of assessing risk of climate change. DOC fluxes transferred from land to the oceans are an important part of these C budgets.
- Understanding the processes regulating DOC concentrations in rivers and lakes/reservoirs is a crucial prerequisite to catchment management in ways beneficial to the water industry to minimise water treatment costs.
- Long-term monitoring of DOC and DON concentrations is a potentially useful indicator of environmental change.
- Monitoring changes in the near u.v. and visible region absorption spectral characteristics of DOM could be considered by monitoring agencies as a method to provide early warning of changes in soil ecological processes at relatively low cost.
- Comparisons of DOC concentrations in comparison with expected concentrations for a site during routine monitoring could provide a useful indication of pollution incidents.
- It is important for environmental managers to understand the links between alkalinity and the neutralisation of organic acidity and acid deposition from the atmosphere if adequate alkalinity is to be maintained in surface fresh waters in acidification-sensitive areas.
- Policy makers need to be aware of how attempting to increase soil carbon store to ameliorate effects of rising concentrations of the greenhouse gas carbon dioxide in the atmosphere could influence DOM concentrations in surface waters.

CHAPTER REVIEW EXERCISES

Exercise 9.1

Intact soil cores were collected from a freely draining acid brown earth soil profile at a site in the British uplands in heavy duty, 150 mm diameter, PVC pipes. They were collected to total soil depths of 120 mm, 210 mm and 500 mm. A plate with a small central drainage hole was attached to the bottom of each core so that drainage water could be collected, and each was subjected to simulated rainfall for three hours. The DOC concentrations in the drainage water were 18.5, 9.30 and 4.2 μg ml^{-1} for the 120 mm, 210 mm and 500 mm deep cores respectively.

Explain the most probable reasons for the differences between the values found. How are such data useful in interpreting changes in DOC with time in upland streams that occur during and following storm events?

Exercise 9.2

A set of 24 river water samples was collected over a year at approximately two-weekly intervals from a stream at the outlet of a small upland catchment in northern Europe, and their TOC concentrations and u.v. absorbance values at 350 nm

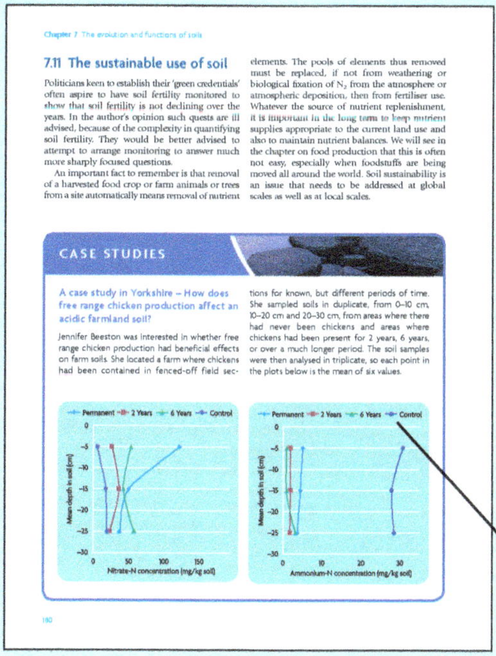

7.11 The sustainable use of soil

Politicians keen to establish their 'green credentials' often aspire to have soil fertility monitored to show that soil fertility is not declining over the years. In the author's opinion such quests are ill advised, because of the complexity in quantifying soil fertility. They would be better advised to attempt to arrange monitoring to answer much more sharply focused questions.

An important fact to remember is that removal of a harvested food crop or farm animals or trees from a site automatically means removal of nutrient elements. The pools of elements thus removed must be replaced, if not from weathering or biological fixation of N_2 from the atmosphere or atmospheric deposition, then from fertiliser use. Whatever the source of nutrient replenishment, it is important in the long term to keep nutrient supplies appropriate to the current land use and also to maintain nutrient balances. We will see in the chapter on food production that this is often not easy, especially when foodstuffs are being moved all around the world. Soil sustainability is an issue that needs to be addressed at global scales as well as at local scales.

CASE STUDIES

A case study in Yorkshire – How does free range chicken production affect an acidic farmland soil?

Jennifer Beeston was interested in whether free range chicken production had beneficial effects on farm soils. She located a farm where chickens had been contained in fenced-off field sections for known, but different periods of time. She sampled soils in duplicate, from 0–10 cm, 10–20 cm and 20–30 cm, from areas where there had never been chickens and areas where chickens had been present for 2 years, 6 years, or over a much longer period. The soil samples were then analysed in triplicate, so each point in the plots below is the mean of six values.

Case studies, usually based on research studies, give a range of examples and illustrations to add a real-world relevance to topics discussed.

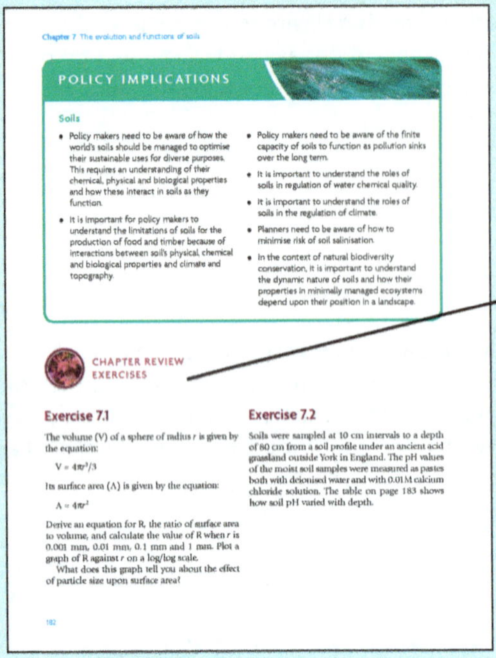

Chapter review exercises provide a range of both analytical questions (often based on data) and reflective questions which stimulate thinking and can be used to test and consolidate learning.

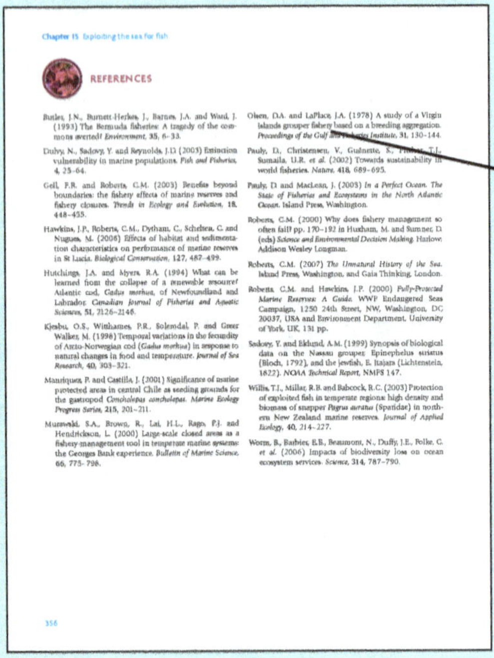

References, provide sources of scientific literature referred to in the text and additional suggestions for further exploration of the topic in question.

Preface

For decades now, academics in the UK and elsewhere in the world have been visited by publishers' representatives, all trying to sell them books, but some also looking for aspiring authors and/or exciting new ideas. If the answer to their 'Have we got the book you need?' question is 'No, afraid not' then the inevitable follow-up question is 'Well what do you need for your course?' The answer to just such an exchange with Andrew Taylor (a particularly interesting Editor from Pearson) prompted a lengthy discussion and, six years later, has resulted in the present volume. What was wanted was a book that introduced Environmental Science as a totally integrated science. It had to embrace fully the complex interactions between chemistry, physics, biology, geology, hydrology, cosmology, archaeology and even the history of science and how science evolves. It shouldn't just cover what we know, but also how we know what we know. Therefore it should include real data drawn from recent or current research studies or from case studies where appropriate, so readers understand how facts are established and where uncertainty lies. More than that, though, it had to cut across the science/social sciences divide; it was important that readers would understand how environmental scientists have to interact with policy makers and the public at large in the formulation of the environmental management policies that are now so desperately needed in the interests of global sustainability. The text had to convey the authors' enthusiasm and passion for their subject, while getting across the message that although we have learned much, we still have much to learn. The subject is not static – it is developing all the time and sometimes at an alarming rate. So readers critically engaging with the subject really can make a difference.

'Authors' had to be plural, since no one person could cover all that was required in the way required. The team was carefully selected, based upon each individual's enthusiasm, internationally recognised expertise in one of the topics that we felt needed to be covered, commitment to engaging and inspiring teaching, and conviction about the need for a highly interdisciplinary approach. As a result we believe that we have produced a book that differs from all those coming before it, and one that is right for current needs. To some, our choice of topics may seem surprisingly diverse initially, but that is all part of the fascination of the subject. Our intended audience is primarily undergraduate students, or postgraduates changing direction, but we hope that policy makers too, and even the general public with an interest in science and the environment, will find the book valuable. Above all, we hope that whoever reads the book will find it fascinating and thought provoking, and at times even worrying.

So if you're a student, how should you use this book? Each chapter starts with a set of intended learning outcomes. This will tell you what you should have learned about by the end of the chapter; but once you've got to the end, you can refer back to see if the content has worked for you as an individual. We have done our best to reinforce the written word with many pictorial images. The photographs are never merely ornamental. Most are everyday images from the lives of the contributing scientists that have been gathered over many years. They are there to help clarify the text, but also to help retention of ideas and mental pictures. For revision purposes there is a lot to be said for browsing and asking why an image was included – what's its message? To help you understand sources of facts and how hypotheses are tested, numerous charts and graphs have been drawn from original research data.

No one who aspires to be a scientist should ever lose sight of the importance of thought. In environmental science the importance of your existing perceptions about how terrestrial and aquatic systems work is very important. It will influence many decisions that you may make on a daily basis. To encourage using your preconceptions as part of the learning process, we have threaded lead-in questions throughout the text. If you try to answer these questions before proceeding, you will often be encouraged by how much you already know if only you stop to think!

Most people who read this book will really care about the world they live in. Many will want to make a difference to how its resources are used and to stop its resources being wastefully abused. To do this means they need to understand how policy makers think and act, whether wisely or unwisely. Therefore, at the end of each chapter, we have included a box summarising the authors' perceptions of the main policy implications of that chapter's contents.

Finally, in recognition of the fact that for many readers their knowledge and understanding will need to be confirmed, at the end of each chapter we have included a modest selection of questions, often based upon real data. We hope that these may also be a useful source of ideas for instructors too.

In assembling this approach we have been helped tremendously by staff from Pearson, and especially Andrew Taylor, Rufus Curnow, Patrick Bond, Helen Leech and Sarah Busby. They put an enormous effort into sending chapters out for review to international referees and getting and giving highly constructive feedback on our ideas, sometimes hopefully stopping us going down the wrong road. We greatly appreciate too help in the final production stages from Philippa Fiszzon, Lynette Miller and Jonathon Price, and their meticulous attention to detail. If you enjoy this book it is in no small way due partly to all their effort, the efforts of the referees, and the feedback from the thousands of students that we have all enjoyed teaching over the years.

Malcolm Cresser
Lesley Batty
Alistair Boxall
Craig Adams

About the authors

Craig Adams earned a PhD in Environmental Health Engineering at the University of Kansas, USA. He is currently Professor and Head of Civil and Environmental Engineering at Utah State University in Logan, Utah, USA. His research focuses on the analysis, treatment, properties, fate and modelling of emerging contaminants (including antibiotics, endocrine disrupting chemicals, estrogens and disinfection byproducts) in drinking water, wastewater, seawater and food, and on appropriate water and sanitation technologies for developing countries. He has over 80 journal publications, is a Fellow of ASCE, and has been recognised with research awards including the WEF Eddy Principles and Processes Medal (2008) and the ASCE State-of-the-Art Civil Engineering Award (2005).

Cumhur Aydinalp obtained a PhD at the University of Aberdeen, UK in 1996 and is now Professor of Soil Science in the Department of Soil Science & Plant Nutrition at Uludag University, Turkey. His research interests include pollution effects on soil/plant/water ecosystems, sustainable use of soil and water resources, soil pollution and remediation, soil chemistry and soil genesis and classification. He has numerous research publications in these areas. He is on the editorial boards of three journals and referees for many international journals. During his career he has attained several awards such as International Scientist of the Year (2005), 2000 Outstanding Intellectuals of the 21st Century (2006) and Leading Educators of the World (2006).

Paul Ayris studied Environmental Science at the University of Lancaster. During a second-year placement at Oregon State University, he secured a volcanic research project and continued this research theme in his MRes degree. These subsequently enabled him to fulfil his childhood dream of becoming a volcanologist, as he then completed a PhD on volcanic ash surface chemistry at the University of York. Paul is currently a postdoctoral researcher at the Université Catholique de Louvain in Belgium. His current research interests include the interaction of eruption plume gases with volcanic ash surfaces and the environmental impacts of volcanic ash emission.

Lesley Batty trained as an Environmental Scientist at the University of Sheffield where she gained her BSc. Following that she completed an MRes at Reading before returning to Sheffield to complete her PhD on Metal Removal Processes in Constructed Wetlands. She spent some time at Newcastle as a postdoctoral fellow in minewater treatment before starting a lectureship at the University of Birmingham where she has been for seven years. Lesley has particular interests in the ecology of industrial pollution and works closely with agencies such as the UK Environment Agency and Coal Authority. She is an active member of the British Ecological Society and is an Associate Editor of the *Journal of Applied Ecology and Mine Water and Environment*.

Monika Böhm graduated with a BSc in Zoology from the University of Aberdeen before relocating to the University of York for postgraduate studies. After the completion of her PhD on livestock–wildlife interaction patterns and their implications for disease transmission, she moved to the Zoological Society of London – via a short stint of travelling. Monika is now working on biodiversity monitoring and indicator development within the Institute of Zoology where she has been closely involved with the IUCN Red List programme, National Red List development and international biodiversity conventions such as the Convention on Migratory Species (CMS).

Alistair Boxall is a Professor in Environmental Science in the Environment Department at the University of York, UK. Alistair's research focuses on understanding emerging and future ecological and health risks posed by chemical contaminants (such as pharmaceuticals, nanomaterials and chemical degradation products) in the natural environment. He is a member of the Defra Advisory Committee on Hazardous Substances and regularly advises national and international organisations on issues relating to chemical impacts on the environment.

Nicola Carslaw has a BSc in Chemistry with Environmental Science, and an MSc in Atmospheric Science from the University of East Anglia. After a year in industry, she returned to UEA to complete a PhD in atmospheric chemistry. Nicola then moved to the University of Leeds as a PDRA before securing a lectureship in Environmental Science at the University of York in 2000. She was promoted to senior lecturer in 2008. Her research spans from the chemistry of air pollution episodes to indoor air quality. Nic uses detailed chemical models to understand the processes that cause high levels of air pollution. She has more than 40 papers in these fields and is on the editorial board of *Atmospheric Environment*.

James Clark is a graduate of Kings College and after postdoctoral work in Canada and the UK, he moved to the University of York where he is now Professor of Chemistry and Director of the Green Chemistry Centre of Excellence. He leads a group of over 70 graduate researchers and support staff working on pure and applied research on developing greener and more sustainable chemical products and processes. James has published over 400 refereed research papers and written or edited over 20 books. He was the founding editor of the world-leading journal *Green Chemistry* and the founding Director, now President, of the international Green Chemistry Network. He has won numerous awards from around the world including the 2011 Royal Society of Chemistry Environment Prize.

After **Malcolm Cresser** completed a BSc and PhD in analytical chemistry at Imperial College, London, he spent a short period at Villanova University in Pennsylvania before moving to Aberdeen University as a lecturer in soil science. During his 29 years there he developed a real interest in the links between soil and water chemistry and quantifying and modelling pollution effects on soils and surface waters. He has sat on numerous national and international research committees and published > 300 papers and several books. Because of the highly interdisciplinary nature of his research and teaching he is a Fellow of the Society of Biology and the Royal Society of Chemistry, as well as an honorary member of the British Society of Soil Science.

Elena Dawkins graduated with a BSc in Environment, Economics and Ecology at the University of York and went on to work for Defra and Government Office for Yorkshire and The Humber. After a number of years Elena returned to the University of York to work as a Research Associate for the Stockholm Environment Institute. There she works on a range of sustainable consumption and production projects, with the aim of linking scientific research and policy making. Elena publishes both academic and policy literature for a wide range of audiences.

Keith Goulding joined Rothamsted in 1974 after completing a Masters in Soil Chemistry at Reading University. He gained his PhD in soil chemistry from Imperial College in 1980. He studies how the plant foods (nutrients) in soils become available to growing plants and the best ways of augmenting these with fertilisers and manures without polluting air and water. Keith is a visiting Professor at the University of Nottingham, a Fellow of the Institute of Professional Soil Scientists and a Chartered Scientist. He was awarded the Royal Agricultural Society of England's (RASE) Research Medal in 2003 for his work on diffuse pollution from agriculture and elected an Honorary Fellow of the RASE in 2010. He received a Nobel Peace Prize certificate for his contribution to the work of the Intergovernmental Panel on Climate Change, for which the Panel and Al Gore were jointly awarded the Peace Prize in 2007.

After graduating with distinction in her BSc in Environment, Economics and Ecology at the University of York, **Sophie Green** worked at the Buildings Research Institute in the UK for a while before returning to York to complete a PhD on the impacts of road salting on soils and waters in UK uplands. Subsequently she has worked at Queen Mary University of London, The Open University at Milton Keynes and, most recently, the University of Leeds on carbon cycling and greenhouse gas emissions from peatlands. Sophie has numerous research publications in these research areas.

Julie Hawkins is a lecturer in Marine Environmental Management in the Environment Department at the University of York, where she runs an MSc in this subject. Her research focuses on how to reduce the problems that human impacts cause to marine ecosystems, in particular how marine protected areas can benefit the oceans. Through her work, Julie has lived in the Middle East and Caribbean and travelled extensively throughout the latter.

After a distinction in his MRes in clean chemical technology, **Andrew Hunt** obtained a PhD from the University of York on the extraction of high value chemicals from British upland plants. His postdoctoral research highlights included innovative work on the use of supercritical carbon dioxide for the extraction of liquid crystals from defunct display devices, work that led to a Rushlight award for innovation in recycling, and other related work on the recovery of waste polyvinyl alcohol. His work received significant press attention, including an ASC press conference at the ASC Green Chemistry Conference in Washington DC in 2010. Andrew currently is the scientific leader of the natural solvent technology platform at the Green Chemistry Centre of Excellence at York. He has numerous research papers in the area of green and sustainable chemistry.

After completing a degree in Biology at the University of Southampton, **Michael Hutchings** moved to the University of Bristol where he developed his interests in wildlife ecology, completing an MSc on the status of brown hares in Britain and a PhD on the role of badgers in the epidemiology of bovine TB. After moving to the Scottish Agricultural College (SAC) he won a five-year Senior Research Fellowship on disease ecology and he now leads the SAC's Disease Systems team.

Professor **Ken Killham**, BSc PhD (Sheffield), FRSE, FAAM, is one of the UK's foremost authorities on the assessment and remediation of contaminated environments. He is director of Science at Remedios Ltd and was formerly the established Professor of Soil Science at Aberdeen University, chaired the UK Soil Science Advisory Committee, is Past-President of the British Soil Science Society and sat on/chaired several UK/international Research Council committees and institute boards, providing expertise on soils and contaminated land issues.

With a BSc in Geography and Mathematics from the University of Sheffield, **Anne Owen** worked as a Secondary School Maths teacher before gaining an MSc in Geographical Information Science from the University of Leeds. She spent the next six years as a Research Associate at the Stockholm Environment Institute at the University of York, working on a variety of projects in the field of Sustainable Consumption and Production and guest lecturing for the Environment Department. Anne has now returned the University of Leeds to start a part-time PhD considering the role Global Trade Models can play in Climate Change Policy.

Graeme Paton combines his role as a technical director at Remedios, an environmental technology company with extensive expertise in the environmental investigation, assessment and remediation industry, successfully concluding a number of high profile projects for clients in the UK and internationally, with a Professorship at the University of Aberdeen in Soil Science. At Remedios he directs projects and works with the team to ensure client satisfaction and that regulatory approval is met.

Dave Raffaelli was trained as a Zoologist and Entomologist at Leeds University and then went on to Bangor in North Wales and to New Zealand to develop his interests in marine ecology. He spent 20 years at Aberdeen's Culterty Field Station studying the complex dynamics of coastal food webs and then moved to York in 2001 to better engage with those in the social and economic sciences, where he is at present. He is currently Director of a major UK programme on Biodiversity and Ecosystem Services (BESS) and advises the UK government on a range of ecological issues.

Callum Roberts is a marine scientist and conservationist at the University of York, UK, and author of *The Unnatural History of the Sea*, an account of the effects of 1000 years of hunting and fishing on ocean life. His research has revealed the extraordinary rise and fall of fisheries over the last 200 years, but also shows how life can make a remarkable comeback after protection is granted. His team at York provided the scientific case for the world's first network of high seas marine reserves in the North Atlantic that in 2010 placed nearly 300,000 km^2 of ocean under protection. His second book, *The Ocean of Life*, explores how the oceans are changing under human influence.

Piran White holds a BSc in Ecology from the University of East Anglia and a PhD from Bristol. Initially a lecturer, he now holds a Personal Chair in the Environment Department at the University of York. He has held visiting research posts at New South Wales Agriculture (Orange, Australia), AgResearch (Lincoln, New Zealand) and the University of Waikato (Hamilton, New Zealand). Piran is Deputy Director of the NERC-funded Biodiversity and Ecosystem Service Sustainability Directorate, a £13m research programme on ecosystem services. He coordinated the ESRC trans-disciplinary seminar series on coastal wetland ecosystem services and leads the University of York's team in the Australian-based Co-operative Research Centre on Invasive Animals. He is Editor of *Wildlife Research* journal, and was an Associate Editor of *Journal of Animal Ecology*.

Clare Wilson gained a BSc in Environmental Science from the University of Aberystwyth, before moving to study MSc Environmental Archaeology at the University of Sheffield. Her PhD from the University of Stirling combined both disciplines with a study of soils buried below archaeological monuments. After working first as a geo-archaeological consultant and then as a postdoctoral research assistant, Clare is now a lecturer in Soil Science at the University of Stirling. Her research focuses on the application of soil science to archaeological questions and problems. Of particular interest to her are the chemical and physical legacies of past land use and human activity that can be preserved in soil.

Acknowledgements

We are grateful to the following for permission to reproduce copyright material:

Cartoons

Cartoon 13.10 reprinted from *Trends in Microbiology*, 16(9), White, P.C.L., Böhm, M., Marion, G. and Hutchins, M.R., Control of bovine tuberculosis in British livestock – there is no 'silver bullet', pp. 420–427, Copyright 2008, with permission from Elsevier.

Figures

Figures on pages 27, 28 based on work by Laura Suddaby; Figure 2.5 Australian Rainfall Analysis (mm) 1st to 17th October 2008, http://www.bom.gov.au, © Commonwealth of Australia 2008, Australian Bureau of Meteorology; Figure 3.3 from Small Comets, http://smallcomets.physics.uiowa.edu/www/faq.htmlx, University of Iowa/NASA; Figure 3.4 from Miller-Urey experiment, http://en.wikipedia.org/wiki/File:Miller-Urey_experiment-en.svg. This file is licensed under the Creative Commons Attribution-Share Alike 3.0 Unported, 2.5 Generic, 2.0 Generic and 1.0 Generic license, Attribution-ShareAlike 2.5 Generic (CC BY-SA 2.5) http://creativecommons.org/licenses/by-sa/2.5/deed.en; Figure 3.6 from http://www.pmel.noaa.gov/vents/gallery/smoker-images.html, NOAA PMEL Vents Program; Figure 3.8 from http://oceanexplorer.noaa.gov/okeanos/explorations/ex1104/background/microbes/media/microbes_sem_tem.html, Image courtesy of Julie Huber at NOAA; Figure 3.11 reprinted by permission from Macmillan Publishers Ltd: The rise of atmospheric oxygen, *Nature*, 451, pp. 277–278 (Kump, L.R. 2008), copyright 2008; Figure 3.12 from SOLVE II Science Implementation http://www.espo.nasa.gov/solveII/implement.html, NASA; Figure 3.13 Phylogenetic Tree of Life, http://nai.arc.nasa.gov/library/images/news_articles/big_274_3.jpg, NASA; Figure 3.14 from Lutgens, Frederick K.; Tarbuck, Edward J., *Atmosphere, The: An Introduction To Meteorology*, 8th, © 2001. Printed and electronically reproduced by permission of Pearson Education, Inc., Upper Saddle River, New Jersey; Figure 3.18 data from NASA Ozone Watch, http://ozonewatch.gsfc.nasa.gov/meteorology/annual_data.html; Figures 3.19a, 3.19b, 3.19c from http://earthobservatory.nasa.gov/Features/WorldOfChange/ozone.php, NASA Earth Observatory; Figures 5.9, 5.12 reprinted from *Science of the Total Environment*, 400, Cresser, M.S., Aitkenhead, M.J. and Mian, I.A., A reappraisal of the terrestrial nitrogen cycle: What can we learn by extracting concepts from Gaia theory?.pp. 344–355, Copyright 2008, with permission from Elsevier; Figure 6.4 reprinted by permission from Macmillan Publishers Ltd: Biodiversity hotspots for conservation priorities, *Nature*, 403(6772), pp. 853–858 (Myers, N., Mittermeier, R.A., Mittermeier, C.G., da Fonseca, G.A.B. and Kent, J. 2000), copyright 2000; Figure 6.5 after *Biology of Freshwater Pollution*, Longman (Mason, C.F. 1996); Figure 6.12 from MacArthur, Robert; *The Theory of Island Biogeography*. © 1967 Princeton University Press, 1995 renewed PUP. Reprinted by permission of Princeton University Press; Figures 7.19, 7.33 adapted from *Soil Chemistry and its Applications*, Cambridge University Press (Cresser, M.S., Killham, K.S. and Edwards, A.C. 1993); Figures on pages 180, 181 by Jennifer Beeston; Figure 8.1 from Observations of Extratropical Variability by James W. Hurrell, http://www.asp.ucar.edu/colloquium/2000/Lectures/hurrell1.html, with permission from Jim Hurrell, NCAR Earth System Lab (NESL); Figure 8.2 from Atmospheric absorption, http://eduspace.esa.int/subtopic/images/07-atmosvindue.gif, image courtesy of the European Space Agency; Figure 8.3 from Earth's energy balance – Solar and Terrestrial, http://apollo.lsc.vsc.edu/classes/met130/notes/chapter2/ebal3.html, with permission from Department of Atmospheric Sciences, Lyndon State College; Figure 8.4 reproduced with the permission of Nelson Thornes Ltd. from *Geography: An Integrated Approach*, David Waugh, 978-1-4085-0407-9, first published in 2009; Figure 8.6 from http://www.windows2universe.org/earth/Water/ocean_currents.html. This image is Windows to the Universe® (http://windows2universe.org) © 2010, National Earth Science Teachers Association. This work is licensed under a Creative Commons Attribution-ShareAlike 3.0 Unported License; Figure 8.12 from Figure 2 Milankovitch cycles, http://skepticalscience.com/co2-lags-temperature-intermediate.htm, with permission from Skeptical Science; Figure 8.13 from *Climate Change 2007: The Physical Science Basis. Working Group I Contribution to the Fourth Assessment Report of the Intergovernmental Panel on Climate Change*, Cambridge University Press, Figure SPM.1, with permission from IPCC; Figure 8.14 from *Climate Change 2007: The Physical Science Basis. Working Group I Contribution to the Fourth Assessment Report of the Intergovernmental Panel on Climate Change*, Cambridge University Press Figure SPM.2, with permission from IPCC; Figure 8.15 from *Climate Change 2007: Synthesis Report. Contribution of Working Groups I, II and III to the Fourth Assessment Report of the Intergovernmental Panel on Climate Change*, Figure 3.2, IPCC, Geneva, Switzerland; Figure 9.2 reprinted from *Environmental Pollution*, 137(1), Evans, C.D., Monteith, D.T. and Cooper, D.M., Long-term increases in surface water dissolved organic carbon: Observations, possible causes and environmental impacts, pp. 55–71, Copyright 2005, with permission from Elsevier; Figure 10.3 adapted from *Gaia: The Practical Science of Planetary Medicine. Reprinted as Gaia: Medicine for an Ailing Planet* (revised 2nd edition 2005), Gaia Books (Lovelock, J. 1991), courtesy of Octopus Publishing Group; Figure 11.1 from Cefic (2011) Chemicals industry profile. Downloaded from http://www.cfec.org/Global/Facts-and-figures-images/Graphs%202011/FF2011-chapters-PDF/Cefic-FF%20Rapport%202011_11_ChemIndProfile.pdf, with permission from Cefic – European Chemical Industry Council; Figures 11.3, 11.4 kindly provided by Dr Igor Dubus; Figure 11.13 from The Voluntary Initiative, http://www.voluntaryinitiative.org.uk/_Attachments/resources/950_S4.pdf, reproduced with permission from the Crop Protection Association UK Ltd; Figure 11.15 from Impacts of Climate Change on Indirect Human Exposure to Pathogens and Chemicals from Agriculture, *Environmental Health Perspectives*, 117, pp. 508–514 (Boxall, A.B.A., Hardy, A., Beulke S., Boucard, T., Burgin, L., Falloon, P.D., Haygarth, P.M., Hutchinson, T., Kovats, R.S., Leonardi, G., Levy, L.S., Nichols, G., Parsons, S.A., Potts, L., Stone, D., Topp, E, Turley, D.B., Walsh, K., Wellington, E.M.H., Williams, R.J. 2009), reproduced with permission from *Environmental Health Perspectives*; Figure 13.4 adapted from The use of mathematical models in the epidemiological study of infectious diseases and in the design of mass immunization programmes, *Epidemiology and Infection*, 101,pp. 1–20 (Nokes, D.J. and Anderson, R.M. 1988), reproduced with permission from Cambridge University Press; Figure 13.6 Graph supplied by kind permission of the Game & Wildlife Conservation Trust; Figure 13.7 from A disease-mediated trophic cascade in the Serengeti and its implications for Ecosystem C, *PLos Biol*, 7(9): e1000210. doi: 10.1371/journal.pbio.1000210 (Holdo, R.M., Sinclair, A.R.E., Dobson, A.P., Metzger, K.L., Bolker, B.M. *et al.* 2009), reproduced under the terms of Creative Commons Attribution Licence 2.5 Generic (CC BY 2.5); Figure 13.8 reprinted from *Biological Conservation*, 131(2), Sterner, R.T. and Smith, G.C., Modelling wildlife rabies: Transmission, economics, and conservation, pp. 163–179, Copyright 2006, with permission from Elsevier; Figure 13.12 reprinted from *Trends in Microbiology*, 16(9), White, P.C.L., Böhm, M., Marion, G. and Hutchins, M.R., Control of bovine tuberculosis in British livestock – there is no 'silver bullet', pp. 420–427, Copyright 2008, with permission from Elsevier; Figure on page 299 from Livestock grazing behavior and inter- versus intra-specific disease risk via the fecal-oral route, *Behavioural Ecology*, 20(2), pp. 426–432 (Smith, L.A., White, P.C.L., Marion, G. and Hutchins, M.R. 2008), by permission of Oxford University Press; Figure on page 306 from Predictive spatial dynamics and strategic planning for raccoon rabies emergence in Ohio, *PLoS Biol* 3(3): e88. doi:10.1371/journal.pbio.0030088 (Russell, C.A., Smith, D.L., Childs, J.E. and Real, L.A. 2005), reproduced under the terms of Creative Commons Attribution Licence 2.5 Generic (CC BY 2.5); Figure on page 310 from Contact networks in a wildlife-livestock host community: Identifying high-risk individuals in the transmission of bovine TB among badgers and cattle, *PLoS ONE* 4(4): e5016. doi:10.1371/journal.pone.0005016 (Böhm, M., Hutchings, M.R. and White, P.C.L.), reproduced under the terms of Creative Commons Attribution Licence 2.5 Generic (CC BY 2.5); Figure on page 312 from The spread of bluetongue virus serotype 8 in Great Britain and its control by vaccination, *PloS ONE 5(2):* e9353 (Szmaragd, C. Wilson, A.J., Carpenter, S., Wood, J.L.N., Mellor, P.S. and Gubbins, S. 2010), reproduced under the terms of Creative Commons Attribution Licence 2.5 Generic (CC BY 2.5); Figure 14.7 reprinted from *Environmental Pollution*, Vol. 136 (1), Smart, R.P., Cresser M.S., Calver, L.J., Chapman, P.J., Clark, J.M., A novel modelling approach for spatial and temporal variations in nitrate concentrations in an N-impacted UK small upland river basin, pp. 63–70, Copyright 2005, with permission from Elsevier; Figure 14.8 adapted from The importance of ammonium mobility in nitrogen-impacted unfertilized grasslands:A critical reassessment, *Environmental Pollution*, 157(4), pp. 1287–1293 (Mian, I.A., Riaz, M. and Cresser, M.S. 2009); Figures 14.10, 14.11, 16.9 from graphs courtesy of Shaheen Begum; Figure 15.1 reprinted by permission from Macmillan Publishers Ltd: Towards sustainability in world fisheries, *Nature*, 418, pp. 689–695 (Pauly, D., Christensen, V., Guénette, S., Pitcher, T.J., Sumaila, U.R. *et al.* 2002), copyright 2002; Figure 15.2 from SeaWiFS image of our world, http://oceancolor.gsfc.nasa.gov/SeaWiFS/TEACHERS/sanctuary_7.html, GeoEye Satellite Image; Figure 15.6 reprinted from *Journal of Sea Research*, 40(3–4), Kjesbu, O.S., P.R. Witthames, P. Solemdal, M. Greer Walker, Temporal variations in the fecundity of Arcto-Norwegian cod (*Gadus morhua*) in response to natural changes in food and temperature, pp. 303–321, Copyright 1998, with permission from Elsevier; Figure 15.8 from http://seawifs.gsfc.nasa.gov/OCEAN_PLANET/IMAGES/I-71.gif, GeoEye Satellite Image; Figure 15.9 reprinted from *Biological Conservation*, 127(4), Hawkins, J.P., C.M. Roberts, C. Dytham, C. Schelten and M. Nugues, Effects of habitat characteristics and sedimentation on performance of marine reserves in St Lucia, pp. 487–499, Copyright 2006, with permission from Elsevier; Figure 15.10 from Impacts of biodiversity loss on ocean ecosystem services, *Science*, 314(5800), pp. 787–790 (Worm, B., Barbier, E.B., Beaumont, N., Duffy, J.E. and Folke, C. *et al.* 2006). Reprinted with permission from AAAS; Figures 16.2, 19.9 from Centre for Ecology & Hydrology (http://www.ceh.ac.uk/); Figure 16.17 reprinted from *Water Research*, 34(6), White, C.C., Smart, R. and Cresser, M.S., Spatial and temporal variations in critical loads for rivers in N.E. Scotland: a validation of approaches, pp. 1912–1918, Copyright 2000, with permission from Elsevier; Figure 16.18 reprinted from *Environmental Pollution*, 66(1), Billett, M.F., Fitzpatrick, E.A. and Cresser, M.S., Changes in carbon and nitrogen status of forest soils organic horizons between 1949/50 and 1987, pp. 67–79, Copyright 1990, with permission from Elsevier; Figure 16.20 chart provided by Felicity Hayes, CEH Bangor; Figure 19.5b with kind permission from Springer Science+Business Media, *Water, Air and Soil Pollution – Focus*, What factors control soil profile nitrogen storage?, 4(6), 2004, pp. 75–84, Crowe, A.M., Sakata, A., McClean, C. and Cresser, M.S., Figure 3; Figure 19.7 with kind permission from Springer Science+Business Media, *Water, Air and Soil Pollution – Focus*, Controls on leaching of N species in upland moorland catchments, 4(6), 2004, pp. 85–95, Cresser, M.S., Smart, R.P., Clark, M., Crowe, A., Holden, D., Chapman, P.J. and Edwards, A.C.; Figure 19.18 reprinted from *Environmental Pollution*, 152(1), Green, S.M., Machin, R. and Cresser, M.S., Effect of long-term changes in soil chemistry induced by road salt applications on N-transformations in roadside soils, pp. 20–31, Copyright 2008, with permission from Elsevier; Figure 19.20 from Long-term salting effects on dispersion of organic matter from roadside soils into drainage water, *Chemistry and Ecology*, 24(3), pp. 221–231 (Green, S.M., Machin, R. and Cresser, M.S. 2008), reprinted by permission of the publisher (Taylor & Francis Ltd., http://www.tandf.co.uk/journals); Figure 21.7 reprinted from *Agricultural Systems*, 99(2–3), Glendining, M.J., Dailey, A.G., Williams, A.G., van Evert, F.K., Goulding, K.W.T and Whitmore, A.P., Is it possible to increase the sustainability of arable and ruminant agriculture by reducing inputs?, pp. 117–125, Copyright 2009, with permission from Elsevier; Figure 21.8 Loddington Farm cropping mosaic, 2005. © Allerton Trust; Figures 21.9, 21.10 based on data from the Game & Wildlife Conservation Trust, http://www.gwct.org.uk/research__surveys/the_allerton_project/gamebird_songbird_counts/default.asp, with permission from Dr Alistair Leake, The Allerton Project, Game & Widlife Conservation Trust; Figures 22.1, 22.2 licensed under Creative Commons License (CC BY-NC-ND 3.0), © Copyright SASI Group (University of Sheffield) and Mark Newman (University of Michigan); Figures 22.4, 22.5 from *The Living Planet Report 2008* (WWF 2008) © 2008 WWF, with permission from WWF International; Figure 22.10 reprinted with permission from The budget approach: A framework for a global transformation toward a low-carbon economy, *Journal of Renewable and Sustainable Energy*, 2(3) (Messner, D., Schellnhuber, J., Rahmstorf, S., Klingenfeld, D. *et al.* 2010), Copyright 2010 American Institute of Physics; Figure 22.11 from *The right to development in a climate constrained world*, Berlin, Germany, Heinrich Boll Foundation, ChristianAid, Stockholm Environment Institute (Baer, P., Athanasiou, T. and Kartha, S. 2007), with permission from Heinrich-Böll-Stiftung e.V.; Figure 23.5 courtesy of Helen Parker; Figure 23.8 Starbon® diagram by Dr. Vitaliy Budarin; Figure 24.6 NCA base map, used under the terms of the Open Government Licence, http://www.naturalengland.org.uk/Images/open-government-licence-NE-OS_tcm6-30743.pdf. © Natural England copyright. Contains Ordnance Survey data © Crown copyright and database right 2012; Figure 24.7 map copyright Natural England. Used under the terms of the Open Government Licence, http://www.naturalengland.org.uk/Images/open-government-licence-NE-OS_tcm6-30743.pdf. © Natural England copyright. Contains Ordnance Survey data © Crown copyright and database right 2012; Figure 24.7 adapted from Figure in *Applying an*

Ecosystem Approach in Yorkshire and Humber, Yorkshire Futures (Raffaelli, D., White, P.C.L. and MacGilvray, A. 2010) (former Yorkshire Forward project number 903609), with permission from The Department for Business Innovation and Skills; Figure 25.5a adapted from The ecological consequences of changes in biodiversity, *Ecology*, 80, pp. 1455–1474 (Tilman, D. 1999), republished with permission of Ecological Society of America; permission conveyed through Copyright Clearance Center, Inc; Figure 25.5b after Plant diversity and productivity experiments in European grasslands, *Science*, 286, pp. 1123–1127 (Hector, A., Schmid, B., Beierkuhnlein, C., Caldeira, M.C., Diemer, M., Dimitrakopoulos, P.G. *et al.* 1999). Reprinted with permission from AAAS; Figure 25.12 from Applications of ecosystem health for the sustainability of managed systems in Costa Rica, *Ecosystem Health* 5, pp. 1–13 (Aguilar, B.J. 1999), Copyright © 1999. Reproduced with permission of John Wiley & Sons Ltd; Figure 25.13 reprinted from *Journal of Environmental Management*, 91(7), Wiegand, J., Raffaelli, D., Smart, J.C.R. and White, P.C.L.W., Assessment of temporal trends in ecosystem health using an holistic indicator, pp. 1446–1455. Copyright 2010, with permission from Elsevier.

Tables

Table 1.1 adapted from *Soil Chemistry and its Applications*, Cambridge University Press (Cresser, M.S., Killham, K.S. and Edwards, A.C. 1993); Table 9.1 reprinted from *Journal of Hydrology*, 257, Dawson, J.J.C., Billett, M.F., Neal, C. and Hill, S., A comparison of particulate, dissolved and gaseous carbon in two contrasting upland streams in the UK, pp. 226–246, Copyright 2002, with permission from Elsevier; Table 11.3 from Royal Society of Chemistry presentation on natural vs man-made chemicals, reproduced by permission from The Royal Society of Chemistry and Penny Le Couteur from http://www.rsc.org/learn-chemistry/resource/res00000140/ready-made-careers-presentations-natural-or-man-made-chemicals; Table on page 301 from Bovine tuberculosis in southern African wildlife: A multi-species host-pathogen system, *Epidemiology and Infection*, 135, pp. 529–540, Table 1, p. 2 (Renwick, A.R., White, P.C.L. and Bengis, R.G. 2006), reproduced with permission from Cambridge University Press; Table 23.1 Starbon® water purification data by Dr Vitaliy Budarin; Table 25.1 from Using Red List Indices to measure progress towards the 2010 target and beyond, *Philosophical Transactions of the Royal Society, B*, 360(1454), pp. 255–268 (Butchart, S.H.M., Stattersfield, A.J., Baillie, J., Bennun, L.A., Stuart, S.N. *et al.* 2005), Copyright © 2005, The Royal Society, by permission of the Royal Society; Table 25.2 from Principles, http://www.cbd.int/ecosystem/principles.shtml, with permission from the Secretariat of the Convention on Biological Diversity (CBD), United Nations Environment Programme.

Text

Extract on page 576 adapted from *Biodiversity: Exploring Values and Priorities in Conservation*, Blackwell Science (Perlman, D.L. and Adelson, G. 1997), with permission from John Wiley & Sons Ltd.

Photographs

(Key: b-bottom; c-centre; l-left; r-right; t-top)

Craig Adams: 424, 430, 433l, 433r, 434, xxiiit; **Gabriel Guitierrez-Alonso:** 31, 34r; **Cumhur Aydinalp:** xxiiic; **K. J. Ayris:** 36, 41b; **Paul Ayris:** xxiiib; **courtesy of Dr. Tom Batey:** 122, 161tl, 161tr, 161br; **Lesley Batty:** 139, 140, 141t, 141b, xxivt; **Monika Böhm:** xxivtc; **Alistair Boxall:** xxivbc; **Paul van den Brink:** 253t, 253b; **(c) Bruno Comby – EFN – Environmentalists For Nuclear Energy –** http://www.ecolo.org: 225, 226; **Nicola Carslaw:** xxivb; **James Clark:** xxvt; **courtesy of Sir Ron Cooke:** 359t, 359b; **Corbis:** 443; **courtesy of Dr John Creasey:** 316, 395; **Malcolm Cresser:** 7l, 7r, 7bl, 9tl, 9bl, 10tr, 10br, 11tl, 11bl, 12tr, 12br, 13tl, 13tr, 13br, 14tl, 14tr, 14br, 15tl, 15bl, 15br, 16tl, 16tr, 16bl, 16br, 17tl, 17bl, 17br, 18tl, 18tr, 18bl, 18br, 19, 20t, 20c, 20b, 21tl, 21tr, 21bl, 21br, 22tl, 22tr, 22bl, 22br, 23tl, 23tr, 23br, 24tl, 24r, 24bl, 25, 26, 29, 30, 33t, 52, 53t, 53b, 54t, 54b, 58tr, 58bl, 99l, 99r, 100tl, 100bl, 100br, 113tr, 114tr, 114l, 114br, 115t, 115c, 115b, 116l, 117, 119, 157l, 157r, 158tl, 158tr, 158br, 159t, 159bl, 159br, 160, 162tl, 162tr, 162b, 163t, 163b, 164, 167, 168l, 168r, 169l, 169r, 170tl, 170cl, 170bl, 170br, 171t, 171b, 175, 208, 212l, 212r, 265l, 265r, 266b, 267t, 267c, 267b, 270, 271t, 271b, 317t, 317b, 327, 328br, 329l, 329r, 330t, 330b, 335, 361, 362, 363, 364, 382tl, 382bl, 382br, 390l, 390r, 391tl, 391tr, 391br, 394, 396l, 396r, 397l, 397r, 398l, 398r, 400t, 400b, 401l, 401r, 402t, 402b, 403, 404t, 404bl, 404br, 405l, 405r, 406tl, 406tr, 406b, 407t, 407b, 408l, 408cl, 408cr, 408r, 409l, 409cl, 409cr, 409r, 452l, 457, 458tl, 458bl, 458br, 466bl, 466br, 467, xxvtc; **Elena Dawkins:** xxvbc; **Ana Deletic:** 466t; **Digital Vision:** 1, v, 27, 44, 56, 65t, 91, 106, 108t, 111, 120, 129t, 154t, 180, 182t, 184, 205, 207, 216, 222t, 233t, 246, 247, 261, 273, 281, 313, 335b, 355t, 358, 385, 408, 410, 439, 441t, 468, 488, 491t, 503t, 524, 526, 527, 530, 534, 548t, 563t, 582t, iii, vii, xxi, xvii, xviii, xxiii, xxix, 1, v, 27, 44, 56, 65t, 91, 106, 108t, 111, 120, 129t, 154t, 180, 182t, 184, 205, 207, 216, 222t, 233t, 246, 247, 261, 273, 281, 313, 335b, 355t, 358, 385, 408, 410, 439, 441t, 468, 488, 491t, 503t, 524, 526, 527, 530, 534, 548t, 563t, 582t, iii, vii, xxi, xvii, xviii, xxiii, xxix, 1, v, 27, 44, 56, 65t, 91, 106, 108t, 111, 120, 129t, 154t, 180, 182t, 184, 205, 207, 216, 222t, 233t, 246, 247, 261, 273, 281, 313, 335b, 355t, 358, 385, 408, 410, 439, 441t, 468, 488, 491t, 503t, 524, 526, 527, 530, 534, 548t, 563t, 582t, iii, vii, xxi, xvii, xviii, xxiii, xxix; **Tony Edwards:** 57; **courtesy of E.A. Fitzpatrick:** 100tr, 166, 365–366; **fossilmall.com:** 132; **Getty Images:** Fox Photos / Hulton Archive 85; **Keith Goulding:** xxvb; **Sophie Green:** xxvit; **Julie Hawkins:** xxvitc; **Felicity Hayes:** 384l, 384r; **courtesy of J. Hilton:** Dr Jason Hilton 152; **Andrew Hunt:** xxvic; **Michael Hutchings:** xxvibc; **Photo by David Iliff:** 86; **Imagestate Media:** 413, John Foxx Collection 485; **Ingram:** 46, 156; **iStockphoto:** 235, 471, 494, 539, 550; **Ken Killham:** 328tl, 472, 476, 478, 479t, 479b, 480t, 480b, 482t, 482b, 486, 488l, 488r, 489, xxvib; **courtesy of Martin Kull:** 536; **Photograph by courtesy of Tim Megginson of GWE Biogas Ltd:** 269t, 269b; **NASA:** 2, 4, 68, image by Jesse Allen 76, Mark Boyle 77; **Anne Owen:** xxviit; **Graeme Paton:** xxviitc; **PhotoDisc:** 45, 65, 66, 92, 108, 110, 129, 130, 131, 154, 155, 182, 183, 206, 222, 224, 233, 234, 262, 263, 264, 282, 314, 315, 336, 338, 355, 356, 357, 386, 388, 411, 412, 441, 442, 469, 491, 492, 503, 504, 531, 532, 548, 549, 563, 564, 582, 583; **Photograph courtesy of Nigel Poole:** 278; **Dave Raffaelli:** 553tl, 553tr, 553br, 556r, 566tl, 566tr, 566bl, 566br, 576, 577, xxviibc; **courtesy of Dr David Rippin:** 82; **Callum Roberts:** xxviib; **reproduced by kind permission of the Save the Tasmanian Devil Program:** Copyright the Tasmanian Government 287br; **Science Photo Library Ltd:** Adam Hart-Davis 556l, Sinclair Stammers 136; **Peter Scott:** 55; **Ute Skiba:** 113bl, 116r; **Richard Smart:** 62l, 454; **Photograph courtesy of Elizabeth Smith:** 266t; **Sozaijiten:** 506, 565; **USGS:** Photograph by Mehmet Celebi 33bl, Photograph by Tom Furnal 34l, Image by J.D. Griggs, courtesy of the USGS Hawaiian Volcano Observatory 37r, Image by C. Newhall, courtesy of the USGS 42, Image courtesy of the USGS Hawaiian Volcano Observatory 38t, 38b, 41t, Image by C. Waythomas, courtesy of the Alaska Volcano Observatory / U.S. Geological Survey 37l, Photograph H.G. Wiltshire, U.S. Geological Survey 33br; **Piran White:** 284, 304, 310, xxviiit; **courtesy of M.A. Whyte:** 153; **Clare Wilson:** xxviiib; **www.imagesource.com:** 67, 93.

In some instances we have been unable to trace the owners of copyright material, and we would appreciate any information that would enable us to do so.

CHAPTER 1

A trip through time on planet Earth

Malcolm Cresser and Paul Ayris

Learning outcomes

By the end of this chapter you should:

- Understand current ideas about how the planet formed.
- Know the elemental composition of the Earth's crust and be aware of the reasons for the relative element abundance.
- Start to understand the structures of primary and secondary minerals.
- Be aware of the nature of major rock types and their significance to landscape evolution.
- Appreciate how we can understand, and reconstruct, the processes that led to current rock type distribution at the Earth's surface.
- Understand how rocks can be dated.
- Be familiar with some key methods used to help identify rocks and minerals.
- Know about the nature and importance of plate tectonics and volcanoes.
- Be aware of the origins of water on the planet, and what regulates the chemical composition of the Earth's oceans.

1.1 How the Earth formed

Even very young children tend to be intrigued by their environment, a fascination which sometimes sends embarrassed parents scurrying to a library or the internet to answer questions such as 'Why is the sky blue?' or 'Why is the sea salty?'. It therefore seems appropriate to start this text with a concise overview of how most scientists currently think that the small planet that we live on originated, and how it changed over many millions of years to the point where it can support the amazing biodiversity that it currently does today. Along the way the authors hope to rekindle the reader's fascination with why, when and how, and to stimulate their curiosity sufficiently to make the fuller accounts that follow in subsequent chapters interesting.

1.1.1 The Big Bang theory

At the time of writing there is general agreement among most scientists that the universe originated with a cosmic explosion some 13.7 billion (13.7 thousand million or 13.7 times 10^9 years) ago, the 'Big Bang'. Prior to that, however, we are confronted with the concept of all matter and energy concentrated into 'a single, inconceivably dense point' (Grotzinger *et al.*, 2007). It was struggling with this concept that led many scientists a few decades ago to believe that although the Universe is known to be expanding now, at one time it must have been contracting. Thus their concept was one of an oscillating universe, and one that probably always was. Perhaps we should not be surprised that some people opt for the concept of a supernatural being creating everything! Fortunately for us we need to concern ourselves only with the formation of our own solar system, and especially that of planet Earth, an event that occurred almost 10 billion years later.

1.1.2 The nebular hypothesis

Developments in astronomy and space exploration over recent decades have led to the identification of many rotating clouds of gas and dust across the universe. The gas mainly consists of two very light elements, hydrogen and helium, whereas the dust particles contain elements found in the solid mass of our own planet. One of the nearest rotating clouds (nebulae) to the Earth, the Andromeda Nebula, is illustrated in Figure 1.1. Andromeda is about 1.8 million light years away from the Earth, and visible (though not in such

Figure 1.1 The Andromeda Nebula.

Source: NASA/courtesy of nasaimages.org

impressive detail!) to the naked eye. This means that light from the galaxy takes about 1.8 million years to reach us, so we see it as it was 1.8 million years ago. Studying many such galaxies allows astronomers to establish the sequences of events that can lead to the evolution of a solar system such as our own, in which a series of planets rotates around a star (the sun in our case).

In the rotating cloud from which our own solar system formed, gravitational forces started to attract matter increasingly inwards and the rotating disk started to contract. As a rotating disk contracts, the rotating matter starts to move progressively faster (distance covered in the time taken for one rotation is greater the further away a particle is from the centre). At the centre of the nebula a protostar was formed (our juvenile sun) as highly energetic particles and atoms collided to be further compressed under the protostar's own accumulating mass. The energy released was sufficient to cause an enormous increase in temperature, sufficient at the high pressure that also existed to cause hydrogen atoms to fuse together to form helium atoms. Hydrogen atoms are the simplest of all atoms, each with a minute, negatively charged, electron orbiting around a larger, positively charged proton as its nucleus. Because each nucleus is surrounded by a cloud of negative charge the atoms naturally tend to repel each other (opposite charges attract, like charges repel). However, under the extreme high temperature and pressure conditions in the protostar, interaction (fusion) becomes possible.

In the fusion process matter is converted to energy. Each hydrogen atom contains one proton and one electron, and has an atomic mass of 1.007947, whereas a helium atom contains two protons, two neutrons and two electrons and has a mass of 4.002602, significantly less than four times the mass of the hydrogen atom. The energy released by the conversion of matter to energy is governed by Einstein's famous equation, $E = mc^2$, where E is energy, m is mass and c is the velocity of light. The energy released helps maintain the sun's high temperature and provides the energy released into space as solar radiation.

1.1.3 The formation of the planets

Solid particles and gas molecules further out in the nebula also underwent energetic collisions, forming progressively larger and larger solid bodies known as planetesimals. These, in turn, underwent further collision sequences, leading ultimately to bodies from the size of our moon (or less) up to the size of our planets, aided as they grew larger by the gravitational fields that they generated. As the conditions were favourable, these then exerted a sufficient gravitational influence upon each other to remain in relatively stable orbits around our recently formed sun, though most stars do not have planets orbiting around them. Those that do outside our own solar system are currently attracting considerable interest, in the context of the possible existence of life on other planets.

Because of the way our solar system emerged, gases remained dominant in the outer reaches of the nebula. As a consequence they remain dominant in the constitution of the outer planets of the solar system, Jupiter, Saturn, Uranus and Neptune. Thus Jupiter is 75 per cent hydrogen and 25 per cent helium on a mass basis, with trace amounts of water, ammonia and methane, but it is thought to possess a rocky core. Its composition is believed to be close to the composition of the original nebula from which our solar system evolved. Saturn is even an oblate spheroid (visibly flattened at the poles) when studied through a telescope because of its fluid state and rapid rotation. Uranus consists of rock and frozen liquids with about 15 per cent hydrogen and some helium, surrounded by an atmosphere of 83 per cent hydrogen, 15 per cent helium and 2 per cent methane. Neptune consists predominantly of liquid water, ammonia and methane, but is surrounded by a thick atmosphere dominated by hydrogen and helium.

The inner planets, Mercury (closest to the sun), Venus, Earth and Mars (furthest away of the four), are all Earth-like, in that they have compositions high in rocks and metals. They formed much closer to the sun at high temperatures, resulting in the loss of gases, especially those of lighter elements, into space. They are also therefore smaller.

Mercury has a large iron core from 3600 to 3800 km in diameter, surrounded by a silicate layer about 500 to 600 km thick. It is currently thought that some of the core is molten metal. Mercury actually has a very thin atmosphere because of the high temperature so close to the sun. In the absence of the types of weathering that occur on Earth its surface is not unlike that of our moon (Figure 1.2), displaying many impact craters. Mercury's orbit is eccentric, its distance from the sun varying between 46 million and 70 million km. Some spectacular views of Mercury (and indeed of the other planets) may be seen on the solarviews.com website or at http://pds.jpl.nasa.gov/planets/.

Venus differs dramatically from the Earth, in spite of its being a near neighbour, formed at about the same time and condensed out of the same nebula. Unlike the Earth it has no oceans and it has a very different atmosphere, composed mainly of carbon dioxide. Its clouds are composed of sulphuric acid droplets rather than water. Its surface temperature is about 482 °C, even hotter than Mercury's, primarily due to the carbon dioxide in the atmosphere trapping infrared emission (a worst case scenario of the greenhouse effect that is discussed later in this book in Chapter 8).

Figure 1.2 A NASA photomosaic of mercury assembled from a set of high-resolution images taken by Mariner 10 in 1974.

Source: NASA/courtesy of nasaimages.org

Mars too has an atmosphere very different from Earth's. It is >95 per cent carbon dioxide with small amounts of nitrogen (2.7 per cent) and argon (1.6 per cent) and traces of neon (0.00025 per cent). It is possible that, in the distant past, the Martian atmosphere may have been more dense. This might have allowed water to flow on the planet, which could explain the physical features resembling shorelines, gorges, riverbeds and islands. The average temperature is −63 °C, with a maximum of 20 and a minimum of −140 °C. Ice occurs at the poles, being either frozen water or frozen carbon dioxide.

Considering that this book is about Earth both before and after human intervention, the perceptive reader might be starting to wonder why Earth hardly gets a mention in the opening pages. However, considering what we know about adjacent planets in our solar system raises some very interesting points about the Earth itself. Why does it have a relatively pleasant temperature for its current biodiversity (except in the most climatically extreme regions), and most importantly, why is its atmosphere not heavily dominated by carbon dioxide like the atmospheres of its two nearest neighbours in the solar system? The answers to these questions will emerge over the coming pages and chapters.

Lead-in question

From our discussion of the formation of the planets in our solar system so far, what would you expect the composition of the Earth to be like?

1.2 What controlled the distribution of elements on Earth?

Bearing in mind that Earth is the third planet away from the sun, we would expect it perhaps to have a metal core, and an outer layer of silicate mineral-rich material. We might anticipate an atmosphere containing a high concentration of carbon dioxide and some nitrogen, argon, sulphur dioxide and water vapour and possibly ammonia and methane. We would expect it to be cooler than Venus as it's further from the sun (148 million km compared with 108 million km), but appreciably warmer than Mars (at 228 million km), and kept hot by the high carbon dioxide concentration. Remember that the surface area of a sphere is $4\pi r^2$, so the solar radiation per unit area at the Earth compared with that at Venus will be reduced by a factor of 108^2 divided by 148^2, or by a factor of 0.533.

When the Earth first formed, 4.56 billion years ago (sometimes written as 4.56 Ga), the carbon dioxide concentration in the atmosphere must indeed have been much higher than it is now, and the oxygen concentration negligible. At the high temperatures prevailing at the time, any free oxygen would have reacted with metallic and other elements present. Much oxygen reacted with hydrogen to form water, H_2O, or possibly OH^- ions in silicate minerals, helping to retain hydrogen that would otherwise have disappeared into space. This helped later to give the planet its characteristic, relatively high, water content. The atmosphere probably would also have contained more nitrogen than those of neighbouring planets.

Although the Earth was (and is) tiny compared to the sun, towards the centre of the planet the temperature and pressure must be extremely high, though not high enough to instigate significant nuclear fusion of the predominantly heavier element atoms present. However, the trace amounts of naturally radioactive elements present from the nebula parent materials liberate substantial amounts of energy as they decay. In the confines of the centre of the planet this contributes to maintaining a very high temperature.

1.2.1 Formation of the moon

We know from studies of other planets and from residual signs of often massive craters on the surface of the Earth itself that over time planets have been subjected to many impacts from meteorites and asteroids from space. This would have been especially true during the early stages after the formation of our solar system. It is now thought that about 4.5 billion years ago, a body about the size of Mars collided with the newly formed Earth. The kinetic energy (energy associated with a mass moving at high velocity) would have been almost instantly turned into a combination of thermal energy and added kinetic energy for the Earth itself. The immense amount of thermal energy would almost certainly have been enough to melt all the rocks on the planet, and the collision would have caused large masses of solid and molten debris from both the Earth and the colliding body to fly off into space, where they would start to orbit the Earth. The orbiting debris, via a series of collisions, is thought to have aggregated to form our moon, which still orbits the Earth today. A secondary effect of the collision was that the Earth's orbital plane was tilted away from its original path, and the Earth started to rotate more rapidly about its axis. We will see later that this is, to a large extent, responsible for the distinctive seasons we have on Earth, especially to the far north or south of the equator, and had a marked effect on how biodiversity evolved.

1.2.2 Differentiation and formation of a layered planet

The consequence of the collision would have been a very deep sea of molten rock or magma at the surface of the planet, which would have allowed fractionation of elements and minerals over a long period of time as the magma slowly cooled. Precisely what happened at any point would depend upon the relative abundance of all elements present, temperature and pressure, but a good overview has been presented by Smith and Pun (2006). They postulate that dense droplets of molten iron formed within the massive silicate melt under a thin, chilled solid crust. These droplets have a high density (mass per unit volume), so sink through the silicate melt towards, and eventually through, the partially molten lower mantle. Their perception is that molten iron/nickel and molten silicate remain immiscible fluids, much like oil and vinegar in a salad dressing. Eventually silicate minerals containing mainly lighter elements would start to crystallise out near the surface,

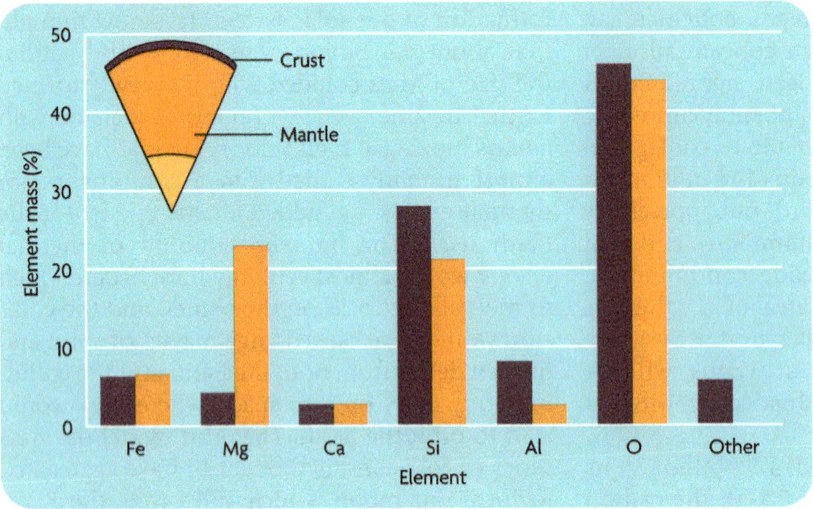

Figure 1.3 Difference in element abundances in the crust and the mantle due to differentiation.

depending upon the relative abundance of the elements left and the melting points of the individual minerals. Thus we would soon have been heading for an Earth structure much as it is today – an outer crust about 40 km thick, over a mantle somewhat enriched in iron and magnesium, around 2900 km thick, surrounding a liquid iron outer core and a solid iron inner core giving the element distribution indicated in Figure 1.3.

Although we cannot see what the iron core is made of directly, we can get a good idea from meteorites that from time to time enter the Earth's atmosphere from space. Small meteorites burn up and become dispersed as they travel through the outer atmosphere at very high speed and are only recognisable as shooting stars. However, if the meteorite is big enough large pieces may survive and crash into the Earth's surface. They will often display evidence of at least partial melting. Figure 1.4 shows two typical fairly large meteorites that were found in the Nullarbor Plain about 630 km east of Kalgoorlie in Australia. They weigh over 800 kg each, and are now exhibited outside the Western Australian museum in Perth, where these photographs were taken.

1.2.3 What would have happened to the early atmosphere?

It is highly likely that massive additional amounts of nitrogen, carbon dioxide and water vapour were released through the fragile outer crust at this time in the planet's evolution, probably through volcanic-style eruptions. Unlike hydrogen, these compounds have a sufficiently high vapour density to have been retained in the atmosphere. It is also possible that the accumulation of chlorine that eventually found its way to the oceans as chloride started at around this time too, though Schilling *et al.* (1978) have concluded that current levels of volcanic activity over a geological timescale are adequate to explain existing concentrations, and no early episode of out-gassing has to be invoked.

The planet surface would have remained hot for a long time with the high carbon dioxide concentration as a consequence of radiated heat being trapped by the gas in the atmosphere, although clouds would have afforded some protection. However, evolution of photosynthetic organisms eventually started to change the atmosphere's composition to one closer to today's, by converting carbon dioxide and water to biomass, liberating oxygen in the process. We shall return to this topic in Chapters 3, 4 and 5, because it is of monumental importance. The evolution of life as we know it was only possible once photosynthesis started to play a role. Only then, for example, could soils start to evolve, let alone other higher organisms.

(a)

(b)

Figure 1.4 Two iron meteorites discovered in Australia.

1.3 Primary and secondary minerals

1.3.1 Primary minerals: framework silicates

In the preceding section we saw that the solid, outer crust was formed from a molten mass containing predominantly the elements oxygen, silicon, aluminium, iron, magnesium, calcium, and smaller amounts of other elements such as sodium and potassium. As this mass cooled, solid minerals would have started to crystallise out within the molten mass. The precise nature of the solid minerals formed would depend upon the ratios in which the elements were present in the melt and the temperature. For example, minerals with a higher melting point would tend to solidify before those with lower melting points. The size to which minerals grew would depend on cooling rate (in this case mostly incredibly slow), and whether the crystals moved through the molten mass. The extent to which they moved would depend upon the relative densities of the hot mineral crystals and molten mass.

With so much silicon and oxygen in the melt, formation of quartz, or silicon dioxide, SiO_2, would be highly probable. Minerals formed under such conditions are known as primary minerals. Figure 1.5 shows what a nicely crystalline specimen of quartz looks like.

Quartz forms rather attractive hexagonal prisms under favourable conditions, but may also be found in massive structures too. It has a melting point of 1710 °C. Being formed from relatively light elements (silicon has an atomic mass of 28.09 and oxygen 16.0), from the top two rows of the periodic table, silica has a quite low specific gravity (SG) of only 2.6. An SG of 2.6 means 2.6 times more dense than water. In other words, as a cubic centimetre of water weighs 1.0 g at 25 °C, a cubic centimetre of quartz would weigh 2.6 g.

Figure 1.5 A specimen of the primary mineral quartz (SiO_2).

It is appropriate at this point to look at the chemical bonding within silica (i.e. how the silicon and oxygen atoms are held together). Oxygen has an atomic number of 8, meaning that an oxygen atom has 8 protons in its atomic nucleus. The nucleus also contains 8 neutrons, (hence the atomic mass of 16) and is surrounded by 8 electrons. Two of these electrons fill the inner 1s orbital, two fill the 2s orbital, and four occupy 2p orbitals. This electronic structure is not especially stable, and the oxygen atom tends to acquire two additional electrons so that it has the stable electronic configuration of the inert gas neon, namely $1s^2 2s^2 2p^6$. Inserting one electron into an oxygen atom, to give an O^- anion, requires a considerable amount of energy, however, and inserting a second requires even more, because a negatively charged particle is being added to an already negatively charged anionic species. Some metal cations do form ionic oxide compounds because the energy needed to form O^{2-} is offset by the energy liberated when a stable crystalline lattice forms (the lattice energy).

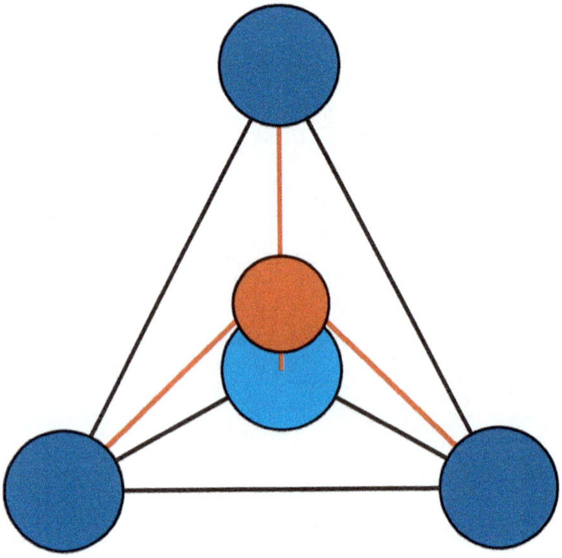

Figure 1.6 A simplified representation of the silica tetrahedron. The silicon atom (red) is bound to four oxygen atoms (blue) so that each is at one apex of a tetrahedron. The oxygen atom to the rear is shaded in a paler blue.

So what happens with silicon atoms, which have the electronic configuration $1s^2 2s^2 2p^6 3s^2 3p^2$? To form the cation Si^{4+} with the inert gas stable electronic configuration of argon, each silicon atom would need to lose 4 electrons. This would require a very large input of energy, as the removal of the second, third and fourth negatively charged electrons would be from a cationic species with increasing positive charge. In energy terms it is much more favourable for silicon and oxygen atoms to share electrons, by what is known as covalent bonding. This happens in the mineral, quartz. Each silicon atom is surrounded by four oxygen atoms in a tetrahedral configuration. Each of these oxygen atoms is, in turn, linked to another silicon atom, and so on, to form a framework of interlinked silicon tetrahedra extending in three-dimensional space. Each of the oxygen atoms thus shares one of its own electrons with two silicon atoms, and each silicon atom acquires an electron from each one of the four oxygen atoms surrounding it in tetrahedral coordination, as indicated in Figure 1.6.

We can ask here what happens when crystalline quartz melts, bearing in mind the extreme formation conditions that we have been discussing. Molten quartz is very viscous (a bit like treacle stored in a fridge!), because it cannot simply break up into conveniently small molecular units the way low molecular mass organic compounds do. Rather it appears to form a glassy mass in which giant molecular fragments can move relative to each other, presumably with some bond breakage and reformation. If this is then cooled very slowly, frameworks reform and crystals can reappear.

Suppose in the framework silicate quartz, one silicon atom from the sub-unit $(SiO_2)_4$ is replaced by an aluminium atom as magma cools. This leaves a deficit of a unit positive charge. If this is then balanced by the incorporation of sodium into the framework, we would have the alkali feldspar mineral albite, with a sub-unit $NaAlSi_3O_8$. If on the other hand the charge deficit was balanced by potassium, we would have $KAlSi_3O_8$. This is the potassium feldspar orthoclase. Depending on the impurities also present, orthoclase may be white, pale pink, yellow or brownish. It has a vitreous lustre, and often shows quite distinct cleavage in two directions. A hand specimen can be seen in Figure 1.7. Often these alkali feldspars contain both sodium and potassium, and a continuous series exists between albite and orthoclase.

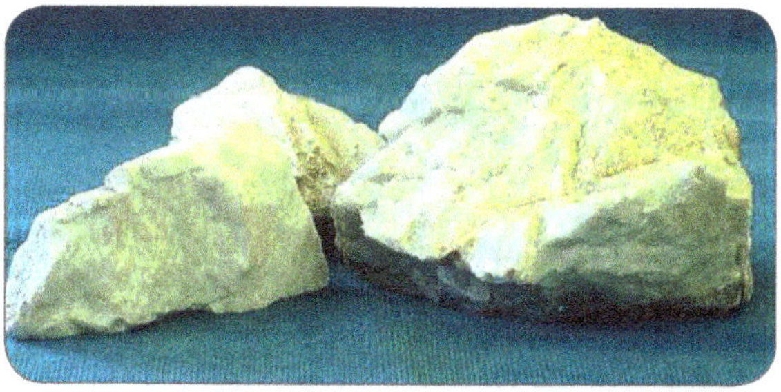

Figure 1.7 A sample of the alkali feldspar orthoclase.

The substitution of one ion in a mineral matrix by another of similar size is known as isomorphous substitution. Isomorphous substitution is very important in silicate mineralogy, especially where the displacing ion or ions possess a different charge. If a silica lattice has twice as much isomorphous substitution as that found in alkali feldspars, and the charge deficit is provided by calcium, Ca^{2+}, we have the structural sub-unit $CaAl_2Si_2O_8$, which is the framework silicate anorthite. In practice there is a continuous series of feldspars with steadily increasing amounts of calcium between albite and anorthite. Feldspars from this series are known as plagioclase feldspars.

Figure 1.8 shows a rather nicely crystalline garnet. Garnets are silicates containing various divalent and trivalent metals, with the formula $M^{2+}{}_3M^{3+}{}_2\,(SiO_4)_3$. M^{2+} is calcium, magnesium, iron or manganese, and M^{3+} is iron, aluminium, chromium or titanium. Thus the colour is very variable, as are the hardness and SG. The crystals may be distinctly dodecahedral (12 sides) or tetragonal (6 rectangular sides) or trisoctahedral (24 sides). They are often found in association with metamorphosed rocks (rocks in which recrystallisation has occurred as a consequence of the rock being subjected to extreme temperature and/or pressure).

Figure 1.8 A garnet attached to a lump of schist (schist as a metamorphosed rock is discussed later).

1.3.2 Primary minerals: sheet silicates

In silica, we saw that the $SiO_4{}^{4-}$ silicate tetrahedrons were linked into a continuous 3-d framework. However, they may also link up in other ways in the formation of primary minerals. One very important group of silicate minerals is the sheet silicates. In these the silica tetrahedrons are linked to form continuous sheets, as shown in Figure 1.9. In this figure two oxygen atoms out of every five (the ones rising vertically upwards from the page) are not joined to another silicon atom, but have a negative charge. The repeat unit in the structure, shown in darker shades in Figure 1.9, is thus $Si_2O_5{}^{2-}$.

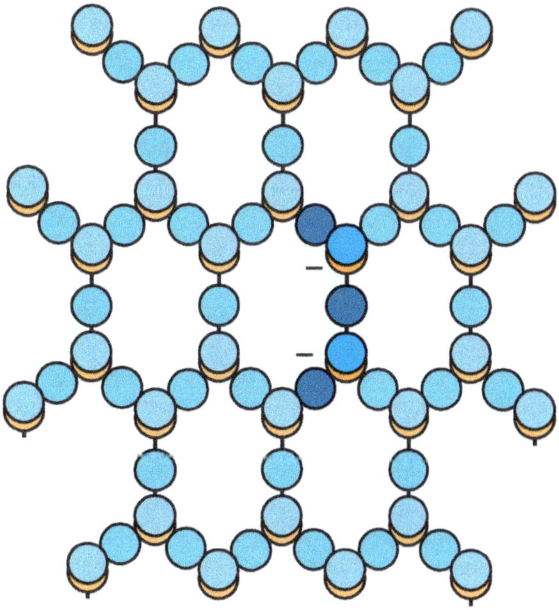

Figure 1.9 A simplified representation of a silicate sheet. Each silicon atom (orange) is surrounded by four tetrahedrally coordinated oxygen atoms (blue).

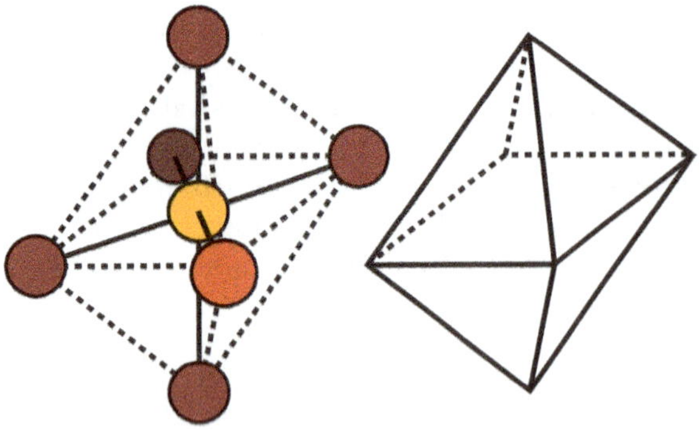

Figure 1.10 Representation of the octahedral coordination of aluminium.

Figure 1.11 The typical platy structure of a muscovite sample.

Aluminium hydroxide (gibbsite, $Al(OH)_3$) also forms extended sheets that are similar in many respects to silicate sheets. However, aluminium favours octahedral coordination, as shown in Figure 1.10. In a gibbsite sheet the octahedra are linked in such a way that the Al atoms all sit in a central plane, sandwiched between two planes of hydroxyls. Magnesium hydroxide (brucite, $Mg(OH)_2$) forms similar sheets via its octahedral coordination.

In the mineral pyrophyllite, a gibbsite sheet is sandwiched between two silicate sheets, and pyrophyllite is therefore known as a 2:1 type mineral. Two thirds of the gibbsite sheet hydroxyls are replaced by apical oxygens from the silicate sheet, which are all therefore directed towards the gibbsite sheet. Thus we should start with two silicate sheet subunits, as described above, $2(Si_2O_5^{2-})$, and one $Al_2(OH)_6$ sub-unit, and get rid of four of the six hydroxyls. This leaves us with $Al_2Si_4O_{10}(OH)_2$, which in turn gives $4[Al_2Si_4O_{10}(OH)_2]$ as a balanced structural sub-unit for pyrophyllite.

The mica mineral muscovite is structurally very similar to pyrophyllite, except that one in four tetravalent silicon atoms is replaced by a trivalent aluminium atom, leaving a shortage of positive charge. This is balanced in muscovite by a layer of potassium ions (K^+) between each 2:1 layer, to maintain electro neutrality. Thus muscovite mica samples have a very clear layered structure. This can be seen in Figure 1.11.

The bonding between the 2:1 layers is weak, and mica can very easily be cleaved along the plane of the sheets to get very thin, flexible sheets. The sheets may be so thin that they become transparent, as in Figure 1.12, and easily cut through with scissors.

Biotite is another common mica 2:1 type mineral found in rocks, but has a brucite sheet in place of the gibbsite sheet of muscovite. However, again it has one in four silicon atoms of the silicate sheet replaced by an aluminium atom, and again potassium in interlayer spaces holds the layers together. In addition, however, about one Mg^{2+} in three in the brucite sheet is replaced by Fe^{2+}. This imparts a characteristic dark brown colour to the mineral (Figure 1.13). This occurs because iron is a transition metal with a more

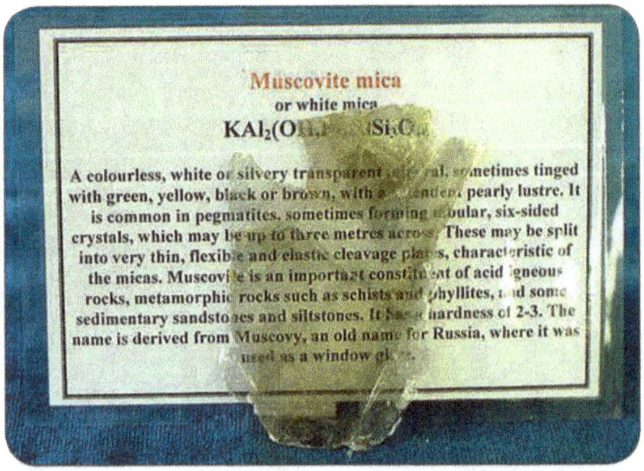

Figure 1.12 A transparent thin sheet of mica. The text underneath may be read quite easily.

Figure 1.13 A sample of the 2:1 sheet silicate mineral, biotite.

complex electronic structure than magnesium. Thus electronic excitation is possible using vacant d orbitals via the absorption of light with an appropriate wavelength (colour). Like muscovite, because of its layered structure and weak bonding between 2:1 units, biotite is a very flaky mineral, and the sheets are quite easily bent when thin. The basal unit is $K(Mg,Fe)_3AlSi_3O_{10}(OH)_2$. The notation $(Mg,Fe)_3$ here denotes that the amounts of Fe^{2+} and Mg^{2+} are variable.

The minerals biotite and muscovite are very common (admittedly not in such large pieces as the specimens illustrated), because of the great abundance of silicon, oxygen, aluminium, magnesium and iron in the Earth's crust (see Figure 1.3). Figure 1.14 shows a much rarer, lilac coloured lithium mica, lepidolite, which again has perfect basal cleavage, but the rather more complex basal unit $4[K_2Li_3Al_4Si_7O_{21}(OH,F)_3]$.

Figure 1.14 A specimen of the lithium mica lepidolite.

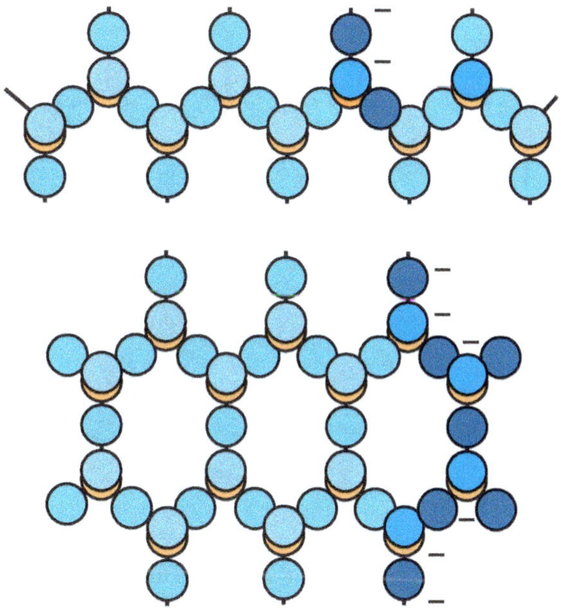

Figure 1.15 Linkages of silica tetrahedral in pyroxene chains (top) and amphibole chains (bottom).

1.3.3 Primary minerals: chain silicates

As well as frameworks and sheets, in some minerals the silicate tetrahedra may be interlinked in chains. Examples are shown in Figure 1.15. In this figure the upper chain structure is the pyroxene chain, for which the repeat unit is SiO_3^{2-}. The pyroxene group includes hypersthene $(Mg,Fe)OSiO_2$, augite (calcium magnesium iron aluminium silicate of variable composition, represented therefore as $(Ca,Mg,Fe,Al_2(AlSi)_2O_6)$, and diopside $(CaO.MgO.2SiO_2)$, as well as several other minerals.

The lower chain structure in Figure 1.15 is the amphibole chain. It has a repeat unit of $Si_4O_{11}^{6-}$. The amphibole group includes hornblende, in which the balancing cations are calcium, magnesium, sodium, iron and aluminium covering a diverse range of compositions, and actinolite $(Ca_2(Mg,Fe)_5Si_8O_{22}(OH)_2)$, in which the amounts of iron and magnesium are variable.

1.3.4 Primary minerals that are not silicates

Up to this point we have concentrated upon the formation of silicate minerals because of the dominance of silicon and oxygen in the Earth's

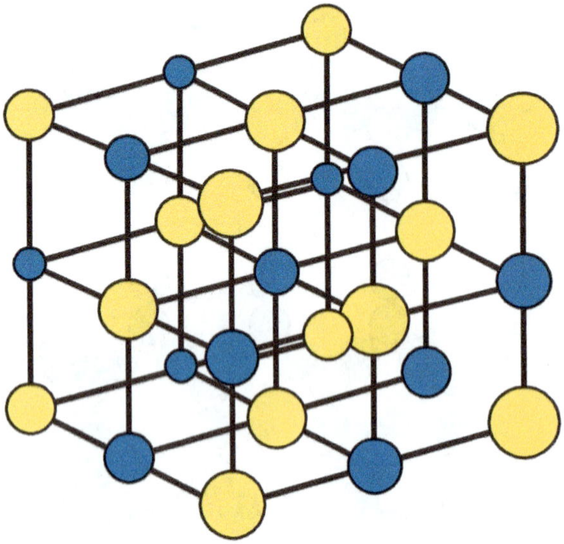

Figure 1.16 A representation of the structure of lead sulphide (galena). Pb^{2+} ions are shown as buff coloured and S^{2-} ions as blue. Ions further back from the front face of the structure are drawn progressively smaller for clarity.

Figure 1.17 An example of the primary mineral galena (lead sulphide).

crust, as discussed earlier. However, there are numerous primary minerals that are not silicates. Several of these primary minerals have structures that are much easier to visualise than that of quartz. Galena, for example, an important ore of the metallic element lead, contains a very regular array of lead Pb^{2+} and sulphide S^{2-} ions. The ions are neatly arranged in a cubic crystalline lattice (Figure 1.16), identical to that of rock salt (sodium chloride). The bonding is termed ionic bonding. The lattice energy is sufficiently large to overcome the energy required to form lead Pb^{2+} cations and sulphide S^{2-} anions from the element atoms. Lead is a heavy element (atomic mass 207.2) and lead atoms are relatively large. The two outermost electrons are shielded from the attraction to the positively charged protons in the nucleus by 80 electrons in intermediate orbitals, making them easier to remove to form the cation Pb^{2+}.

Sulphides of many lighter metals tend to have more complex structures and bonding which is neither purely ionic nor purely covalent. Like rock salt, galena forms very attractive cubic structures (Figure 1.17), but it has a much higher SG (7.58).

Pyrite or iron pyrites has a similar cubic structure, and can form often quite sizeable crystals with a brassy, metallic lustre (Figure 1.18), and a SG of about 5.0. Because of its often golden colour it is commonly known too as fool's gold, for obvious reasons. It has the formula FeS_2, which at first glance makes the iron look tetravalent rather than being in its usual divalent or trivalent state. However, this is not the case. The sulphur is in the form of S_2^{2-} units, or as $^-S—S^-$ ions. These and Fe^{2+} ions are arranged in a slightly distorted cubic lattice.

Figure 1.18 An attractive specimen of the mineral pyrite (FeS_2), showing its cubic structure.

Figure 1.19 An example of barytes (4[$BaSO_4$]).

Pyrite may occur as a secondary mineral. This is the name applied to any mineral formed as a consequence of a chemical reaction between elements released by chemical weathering (chemical decomposition) of primary minerals. It forms in anaerobic soils in which Fe^{3+} has been reduced to Fe^{2+}, and sulphate to S^{2-} and S_2^{2-}.

Figure 1.19 is a sample of barytes (barium sulphate) picked up at an old lead mining site in England. The mineral takes several crystalline forms, but is generally whitish, but often tinged with yellow/brown. Barium is one of the heavier alkaline earth metals, with an atomic mass of 137.3, and although sulphur and oxygen are lighter elements, the mineral still has an SG of about 4.5. It commonly occurs in mineral veins in rock, accompanied by galena (which we looked at earlier), zinc blende, fluorite and quartz.

Figure 1.20 shows a specimen of fluorite (also known as fluorspar), CaF_2. Fluorite forms often quite distinctive cubic crystals, in a range of colours from colourless to almost black, and various other shades including violet as in Figure 1.20 through greeny yellow and brown. Because of the stoichiometry of CaF_2, the arrangement of ions is a little more complex than that shown in Figure 1.16. It may be regarded as parallel layers of fluoride anions with half the interstitial spaces containing Ca^{2+} cations. The crystal lattice energy is high, so the mineral has a very low solubility.

Figure 1.20 A specimen of fluorite (CaF_2), displaying the mineral's distinctive cubic structure.

We saw earlier that although iron is particularly concentrated in the Earth's core, it is also abundant in the mantle and crust. Bearing in mind the high oxygen content too, we might expect to find plenty of iron oxide, and indeed we do. Hematite (2[Fe_2O_3]) occurs world wide, and is an important ore of iron. It can occur in various crystalline forms, or sometimes in the form shown in Figure 1.21, when it is known as kidney ore.

Figure 1.21 also shows the mark made by the kidney ore on a streak plate. The colours of minerals by reflected light may be highly variable, as in the

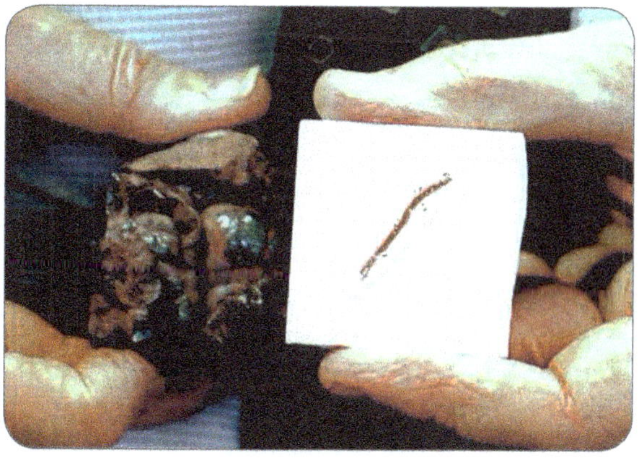

Figure 1.21 A specimen of the iron oxide mineral hematite (Fe_2O_3) in the form known as kidney ore, and the mark it makes on a streak plate.

Figure 1.22 A sample of black velvet tourmaline, an important boron containing mineral.

Figure 1.23 A specimen of gypsum, showing clear cleavage in two directions.

specimen shown. Sections of kidney ore may be almost black, and have a metallic lustre. However, when the mineral is scraped along a plate of mildly abrasive ceramic material, a streak of much more uniform colour is left. Doing this allows geologists to describe colour more reproducibly.

An important mineral that crystallises out occasionally as some igneous or metamorphic rocks form is tourmaline, a complex alkali metal aluminium magnesium and iron borosilicate with the long formula: $(NaCa)(Li,Mg,Fe^{2+},Al)_3(al,Fe^{3+})_6B_3Si_6O_{27}(O,OH,F)_4$. As we will see later in Chapter 5, boron is an essential element for plants and also occurs in sea water, so identifying mineral sources that may release the element by weathering is important. Figure 1.22 shows a form of tourmaline known as black velvet tourmaline. The mineral is invariably very dark to black in colour.

1.3.5 Secondary minerals

The minerals that we have considered so far are mostly primary minerals; this means that they have been formed by crystallisation from magma or veins of molten rock. When these minerals weather on exposure to the atmosphere they release ions into solution which may then react with each other to form precipitates, often over many years and far removed from their original parent materials. Such minerals are known as secondary minerals. One such mineral is gypsum, which is hydrated calcium sulphate, with a structural unit $8[CaSO_4.2H_2O]$. It deposits as colourless to white tabular or prismatic crystals, sometimes tinged with yellow or grey. It often shows distinct cleavage in two directions, as for example in Figure 1.23.

Quartz can also reappear as a secondary mineral. One very well known form is flint, which shows conchoidal fracture patterns. It is quite common in many chalk deposits, where it has filled voids over very long periods of time. It sometimes contains traces of organic remains. The white chalk can still be seen on the surface of the flint fragment in Figure 1.24.

Sometimes quartz deposits as a secondary mineral inside holes left in sediments after the original organic contents have long since disappeared via decomposition, thereby recreating

Figure 1.24 A broken flint from a chalk deposit.

(a)

(b)

Figure 1.25 A fossil ammonite carefully removed from a sedimentary rock (a), and (b) a similar ammonite after bisection and polishing.

the shape of the original occupant. Figure 1.25(a) shows a fossil ammonite produced in this way. Carefully removed, bisected and polished, such items often end up as ornaments in souvenir shops, as shown in Figure 1.25(b).

Another common secondary mineral is calcite, calcium carbonate. Calcium hydrogen carbonate is released by the biogeochemical weathering of many minerals, and may then react in the air or in the soil atmosphere under alkaline conditions to produce calcium carbonate:

$$Ca(HCO_3)_2 \rightarrow CaCO_3 + H_2O + CO_2$$

Readers who live in hard water areas will be familiar with the scale that forms in kettles and sometimes under taps, and occasionally wrecks washing machines by coating their heating elements. Hard water contains calcium and magnesium hydrogen carbonates, and the above reaction is responsible for the scale formation problems. More impressive though are the stalactites (the spiky features hanging from cave roofs) and stalagmites (the features rising up from the floors) in limestone caves, such as the Caves of Drach in the Spanish island of Majorca (Figure 1.26). These caves are carefully and thoughtfully illuminated, so that the overall effect for visitors is quite spectacular. More entertaining than spectacular is Mother Shipton's Cave in Yorkshire, where the tradition has emerged of hanging assorted everyday items in the dripping waters from the overhanging rock, so that they become coated

Figure 1.26 The stalactites and stalagmites in the Caves of Drach, Majorca.

with calcium carbonate (Figure 1.27). See what you can identify being 'turned to stone'!

The mineral may be found in a number of crystalline forms with varying degrees of transparency, from course granular to spiky hexagonal crystals. Figure 1.28 is just one of several types.

In the presence of high magnesium concentrations, dolomite, calcium magnesium carbonate, may form rather than calcite, and sometimes the two may be found side by side. Its colour depends upon the impurities present, going from colourless when pure through to yellow/brown (Figure 1.29) or even greenish with iron impurities, and pink with manganese impurities (Figure 1.30).

Figure 1.27 Coated artefacts in Mother Shipton's cave, Yorkshire.

Figure 1.28 Calcite crystals, showing one common crystal form.

Figure 1.29 Dolomite crystals ($CaMg(CO_3)_2$).

Figure 1.30 Rose dolomite, with a pink discolouration associated with the presence of manganese.

1.4 Major rock types and their significance to landscape evolution

1.4.1 Igneous rocks

As might be expected from the discussion of the evolution of the Earth, the original molten magma would have had a fairly variable composition on moving across the Earth's surface and cooling conditions would also have varied. Thus igneous

Table 1.1 Classification of igneous rocks, after Cresser *et al.* (1993). Serpentinite is commonly classified as a metamorphosed ultrabasic igneous rock.

	Acid	Intermediate	Basic	Ultrabasic
Very coarse-grained	Pegmatite	Pegmatite		
Coarse-grained	Granite	Diorite	Gabbro	Peridotite
Medium grained	Micro granite	Micro diorite	Dolerite	Serpentinite (metamorphosed)
Fine-grained	Rhyolite, pumice	Andesite	Basalt	
Glassy	Obsidian, pitchstone	Obsidian, pitchstone		
Colour	Light	Variable	Dark	Dark

Source: Adapted from Cresser *et al.*, 1993.

rocks formed on cooling and solidification of magma might be expected to display substantial variation in composition, and this is what is observed in practice. Igneous rocks are classified according to how rich they are in base cations (primarily calcium, magnesium, sodium and potassium) and the size of the grains or crystals that constitute the matrix, as in Table 1.1. Thus base cation contents are higher moving to the right-hand side of the table, while quartz contents are lower. This table is slightly simplified and not fully comprehensive, but adequate to give an understanding of general principles.

How this classification works in practice should become clearer if we consider a few selected examples. Figure 1.31, for example, shows two samples of granite, and Figure 1.32 shows a magnified section of the second of the two specimens. The coarse grain size is readily apparent.

In the close-up shot of the granite (i.e. magnified), again the coarse grain size is readily apparent, as are the individual pink orthoclase crystals, colourless quartz crystals, white albite crystals and flaky

Figure 1.31 Examples of (a) grey and (b) pink granite rock. The potassium feldspar orthoclase in (b) gives it its pinkish colour.

Figure 1.32 A close up view of the pink granite in Figure 1.31, clearly showing the individual mineral grains.

plates of the darker biotite. The overall appearance of the dry hand specimens is light in colour, as in the table.

Figure 1.33 is a rather less common granite, an orbital granite, from the Murchiston District in Western Australia. The photograph was taken of a small section of a large polished slab. The rounded, concentric 'orbs' sometimes form as molten magma starts to crystallise slowly deep below the Earth's surface. Small single crystals form first and act as nuclei around which other minerals start to attach themselves in sequence. The small vein of granite cutting across the picture is a later intrusion.

Figure 1.34 shows specimens of another igneous rock, basalt, from the Isle of Skye, off western Scotland. Whereas granite is an intrusive igneous rock, meaning it has crystallised slowly below the surface (which allows quite coarse grain size formation), basalt is extrusive. This means that it has penetrated the surface of the crust and cooled much more rapidly. It therefore contains very fine grains, as indicated in the table, and it is dark in colour. Basalts contain a ferromagnesian mineral, usually the dark grey to black augite, $(Ca,Mg,Al,Fe)_2(Al,Si)_2O_6$, but sometimes hypersthene (hornblende) $16[(mg,Fe)SiO_3]$. Hypersthene is dark geeny-brown to black. They also contain plagioclase and may contain some quartz or orthoclase.

The intermediate rock diorite sample in Figure 1.35 is coarse-grained as Table 1.1 anticipates and towards the darkish end of the 'variable' colour range. It contains plagioclase but no quartz, and as for basalt, augite and hypersthene contribute to its colour. Note the rusty brown colour of the weathering crust that may be seen on some surfaces, due to the formation of some iron oxides.

The final igneous rock included is a coarse-grained basic rock, gabbro (Figure 1.36). The

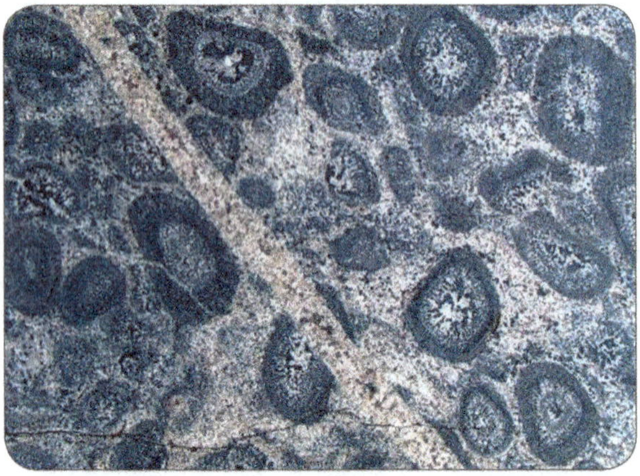

Figure 1.33 A polished section of an orbicular granite from Western Australia.

Figure 1.34 Specimens of the extrusive basic igneous rock basalt, from the Isle of Skye.

Figure 1.35 Samples of the intermediate rock diorite.

Figure 1.36 A sample of gabbro, dominated by large feldspars.

calcium-rich feldspars in this particular specimen would be classified as coarse, and in the photograph tend to somewhat lighten the dark green of the coarse-grained iron magnesium calcium silicate pyroxene minerals, so some care is needed when using simple tables such as Table 1.1. The overall impression is still dark though.

Lead-in question

Are all rocks igneous, or are there other mechanisms for rock formation?

1.4.2 Sedimentary rocks

We saw that the crust initially would have been thin, and constituted from igneous rocks and at high temperature. The atmosphere would have been hot, and very rich in carbon dioxide. The early crust would have been subject to some incredible stresses and strains as a consequence of changes in the molten magma mass, and undoubtedly eruptions would have occurred to relieve some of the very high pressures that would have built up, leading to a very uneven rugged landscape. Physical and chemical weathering would have been severe, with chemical attack of surface minerals by acids such as sulphuric acid and carbonic acid, and possibly hydrochloric acid, present in rainfall, aided by the high temperatures. Consider, for example, acid attack on the framework silicate anorthite, $CaAl_2Si_2O_8$. We can write a balanced equation as follows (see Lindsay, 1979):

$$CaAl_2Si_2O_8 + 8H^+ \rightarrow Ca^{2+} + 2Al^{3+} + 2H_4SiO_4$$

Some of the weathering products (the aluminium cations and silicic acid) would have reacted to produce a secondary un-substituted mineral, kaolinite, $Al_2Si_2O_5$, as indicated below:

$$H_2O + 2Al^{3+} + 2H_4SiO_4 \rightarrow Al_2Si_2O_5 + 6H^+$$

Kaolinite is a stable end product of weathering, and as we will see in Chapter 7 it is a common mineral in soils and can be detected in weathered rock surfaces too. The Ca^{2+} would eventually start to precipitate out in surface waters as calcium carbonate. Thick deposits of chalk could form under appropriate conditions, and these would eventually be subjected to sufficient over-burden pressure to be consolidated into rock. Such a specimen can be seen in Figure 1.37.

Chalk deposits could also contain other mineral grains, many of which would have been released by a combination of physical and chemical weathering and washed down slope (erosion) by rain and surface water, or moved by wind. Not all samples therefore look as white as the one in Figure 1.37. In places eventually sediments would also form in which carbonates were virtually absent. Such sediments may originate from rock surfaces that crumble more readily by physical weathering once the most readily chemically weathered mineral grains have been totally or partially removed. This would leave separated grains of the more resistant minerals. These too could be transported by natural processes to form aquatic sediments or deep terrestrial sediments, that could again eventually be consolidated to form rocks such as sandstone (Figure 1.38), or shale. In shales the particles forming the rock are very much finer grained (clay-sized – see Chapter 7). Figure 1.39 shows an example of a shale.

Sometimes the layered nature of such rocks can still be seen at the surface at rock faces, and there may be planes of weakness between beds

Figure 1.37 A specimen of chalk, containing grains of some other minerals.

Figure 1.38 Lumps of red sandstone, in which the gritty parent material is still visible.

Figure 1.39 A specimen of shale, showing typical platy structure.

where deposition conditions have suddenly changed for a period of time. This is the case beside the river shown in Figure 1.40. Note that in this figure the original planes are tilted compared to the current horizon, reflecting past movement of bedrock or crustal rocks when subjected to huge natural pressure.

Figure 1.40 Layered siltstone rock exposure beside a river in NW England.

The observant reader may think that we have rather digressed occasionally from a smooth passage through time in the Earth's geological history by including pictures with riverside vegetation, fossils from secondary minerals, and even teddy bears and a slipper at Mother Shipton's tourist attraction! And indeed we have, as we still have four billion years to go, there is no life on Earth as yet, and the atmosphere is still hostile and full of carbon dioxide and nitrogen and possibly sulphur dioxide and hydrogen chloride at significant concentrations. However, if anything, chemical and physical weathering, erosion, volcanic eruptions and faulting and folding of rocks (discussed later) probably occurred at an even greater rate than they do today, possibly aided by more frequent and more substantial meteorite collisions in the early stages. But before considering formation of sedimentary rocks any further, we need to consider the start of life on Earth.

Fossils of cyanobacteria have been detected in Archaean (2.8 to 3.8 billion years BP) rocks from Western Australia that have been accurately dated, and found to be 3.5 billion years old. They are easily recognisable by electron microscopy. An interesting account of their discovery and characteristics may be found on the University of California Museum of Paleontology website. The fact that these organisms could photosynthesise would have started to lower the carbon dioxide concentrations in the atmosphere, and generate oxygen instead. Carbon would have been retained in the cells' biomass, predominantly in the oceans. The possible origins of life are discussed in Chapter 3 and the carbon cycle and photosynthesis are discussed in Chapter 4 in detail. This process started in the Archaean Era, and continued to grow in importance throughout the Proterozoic Era.

Eventually higher forms of life started to form in the oceans, and fossil evidence for many of these is now very common in sedimentary carbonate rocks that were once deposits on the ocean floor. Many of them are readily identifiable today as being of marine origins, even if the sedimentary rocks now are many km from the seashore. Figure 1.41, for example is a crinoidal limestone, containing the fossilised remains of these early creatures. They are sometimes known as sea lilies because of their plant-like appearance. The organic matter

Figure 1.41 A specimen of crinoidal limestone, with clearly visible crinoid fossil fragments. Some excellent photographs and more information about crinoids may be found on the Virtual Fossil Museum website at www.fossilmuseum.net.

Figure 1.42 A close up view of the specimen in Figure 1.41.

has long since gone, but the cavities left have been filled by a perfect calcite replica that shows us the original shape. These can be seen better in close up in Figure 1.42, or via the ring features in the beach pebble in Figure 1.43.

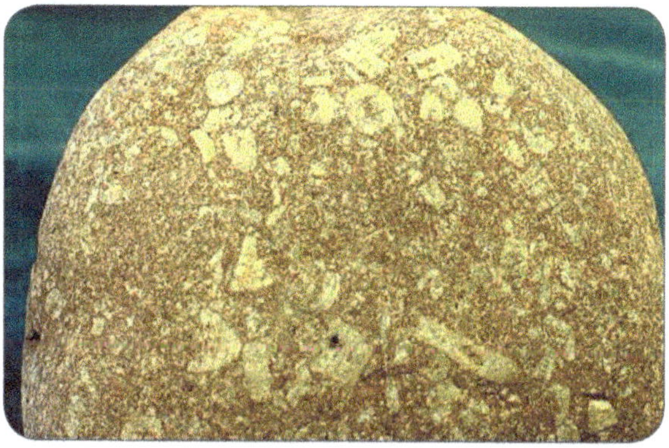

Figure 1.43 Part of a beach pebble displaying crinoid cross-sections. The precise shape of cross-sections depends upon the angle at which they have been broken during weathering, so not all are circular.

Figure 1.44 A dark coloured shelly limestone specimen.

Fossils of a wide range of shells may also be found in limestone deposits. Two examples have been included here, from different parts of England, as a reminder that limestones may be found in diverse colours (Figures 1.44, 1.45). Traces of shelly deposits may be found in non-calcareous matrices too, such as shale formed from clay deposits (Figure 1.46).

In Figures 1.41 to 1.46 the fossils are readily identifiable by the unaided eye. Oolitic limestone is formed from ooliths, and it is easier to identify its origins at slight magnification, as in Figure 1.47.

Limestones are quite soft, and for that reason relatively easy to quarry as building stones, especially if ornate carving is required. Sadly the low lattice energy of calcium carbonate makes it readily

Figure 1.45 A limestone specimen containing paler coloured fossil shells in a brownish matrix. The matrix also contains some calcite crystals.

Figure 1.46 A closer view of the shale seen in Figure 1.39.

convertible back to Ca^{2+} ions. So the stonework of buildings created tends to weather rather rapidly. We will see later in the book that this effect has been compounded by the ravages of acid rain, as in the example shown for the walled garden at the University of York (Figure 1.48). However, the problem is not just one that came after the Industrial Revolution. Natural weathering of the limestone and dolomitic rock (dolorock) used in the construction of York Minster, the famous medieval cathedral in York has been a documented problem for centuries (see Chapter 16).

Figure 1.47 A specimen of oolitic limestone, photographed at approximately five-fold magnification. The ooliths are formed by slow growths of calcium carbonate around small existing nuclei.

Figure 1.48 A heavily weathered ornamental garden feature made from limestone in the grounds of the University of York in England.

1.4.3 Metamorphic rocks

When solid rocks are exposed to high temperatures and pressures over extended periods of time (a million years or longer), although still below their

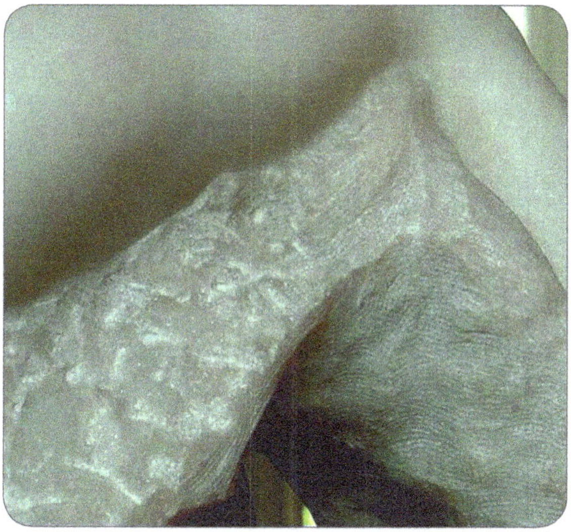

Figure 1.49 A close up of part of 'The Earth and the Moon', by Rodin, in the Rodin Museum in Paris. The small calcite crystals are more visible in the roughly hewn base than in the more carefully worked leg section.

Figure 1.50 Rodin's 'The man with a broken nose', in the Rodin Museum in Paris. Rodin created an earlier clay version before this one in white marble, which won greater acclaim.

melting point, the mineral compositions and crystal textures (grain size) and alignments may change. The rocks formed in this way are known as metamorphic rocks. Perhaps the best-known example is marble, which may be produced from fossil-rich sedimentary limestone, eliminating any visible trace of its initial origins in the process. When the 3-d mosaic of large calcite crystals fits together so well that no individual crystals may be discernible, it becomes a very attractive medium for sculptors and stonemasons (Figures 1.49, 1.50).

Slate is another widely recognisable metamorphic rock, formed at a regional scale, most commonly from shales (we saw a shale in Figure 1.39). It is enriched in mica and exceptionally fine grained, with perfect planar cleavage making it excellent as a durable roofing material. Evidence of metamorphism is perhaps more immediately apparent in the chiastolite-slate in Figure 1.51; formed from carbonaceous sediments, it contains needle-like crystals of chiastolite in a cryptocrystalline matrix. As for the crinoid fossils the shape actually seen reflects the angle at which the crystal has been broken.

Figure 1.51 Chiastolite slate, clearly showing white chiastolite crystals.

Figure 1.8 showed another common regional metamorphic rock, mica schist. It contains micas, quartz and feldspars, and has a very sparkly appearance as light is reflected off of the mica crystal surfaces. Schists may sometimes contain garnets, and we saw an example of this in Figure 1.8. The schist in that figure also has a somewhat banded appearance. Such banding is a common feature of schists if they have formed over a very

Figure 1.52 A specimen of biotite gneiss from southern Norway. Note the banded appearance.

long time and the crystals have grown to a significant size.

Figure 1.52 is a lump of biotite gneiss, a coarse-grained metamorphic rock with some indication of folding. The bulk of the matrix contains lighter coloured quartz and feldspars, and biotite and muscovite micas and hornblende form the darker bands.

Our final metamorphic rock specimen, forsterite marble from the Isle of Skye, contains the magnesium silicate mineral forsterite (Figure 1.53). The greenish banding is readily visible in this example.

Figure 1.53 A specimen of forsterite marble.

A very readable and more extensive account of the formation of metamorphic rocks may be found in Grotzinger, Jordan, Press and Siever's, *Understanding Earth* (2007).

1.5 Properties of rocks and minerals used for identification

Experienced geologists soon learn to identify many rocks and minerals by examination of hand specimens, often with the help of a hand lens or microscope. We saw earlier in this chapter that cleavage can be very characteristic of some minerals and is thus diagnostically a useful property to aid with identification. Colour is sometimes useful, though many minerals may assume different colours, depending upon the trace impurities present. For example, fluorite may be mauve as in Figure 1.20, but also olive-green, orange, pale yellow or brown. The dolomite in Figure 1.30 was pink due to the presence of manganese traces. It is also important to remember the need to use streak plates (as seen in Figure 1.21) when assessing the colour of some minerals.

Chemical tests are not used that often, a notable exception being the use of dilute mineral acid to test for the presence of carbonates. Effervescence from the rapid evolution of carbon dioxide is readily seen (Figure 1.54). Full spectrochemical analysis is possible though.

Figure 1.54 Effervescence on adding dilute hydrochloric acid to a chalk sample.

One drawback of trying to teach rock and mineral identification from photographs, rather than by handling actual specimens, is that you do not get any idea about the density of the samples. Most minerals have densities in well known and often narrow ranges. If you pick up the samples in Figures 1.17 (galena or lead sulphide) or 1.19 (barytes or barium sulphate), it is immediately obvious that the mass of mineral per unit volume is very high. Density may be measured quite easily, so it is another useful determinant for mineral identification.

The photographs of minerals included earlier in this chapter show that some, such as galena (Figure 1.17) or pyrites (Figure 1.18), have a metallic lustre. Others may be glass-like (vitreous) or silky, like talc. Others are pearly, with a whitish iridescence. Yet others have the brilliant lustre of diamonds (known as adamantine). Lustre too therefore may be diagnostically useful.

For some minerals, the way in which they fracture may be informative, e.g. Figures 1.23 and 1.24.

The final useful property of minerals is their hardness, expressed on a scale known as Mohs scale of hardness. Diamond is hardest, and has a scale value of 10, while talc is softest with a value of 1. Moving upscale from talc come gypsum (2), calcite (3), fluorite (4), apatite (5), orthoclase (6), quartz (7), topaz (8) and corundum (9). Sets of small tools are available fitted at one end with chips of diamond, corundum, topaz, quartz, etc. If a mineral can be scratched by quartz, but not by orthoclase, for example, it has a scale value of 6–7. Sheet silicates are softer (1–3) than framework silicates (5–7).

1.6 How fast do rocks weather?

How fast rocks weather is an issue of paramount importance to landscape evolution, maintaining natural soil fertility, and the quality of surface waters and groundwaters, yet it is something that is difficult to quantify reliably for several reasons. Rock weathering is a result of physical, chemical and biological processes, all of which may be interactive. Most rocks are, as we have seen, matrices containing interlocked crystals of a diverse range of minerals (the exceptions being simpler rocks such as quartzite and limestone). Weathering is also slow, and rates will change over time in the long term, making the conclusions questionable when the results of simple laboratory experiments are interpolated to predict behaviour under field conditions that are often very variable.

Opportunist scientists have sometimes exploited information in cemeteries on dated gravestones, at least semi-quantitatively. For example, the weathering-resistant granite gravestone in Figure 1.55 in a cemetery in York has hardly weathered at all since 1888; however, an adjacent gravestone from only two years earlier made from sandstone is much harder to read because of weathering (Figure 1.56).

Even for such basic studies caution is needed however. For example, the stone in Figure 1.55 could have been replaced in 1891, when Elizabeth's husband William (a survivor of the famous Charge of the Light Brigade) was eventually laid to rest at the same spot. So one has to be sure that the additional inscription was added in 1891, rather than a new tombstone. Ironically, the low weathering rate makes it difficult now to identify addenda

Figure 1.55 A gravestone apparently from 1888 made from granite, showing little sign of any weathering.

Figure 1.56 A gravestone more or less contemporary to that in the previous figure, showing a much greater degree of weathering.

to inscriptions. However, gravestones have been used with a reasonable degree of success to compare weathering under contrasting climatic conditions.

For minerals at least we can place them into a weathering sequence with respect to chemical weathering. As we have seen, resistance to dissolution depends upon the stability of the mineral crystal lattice, or the amount by which the lattice energy exceeds the energy that is required to produce individual cationic and anionic species. Thus, for example, the macromolecular matrix of quartz, SiO_2, makes it highly resistant to chemical weathering. So does the silicate sheet/gibbsite sheet/silicate sheet layer structure of muscovite when the layers are permanently held together by potassium (K^+) cations balancing isomorphous substitution of aluminium in the silicate layers. Framework feldspar silicates such as orthoclase, albite and anorthite weather slightly more readily, and other layer silicates such as biotite, magnesium chlorite, glauconite, antigorite and nontronite more readily still. Orthosilicates and chain silicates come next in order of weatherability, followed by calcite, dolomite and apatite, and finally simple salts such as halite and gypsum head the list. These ideas came from Jackson and Sherman back in 1953, and were extended by Sposito in 1989. Their full lists also include a wide range of secondary minerals commonly found in soils often as stable end members of weathering sequences, but we will consider mineral weathering again in Chapter 7 (section 7.4.5) and when we consider buffering against acid deposition in Chapter 16.

A more quantitative approach to quantifying weathering rates is to measure input–output balances for elements of interest for water-tight drainage basins or catchments. Suppose, for example, we measure all the calcium (i.e. the annual amount or flux, and not just the concentration) deposited on a catchment from the atmosphere, and all the calcium exiting the catchment in a river draining the catchment. The amount by which the output exceeds the input is a measure of the rate of release of calcium from the minerals in the soils and rocks within the catchment. How such measurements are made in practice is discussed in Chapter 17.

The timescales for depletion of key nutrient base cations are sometimes surprisingly short in the context of long-term sustainability and geological timescales. For example, Reid and colleagues (1981) showed that the Glen Dye catchment in N.E. Scotland, with soils derived from weathered granite, is losing calcium at a rate of about 17 kg per hectare per year (17 kg ha^{-1} yr^{-1}). This does not sound like a massive amount. However, it was later shown that it takes 5733 kg of the soil's sand and silt to lose all its calcium per year to provide this output of calcium each year (Creasey *et al.*, 1986). If we assume that the density of the sand and silt is about 1.6 g per cubic cm (1.6 g cm^{-3}), and we know a hectare is 10^8 cm^2, then a 1 cm depth of soil over a hectare weighs 1.6 times 10^8 g, or 1.6×10^5 kg. Dividing 1.6×10^5 by 5733 gives about 28, and therefore it would take 28 years to strip all the calcium from 1 cm depth of soil over one hectare, or 280 years to strip the calcium from 10 cm of soil.

CASE STUDIES

A case study at York – Release of base cations from weathering biotite

Laura Suddaby and Malcolm Cresser in the Environment Department at the University of York studied the effects of adding the mineral biotite, either cut to pieces about 1 mm by 1 mm or finely ground in a ball mill, to an acidic soil. The mineral and soil samples were thoroughly mixed and added to covered leaching tubes. Each tube was watered once every three days with artificial rain, and the drainage water was collected and analysed for base cations. Glucose at one of two concentrations was also added to some of the rain to see if stimulating microbial activity increased production of carbonic acid, H_2CO_3, and hence increased weathering rate. From the known structure of the 2:1 mineral, which was discussed in section 1.3.2, it was hypothesised (1) that potassium, K^+, and magnesium, Mg^{2+}, concentrations would show the biggest increases; (2) that the increases would be greater from the finely ground biotite because of the increase in mineral particle surface area exposed to weathering; (3) that addition of the glucose organic substrate would stimulate weathering.

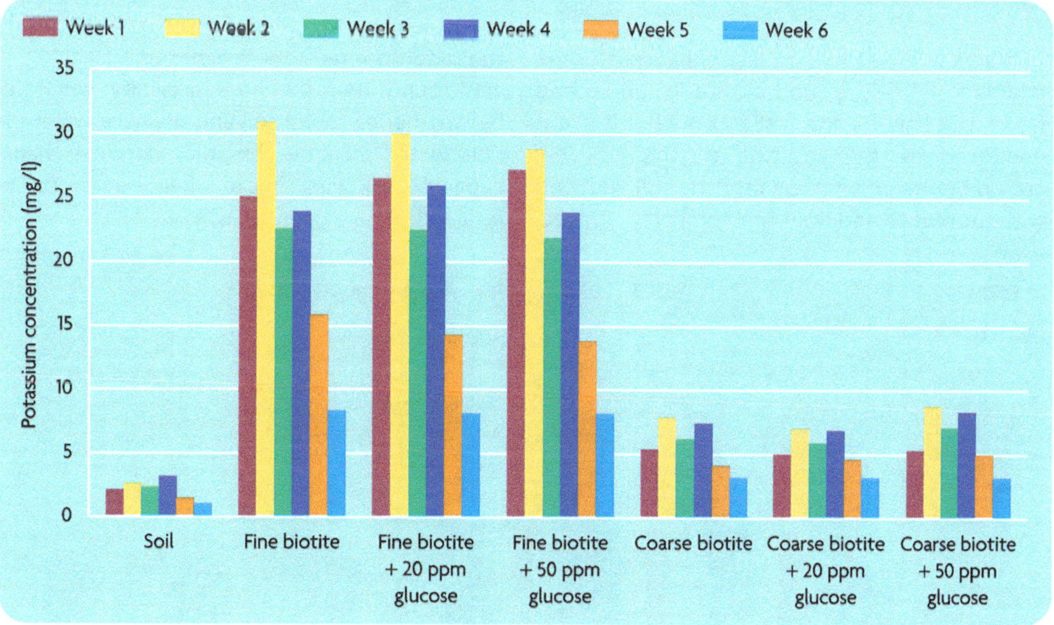

The figure above for potassium concentration confirms that potassium is released by weathering of biotite (hypothesis 1) and that the release is greater when the mineral is finely ground (hypothesis 2). However, the effect is greatest over the first four weeks of the experiment. The effects of glucose substrate addition, however, are not consistent.

➤

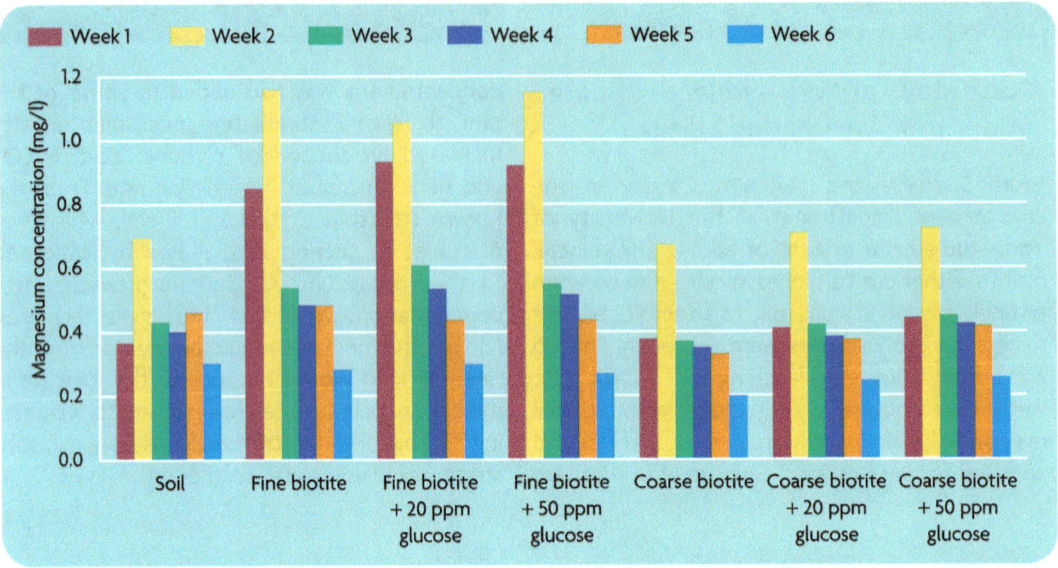

Magnesium leaching is also enhanced by the addition of finely ground biotite in the second figure, but only for the first four weeks. It is not enhanced by the addition of the coarsely cut up biotite when results for the soil with and without biotite additions are compared. Indeed, the addition reduces magnesium leaching. It was found that the base alkalinity generated by weathering raised soil and drainage water pH. This would increase the soil's cation exchange capacity, allowing more magnesium to be retained.

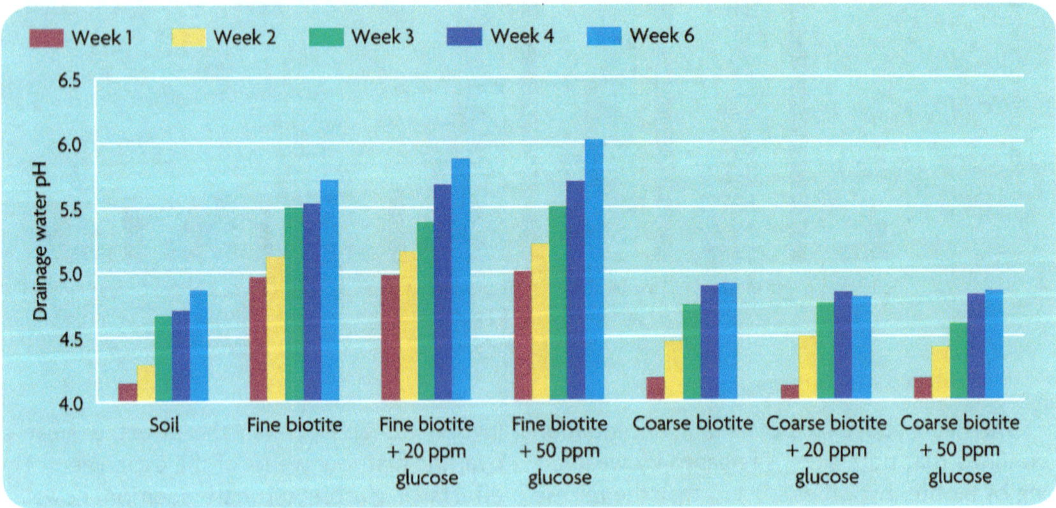

An interesting unanswered question remains however: what is the arrangement of atoms at the broken edges of biotite crystals, and what does this arrangement mean for the weathering rate? We do not know the answer to this question yet.

1.7 How do we know what led to current rock type distribution at the Earth's surface?

In section 1.10 we will see how the presence of a particular fossil species only on two separate continental land masses suggested that, at the time the species were alive, the two land masses were joined. This, in fact, is just one example of how matching fossil communities may be used across a landscape to factor out how the landscape evolved over geological time.

Where sedimentary rocks are stratified and close to horizontal, it generally may be assumed that the youngest rocks are those closest to the surface. Tectonic activity may cause localised vertical displacement of strata, and subsequent erosion may remove one or more strata at some outcropping rock sites. It will become clearer later in the chapter from Figure 1.62 how this could happen.

Matching of fossil types (Figure 1.57) and communities in strata of spatially separated rock outcrops, especially if matches occur over more than one stratum, often allows elucidation of the sequences of geological sedimentation and erosion processes that have occurred in the geological past. Coupled with observed similarities in stratification between sites, this often allows evolutionary sequences to be worked out at the landscape scale, even where some strata are missing.

The problem with the above approach is that while it may be invaluable for assessing relative age of strata, it tells us nothing about absolute age. For that we need to use appropriate rock dating techniques that are capable of indicating rock ages lying between several thousands and many millions of years.

Figure 1.57 Fossilised trilobites in sedimentary rock. Many visible features are well preserved, facilitating their use in matching communities in spatially separated rock outcrops.

1.8 How can rocks be dated?

A number of elements found in minerals are naturally radioactive, transforming as a consequence of radioactive decay to different elements. For example rubidium-87 decays by converting a neutral neutron in the rubidium atom nucleus to a positively charged proton and an electron, thereby converting the rubidium-87 atom to a strontium-87 atom. Note that in this instance the parent element (^{87}Rb) and daughter element (^{87}Sr) have the same atomic mass. The conversion is a very rare event, and it is known to take 47 billion years for half of the rubidium to be converted by this process. Figure 1.58 is a simplified representation of the decay process. For the above conversion, the five sketches would represent the initial state, and the states after 47 billion, 94 billion, 141 billion and 188 billion years respectively.

If we know the half life, and the amounts of parent and daughter element, and assume for igneous rock minerals that only the parent element was present initially, it is possible to estimate how long ago the mineral crystallised for the first time, and hence the age of the rock formation. Rubidium-87 occurs naturally in biotite, muscovite and potassium feldspars, so this

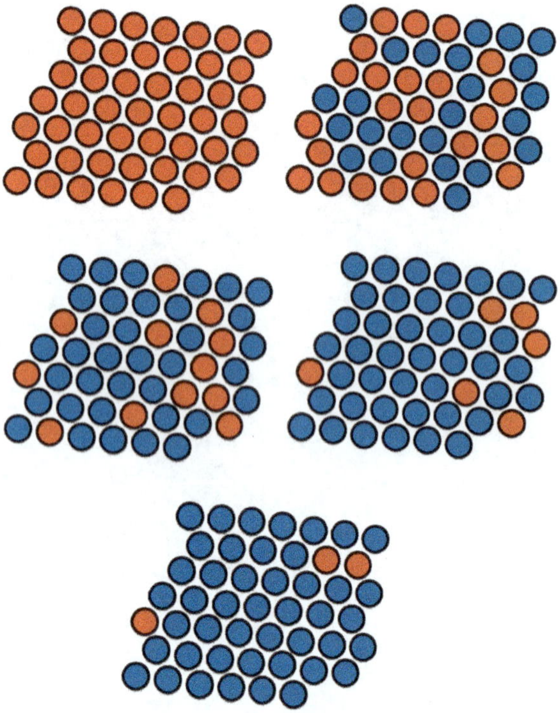

Figure 1.58 Simple schematic representation of radioactive decay, as used for mineral dating. At top right, crystal forms containing atoms (red) of a parent element, which decays to atoms of a daughter element (blue). The sequence shows how many red atoms would be left, and daughter atoms formed, at the half life, twice the half life, three times the half life, four times the half life, etc. In reality, of course, the numbers of parent and daughter element atoms would be many orders of magnitude larger than the simplified figure suggests.

particular transformation is very useful for dating rocks between 10 million and 4.6 billion years old (Grotzinger *et al.*, 2007).

Most of the radioactive decay processes used for dating involve more than a single transition, but this is not a problem as long as the half life for the series is known. Decay of uranium (^{238}U or ^{235}U) in apatite or zircon is often used, the former to ^{206}Pb and the latter to ^{207}Pb, as is the decay of potassium-40 in micas or horneblende to Argon 40.

The above group of transitions is useful for dating rocks over the range 50 thousand years to 4.7 billion years. Radiocarbon dating, using the transition of ^{14}C to ^{14}N, is useful over a more recent timescale, from 100 to 70 000 years b.p.; thus it may be used for dating wood, charcoal, bone, shell, etc., or even glacier ice provided it contains trapped carbon dioxide. The formation of carbon dioxide effectively resets the clock, so trapped organic material could not be used in this way.

1.9 Rock movements in the crust and their significance

1.9.1 Folds and faults

Crustal rocks were, and still are, continuously being exposed to extreme heat and pressure because of the processes occurring at depth in or below the crust and mantle. This can result in folding of rocks (Figure 1.59) as well as tilting of previously horizontal strata. The concept of stratification and the idea that all layers of sediment would initially have been horizontal are important starting points when attempting to interpret visible geological features. It is generally assumed that upper strata are the younger rocks.

Sometimes a vertical or angular fracture occurs, known as a fault, with rock strata often being displaced relatively by considerable vertical

Figure 1.59 A simple example of folding of sedimentary rock strata on a coastal outcrop in Victoria, Australia. The ocean has undercut the rock outcrop, forming a clearly visible bridge structure.

distances. Two possible fault types are illustrated in Figure 1.60. If the crust is being stretched (pulled apart), strata on one side of the fault line tend to drop relative to those on the other side. This is known as a normal dip-slip fault. If however the crust is being compressed from two opposite directions, strata on one side of the fault lift above those on the other side and the fault is known as a reverse dip-slip fault. A ridge is formed across the landscape.

Where strata overlay each other at obviously different angles, the system is described as an angular unconformity. Figure 1.61, kindly provided by Gabriel Gutierrez-Alonso of the Universidad

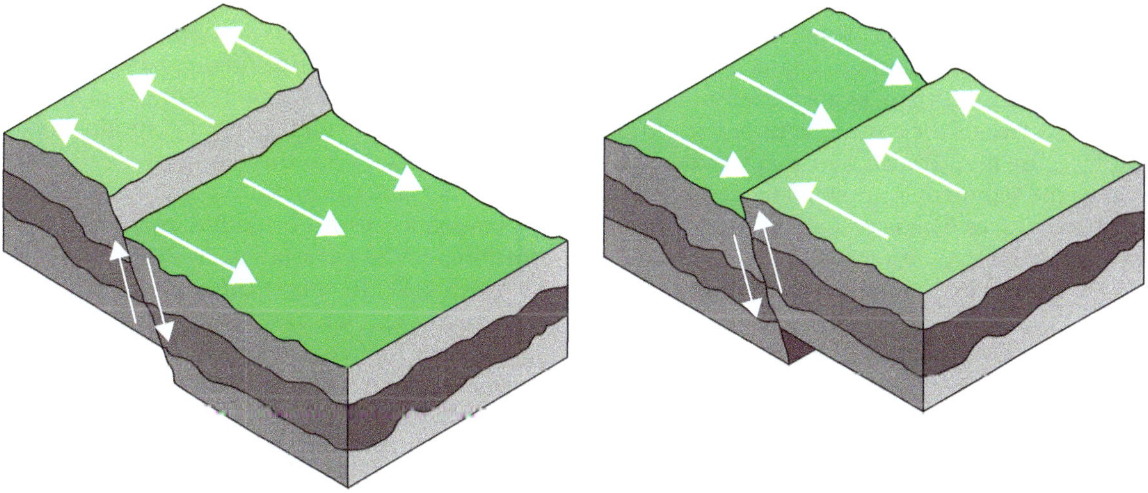

Figure 1.60 Simple schematic representation of faults involving vertical displacement. In the reverse fault on the right, the block of rock on the right-hand side is pushed upwards under the massive compressive forces.

Figure 1.61 Photograph of folding and angular unconformities at Telheiro Beach, Portugal.

Source: Copyright of Gabriel Gutierrez-Alonso.

de Salamanca in Spain, shows angular unconformities formed after 320 million year old Carboniferous rocks were folded during the Varscan-Alleghenian orogeny, eroded and buried over the next million years. The site of this outcrop is in south-western Portugal. Sedimentary rock from the crest of the original folds has been eroded away, and fresh layers of sediment (closer to the horizontal) deposited later. This concept is perhaps easier to follow however in the simplified schematic representation in Figure 1.62.

In Figure 1.62(a), rocks are folded and uplifted at the surface of the crust. The upper, protruding surfaces are then eroded over many thousands of years, and sediment is horizontally deposited over the new surface. This is eventually consolidated to form new rock.

In Figure 1.63 it appears that the almost horizontal strata of light brown, coarse-grained sandstone have been uplifted, sliding across the lower layers of grey, finer grained sandstone at an angle of about 45°. It might be thought that the lower, grey-coloured strata were already tilted before the fault occurred by previous uplift in the region, and the enormous pressure at the interface between the two rocks when the fault was occurring slightly masks the clarity of the boundary. However, at this particular site the red sandstone strata were simply deposited after folding had tilted the older, grey sandstone. This example was included to show that great care is needed when interpreting visible features of sedimentary rocks!

The faults we have considered so far have involved primarily vertical movement of massive blocks of rock in the Earth's crust. Lateral strike slip faults also occur, in which masses of rock are displaced sideways relative to each other. The best known major fault in the world is probably the San Andreas Fault in southern California. The USGS Western Earth Surface Processes Team has an excellent website discussing the major fault lines in California, with the following URL: http://geomaps.wr.usgs.gov/socal/geology/inland_empire/socal_faults.html. The site also provides a useful introduction to plate tectonics.

Whether faults are strike-slip, normal or reverse, the consequences for buildings and urban infrastructure may be devastating, as shown in Figures 1.64 and 1.65. In the Michoacan Earthquake in Mexico, more than 9500 lost their lives and 30,000 were injured. Eventually 100,000 were left homeless. The Loma Prieta Earthquake (Figure 1.65) was a consequence of a crustal slip of as much as 2 m at a depth of 18 km along a distance of 35 km. However, the fault did not break the land surface.

In more rural areas the impacts may appear less traumatic, even when the rupture occurs over long distances. Figure 1.66 shows surface damage in a field following the 1999 Izmet Earthquake in Turkey in 1999. On this occasion the rupture was generally 2.5 to 3 m, along a distance of about 110 km.

(a) (b)

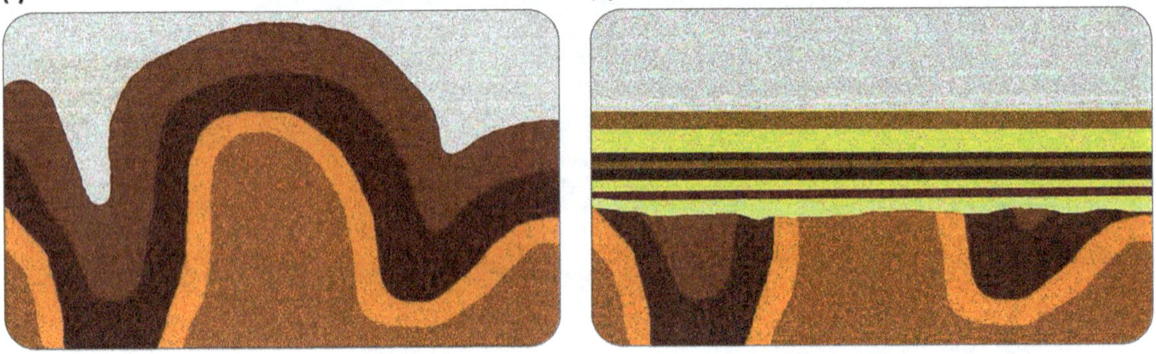

Figure 1.62 Simplified schematic representation of the formation of angular unconformities. For simplicity effects of heating and crushed rock are excluded from the left-hand image (a).

Figure 1.63 An angular unconformity. The browner sandstone strata are close to the horizontal, and overlay much more steeply inclined strata of grey sandstone. The brown material was deposited at this site after the grey material had been uplifted and tilted.

Figure 1.64 Wreckage of a 21-storey steel-constructed building after the Michoacan Earthquake in Mexico City in 1985.
Source: Photograph by Mehmet Celebi/USGS.

Figure 1.65 Collapse of support columns and upper deck of the Cypress Viaduct of Interstate Highway 880, in the USA, following the Loma Prieta Earthquake in San Francisco, October 1989.
Source: Photograph by H.G. Wiltshire, US Geological Survey.

Figure 1.66 Damage in a field as a consequence of a fault line following the 1999 Izmit Earthquake in Turkey.
Source: Photograph by Tom Furnal, US Geological Survey.

Figure 1.67 An example of mountains formed as a consequence of plates colliding at the time of the Variscan Orogeny, 310 to 390 million years ago.
Source: Photograph copyright of Gabriel Gutierrez-Alonso.

1.10 The nature and importance of plate tectonics

At any particular moment of time, standing looking at a landscape it is not often immediately obvious that land masses move around the surface of our planet on a massive scale and have done so for thousands of millions of years. But if you look carefully at Figure 1.67, in the bottom left-hand corner you can see a road and road tunnel beside the lake. Once such a scale is established, the picture then provides a very clear indication of the massive forces that would have been involved in folding the rock strata and causing the mountain ranges to form. Our massive continental land masses are still slowly moving around at the surface of the planet. When continents collide, one must slide under or over the other, causing the formation of mountain ridges. Just such a collision formed the mountains in Figure 1.67. At the same time rocks that were once close to the surface start to become buried at great depth. At other parts of the Earth's surface, continental land masses or plates are drifting apart. The study of the movement of plates is known as plate tectonics.

The crust and cooler upper part of the mantle, together known as the lithosphere, are not fluid in nature, so the lithosphere bends, flexes or breaks when subjected to great force. It is immediately underlain by the asthenosphere, which can flow, but only at very slow rates. In effect the lithosphere, which varies greatly in thickness between continental land masses and the oceans, is floating over the asthenosphere. Thus South America and Africa are drifting apart at about 1.7 cm each year (cm y^{-1}) (Marshak, 2008).

In the past, over a geological timescale, the relative positions of continents were very different from what they are today. It was the observation that current continental shapes fit together so well, almost like the bits of a giant, three-dimensional jigsaw puzzle, that provided one of the first clues that continental land masses are mobile. Figure 1.68 provides an indication of what we now think our planet's surface looked like around 225 and

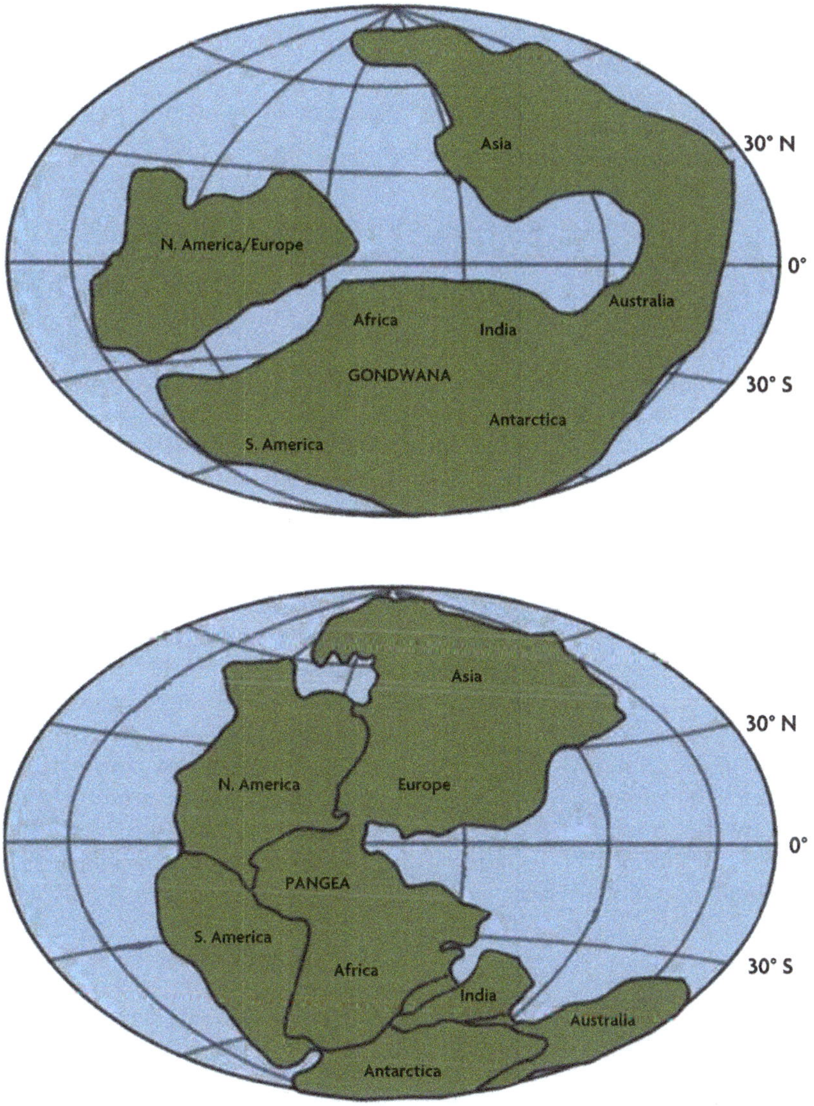

Figure 1.68 Sketch showing the origins of the current continental land masses before continental plates were forced apart. The upper image shows the estimated locations of current continental land masses in the Carboniferous (circa 340 million years ago) relative to the supercontinent Gondwana; the lower image indicates where the supercontinent Pangea started to split up in the Permian/Triassic 225 million years ago to form current continents. A more complete chronosequence may be seen in Grotzinger *et al.* (2007).

340 million years ago. Note that some continents that were once joined are now isolated land masses, Australia being a good example.

It should be stressed that it is not only the apparently snug fit of interlocking continental shapes that supports the concepts of plate tectonics. If the spatial distribution of ancient rock types is superimposed upon the continental outline maps, then they too fit remarkably well when the land masses are rejoined. Moreover, Mesosaurus fossils from 3 million years ago have been recorded in South America and Africa, but nowhere else on the planet, suggesting that at that time those two land masses were connected.

If continental land masses are slowly moving apart, it follows that, in the oceans, new ocean floor must slowly be being created all the time. By the 1970s, plate tectonics had been more or less universally accepted by informed scientists. The concept of seafloor spreading as a consequence of material from the intensely hot mantle drifting up by convection through the asthenosphere to create zones of intense volcanic and earthquake activity became well recognised following thorough investigations of the nature of the Mid-Atlantic Ridge. These processes are much more fully described in recent excellent geology texts such as those by Grotzinger *et al.* (2007) or Marshak (2008).

1.11 Volcanoes

Volcanoes are perhaps the best, and among the most unwelcome, reminders that the world beneath our feet is both complex and dynamic. Volcanic eruptions can reshape our lives, our world and even our entire existence. We cannot avoid them and we cannot prevent them; volcanoes are distributed across the world and as our population expands, many of our cities and towns now lie in the shadow of an active volcano. Even where the volcano is only a distant peak, you may not need to dig far into the soil to find signs of geologically recent volcanic activity in the area.

Since we cannot prevent volcanic eruptions, we have to learn to live with them. By studying past and present volcanoes, and through modelling and experimental research, volcanologists seek to determine how local, regional and global communities could survive, mitigate and where possible, remediate the impacts of volcanic eruptions.

1.11.1 How does a volcano start?

Contrary to the mental picture we may have acquired from our childhood, a volcano does not start life as a conical flat-topped mountain (Figure 1.69). A volcano begins with a patch of bare ground or ocean floor and a buoyant plume of molten rock somewhere deep beneath the Earth's surface. The magma pushes its way through the crust and mantle, providing its density is less than that of its surroundings. When its density is equal to that of the rocks around it, or where the overlying rock impedes its ascent, a reservoir or *magma chamber* may form.

Magma will continue its ascent towards the surface from the magma chamber if the magma buoyancy is altered, perhaps by a new injection of fresh magma, or if there is a decrease in the overlying pressures constraining it. Renewed ascent may carry molten rock towards the surface until it has sufficient energy to break violently through any overlying rock remaining and to breach the surface. Where once there was an empty patch of ground or ocean floor, an opening or volcanic vent now emits molten rock and volcanic gases. A volcano is born!

Figure 1.69 The classic view of a volcano, the summit of Mt Etna, Sicily. Note the steam gently rising from the collapsed crater wall on the right.
Source: Image courtesy of K.J. Ayris.

1.11.2 The constituents of an eruption

If you asked someone to describe a volcanic eruption you would probably get an amalgam of various phenomena from half-remembered documentaries and the half-truths of Hollywood. In fact, volcanoes have many forms, ranging from the classic flat-topped conical mountain of our childhood to gently sloping mountains, and even to simple fissures in the ground. Each volcano features a specific combination of eruptive phenomena and environmental effects, but all eruptions feature the same basic constituents.

All volcanic eruptions begin, however briefly, by ejecting mixtures of gas and fragments of molten rock, collectively referred to as *tephra*. The larger fragments, which can be metres across, are called *bombs*, while those less than 2 mm across are called *ash* (Figure 1.70). After the gas and tephra producing phase, it is then possible for molten rock to be erupted directly as a continuous stream of lava from the vent.

There remains considerable variation in each of the above eruptive phases. During the gas and tephra producing phase, some volcanoes may produce fountains of molten rock and gas which can reach heights of 10s or even 100s of metres above the vent (Figure 1.71). These cascades produce droplets and strands of fluid magma which are quenched to form glassy, aerodynamic tephra. Other volcanoes erupt more dramatically, producing an explosion sending a jet of sharp, angular molten rock fragments into the atmosphere, entrained within a turbulent convecting plume of volcanic gas. The plume cools, becomes diluted and the heavier rock fragments are deposited from it. The ascending plume can reach heights of up to 50 km before it becomes neutrally buoyant. The plume then spreads laterally, forming a volcanic cloud which may circle the Earth many times before deposition.

Figure 1.70 Ash-bearing eruption plume at Redoubt volcano in Alaska on 31 March 2009.

Source: Image by C. Waythomas, courtesy of the USGS/Alaska Volcano Observatory.

Figure 1.71 Lava fountain from the 1983 eruption of Kilauea volcano, rising approximately 10–15 metres into the air before crashing back down.

Source: Image by J.D. Griggs, courtesy of the USGS Hawaiian Volcano Observatory.

Lava produced during a volcanic eruption can also vary dramatically in its behaviour. Some lavas are highly fluid and can cascade down the volcano flanks, or spread smoothly outwards along flatter ground. Their surface cools to form solid crusts which restrict the flow, and these crusts can be blocky masses of rubble and debris (*'aa' flows*, Figure 1.72) or smooth skins with ripples and folds, forming all manner of bulbous shapes (*'pahoehoe' flows*, Figure 1.73). A more viscous lava is more rapidly constrained by its crust, as it tends to be cooler and does not flow as easily. Viscous lavas can form bizarre edifices depending on the geographic barriers that stand in their way.

1.11.3 How does an eruption work?

The different eruptive styles at different volcanoes result from the presence of dissolved volatile gases within the molten rock and the latter's variable

Figure 1.72 The rubble flow of an 'aa' lava from the 29 January 2008 eruption at Kilauea volcano.
Source: Image courtesy of the USGS Hawaiian Volcano Observatory.

Figure 1.73 A crack in a fresh, rippled crust of a 'pahoehoe' from 22 April 2010 eruption of Kilauea volcano. The crack reveals the high temperature lava still present beneath the crust.
Source: Image courtesy of the USGS Hawaiian Volcano Observatory.

fluidity (*viscosity*). In the high pressure and high temperature world beneath our feet, H_2O, CO_2, SO_2, H_2S and various other volatiles are all dissolved within the magma. The concentrations of these volatiles in solution are limited by their solubilities. Excess volatiles therefore nucleate to form gas bubbles. Gas solubility varies with pressure and temperature, decreasing at higher temperature and lower pressure. You can observe the latter when you open a carbonated drink for the first time. Initially, the liquid in the pressurised bottle shows no sign of the dissolved CO_2 within it. As soon as you open the bottle, the pressure is released, CO_2 solubility drops and bubbles rapidly form within the liquid. This process occurring within the magma is fundamental to volcanic activity.

In both fluid and viscous magmas, bubble formation lowers the overall density, and renders the magma more buoyant within the crust. As the magma rises through the crust, gas bubbles expand as the pressure decreases (an application of Boyles' Law) and the solubility of more volatiles is lowered, leading to further nucleation of additional volatiles.

Within fluid magmas, gas bubbles may percolate through the molten rock and collect at its surface as a low density foam. This foam undergoes rapid acceleration up the volcanic conduit, and may be reduced to a mixture of molten rock fragments and volatile gases within the vent. Disruption of the foam may be due to shearing of gas bubble walls by the rapid acceleration, or due to gas bubble bursting when the internal of the bubble pressure exceeds the ever-decreasing overlying pressure.

More viscous magma constrains gas bubbles and reduces their expansion during magma ascent, preventing them from escaping. This is the reason why the pumice stone in your bathroom floats; it was initially a viscous magma filled with gas bubbles. In the magma chamber and conduit, gas bubbles attempt to expand as the magma ascends. The viscous magma inhibits the expansion, and the pressure within the bubbles increases until the bubble explodes. The energy required to disrupt the overlying viscous magma is much greater than that in a less viscous magma, as anyone trying to blow bubbles in a thick milkshake and in a glass of water can tell you. When the pressure does eventually reach the critical level, a high energy explosive fragmentation occurs, shattering the viscous bubble walls into angular, broken fragments which are then entrained in the released gas. Explosive eruptions driven by this mechanism can have tremendous energy. The eruption of Krakatau in 1883, for example, had an explosive yield of 150–200 million tonnes (megatonnes) of TNT, three times larger than the most powerful thermonuclear device ever tested.

When both viscous and fluid magmas are reduced to mixtures of gas and molten rock in the conduit, their rapid acceleration upwards leaves a void behind them. Molten rock rises to fill this void, and when still charged with volatiles, this can result in further fragmentation in a propagating 'chain reaction'. When the volatile content is too low to maintain the fragmentation, the magma may continue to rise up the now open volcanic conduit. At the surface, the degassed magma may then be erupted as flowing lava rather than as the explosive or effusive phase. This may continue until the volume of magma is no longer sufficient to hold the conduit open, which then collapses, effectively halting the eruption. In the case of very large eruptions, the emptying of the magma chamber can lead to its collapse and the formation of a deep basin, or *'caldera'* (Figure 1.74).

Figure 1.74 'Wizard Island', the remnants of a lava dome within the flooded caldera lake known as Crater Lake. The caldera was formed by the eruption of Mt Mazama, Oregon approximately 7700 years ago. Source: Photo from S. McKeown/ USGS.

1.11.4 Where do different eruptions occur?

The silicic nature of magma is the primary control on gas solubility and magma viscosity, and so is a primary control on the eruption itself. Magma viscosity is dictated primarily by the strong silicon–oxygen bond. To break the silicon–oxygen bond requires very high energy. To begin to break down the silicon–oxygen bonds in macromolecular lattice of quartz, for example, needs temperatures in excess of 2000 °C at atmospheric pressure. Magma, as a multi-component aluminosilicate, melts at much lower temperatures. Basaltic magma, for example, is fluid at around 1200 °C, which is due to the presence of additional major elements and dissolved volatiles. Some of these major elements (i.e. Na, Ca, K, Mg) and volatiles (i.e. H_2O) can disrupt the Si-O-Si bond. Reducing the number of Si-O-Si bonds therefore reduces the energy required to initiate melting within the rock. In addition, the presence of non-Si cations enables dissolved volatiles to be better incorporated into the magma, increasing their solubility. In a silicon-rich magma, volatile solubility will be therefore be low, and gas bubble formation will be extensive, but the magma viscosity will constrain these bubbles. The potential for more Si-rich magmas to build to an explosive eruption to a greater degree than lower Si magma should therefore be apparent.

It can therefore be concluded that the reason why different volcanoes and eruptive styles occur in different regions of the world comes down to differences in the magma that supplies them.

At divergent plate margins (Figure 1.75, A), where ocean or continental crust is pulled apart, shallow mantle rock rises to fill the gap between the plates. You may recall from the discussion in section 1.2 that mantle rock tends to be lower in Si content than crustal rock, having a more basaltic

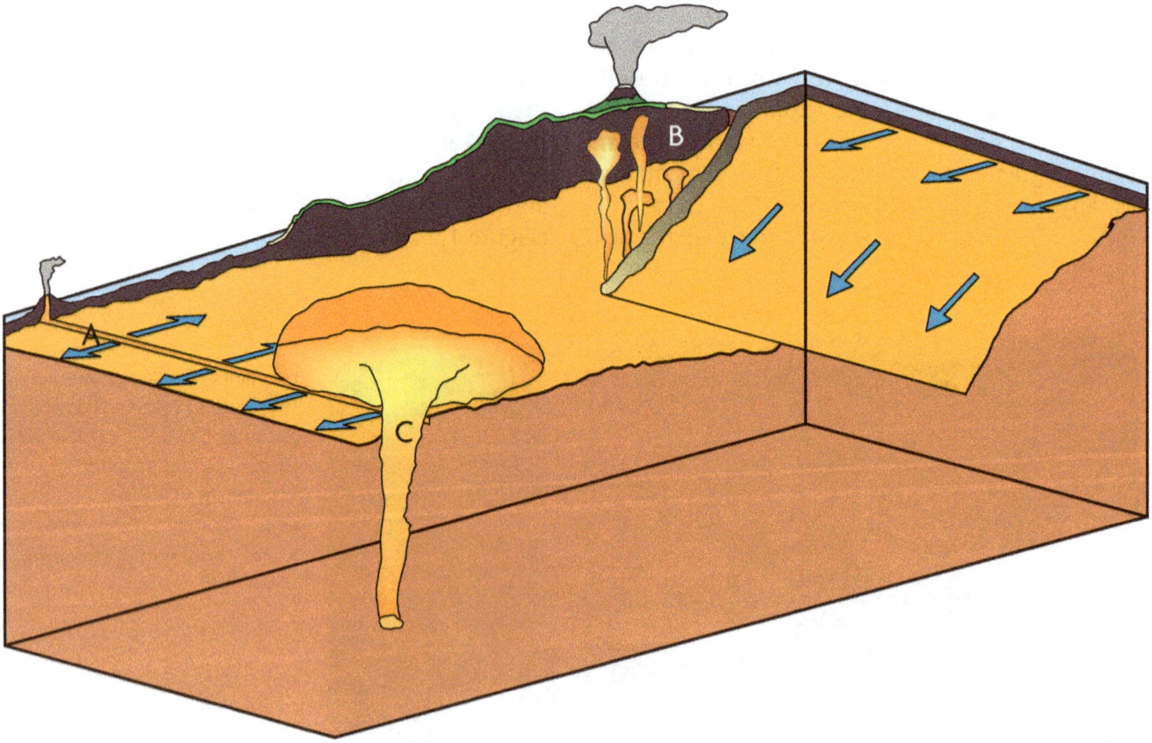

Figure 1.75 Cartoon showing the different volcanic eruptions and their origins, albeit not to scale. At point A, the divergent plate results in upwelling magma. At point B, the subducting plate induces mantle melting, and point C, a deep mantle plume rises ominously to the surface.

composition. Eruptions at divergent margins may therefore tend to be more effusive, producing 'fire-fountains' and erupting more fluid lava.

At convergent plate margins (Figure 1.75, B), where a water-rich oceanic plate subducts beneath a continental plate, the former slowly melts. The water within the melting rocks is added to the overlying mantle, depolymerising it and triggering melting. Mantle rock may then rise through the thick continental crust, assimilating the more Si rich rock and becoming more viscous. Accordingly, volcanoes along subduction zones tend to be more explosive.

It is also possible for volcanoes to form in the middle of plates, far from the spreading centres and subduction zones. This volcanic activity is triggered by infrequent and thankfully rare upwellings of hot magma from the lower mantle (Figure 1.75, C). Huge plume heads, thousands of km wide, rise towards the surface, too large to be affected by the overlying tectonic setting. Upon their eruption, these plumes can pour low Si magma over wide areas for millions of years. Even when the plume head is spent, the remnant tail may feed volcanic activity in the plates tracking over the hotspot for millennia to follow.

1.11.5 What are the impacts of a volcanic eruption?

The most obvious impacts of volcanic activity upon our environment and upon our existence are those which physically intrude upon them. These include lava flows, ashfall deposits and pyroclastic flows, but can also extend to water and mudflows produced by the heat of the eruption. In addition to these, the emissions of volcanic gases from the volcano may alter the chemistry of the atmosphere, resulting in potential climate impacts.

Lava

Lava is perhaps the most obvious and well-recognised physical impact of volcanic activity. The most spectacular impacts are associated with lava flows from low-Si magmas. Eruptions fed by high-Si, more viscous, magma sources are more explosive, and any lava produced flows slowly and cools quickly. On the geologic timescale, the impacts of flowing lava can be extensive. Mantle plume volcanism, for example, poured fluid lava across entire continents over millions of years. The impacts of such volcanism on the world may be profound; in the past, global mass extinction events frequently coincide with mantle plume volcanism, though a formal link has yet to be unequivocally demonstrated.

On the human timescale, lava flows burn and bury forests, arable land, roads and buildings (Figures 1.76 and 1.77). In Hawaii, this happens so frequently that recovery efforts in the affected

Figure 1.76 Lava flows covering the roads beneath Kilauea volcano on 5 May 2010. Note the flames licking around the edge of the flow as the asphalt burns.

Source: Image courtesy of the USGS Hawaiian Volcano Observatory.

Figure 1.77 The plaster casts of the bodies buried at Pompeii by ashfall after the AD 79 eruption of Mt Vesuvius.

Source: Image courtesy of K.J. Ayris.

regions are no longer cost effective. In general, lava flows are a threat to physical environments but not normally to human life. Flowing lava is avoidable and evacuation is usually possible in advance. This is not always the case. At Nyiragongo volcano in Zaire in 1977, a rapidly drained lava lake sent fluid lava raging down the mountainside at speeds of up to 100 km h^{-1}. Between 60–300 people died when their villages were engulfed.

Ashfall

As the eruption plume ascends into the atmosphere, the heaviest ash particles rain from it in a continuing cascade of rock fragments. The remaining particle mass is carried over tens, hundreds and even thousands of km before being deposited. The impacts associated with volcanic ashfall are arguably the most far-reaching of any of the manifested physical impacts. The range and severity of ashfall depends upon the size and explosivity of the eruption. It can therefore be instantly recognised that the less explosive, more effusive eruptions will generally produce smaller and more localised ashfall in comparison to larger, explosive eruptions. In the case of the largest eruptions, the so-called 'supervolcanoes', ashfall deposits have been recorded at distances of thousands of km away. The ash deposits of Yellowstone Caldera, erupting approximately two million years ago, can be found throughout almost the entire continental United States.

Ash deposits may range from mm to several metres in thickness. The latter are confined to near-vent regions by rapid sedimentation from the plume. Within the near-vent regions, ash deposits which are dozens of cm or even metres thick can be devastating. Anything natural or manmade which is not buried may be overloaded to the point of collapse by the mass of ash deposited. The geography of entire regions may be altered as ashfall reshapes hills and valleys, chokes rivers and fills lakes. Any living creature in the region which cannot escape may be smothered, as the ruins of Pompeii after the AD 79 Vesuvius eruption remind us (Figure 1.77).

Where the deposits are thin, the devastation decreases, but severe effects can still result. It is well known that thin crusts of ash can impede the exchange of heat, light and water between both soil and vegetation surfaces and the atmosphere. On a global scale, ash within the atmosphere is known to scatter solar radiation, resulting in climatic cooling effects. It has also recently been hypothesised that large ash blankets may also induce localised cooling and perturb the global climate.

The effects of ashfall persist as long as it remains uneroded by wind or water. Erosion may cause delayed secondary impacts after the eruption, extending even into areas which had escaped damage during the event. Glass shards being blown or washed along the ground can shred new plant growth and erode soil, while airborne ash damages canopy vegetation and poses a respiratory hazard to both animal and human life. High volumes of ash suspended in river water can erode the channel downstream, potentially propagating changes down the river system and influencing the entire ecosystem as a result.

Pyroclastic flows

A further hazard associated with volcanic eruption plumes, and hence with explosive volcanism, is the potential for pyroclastic flows. These are mixtures of hot lava, ash and gas in the plume which become too heavy to remain aloft. The plume collapses under gravity down the flanks of the volcano (Figure 1.78). In comparison to ashfall and lava, pyroclastic flows are generally the most dangerous.

Figure 1.78 Pyroclastic flow surging down the flanks of the Mayon volcano in the Philippines 15 September 1984. Note the multiple flows travelling down different areas of the mountainside.

Source: Image by C. Newhall, courtesy of the USGS.

Ashfall builds slowly over the course of the eruption and lava, in most cases, can be avoided. Pyroclastic flows, on the other hand, travel very fast (up to 300 km h^{-1}) and can travel for dozens of km, even over water. As a fast moving suspension of rock and gas, pyroclastic flows have tremendous kinetic energy and can demolish both trees and manmade structures alike. Once that kinetic energy is expended, a thick deposit, which can be metres, 10s of metres or even 100s of metres thick, is left behind, filling valleys and swamping cities. In the 1902 eruption of Mt Pelée in Martinique, 28,000 were killed when the city of St Pierre was almost totally destroyed by pyroclastic flows which engulfed the city.

Mudflows and floodwaters

Although fire and heat are the products of the volcano, vast quantities of water can be the result! Heat from the eruption, lava flows or pyroclastic debris may melt icecaps or snow cover in high altitude and high latitude locations. Meltwater generated from this heat may collect behind a geographic barrier. If that barrier is breached, a sudden flood of water can be unleashed, and this is known as a *jökulhlaup*. Where flood water is mixed with ash or pyroclastic deposits, a *lahar* is formed, a muddy slurry of volcanic debris and water. *Lahars* can flow at speeds of between 15–30 km h^{-1} but with the right combination of flow conditions and topography, have been observed at up to 180 km h^{-1}. Like pyroclastic flows, the kinetic energy of a *lahar* can tear down most obstacles within its path, and once abated, it leaves behind massive quantities of mud and debris. The eruption of Nevado Del Ruiz in 1985 produced lahars which distributed 90 million m^3 of slurry across 2100 km^2 of land, including the town of Amero. The death toll from the destruction of Amero was 23,000, making the event one of the worst volcanic disasters of the twentieth century.

Volcanic gases

Volcanic gas emissions are a further source for environmental impacts, and these may have local, regional or even global implications. On local scales, the impacts of gas emissions may be primarily concerned with the deposition of acids into the environment. Acid deposition may occur either via wet and dry deposition, or via condensation on ash particles. The former of these may occur during both eruptive and passive degassing, but the latter is confined to eruption plumes. These emissions may be a significant source of acid delivered into the local environment, causing vegetation damage and potentially contaminating human or animal drinking water. Although human health impacts are obviously of concern, the impacts on animals and agriculture should not be overlooked. The loss of a key crop can create a ripple-effect of economic repercussions for areas far beyond the reach of the volcano.

On regional scales, large eruptions cause direct chemical effects over much larger areas, and may also perturb the global climate. In 1783, a fissure eruption at Laki in Iceland which lasted more than 8 months produced enough S, Cl and F acids to create a haze of aerosol droplets which persisted over Europe and parts of Asia for several months. More recently, the eruption of Mt Pinatubo in the Philippines in 1991 sent almost 19 million tonnes of SO_2 30 km into the atmosphere, and the sulphate aerosol produced caused a 0.1 °C decrease in the average mean temperature of the northern hemisphere for the following year.

Finally, on the global scale, attention must be given to the 'supervolcanoes', as these may have calamitous consequences. The amount of sulphate aerosol produced from the eruption of Lake Toba, approximately 74,000 years ago, has been estimated to be more than 100 times larger than that of Mt Pinatubo. The impact on climate of such large-scale sulphur emissions has been estimated to result in global temperature decreases of several degrees for more than a decade. The effect of such disastrous climatic disturbance cannot be understated; the impacts of the Lake Toba eruption are thought to have brought the human race to the brink of extinction.

In the previous sections, we have discussed how different volcanic eruptions are produced at different locations. We have also detailed some of the impacts which can be expected to occur at each of these sites. By combining our knowledge, we can therefore attribute the impacts and products of every volcanic eruption across the world to the unique and dynamic forces acting deep beneath it. The interactions occurring within that subterranean realm, the motion of plates, the convecting mantle and the melting of rock under high temperature and pressure are all fundamentally intertwined in each and every eruption. At the heart of these complex interactions, we are brought right back to the first billion years of our planet's existence. In sections 1.1 and 1.2, we considered the 'birth' of the planet, condensing from the solar nebula, and how it was shaped by fractionation, cooling and titanic impact events. These events were fundamental to the composition and structure of the planet we live on today. With every volcanic eruption, we are presented with a reminder of the importance of the dynamic, underground world beneath us. With every eruption, we are reminded of the multiple chance events which have brought our planet to this point in time, and of our unique position within the solar system.

POLICY IMPLICATIONS

Understanding how Earth evolved

- Policy makers need to be aware of the roles of geology in landscape evolution so that they can make international, national and regional assessments of risks of catastrophic events. Such risk assessments should cover both probabilities and probable extents of impacts of such events.
- Knowledge of weathering and erosion rates is necessary to formulate policies associated with assessing long-term sustainability of ecosystems.
- Understanding the importance of rock chemical and mineralogical composition is important to understanding how geology impacts upon soil evolution and fresh water chemical quality and upon the probable availability of natural geochemical resources.
- It is important that policy makers understand the dynamics and impacts of natural processes that shape the planet and how it functions, before they start attempting to assess impacts of anthropogenic activities that generate pollution upon sustainability.
- Understanding the processes and importance of natural biogeochemical cycling is important before attempting to assess the sustainability of methods of food production to meet global food market aspirations.
- Policy makers need to be aware of the nature and impacts of past volcanic eruptions so that they can plan how to best deal with future eruptions.

Some of these policy implications should become even clearer after reading later chapters.

CHAPTER REVIEW EXERCISES

Exercise 1.1

Why does a framework silicate mineral such as orthoclase feldspar show cleavage in two directions whereas micaceous minerals, such as muscovite and biotite, may be cleaved in a single direction into sheets that are so thin that they become transparent?

Exercise 1.2

Mineral crystals in a hand specimen may be scratched by tools tipped with diamond, topaz or corundum, but not by a tool tipped with an orthoclase crystal. What is the hardness of the specimen on Mohs scale of hardness? The crystals in the specimen are colourless hexagonal prisms. What do you think the mineral might be? What additional physical or chemical tests could you perform to help confirm your identification?

Exercise 1.3

Draw a simple, two dimensional, sketch to illustrate how differing fossil communities in strata of sedimentary rocks may be used to establish links between depositional sequences in three spatially separated rock outcrops, even when erosion may have removed one or more strata at one rock outcrop site.

Exercise 1.4

Why would you not expect to find fossils in a metamorphic rock?

Exercise 1.5

Why are pyroclastic flows from volcanoes especially dangerous?

REFERENCES

Creasey, J., Edwards, A.C., Reid, J.M., MacLeod, D.A. and Cresser, M.S. (1986) The use of catchment studies for assessing chemical weathering rates in two contrasting upland areas in north-east Scotland, in *Rates of Chemical Weathering of Rocks and Minerals*, S.M. Colman and D.P. Dethier (eds), Academic Press, 467–502.

Cresser, M., Killham, K. and Edwards, T. (1993) *Soil Chemistry and its Applications*. Cambridge University Press, Cambridge, 192 pp.

Grotzinger, J., Jordan, T.H., Press, F. and Siever, R. (2007) *Understanding Earth*, 5th Edn. W.H. Freeman and Co., New York, 579 pp. plus appendices.

Jackson, M.L. and Sherman, D.G. (1953) Chemical Weathering of Minerals in Soils. *Advances in Agronomy*, **5**, 219–318.

Lindsay, W.L. (1979) *Chemical Equilibria in Soils*. Wiley Interscience, New York, 449 pp.

Marshak, S. (2008) *Earth: Portrait of a Planet*, 3rd edn. W.W. Norton & Co., New York, 832 pp. plus index and appendices.

Reid, J.M., MacLeod, D.A. and Cresser, M.S. (1981) Assessment of weathering rates in an upland catchment. *Earth Surface Processes and Landforms*, **6**, 447–457.

Schilling, J.-G., Unni, C.K. and Bender, M.L. (1978) Origins of chlorine and bromine in the oceans. *Nature*, **273**, 631–636.

Smith, G.A. and Pun, A. (2006) *How Does Earth Work? Physical Geology and the Process of Science*. Pearson Prentice Hall, Upper Saddle River, NJ, 642 pp. plus appendices.

Sposito, G. (1989) *The Chemistry of Soils*. Oxford University Press, Oxford, 304 pp.

CHAPTER 2

The global cycling and functions of water

Malcolm Cresser and Paul Ayris

Learning outcomes

By the end of this chapter you should:

- Recognise that water is essential to all life on Earth.
- Understand the processes and extent of transfers of water between the atmosphere, the oceans and terrestrial surfaces.
- Be aware of what limits the amount of natural fresh water available to the growing world population.
- Start to be more aware of some problems associated with over exploitation of fresh water resources.
- Start to appreciate the contributions of river systems in element cycling and the rock cycle.
- Start to understand the links between water cycling and landscape evolution and land use potential.
- Better appreciate the many demands humans make on river systems.

2.1 The importance of water

In many countries of the developed world, a regular supply of clean, potable fresh water has been rather taken for granted by the general population, at least until recent decades. Indeed, human need for water to drink, and desire to use even more for sanitary purposes, has, in the past, generally dictated where settlement has occurred. Today, however, constraints on potential size of sustainable water supply often dictate the limits of expansion to those same settlements. As the global population has grown, as indicated in Figure 2.1, so too has the demand for water, not just for drinking, leisure activities, washing and waste disposal, but also for food production and food processing and for industrial consumption. Thus need for sustainable water supplies which are adequate both in terms of quantity and chemical and biological quality is now close to the top of the global environmental management agenda.

Lead-in question

What is the importance of water to the functioning of planet Earth apart from meeting the needs of humans?

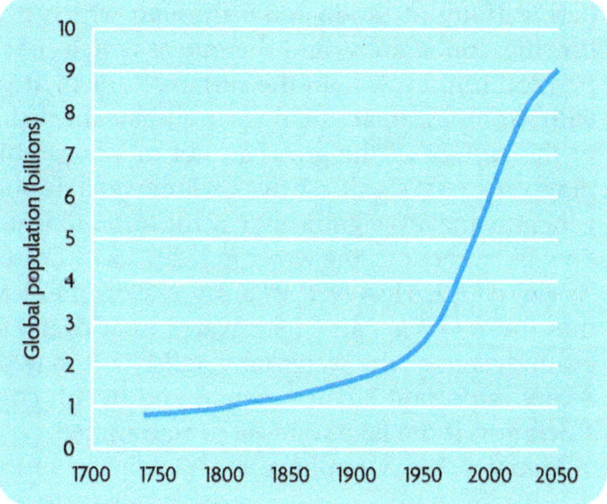

Figure 2.1 Growth of the global population, based upon UN World Population Prospects.

Aside from the obvious needs of humans, and, of course, other animals, water has a far greater importance to the functioning of planet Earth. When we look at the global carbon cycle in Chapter 4, we will see that water is an essential reactant in the photosynthetic process. *Via* photosynthesis and biomass production, water is crucial to the local and global cycling of carbon, and hence to the control of our planet's climate. We also will see later that microbial decomposition of plant litter and other organic materials produces carbon dioxide, which in turn dissolves in soil water to produce carbonic acid (H_2CO_3), a key agent in biogeochemical weathering of soil and rock minerals. Some of this soil water may drain to rivers, transporting weathering-product base cation elements such as calcium (Ca) and magnesium (Mg) as their hydrogen carbonates, $Ca(HCO_3)_2$ and $Mg(HCO_3)_2$, to rivers, and potentially on into ocean waters. There they may participate in the formation of sedimentary rocks, as discussed in Chapter 1. Many millennia later these may be uplifted as a consequence of tectonic activity, and thus water contributed to rock cycling. Similarly water transports sediments from physical erosion to the oceans, and these may undergo a similar fate.

We shouldn't forget that the microorganisms mentioned above also require water to function, and they need water of adequate chemical quality as well as an appropriate amount of water. It's no wonder then that the first question scientists often ask when considering the possibility of life on other planets in the solar system is: 'Is there water there?'

Lead-in question

Where do the terrestrial biota on Earth get their fresh water from?

2.2 The global water cycle

The obvious answer to the above lead-in question has to be from the atmosphere initially; 'initially' because, in days following precipitation events, water rising from depth in the soil by capillary action and/or drainage water from upslope may also help provide water needs of biota.

Precipitation from the atmosphere may be in the form of rain or snow. Rainfall is usually measured in mm, and snowfall generally converted to a rainfall equivalent in mm. The field methods by which rainfall and snowfall are measured in practice will be discussed later in Chapter 17, so need not be considered further here. Water may also be deposited as mist or dew (when water vapour in moist air condenses to form fine droplets as warm, moisture-laden air cools) or as frost when water vapour is deposited as ice when deposition is onto surfaces at sub-zero Celsius temperatures.

If a fairly dry region has 500 mm of precipitation each year on average, this means that the volume of water falling per hectare per year would be 50 cm by 10^8 cm^2 (as a ha = 10^8 cm^2), or 5 by 10^9 cm^3 or 5 by 10^6 litres (i.e. 5 million litres). This sounds like a lot of water, but much of this may return to the atmosphere as water vapour, either by evaporation from the surface or via transpiration losses from plant foliage. In the UK, for example, evapo-transpiration would typically be around 450 mm, so only about 10 per cent of the 500 mm of precipitation might pass on to rivers or the recharge of groundwaters held below the terrestrial surface. This is why the region could be described as 'fairly dry'. So humans in the region only potentially acquire around 5.10^5 litres of water a year, assuming (wrongly if you don't want rivers to dry up all year!) that all of the excess of precipitation over evapo-transpiration is 'available' for abstraction.

Five hundred thousand litres per hectare per year still sounds like a lot of water, but let's consider this from the perspective of human use. Suppose an individual human uses 100 litres of water a day for drinking, washing clothes and dishes, showering, flushing toilets, etc. They would use 36,500 litres of water a year. At this rate of consumption in this area each hectare could only meet the aspirations of approximately 13 individuals. Worryingly, this is an over-estimate, as it takes no account of water use in other aspects of the individuals' lives, such as food production and processing or watering the garden.

2.2.1 How much fresh water is there on Earth?

Viewed from space, the Earth is often described as the blue planet because of the high proportion of the surface which is covered by water. It is estimated that the total volume of water is about 1.4×10^9 km^3. However, the bulk of this, almost 96 per cent, is salt water in seas and oceans, and only just over 4 per cent is fresh water. Moreover, almost three quarters of this fresh water is currently locked up in polar ice and glaciers, and most of the fourth quarter is in underground water (1.05 per cent of the total water on earth, or 1.54×10^7 km^3) (Grotzinger *et al.*, 2007). All the world's lakes and rivers together contain only about 0.009 per cent of the planet's total water, and that is nine times more than the atmosphere contains at any moment in time. But it is the part of this tiny fraction of the total that is deposited on land that we depend upon, via recycling, to maintain the needs of terrestrial and fresh water biota.

Water from below the ground surface (groundwater) has been very extensively used to provide potable water or water for irrigation of crops for many centuries. For example, the remains of an early pumping system used in Roman London were discovered not that long ago in England. Figure 2.2 is a simple schematic representation of a pumped groundwater. It should be realised from the outset that groundwater is rarely static. The speed at which it flows depends upon a combination of the permeability of the material through which it is flowing, topography, and the rate at which precipitation infiltrates from the surface. Thus in areas with strong seasonal variations in climate the depth to the surface of the groundwater or water table may vary very significantly. Groundwater inputs generally make a significant contribution to water flow in rivers or to the water in lakes, as indicated in Figure 2.2. However, in a dry season the flow may be reversed, and lake water may drain to groundwater as the water table falls. This may be a very important environmental and health consideration if the lake water is contaminated.

Figure 2.3 is a graphical representation of the relative amounts of water in the water cycle at a global scale, expressed as percentages. Of the 5.05 by 10^5 km^3 of water that pass into the atmosphere

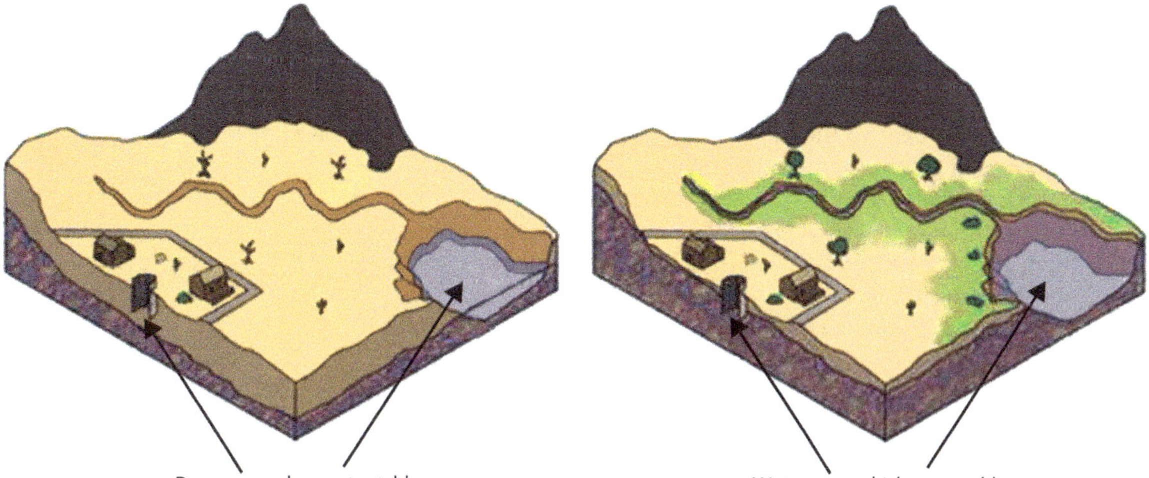

Figure 2.2 Diagram showing the seasonal change in the position of the groundwater table in an excessively dry summer (groundwater is represented here by mauve mottled areas). In the wet season, fresh water starts to refill the lake from the river and from near-surface groundwater. To maintain a water supply in summer, wells must be appreciably deeper than the lowest summer water table.

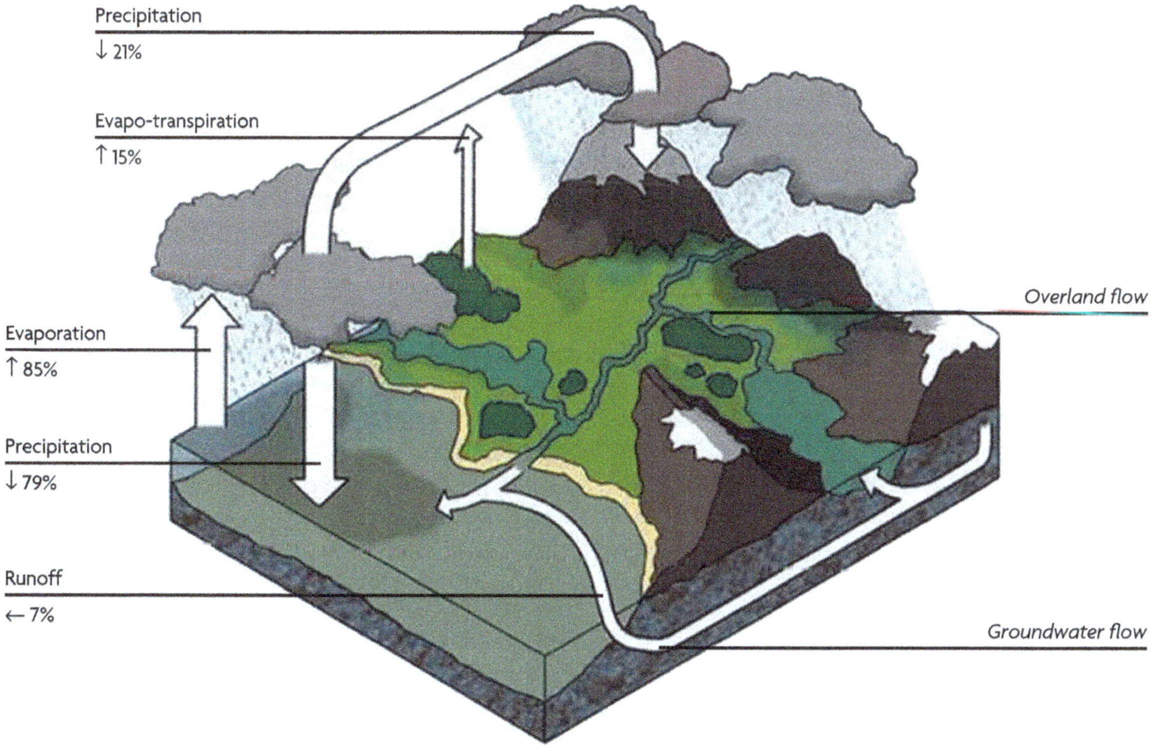

Figure 2.3 The global water cycle, showing relative importance of water pathways, based upon estimates in Grotzinger *et al.* (2007).

each year, 85 per cent is from evaporation from the oceans and 15 per cent from terrestrial surfaces. The same total volume of water returns to the oceans and land surfaces each year as precipitation (rain, snow, etc.), 79 per cent to the oceans and 21 per cent to terrestrial surfaces. At first glance it looks as if land is accumulating water (i.e. precipitation minus evapo-transpiration = 21% – 15%), but the 'missing' 6 to 7 per cent is the water that runs off from and/or through terrestrial surfaces into the oceans each year, transporting both solute chemical species and suspended sediment as it does so.

2.2.2 Do solute species in sea water cycle too?

One aspect of water cycling not accurately represented in Figure 2.3 is the transfer of ocean water spray up into the atmosphere. At the ocean surface, especially during stormy weather, droplets of sea water aerosol may be produced as waves break at the surface. Much of this aerosol is transported up into the atmosphere by wind. This process transfers both water and the ions it contains up into the atmosphere. Thus invariably precipitation contains significant concentrations (at concentration levels of a few $\mu g\ ml^{-1}$) of sodium, magnesium, chloride and sulphate especially, and lower concentrations of other species such as boron. The effects and importance of this on terrestrial ecosystems will be discussed later in Chapters 5 and 7. It is mentioned here because such salt deposition from the atmosphere can have major impacts upon the conditions in the soils upon which they are deposited and their associated vegetation, especially in regions of the world where evapo-transpiration exceeds precipitation. Where this happens regularly the soluble salts start to accumulate and soils may become strongly alkaline as there is insufficient drainage water to flush them out of the soil.

2.3 Topographic effects on precipitation

Although we have considered how large masses of water are transferred from ocean and terrestrial surfaces to the atmosphere, so far we have not considered why this is eventually returned as precipitation. If you carefully observe the space over water boiling in a beaker, you will clearly see steam forming as the air cools immediately above the top of the beaker. What you are observing is the condensation of water vapour from a gaseous phase to suspended water droplets, effectively cloud formation. If you don't have a beaker and hotplate to hand, think of the contrails behind an aircraft as warm water vapour emitted as a combustion product from engines cools rapidly, forming water (and then often ice crystals) or, closer to home, of car exhausts on a cold morning. This process will occur eventually wherever air becomes super-saturated (i.e. the water vapour partial pressure in the air mass exceeds the saturation water vapour pressure at the temperature of the air mass). This may happen in rising air masses as temperature falls with altitude, or wherever warm air meets a mass of colder air.

The decrease in temperature with altitude comes into play very regularly when prevailing wind direction causes moist air to rise over hills or mountains, as indicated in Figure 2.4. If the altitude is sufficient, by the time the air mass has passed over the mountain range much of its condensed water vapour may have been deposited as mist, rain or snow. Wherever this happens, precipitation levels may consistently be much lower on the far side of the mountain range. In extreme cases luscious vegetation on the windward side may give way to almost desert on the leeward side.

The effect described above, known as a rain shadow effect, has an important impact upon the climates in Ireland and Great Britain, where prevailing winds come in from the west across the Atlantic Ocean. Thus western Scotland, Wales and England tend to have substantially more precipitation each year than eastern Britain.

The above effect is relatively more important than it might at first look, because effects of climate upon water quality are more dependent upon runoff than on precipitation amount. For example, if an area to the west of a country has annual precipitation of 1100 mm, double the precipitation of an area to the east with 550 mm, but evapo-transpiration is 450 mm in both areas, then runoff would be 650 mm to the west but only 100 mm to the east, i.e. six-and-a-half times higher rather than two. Other factors being equal, therefore, dilution of solute species such as nitrate from agriculture in river water will be substantially

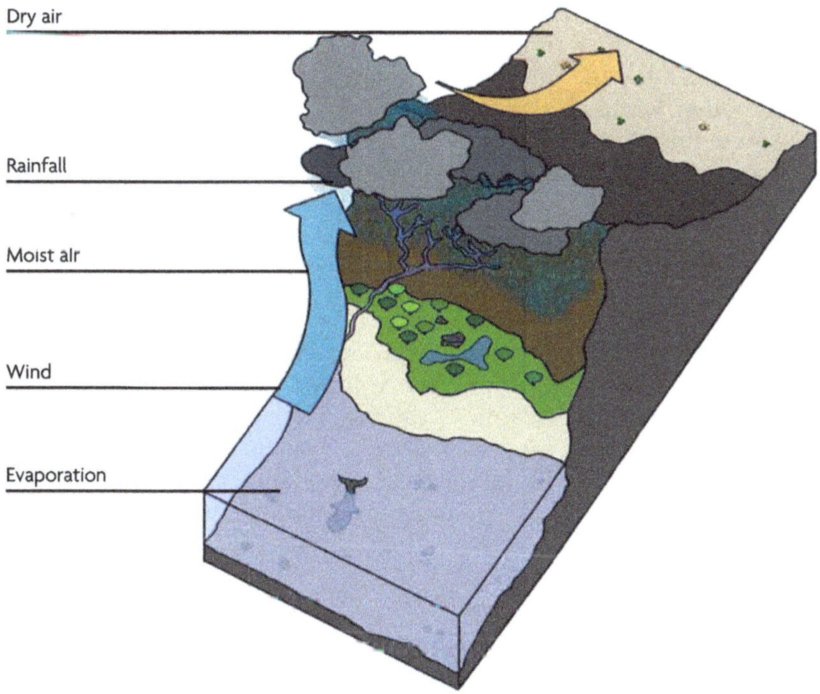

Figure 2.4 Schematic representation of the rain shadow effect. Water vapour from evaporation from the ocean is carried in by the prevailing wind and forms clouds as the rising air mass cools. These lead to precipitation over higher ground. Air passing over the top of the mountain range is much drier, so less precipitation falls on the far side of the range.

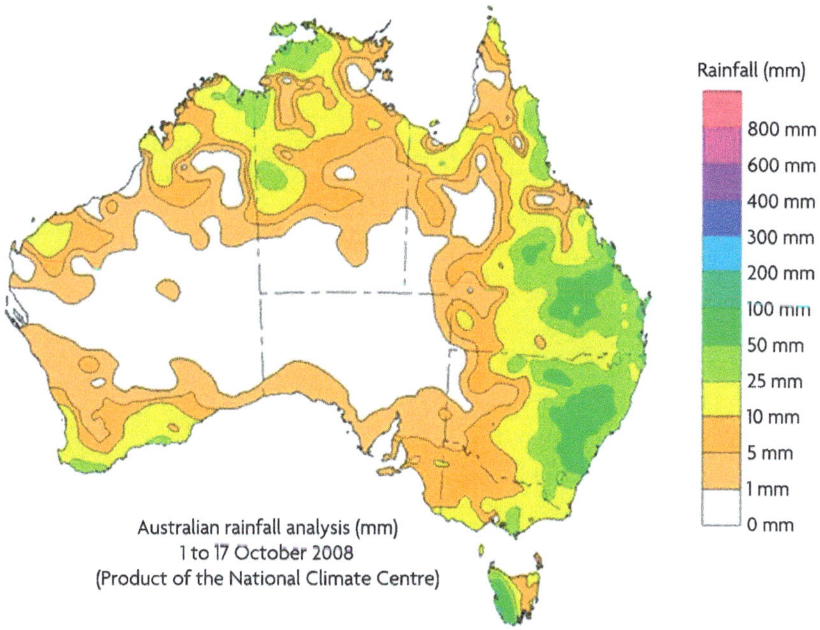

Figure 2.5 An example of a precipitation distribution map as downloadable from the Australian Government Bureau of Meteorology website, showing rainfall between 1 October 2008 and 17 October 2008.
Source: www.bom.gov.au

higher to the west. This effect can be very clearly seen in Walling and Webb's informative chapter on water quality in John Lewin's excellent early book on British rivers (Lewin, 1981).

Many countries produce very detailed precipitation distribution maps of their entire country and/or individual regions. An excellent example is the Australian Government Bureau of Meteorology website, which allows you to look at rainfall over 1 d, 1 w, or the past 1, 3, 6, 9, 12, 18, 24 or 36 months, or over the month to date or year to date. Such maps (e.g. Figure 2.5) are prepared taking local knowledge about impacts of features such as topography and wind direction effects into account. This is done by careful plotting of isohyets, which are lines (precipitation contours) joining points where total precipitation over the assessment period is predicted to have corresponded to the precipitation amount being plotted.

Lead-in question

Who would find such information useful?

Such detailed information is invaluable to environmental managers responsible for water supplies, but also to farmers who need to assess the need for irrigation and the level of irrigation required to meet crop requirements while minimising water waste. Comparisons with trends in previous years (and their consequences!) may also be invaluable.

2.4 Water cycling on a volcanic island

The significance of the points raised so far should become much clearer if we look at a specific example. The volcanic island of Tenerife, in the Atlantic Ocean due west of North Africa and at the same latitude as the Sahara Desert, is near perfect for this purpose. The altitude ranges from sea level to 3718 m at the summit of the largest volcano (Teide, Figure 2.6). In summer, temperature

Figure 2.6 Students setting off for the summit of Teide in the National Park in Tenerife. (Prior permission to approach the summit is required for safety reasons.)

Figure 2.7 Assessing the biodiversity of drought- and salt-tolerant species at low altitude in Eastern Tenerife.

ranges from 26–28 °C in coastal areas to close to freezing near the Teide summit. The windward half of the island receives about three quarters of the island's total precipitation.

At low altitude the climate is extremely dry, and evaporation substantially exceeds precipitation. In some months there may be no precipitation at all. Salts accumulate in the alkaline soils as there is negligible leaching of the salts of oceanic origins. Thus the vegetation community consists of species able to tolerate drought and salinity (Figure 2.7).

Less salt- and drought-tolerant species may be grown, but in coastal regions this is generally in well-watered tubs for shade plants or using trickle irrigation to flush out deposited salts, as in Figure 2.8.

As the warm, moist Atlantic air is forced to rise as the prevailing trade winds move inland, the cooling effect produces a distinctive layer of mist and cloud between a few hundred metres above sea level and about 1800 m (Figure 2.9; the exact

Figure 2.8 Trickle (irrigation via the black piping) being used to overcome the salinity problem. Suitable wastewater may be used for this purpose.

Figure 2.9 Typical cloud layer at Tenerife viewed from about 2000 m.

altitude range of the cloud layer varies seasonally across the island). Precipitation is sufficient at these altitudes to exceed evaporation, so leaching of deposited salts occurs and conditions become much more favourable (less saline) for soil evolution and the establishment of vegetation. Thus extensive zones of Canary Pine forest occur in this altitude range (Figure 2.10).

At higher altitudes (>2000 m) the rising air mass meets denser, colder air coming in from over the ocean, forming a barrier to further rise, so although some rain and snow fall, precipitation is lower. Conditions for the early stages of soil formation from the recently erupted rock *via* lichen colonisation (see also Chapter 5) are poor because of a combination of limited moisture supply, consistently low temperature, and possibly adverse effects of toxic hydrogen sulphide in the atmosphere locally and the intensity of potentially damaging ultra-violet radiation. Phosphorus supply is limited too, because of the fixation of phosphate by the hydrous oxides of iron and aluminium formed when the volcanic parent materials do start to weather (see also Chapter 7).

2.5 What happens during storms?

So far we have mostly been considering the cycling and fate of water in rather general terms, often with values quoted representing mean values over a year. However, we also need to consider what happens during individual precipitation

Figure 2.10 Canary Pine forest in Tenerife at around 1200 m above sea level. The pines show blackening of trunks from previous forest fires, but Canary Pines are very fire resistant because of their extremely thick bark. The yellow flowers of the ground vegetation are *Lotus lancerotennsis*, a nitrogen-fixing legume that grows well if nutrient nitrogen is lost from the litter layer in the fires (see also Chapter 5).

Figure 2.11 Typical upland landscape of central Scotland. Note the river meandering through the valley floor.
Source: Peter Scott.

events, especially when the climatic conditions are extreme. What might be expected to happen, for example if the upland Scottish landscape in Figure 2.11 was subjected to a prolonged period of heavy rainfall?

The fate of rainfall would clearly depend upon where it lands. On bare rock, and especially on steep slopes, runoff would be more or less instantaneous. Rain landing upon soil might initially infiltrate the soil, depending on the extent to which infiltration rate exceeds the rate at which the precipitation falls on the soil surface. Infiltration rate would depend upon the prior soil moisture status and upon soil texture or particle size distribution. We will see in Chapter 7 that water can pass much more rapidly through soils where the particle size is coarser than where particle size is much finer. On steeper slopes, especially on convex slopes, mineral particles are prone to physical erosion, so soils often tend to be thinner over underlying rock. Thus soils may become saturated very rapidly during heavy rain, to the extent that overland flow of water may occur. Even if it does not, water may start to flow laterally down slope through the soil.

The discharge in the river flowing through the valley would start to rise quite rapidly, typically within a few hours (e.g. 2 h or so). Clearly the precise time to peak discharge would depend upon a combination of topography, catchment size, antecedent soil moisture conditions, and precipitation intensity and duration. Once the precipitation ceases, the river discharge starts to fall quite quickly, especially in smaller catchments, though the storm hydrograph will not be symmetrical (Figure 2.12). Reliable modelling of the shape of storm hydrographs is achievable, but complex, and beyond the scope of this text, so interested readers should consult a more specialised monograph on hydrology (e.g. Beven, 2004).

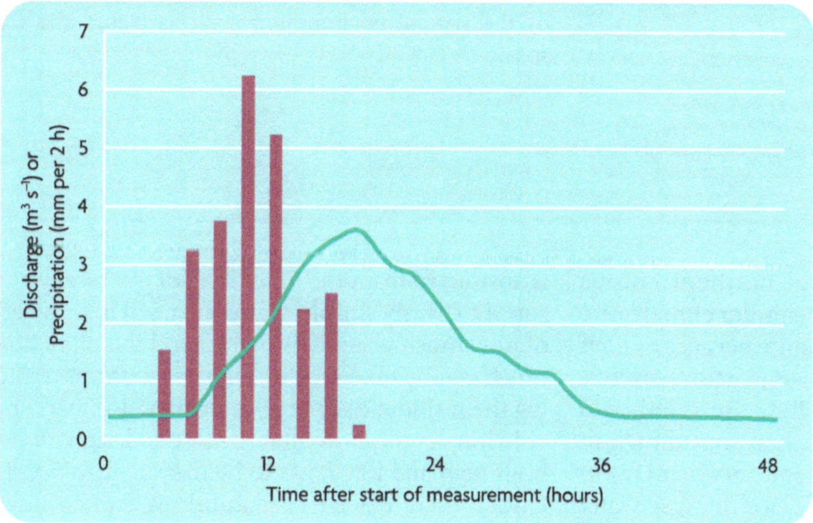

Figure 2.12 A typical storm hydrograph for an upland area in northern Europe. Red bars show precipitation.

CASE STUDIES

A case study in Scotland – How does rainfall composition change during storms?

When investigating the environmental impacts of acid deposition in the United Kingdom, Edwards, Creasey and Cresser (1984) became interested in how precipitation chemical composition changed over time throughout rain storm events. They hypothesised that precipitation early on in the storm would be more acidic as it removed acidic pollutants from the atmosphere. They therefore designed a simple mechanical fractionating rain gauge to test their hypotheseis. An example of what they found is shown in the graph below.

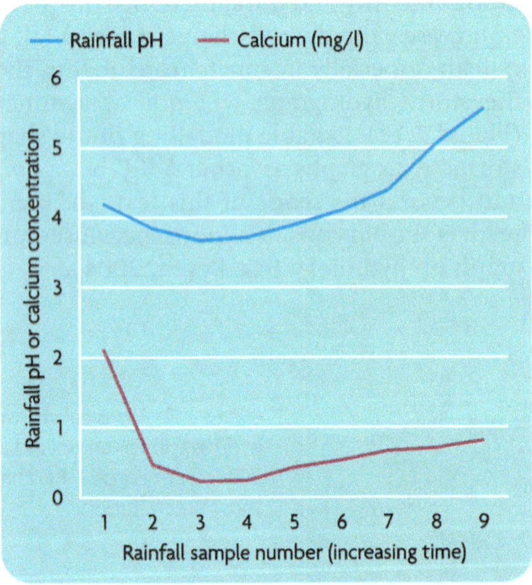

In practice, they found that the earliest rainfall, right at the start of the storm, was not the most acidic. They suggested that this was because terrestrial dust was scrubbed out of the atmosphere, as well as acidifying gases and acidic aerosols. Terrestrial dust tends to have a fairly high concentration of base cation elements such as calcium, magnesium, sodium and potassium, and may have a neutral or slightly alkaline pH. The dust is scrubbed out of the sub-cloud atmosphere even more rapidly than gases, as can be seen from the plot of calcium concentration over time that is also shown. Potassium and sodium gave similar trends with time (Edwards *et al.*, 1984).

Towards the end of a storm, rainfall generally becomes less intense, and under these conditions the base cation concentrations tend to increase slightly as a consequence of evaporation. The rainfall pH becomes higher because there is less acidic pollution left in the atmosphere for the rainfall to remove.

The results are interesting because the rainfall early on in a storm tends to be what is absorbed into porous building materials. Also consideration should be given as to whether the variation in precipitation chemistry over time during storms, significantly affects its impacts upon components of the natural environment such as plants and soils.

Rivers in upland regions of northern Europe and regions of the world with similar climates are often described as 'flashy' in nature because of their rapid (but reversible) response to storm events. In many rivers discharge or flow are monitored continuously using flumes such as that shown in Figure 2.13. This particular flume, at Glen Dye in north eastern Scotland, is in place because water is abstracted from the River Dye for the local water supply. Clearly supply managers need to be aware of how much water is available, and therefore how much can be taken for human use without reducing the residual discharge in the river to unacceptably low levels. Note that the flume design allows both high and low flows to be measured accurately. Under high discharge conditions, water flows

Figure 2.13 A concrete flume at Glen Dye in NE Scotland. Under high flow conditions, as here, flow passes over the outer limbs, so it can still be accurately measured. Under low flow conditions the water is confined to the narrower, central channel.
Source: Tony Edwards.

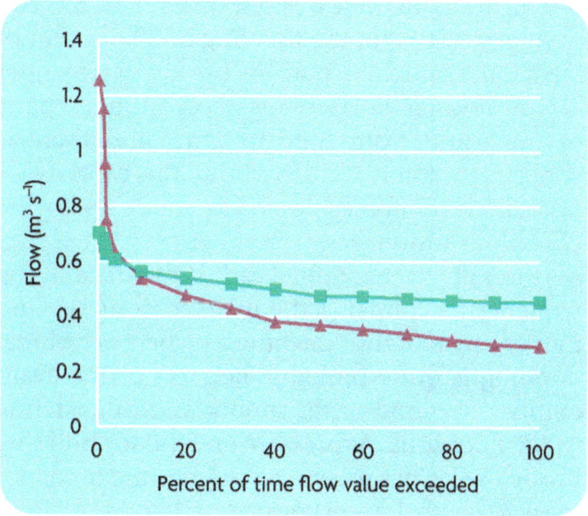

Figure 2.14 Flow duration curves for two drainage basins in the UK uplands: the green line is for a catchment with more shallow slopes.

through the outer channels of the flume, as well as through the narrow central channel which becomes important at low flows.

Rivers draining upland areas supply a substantial proportion of the water supply in Scotland, but in some other parts of the UK and of the world reservoirs or groundwater supplies may be relatively more important. It is important to realise that for each of these sources the precipitation component of the water cycle is what maintains the sustainability of the supply. If abstraction regularly exceeds runoff (i.e. exceeds precipitation minus evapo-transpiration) we will see later in Chapter 14 that problems of declining water chemical quality invariably will ensue.

Hydrologists often illustrate the flashy nature of river systems using flow duration curves. These are graphs illustrating how the percentage of total time a specified discharge value is equalled or exceeded varies with discharge. Figure 2.14 shows such curves for two quite small upland drainage basins. The catchment of the more flashy river has steeper slopes and shallow soils, so water during storms is transferred very rapidly to the river, and discharge subsides correspondingly rapidly very soon after precipitation ceases. Thus very high discharge occurs only for short intervals during a small number of particularly severe storm events.

Lead-in question

What changes do you think might happen in rivers during brief periods of exceptionally high discharge?

It is appropriate now to consider what happens during periods of abnormally high discharge. When rivers were left to their own devices, river water would tend to flow over the river's natural banks and inundate the river's natural flood plains, especially in lower lying and flatter parts of a catchment. When the river water velocity is high at high discharge, the turbulent water may move large amounts of suspended sediment. When the river floods over its banks, the water velocity slows dramatically, and sedimentation starts to occur. Larger particles tend to deposit first, often close to the original river channel. This tends to create natural banks of sediment known as levees. Over several decades these can build up and help minimise subsequent flood risk. However, if a storm event is very extreme then the levees may be breached and partially washed away, resulting in sudden and severe local flood events. The collapse of levees at several points as a consequence of the severity of Hurricane Katrina was responsible, at the end of August 2005, for the flooding of 80 per cent of New Orleans, where parts of the city were under 4.5 m of water. Such events occur in many parts of the world from time to time, and there is concern in some regions about the capacity of natural or reinforced levees to stand up to the effects of earthquakes.

Ironically, as mentioned earlier, there is a strong tendency for human settlements to occur around the very areas of river catchments where occasional flooding is quite probable because of the availability of a good water supply and often fertile valley-floor soils. As pressure on land for building in such areas has grown, so too has the tendency to try to control river channels and build on flood plains. Flood protection schemes are not always adequate however, and in any case often tend to increase flood risk further down-river. For these reasons flooding is not uncommon in cities such as York in England, for example (Figures 2.15 and 2.16), but improved flood protection in York may increase the flood risk further down-river at Selby.

Even quite a cursory glance at Figures 2.15 and 2.16 makes it very obvious why insurance companies are so interested in flood prediction, and particularly in the ways in which climate changes are likely to influence flood risk distribution. It is fairly clear from the colour of the water in Figure 2.15 that it has a fairly high concentration of fine suspended sediment in the relatively slowly moving flood water.

Figure 2.15 A low-lying car park in York beside the River Ouse that is flooded relatively frequently. The need for flood warnings here is fairly obvious, though warnings are not always heeded in time!

Figure 2.16 A further example of local flooding in York. The flood waters usually subside quite rapidly, but may still do a lot of damage to buildings.

2.6 The roles of groundwater

Lead-in question

Is all groundwater fresh water?

2.6.1 Sea water incursion

As seas and oceans form, or change shape over geological timescales, the water infiltrating to groundwater is clearly likely to become highly saline. Thus groundwaters under the ocean floor are saline. Therefore in coastal zones, fresh water groundwater flowing from higher terrestrial regions

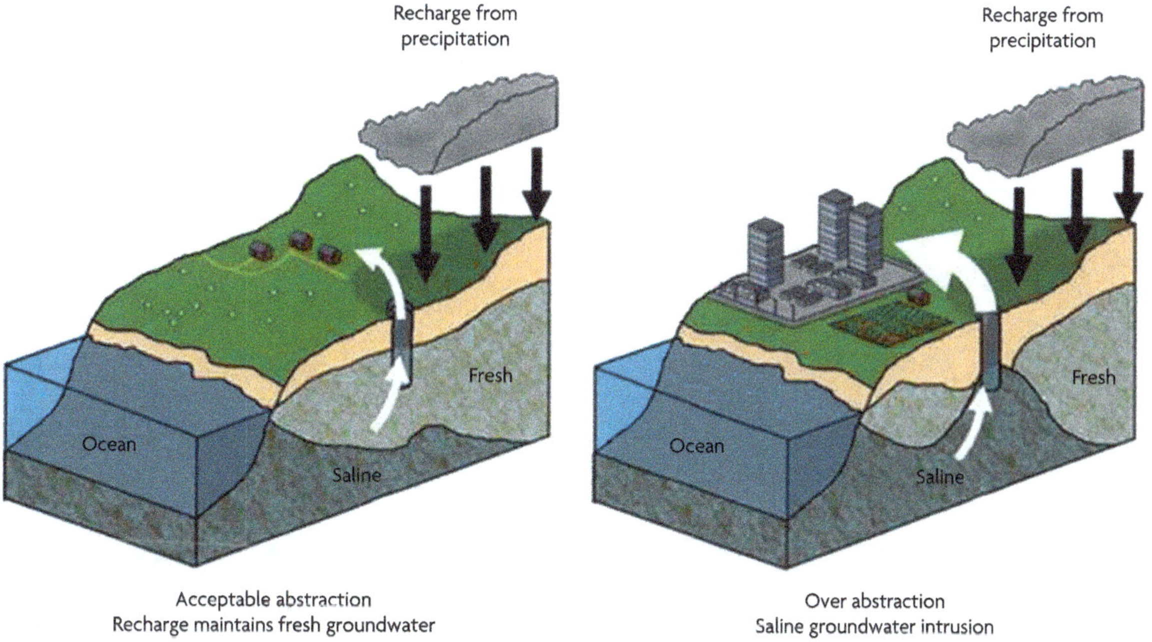

Figure 2.17 Illustration of what happens when fresh groundwater meets saline groundwater in coastal zones. Excessive abstraction, for example due to over-development of the coastal region for urban land use and agriculture, may cause the fresh water/groundwater boundary to move inland, until eventually saline water is being pumped.

inland meets saline groundwater, as indicated in Figure 2.17. Where the annual amount of precipitation exceeds the actual amount of evapotranspiration, under natural conditions such a boundary generally would be retained by the slightly positive pressure head of the fresh water coming from the land.

We will see later, in Chapter 14, that if terrestrial groundwaters are over exploited, especially if too much groundwater is pumped back to the surface for human use in low lying coastal areas, the barrier between the salt water and the fresh water may start to move inland. Ultimately this sea water incursion results in salinisation of the pumped water. Eventually the water may become unsuitable for human consumption and, eventually, even for irrigation purposes. This is another example of the reason why the rate of groundwater abstraction should never normally exceed the rate of recharge by precipitation and drainage. The recharge sometimes may come from precipitation in mountainous regions at an appreciable distance from where the groundwater is being pumped. It is therefore very important for those managing water supply to be aware of the extent and any flow direction of groundwater bodies.

2.6.2 Perched water tables

The author can still remember going on a field trip as a newly appointed lecturer in soil science to look at some soil profiles on a hill slope in Scotland. It was raining lightly at the time, so when the half-dug soil pit started to fill up with water (and wishing to make a good impression on new colleagues!) he readily agreed to lie down beside the pit and bale it out as the digging continued. However, it soon became clear that the water was winning. When the digging stopped momentarily, it became immediately obvious that there was a source of water spouting up gently from the bottom of the pit. The water had been flowing down slope through a layer of permeable, coarse sandy material sandwiched between two very much less permeable soil layers. Breaking through the upper impervious layer with a pick

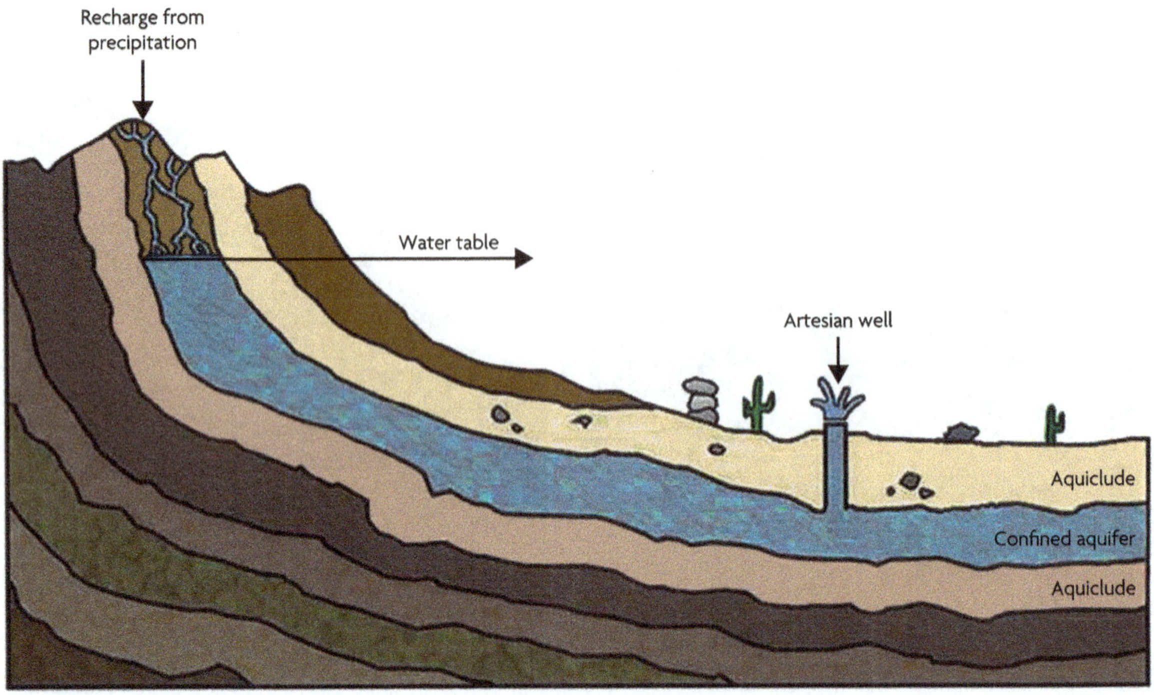

Figure 2.18 Illustration of how an artesian well functions. A trapped layer of groundwater flows down slope through a layer of permeable material sandwiched between two aquicludes (layers of very low permeability or impermeable material). The head of water can force water up to the surface down slope if the upper aquiclude layer is penetrated.

unleashed a small spring, effectively creating an artesian well (Figure 2.18).

In fact, the stopping (or substantial slowing down) of water infiltration because of the presence of a local impermeable layer (an aquiclude) between the soil surface and underlying bedrock is not that uncommon a feature in uplands of northern Europe. Such layers may create a perched water table, as illustrated in Figure 2.19. The presence of such perched water tables may result in the development of springs on hillsides, as also illustrated in the figure. Sometimes these may only flow intermittently, drying up after extended periods of dry weather.

Where the topography, climate and the nature of the underlying strata are all appropriate, artesian wells may provide an excellent supply of clean, potable water throughout the year. This would have been recognised from pre-historic through to more modern times when choosing a good place to settle.

2.7 Discontinuities in river flow

Lead-in question

Do you think all river channels are water-tight?

2.7.1 Sink holes and Karst topography

In Chapter 1 we saw that calcareous rocks such as calcite and dolomite are readily attacked by carbonic acid in rain formed by the dissolution of atmospheric carbon dioxide:

$$CO_2 + H_2O \rightarrow H_2CO_3$$

$$CaCO_3 + H_2CO_3 \rightarrow Ca(HCO_3)_2$$

The carbonic acid concentration in soil solution is even more concentrated, because the concentrations of carbon dioxide (CO_2) in the soil atmosphere are much higher than those in air. This

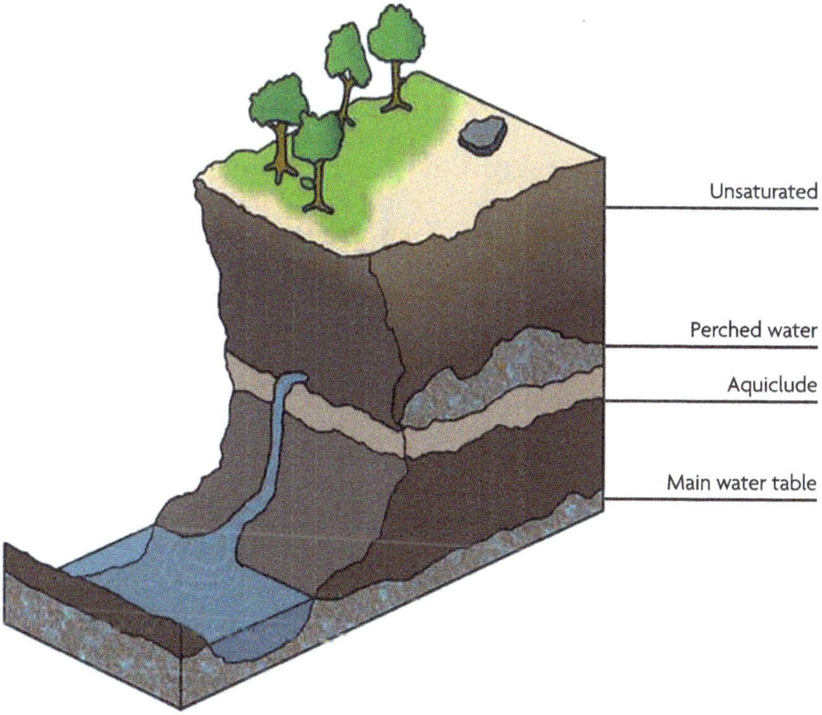

Figure 2.19 Diagram illustrating how the presence of an aquiclude may lead to the development of a perched water table at a sloping site in upland areas. Surfaces of aquicludes are often flatter and more parallel than indicated here.

process, over long periods of time, may produce substantial cave systems underground through limestone rock that then can act as conduits for the rapid transport of water. What happens to this water subsequently depends upon the interconnectivity of the channels formed and upon local topographic features. It often re-emerges at the surface as a new river source or it may join an existing river. However, the channels may also act as a rapid sink for water that was previously flowing over the land surface. Such terrain, known as Karst topography, is illustrated in Figure 2.20.

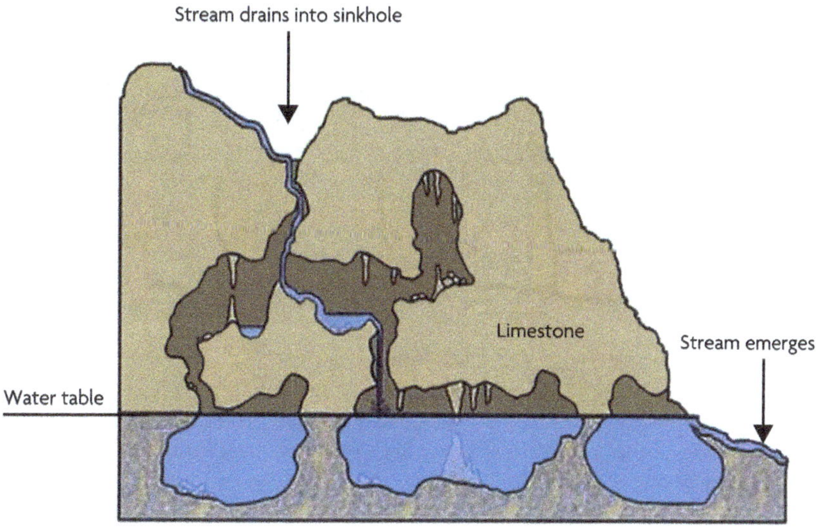

Figure 2.20 Illustration of how dissolution of readily soluble rock such as limestone may lead to substantial underground channels and caves (Karst topography) that allow rapid flow and routing of water below the surface. Water may drain rapidly into such spaces from both terrestrial surfaces and from river beds via sink holes. The water may then resurface down slope forming a new river channel. Sometimes water apparently in the same river channel may originate from substantially different land drainage areas.

Often the presence of such underground features may not be immediately obvious at the surface. In rivers, however, the effects of sink holes or underground water inputs to rivers often may be detected by sudden changes in the chemical composition of the water along a relatively short stretch of a river where there is no other obvious potential cause. Sometimes sudden changes in temperature by a few °C may similarly be indicative of a groundwater input.

2.8 Rivers in landscape evolution

Lead-in question

How may rivers influence landscape evolution over geological timescales?

2.8.1 Erosion and transport of sediment

Recent public concerns over global warming have stirred much interest in the melting of polar and glacier ice. Most readers will be aware already of the fact that glaciers may scour out channels at a dramatic spatial scale as they migrate across the land surface, both creating and depositing sediment as they progress. They leave distinctive landscape features many thousands of years later that are still recognisable as being of glacial origins.

Major rivers systems too may move large quantities of sediment eroded from the land surface, to deposit it eventually down-river in estuaries and the ocean. The size of particles being transported and the speed of their transport increase with increasing discharge, because the water flow velocity increases at higher flows. Comparison of the two photographs in Figures 2.21 and 2.22, which show a view of the River Dee in Scotland at low flow in winter and at moderately high flow in summer, gives some idea of the size of the granite boulders that can be rolled along the river bed by the water in severe storm episodes.

The deposition of sediments with different size ranges by waters flowing at different velocities often explains the size distribution of soil particles found in soils (discussed in Chapter 7) at sites where the soil has evolved from ancient river sediments.

Figure 2.21 The River Dee at low discharge in winter, exposing the size of boulders that can be slowly transported along the river bed by high water velocities in extreme events. The slower the flow velocity, the smaller the size of the particles than can be re-deposited on the river bed.
Source: Photographed by Richard Smart Ph.D.

Figure 2.22 A similar view of the River Dee at high discharge in summer. Quite large particles can be rolled along the river bed and moved in suspension by the high water velocity. Turbulent flow can re-suspend particles from the bed.

2.8.2 Transport of solute species in rivers

In Chapter 1 and section 2.7.1 in this chapter, we saw that soil or rock minerals are subject to slow dissolution by biogeochemical weathering processes involving reaction with carbonic acid. As a consequence, rivers transport substantial amounts of solute species such as $Ca(HCO_3)_2$ and $Mg(HCO_3)_2$ down-river and on into the ocean. Prior to the start of a precipitation event, water drains into the river through riparian zone soils immediately adjacent to the river channel. Typically the water will drain predominantly from near-surface groundwater. Such water has had plenty of time to equilibrate with the soil atmosphere and usually contains significant concentrations of base cations such as calcium (Ca^{2+}) and magnesium (Mg^{2+}). In Figure 2.23, which shows how flow pathways of drainage water tend to change during storms in upland areas, this is represented by pathway C.

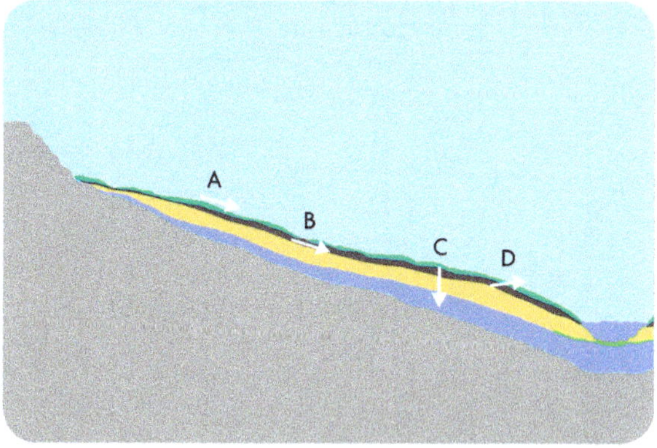

Figure 2.23 Schematic representation of drainage water flow pathways during storms; A, overland flow or surface runoff; B, throughflow through near-surface soil horizons; C, water draining to join groundwater flow and D, return flow (throughflow returning to the surface on lower slopes) (Based on Cresser, Killham and Edwards, 1993).

During a storm event, precipitation rate may eventually start to exceed infiltration rate (often in local areas of the catchment rather than everywhere). Overland flow (pathway A) and return flow (pathway D) may then start to occur, and also water often starts to flow laterally down slopes through near surface-soils (pathway B). These soils are generally more organic matter-rich than the underlying mineral soils containing groundwater. Therefore, soon after the river discharge starts to increase, as water draining through or over surface soil layers becomes relatively more important, the total organic carbon (TOC) concentration of the river water increases; this is indicated in Figure 2.24. Conversely, the concentration of calcium from weathering of minerals in

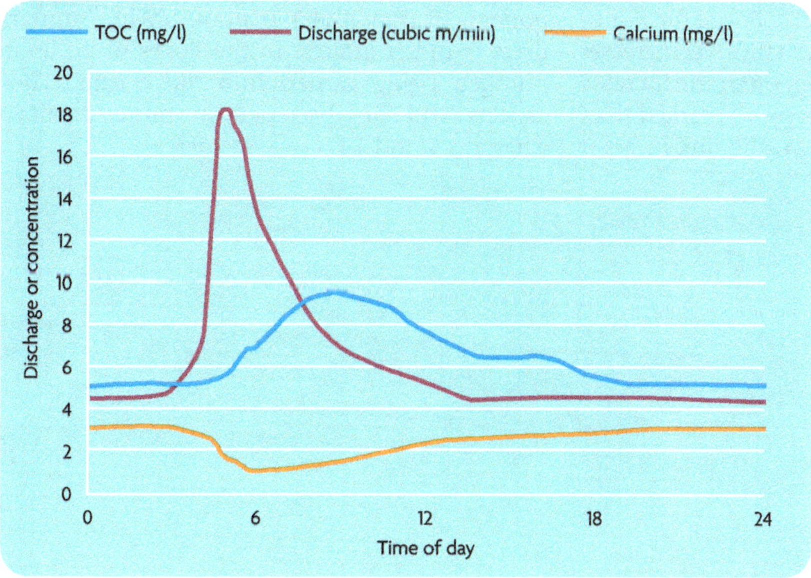

Figure 2.24 Graphs showing how calcium concentration falls but TOC concentration rises during periods of rising discharge in a small upland catchment in northern Europe. Base flow calcium concentrations may be much higher in a limestone catchment.

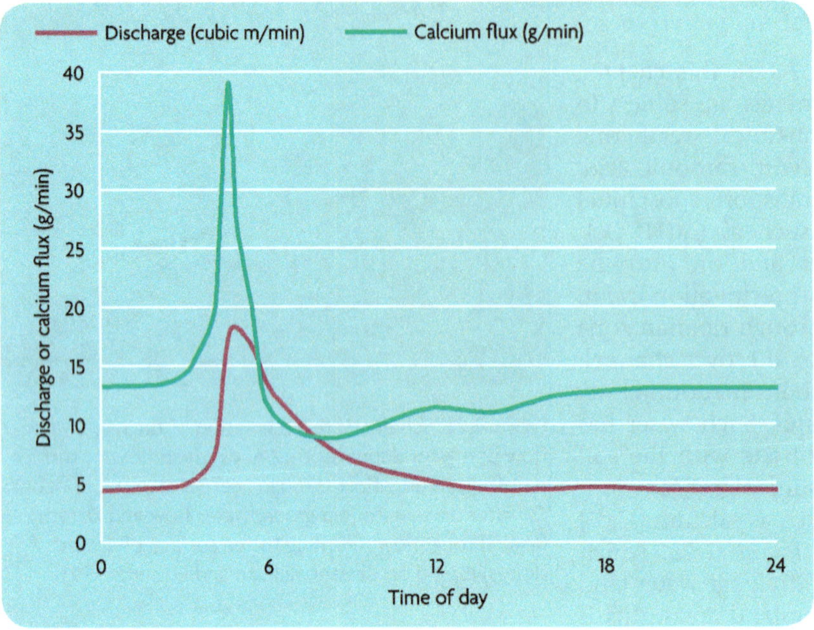

Figure 2.25 Graphs showing how calcium flux increases during periods of rising discharge in the same small upland catchment, even though the calcium concentration falls.

the more mineral-rich subsoil (which was relatively high when groundwater dominated flow) starts to fall (Figure 2.24). Alkalinity, which is a measure of HCO_3^-, falls in the same way as calcium concentration. This is to be expected as HCO_3^- is often the dominant anion that balances the positive charge of calcium cations (Ca^{2+}).

Note that, relatively, the rise in discharge (4-fold) is higher than the decrease in calcium concentration (3-fold) in this instance. Therefore the *flux* of calcium leaving the catchment in the river increases during storm events. Moreover, often the increase in flux during storms is much more marked than in this example. In Figure 2.24, the flux in g per minute at any chosen time can be calculated by multiplying the concentration, after conversion to g per cubic metre, by the flow in cubic metres per minute. This has been done to produce Figure 2.25. Measurements of fluxes of solute species lost in rivers are very important because they provide information about the natural rates of decline of catchment soils. For carbon, fluxes transported in rivers to the oceans are an important component of the global carbon cycle. Similarly there is much interest in the fluxes of pollutant nitrogen being carried into the oceans. These aspects will be discussed in more detail in Chapters 9 and 14.

POLICY IMPLICATIONS

Global water cycling

- Policy makers need to know how national- and global-scale water budgets are changing, especially in the context of assessing risk from climate change as a consequence of rising carbon dioxide, methane and other greenhouse gas concentrations in the atmosphere.
- Understanding of the need for groundwater recharge, and of what limits the amounts of water that can be abstracted for human use, is essential when formulating policies to protect groundwater resources over the long term.
- Detailed spatial and temporal data on precipitation amounts are essential for sustainable use of water for irrigation purposes and for management of water supplies.
- An understanding of the relationships between climate and river hydrographs is crucial for flood prediction.
- In more arid climates especially, an understanding of the risk to groundwater quality of sea water incursion as a consequence of over abstraction of groundwaters is essential.
- A sound understanding is necessary of risks of groundwater contamination associated with groundwater flow.
- Planners need to keep water supply limitations at the forefront of planning how to deal with population increases.
- Knowing the input/output balances for base cations and nutrient elements allows assessment of sustainability of natural ecosystems and of soil use and management practices over mid- to long-term timescales.
- Policy makers need to be aware of the changes in water chemical quality that occur during storm events or rapid snow melt because of potential adverse effects on aquatic biodiversity.

CHAPTER REVIEW EXERCISES

Exercise 2.1

If the world's population is 8 billion people, and each aspired to use 100 litres of water each day, calculate the total volume of water that would be required to meet this aspiration in cubic kilometres. How does your result compare with the total volume of water present in the world's lakes and rivers (see section 2.2.1)?

(A litre is 1000 cm^3; a billion is 10^9; 1 km^3 is 10^{15} cm^3.)

Exercise 2.2

If 1000 mm of precipitation falls on a watertight catchment with an area of 1000 hectares (1 ha = 10^4 m^2), in an area where annual actual evapotranspiration is 400 mm, calculate the volume of runoff leaving the catchment each year in its river system. Express your answer in both cubic metres (m^3) and litres.

What assumption is being made about the total volume of groundwater in the catchment in this calculation?

Exercise 2.3

We saw in this chapter that maps showing spatial distributions of precipitation at regional or national scales are often produced by careful plotting of isohyets, which are lines (precipitation contours) joining points where total precipitation over the assessment period is predicted to correspond to the precipitation value of the contour being plotted. What topographic and climatic features should be considered when plotting isohyets and why?

Exercise 2.4

Use the information provided in section 2.2.1 to calculate the total volume of water per year, in thousands of cubic km:

1 in precipitation falling over the oceans.

2 in precipitation falling over land surfaces.

3 passing from the oceans up into the atmosphere.

In each case, also calculate what percentage your result is of the total amount of fresh water on Earth.

REFERENCES

Beven, K.J. (2004) *Rainfall-Runoff Modelling: The Primer* (Paperback). John Wiley & Sons Inc., 372 pp.

Cresser, M., Killham, K. and Edwards, A. (1993) *Soil Chemistry and its Applications*. Cambridge University Press, Cambridge, 192 pp.

Edwards, A.C., Creasey, J. and Cresser, M.S. (1984) The conditions and frequency of sampling for elucidation of transport mechanisms and element budgets in upland catchments. *Proceedings of the International Symposium on Hydrochemical Balances of Freshwater Systems*, Eriksson, E. (ed.), Uppsala, Sweden, IAHS Publication No. 150, 187–202.

Grotzinger, J., Jordan, T.H., Press, F. and Siever, R. (2007) *Understanding Earth*, 5th edn. W.H. Freeman and Co., New York, 579 pp. plus appendices.

Lewin, J. (ed.) (1981) *British Rivers*. George Allen & Unwin London, Boston, Sydney, 216 pp. + xii.

CHAPTER 3

The origins of the atmosphere

Nicola Carslaw

Learning outcomes

By the end of this chapter you should:

- Understand current ideas about how the atmosphere formed.
- Appreciate the composition of the three atmospheres of Earth and what caused each of them to form.
- Appreciate the symbiosis between evolution of life and development of the atmosphere.
- Understand the main theories for the creation of the first life forms.
- Be aware of the forms of early life and how these developed into more complex forms.
- Understand what photosynthesis is and how it dramatically changed the composition of the atmosphere.
- Be aware how and why the ozone layer formed and the implications it has for life on Earth.
- Know the current composition of the atmosphere and the major perturbations caused by the advent of industrialisation.

3.1 Introduction

The atmosphere is often referred to as the thin blue veil or envelope. When viewed from space, it is easy to see why (Figure 3.1). Our atmosphere, only about 100 km thick, forms a relatively thin layer around the Earth with a diameter of ~12,750 km. Yet this thin veneer protects us from the harshest extremes of the sun, in terms of both UV light and also temperature. More importantly, it provides exactly the right conditions for all life in its many forms on both land and in the oceans to thrive. Without the atmosphere we simply would not exist.

We tend to take the presence of our atmosphere very much for granted. Few of us give much thought to the fact that the composition of the atmosphere is exactly as we need it. However, the composition of the atmosphere today, largely nitrogen molecules (78 per cent) and oxygen molecules (21 per cent), came about through a remarkably symbiotic relationship with first the presence of very basic life forms and then with increasingly complex forms of life. This remarkable relationship is put into even sharper focus when one considers the very different fates of our nearest neighbours in our solar system, Venus and Mars. Despite their relative proximity to the Earth (at least in terms of the vast realms of space), the atmospheres of these two planets are very different to that of Earth, as discussed briefly in Chapter 1 and in more detail later in section 3.10.

Figure 3.1 The thin blue veil of the Earth's atmosphere around the planet.

Sources: Credit NASA Goddard Space Flight Center Image by Reto Stöckli (land surface, shallow water, clouds). Enhancements by Robert Simmon (ocean colour, compositing, 3D globes, animation). Data and technical support: MODIS Land Group; MODIS Science Data Support Team; MODIS Atmosphere Group; MODIS Ocean Group. Additional data: USGS EROS Data Center (topography); USGS Terrestrial Remote Sensing Flagstaff Field Center (Antarctica); Defense Meteorological Satellite Program (city lights).

3.2 The primitive atmosphere of Earth

As discussed in Chapter 1, section 1.2, the Earth was formed ~4.56 billion years ago through a process of condensation and accretion of solid particulates and gases in the cosmos. The Earth at this point was a ball of molten rock surrounded by an atmosphere that was heavily influenced by the gases present in the initial generating nebula, namely hydrogen and helium.

3.3 The second atmosphere of Earth

With time, the Earth began to cool and a number of processes changed the original composition of the Earth's atmosphere:

- The relatively highly volatile components, hydrogen and helium gases, boiled away into space.
- Volcanic eruptions added new gases to the atmosphere.
- Comets entering the Earth's atmosphere brought new material.

Earth and the other inner planets are all relatively close to the sun (Figure 3.2). The absence

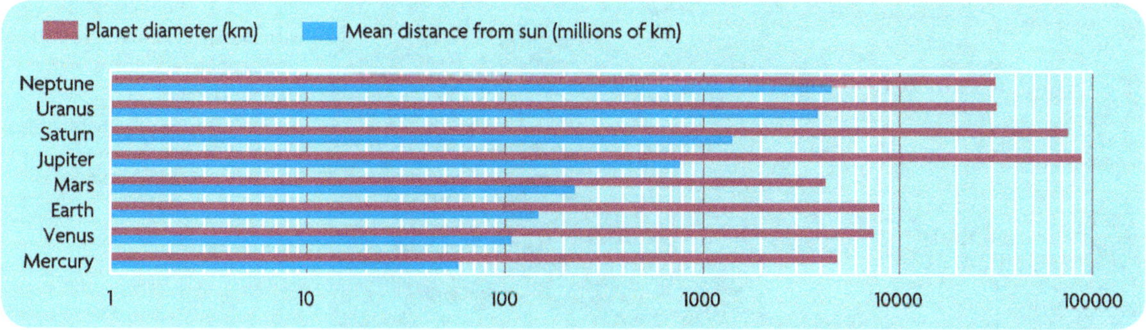

Figure 3.2 The mean distances of the planets in our solar system from the sun, and planet diameters. The hot inner planets lost their volatile material which went on to help form the outer planets in the colder reaches of space.

of a differentiated core (see section 1.2.2) meant that there was no magnetic field; consequently, the intense solar wind blew components of the early atmospheres on Earth, especially those with low atomic and molecular masses such as hydrogen and helium, away into space. This volatile material moved to the colder reaches of space to form the outer planets, or *gas giants*.

Consequently, the inner planets are small, made of rocks and metals, with low concentrations of very low atomic mass elements and very low molecular mass compounds in their atmospheres. The outer planets on the other hand contain rocky cores, but consist mainly of ice and gases captured by their gravitational fields from space and the outgassing of the inner planets.

A traditional view of the modification of the early atmosphere is that as the early Earth was very hot, a dense atmosphere soon formed from gases expelled from the interior through volcanic-like activity. Obviously the composition of the underlying rocks was critical in determining the composition of the expelled gases. It is likely that the major components in the atmosphere from such activity were methane (CH_4), hydrogen gas (H_2), nitrogen gas (N_2), water vapour (H_2O), carbon monoxide (CO) and carbon dioxide (CO_2). Sulphurous and halogenated compounds may also have been present, as well as ammonia. There is still much debate over the exact composition of the second atmosphere of Earth. However, more important than the exact composition of the expelled gases, was the absence of free oxygen. Oxygen levels in the atmosphere were very low and certainly not able to sustain life as we now know it.

A more recent theory on the modification of the early atmosphere was that water vapour was provided through the bombardment of Earth by comets (Frank *et al.*, 1986). Dark spots observed on satellite images, based upon the wavelength of light they absorbed, were postulated to correspond to clouds of water vapour. The theory is that every few seconds these so-called 'cosmic snowballs' (of $\sim 10^5$ kg) approach the Earth and break up in the atmosphere, leaving behind a large trail of water vapour (Figure 3.3).

Over geological time, these snowballs could have provided all of the water in our oceans and in our atmosphere. Some scientists believe that the comets also may have bought other geochemically and biologically important compounds into the Earth's atmosphere (see section 3.4.3).

The presence of water at the surface of the Earth permitted some of the atmospheric gases to dissolve, providing another means of altering the atmospheric composition. For instance, large quantities of the CO_2 in the early atmosphere are likely to have dissolved in the newly forming oceans, so keeping the atmospheric concentration of this gas much lower than it otherwise would have been. Liquid water is also vital for life as we now know it.

The second atmosphere of Earth was therefore very different to the primeval atmosphere. Much of the volatile material had been lost to space, and

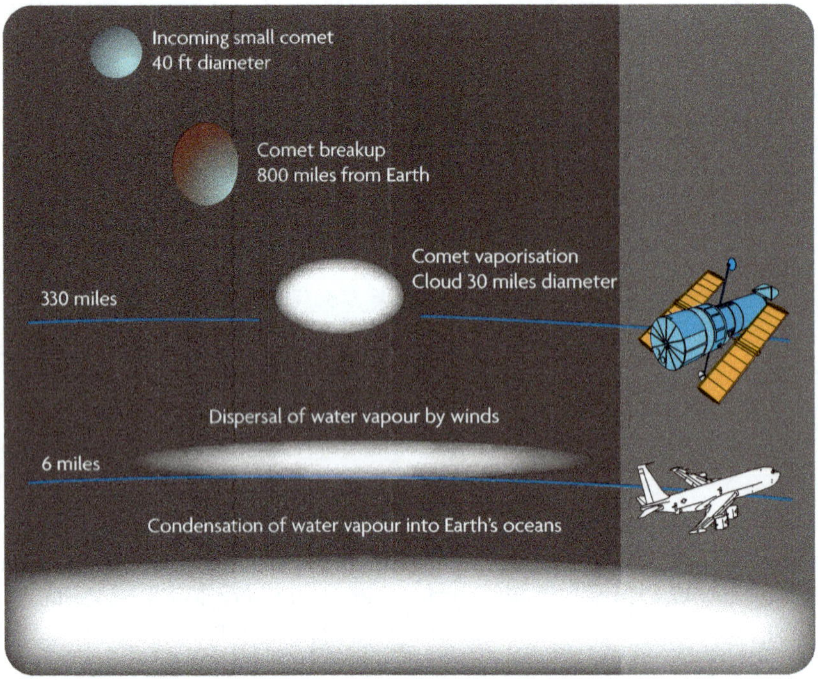

Figure 3.3 The process by which comets enter the Earth's atmosphere and leave water vapour behind.
Source: http://smallcomets.physics.uiowa.edu/www/faq.htmlx University of Iowa/NASA

what remained was very much a reducing atmosphere, with very low concentrations of free oxygen.

3.4 The evolution of life and the modern atmosphere

We know that sometime around 3.5 billion years ago life evolved on Earth, but for a long time, scientists wondered how this might have happened. A conundrum existed known as the *Oxygen Catch*-22. From experiments carried out in laboratories, scientists knew that the simple amino acids necessary to form the building blocks of life (proteins) had to be formed under reducing conditions (amino acids react with oxygen to form CO_2 and water). However, the low concentration of oxygen in the atmosphere at that time meant that there was no ozone layer. Life on Earth is currently protected from the strongest UV radiation emitted by the sun through the presence of the ozone layer. Consequently, many believe that any life that managed to form on the surface would have been subjected to very high levels of UV radiation (although in a reducing atmosphere high concentrations of gases such as ammonia and hydrogen sulphide could absorb much of the UV radiation). Although the reducing conditions potentially allowed life to form, many argue that it is hard to see how it could be sustained and develop in the absence of an ozone layer. In recent years, however, suggestions have been put forward for how life on Earth could have began.

3.4.1 Prebiotic soup theory

In the 1950s, Stanley Miller (1953) synthesised amino acids in a laboratory experiment as illustrated in Figure 3.4.

He sent electrical sparks (as a proxy for lightning) through a sealed flask with 'an atmosphere' containing gases that were representative of the second atmosphere of Earth, namely ammonia, methane, hydrogen and steam (Miller, 1953). The chemicals formed by the electrical sparks rained down into the 'ocean' below. After about a week of these conditions, the solution in the flask went red and murky and was found to contain amino acids as well as cyanide and formaldehyde. Thus the popular theory of life evolving in some sort of prebiotic 'soup' began to gain much credence.

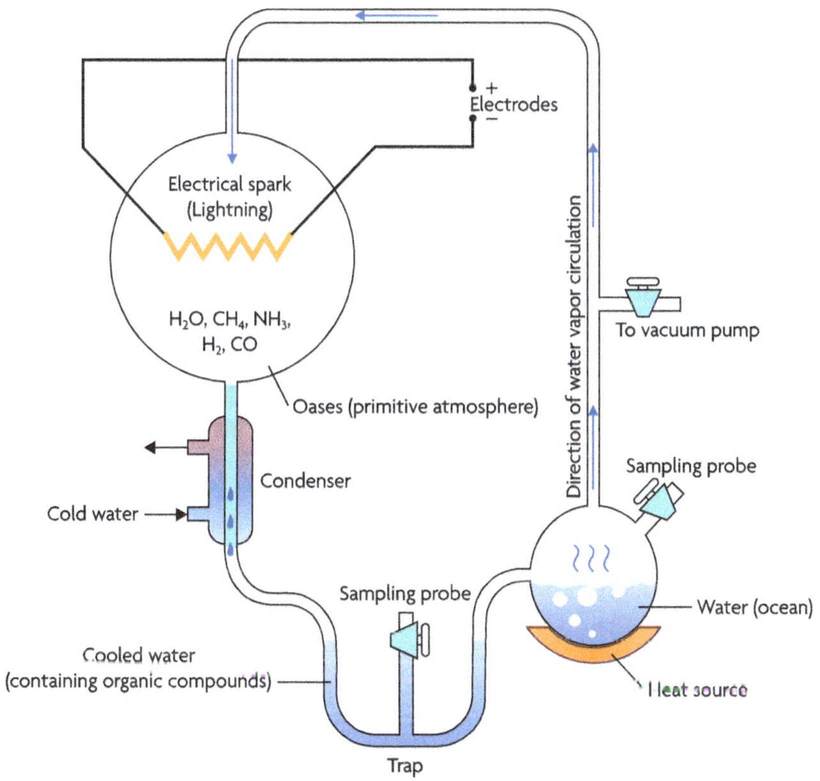

Figure 3.4 Stanley Miller's experimental set-up of 1953 to try to simulate the creation of early life.
Source: http://en.wikipedia.org/wiki/Miller–Urey_experiment

Although Miller synthesised amino acids in this way, his theory that life began in such a manner was criticised by many. Amino acids are the building blocks of proteins which are the building blocks of life. Although Miller produced amino acids, there are still a number of steps that need to occur before proteins are made. For instance, 22 different amino acids need to form into a long chain-like structure to produce a protein (Figure 3.5). Miller managed to produce the building blocks, but not the final assembled structure.

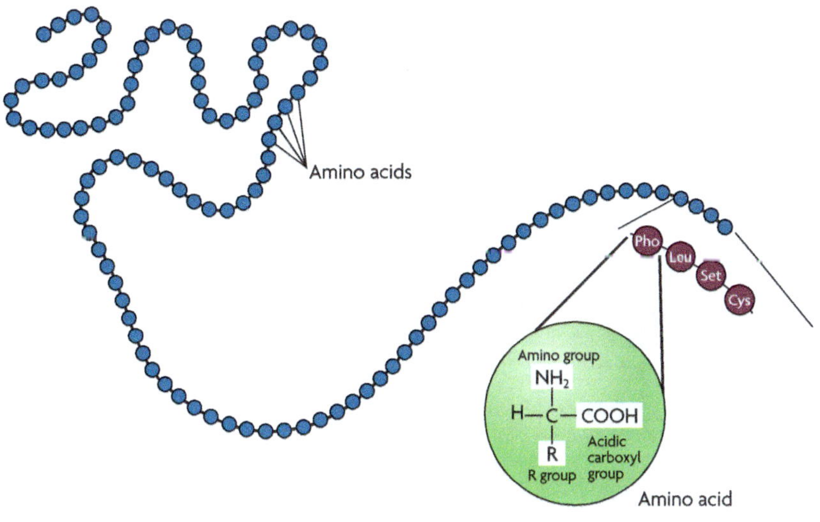

Figure 3.5 Protein primary structure showing a chain of amino acids.

A second problem with the theory was that scientists increasingly came to believe that the composition of gases used by Miller in his original experiment did not represent the atmosphere of the Earth at the time life evolved. Many now think the atmosphere was much more CO_2 rich than Miller assumed and that gases such as methane and ammonia would have been photolysed under the high UV light conditions. Finally, many scientists thought that the 'soup' was simply too dilute to enable the chemicals to come together to react in the first place.

Interestingly, in 2008, Johnson *et al.* (2008) reanalysed the extracts from one of Miller's less well-known experiments, and found a much wider variety of amino acids than reported for the more famous experiment. They speculated that given that lightning and volcanically derived gases were common in the early atmosphere, it was possible that prebiotic compounds could have been synthesised and then further processed into proteins.

Had Miller's amino acids gone on to form proteins and early life forms, they would have needed protection from the strong UV light levels. This may have been possible in rocky crevices, sheltered tidal pools or within sediments but not at the surface.

3.4.2 Pioneer metabolic theory

In 1977, an oceanographer by the name of Jack Corliss made an incredible discovery while exploring a volcanic ridge at the bottom of the Pacific Ocean. Oceanographers had considered the ocean floor to be mostly cold and barren, but Corliss observed clams, tube worms, and strange microbes thriving at the vents. Given that these creatures were able to exist in such hostile conditions (high temperature and pressure and also extremely sulphurous), he began to wonder if life could have evolved in such locations.

A submarine hydrothermal vent (SHV) is basically an opening in the ocean floor, where superheated water from deep below the Earth's crust can escape. This water contains crustal material in the form of minerals, such as sulphides, and as it passes through the vent, the minerals within the water precipitate out to form chimney-like structures on the sea bed (Figure 3.6).

A major advantage of the proposal put forward by Corliss was that any life that evolved at SHVs would be protected from the sun by the sheer volume of water above it. However, despite the undoubted promise of this theory, for the next

Figure 3.6 Sully Vent in the Main Endeavour Vent Field in the North-East Pacific. The base of the black smoker is covered with a bed of tube worms, while an acoustic hydrophone and resistivity-temperature hydrogen probe are also visible in the image from the dive in 2004.

Source: Image from http://www.pmel.noaa.gov/vents/gallery/smoker-images.html courtesy of the NOAA PMEL Vents Program.

20 years scientists remained absorbed with Miller's prebiotic soup theory.

The next development for the pioneer metabolic theory came when Günter Wächtershäuser proposed a possible mechanism for life to form under such conditions. He noted the dependence of respiring life forms today on the citric acid cycle (a series of chemical reactions catalysed by enzymes that cells use to extract energy from food) and speculated that this cycle could be the basis for the early evolution of life. He considered the basic starting materials present at the time – high temperatures and pressures, volcanic gases and sulphide deposits – and proposed that life on Earth evolved through a series of reactions beginning with the formation of acetic acid. He got over the dilution problem of the prebiotic soup theory by proposing that the surface of the sulphide deposits provided a reaction centre, effectively concentrating the reactants.

In the absence of enzymes, metal ions associated with the sulphur deposits were the necessary catalytic species to enable the carbon, hydrogen and oxygen atoms to combine to produce acetic acid, and then more complex molecules such as peptides through further reactions. The key reactions are summarised in Figure 3.7 and the individual steps have since been demonstrated through experiment (Wächtershäuser, 2000). Wächtershäuser believed that once this early metabolism evolved, it became self-sustaining and it was only much later that other basic elements were added in, such as genetic code.

What is still unclear, however, is which element is the right metal catalyst that will cause the peptide chains to lengthen and reproduce themselves and effectively autocatalyse this cycle. In addition, any proposal for a likely self-replicating peptide or RNA strand, would have to be supported by an explanation of where it originated given the primitive conditions.

Critics argue that there are other problems with this theory. For instance, the temperature near the centre of a SHV (~350 °C) is much higher than the temperature that Wächtershäuser used in the laboratory to generate acetic acid, which at 100 °C, was closer to the temperature at the fringe of the SHV. At higher temperatures, it was argued that the acetic acid would be destroyed after only a few minutes, leaving little time for it to go on and form more complex material. Some critics also argue that the same peptide links could have been made in the primordial soup mixture, just more slowly at the lower temperatures. However, Wächtershäuser argues that given the molecular links are broken nearly as fast as they're created in both situations, life has more chances to evolve given the higher activity rates at the SHVs.

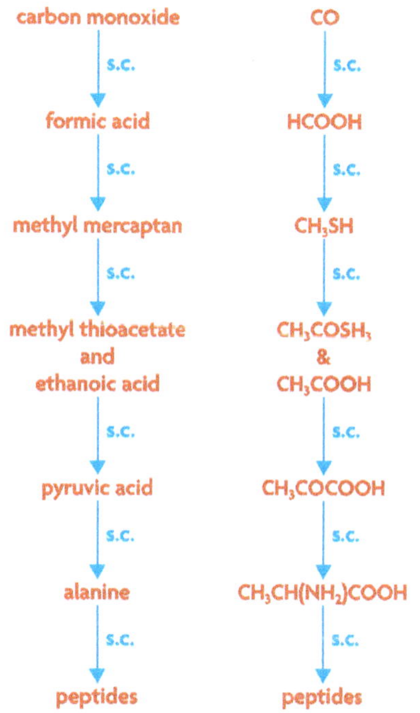

Figure 3.7 Sequence of reactions, based upon the work of Wächtershäuser (2000), thought to occur catalytically on Ni/Fe/Co sulphide surfaces, that convert carbon monoxide to peptides.

The pioneer metabolic theory opens up some other interesting possibilities. There are other planets in the solar system that are subject to geothermal forces. There has been much speculation in the past that life may have once existed on Mars for instance. In addition, many scientists believe that Jupiter's ice-covered moon Europa may hide a liquid ocean which contains similar SHV systems to those found on Earth. Assuming the Europa Jupiter System Mission goes ahead in 2020, some scientists believe that we will discover that we are not the only life forms in our solar system.

3.4.3 Panspermia

A much more controversial theory at present suggests that life on Earth may have been seeded from another planet. It is well established that Martian crustal material (from meteorites) has been found on Earth. Further, biologists believe that some organisms exist that potentially could survive short trips inside meteorites. Therefore, in theory at least, some organisms may have originated on another planet and then arrived on Earth. Such organisms may have provided the early seeds of life, which continued their evolution once on Earth.

Recent scientific evidence appears to add weight to some parts of this theory as summarised by Warmflash and Weiss (2005). For instance, it has now been shown that meteoritic material is exchanged between planets, and that some of the microorganisms within this material could survive ejection into space (e.g. from Mars) and entry into the Earth's atmosphere. However, whether such processes occurred and if so, how important they were in creating life on Earth is impossible to say at present. One fascinating prospect lies with future missions to Mars. As Warmflash and Weiss (2005) point out, should the biochemistry on Mars show similarities to our own, the likelihood of panspermia having had an influence on life on Earth becomes much greater. In a sense though the question would then become: 'How did life first form on Mars?'.

3.4.4 Summary

This section has hopefully demonstrated that there is still much that we do not understand about how life on Earth evolved. What we know for sure is that life evolved at some point from non-living matter, and then evolved over time into the rich complexity of life forms we see on the planet today. There is still much debate over this topic, which is often hostile, particularly between the prebiotic soup and the pioneer metabolic camps (see for instance the Letters page in *Science*, Volume 315, pp. 937–939 in 2007).

We will never know for sure how life began, but scientists are hopeful that in the next few years, we will at least be able to put forward some reasonable accounts of how it may have happened. No doubt, faced with continued uncertainty, creationists will continue to faithfully believe in some superpower as the creator of life at some point in time.

3.5 Early life forms

There is much debate in the literature about the time that the first life appeared on Earth, with some estimates putting life on Earth as early as ~3.85 Ga ago. However, this is very close to the end of the late heavy bombardment period and there is uncertainty over interpretation of the fossil records (Brasier *et al.*, 2006). General consensus currently is that early life forms were probably established by around 3.5 Ga ago (López-García and Moreira, 1999; Brasier *et al.*, 2006).

Based on what we know about species that live around SHVs today, it is believed that the earliest life forms were Archaea (from the ancient Greek meaning 'ancient things'), which are single-celled microorganisms (Figure 3.8). Although many Archaea live in mild environments, some prefer extreme environments such as SHVs, geysers and inside volcanoes (thermophiles). Such species are likely to have been among the first living species on Earth.

Early Archaea were autotrophs, which meant that they had to generate energy to produce food inside their cells, rather than consuming food directly from the environment like heterotrophs. They are likely to have used lithotrophy, which is the process where chemical energy is extracted from inorganic molecules. For instance, microorganisms could gain energy for chemosynthesis from the process represented in reaction R1:

$$4H_2 + CO_2 \rightarrow CH_4 + 2H_2O + \text{energy} \qquad \text{(R1)}$$

Moreover, the greenhouse warming produced by the CH_4 formed in this reaction would have more than compensated for the cooling produced when CO_2 (also a greenhouse gas) was removed. Methane is a much more potent greenhouse gas than CO_2 as we will see in Chapter 8, section 8.8. This conversion between CO_2 and CH_4 would have kept conditions warmer during the Archaean aeon than they otherwise would have been. Indeed, there are many such intricate links between the

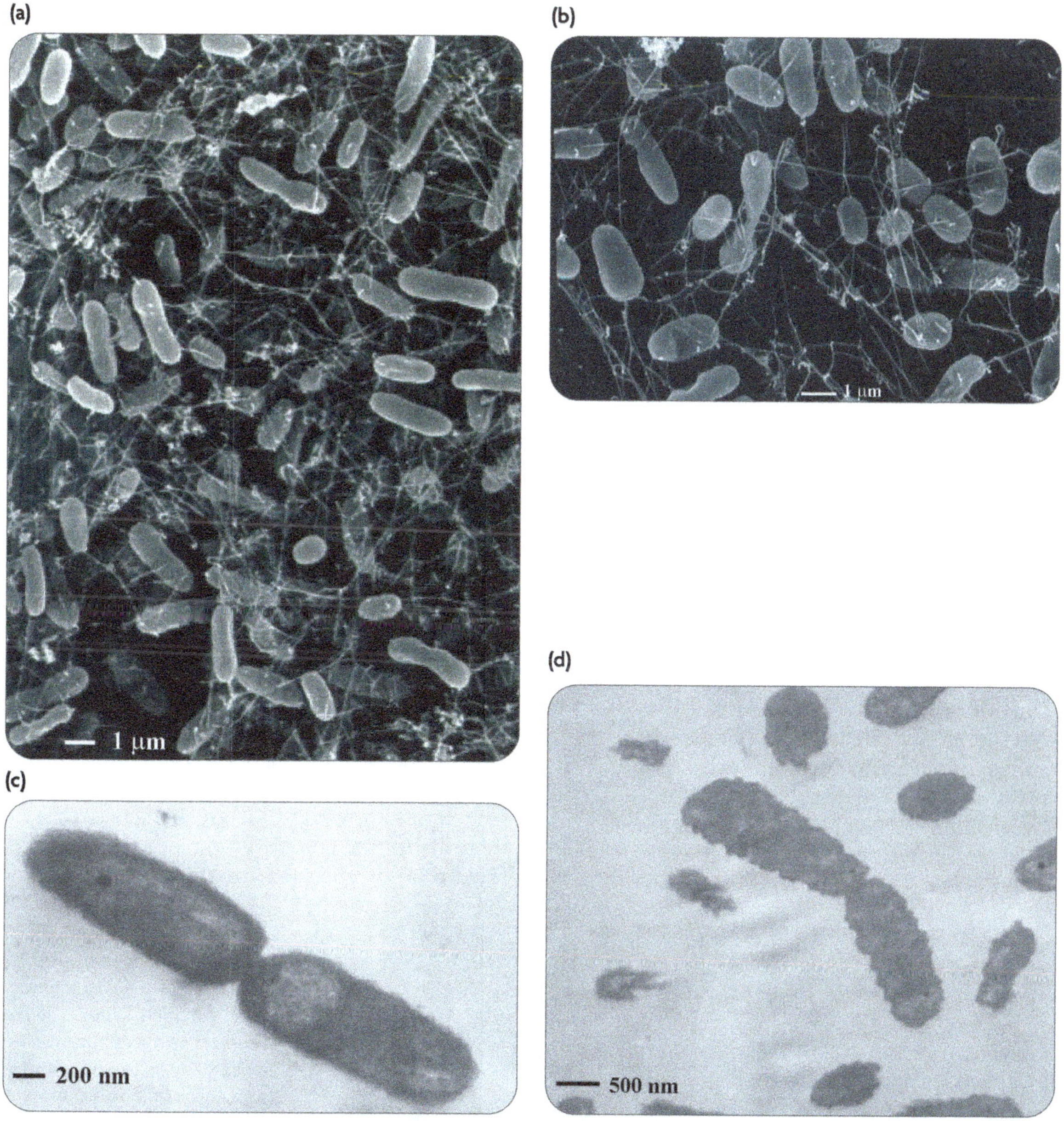

Figure 3.8 Images of a sub-seafloor thermophile isolated from a deep-sea hydrothermal vent. Images A and B show scanning electron micrographs while C and D show transmission electron micrograph thin sections.

Source: Image courtesy of Julie Huber at NOAA: http://oceanexplorer.noaa.gov/okeanos/explorations/ex1104/background/microbes/media/microbes_sem_tem.html

composition of the atmosphere and the development of life forms. We will see this throughout this chapter and later in Chapter 10 when we consider the Gaia Hypothesis and how living organisms regulate their environmental conditions.

Lead-in question

What happened to increase the oxygen concentration?

Some early microorganisms contained molecules that could capture light energy, and an early form of photosynthesis became possible (see also Chapter 4). Sunlight was generally a reliable source of energy and as the photosynthetic microorganisms that utilised solar energy died, their cell contents became available for other microorganisms (such as heterotrophs) to consume. In effect, such a process provided one of the Earth's first food webs.

The earliest photosynthetic reactions could probably have looked very different to the standard reaction we recognise today (see section 3.6). For instance, based on the chemical products found by hydrothermal vents in the deep oceans, we could suggest a chemosynthetic reaction or a photosynthetic reaction such as R2:

$$3H_2S + 2CO_2 + h\nu \rightarrow (CH_2O) + CH_4 + SO_3 + 2S \quad \text{(R2)}$$

Such a reaction would have enabled solar energy to be harnessed to form organic material (represented by the basic unit, CH_2O). The sulphur trioxide would dissolve to form sulphuric acid. The term 'hv' denotes a photon of light of sufficient energy to break the relevant bond (the process of *photolysis*). This relationship is defined by the Plank–Einstein equation whereby:

$$E = h\nu$$

where E is the energy of the radiation in Joules, h is Plank's constant (6.626×10^{-34} J s) and ν is the frequency of light in s^{-1}. Stronger bonds need light with more energy to break them apart than weaker bonds.

What about the production of oxygen at this time? Only small amounts of O_2 were being emitted into the atmosphere via the photo-dissociation of water:

$$2H_2O + h\nu \rightarrow 2H_2 + O_2 \quad \text{(R3)}$$

Such a process was possible owing to the high levels of UV that were able to reach the surface of the planet in the absence of an ozone layer. However, it is believed that this process provided perhaps 10^{-14} of the present atmospheric level of O_2 (Buick, 2008). Luckily for us, some species developed the ability to produce oxygen via photosynthesis using CO_2 and water, and the atmosphere entered a completely new stage in its composition. If that had not happened there'd be no 'us'!

3.6 The rise of photosynthesis

The earliest producers of oxygen via photosynthesis of water were probably cyanobacteria, sometimes known as blue-green algae because of their distinctive colour. Blooms of these bacteria (Figure 3.9) in enclosed water systems can be toxic,

Figure 3.9 A large bloom of aquatic cyanobacteria spread across Lake Atitlán, Guatemala in the autumn of 2009. The lake is subject to agricultural runoff, sewage, and increased runoff from deforestation around the lake basin.

Source: Image by Jesse Allen from NASA Earth-Observatory, http://earthobservatory.nasa.gov/IOTD/view.php?id=41385, based on data from the NASA/GSFC/METI/ERSDAC/JAROS, and US/Japan ASTER Science Team.

and sometimes lead to such water bodies being closed for recreational purposes. Cyanobacteria are ubiquitous on the planet even now, and come in many different shapes and sizes. They are small and usually unicellular, but often grow in colonies that are big enough to see.

In terms of the fossil record, there is evidence for 'microbial mats' dating back around 3.5 Ga (López-García and Moreira, 1999). Such mats are multi-layer sheets of microorganisms (mainly Archaea and Bacteria) that thrive at the interface between different types of materials, particularly on submerged or moist surfaces. They can survive at a range of temperatures, pressures and altitudes and as such, are ubiquitous on the planet. The best known forms of microbial mats are stromatolites (Figure 3.10), still common in the tidal flat areas of Australia today, as well as in the fossil records. Such features are formed when sediment produced by waves adheres to the slime secreted by the bacteria and is laid down, layer upon layer. When this process is repeated numerous times, large pebble like structures are able to form. It is thought that ancient stromatolites were formed through similar processes.

Cyanobacteria (unlike Archaea) were able to use the sun's energy to split water and carbon dioxide to produce organics and oxygen through photosynthesis:

$$nCO_2 + nH_2O + h\nu \rightarrow n(CH_2O) + nO_2 \qquad \text{(R4)}$$

Figure 3.10 Stromatolites growing in Hamelin Pool Marine Nature Reserve, Shark Bay in Western Australia.

Source: Image from NASA/courtesy of nasaimages.org. Photo by Mark Boyle. http://fettss.arc.nasa.gov/collection/details/shark-bay/

The organic material in this equation (units of CH_2O) could go on to produce the complex molecules required for living tissue, such as starch and cellulose.

To begin with, most of the oxygen produced by cyanobacteria would have been almost instantly removed through reaction with the predominantly reduced volcanic gases and weathering of crustal minerals present at the time, notably those containing iron, but also sulphides. Indeed, the oxidation of iron (or rusting) left behind banded iron formations, suggesting that for millions of years any oxygen released into the atmosphere was rapidly removed. The first evidence that oxygen began to accumulate in the atmosphere is from around 2.3 Ga, when sediment records suggest that rivers ran red with dissolved iron (Cowen, 2005). Such an observation is consistent with the presence of enough oxygen in the atmosphere to oxidise iron minerals at the surface, which were then washed into the rivers. This period of time is sometimes referred to as the *Great Oxidation Event*, as it heralded the start of the accumulation of free oxygen in the atmosphere (Figure 3.11). Estimates of how long ago exactly this event happened vary between 2.5 Ga to 2 Ga, but we know that somewhere in this time range, the atmosphere shifted from an anoxic to an oxic environment (Kump, 2008). Most importantly, the oxygenation of the atmosphere permitted aerobic respiration and the subsequent development of more complex life forms.

Following the Great Oxidation Event, scientists believe that oxygen concentrations stayed fairly constant until about 600 million years ago. At around this time, the fossil record shows that a variety of different organisms became much bigger, consistent with higher O_2 levels. Nobody is sure what made the oxygen concentration increase at this time, although it does seem to be associated with extreme glacial periods, so-called *'snowball Earth'* episodes.

Lead-in question

Does oxygen in the atmosphere do anything else useful apart from supporting life?

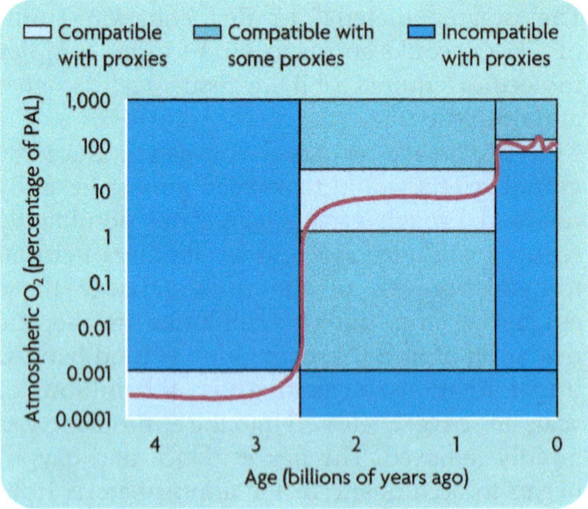

Figure 3.11 Change in atmospheric O_2 concentrations expressed as a percentage of present atmospheric level (PAL) over geological time. Oxygen concentrations began to increase around 2.5 Ga ago (although the exact timing is unclear) and are thought to have reached present levels around 600 million years ago. The figure, reproduced from Kump (2008), shows that much of what we know about past oxygen levels is inferred from proxies (and can often only provide upper and lower limits). There is a wide range of possible O_2 concentrations over much of geological time.
Source: Kump (2008).

3.7 Formation of the ozone layer

As oxygen accumulated in the atmosphere, an ozone (O_3) layer formed around the planet and was able to protect developing life forms from harmful UV light. Ozone molecules form when oxygen atoms react with oxygen molecules. The oxygen atoms are formed when high energy UV light strikes an oxygen molecule splitting it in two atoms. These oxygen atoms can then recombine with oxygen molecules to form ozone molecules, through a series of reactions (R5–R8) known as the Chapman mechanism:

$$O_2 + h\nu \rightarrow O + O \quad \text{SLOW} \quad \text{(R5)}$$

$$O + O_2 + M \rightarrow O_3 + M \quad \text{FAST} \quad \text{(R6)}$$

$$O_3 + h\nu \rightarrow O + O_2 \quad \text{FAST} \quad \text{(R7)}$$

$$O + O_3 \rightarrow 2O_2 \quad \text{SLOW} \quad \text{(R8)}$$

Note that 'M' in reaction R6 is a third body (usually a dinitrogen or oxygen molecule) that needs to be present to remove excess energy from the reaction.

There is a balance between the reactions in the Chapman mechanism, particularly R5 and R6, which determines the altitude at which the maximum ozone concentration occurs. At high altitudes, the photolysis rate of oxygen molecules is high but the concentration of oxygen molecules is relatively low, as the density of air decreases with height. At lower altitudes, there is a much higher concentration of oxygen molecules, but less UV radiation to dissociate them. Consequently, the maximum ozone concentration, or the 'ozone layer', is found between altitudes of 15–30 km, where the net formation of ozone is greatest.

The Chapman mechanism also demonstrates how the ozone layer protects us. Despite the concentration of ozone in the stratosphere being relatively low (a maximum of only a few parts per million), it can absorb harmful UV radiation from the sun (through reaction R7) that would otherwise cause significant damage to biological tissue.

Figure 3.12 shows that UV radiation is classified into three wavelength ranges.

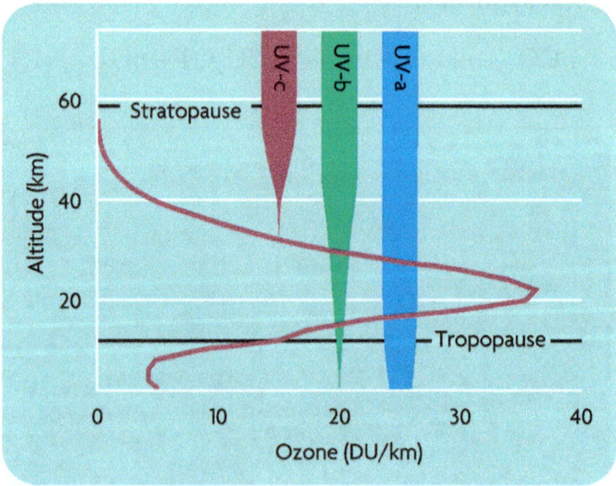

Figure 3.12 The three types of UV radiation. UV-C (red) from 100–280 nm, UV-B (green) from 280–315 nm and UV-A (blue) from 315–400 nm. Stratospheric ozone absorbs much of the high energy radiation allowing life to flourish below.
Source: www.espo.nasa.gov/solveII/implement.html/NASA

The most potentially dangerous UV radiation is known as UV-C. It covers the wavelength range from 100–280 nm and would be extremely harmful to life if it reached the ground. Why is that? Well that's because the lower the wavelength of light, the higher the energy associated with it. Frequency, wavelength and speed of light are related as:

$$c = \lambda\nu$$

where c is the speed of light in m s^{-1} and λ is the wavelength of light in m. By rearranging the Plank–Einstein equation we get:

$$E = (hc)/\lambda$$

showing that energy and wavelength are inversely related. Therefore, lower wavelengths are associated with higher energies and vice versa.

Luckily for us, ozone in the higher reaches of the stratosphere at around 35 km absorbs UV-C radiation and prevents it from reaching the ground. UV-B radiation covers the range from 280–315 nm. The ozone layer is quite effective at absorbing radiation within this wavelength band, and consequently, the intensity of radiation with a wavelength of 290 nm is 350 million times stronger at the top of the atmosphere than at the Earth's surface. Nevertheless, some UV-B still reaches the surface and can cause sunburn and potentially genetic damage in the form of skin cancer if exposure is prolonged. It is for this reason that there was so much concern when a thinning of the ozone layer was observed in the mid-1980s, a subject we return to later in section 3.11.3. Finally, most of the UV-A reaches the surface, but as it is not as genetically damaging as the other two forms of UV, it tends to be of less concern.

A footnote to this section should make the distinction between the stratospheric ozone we have been discussing in this chapter and that existing in the troposphere. While stratospheric ozone protects us from UV and is an essential element for life on Earth, ozone in the troposphere is a greenhouse gas and a harmful air pollutant. For this reason, stratospheric ozone is sometimes referred to as 'good' ozone, while tropospheric ozone is referred to as 'bad' ozone. While this distinction is somewhat simplistic, it serves to remind us of the very different role that ozone can play, depending on its altitude in the atmosphere.

3.8 More advanced life forms

Following the development of aquatic photosynthetic bacteria, the atmosphere began to slowly accumulate biologically produced oxygen. Coupled with the formation of the protective ozone layer, conditions at the surface of the planet became increasingly conducive to the evolution of more complex life forms.

It is likely that Eukaryotes (organisms that consist of one or more cells, containing a nucleus enclosed by a membrane and with well-developed intracellular compartments) developed through a metabolic symbiosis between bacteria and the early methanogenic Archaea. Such an idea helps to explain why present day eukaryotes have features of Archaea (their genetic machinery) as well as of bacteria (such as in their metabolisms) as well as distinct characteristics which have evolved through the symbiosis process (López-García and Moreira, 1999). Indeed, the life that was able to flourish following this early symbiosis has led to the remarkable diversity of life forms we see today (Figure 3.13).

Table 3.1 shows when the major organism groups are likely to have appeared on the planet. It is hard to be precise about when the different groups first appeared, as the fossil record is notoriously difficult to interpret. It is easier to find bigger fossils than smaller ones and it is often difficult to say exactly when one organism group becomes distinct from another. Therefore, for some classes, a range of dates or geological age is given whereas for others, it is possible to date more precisely their first appearance.

The first land animals were likely to be marine creatures that had left the sea, such as some sort of arthropod (e.g. foraging crabs) attracted by organic debris on the beach (Cowen, 2005). For further detail on the history of life and a detailed description of the development of the different organism groups, the reader is directed to the *History of Life* by Richard Cowen.

An excellent educational resource from the British Geological Survey is a Geological Timeline,

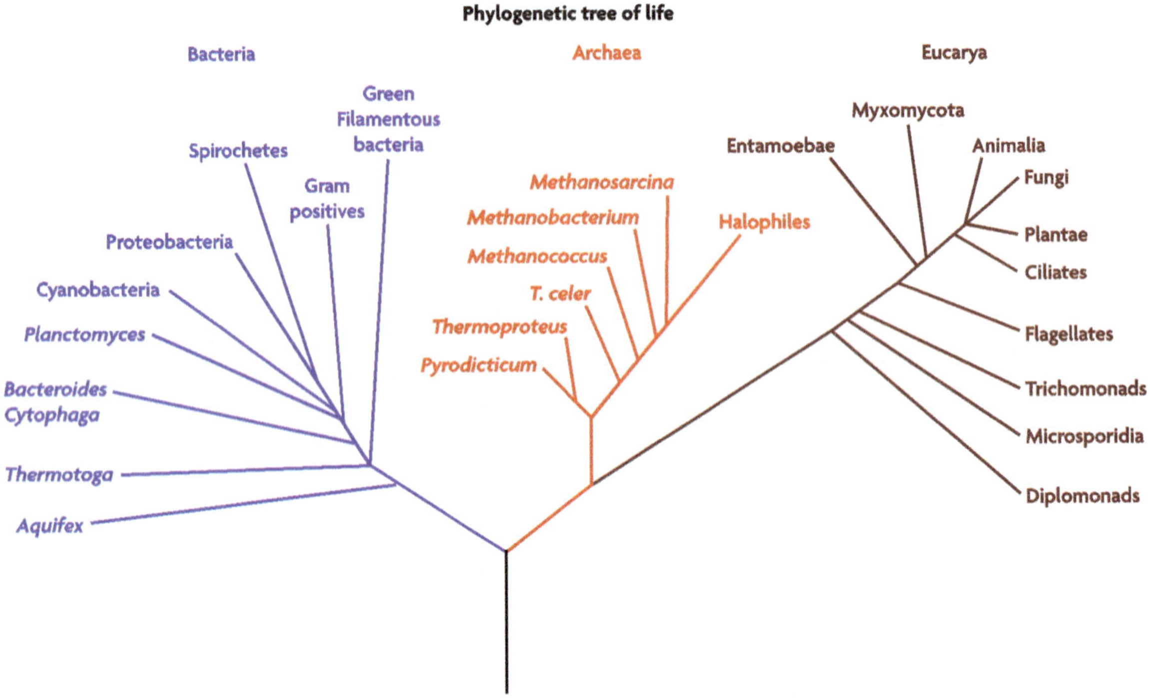

Figure 3.13 Tree of life showing the connectedness between species.
Source: Internet image based on a NASA image. http://nai.arc.nasa.gov/library/images/news_articles/big_274_3.jpg/NASA

Table 3.1 Appearance of major organism groups on Earth based on oldest known fossils for each group (Based on Cowen, 2005).

Organism group	Time of origin
Marine invertebrates	575 Ma
Fish	Ordovician[1]
Land plants	Middle Ordovician[1]
First land animals	Late Silurian[2]
First tetrapods	368 Ma
Dinosaurs	225 Ma
Mammals	Triassic/Jurassic boundary[3]
Flowering plants	Jurassic/Cretaceous boundary[4]

[1] The Ordovician ranged from 488–443 Ma.
[2] The Silurian covered the period from 443–416 Ma.
[3] The Triassic/Jurassic boundary was 199 Ma B.P.
[4] The Jurassic/Cretaceous boundary was 145 Ma B.P.

available at http://www.bgs.ac.uk/education/timeline/home.html. This tool rescales time to illustrate relative age of major biological changes. The Earth is taken to be 46 years old instead of 4.6 billion years and significant birthdays are then highlighted on this scale. So on the planet's 7th birthday, the first living organisms developed. Cyanobacteria appeared for the Earth's 16th birthday, but fish did not arrive until the planet was 41. Animals began to colonise the land when the planet reached 43 years old and the dinosaurs and first mammals at 44. The first birds and flowers didn't appear until the Earth was 45 years old. Four months later the dinosaurs became extinct, followed another 4 months later by the appearance of primates. The first people appeared a mere 22 hours ago, with modern humans 18 hours later. Complex life really has existed for a relatively short amount of time, and the changing composition of the atmosphere has allowed life to adapt and develop.

3.9 Present day atmosphere

The present day atmosphere is a relatively stable mixture of hundreds of different gases which combine to form the thin veil around the Earth referred to in section 3.1. The major components up to about 80 km are O_2 (21%), N_2 (78%) and Argon (1%). There are also smaller amounts of the so-called 'trace gases'. However, despite the low abundance of the latter, we will see in section 3.11 and in Chapter 6 that these trace gases have a profound effect on the planet.

The atmosphere is divided into several layers, which are defined by changes in temperature (Figure 3.14).

Each layer is a *'sphere'* and the break between two different layers a *'pause'*. The layer nearest to the surface is called the troposphere, where temperature falls away gradually with height as you leave the surface. This fall off in temperature occurs as the effect of warming at the surface by the sun becomes less pronounced as you move away from the ground. The tropopause is where our weather occurs and most pollution is emitted and is consequently the area of the atmosphere that has been studied most frequently.

Above the tropopause we have the stratosphere. Throughout this layer of the atmosphere, temperature increases with height, and this is because of the heat given out by the Chapman mechanism as ozone absorbs radiation (section 3.7). The stratosphere has been the subject of considerable research following the discovery of the ozone hole in the 1980s (section 3.11.3).

The mesosphere lies above the stratosphere and away from ozone chemistry, and temperature falls with height. Finally, temperature increases with height again in the thermosphere. At such altitudes, there is a very low air density, so the heat capacity is small and temperature increases with proximity to the sun.

Of note in the northern hemisphere thermosphere is the existence of aurora borealis (or aurora australis in the southern hemisphere). This phenomenon arises because of the interaction of the Earth's magnetic field with the solar wind, which

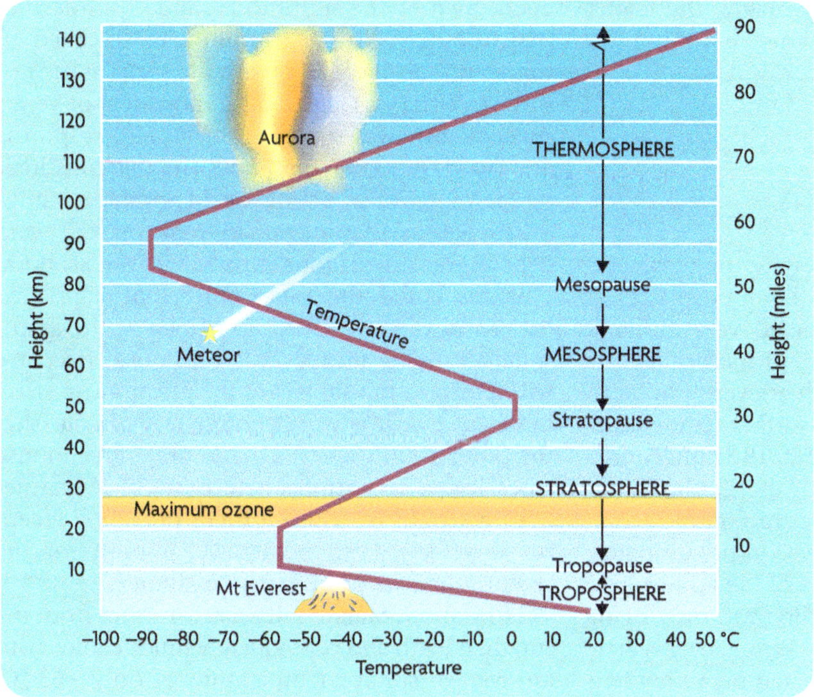

Figure 3.14 The layers of the atmosphere and the temperature variation with altitude.

Source: Lutgens, F.K. and Tarbuck, E.J. (2001), Fig. 1–19, p. 20.

Figure 3.15 Aurora Borealis over the valley Kårsavagge in Northern Sweden, from near the mountain hut Kårsavaggestugan.

Source: Courtesy of Dr David Rippin, Environment Department, University of York.

Table 3.2 A comparison of the key properties of Earth, Venus and Mars, as well as the major constituents of their atmospheres.

	Earth	Venus	Mars
Radius (km)	6371	6052	3396
Distance to sun (millions of km)	150	108	228
Solar constant (W m^{-2})	1366	2611	589
Surface temperature (°C)	15	460	−63
Oxygen (%)	21	<0.0001	0.13
CO_2 (%)	0.038	97	95
H_2O(g) (%)	1*	0.002	0.021

* Varies according to location and climate

produces large electrical currents. These currents cause gases in the upper atmosphere to glow, leading to some beautiful displays of colours (Figure 3.15).

The colour of the light varies depending on the height of the interaction and the composition of gases in that part of the atmosphere. For an excellent overview of auroras and images, the reader is guided to the very informative and stimulating resource at: http://www.exploratorium.edu/learning_studio/auroras/index.html.

3.10 Goldilocks Earth

We mentioned earlier that despite the proximity of Earth to Mars and Venus, the atmospheres of these three planets have suffered very different fates. Table 3.2 shows some properties of the three planets. Venus and Earth are very similar in terms of size, but the average surface temperature of Venus is 460 °C compared to the rather more benign 15 °C at the surface of the Earth. This is despite the solar constant (the amount of incoming solar electromagnetic radiation per unit area, measured on the outer surface of the atmosphere in a plane perpendicular to the rays) only being approximately twice as large for Venus. Why is there such a big difference in surface temperature then?

Well for a start, Venus is much closer to the sun than the Earth (Table 3.2), so Earth has always been cooler. Whereas oceans were able to form on Earth, it was too hot on Venus. On Venus, it is possible that the water vapour expelled by volcanic activity was never able to condense at the surface, but was dissociated by the sun into H_2 and O_2 molecules (reaction R3). The H_2 would have escaped to space and the O_2 reacted with the rocks at the hot surface. What was left in the atmosphere of Venus consequently was largely CO_2, but also sulphuric acid clouds from the sulphurous volcanic emissions.

The presence of oceans on Earth made a huge difference, as they began to soak up CO_2 and deposit it into carbonate rocks on the seafloor. This prevented the catastrophe that happened on Venus, where the CO_2 remained in the atmosphere, causing a runaway 'greenhouse effect'. We will return to the subject of the greenhouse effect in Chapter 8. It is worth noting at this stage that the presence of water, CO_2 and other greenhouse gases keeps our atmosphere over 30 K warmer than it would be without them (a *natural* greenhouse effect), even before we account for the recent influence of humans on climate.

How about Mars? Well Mars is further from the sun than Earth and consequently much cooler, with an average surface temperature of only −63 °C, and certainly too cool for advanced life forms to exist. Mars, being smaller than Earth, cooled more quickly and its volcanic activity therefore ceased more quickly. Volcanic eruptions were no longer able to spew greenhouse gases into the atmosphere and the planet cooled even further. The low temperatures meant that water was locked up in ice rather than able to act as a greenhouse gas in the atmosphere. In addition, as Mars is quite a lot smaller than the Earth and Venus, and as gravitational attraction is proportional to mass, the heavier a planet, the better able it is to hold onto an atmosphere. Mars has a very thin atmosphere, composed mostly of CO_2. Mars had surface water in the distant past, although it is debatable whether it ever had oceans. The surface of this planet, a long way distant from the sun, is now cold and frozen.

Remarkably then, while Venus is too hot to sustain life and Mars is too cold, Earth is just right.

3.11 Human influence on the atmosphere

The discussion in this chapter has focused so far on the natural atmosphere. However, as most readers will know, the state of the atmosphere is far from natural at present. Into the thin envelope that surrounds the planet, we increasingly emit tonnes of pollutants from activities such as fossil fuel combustion for transport and electricity generation, biomass burning for land clearance and intensive agriculture.

Since the Industrial Revolution, human activities have had a particularly profound effect on the atmosphere. Some of these major impacts will now be briefly discussed here. Climate change is the major example of how humans have perturbed the atmosphere, but given the global importance of this issue, Chapter 8 will cover this topic in detail. For readers who would like more information on atmospheric pollution, it is suggested they consult the book of the same title by Mark Jacobson (2002).

3.11.1 Historical air pollution

We tend to think of air quality issues as being a modern-day problem related to the large amounts of pollutants emitted from urban areas; industrial, domestic and vehicular emissions are concentrated in these areas. Given that most of the world's population lives and works in urban areas and that many air pollutants have deleterious health effects at high concentrations, there are good reasons why we should be concerned about the composition of the air that we breathe.

Such concerns are not a new phenomenon however, as can be seen by looking back through the history of air pollution, using London as an example (for a really excellent review of this topic, the reader is encouraged to read *The Big Smoke* by Peter Brimblecombe (1987)). Some of the first complaints about poor air quality date back to the thirteenth century and were associated with lime production for castle building (Brimblecombe, 1987). Limestone was heated in kilns to produce calcium oxide (R9) and the calcium oxide mixed with water to produce the calcium hydroxide needed for building (R10):

$$CaCO_3 \rightarrow CaO + CO_2 \quad \text{(R9)}$$

$$CaO + H_2O \rightarrow Ca(OH)_2 \quad \text{(R10)}$$

The combustion process used to convert $CaCO_3$ to CaO (reaction R9) also released nitrogen oxides and particles as well as CO_2, and would have been a messy business. Lime production required vast quantities of lime, so the kilns needed to be near to the settlements where it was needed. Industry was typically based in the forests before lime production commenced (e.g. blacksmiths), close to a source of wood as a fuel. Therefore, the pollution caused by this process would have been quite a new and unpleasant experience for many people, though they already had to live with inadequate disposal of human excretia.

The changing dynamics between wood and coal use as fuels had a large impact on pollution over the next few hundred years (Brimblecombe, 1987). There were a number of factors that determined which of these fuels dominated between the thirteenth and seventeenth centuries, including relative prices, meteorology, population growth (and collapse after the Black Death), politics, and even societal pressures (the rich took up coal as a fuel after King James I began to use it in the English court following his move from Scotland, where it was used more commonly). However, by the end of this period, coal was firmly established as the major fuel both for industrial and domestic purposes.

By the late 1800s and into the early 1900s, air quality in London and many other cities was appalling. The advent of the steam engine (powered by coal combustion) in the eighteenth century meant that industry and transport could become mechanised. Engines were widely used, factories set up and most of this activity was centred in urban areas, where coal was also used for domestic purposes. There was a hundred-fold increase globally in the use of coal for combustion between 1800 and 1900, but in the UK, the resulting air pollution has been estimated to have killed 4–7 times more people than in other countries worldwide (Jacobson, 2002).

The foul smelling air in London had its own nickname – the *Great Stinking Fog* or the *London Particular*. Imagine if you will a mix of emissions from coal burning fires in the city, manufacturing processes (with no regulations regarding waste management) and imperfect (putting it mildly) sewage systems. Coupled with the fogs that were much more common at this time, the odorous vapours hung over the city blocking out sunlight and causing much distress. Such effects are reflected in the writings of the time by novelists such as Charles Dickens, Arthur Conan Doyle and Wilkie Collins. The smoggy conditions were used to great effect to add to the air of murderous intent in many of the Sherlock Holmes stories for instance.

The turning point for air quality in London was reached with the Great Smog of December 1952 (Smog from SMoke and fOG), when the acidic conditions within the air killed approximately 4000 people. Although smoggy conditions in London had led to excess deaths before, none had killed so many over such a short period. Many of those who died were the more vulnerable members of society – the young, the elderly and those already suffering from respiratory problems. Some experts think that in the weeks that followed, a further 8000 people may have died as a result of this event.

So why was it so bad? The emissions from London coincided with a period of cold, stagnant weather associated with an anticyclonic weather system. The stagnant conditions brought low wind speeds so pollution was not dispersed. In addition, the cold conditions caused a temperature inversion close to the ground, effectively trapping the pollutants in a shallow layer (Figure 3.16). The fog that formed was the thickest many had seen and lasted for several days (see the excellent book by William Wise (1968) called *Killer Smog: The World's Worst Air Pollution Disaster*. It contains the facts, but also anecdotal evidence from people who lived through the episode.

The alarm was only raised when livestock starting dying at Smithfield Market and undertakers began to run out of coffins. It wasn't until mortality statistics were assembled later that the large number of fatalities was revealed. In fact, it was this event that finally precipitated the UK government into introducing the Clean Air Act of 1956, which has been followed up by various initiatives since. This policy had a profound effect on cleaning up the air by forcing industry out of the city and introducing smokeless fuel zones. At the same time, many people began to convert their coal burning fireplaces to gas or electric heating systems and the air quality in London started to improve.

So what caused the damage? The SO_2 formed from burning sulphur-containing coal was converted to sulphuric acid (H_2SO_4) in the fog droplets through a series of reactions:

$$S_{(s)} + O_{2(g)} \rightarrow SO_{2(g)} \qquad \text{(R11)}$$

$$SO_{2(g)} + H_2O_{(aq)} \rightarrow H^+_{(aq)} + HSO_3^-{}_{(aq)} \qquad \text{(R12)}$$

$$2HSO_3^-{}_{(aq)} + O_{2(aq)} \leftrightarrow 2H^+_{(aq)} + 2SO_4^{2-}{}_{(aq)} \qquad \text{(R13)}$$

$$2H^+_{(aq)} + SO_4^{2-}{}_{(aq)} \leftrightarrow H_2SO_{4(aq)} \qquad \text{(R14)}$$

Indeed, acid deposition became a cause for international concern during the 1950s and 1960s when it was found that emissions of sulphur (and also nitrogen containing species) from the UK and other parts of the EU were causing acidification of Scandinavian lakes (see Chapter 16).

Despite such air pollution events being consigned to history for much of the developed world, there are still some parts of the world where such conditions persist. For instance, China is industrialising at a rapid rate and has plentiful supplies of coal (with relatively high sulphur content). Therefore, in many of China's cities, acidic smogs are still a major issue.

Figure 3.16 The smog in London, 1952.
Source: Fox Photos/Hulton Archive/Getty Images.

3.11.2 Urban air quality issues today

The major type of pollution experienced in major cities around the world today is photochemical in origin. Such pollution was first noted in Los Angeles in the late 1930s/1940s, when people began to report eye irritation and respiratory problems. There was also a significant reduction in visibility (Figure 3.17), increased crop damage and material (such as rubber) deterioration. However, the conditions for forming pollution were very different to those in London at the same time, nominally relatively low humidity, plenty of sunshine and low wind speeds.

The cause of the problem was unclear until the seminal work of Haagen-Smit (1952) who concluded that ozone could be produced in the presence of both nitrogen oxides and hydrocarbons (both emitted from motor vehicles) in a sunlit atmosphere. The problem was compounded by the topographical conditions in LA, which is bounded by mountains downwind of the city, which limit the dispersion of pollutants. Such conditions are ideal for the formation of high concentrations of ozone and other secondary pollutants.

The chemical reactions that lead to photochemical pollution episodes are complex. Nitric oxide (NO) is formed during combustion processes and once released into the atmosphere its main fate is reaction with O_3 (R15):

$$NO + O_3 \rightarrow NO_2 + O_2 \quad \text{(R15)}$$

Figure 3.17 Los Angeles and Griffith Observatory, as viewed from the Hollywood Hills.
Source: Photograph by David Iliff.

The NO_2 (nitrogen dioxide) formed in reaction R15 can be photolysed back to NO (R16) and the O atom can recombine with oxygen molecules to reform O_3 (R6):

$$NO_2 + h\nu \rightarrow NO + O \qquad \text{(R16)}$$

$$O + O_2 + M \rightarrow O_3 + M \qquad \text{(R6)}$$

If reactions R15, R16 and R6 were the only ones that mattered, there would be no net ozone production. Each time an ozone molecule is produced (e.g. R6), it is destroyed again by NO (R15). However, measurements from polluted atmospheres show that the concentration of O_3 typically increases over the day, building to a peak in late afternoon. Clearly then, there are species present that convert NO to NO_2 other than O_3. Such species are called peroxy radicals, and are formed when the hydroxyl radical (OH) reacts with volatile organic compounds (VOCs) such as hydrocarbons.

Lead-in question

Does anything get rid of pollutants once they've entered the atmosphere?

The OH radical controls the atmospheric lifetimes of many species such as CO, CH_4 and hydrocarbons, and is consequently often referred to as the atmospheric vacuum cleaner. It is formed by the photolysis of O_3 by high energy sunlight (wavelengths less than about 320 nm) and the subsequent reaction of excited oxygen atoms (O*) with water:

$$O_3 + h\nu \rightarrow O^* + O_2 \qquad \text{(R17)}$$

$$O^* + H_2O \rightarrow 2OH \qquad \text{(R18)}$$

OH is very reactive and, once formed (it has a typical lifetime in the atmosphere of less than one second), can react with a variety of atmospheric species. Although it exists at very low concentrations (usually around 1 molecule in 10^{13}), it is hard to over-emphasise the importance of this radical. It controls the removal of most species in the atmosphere, species that can go on to contribute to global warming, to form acid rain, or to lead to poor urban air quality. Without this radical, the lifetime of many pollutants in the atmosphere would be significantly longer.

Let's consider the atmospheric oxidation of the simplest hydrocarbon, methane. The initial step occurs when the OH radical abstracts a hydrogen atom from methane to form water and the methyl peroxy radical, CH_3O_2 (R19):

$$OH + CH_4\ (+O_2) \rightarrow CH_3O_2 + H_2O \qquad \text{(R19)}$$

The $(+O_2)$ indicates that reaction R19 requires the presence of oxygen, but given 21 per cent of the atmosphere is composed of oxygen this requirement is obviously fulfilled.

In polluted atmospheres, the main fate of CH_3O_2 is to react with NO to form the methoxy radical (CH_3O) and NO_2:

$$CH_3O_2 + NO \rightarrow CH_3O + NO_2 \qquad \text{(R20)}$$

The methoxy radical then reacts with O_2 to form formaldehyde (HCHO) and the hydroperoxy radical (HO_2):

$$CH_3O + O_2 \rightarrow HCHO + HO_2 \quad (R21)$$

As with CH_3O_2, the major fate of the HO_2 radical is to react with NO in polluted atmospheres and hence regenerate OH and form another molecule of NO_2:

$$HO_2 + NO \rightarrow OH + NO_2 \quad (R22)$$

The sum of reactions R19–R22 is:

$$CH_4 + 2O_2 + 2NO \rightarrow H_2O + HCHO + 2NO_2 \quad (R23)$$

It can be seen that CH_4 has converted two NO molecules to two NO_2 molecules, which can then produce two O_3 molecules through R16 then R6. In addition, the OH regenerated through reaction R22 can start the whole process over again by oxidising another VOC. In this way, the ozone concentration is able to increase. Because this process is limited by the initial oxidation step (e.g. reaction R19) which typically takes a few hours for many hydrocarbons, maximum ozone concentrations in polluted atmospheres tend to occur mid to late afternoon.

There are hundreds of different VOC molecules in polluted atmospheres and each is oxidised by OH to a greater or lesser extent and can go on to produce ozone. Hydrocarbons with more carbon atoms have the potential to achieve more NO to NO_2 conversions than simpler ones. The overall effect is that with many peroxy radicals converting NO to NO_2, any O_3 that is lost through reaction with NO in R15 is more than adequately compensated through peroxy radical conversions of NO to NO_2 (e.g. reactions R20 and R22).

Concentrations of ozone are typically 30–40 ppb in the background atmosphere, but can reach 500 ppb in polluted regions (note that a ppb is a part per billion or 1 part in 10^9). During a moderately polluted afternoon, the concentration may be around 150 ppb (Jacobson, 2002). Such concentrations are a concern as ozone can cause headaches at concentrations above 150 ppb, chest pains above 250 ppb and a sore throat and cough above 300 ppb (Jacobson, 2002). Ozone can also cause damage to materials (cracked tyres in LA were an early indication that high concentrations of ozone were damaging rubber) and vegetation. The latter is of particular concern in developing parts of the world, where the increased ozone concentrations predicted for the future will undoubtedly impact on crop yields, crops that are the major income for many subsistence farmers; this is discussed further in Chapter 16.

We also mentioned that eye irritation was reported in Los Angeles in the 1950s. In the presence of oxygen, acetaldehyde (CH_3CHO) reacts with OH as follows:

$$CH_3CHO + OH + (O_2) \rightarrow CH_3CO.O_2 + H_2O \quad (R24)$$

Acetaldehyde can be directly emitted as a pollutant, or formed as a secondary product through other reactions in polluted atmospheres. The acetyl peroxy radical ($CH_3CO.O_2$) can then react with NO_2 to produce peroxyacetylnitrate (PAN):

$$CH_3CO.O_2 + NO_2 \leftrightarrow CH_3CO.O_2NO_2 \quad (R25)$$

Note that reaction R25 is reversible; at high temperatures, PAN will decompose back to the acetyl peroxy radical and NO_2. PAN is known to be an eye irritant and is one of the causes of human discomfort during photochemical pollution episodes.

The discussion so far has focused on gas-phase chemistry. However, also of concern in urban areas is the presence of particulate matter. Oxidation of VOCs (as well as of NO_2 and SO_2) can produce particulate matter in addition to the gas-phase species discussed above. Particulate matter also enters the atmosphere through combustion processes (such as fossil-fuel burning) as well as through a range of natural processes such as volcanic eruptions, tidal action (sea spray) and sand storms.

This particulate matter tends to be considered in two main size classes: PM_{10} (particulate matter with an aerodynamic diameter less than 10 μm) and $PM_{2.5}$ (particulate matter with an aerodynamic diameter less than 2.5 μm). As a general distinction, natural material tends to fall into the PM_{10} fraction, while manmade particles (oxidised hydrocarbons, combustion and smelting products) fall within the $PM_{2.5}$ fraction.

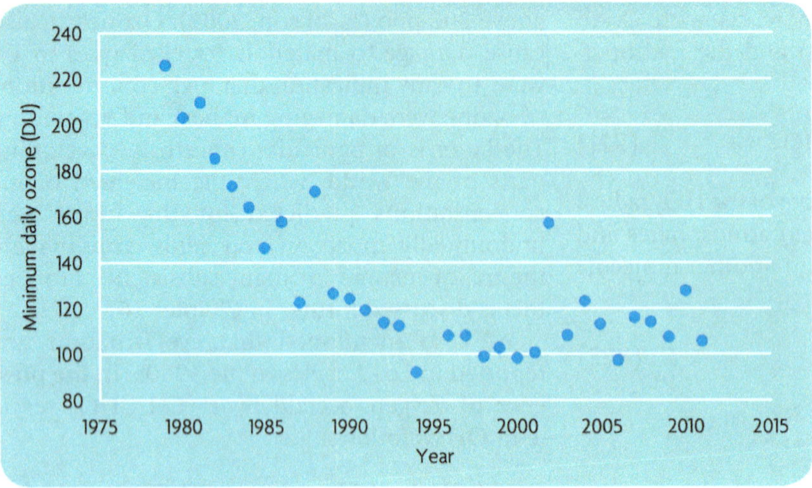

Figure 3.18 Variation in the minimum of the mean ozone (total column density) for the period from 21 September–16 October since 1979.
Source: Data taken from the NASA Ozone Watch site at http://ozonewatch.gsfc.nasa.gov/meteorology/annual_data.html

The health concerns related to particles centre on the fact that such material can effectively penetrate the respiratory system. Particles within the 5–30 μm range tend to impact the throat or nasal passages and be removed there. Particles in the size range from 1–5 μm can make it as far as the trachea and larger branches of the lungs. However, particles smaller than 1 μm can enter the alveolar region, the smaller branches of the lungs and the air exchange area.

So the smaller material can make it further into the respiratory system. Moreover, it tends to be such particles that contain more toxic material, such as heavy metals owing to the processes that produced them (e.g. smelting, diesel fuel combustion). Pope (2000) conducted a review of several studies of the health effects of PM and surmised that short-term increases of PM_{10} concentrations of 10 μg m^{-3} were associated with a 0.5–1.5 per cent increase in daily mortality (usually 1–5 days after the air pollution episode), as well as increased incidence of other respiratory and cardiovascular problems. Much of the focus nowadays in urban areas is on how concentrations of particulate matter can be further reduced.

3.11.3 Ozone hole

In the 1970s, two groups of independent scientists predicted that oxides of nitrogen and chlorine from CFCs emitted at ground level would lead to stratospheric ozone depletion (Paul Crutzen, Mario Molina and Sherwood Rowland were later awarded the Nobel Prize for Chemistry in 1995 for this seminal work). However, the first evidence from measurements didn't arrive until 1985, when British scientists working at Halley Bay in Antarctica discovered the so-called ozone 'hole' (it's really a thinning of the ozone layer rather than a hole) using ground-based measurements and published their results in Nature (Farman *et al.*, 1985). Additional data from satellites confirmed that the ozone concentration above Antarctica was indeed decreasing rapidly (Figure 3.18).

Note that the instruments measuring ozone depletion in Antarctica measure the total column amount of ozone in Dobson Units (DU). If all of the O_3 molecules in the stratosphere could be brought down to the surface at 0 °C and 1 atmosphere pressure, the ozone layer would be only 3 mm thick, which corresponds to 300 DU.

Obviously this observation was worrying, given the protective effect of the ozone layer (section 3.7). Depleting stratospheric ozone increases the amount of UV radiation reaching the surface.

We discussed the Chapman mechanism for ozone chemistry in the stratosphere earlier (section 3.7, reactions R5–R8), in the context of how the stratosphere is warmed. However, what was postulated in the 1970s, and confirmed by measurements in the 1980s, was that chemicals released at the ground were sufficiently long-lived to pass through the tropopause into the stratosphere, where they disrupted the Chapman mechanism and led to ozone destruction.

Chlorofluorocarbons (CFCs) were widely regarded as very useful chemicals, being used extensively in industrialised nations as propellants, refrigerants and solvents. These CFCs, which contain chlorine, fluorine and carbon, are non-toxic and chemically inert, making them particularly attractive for use in the home. As refrigerants they were particularly attractive, replacing the more toxic substances used previously such as ammonia, chloromethane and sulphur dioxide. The principal CFCs are CFC-11 (CCl_3F) and CFC-12 (CCl_2F_2), which have lifetimes in the atmosphere of around 50 and 100 years respectively. Their brominated counterparts, bromofluorocarbons, are made of bromine, fluorine and carbon atoms and have applications in fire retardants, agriculture and dry cleaning.

Once in the upper stratosphere beyond the protection of the ozone layer, the stronger UV radiation can break the bonds in these molecules and liberate halogen atoms (e.g. R26):

$$CCl_3F + h\nu \rightarrow CCl_2F + Cl \qquad (R26)$$

The Cl atoms can then react with ozone and a series of catalytic reactions are initiated:

$$Cl + O_3 \rightarrow ClO + O_2 \qquad (R27)$$

$$ClO + O_3 \rightarrow Cl + 2\ O_2 \qquad (R28)$$

Note that the ClO molecule formed in reaction R27 can also react with O_3, while regenerating the Cl atom. Bromine (Br) atoms can undergo similar cycles once liberated from their parent bromofluorocarbon in a process analogous to that in R26. Although there is less bromine in the atmosphere, it is more efficient at destroying ozone so both halogens are important. The overall effect of these reactions is to deplete the ozone concentration in the stratosphere.

Such O_3 destruction, however, can be short-circuited as the catalytic species Cl, ClO, Br and BrO can be effectively removed by the following reactions:

$$Cl + CH_4 \rightarrow HCl + CH_3^{\cdot} \qquad (R29)$$

$$Cl + HO_2 \rightarrow HCl + O_2 \qquad (R30)$$

$$Br + HO_2 \rightarrow HBr + O_2 \qquad (R31)$$

$$ClO + NO_2 + M \rightarrow ClONO_2 + M \qquad (R32)$$

$$BrO + NO_2 + M \rightarrow BrONO_2 + M \qquad (R33)$$

Hydrogen chloride (HCl), hydrogen bromide (HBr), chlorine nitrate ($ClONO_2$) and bromine nitrate ($BrONO_2$) are all reservoir species and effectively prevent the halogen atoms from destroying more ozone. On their own, these reactions would cause some ozone destruction, but it would be tempered through the formation of the reservoir species.

However, peculiar meteorological conditions in the Antarctic also play a role. The onset of winter brings the air temperature to as low as 185K, at which point, the water in the air forms ice crystals or *polar stratospheric clouds* (PSCs). A strong westerly vortex is set up by the Coriolis force at this latitude (see Chapter 8). This vortex effectively traps and cools the air and any pollutants within it, including the reservoir compounds formed in reactions R29–R33. The PSCs within the vortex form a very efficient surface on which chemical reactions can take place.

The chlorine and bromine reservoir compounds can then be converted, e.g. to Cl_2 and HOCl on the surface of PSCs (either at the water–ice surface or to adsorbed species $_{(ads)}$ at the surface) as demonstrated in reactions R34 and R35:

$$HCl_{(ads)} + ClONO_{2(g)} \rightarrow Cl_2 + HNO_{3(ads)} \qquad (R34)$$

$$H_2O_{(s)} + ClONO_{2(g)} \rightarrow HOCl_{(g)} + HNO_{3(ads)} \qquad (R35)$$

Reactions R34, R35 and their analogues also ensure that NO_X is removed from the system, so reducing the ability of reactions R32 and R33 to reoccur, which would otherwise tie up BrO and ClO.

Effectively, these reactions activate chlorine from a reservoir species to a photo-labile form. Consequently, at the end of the polar winter when sunlight returns, these activated forms are photolysed to form chlorine (and their brominated analogues form bromine) atoms, e.g.:

$$Cl_2 + h\nu \rightarrow Cl \qquad (R36)$$

$$HOCl + h\nu \rightarrow Cl + OH \qquad (R37)$$

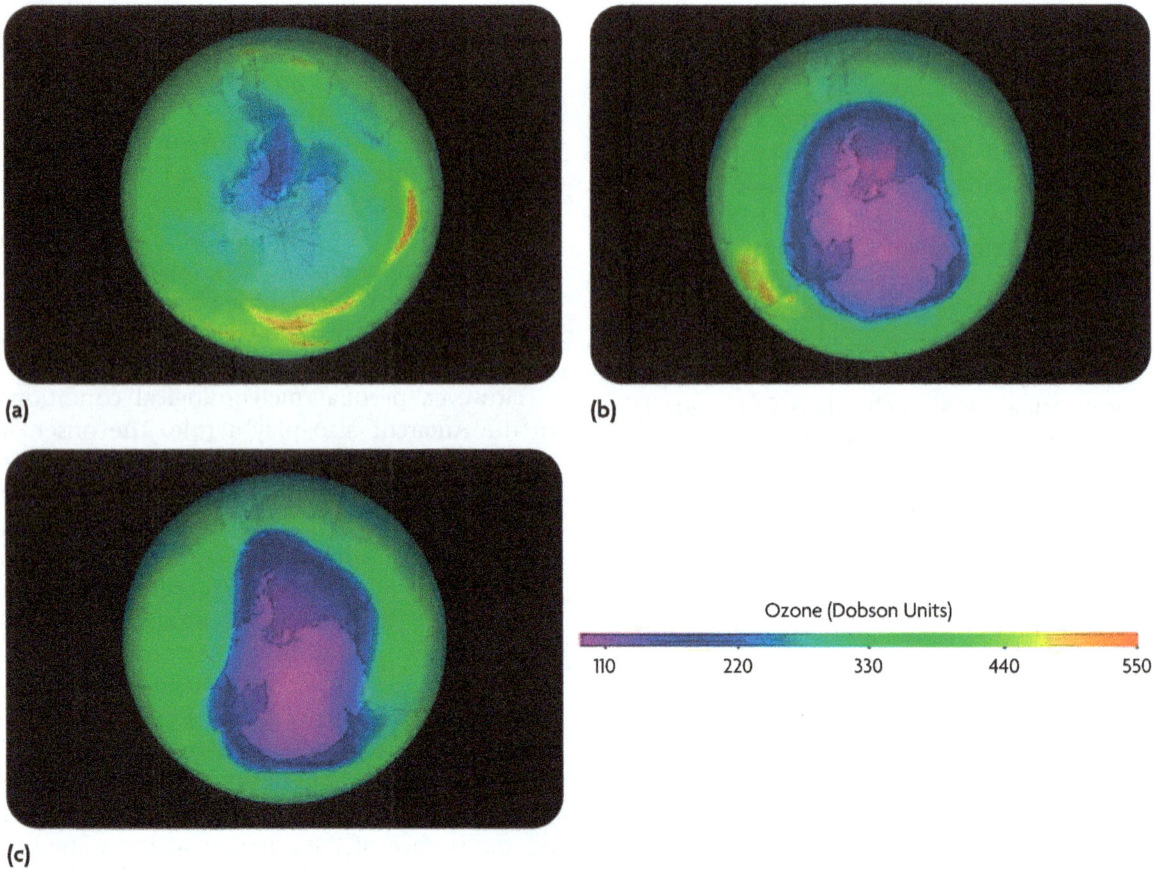

Figure 3.19 (a) Ozone hole in 1979 (first image available); (b) 1994 (the deepest ever recorded ozone hole) and (c) 2006 (the largest ozone hole at the time of writing).
Source: Courtesy of http://earthobservatory.nasa.gov/Features/WorldOfChange/ozone.php/NASA

followed by reactions R27 and R28 to destroy ozone. The newly formed ClO through R27 can no longer react with NO_2 through R32 as NO_X has been removed from the vortex. The catalytic cycles can therefore begin again.

The O_3 destruction reactions reach a maximum in Antarctic spring (September to December) when the sunlight returns, and this is when maximum ozone loss in the lower stratosphere (the ozone hole) is observed. At this point, the PSCs melt and the trapped compounds are released. Once the vortex breaks up around mid-December, ozone-rich air from lower altitudes returns to the Antarctic and the ozone hole disappears.

NASA have an excellent web resource for those interested in learning more about the ozone hole at http://earthobservatory.nasa.gov/Features/WorldOfChange/ozone.php. In particular, they have a time series of observations from 1979 to the present that show the maximum ozone hole depth (minimum concentration of ozone) in the Antarctic each year. Figure 3.19 shows a series of images from this website.

The maximum depth of the hole in 1979 was 194 DU, and the minimum concentration stayed at around this level until 1983, when it began to decline quite rapidly. The deepest ozone hole ever recorded was in 1994, when concentrations fell to 73 DU at the end of September. The largest hole was recorded in 2006, showing that the hole doesn't necessarily peak in terms of depth and size at the same time. However, generally there was a consistent trend in growth of the hole in terms of depth and size through to the mid-nineties when a stabilisation occurred. The variation from year-to-year is caused by changes in the stratospheric temperature and the circulation patterns of the vortex. The colder it is, the more ozone can be depleted as the PSCs are formed more efficiently.

Why then is there no hole in the Arctic during northern hemisphere spring? Well meteorological conditions are very different there. The land masses surrounding the Arctic prevent such a strong and sustained vortex forming there and temperatures are also about 10 K warmer. So although ozone depletion is noted at this time of year, there is no analogous 'hole'.

In 1989, the Montreal Protocol banned the use of CFCs in recognition of their destructive effect on the ozone layer. Owing to the long lifetimes of these substances in the atmosphere, it is unlikely that the Antarctic ozone layer will recover before the middle of this century. In the meantime, HCFCs (hydrochlorofluorocarbons) have been introduced as a short-term replacement. They are much shorter lived and better than the CFCs in that sense, but like the CFCs, are potent greenhouse gases. The search for the ultimate CFC replacement continues, but the fact that the world came together to solve this issue at least gives some hope for the future in terms of finding a solution for other global environmental problems such as climate change. Those readers who are inspired by Doomsday Scenario type stories may like to read a version of what might have happened had we not signed up to the Montreal Protocol at http://earthobservatory.nasa.gov/Features/WorldWithoutOzone/.

3.12 The future of the atmosphere

So how about the future of the atmosphere? This chapter has shown that the atmosphere has undergone massive changes throughout the history of the Earth, culminating in perfect conditions for the diverse life we experience today. However, we stand to lose much if we continue to pollute our atmosphere at the rate we are currently doing. If we can't, as a race, put a stop to human-induced climate change, the atmosphere could once again render the planet inhospitable to most forms of life.

POLICY IMPLICATIONS

The origins of the atmosphere

- It is crucial that policy makers fully appreciate how biota influence the composition of the modern-day atmosphere.
- Policy makers need to fully appreciate the effects of all gaseous components in the atmosphere and what regulates their concentrations.
- Policy makers need to be aware of how atmospheric pollution can modify the atmosphere, of what causes smog and elevated ground level ozone concentrations and of the associated potential adverse impacts on human health.
- Policy makers need to be aware of the past causes of stratospheric ozone depletion and to continue to monitor stratospheric ozone concentrations because of the risks to human health.
- Policy makers should understand the significance of classification of ultraviolet radiation into UVA, UVB and UVC and what controls the exposure of biota on Earth to UV in the three wavelength ranges.
- Policy makers need to be aware of the importance of particulate pollutants in the atmosphere and the associated health risks for different size ranges.

CHAPTER REVIEW EXERCISES

Exercise 3.1

Calculate the energy associated with light with a wavelength of 300 nm and 700 nm. What does your answer tell you about the relationship between the wavelength of light and the associated energy?

Exercise 3.2

What are the two main theories for the creation of life? What are the strengths and weaknesses for each theory?

Exercise 3.3

Explain how the ozone layer formed and why it's so important. How was the ozone layer perturbed by manmade pollution?

REFERENCES

Brasier, M., McLoughlin, N., Green, O. and Wacey, D. (2006) A fresh look at the fossil evidence for early Archaean cellular life, *Philosophical Transactions of the Royal Society B*, **361**, 887–902.

Brimblecombe, P. (1987) *The Big Smoke: A History of Air Pollution in London Since Medieval Times*. Methuen, London, UK.

Buick, R. (2008) When did oxygenic photosynthesis evolve? *Philosophical Transactions of the Royal Society B*, **363**, 2731–2743.

Cowen, R. (2005) *History of Life*, 4th edn. Blackwell Publishing Limited.

Farman, J.C., Gardiner, B.G. and Shanklin, J.D. (1985) Large losses of total ozone in Antarctica reveal seasonal ClO_X/NO_X interaction. *Nature*, **315**, 207–210.

Frank, L.A., Sigwarth, J.B. and Craven, J.D. (1986) On the influx of small comets into the Earth's upper atmosphere II. Interpretation. *Geophysical Research Letters*, **13**, 307–310.

Haagen-Smit, A.J. (1952) Chemistry and Physiology of Los Angeles Smog. *Industrial and Engineering Chemistry*, **44**, 1342–1346.

Jacobson, M.Z. (2002) *Atmospheric Pollution: History, Science and Regulation*. Cambridge University Press, Cambridge, UK.

Johnson, A.P., Cleaves, H.J., Dworkin, J.P., Glavin, D.P., Lazcano, A., and Bada, J.L. (2008) The Miller Volcanic Spark Discharge Experiment. *Science*, **322**, 404.

Kump, L.R. (2008) The rise of atmospheric oxygen. *Nature*, **451**, 277–278.

López-García, P. and Moreira, D. (1999) Metabolic symbiosis at the origin of eukaryotes. *Trends in Biochemical Sciences*, **24**, 88–93.

Miller, S.L. (1953) A production of amino acids under possible primitive Earth conditions. *Science*, **117**, 528–529.

Pope, C.A. (2000) Review: Epidemiological basis for particulate air pollution health standards. *Aerosol Science and Technology*, **32**, 4–14.

Wächtershäuser, G. (2000) Origin of life: life as we don't know it. *Science*, **289**, 1307–1308.

Warmflash, D. and Weiss, B. (2005) Did life come from another world? *Scientific American*, **293**, 64–71.

Wise, W. (1968) *Killer Smog: The World's Worst Air Pollution Disaster*. Rand McNally & Company.

CHAPTER 4

The natural carbon cycle

Malcolm Cresser

Learning outcomes

By the end of this chapter you should:

- Be more aware of the importance of each step in the C cycle to atmospheric chemistry and hence to life on Earth as we know it.
- Know how our understanding of photosynthesis evolved.
- Be more aware of the importance of setting up and testing hypotheses in scientific investigations.
- Know where the pools of organic carbon (C) in soils and plants originate.
- Realise the importance of microorganisms to the cycling of C at local and global scales.
- Appreciate the crucial importance of processes in the oceans to the global C cycle.
- Appreciate the importance of C cycling to the buffering of water pH.
- Start to understand the direct effects of atmospheric CO_2 enrichment on plants and the reasons for uncertainties about future predictions.
- Better understand the link between cycling of C and cycling of other plant nutrient elements.

4.1 Introduction

Nowadays most people, even those with limited education, are aware of the importance of photosynthesis to the production of plant biomass, and of the role of carbon dioxide (CO_2) in the atmosphere to this process. Many, having been made more aware by the media, would appreciate the importance of carbon dioxide and methane as greenhouse gases too. Far fewer though would be able, if asked, to elaborate on how important the carbon cycle is to the control of the oxygen content of the atmosphere, the cycling of plant nutrients, soil stability and water retention, and water chemical quality.

Lead-in question

Why is there such limited public awareness of the importance of the carbon cycle?

Try to imagine going back in time for three or four thousand years. If you saw an oak tree growing from a tiny acorn, or even grasses growing from seed, what do you think you would have deduced about the origins of the plant material springing up before your eyes? The chances are you would conclude that, somehow, material in the soil was converted to green plant material. Early man, even at the time of the late stone-age, would soon have discovered that if crops were harvested from soil year after year, growth would get progressively poorer, suggesting something (or some group of things) in the soil was present in limited amounts, and eventually ran out or had to be replaced. Such an observation would reinforce the idea that everything the plant needed to grow came from the soil. Effects of drought would become apparent in dry summers, however, and the need for water soon realised, leading to early irrigation schemes. A further piece in the jigsaw would come from looking at the small amount of ash left after large amounts of solid plant material were burnt. The bulk of the material would apparently vanish into the air. I think, under the circumstances, I might have finished up like the ancient Greeks believing that there were four factors involved in supporting all living things: earth, air, fire and water.

4.2 Setting up and testing hypotheses

4.2.1 How scientific progress is made

Lead-in question

Would you have been inquisitive enough to want to know how the four factors, earth, air, fire and water, interacted? If so, what experiments would you have done?

I was once told by a celebrated Czech chemist that we all make discoveries. The difference between a good scientist and a poor scientist, he said, is that when the good scientist sees something unexpected, he or she has an insatiable desire to find out what caused the surprising event.

The example he gave was complexometric analysis, which was allegedly discovered as a result of a washing up session in a chemical laboratory. Contents of a flask containing a solution of a strong organic complexing reagent, EDTA (ethylenediaminetetraacetic acid), were emptied down a sink. Fortuitously this happened just after the contents of another flask, containing a solution of a coloured organic compound of the element calcium, had also been poured into the sink. Apparently the sink went from pink to blue.

An uninspired chemist might have thought nothing of it at all, or perhaps 'oh, that's pretty!' However, the inquisitive chemist wanted to know what was going on, and made up all the reagents he had just thrown away afresh. A few days later he had worked out that EDTA solution could be used in a titration to measure the calcium concentration of a solution in the presence of a suitable organic complexing agent (indicator). The indicator forms a coloured complex with calcium ions, but EDTA forms an even stronger, but colourless, complex. As long as calcium is present in excess, the colour of the indicator-calcium compound will be seen. Once, however, enough EDTA has been added to react with all the calcium present, the colour of the indicator-calcium compound disappears, leaving the colour of the free, un-bound indicator. Thus a new analytical method was born from a chance observation.

In the context of this chapter, chance observations over many years were probably incredibly important. We know, for example, that shell sand was used as a liming material back in the stone age. Almost certainly this would have resulted from seeing growth improvement around areas where shells from shell-fish meals had been dumped on acidic soils. Primitive fertiliser trials would have followed and, even though there would have been no understanding of the neutralising process:

$$CaCO_3 + 2H^+ \rightarrow Ca^{2+} + H_2O + CO_2$$

the practice of adding shell sand would have been adopted on a trial and error basis. Early man was setting up a hypothesis to test a specific question: 'Application of ground up shells to soil can improve the growth of plants'. He/she may not have been wearing the stereotypical scientist's white lab coat when testing the hypothesis experimentally, but test it he/she did. Similarly it's not that hard to imagine how the fertiliser potential of faeces, manures and urine was recognised thousands of years ago!

I have sometimes wondered, however, what observations prompted experiments leading to the discovery of oxygen and its properties and importance. It doesn't seem intuitively obvious, for example, to blow on glowing embers to get them to burst into flame unless you already know that you're adding extra oxygen to aid combustion. But if you noticed a sudden accidental draft, perhaps from shaking a sheet of animal skin, caused an extra glow in embers of a fire, you might well try to create one deliberately by blowing to see what happened. When blowing worked, you'd be half way to inventing bellows (well part way, anyway). From there it's a short step to realising there must be something in air that helps combustion, and to wanting to know if air is a simple single substance or a mixture of something that supports combustion (oxygen, O_2) with something inert (mainly nitrogen, N_2).

Back at the start of the eighteenth century, however, early chemists for a while took a different view, based on the ideas put forward by Becher and developed by Stahl. They concluded that flammable substances or substances that reacted with air on heating contained a curious colourless, odourless component that they termed 'phlogiston'. During combustion this component was removed into the air, but because they knew from experiments that a particular volume of air could only support a limited amount of combustion, they concluded that air had a specific capacity for absorbing phlogiston. When nitrogen was discovered as an inert gas incapable of supporting combustion, it was simply regarded as air in which the capacity for phlogiston absorption was spent.

It's not that hard to follow the probable logic of these early chemists. Burning wax or oil, for example lost weight during partial combustion, so something was lost to the air. Now of course we know that the loss was primarily of carbon and hydrogen as carbon dioxide (CO_2) and water (H_2O), products of combustion of hydrocarbons and other organic materials, but we have the advantage of another three centuries of scientific hypothesis testing.

Where the phlogiston theory of the late seventeenth century ran into trouble was when chemists started to 'burn' metals in limited amounts of air by heating them using solar energy focused through a lens. Metals such as magnesium formed the oxide:

$$2Mg + O_2 \rightarrow 2MgO$$

In doing so it gained weight. Proponents of the phlogiston theory concluded this must mean that phlogiston had negative weight! They would perhaps have been better to accept that the initial hypothesis that phlogiston was lost to the air had been disproved sooner than they actually did.

Lavoisier (before sadly going to the guillotine) showed conclusively that combustion requires the presence of oxygen, suggesting oxygen was a key reactant in combustion or oxidation processes.

Testing hypotheses by well-designed experiments that lead to unequivocal conclusions is how science advances. Even if a hypothesis is disproved, progress has been made, because a new, alternative hypothesis must then be set up and tested.

4.3 Respiration and photosynthesis

4.3.1 Some key historical experiments

Chance observations may well have played an important role in our understanding of photosynthesis and respiration too. Perhaps a few pets or farm animals (or even humans!) died after being shut in air-tight containers for a few hours and someone wondered why. It is a short step from there to running an experiment with a mouse in a bell jar over a trough of water (to get an air-tight seal) to see how long it lived, similar to experiments with a candle to see how long it burned. It is another short step to showing the respired gas eventually would not support combustion, thus establishing that oxygen had been consumed.

Joseph Priestly, an eighteenth century clergyman, made an important contribution towards our understanding of photosynthesis. He developed a strong interest in what turned out to be carbon dioxide that may well have stemmed from his living close to a brewery, and observing difficulty in breathing freely in the air within the space at the top of the beer fermentation vats. From the experiments he performed, Priestly must have wondered if, and how, plants and animals interacted with the atmosphere.

Even though Priestly was still an advocate of the phlogiston theory towards the end of the eighteenth century (Priestly, 1796), he had earlier found that growing a plant in air 'injured by the burning of candles' restored the capacity of the air to support combustion and allow a mouse to breathe. Thus although he didn't offer an explanation in modern scientific parlance, he had discovered that photosynthesis converts carbon dioxide to oxygen.

Stimulated perhaps by his awareness of Priestly's experimental observations, a Dutch physician, Jan Ingenhousz, showed that sunlight was crucial to the conversion of bad air back to good (i.e. for photosynthesis). The present author strongly recommends reading exerts from his inspirational work on 'Experiments upon vegetables, discovering their great power of purifying the common air in the sun-shine, and of injuring it in the shade and at night, to which is joined, a new method of examining the accurate degree of salubrity of the atmosphere' (Ingenhousz, 1779).

The work discusses, among other things, the adverse effect of partial defoliation of fruit trees on fruit quality, and how complete leaf removal caused fruit to die back (call me cynical if you like, but I often wonder if the leaf stripping was done with the hope of improving fruit size if leaves didn't have to grow too! Sometimes advances are made in environmental science when the expected does not happen and a new hypothesis has to be set up and tested). The work really is an announcement of how important leaves are to plant growth. It goes on to talk about production of bubbles (of oxygen) when leaves are submerged under water in the presence of sunlight, and how leaves align themselves to optimise the amount of light that they receive. Dephlogisticated air (oxygen in reality) was emitted for the benefit of animals to breathe. He goes on to suggest that somehow leaves manage to offset the influence of leaf decay. He also realises the role of algal growth in water as a source of dephlogisticated air (oxygen). If you read Ingenhousz's original work you will see that he was not only introducing us to the carbon cycle, but was even pointing towards the Gaia hypothesis discussed in some detail in Chapter 10.

Sometimes when hypotheses are tested in science, incorrect conclusions may be drawn if not all possible mechanisms have been adequately considered. A good example of this is the research of Jan Baptista Van Helmont back in 1649. He transported a young willow tree shoot (a great choice – they are easy to grow even from simple cuttings and grow very quickly!) into a tub containing a known (carefully weighed) amount of soil. He grew the tree for five years, and then weighed both tree and soil. The tree weighed 164 pounds, but the soil had lost only a couple of ounces. He deduced that the plant had derived virtually everything it needed from the water added over five years. Nothing he had seen had given him any reason to consider direct inputs from the atmosphere.

It was not until the start of the nineteenth century that Nicolas Theodore de Saussure performed a more sophisticated version of Van Helmont's experiment, in which he measured the amounts of carbon dioxide and water that the plant received. He showed that the plant carbon came from carbon dioxide and its hydrogen from added water. In the following decades the crucial importance of inputs of solar energy were confirmed, and the central role of photosynthesis in the carbon cycle was established beyond any doubt.

4.4 The carbon cycle

4.4.1 The main features of the C cycle

Figure 4.1 is a simplified representation of the carbon cycle as we consider it today. The absorption of solar light by plant pigments such as chlorophyll *a* and chlorophyll *b* provides the energy required to produce carbohydrate:

$$6CO_2 + 6H_2O \xrightarrow{\text{sunlight}} C_6H_{12}O_6 + 6O_2$$

Carbon dioxide here is chemically reduced to produce simple sugars using hydrogen produced from the photolysis of water.

All elements participate in biogeochemical cycles to a greater or lesser extent, as we will see in the following chapter. However, it is very appropriate to consider the carbon cycle first; the growth of biomass, and the subsequent decay of live and dead components of biomass, are central to the cycling of other elements. If there's no photosynthesis, then the other elements being cycled have no role in plant growth to fulfil. Nor will other elements be recycled effectively if plant and animal residues and waste products are not decomposed by microorganisms.

The legend to Figure 4.1 describes the portrayal as 'simplified'. This is because a number of components and processes of lesser importance have been omitted. However, before considering what's missing, let's consider what is shown. Sunlight

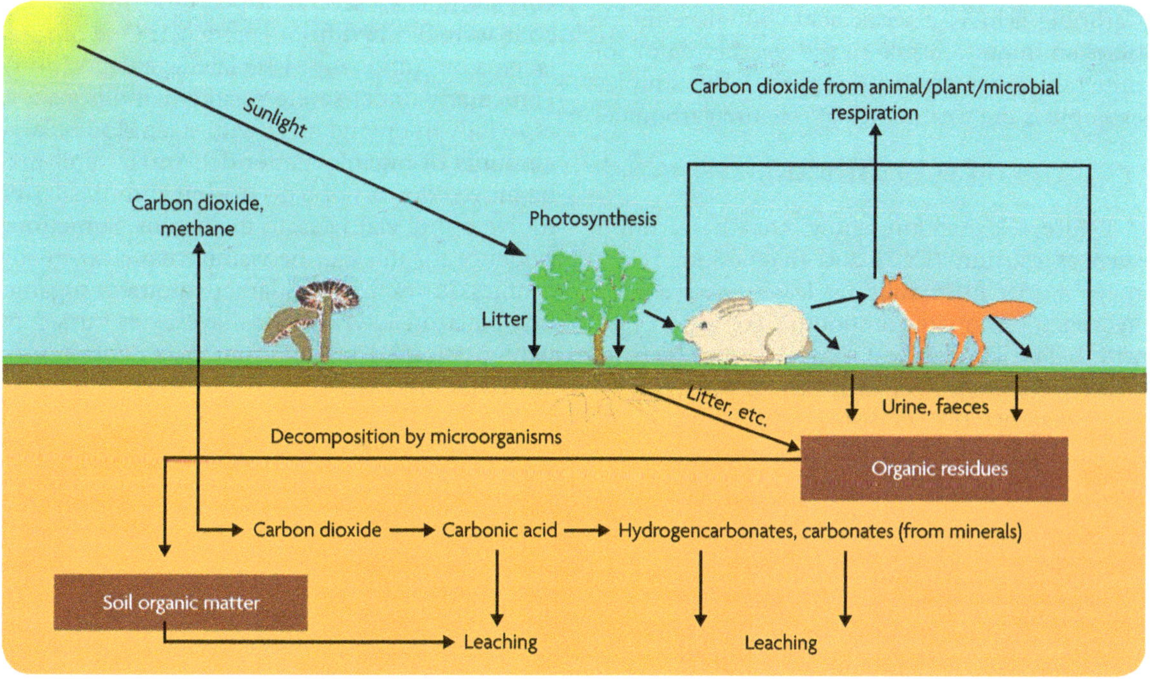

Figure 4.1 A simplified schematic representation of the carbon cycle.

provides the energy needed for photosynthesis to produce sugars for plant growth from carbon dioxide in the atmosphere and water. Plant tissue may be digested by herbivores, providing them with energy from food. Some of the carbon from the food will be deposited on the soil in animal faeces, etc., and some returned to the atmosphere via animal respiration. The herbivore animal itself may become part of a food chain. Animal waste, along with plant litter from falling leaves, twigs, branches, or dying roots, and organic compounds exuded by roots into the soil provide a substrate to support soil microorganisms. Some of the carbon in these materials will be lost back to the soil atmosphere via microbial respiration, mainly as carbon dioxide, some will become part of the microbial biomass and some will become part of the soil organic matter pool, as discussed later. We rarely think about soil microbes, unless we see fungal fruiting bodies such as mushrooms and toadstools.

So far we have considered carbon dioxide only in the gaseous form, but the gas dissolves in water to produce carbonic acid:

$$CO_2 + H_2O \leftrightarrow H_2CO_3$$

Carbonic acid is a weak acid, but nevertheless plays an important role in the weathering of soil minerals, as we saw in Chapter 1. The simplest example is the weathering of calcium carbonate:

$$CaCO_3 + H_2CO_3 \rightarrow Ca(HCO_3)_2$$

Figure 4.1 therefore shows that base cations such as calcium (Ca^{2+}) and magnesium (Mg^{2+}) may be lost from soils by leaching as soluble hydrogen carbonate compounds (or as ion pairs with balancing electrical charge, so that there are two HCO_3^- anions for each Ca^{2+} or Mg^{2+} cation). At higher pH (i.e. under more alkaline conditions), carbonates may be leached, for example with two sodium (Na^+) or potassium (K^+) cations accompanying each carbonate (CO_3^-) anion.

The leaching of calcium and magnesium hydrogen carbonates from soils to rivers is a very important component of the carbon cycle for two main reasons. The first is that the soluble salts may be transferred to lakes or the oceans, and ultimately the transported ions may end up in shells or ocean floor deposits. On the geological timescale of course such deposits may eventually be returned to the Earth's surface by uplift. The second reason is that the negatively charged HCO_3^- ions balancing the positive charges associated with cations have to remain in solution to maintain electro-neutrality or charge balance, and this has an important consequence for the buffering of the acidity of river water closer to biologically favourable, circum-neutral values. This is considered further in section 4.5.

Figure 4.1 indicates that methane gas may be a significant product from the decomposition of organic materials in soil. Until the last decade or so, methane emissions from soils attracted little or no attention, but this has changed recently because methane is important in the context of global warming, being a potent greenhouse gas. If soils become anaerobic (because virtually all the oxygen has been depleted by respiration, and not replaced by diffusion from the atmosphere), methanogenesis may occur as carbon dioxide is reduced to methane by *methanobacterium*.

4.4.2 Other aspects of the C cycle: erosion

For simplicity, several aspects of the carbon cycle were omitted from Figure 4.1. One of these is erosion (and any associated re-deposition). Particularly under severe weather conditions and especially in upland areas with steep slopes, large amounts of organic-matter-rich surface soil may be moved down slope by erosion and transferred to rivers and, via rivers, to the oceans. Sometimes evidence for this can be visibly very clear in soil profiles, for example if darker coloured, organic-rich soils on lower slopes have been buried by lighter coloured mineral materials deposited by erosional processes.

An example of this may be seen in Figure 4.2. Large amounts of plant material may be moved in this way too, often, but by no means always, as a consequence of river bank erosion. Figure 4.3, for example, shows mainly plant detritus from storm damage trapped under a small bridge on the Bobby Jones Golf Course in Atlanta, Georgia, after only a moderately severe storm.

Quantifying the removal of eroded material in rivers is difficult, because the amount being

Figure 4.2 At some time in the past, each of the darker layers in the profile of this river-bank soil was at the surface. Past erosion by down-slope overland flow (or by high stream flow) has stripped off the older, darker, organic-rich surface layers. In subsequent periods, deposition of lighter mineral materials has buried each of the older soil layers.

Figure 4.3 During periods of high river discharge during severe storms, large amounts of carbon may be transported in dislodged plant materials, as shown here on an urban golf course in the USA.

moved depends strongly, at any moment in time, upon the severity of recent and current climatic conditions, and upon drainage basin characteristics such as soil types, land use and management, grazing, slope, the presence or absence of drainage system(s), etc. The flux of carbon per hour may change dramatically over periods of a few hours. Intensive sampling is therefore necessary to accurately estimate erosion losses of carbon under high discharge conditions. Sediment traps are fine under moderate flow conditions, but don't cope well with large branches and tree trunks! Under really severe conditions use of auto-samplers such as those described later in Chapter 17 is not really an option.

4.4.3 Other aspects of the C cycle: mixing by soil fauna and mammals

Another important aspect of the carbon cycle not indicated in Figure 4.1 is the mixing influence of earthworms, soil invertebrates and burrowing mammals. Sometimes the impact of these is very obvious. For example, moles leave their characteristic trademark molehills, as in Figure 4.4. Earthworms too may play an important role in mixing soil layers (Figure 4.5), moving organic litter lower in the profile where it is more readily decomposed. In fact the gut of earthworms is an excellent source of microorganisms, so they not only move organic litter and break it up into smaller fragments; they also apply a liberal input of microorganisms to it, facilitating subsequent decomposition.

Lead-in question

If earthworms and burrowing mammals do such a thorough mixing job, why are the boundaries between the individual soil layers in Figure 4.2 so sharp and clear?

Figure 4.4 Typical molehills (and the author's daughter, at the age of 5, obligingly doing her best to perform a mole impression for purposes of establishing approximate scale!).

Figure 4.5 Earthworms make a major contribution to mixing carbon-rich soil residues within a soil profile. Earthworm casts provide visual evidence of their activities.

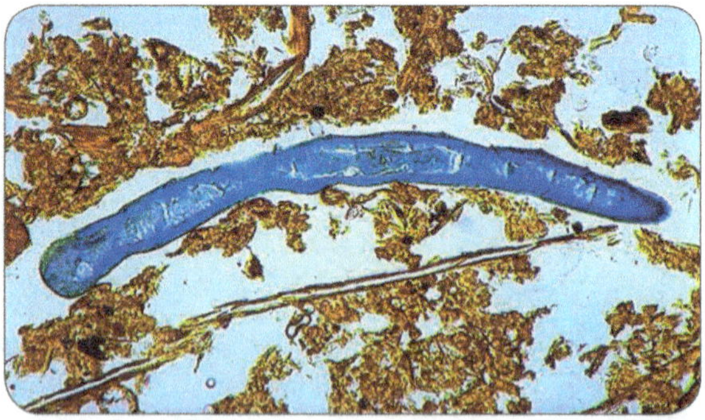

Figure 4.6 Microscopically enlarged image of an Enchytraeid worm (stained blue here) in a soil thin section. Enchytraeids contribute to comminution and mixing carbon-rich soil residues within the upper 5–10 cm of acidic moorland soils.

Source: Photograph courtesy of E.A. Fitzpatrick.

Figure 4.7 A typical agricultural soil. The earthworm activity works as effectively as cultivation here to mix the organic matter in the surface soil thoroughly to a depth of about 30 cm.

In very acidic soils, the activity of earthworms may be negligible, and indeed they may be totally absent. Very small (a few mm long) enchytraeid worms (Figure 4.6) may contribute to mixing within the surface organic layer but have little effect at depth. The soils in Figure 4.2 have a very low pH value below 4 (i.e. they are very acidic), so the soil layers remain relatively unmixed. Compare this soil profile with that of a more typical agricultural soil (Figure 4.7). In the latter the organic matter and its carbon are mixed to a depth of more than 15 cm. Even so the darkening effect of the presence of organic matter is still visible in the upper soil layer, even though the soil organic matter content is only around 6–8 per cent. The redistribution of organic detritus over 15 to 25 cm depth means that the recycled nutrient elements are released to a greater depth in the soil, from where they may be taken up again by plant roots.

4.4.4 Carbon dioxide and the oceans

As already stated, several aspects of the carbon cycle were omitted in the sketch of the carbon cycle in Figure 4.1 for simplicity. In essence, Figure 4.1 is the sort of sketch that a terrestrial ecologist might typically draw, because although it acknowledges that both organic carbon and inorganic carbon species may be lost in leached drainage water, implying that they may eventually pass on to the oceans, it makes no mention of ocean processes. Over recent years the capacity of the oceans to act as a sink for carbon dioxide has been increasingly widely recognised because of concerns over the sharply rising concentration of the gas in the global atmosphere and its potential impact upon climate. The growth of photosynthesising algae and their subsequent fate, and the fate of inorganic salts of carbonic acid entering the oceans are clearly very important to the global carbon cycle, especially considering what a high proportion of the planet's surface is covered in ocean waters.

Data summarised in a recent review by Lal (2008) highlight very effectively why oceans are so important to the global carbon cycle. The review points out that the total pool of carbon in near-surface ocean water globally is about 900 Pg (petagrams, one petagram equals 10^{15} g or 10^{12} kg or 10^9 metric tonnes), while that in intermediate and deep ocean water is about 37,100 Pg. Compared to these values, the crucial mass in the atmosphere is only around 780 Pg, and each year around 90 Pg is transferred between oceans and the atmosphere and vice versa. However, a cause for concern is the observation that the flux from the oceans to the atmosphere slightly exceeds the flux in the reverse direction (by about 2–3 Pg y^{-1}). This has led to concerns about significant acidification of sea water before the end of the century (Feely *et al.*, 2008; Cao and Caldeira, 2010). Terrestrial vegetation, by comparison, holds only about 500–650 Pg of carbon, 400–500 Pg of that being in forest. Soils and litter contain 2500 and 40–80 Pg respectively, with ca. 150 Pg being in peat. What happens to carbon cycling in the Earth's oceans is therefore crucial to the regulation of the air we breathe and to everything that happens in terrestrial ecosystems, because they hold the greatest carbon pool. Moreover, it is thought that the oceans have absorbed around half of the extra CO_2 that has entered the atmosphere as a consequence of human activities since the onset of the Industrial Revolution around 1800 (Sabine *et al.*, 2004).

Phytoplankton and the species that graze on them play a crucial role in removal of CO_2 from the atmosphere by ocean waters. Algal blooms tend to respond more to inorganic nitrogen species than to phosphorus in ocean waters. However, for more than two decades it has been recognised that the very low concentrations of iron in ocean waters may significantly restrict growth, and experiments have been conducted to see if iron fertilisation of ocean waters can increase removal of CO_2 from the atmosphere (Watson *et al.*, 2001).

Many have criticised small-scale simulation experiments and even experiments conducted in open ocean waters because of their lack of realistic conditions. For this reason a study that exploited a naturally occurring large phytoplankton bloom which was supplied with iron and other nutrients by upwelling deep water from below in the Southern Ocean appears to be especially attractive (Blain *et al.*, 2007). The results suggested that such upwelling may have played key roles in the past, and may do again in the future. Because of the complexity and scale of ocean systems, there no doubt will continue to be concern about the possible use of iron fertilisation of ocean waters as a pollution management technique for many years yet to come.

4.4.5 Carbon dioxide and plants

When concerns about climate change and elevated CO_2 concentrations in the atmosphere were first raised, it caused a number of scientists to consider what direct effects the increasing concentration might have upon plant growth (e.g. Warrick *et al.*, 1986; Eamus and Jarvis, 1989). Many experiments were carried out, mainly on agricultural crops, which showed growth stimulation.

This was as might be expected, since even then elevated CO_2 concentrations were being used in crop production in glass houses to accelerate growth. The author can remember being at a meeting at the time where most of the scientists present saw the increase as a free fertiliser rather than a potential problem in this respect. This was, in some respects, naïve, because sustaining any increased crop yields would obviously mean increased inputs of other nutrient fertilisers over the longer term. Nevertheless, the direct effects on non-agricultural systems received relatively little attention at the time as a result. This is important, because we need to know whether such natural and minimally managed systems are sustainable when nutrient demands are increased by plants often growing on nutrient-poor soils, litter inputs from perennial crops may be increasing, and water use efficiency is changing because of changes in stomatal distributions on foliage (Woodin *et al.*, 1992).

Experiments with heather plants (*Calluna vulgaris*) growing in natural, nutrient-deficient soils at ambient and ambient plus 100 or plus 200 ppm CO_2 showed suppression of growth in the first year, but enhancement in the second year (Woodin *et al.*, 1992). This really highlights the problems with much of the experimental work in this area, which has been done for relatively short periods and with annual crops, and with sudden jumps in CO_2 concentration rather than a gradual increase over a few years or decades. This makes model extrapolations into the future much less reliable, especially as we have little idea how soil process will change over the longer term.

One fact that we can be fairly sure of from horticultural experiences is that physiological plant growth stages will be speeded up, and flowering will occur earlier. However we know less about the consequences of this for insect populations and higher species of food chains.

4.5 Carbon cycling and water pH

As we saw earlier, carbon dioxide dissolves in water to produce a weak acid called carbonic acid, H_2CO_3. The amount of carbon dioxide that dissolves depends upon the concentration of gas in the atmosphere with which the water is equilibrating. The higher the atmospheric concentration of the gas, the higher the concentration of carbonic acid will be. We write the equation for its dissolution as a reversible equilibrium:

$$CO_2 + H_2O \leftrightarrow H_2CO_3$$

The more carbon dioxide there is in the air equilibrating with the water, the higher the concentration of carbonic acid formed. Double the CO_2 concentration and you double the H_2CO_3 concentration. We can express this as a simple mathematical equation:

$$pCO_2/\{H_2CO_3\} = K$$

where K is a constant, pCO_2 represents the partial pressure of carbon dioxide in the atmosphere and $\{H_2CO_3\}$ is the concentration of carbonic acid in the equilibrating solution.

We can also write an equation for the dissociation (i.e. for the release of hydrogen ions (H^+) or protons) of carbonic acid:

$$H_2CO_3 \leftrightarrow H^+ + HCO_3^-$$

For this reversible reaction, the probability of the free, un-dissociated acid re-forming increases in proportion to the concentration of H^+ cations and to the concentration of HCO_3^- anions present in the solution phase. This is intuitively obvious, as the probability of the cation and anion species coming close enough together to interact depends on the concentrations of both species. We can express this in simple arithmetic terms as follows:

$$\{H^+\}\{HCO_3^-\}/\{H_2CO_3\} = k$$

where a set of curly parentheses, $\{\}$, denotes concentration of the species in the parentheses. The constant, k, is the dissociation constant of the acid. Thus if we double the amount of H^+ present without changing the concentration of HCO_3^-, we double the concentration of un-dissociated carbonic acid, and so on.

We will see in Chapter 7 that the atmosphere in soils, because of all the microbial and root respiration, contains a much higher concentration of carbon dioxide than the air that we breathe, typically 1 to 2 per cent compared to 385 ppm (parts per million) or 0.0385 per cent. Thus when water drains out of a soil it re-equilibrates with an atmosphere with a much lower carbon dioxide concentration, typically around 30 to 60 times lower. The consequence of this for a solution of carbon dioxide in pure water would be out-gassing (release to the atmosphere) of carbon dioxide to attain a new equilibrium. At the new equilibrium, the concentration of H_2CO_3 would be 30 to 60 times lower in solution, which means, according to the acid dissociation equation given earlier, that the product of $\{H^+\}$ and $\{HCO_3^-\}$ would also decrease 30- to 60-fold. This means that the H^+ ion concentration would be decreased by ca. 5.5- to 7.8-fold (i.e. by $\sqrt{30}$ to $\sqrt{60}$). Thus carbon dioxide out-gassing raises the pH of water as it drains from soil into a river.

In soil solution, as opposed to a pure, aqueous solution of dissolved carbon dioxide, any HCO_3^- ions present to balance the charge on cations such as Ca^{2+} or Mg^{2+} have to remain in solution. Thus if the product of $\{H^+\}$ and $\{HCO_3^-\}$ falls by 30- to 60-fold as a consequence of CO_2 out-gassing, as in the above example, but the $\{HCO_3^-\}$ concentration can only fall by a factor of 2-fold, then the $\{H^+\}$ would have to fall by 15- to 30-fold. The resulting river water would thus be much less acidic than the soil solution from which it originated.

In natural ecosystems, hydrogen ion concentration varies over a very wide range, from around 0.01 molar (i.e. 0.01 moles per litre or 10^{-2} M) right down to around 10^{-10} M (0.000 000 000 1 molar). For convenience, scientists define the acidity of solutions on a logarithmic scale known as the pH scale, as follows:

$$pH = -\log_{10} \{H^+\}$$

On this scale a solution containing 10^{-2} moles of H^+ per litre has a pH value of 2, and one with 10^{-10} moles per litre has a pH of 10. Very pure, deionised water at 25 °C has a pH of 7. This is because pure water, a very poor conductor of electricity because it contains so few ions, does nevertheless dissociate slightly:

$$H_2O \leftrightarrow H^+ + OH^-$$

As before we can write an equation to describe this equilibrium:

$$K_w = \{H^+\}\{OH^-\}/\{H_2O\}$$

The concentration of water (in water) in this equation, $\{H_2O\}$, never changes significantly so it is assigned a fixed value of 1. Thus we can simplify the equation for the dissociation of water to:

$$K_w = \{H^+\}\{OH^-\}$$

The value of K_w is 10^{-14}, so for charge neutrality both $\{H^+\}$ and $\{OH^-\}$ must equal 10^{-7}. A 10-fold shift in the value of $\{H^+\}$ corresponds to a pH shift of 1 unit, and a 100-fold shift in $\{H^+\}$ to a pH change of 2 units. Thus CO_2 out-gassing of soil solutions as they drain into rivers may cause the water pH to increase by 1 to 2 pH units. This may not look very important, but it might be a matter of life or death to an acid-sensitive fish species!

4.6 The importance of organic matter in soil

We have seen that plant biomass production depends upon solar energy providing the energy needed for photosynthesis to proceed. Organic matter of plant origins subsequently decomposes in near-surface soils as a consequence of microbial degradation, sometimes aided by soil animal activity. This releases carbon dioxide (CO_2) back into the atmosphere, where it may again participate in photosynthesis. The decomposition process occurs at a rate that depends upon the nature of the organic material being decomposed and the environmental conditions such as soil moisture content, acidity and temperature. The process is not instantaneous, and not all of the carbon is converted to CO_2. So not all passes back to the atmosphere or is converted to carbonic acid so that it can participate in soil mineral weathering. Some is incorporated into soil organic matter (often referred to as soil humus). We saw the visible effect that soil humus has on the darkening of soil colour in Figure 4.2.

In temperate regions, agricultural soils typically contain about 5 to 6 per cent organic matter formed in this way. In hotter climates the figure may be lower, and in wetter and/or colder regions and in more acidic, minimally managed soils, where decomposition is slower, organic matter concentrations may be much higher.

At the present time, the amounts of organic carbon stored in soils are attracting considerable attention because of growing concerns about human disruption of the natural carbon cycle, especially through excessive exploitation of fossil fuel resources. There are two main issues that are causes for concern. The first is the extent to which the pools of carbon stored in soils are likely to change in response to temperature change (e.g. Thornley and Cannell, 2001). This is a cause for concern because a positive feedback occurs if pools decline and more CO_2 passes to the atmosphere, further enhancing warming. However, we also need to consider impacts of changes in precipitation patterns, because soil moisture content has a large impact upon turnover of organic matter. The second issue is the potential effect of land use change. Most concern in this respect has centred around impacts of deforestation (Lorenz and Lal, 2010), but a recent review has pointed out that rapid changes in carbon storage may also occur in peatlands and grasslands (Ostle *et al.*, 2009). The authors suggested that carbon sequestration in soils should become a routine aspect of ecosystem management alongside food or timber production.

We will return to this point later in the book, but there are several other reasons why soil organic matter is important to sustaining ecosystems. They include:

- Build up of cation exchange capacity (CEC).
- Contribution to soil moisture retention.
- Contribution to the development of soil structure.
- Contribution to soil colour and radiative properties.
- Contribution to sustaining microbial biomass.
- Contribution to soil buffering capacity.
- Contribution to the supply of plant nutrients.

These aspects are briefly considered here, though some aspects will be revisited in more detail in the chapter on soil evolution.

4.6.1 Contribution of organic matter to cation exchange properties of soils

The chemical structure of soil organic matter is complex, but it contains some readily identifiable functional groups, including carboxylic acid groups (–COOH) and phenolic –OH groups (hydroxyl groups attached to aromatic ring structures). These groups may be partially dissociated in soils, to produce $-COO^-$ and H^+ or aromatic-O^- and H^+. The negative charge on these functional groups may then be partially balanced by cations that exchange for the initial H^+ cations. In fertile agricultural soils Ca^{2+} is often predominant on these cation exchange sites, typically followed by the other major plant nutrient element cations Mg^{2+} and K^+. Other cations may also be present, such as sodium (Na^+), ammonium (NH_4^+), and various trace metal nutrients such as zinc (Zn^{2+}) and manganese (Mn^{2+}). Thus these cation exchange sites are valuable for retention of metal cations released by mineral biogeochemical weathering and retention of ammonium released during decomposition of biomass residues. In this way they minimise nutrient element cationic leaching losses from soils. We shall see later that soil minerals may also contribute to CEC, but in coarse-textured sandy soils especially the contribution of organic matter is very important.

4.6.2 Contribution of organic matter to buffering capacity of soils

The high CEC of soil organic matter contributes to protecting soils against sudden sharp changes in their chemistry. For example, Figure 4.8 shows what happens when zinc is added to a soil under five different vegetation types. The zinc is initially retained on the cation exchange sites. However, over several days it changes to other forms. In this experiment the rate of change was much slower in the most acidic soils that were present under heather (*Calluna vulgaris*) or bracken (*Pteridium aquilinum*), and much faster in thoroughly limed, reseeded soils.

Lead-in question

How was it possible to distinguish between added zinc and native zinc and between exchangeable zinc and other zinc forms to produce Figure 4.8?

In the experiments used to produce Figure 4.8, the added zinc was spiked with a radioactive isotope of the element, Zn^{65}. If this is the only radioactivity present in the system, then the exchangeable zinc originating from the added spike of zinc can be determined by leaching with another cation, Mg^{2+}, added at large excess, and measuring the radioactivity of the leachate. Soil scientists have developed series of sequential extraction procedures that allow them to quantify trace metals bound to exchange sites, to organic matter, in carbonates and to amorphous and crystalline oxides of iron and aluminium. Users of such procedures have to follow detailed step-by-step instructions very carefully to get meaningful results. The procedures are said to be 'operationally-defined procedures'.

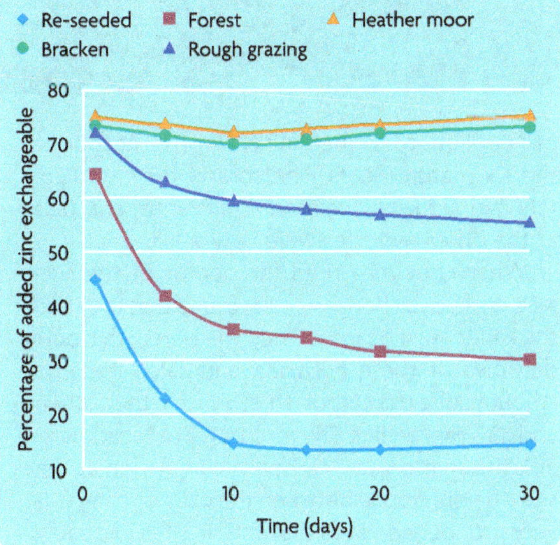

Figure 4.8 Effect of land use and vegetation type upon changes in the amounts of added zinc remaining on the cation exchange sites in soils over 30 days.

The pools of cations bound to cation exchange sites are invariably very large compared to the pools present as free cations in soil solution. If a sudden input of any one cation occurs, the CEC acts as a temporary sink very rapidly, reducing the concentration in soil solution. In this way organic matter can help buffer against sudden changes in bio-availability of metals or in concentrations leached to surface waters or drainage waters.

4.6.3 Contribution of organic matter to plant nutrient supplies and sustaining microbial biomass

Organic matter contributes to biomass nutrient supplies in two main ways. When plant litter decomposes as part of the carbon cycle, other nutrients from the litter are also released into the near-surface soil. If they are not taken up by plants or microorganisms immediately, then the high CEC of soil organic matter contributes to the retention of most of these nutrients within the rooting zone. Nutrient leaching losses are minimised, but nutrient cations held on cation exchange sites are still generally readily available to plants and microbes. In addition, organic matter may help retain soil moisture status during extended dry periods, keeping conditions more favourable for litter decomposition and sustaining a viable microbial population. Figure 4.9 shows further results from the experiment described earlier on the fate of zinc added to soils. Here an operationally defined procedure has been used to quantify zinc passing to the organically bound fraction, which includes microbial and other biomass zinc. The sharp decline in exchangeable zinc seen in the limed improved grassland in Figure 4.8 is accumulated with rapid transfer to the organically-bound phase, associated with the high biological activity in the limed soil.

4.6.4 Contribution of organic matter to soil structure and water retention

Gardeners living in dry regions with very freely draining soils often add large amounts of compost to their soils to improve soil moisture retention. Unless dried out over very extensive periods

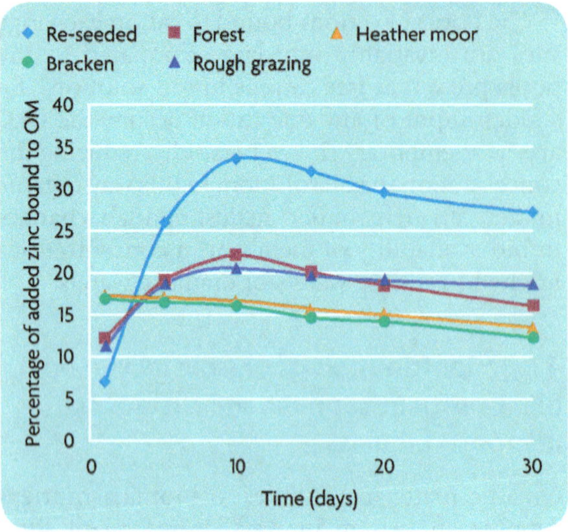

Figure 4.9 Change in organically bound zinc over 30 days under various types of vegetation/land use; the reseed soil has the highest pH (is least acidic), so has high bacterial activity.

(in which case organic matter may become very hydrophobic), humus has a high affinity for water. Peaty soils, which typically contain 95 per cent organic matter when oven dry, typically contain 80 per cent water in the field. Thus 5 g of wet peat dried in an oven at 105 °C might give only 1 g of dry peat. Its moisture content is thus 400 per cent, since moisture contents of soils are always expressed on an oven dry mass basis. Such high values sometimes surprise those not used to dealing with highly organic soils.

Part of the effect of organic matter on soil moisture retention is due to the strong affinity of the organic matter itself for water, but part is due to its beneficial effect upon soil structure. Acting in combination with soil microorganisms (which also contribute organic matter directly), organic matter helps bind soil mineral particles into aggregates. The latter may contain additional small pores that hold water very effectively, reducing the tendency to rapid free drainage.

CASE STUDIES

A case study – Why are changes in soil carbon pools important other than when investigating C budgets?

When conducting research on the effects of atmospheric nitrogen and sulphur deposition on soils, Billett *et al.* (1990) decided to revisit sites in Scottish forests where soils had been sampled for analysis 38 years earlier. The original researcher, E.A. Fitzpatrick, had kept both his original samples and meticulous details of where the original soil pits were, so they could be very accurately relocated. The researchers wanted to test the hypothesis that the near-surface soil C:N ratio would have fallen markedly because of decades of N pollution. To their initial surprise, at four of the six mature Scots Pine forests they studied, C:N had actually increased significantly, and falls at the other two sites were very small.

When they measured the depths of the surface organic soil horizons, they found that they had thickened significantly. They measured the bulk densities of these horizons, and calculated the amounts of extra carbon stored within them, then and 38 years earlier. The results, reproduced here, indicate that organic carbon accumulation at the site (the difference between the pairs of bars) was very pronounced at five of the six sites, probably because acidifying pollution deposition had slowed down the decomposition of organic matter.

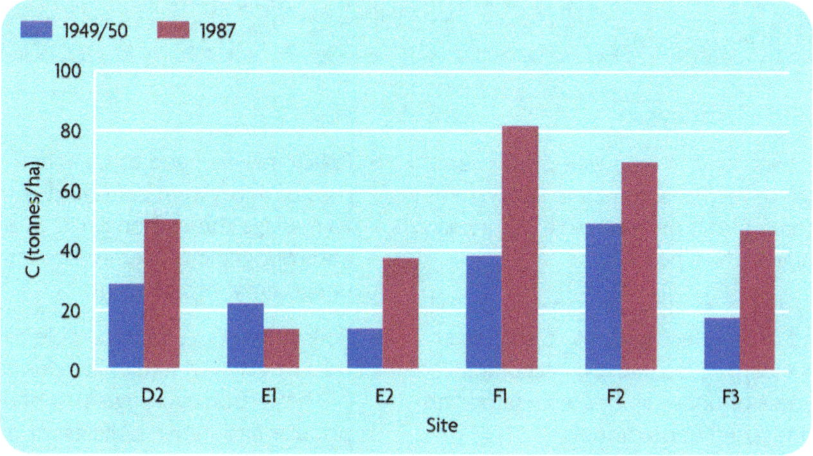

They then calculated the pools of nitrogen stored at the site in a similar way. At five of the six sites, the pool of organic nitrogen had also increased, showing that the N deposition had indeed accumulated at the site over the decades. In the chart below, the values above the bars show the accumulation rates in kg N ha^{-1} y^{-1}. These values are quite close to the N deposition fluxes for the area. A perfect match would not be expected, because position on slope is also important to forest soil processes.

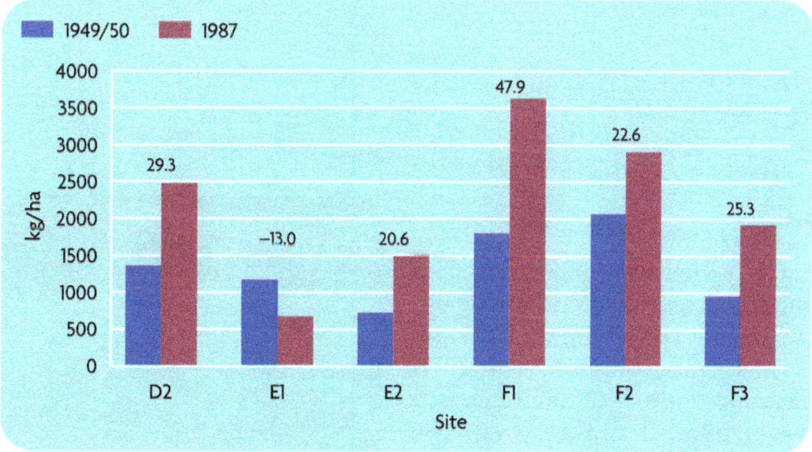

Whether analysing soils or plant tissue materials, it is important to realise that growth rate (or organic matter accumulation rate for soils) will influence the concentrations of elements present in the samples. Faster growth has a dilution effect in plant tissue if the uptake from soil is constant. For example, the extra organic matter accumulating in these soils would have influenced the heavy metal concentrations that were also measured later at the sites (Billett *et al.*, 1991), just as it influenced nitrogen concentrations. Therefore when looking for changes over time, pools are often more informative than concentrations. The results thus provide a cautionary warning to those who attempt to use parameters such as C:N ratio as an index of pollution.

POLICY IMPLICATIONS

The global carbon cycle

- Policy makers need to be aware of the need to manage terrestrial and marine ecosystems in ways that at least maintain soil, vegetation and oceanic carbon pools.
- Knowledge of the long-term capacity of soil, forest and oceanic ecosystems to sequester carbon is essential, but must be accompanied by awareness of uncertainties associated with predictions based upon models.
- Understanding the importance of atmospheric carbon dioxide and methane concentrations to climate change, and vice versa, is important to managing uses of natural resources sustainably.
- Policy makers need to be able to assess risks of ecosystem damage due to attempts to manage the carbon cycle to ameliorate anthropogenic CO_2 emissions, properly taking uncertainty into account.
- Policy makers need to be able to balance risks of ecosystem damage due to disruption of the carbon cycle against potentially less popular alternative policies for managing CO_2 emissions.

Some of these policy implications should become clearer after reading later chapters.

CHAPTER REVIEW EXERCISES

Exercise 4.1

Columns of moist garden topsoil were flushed with either ambient air, or air containing an elevated carbon dioxide concentration of 3 per cent, for 24 hours. The columns were then leached with the same volumes of rainwater, and the leachates were collected and analysed. The following results were obtained for the base cation concentrations, all measured in units of mg l^{-1}.

Element	Ambient air	Air with 3% carbon dioxide
Calcium	0.11	0.15
Magnesium	0.27	0.41
Potassium	0.34	0.47
Sodium	0.35	0.27

Explain the processes whereby the base cation concentrations were increased at the elevated carbon dioxide concentrations.

Exercise 4.2

A soil pit was dug in an agricultural field and the exposed soil profile was sampled to a depth of 100 cm in five 20-cm increments. The samples were analysed for their total carbon concentrations, to give the following results.

Depth range (cm)	Density (g cm^{-3})	C concentration (%)
0–20	1.22	3.60
20–40	1.38	1.95
40–60	1.60	0.58
60–80	1.75	0.25
80–100	1.78	0.04

Calculate the mass of carbon in kg per square metre of the field to a depth of 100 cm and the mass of carbon in metric tonnes per hectare (1 hectare = 10^4 m^2).

Exercise 4.3

Use the data in Exercise 4.2 to plot a graph of density (on the y or vertical axis) against concentration (on the x or horizontal axis). Similarly plot a graph of density versus depth. Give possible reasons for the relationships that emerge.

Exercise 4.4

We saw in this chapter that, as a consequence of photosynthesis, aquatic plants and plankton may liberate oxygen, oxygenating the water in which they are growing. The dissolved oxygen concentration was measured in the water in a shallow lake in northern Britain in Summer, every 2 hours over a 24-hour period from midnight to midnight. The results are shown in the following table.

Time of Day	Dissolved Oxygen Concentration (mg l^{-1})
00.00	4.00
02.00	3.05
04.00	3.00
06.00	3.52
08.00	4.86
10.00	7.02
12.00	9.75
14.00	12.03
16.00	9.85
18.00	7.45
20.00	5.18
22.00	4.95
24.00	4.03

Plot a graph of dissolved oxygen concentration (vertical axis) against time of day (horizontal axis), and explain the changes seen over the 24-hour period. What do you think the corresponding changes would be in carbon dioxide concentration over the same time period and why? What do you think the relevance of the trends seen might be to aquatic biota?

REFERENCES

Billett, M.F., Fitzpatrick, E.A. and Cresser, M.S. (1991) Changes in the carbon and nitrogen status of forest soil organic horizons between 1949/50 and 1987. *Environmental Pollution*, **66**, 67–79.

Billett, M.F., Fitzpatrick, E.A. and Cresser, M.S. (1991) Long-term changes in the Cu, Pb and Zn content of forest soil organic horizons from North East Scotland. *Water, Air and Soil Pollution*, **59**, 179–191.

Blain, S., Quéguiner, B., Armand, L. and 44 others (2007) Effect of natural iron fertilization on carbon sequestration in the Southern Ocean. *Nature*, **446**, 1070–1074.

Cao, L. and Caldeira, K. (2010) Can ocean iron fertilization mitigate ocean acidification? *Climate Change*, DOI 10.1007/s10584-010-9799-4.

Eamus, D. and Jarvis, P.G. (1989) The direct effect of the global increase in atmospheric CO_2 concentration on natural and commercial temperate trees and forests. *Advances in Ecological Research*, **19**, 1–55.

Feely, R.A., Sabine, C.L., Hernandez-Ayon, J.M., Ianson, D. and Hales, B. (2008) Evidence of upwelling of corrosive 'acidified' water onto the continental shelf. *Science*, **320**, 1490–1492.

Ingenhousz, J. (1779) *Experiments upon Vegetables, Discovering their Great Power of Purifying the Common Air in the Sun-shine and of Injuring it in the Shade at Night, to which is Joined a New Method for Examining the Accurate Degree of Salubrity of the Atmosphere.* Printed for P. Elmsley in the Strand and H. Payne in Pall Mall, London. Viewable *via* Google Books.

Lal, R. (2008) Sequestration of atmospheric CO_2 in global carbon pools. *Energy & Environmental Science*, **1**, 86–100.

Lorenz, K. and Lal, R. (2010) *Carbon Sequestration in Forest Ecosystems*. Springer, pp. 277 + xix.

Ostle, N.J., Levy, P.E., Evans, C.D. and Smith, P. (2009) U.K., land use and soil carbon sequestration. *Land Use Policy*, **265**, S274–S283.

Priestley, J. (1796) *Considerations of the Doctrine of Phlogiston and the Decomposition of Water.* Printed by Thomas Dobson, 41 South Second Street, Philadelphia. Viewable via Selected Classic Papers, http://web.lemoyne.edu/~giunta/papers.html.

Sabine, C.L., Feely, R.A., Gruber, N. and 12 others (2004) The oceanic sink for anthropogenic CO_2. *Science*, **305**, 367–371.

Thornley, J.H.M. and Cannell, M.G.R. (2001) Soil carbon storage response to temperature: an hypothesis. *Annals of Botany*, **87**, 591–598.

Warrick, R.A., Gifford, R.M. and Parry, M.L. (1986) CO_2, climatic change and agriculture, in: *The Greenhouse Effect, Climatic Change and Ecosystems* (eds Bolin, B., Doos, B.R., Jager, J. and Warrick, R.A.). *SCOPE*, **29**, 393–473, John Wiley & Sons, Chichester.

Watson, A.J., Liss, P.S. and Duce, R. (1991) Design of a small-scale *in-situ* iron fertilization experiment. *Limnology and Oceanography*, **36**, 1960–1965.

Woodin, S., Graham, B., Killick, A., Skiba and Cresser, M. (1992) Nutrient limitation of the long-term response of heather [*Calluna vulgaris* (L.) Hull] to CO_2 enrichment. *New Phytologist*, **122**, 635–642.

CHAPTER 5

The cycling of nitrogen and selected other elements

Malcolm Cresser

Learning outcomes

By the end of this chapter you should:

- Know where all the nitrogen (N) in soils and plants comes from.
- Be aware of the importance of microorganisms to the cycling of N and other nutrient elements.
- Be aware of what regulates the leaching of N from soils into rivers and lakes and on to the oceans.
- Appreciate the merits of using isotopic tracers in experiments to study element cycling.
- Understand the differences between cycling of N and cycling of other nutrient elements.
- Start to have a better insight into how soils start to form from rocks or sediments.

5.1 Introduction

When I first started to learn inorganic chemistry at high school back in the 1960s, in spite of the best efforts of my chemistry teacher to make it sound interesting, nitrogen struck me as a particularly boring element. Here was an element that made up four-fifths of the air we breathe, so there was a lot of it around, but it really didn't seem to do very much or to be very important. It was years later before it dawned on me in the natural world, nothing could be further from the truth.

Nitrogen's inertness is a direct result of the electronic configuration of N atoms. Each has two 2*s* electrons and three 2*p* electrons orbiting round outside a filled inner 1 s orbital containing another pair of electrons. The group of three 2*p* orbitals, each with a single, unpaired electron, made free N atoms very reactive at the time of the atmosphere's formation, and, in my chemistry teacher's words, 'desperate to pick up three extra electrons from anywhere they could to give them the nice stable electronic configuration of the inert gas neon'. A few years later I would have strongly criticised my students for using such flowery and colloquial English (atoms can't be desperate and electronic configurations can't really be nice). What he meant was that it is energetically more favourable for nitrogen atoms to combine with molecules by covalent bond formation than to form ionic compounds containing the N^{3-} nitride anion. Thus the element in the atmosphere is found predominantly as gaseous dinitrogen, or N_2, molecules, each with a triple covalent bond binding its two N atoms together. Each N atom in N_2 has 10 electrons, 7 of its own and 3 shared ones from its partner.

This triple bond is very strong, making it difficult to get N_2 to react with anything unless you supply a lot of energy to the reactants. An example in the natural world is the generation of nitrogen oxides during flashes of lightning in the atmosphere. My old chemistry teacher did this by putting burning magnesium ribbon into a gas jar filled with N_2 (a particularly memorable experiment, especially when he nearly set fire to his trousers at the first attempt!). The product was a modest amount of magnesium nitride, Mg_3N_2. In this the N atom has acquired 3 electrons, each with a negative charge, to form the nitride anion N^{3-}, but only under very energetically extreme conditions. Hence I dismissed N_2 itself as probably not very important.

Once you start to dabble in organic chemistry of course you soon start to realise that organic N compounds are extremely important to all forms of life on earth, and that nitrogen must somehow be cycled from soils to plants and back again. It is intuitively obvious too that, in the modern world, if too much plant material is taken away, for example in harvested crops, it soon will be necessary to start applying nitrogen fertiliser to avoid nitrogen deficiency problems in plants. But in the absence of man, plants grew, and went on growing, perfectly well. This chapter explains how this happened, not just in the case of nitrogen, but for the other elements essential to plant growth too.

Lead-in question

Where does all the nitrogen in soils come from?

5.2 The origins of soil nitrogen

The bulk of the nitrogen present in soil is associated with the soil organic matter, which was considered in Chapter 4. If a soil has 5 per cent organic matter, and a bulk density of 1.5 g per cubic cm (1.5 g cm^{-3}), then to 10 cm depth there would be, in one hectare (or in 10^8 cm^2):

$$(5/100) \times 10 \times 10^8 \times 1.5 \text{ g of organic matter}$$

This works out at 0.75×10^8 g of organic matter, or 0.75×10^5 kg of organic matter or 75 tonnes of organic matter. Typically this organic matter might contain 5 per cent nitrogen, so we would have 3.75 tonnes of organic N per hectare to the specified depth of 10 cm. This is only a crude estimate, but agrees with the results of direct measurements of the total amounts of nitrogen stored in soils. It's a lot of nitrogen! Such simple sums are useful if we want to estimate how long it would have taken a soil to accumulate all this nitrogen, provided we know much nitrogen an evolving soil could acquire each year.

Every hectare of soil may in fact contain around 10 tonnes (10,000 kg) or more of nitrogen to 1 m depth, mainly in the form of organic N. Back at school I never thought to ask the obvious (with the wisdom of hindsight) question: 'If soil is originally derived from rocks containing only tiny traces of inorganic N compounds, where has all its organic N come from?'

5.2.1 Biological nitrogen fixation

Readers partial to walking in upland areas may well be very familiar with the type of terrain featured in Figure 5.1. They may also be familiar with the common occurrence of lichens and mosses colonising the weathering rock surfaces in such landscapes, as in Figure 5.2.

Lichens are constituted from algae and fungi growing in a symbiotic (mutually beneficial) association. In spite of the strength of the triple covalent bond in N_2, algae are simple organisms capable of converting dinitrogen (N_2) to organic N and of photosynthesising. Thus they can produce organic matter primarily using water and gases present in the atmosphere. However, to multiply they require mineral nutrient elements such as phosphorus (P), sulphur (S), base cations and trace metal elements. They can obtain these from weathering of minerals in rocks. The symbiotic fungal partner facilitates the release of these elements from rock surfaces, making them available to the algae. In return the algae provide the organic matter that the fungi require as a substrate for growth.

Figure 5.1 A typical upland landscape with sparse vegetation colonising weathered rock surfaces.
Source: Photograph courtesy of Ute Skiba.

Figure 5.2 Closer view of lichens and moss colonising weathered rock surfaces.

As lichens grow and eventually start to decompose, organic matter starts to form between weathered mineral rock particles. This increases the water retention of the material, and before too long mosses may start to grow, providing a habitat for some insects. Further development is possibly aided by inputs from birds and ongoing decomposition, until eventually higher plants can start to become established. This can be seen in Figure 5.3. Over the years, the plant material in turn starts to die back and its organic matter too is subject to decomposition and formation of soil organic matter. Over the centuries the process continues, aided by other potential N inputs as discussed later, until soils as we more typically think of them can be found. On lower slopes erosion from upper slopes often will speed up the formation of thicker soils down slope.

Figure 5.3 Organic matter evolved from lichens and moss colonising weathered rock surfaces here has allowed grasses to start to establish.

The first stages of the above process sometimes can be seen occurring in unexpected places. For example Figure 5.4 shows two flat roofs in Stirling, Scotland, both covered in tar. However, to one (top right) granite chips have been added. Mineral weathering from these releases nutrient elements; they also help retain inputs from atmospheric dust and reduce evaporation losses of water. Thus with the chips soil formation is more advanced and higher plants are already starting to establish themselves on the roof. Pockets of trapped terrestrial dust have almost certainly played an important role on the roof in image b, bottom right (no chips added).

(a)

(b)

Figure 5.4 View of lichens and moss colonising two flat roofs, one with (a) and one without (b) a layer of granite chippings over the tarred surface. Note the higher plants starting to colonise the former, only a few decades after the roof was constructed.

The nitrogen-fixing algae in lichens live in a symbiotic association with fungi, but free-living algae also occur in soil and fix nitrogen from the atmosphere. They include blue-green algae (members of the division *Cyanochloronta*), green algae (*Chlorophycophyta*), diatoms and yellow-green algae (*Chrysophycophyta*), and red algae (*Rhodophycophyta*) (Lynch, 1983). However the members of the first two divisions predominate. Free-living algae may fix around 1 kg N per hectare per year (kg N ha^{-1} y^{-1}), so to acquire 10,000 kg ha^{-1} would take *circa* 10,000 years if soils had to depend just upon free-living algae. Many fully developed soils are younger than that as we will see in Chapter 7, so there must be other inputs of nitrogen during soil formation.

A genera of bacteria known as *Rhizobium* can play an important role in the fixation of dinitrogen; they form nodules with the roots of legumes. Figure 5.5 shows the *rhizobium* nodules associated with the roots of clover (*Trifolium* spp.). This is a further example of symbiosis. The plants supply organic substrate to the bacterium, but obtain fixed nitrogen in return. Clover can fix around 20 kg of nitrogen per hectare per year (kg N ha^{-1} y^{-1}). Plants living symbiotically with N-fixing bacteria clearly have a competitive edge over non-legumes in nitrogen deficient soils. A genus of the actinomycetes bacterial group, *Frankia*, also form nitrogen-fixing root nodules with a range of woody plant species, including Alder (*Alnus* species) and sea buckthorn (*Hippophae rhamnoides*). Shrubs such as broom and gorse are often found in road cuttings or anywhere the fertile organic matter-rich topsoil has been stripped away because of the benefits they derive from symbiotic nitrogen fixation. For example Figure 5.6 show opportunist gorse plants growing well in the bunker of a former abandoned golf course.

We considered algae, lichens and legumes, peas, beans, alfalfa, clover, soybeans and peanuts first because they are especially important to the build up of soil nitrogen pools, but there are also aerobic (*Azotobacter*, *Beijerinckia*, *Klebsiella* and some

(a)
(b)
(c)

Figure 5.5 Rhizobia nodules on a clover plant. The nodules are the small protuberances attached to roots, some marked with white arrows. This particular plant is unusual in that it has some four-leaved foliage ((a) – centre right).

Figure 5.6 Nitrogen-fixing rhizobia on the roots of this gorse bush helped it establish in the nitrogen-deficient soil of an old bunker on an abandoned golf course. The brilliant yellow flowers have faded here as the plant is forming seeds which help it colonise more ground.

Cyanobacteria) and anaerobic (some *Clostridium*, *Desulfovibrio*) free-living nitrogen-fixing bacteria in soils (Deacon, 2006). However we will not consider these further here. Cyanobacteria can also be found as symbionts with lichens.

Lead-in question

If it's so difficult to split the double bond in dinitrogen, how do bacteria manage to do it?

The equation generally used to represent fixation of dinitrogen is:

$$N_2 + 8H^+ + 8e^- + 16ATP \rightarrow 2NH_3 + H_2 + 16ADP + 16Pi$$

Two moles of ammonia are produced from each mole of dinitrogen, consuming protons (H^+), electrons (e^-) and adenosine triphosphate (ATP) in the process. ATP is a fascinating molecule that effectively allows energy to be stored in such a way whereby it can be readily exploited by organisms. It can interact with an enzyme to release a substantial amount of energy by loss of a phosphate group to form adenosine diphosphate (ADP). The pivotal role of ATP as an energy store in biochemical processes is one of the main reasons for phosphorus being such an essential element to biological systems. The prokaryotes (bacteria and related microorganisms) in which the reaction occurs use the nitrogenase enzyme complex (which involves an iron protein and a molybdenum-iron protein) to trigger the reduction of dinitrogen. For more information the excellent website by Deacon (2006) is a good starting point.

An adequate supply of available water is essential for microorganisms to function effectively, so moisture and moisture retention are very important to early soil formation and weathering processes. Upper slopes are often shrouded in mist (Figure 5.7), which facilitates the colonisation by mosses and the whole weathering process. Clearly it also facilitates the establishment of higher plants. Often in high uplands root establishment is via small cracks and crevices in weathering rocks, which become filled with fine mineral matter and organic detritus, and thus acquire a capacity for water retention.

Figure 5.7 Upper slopes of mountains are often shrouded in mist.

Source: Photograph courtesy of Ute Skiba.

5.2.2 Other nitrogen inputs to skeletal soils

There were inputs of inorganic nitrogen compounds to soils that would have aided the build up of soil nitrogen over the centuries. As mentioned earlier, the high available energy in lightning, for example, can produce oxides of nitrogen and nitric acid that can be deposited directly as gaseous species or as nitrate in precipitation (rain, snow, cloud water). Ammonia can enter the atmosphere from animal faeces and urine (remember the 'fragrance' of poorly serviced public toilets). Traces are also formed during lighting strikes. This can be taken up directly by vegetation or after deposition as ammonium compounds in rain or as aerosol particles. These could be as ammonium nitrate or ammonium sulphate for example. Sulphate in the atmosphere can originate from volcanic emissions or from natural fires. We will consider the influence of man, and especially man's excessive fossil fuel combustion and intensive animal rearing impacts, later in this book, but for now we should not forget that fires can, and do, occur quite naturally, for example as a result of lightning strikes. The odd insect or small mammal dying in a remote area can also be a useful source of soil nitrogen eventually.

Animals entering an area can contribute waste material even if they don't die of course. In the case of birds this can be a considerable input, especially in the vicinity of a bird colony (Figure 5.8).

Figure 5.8 Excrement from birds can be an important nutrient input to bare rock. This shows droppings down a cliff face in Tenerife under an overhanging ledge.

If you consider all these inputs alongside trapped terrestrial dust, the build up of a skeletal soil such as that in Figure 5.4 over a few decades becomes a lot less surprising.

5.3 The nitrogen cycle

We are now in a position to consider the whole nitrogen cycle as it must have been in the absence of man once a soil had evolved, or at least a simplified schematic representation of it, as in Figure 5.9. It will be apparent that although many of the individual *inputs* have been discussed already, several *outputs* have not. Nor in this chapter have storage pools such as cation exchange sites or secondary mineral fixation of ammonium ions been discussed so far.

5.3.1 Outputs in the nitrogen cycle

We tend automatically, if we see a phrase like 'crop removal', to associate it with harvesting activities of humans, but any plant species taken away from a site as a consequence of any animal activity (denoted generically as 'crop removal' in Figure 5.9) clearly represents a loss of nitrogen and of other plant nutrient elements. This may be material removed by grazing animals, or by birds, etc. Losses from the ecosystem may also occur either by loss of soil particulate material (erosion), or of inorganic or organic nitrogen species in drainage water, or by denitrification, which is considered later in the chapter. Erosion losses may occur steadily over time, or in extreme, and sometimes catastrophic, events as considered in the previous chapter when the carbon cycle was being discussed.

Natural, unmanaged and unpolluted soil ecosystems tend eventually to attain an equilibrium pool of stored N, at which point the mean annual input and output are in approximate balance with respect to elemental nitrogen. As inputs are low, outputs will also be low. Analysis of river water from upland streams in areas where pollution levels are low suggests negligible nitrate and ammonium leaching from soil, at least for most of the year.

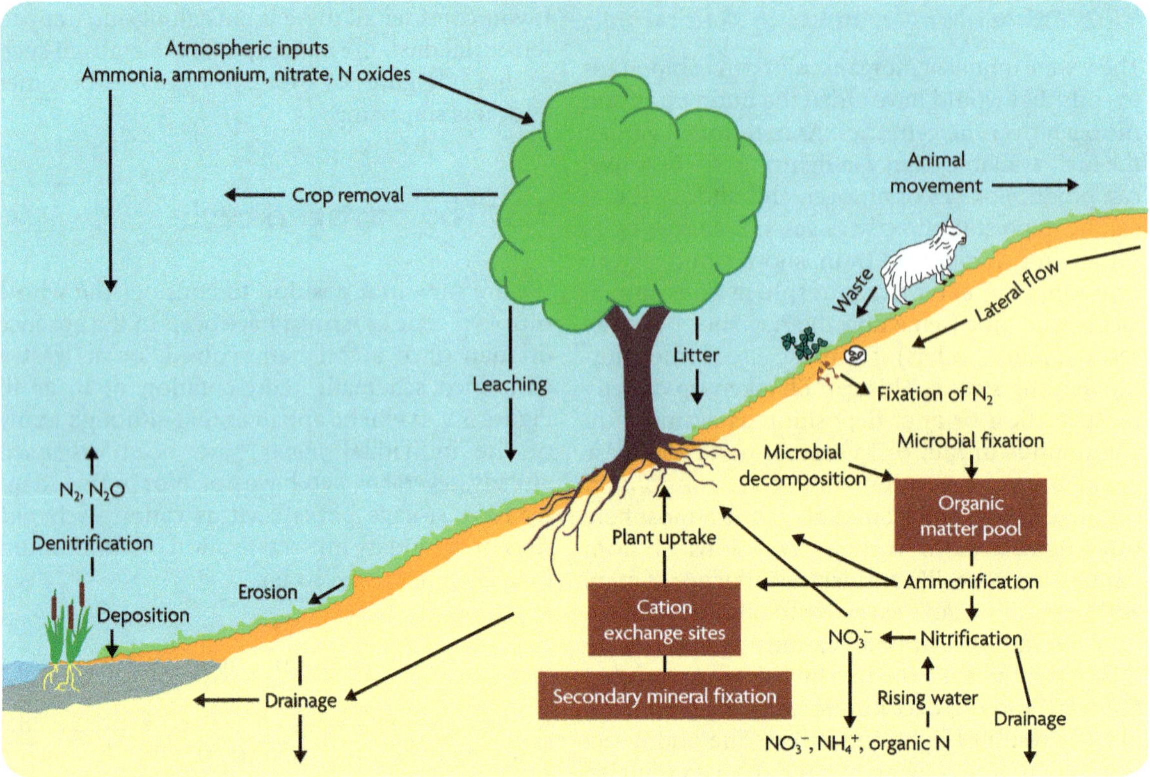

Figure 5.9 Schematic representation of the nitrogen cycle in the absence of pollution.
Source: Cresser, M. *et al.* (2008).

Lead-in question

How are ammonium and nitrate inputs retained so effectively in unpolluted soils?

5.3.2 N storage mechanisms in the N cycle

When considering the nature of soil organic matter in the previous chapter, it was noted that it contained carboxylic acid (–COOH) groups and phenolic hydroxy (–OH) groups. The protons associated with these groups can be displaced (dissociation), to give negatively charged functional groups, $-COO^-$ and $-O^-$. Soil is not electrically charged overall, so other positively charged cations can be absorbed onto the negatively charged carboxylate or phenolate groups as H^+ is displaced. Thus we say the H^+ ion is an exchangeable cation, and the total amount of negative charge associated with soil organic matter is its cation exchange capacity, or CEC. We will see in Chapter 7 that soil minerals also contribute to the CEC of soils.

If ammonia reaches the soil and forms ammonium (NH_4^+) cations, or ammonium is deposited as a salt as aerosol or in rain, the ammonium will initially be held on cation exchange sites. The amount of ammonium ions staying on the exchange sites depends upon the amount already there, and on the relative amounts of all other competing cations entering the soil system, whatever their source. If none of the ammonium-N input was changed to other N chemical species, eventually some ammonium-N would start to leak out. However, ammonium-N may be taken up and used by soil microbes and plants, and thus converted to organic N. Some will be incorporated into non-biotic soil organic matter. Some will be converted to nitrate by microorganisms. These processes deplete the amount of ammonium stored on the CEC, making it more likely to absorb any subsequent ammonium input.

Nitrate inputs too may be taken up by plants and soil microbial biomass. Indeed, nitrate may be removed from rainfall when it is intercepted by plant foliage. Biotic uptake is especially marked in soils with a very limited nitrogen supply, so in unpolluted natural ecosystems, nitrate leaching from soils to rivers and lakes is minimal, as is ammonium leaching. Clearly inorganic nitrogen inputs are likely to end up contributing to the large pool of organic-N stored in soils.

5.3.3 What ultimately limits soil N storage?

Although inorganic N inputs are very effectively retained in natural ecosystems, and only very small amounts of inorganic N ionic species are lost, some N is also lost by leaching of organic-N in drainage water from many systems. This loss eventually helps to maintain the ultimate N-input/N-output balance. So when you see water gushing from an upland catchment, and it looks the colour of that in Figure 5.10, you are watching an outpouring of organically bound nitrogen as well as an output of organic carbon. Fluxes of organic-N leaving a minimally managed (negligible agriculture or forestry) drainage basin often exceed those of inorganic N (ammonium-N and nitrate-N), sometimes by a substantial margin (Chapman *et al.*, 2001). It should be obvious that there must be some potential outlet for nitrogen, and that it cannot continue to accumulate in soils indefinitely with no balancing loss starting to play a role. The following case study also considers the fact that inputs of organic N from the atmosphere can and do occur too, although their exact nature and origins are not yet well understood.

Figure 5.10 The organic matter discolouring river water, especially at high flow, in upland drainage basins may be a major output of N from the catchment.

CASE STUDIES

A case study in the UK – Quantifying atmospheric organic-N inputs

Recognising the potentially damaging effects that the deposition of excessive amounts of inorganic nitrogen species from the atmosphere might have on sensitive natural terrestrial and aquatic ecosystems, the Natural Environmental Research Council in the UK, at the end of the 1990s, set up a very substantial interdisciplinary research programme to identify risks, and quantify key input and output fluxes in the N cycle.

The 'Global Atmospheric Nitrogen Enrichment' programme (GANE) brought together teams of scientists from diverse academic and research institutes. One such consortium was led by Neil Cape, from the Centre for Ecology and Hydrology in Edinburgh, who had found measurable quantities of organically bound nitrogen in precipitation. Cape, along with Pippa Chapman (University of Leeds), Tony Edwards (Macaualy Institute, Aberdeen) and Malcolm Cresser (University of York) measured the inputs of dissolved organic nitrogen (DON) across Britain.

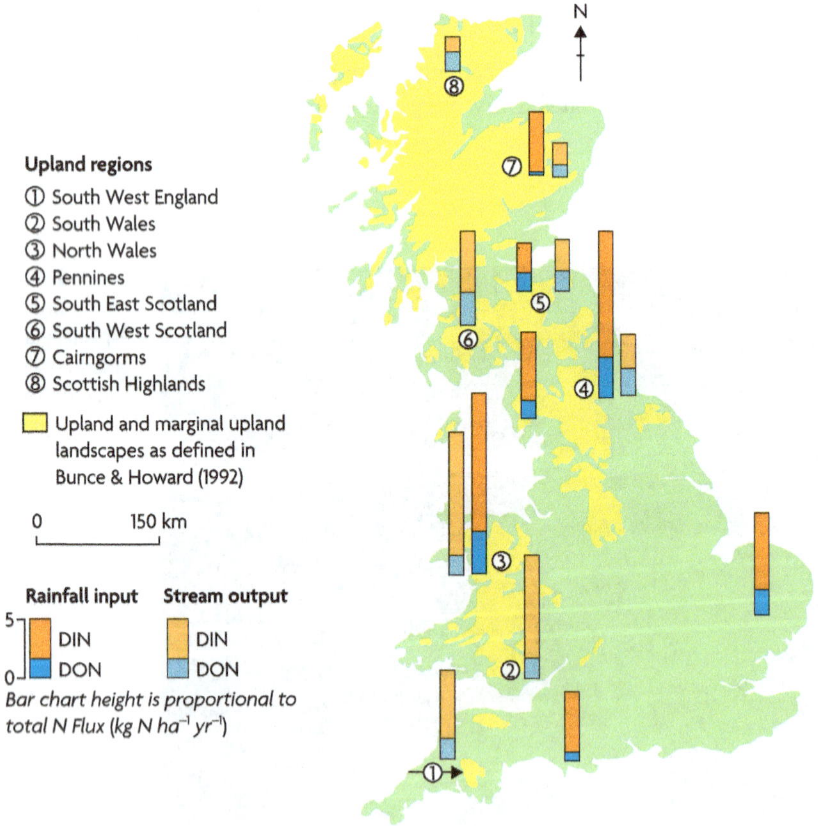

Source: Poster prepared for a NERC meeting by Pippa Chapman, Geography Department, University of Leeds.

It was found that the inputs of DON in precipitation across Britain ranged from 0.3 to as much as 4.5 kilograms of nitrogen per hectare per year (kg N ha^{-1} yr^{-1}). The mean deposition flux of DON was 2.4 kg N ha^{-1} yr^{-1}. The group compared the values found with the output fluxes of DON in upland streams in Great Britain. The latter ranged from 1.0 kg N ha^{-1} yr^{-1} to 3.5 kg N ha^{-1} yr^{-1}. It was concluded that at several upland catchments in GB, organic N inputs exceeded organic-N outputs. The results are summarised in the map presented on page 120. Except in N.W. Scotland, which is relatively unpolluted compared with the rest of GB, inorganic N inputs (DIN) exceeded organic N inputs (DON) from rain. The consortium concluded that more research was needed on the nature, origins and fate of organic N inputs from the atmosphere to terrestrial ecosystems.

5.3.4 Nitrogen cycling *within* soils

We have seen that there are several sources of organic nitrogen inputs to soils. They include plant litter (falling leaves, twigs, branches and dying roots), living and dead soil microoganisms, waste from animals, compounds exuded from roots, small amounts of organic-N compounds in precipitation, and components of terrestrial dust. However, before most of this N can used by plants it has to be converted to the inorganic ions (ammonium and/or nitrate) that plants can readily take up through their roots and use for growth (Fitter and Hay, 2001). The overall processes by which this occurs are collectively known as mineralisation, or N mineralisation. The two key stages are conversion to ammonium (NH_4^+) by ammonification and conversion (oxidation) of ammonium to nitrate (NO_3^-) by nitrification.

Many kinds of microbes contribute to ammonification in soils and sediments, aided by deaminating enzymes. The amount of ammonia liberated from the microbes to soil or sediment depends upon the carbon-to-nitrogen ratio of the organic substrate that is being mineralised. More is released generally when the substrate is nitrogen-rich.

Nitrifying bacteria of the genus *Nitrosomonas* oxidise ammonia to nitrite. The nitrite, in turn, is usually further oxidised to nitrate by *Nitrobacter*. Nitrite is quite toxic to many organisms, but is usually further oxidised quite rapidly, so accumulation to concentrations that exceed toxicity thresholds is not common.

5.3.5 Denitrification in soils and sediments

It is clear from the above discussion that microbes are every bit as important to the nitrogen cycle as they were in the carbon cycle discussed in Chapter 4. Moreover, there is another microbial process that we need to consider here, as indicated in Figure 5.9, namely denitrification. Quite a few species of bacteria are capable of reducing nitrate to gaseous nitrogenous species, most commonly to dinitrogen but also to nitrous oxide (N_2O) or nitric oxide (NO). The range includes organisms from the genera *Pseudomonas*, *Thiobacillus*, *Archromobacter*, *Bacillus* and *Micrococus*.

Denitrification occurs under anaerobic conditions, so for example is common in very poorly drained soils and anaerobic sediments. In the latter, clearly the amount of nitrate (substrate) in the water column is important, as is the dissolved oxygen concentration. The latter depends upon the organic substrate load (which obviously influences the biological oxygen demand or BOD) and temperature. Water column depth and the turbulence of the water surface (more turbulent water entrains more oxygen from the atmosphere) may also be important.

Denitrification in soils can reduce the amount of nitrate reaching surface fresh waters, and help

Figure 5.11 The visible impact of denitrification on a cereal crop along a line of heavily compacted and anaerobic soil.
Source: Photograph courtesy of Dr Tom Batey.

maintain their naturally low nutrient status if the conditions are right. We will see later in the book that this is sometimes exploited in water treatment. However, it is not always beneficial to man. Figure 5.11, for example, shows how compaction of soil caused during the laying of a pipeline across a field has produced anaerobic conditions in the soil, resulting in nitrogen deficiency in a subsequently planted cereal crop. The deficiency manifests itself via the chlorosis (yellowing) of the crop foliage along the route of the underground pipeline.

5.3.6 Estimating transformation rates of N species: stable isotope mass spectrometry

It is appropriate to consider at this point how we know that the microbially mediated processes discussed in the previous few pages actually happen. Much has been learned by studying the behaviour of colonies of individual species of microorganisms isolated from soils and sediments. Provided they are added to an appropriate substrate, for example, we can assess whether processes such as ammonification or nitrification occur. We can study the rates at which any particular organic N compound of interest is decomposed and what the decomposition products are. However, the problem then is knowing how to relate what happens in culture experiments to what will happen in a much more complex soil system.

Nitrogen exists as a number of naturally occurring stable isotopes, and one of these, ^{15}N, is very useful as a tracer in environmental scientific experiments. The isotope ^{14}N is much more common, and constitutes most of the nitrogen in the natural environment. In fact the ratio of ^{15}N to ^{14}N is around 0.0036765 in the atmosphere. Nowadays it is possible to buy almost pure ^{15}N compounds or compounds with an accurately known level of ^{15}N enrichment. If we add a known amount $^{15}NH_4^+$ to a soil, we can then see how much of the ^{15}N added ends up as nitrate or in soil organic matter or biomass, so we can then trace the fate of added ammonium-N and measure how its fate changes over time. Similarly it is also thus possible to study what happens, for example to added ^{15}N-enriched nitrate or to ^{15}N-enriched plant litter. Scrimgeour and Robinson (2004) have written a useful review of the use of stable isotope tracers in environmental research.

5.4 The cycling of base cations

We are now in a position where we should be able to consider the cycling of base cation elements. The cycling of calcium (Ca^{2+}) for example is represented in Figure 5.12.

Lead-in question

What do you think the main difference is likely to be between calcium cycling and nitrogen cycling?

The element calcium is in Group 2 of the periodic table (the alkali-earth elements, Be, Mg, Ca, Sr, Ba and Ra). These elements gain inert gas configurations by losing their two outer electrons, to form cations with two positive charges such as Ca^{2+}. This makes calcium very different from nitrogen, which is in Group 5. Nitrogen most commonly acquires extra electrons by entering into covalent bond formation (shared electrons). Thus nitrogen forms gases such as N_2, N_2O, NH_3, NO, NO_2 rather than ionic solids with massive crystal lattices, and is a constituent element of

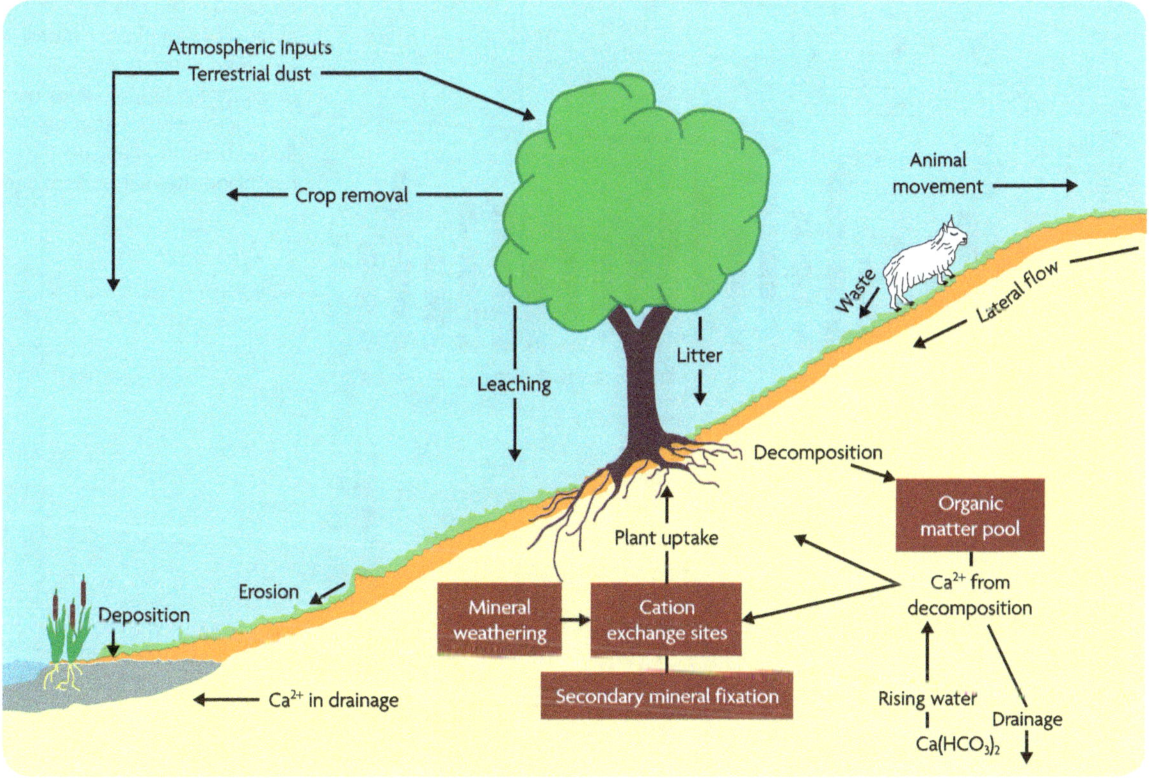

Figure 5.12 The cycling of calcium.
Source: Cresser, M. *et al.* (2008).

many organic compounds of biological importance. Moreover, whereas nitrogen exists in the environment in several oxidation states, calcium exists in just one, so no microbially mediated element species transformations are involved in the calcium cycle. Thus for calcium there are no processes equivalent to biological dinitrogen fixation, ammonification, nitrification or denitrification. As a result the calcium cycle shown in Figure 5.12 is much simpler than the nitrogen cycle that we looked at in Figure 5.9. However, biological cells contain calcium, so calcium cations may be released for recycling during the decomposition of organic residues.

If you compare Figures 5.9 and 5.12 carefully you will notice an extra 'pool' box in Figure 5.12, namely 'mineral weathering'. Carbonic acid is formed in soils when carbon dioxide (CO_2) from decomposition of organic matter dissolves in soil water.

$$H_2O + CO_2 \leftrightarrow H_2CO_3$$

Carbonic acid slowly attacks minerals to form hydrogen carbonates in solution. The simplest example would be the dissolution of calcium carbonate in a calcareous soil:

$$CaCO_3 + H_2CO_3 = Ca(HCO_3)_2$$

The overall process is known as biogeochemical weathering. Calcium in soils then is derived predominantly from the rocks from which the soils evolved, again making it very different from nitrogen in soils. The calcium in plants and animals was originally released from soil minerals by weathering.

The rate at which minerals weather depends upon mineral structure and stability. Minerals may be listed in order of their weatherability, and we will see the importance of this in Chapter 7 when soil evolution is discussed.

The cycling of other base cations is very similar to that of calcium, although the relative importance of different inputs varies. In the case of

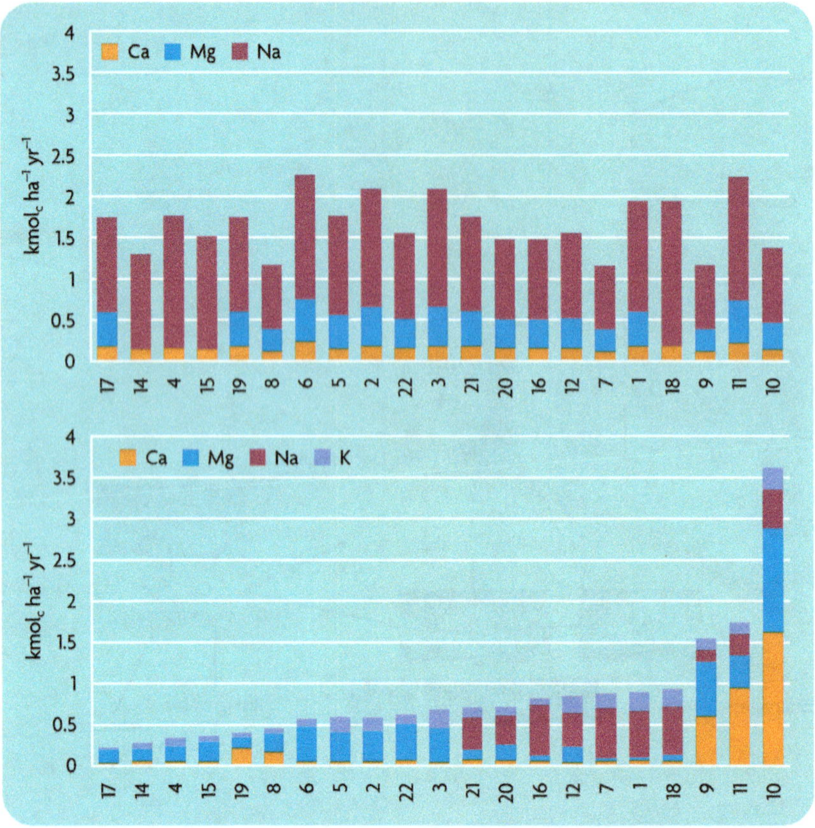

Figure 5.13 The relative amounts of the base cations indicated entering soils via precipitation (upper set of bars) and via mineral weathering (lower bars), at 21 upland heathland sites across Scotland.

sodium, for example, inputs of aerosol that reaches the atmosphere from oceanic spray are often substantial. This is especially true in countries like the UK which have maritime climates, or Ireland where winds sweep inland from across the Atlantic Ocean. Often the inputs from oceanic origins far exceed those derived from biogeochemical weathering, and not just in coastal regions. Peaty soils, where the upper 50 cm or more may be almost pure organic matter, have negligible inputs from biogeochemical weathering. Their chemistry, and that of waters draining out of them, is therefore regulated by the chemical composition of atmospheric inputs. A second cation present at quite high concentration in sea water is magnesium, and for magnesium too atmospheric inputs may exceed those from weathering of soil minerals.

Figure 5.13 compares the inputs of sodium, magnesium and calcium in precipitation (rain, snow, dew) at 21 upland heath sites distributed across Scotland in the UK with the inputs of the same ions, plus potassium, estimated to come from weathering of soil minerals. Weathering rates were estimated using a model known as the PROFILE model, which makes predictions based upon site characteristics such as the type and abundance of soil minerals present, particle size and precipitation. Sites have been arranged in order of increasing weathering rate. The greater relative importance of atmospheric inputs to soils with low weathering rates is very obvious.

5.5 Other element cycles influenced by oceanic aerosol

Lead-in question

If sea salts substantially influence the cycling of sodium and magnesium, what other element cycles would probably be influenced?

5.5.1 The chlorine cycle

Ocean water contains very high concentrations of chloride (Cl^-) and sulphate (SO_4^{2-}) as counter anions to its Na^+ and Mg^{2+} cations to maintain charge balance. Therefore oceanic aerosol is bound to play a significant role in the sulphur and chlorine cycles. This is especially true for chlorine, because, in the natural environment, its chemistry is dominated by the occurrence of chloride anions. The element chlorine is in Group 7 of the periodic table. Chlorine atoms have seven outer electrons, and only need to acquire one extra to get the electronic configuration of the inert gas argon. Thus chloride ions are readily formed and very stable. Chlorine atoms do enter into covalent bonding however, for example in chlorine gas (Cl_2, in which there is one shared pair of electrons, one from each atom) and in organic compounds. In the absence of pollution, however, such molecules are not common, so the cycling of chloride is relatively very simple.

We saw that the calcium cycle that we looked at in Figure 5.12 was simpler than the nitrogen cycle in Figure 5.9 because of the lack of gaseous components. A sketch of the chlorine cycle would be even simpler. The cation exchange process that was so important in the cycling of base *cations* plays no important role in the cycling of chloride *anions*. There is an equivalent process for anions, anion exchange, but it is really only important for polyvalent anions such as sulphate (SO_4^{2-}) and phosphate (PO_4^{3-}). These anions, because of their multiple negative charge, are strongly adsorbed and can displace less strongly held OH^- anions from the surfaces of hydrous iron and aluminium oxides and some other minerals. We will see in Chapter 7 that anion exchange is a very important soil property, but not for Cl^-. Nor was it important for NO_3^-, which is also not significantly held by anion exchange, so we did not need to consider anion exchange when looking at the nitrogen cycle.

In the author's experience, although Cl^- does not participate significantly in anion exchange reactions, its passage through soils at the catchment scale is often slower than might be expected. Thus road salting impacts upon water in adjacent streams may often be observed many months after road salting has ceased, as discussed in Chapter 19. It appears that interim biotic uptake contributes to the delay.

Lead-in question

If chloride enters the atmosphere from the oceans, is deposited on land but then leaves in drainage water via river back to the ocean (perhaps after a few cycles through vegetation), why should we even bother to consider it?

5.5.2 The importance of mobile anions

It is important to realise that cations can't move naturally through a soil or in river water unless they do so in association with a mobile anion to balance their charge. After biogeochemical weathering of minerals, we saw earlier that calcium released from the solid mineral matrix may move as its hydrogencarbonate, $Ca(HCO_3)_2$. We have also seen that calcium cations (Ca^{2+}) may be held on cation exchange sites in soils (or on plant root surfaces). If a storm sweeps in from the ocean, and a lot of sodium chloride aerosol is deposited directly or via rain, the sodium may displace cations such as Ca^{2+}, NH_4^+, H^+ or K^+ from cation exchange sites. These ions may then be lost from the soil accompanied by mobile chloride anions. This mobile anion effect can be very important. We will see later that it may disrupt the nitrogen cycle and may result in acid flushes in soil solution and drainage water if H^+ ions are displaced from acidic soils. It is very important to remember that element cycles are interactive to a greater or lesser extent.

There is another reason that the natural cycling of chloride is important. A number of potentially toxic metal elements such as mercury (Hg) and cadmium (Cd) will form ion pairs or complexes with chloride anions. For example, as the concentration of chloride increases in solution:

$$Hg^{2+} + Cl^- \rightarrow HgCl^+$$

$$HgCl^+ + Cl^- \rightarrow HgCl_2^0$$

$$HgCl_2^0 + Cl^- \rightarrow HgCl_3^-$$

$$HgCl_3^- + Cl^- \rightarrow HgCl_4^{2-}$$

The toxicity of the four mercury species to microorganisms is quite different (Sarin *et al.*, 2000), so it is important to know how much chloride is cycling through soil and sediment systems.

5.5.3 The sulphur cycle

In section 5.5.1 the substantial amount of sulphate in sea water was mentioned, so clearly marine-derived aerosol will be important in the sulphur cycle. We also saw that sulphate anions (SO_4^{2-}), with their double negative charge, participate in anion exchange reactions in soils. Anion exchange therefore can delay the passage of sulphate deposited from the atmosphere into drainage water to rivers and its subsequent journey back to the oceans.

Lead-in question

Will the sulphur cycle be more like the nitrogen cycle or more like the calcium cycle?

Earlier in this chapter we saw that because nitrogen is in Group 5 of the periodic table, it enters into covalent bond formation, rather than readily forming the nitride anion. To understand why it's not easy to form nitride in the natural environment, consider the process of adding three negatively charged electrons, one at a time, to a nitrogen atom so that it can acquire the favoured inert gas electron configuration. By the time we're ready to try to add the third electron, the nitrogen intermediate anion would already have a double negative charge and, whereas opposite (+ and –) charges attract each other, like charges repel each other. Forcing the third electron into the N atom is therefore very difficult without a big input of energy. Sulphur is in Group 6 of the periodic table, and only requires two extra electrons to form sulphide (S^{2-}) anions. This requires much less energy than nitride formation. Thus sulphide occurs quite naturally, especially when oxygen supply is very low. Therefore sulphur can occur naturally in the –2 oxidation state as an anion, for example in iron sulphide (FeS). It can also, like nitrogen, enter into covalent bond formation, so just as nitrogen in Group 5 forms ammonia gas, NH_3, sulphur in Group 6 forms hydrogen sulphide, H_2S. This is analogous to oxygen forming water, H_2O, which should come as no surprise because oxygen (O) lies immediately above sulphur (S) in Group 6 of the periodic table.

Sulphur chemistry therefore is more complicated than the chemistry of a base cation like calcium, which occurs only as Ca^{2+} in nature. Sulphur exists in a range of oxidation states, from –2 in sulphides through 0 in elemental sulphur and +4 in sulphite (SO_4^{2-}) up to +6 in sulphate (SO_4^{2-}). The oxidation state of sulphur is +6 in sulphate, because it has two covalent double bonds to oxygen atoms, giving it four extra electrons, and two single covalent bonds to O^-, giving it an extra two electrons, making six electrons in total (see Figure 5.14). Thus with its range of

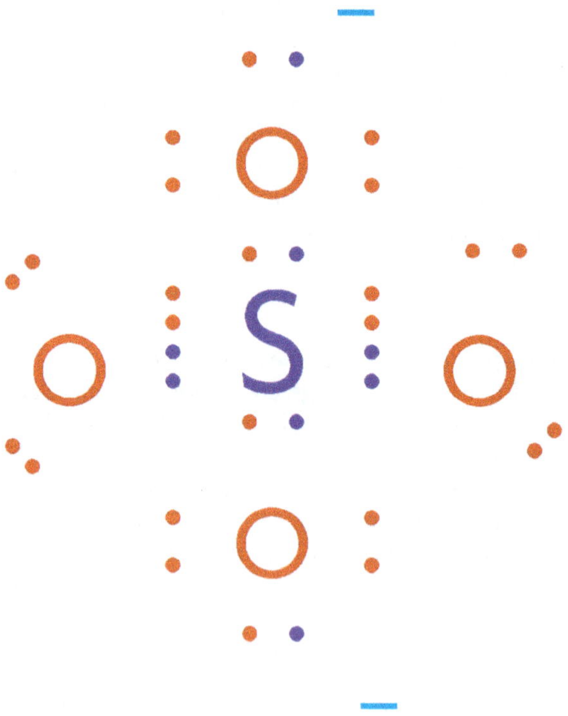

Figure 5.14 The bonding in the sulphate anion. Each oxygen has acquired two extra electrons, to attain eight outer electrons, giving it the electronic structure of the inert gas neon. In the +6 oxidation state, the sulphur atom itself does not have an inert gas configuration.

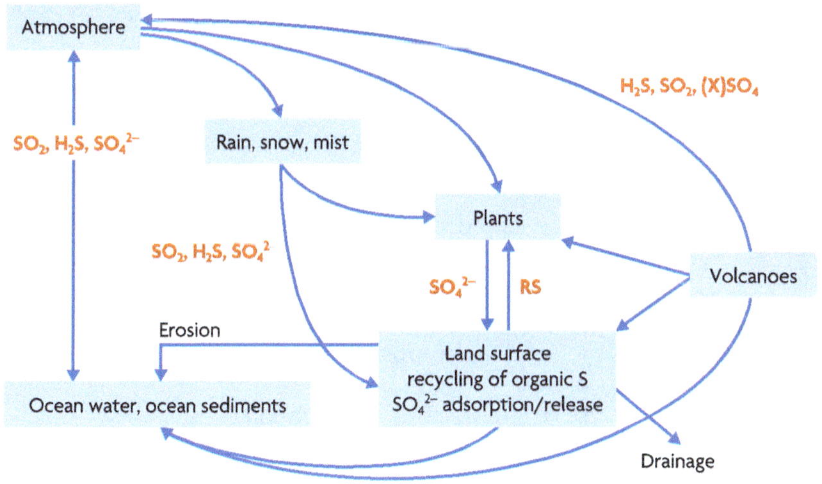

Figure 5.15 Simplified schematic representation of the sulphur cycle. RS denotes organically bound sulphur. In the interests of clarity, volatilisation of organic S compounds is not shown, nor are sulphur mobilisation from natural fires, sulphate from weathering of minerals (which may be locally important in some soils) or effects of animal grazing, movement and waste. Sulphur may also be found in nature as the free element. It is clear that the sulphur cycle is more complex than the calcium cycle.

possible oxidation states and its participation in covalent bond formation (which results in gaseous sulphur compounds being common at room temperature), the sulphur cycle has more features in common with the nitrogen cycle than it does with the simpler calcium cycle or other base cation cycles.

Figure 5.15 is a simple block diagram representation of the sulphur cycle as it would have been in the absence of man. We will see later in this book in Chapter 16, where the influences of human activities are considered, that fossil fuel combustion and land use and soil management changes over recent millennia have had a major disruptive influence on the cycling of sulphur.

It can be seen in Figure 5.15 that, as for nitrogen, a number of gaseous compounds are important in the sulphur cycle, including hydrogen sulphide (H_2S) and sulphur dioxide (SO_2). In most natural ecosystems prior to the ascent of man, sulphate of oceanic origins would have been the predominant input of sulphur, although in the Earth's history during periods of intense volcanic activity this situation could have been very different.

5.5.4 The boron cycle

Boron also occurs in sea water at significant concentrations, the ratio of Na:B in ocean water being around 242:1 (De Mora, 1992). Thus if the precipitation in an area contained 2.42 ppm or 2.42 μg per ml of sodium, it might be expected also to contain 0.01 ppm or 0.01 μg per ml of boron of sea water origin. It has been shown for Scotland that marine-derived inputs of boron can make a major contribution to the boron concentration in soil solution and drainage waters in catchments where soil parent minerals have very low boron concentrations (Jahiruddin *et al.*, 1998). In such catchments, oceanic aerosol is probably the main source of this essential plant nutrient element. Statistical analysis showed that boron concentration in river water throughout a major river network was statistically highly significantly correlated to the concentration of sodium and to the concentration of chloride. This is strong circumstantial evidence that the three species have a common origin. Before we could say the link is proved unequivocally we would have to show that the concentrations of boron and say sodium are not simply both regulated by another common factor, for example the amount of rainfall.

We will see in Chapter 7 that boron is the plant nutrient element with the narrowest range between deficiency and toxicity. As boron concentrations have probably remained little changed in sea water for many millennia, it seems likely that deposition of boron from the atmosphere may have played a role in plant evolution. Plants would have to be able to tolerate the concentrations of boron deposited to survive. It could be that they have evolved to require at least this concentration. In soils where the parent material contains higher boron concentrations, plants also have to be able to tolerate additional boron released from the weathering of boron-containing minerals such as tourmaline.

5.5.5 The phosphorus cycle

Although phosphorus lies just below nitrogen in Group 5 of the periodic table, its chemistry is very different to that of nitrogen. Resemblance is, at best, superficial and not really of environmental relevance. Thus although phosphorus forms phosphine PH_3, analogous to nitrogen forming ammonia, NH_3, phosphine is of no significance in nature, being highly toxic and very readily oxidised. There is no phosphorus molecule analogous to dinitrogen, and elemental phosphorus can be produced in a number of forms; all are solid and reactive. Like nitrogen, it has a strong tendency to form covalent bonds, especially with oxygen, but also in organic molecules. Its most common and important form is the orthophosphate anion, PO_4^{3-}, although in natural ecosystems this is often associated with one or two protons, as HPO_4^{2-}, or $H_2PO_4^-$. The more acidic the solution, the greater the tendency for protonation. Thus in slightly acidic soil solutions, the HPO_4^{2-} form is most abundant, and the predominant form taken up by vegetation. The dominance of the oxyanion stems from the fact that the formation of multiple bonds to oxygen using hybrid orbitals encompassing 3*d* orbitals close in energy to the P atom's outer $3s^2$, $3p^3$ electron configuration is energetically favourable. Thus, unlike nitrogen, phosphorus occurs widely in nature in the +5 oxidation state.

Lead-in question

What will the key stages be in the phosphorus cycle?

Based on the above discussion, and drawing upon our knowledge of the general principles of element cycling from this and earlier chapters, it is a relatively simple matter to list the key features of the phosphorus cycle:

- No gaseous components.
- No significant atmospheric inputs apart from dust.
- Organic P from litter is recycled to phosphate for plant uptake.
- Soil phosphorus must be derived from weathering of phosphate minerals.
- Animal waste and movement can relocate phosphorus.
- Phosphate is retained in soils by anion exchange, so leaching losses in drainage will be small.
- Water erosion may transport phosphorus associated with particulates, via rivers, to the oceans.
- Over a geological timescale, uplift may bring ocean sediments back to the terrestrial surface.

Adequate supplies of phosphorus are crucial at almost every stage of plant growth, so we will return to learn more about this important element in Chapter 7 where the natural evolution of soils and controls on soil fertility are discussed.

POLICY IMPLICATIONS

Biogeochemical cycling of N and other elements

- Policy makers need to be aware of how careful management of biogeochemical cycling of elements is crucial to sustainability of both natural and managed ecosystems.
- Policy makers need to be aware that all element cycles at regional through to global scales are interactive.
- Policy makers need to consider carefully how human activities such as those producing pollution, waste generation (including sewage production) and participation in global food webs disrupt natural biogeochemical cycling (these important aspects will reappear in subsequent chapters).
- Policy makers need to understand the timescales of natural change of natural ecosystems compared to timescales of human driven change.
- Policy makers need to produce strategies for pollution management that are based upon a correct and quantitative understanding of natural element cycling processes.

CHAPTER REVIEW EXERCISES

Exercise 5.1

A pit was dug in a permanent grassland soil and representative samples were taken for each 10 cm increment to a total depth of 50 cm. Samples were taken to the laboratory, oven dried, and analysed for their total nitrogen content. Small cores of known volume from each depth increment were also collected, returned to the laboratory and dried, and the density (dry mass per unit volume) was calculated for each depth increment. The results are summarised in the table opposite.

Depth in soil profile	Soil N %	Soil density g cm^{-3}
0–10	0.45	1.45
10–20	0.26	1.49
20–30	0.10	1.62
30–40	0.03	1.64
40–50	0.02	1.70

Calculate the mass of soil per hectare (10^4 m^2 or 10^8 cm^2) in each depth increment, and then the mass of nitrogen in each depth increment. Express your results in metric tonnes (thousands of kg). Finally, calculate the total amount of nitrogen stored in the soil profile to a depth of 50 cm.

Exercise 5.2

What are the possible origins of the nitrogen found in the soil profile in Exercise 5.1? If free-living algae and lichens could fix 1 kg of dinitrogen (N_2) from the atmosphere each year in physically weathered rock fragments at the surface, how many years would it take them to accumulate the total amount of N found in the soil in Exercise 5.1, assuming no N losses? If it is assumed that the early soil was colonised by clover capable of fixing 10 kg of N_2 per year on average after the first 200 years, how long would it then take to accumulate the total nitrogen amount found in the soil?

Exercise 5.3

If precipitation contains 0.01 mg of boron per litre, calculate the annual flux of boron deposition per hectare (1 ha = 10^4 m^2) in an area that receives 1000 mm of precipitation per year. What do you think the fate of this boron would be?

REFERENCES

Chapman, P.J., Edwards, A.C. and Cresser, M.S. (2001) The nitrogen composition of streams in upland Scotland: Some regional and seasonal differences. *Science of the Total Environment*, **265**, 65–83.

Cresser, M.S., Aitkenhead M.J. and Mian, I.A. (2008) A reappraisal of the terrestrial nitrogen cycle – What can we learn by extracting concepts from Gaia theory? *Science of the Total Environment*, **400**, 344–355.

Deacon, J. (2006) The microbial world: The nitrogen cycle and nitrogen fixation. http://helios.bto.ed.ac.uk/bto/microbes/nitrogen.htm, accessed on 14/06/2006.

De Mora, S.J. (1992) The oceans, in: *Understanding Our Environment: An Introduction to Environmental Chemistry and Pollution* (ed.) Harrison, R.M. Royal Society of Chemistry, Cambridge, pp. 93–136.

Fitter, A.H. and Hay, R.K.M. (2001) *Environmental Physiology of Plants*, 3rd edn. Academic Press, London, 367 pp.

Jahiruddin, M., Smart, R., Wade, A.J., Cresser, M.S. and Neal, C. (1998) Factors regulating the distribution of boron in water in the River Dee catchment in north east Scotland. *Science of the Total Environment*, **210/211**, 53–62.

Lynch, J.M. (1983) *Soil Biotechnology. Microbiological Factors in Crop Productivity*. Blackwell Scientific Publications, Oxford.

Sarin, C., Hall, J.M., Cotter-Howells, J., Killham, K. and Cresser, M.S. (2000) Influence of complexation with chloride on the response of a *lux-marked* bacteria bioassay to cadmium, copper, lead and mercury. *Environmental Toxicology and Chemistry*, **19**, 259–264.

Scrimgeour, C.M. and Robinson, D. (2004) Stable isotope analysis and applications, Chapter 9 in: *Soil and Environmental Analysis: Modern Instrumental Techniques*, 3rd edn. Dekker, New York, pp. 381–431.

CHAPTER 6

Ecology and biodiversity on Earth

Lesley Batty

Learning outcomes

By the end of this chapter you should:

- Understand the basic concepts of ecology and the levels at which it is studied.
- Recognise the difficulties in measuring biodiversity.
- Understand how and why biodiversity varies over spatial and temporal scales.
- Have knowledge of the key abiotic factors that affect organisms.
- Understand how biological organisms can be used as environmental indicators.
- Appreciate how organisms play a key role in maintaining ecosystem function.

6.1 Introduction

We saw in Chapter 1 that for many millennia the Earth was deprived of life, and the forces shaping the planet and its atmosphere were purely chemical and physical. As soon as life appeared, which we currently think was about 3500 million years ago (Grotzinger *et al.*, 2007), it started fundamentally to affect the processes that occur on Earth (see Chapter 3). The first organisms to appear were very simple, single-celled, bacteria, most of which were thermophilic anaerobes. These are organisms that live in an environment without oxygen and at higher temperatures.

A major advance came when some of these bacteria evolved to contain complex organic molecules of chlorophyll that enabled them to use light energy to split carbon dioxide into its two constituent parts. Some of the carbon was assimilated by the bacteria in biomass growth and oxygen was released as a by-product. The presence of trillions of these organisms across the planet over many millennia gradually increased the levels of oxygen in the atmosphere, essentially changing the world.

Figure 6.1 An example of fossil ancient cyanobacteria from Bolivia.

Source: Courtesy John P. Adamek, Fossilmall.com

The first evidence of these organisms can be found in the fossil record in rocks of more than 3000 million years old, where they formed structures known as stromatolites. Sromatolite formation was a consequence of the development of mats of these photosynthesising cyanobacteria (Figure 6.1). Layers of calcium carbonate deposited over the growing bacterial mat as photosynthesis in the bacteria depleted the carbon dioxide concentration in the surrounding water. The carbonate and sediment grains became trapped by adhering to the coating surrounding the bacterial colonies, which then re-established at the surface to form a new bacterial layer. This process repeated many times, forming distinctive multi-layered structures still visible in the fossil record. Living examples can still be found in areas of the world such as Australia today.

It is thought that capture of bacteria within organic membranes eventually led to the development of eukaryotic cells, evidence for which is found as fibrous and spherical fossil cells. Organisms then evolved over many millennia, providing the diversity of life that we see today, although the basic origins of life can be seen still in all organisms at a cellular level. In the greater scheme of things, humans did not evolve until very late on in the history of the Earth, but other organisms, many of which are now extinct, have been interacting with the environment for millennia and it is this interaction that comprises the study of ecology.

Our concepts and understanding of ecology have largely been generated by modern day investigations of existing organisms, although there is some effort to understand ecology in the geological past, before the evolution of humans, which we will also consider. However, first we need to consider the different components of ecological study. Ecological studies can essentially be made on a number of different levels: the single organism, a population, a community, or an ecosystem.

6.2 Organisms and species

Lead-in question

How many different species currently exist on earth?

An organism is a living biological individual and we can examine how it interacts with both its environment and with other organisms. However, ecologists do not usually study one individual but rather a set of individual organisms belonging to the same species group. Therefore the first thing we need to establish is 'what is a species'? The classic definition of a biologically defined species is 'a group of organisms (or individuals) that have the potential to interbreed and produce fertile offspring'. However, there are a number of issues with this definition, the most notable being that it only really applies to organisms that use sexual reproduction. The more general definition of 'organisms that have a high level of genetic similarity' is more appropriate. It should be recognised that no one clear and universally agreed definition of a species exists and the definition employed often varies according to the type of organism that is being studied. When considering any ecological study at the species level therefore, we should recognise any potential limitations associated with this lack of definition.

When a species is identified it is named according to the Linnaean classification system. Under this system organisms are essentially split into six main divisions, referred to as phyla. Phyla are further split into class, order, family, genus and species. However, major revisions to this classification have been made in recent years and the Linnaean classification system should be considered a 'work in progress'. Classification was originally based upon separating out organisms according to their physiological traits. Some scientists have considered the classification of organisms based on physiological traits to be not very useful and instead propose the classification of organisms based upon other properties. For example, organisms may be divided according to their mode of nutrition (Table 6.1).

Organisms are also sometimes referred to as *r* or *k* strategists. This division is based upon the idea that pressures within the environment act to drive evolution. In response to these driving forces, *r* strategists produce many offspring in densely populated niches, whereas *k* strategists invest more resources in producing a small number of offspring, most of which will survive into adulthood, in sparsely populated niches. More recently, methods using genetic based information from DNA have been applied to provide a more evolutionary-based classification system.

In science it is normal to refer to organisms by their Latin name to avoid confusion potentially caused by diverse local names and language differences; for example, *Campanula rotundifolia* is known as harebell, Scotch bluebell, Californian harebell, witches thimbles, old man's bells and fairy bells. For every species that has been officially named there is an originally described 'type specimen' from which the species name

Table 6.1 Division of organisms according to their mode of nutrition.

Mode of nutrition	Energy source	Carbon source	Organisms
Photoautotroph	Light	CO_2	Cyanobacteria, plants
Chemoautotroph	Inorganic molecules	CO_2	Prokaryotes and protists
Photoheteroautotroph	Light	Organics	Prokaryotes
Chemoheteroautotroph	Organic molecules	Organics	Prokaryotes, protists, fungi, animals, some plants

comes and this is also true for fossil species. These type specimens are usually archived in museums for future reference.

Our knowledge of organisms and species in the past is largely constrained by the quality of the fossil record, but it may surprise you to realise that our knowledge of organisms alive at present is also very limited.

There are many problems associated with establishing just how many species currently exist that are just as fundamental as those faced when looking at life in the geological past. This is reflected in the number of species that are recorded in the literature. Estimates of how many species there are at present range from 5 to 150 million, but general consensus is around 15 million. According to the United Nations Environment Programme (www.unep.org) only about 1.75 million species have been described out of an estimated 14 million species on Earth; therefore potentially there is a large number of species that we know absolutely nothing about. We should also remember that many species that we do know about are only sketchily described and we know little about their behaviour or role within ecosystems. Moreover, there is a great disparity in our knowledge of different groups of organisms. For example, it is thought that we have described about 80 per cent of plants and 90 per cent of vertebrates but potentially only about 7 per cent of fungi and 0.1 per cent of bacteria (Stork, 1997).

So why is our knowledge of the number of organisms on the planet so poor? Well firstly it is a question of organism size. It is no accident that bacteria have the largest number of un-described organisms. The use of magnification didn't allow the observation of the basic cell until 1665, and even then bacteria weren't observed until 1673 (remember that bacteria are single-celled organisms). However, it is only relatively recently that we have had the technology to differentiate between different bacterial species; as advances in these techniques are continually being made it leads to new species being described and re-categorisation of existing species. The major developing techniques rely largely upon identification of genetic differences. Such developments led to the identification of a separate group of organisms known as the archaea. These were previously considered to be a type of bacteria but have been shown to have an entirely separate evolutionary history through the use of genetic information.

Secondly there is also a question of access. Although we may think that we have explored every part of this world, it is not true. There are many regions of the world, most notably the deepest oceans, which remain outside our current capabilities of detailed exploration. Lastly it is the sheer difficulty in describing a new species which relies upon a very high level of expertise, and there just aren't enough specialists or resources in taxonomy to carry out the descriptions of species. We should also note the issues of what is meant by a species as discussed earlier. So given our very incomplete knowledge of individual species it is unsurprising that we also have a limited understanding of how all these organisms and species function and interact with their environment.

6.3 Evolution and adaptation

We have seen just how diverse life is on Earth but this represents only a small proportion of the organisms that have existed over geological history, many of which have become extinct. At this point we need to consider the mechanisms by which life forms develop as this is important for the ecological processes discussed later in this chapter.

Evolution is a very familiar term to most people and of course is synonymous with Charles Darwin. In fact, Darwin wasn't the first to propose this idea but was the first to suggest that the process of natural selection was important in the evolution of life. Before we consider evolution we need to examine some basic genetics.

During the process of reproduction, genetic information is passed on to offspring through the transmission of genes. Within any population (which is what we call a group of organisms of the same species) the total genes, or in other words the genetic information, is known as the gene pool. This gene pool should stay constant unless other forces act to change the distribution of genes and it is such a change that constitutes evolution.

There are a number of ways in which the gene pool can change:

1 Genes can change due to mutation. In most cases the mutated genes will result in offspring that are either unable to survive or may not be able to reproduce themselves and therefore the genetic mutation does not persist into future generations. However, in some cases the mutation may not result in this outcome and can be transferred to successive generations. Over very long periods of time these genetic mutations may result in major changes in the gene pool and therefore in the characteristics of the organisms within the population.

2 Mating is not random and due to selection of sexual partners certain genes may be preferentially passed on through subsequent generations.

3 Individuals are not confined to a population and there may be movement into, and out of, a population, changing the composition of the gene pool.

As a result of these forces the inherited physiological, behavioural or morphological traits that result may actually allow the organisms to be better adapted to local environmental conditions and therefore more likely to persist due to the increased survival of individuals. This is *adaptation*. Evolution and natural selection are the processes by which organisms are able to colonise the huge variety of habitats over the Earth from the cold deserts of the Antarctic, through the hot springs of volcanic regions to the deep seas.

It is easy to look at very complex organisms or parts of organisms such as the compound eye and question whether natural processes could be the cause, but it is also difficult for humans to comprehend the lengths of time that are involved in evolution. Remember that the Earth is billions of years old and many organisms can have several hundred generations within a single year.

In addition to the forces of change listed above, there is an additional natural process called genetic drift where, completely by chance, some individuals have more offspring than others and therefore pass on more genetic information. This occurs in all populations, but in larger populations it does not normally change the gene pool significantly. However, in smaller populations it can be far more important and can actually result in populations becoming less well adapted to their environment and make them more susceptible to extinction. There is also a specific phenomenon that occurs when a population is small as a result of either dramatic loss of individuals or it being a 'founder' population of organisms that are colonising a new area. Within these small populations there is very limited genetic diversity and thus there are limited opportunities for evolution and adaptation. This is called a genetic bottleneck and can also make the population prone to extinction (see Chapter 25).

Where a species cannot adapt and change in response to variations in environmental conditions then it is likely to become extinct and this is a process that has occurred throughout the history of the Earth. We shall come back to the fossil record later, but there is evidence of this continuous turnover of species with examples of many species (and whole families) that have become extinct, including the trilobites (Figure 6.2) and graptolites. This has given rise to the concept of the *background rate of extinction* which is estimated to be one extinction per million species per year (Wilson, 1992). However, there have been occasions in the geological past where the extinction rate has dramatically increased and these are referred to as mass extinctions. There have been five mass extinctions during the Earth's history (Figure 6.3). Some scientists now think that we are within a sixth mass extinction, but as a result of human activity rather than 'natural' processes (see Chapter 25).

6.4 Distribution of organisms

Strictly speaking, the global distribution of organisms, why they live where they do and how they got there, is the study of biogeography rather than ecology, but it overlaps with ecology on many levels. The current distribution of organisms is very difficult to interpret because of the major impact that humans have had upon the environment and organisms. However, some major patterns have been observed and a number of

Figure 6.2 A fossil specimen that provides an indication of the high density of trilobites at this site in the geological past.
Source: Science Photo Library.

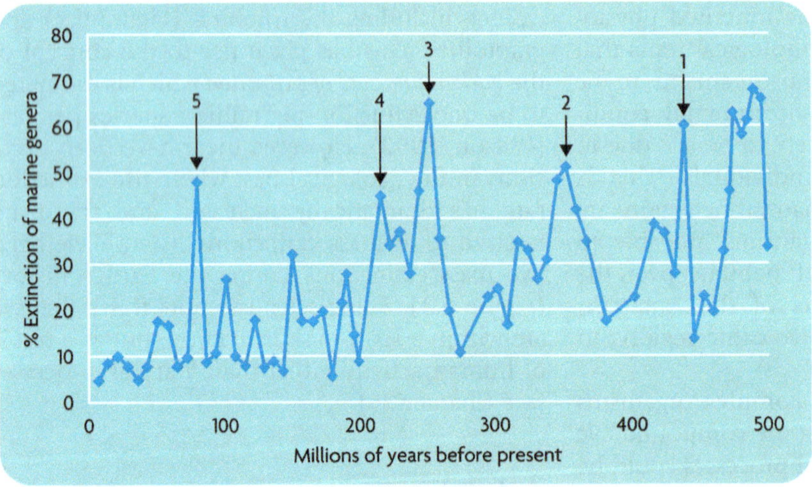

Figure 6.3 The occurrence of mass extinctions, indicated by arrows, from 1 (earliest) to 5 (most recent) in Earth's history.
Source: Based on GCUK ESO-S Project, 2012.

theories put forward to explain their distribution. Over the entire planet it has been noted that there tends to be greater species diversity towards the equator than at the poles. There are two main groups of theories to explain this: (1) Past environmental conditions have controlled the distribution (historical theories) or (2) current environmental conditions are the dominant factor (equilibrium theories). Neither theory has been accepted in preference to the other and, as always with science, it is probably a complicated combination of factors that has led to past and current patterns.

There also appear to be certain areas on the planet that have unusually high numbers of species. These are referred to as biodiversity hotspots (Figure 6.4). They include the tropical rainforests, coral reefs and large tropical lakes. A study by

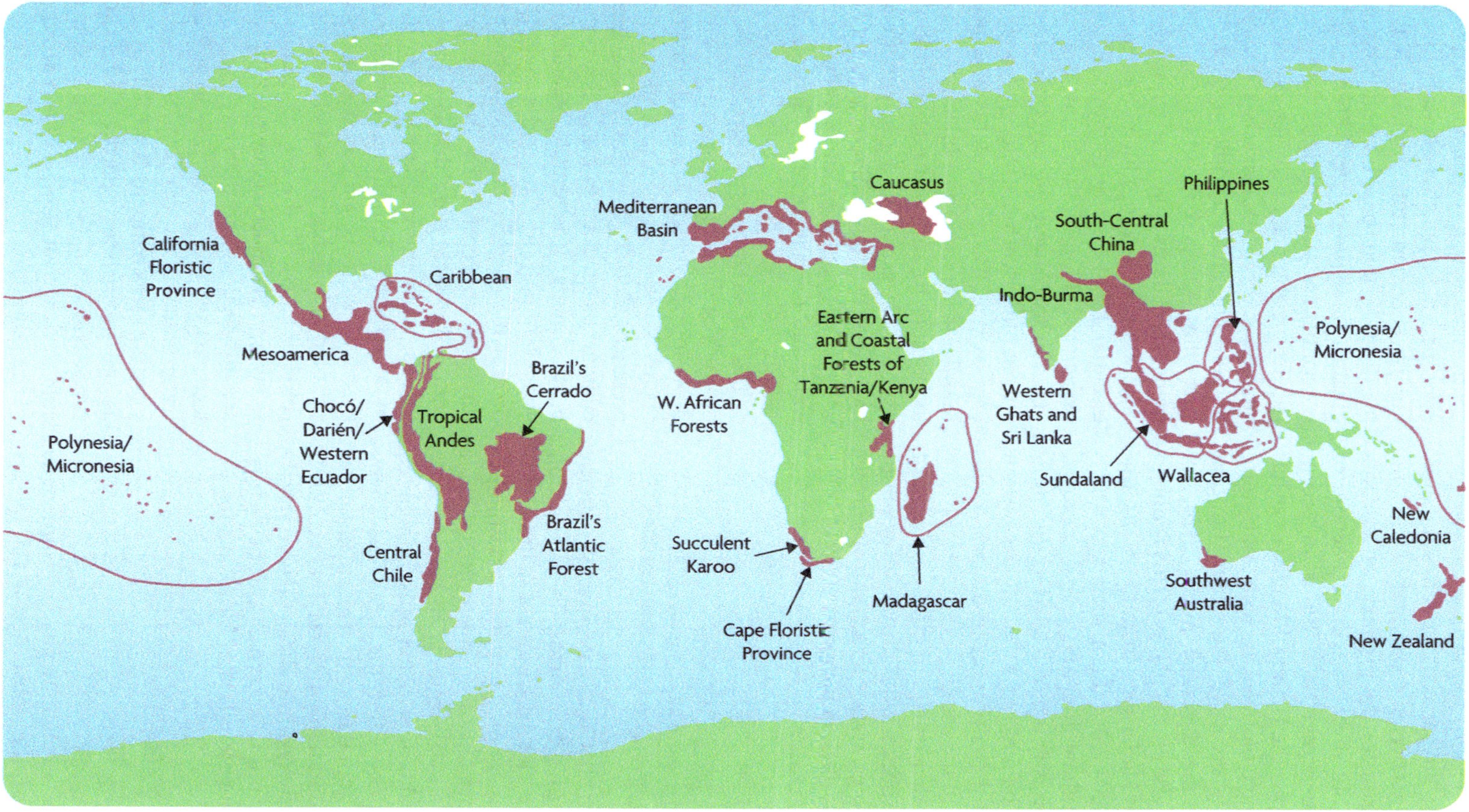

Figure 6.4 Map of the 25 biodiversity hotspots from Myers *et al.* (2000).
Source: Myers *et al.* (2000).

Mittemeier *et al.* (1998) found that 25 hotspots contained about 50 per cent of terrestrial biodiversity but only covered 2 per cent of the land surface. The reason for the presence of hotspots is also unclear but these areas have become of great interest in conservation terms in recent years. For example, the island of Madagascar has an unusually high biodiversity with many rare species that only occur on the island (around 90 per cent of the terrestrial plants there are found nowhere else on the planet). This is due to the separation of Madagascar from the main continents around 80–100 million years ago which has allowed a unique flora and fauna to evolve. However, the diversity is also related to the number of different habitat types that are present on the island, from the rainforests in the east to the arid scrublands in the south.

So, given the deficiencies in our knowledge of the number of species, what do we know about how individual organisms interact with, and meet the challenges of, their environment?

Lead-in question

What governs the distribution of organisms both on a global and a local scale?

We have seen that there are overall patterns in the number of species across the globe but, if we look more closely, we can also note that particular species are only found in certain areas and environments. Organisms are limited in their distribution on a local, regional and global scale by environmental factors as they must be adapted to the environment within the area. This is why bananas are not found in arctic regions and polar bears are not found in the tropics! These controlling factors are divided into those that are strictly physical and chemical, known as abiotic factors, and those that are biological or biotic factors.

6.4.1 Abiotic factors

Oxygen

The presence of oxygen in an environment is probably one of the most important factors controlling which organisms can survive. We examined already different ways that organisms can be classified, but one way that we didn't mention is that based upon whether or not organisms require oxygen. Those to which oxygen is essential are called obligate aerobes; those that require an absence of oxygen are called obligate anaerobes; those that can exist in anaerobic environment but prefer aerobic are called facultative anaerobes. Oxygen is, of course, essential for many animals in that it is required for cellular respiration which produces the energy that is essential for the body to work. Plants produce oxygen via photosynthesis, but they also respire in order to convert the carbohydrates produced during photosynthesis into energy, and this also uses oxygen. Therefore plants also need oxygen in order to survive. Some organisms have adapted to be able to survive under conditions of low oxygen; for example, some aquatic organisms have a very high gill surface area which is where the exchange of oxygen between the water and the gills occurs (Childress and Seibel, 1998).

The importance of oxygen can be illustrated by the response of organisms in a river to the depletion in oxygen as a result of pollution. In Figure 6.5 you can see that in response to the presence of sewage which has a high organic content, bacteria act to break down the complex organic matter into simple organic compounds, and eventually to CO_2 and water. During this process the bacteria consume oxygen and therefore the concentrations of dissolved oxygen within the water drop dramatically. In response to this, those invertebrates that are not tolerant of low oxygen conditions cannot survive and therefore there is an increase in the numbers of tolerant organisms that are largely tubificids (bloodworms). As the organic matter is progressively more consumed downstream, the bacterial activity declines and therefore the organisms use less oxygen. Oxygen is able to diffuse into the water and so its concentration starts to rise slowly. As this happens less tolerant species, such as the invertebrate species *Asellus*, are able to survive until, finally, very sensitive organisms are able to survive at some distance from the source of the pollution.

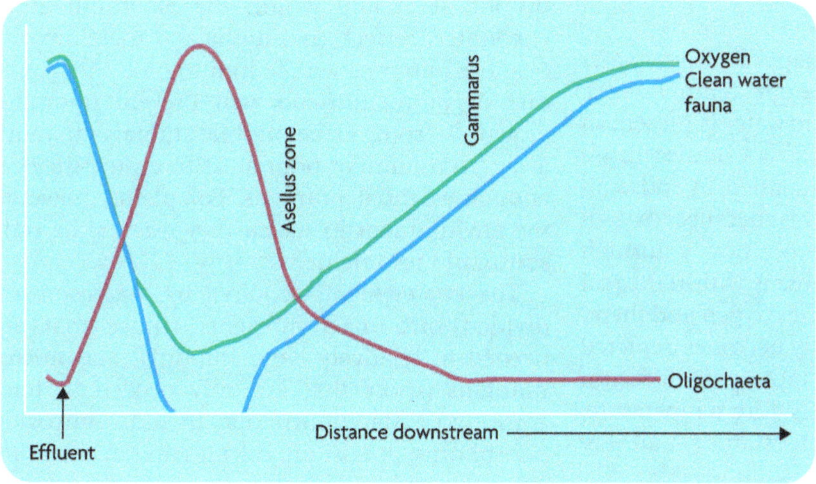

Figure 6.5 Changes in oxygen concentration and associated macroinvertebrates in response to entry of a point source of organic pollution within a river.
Source: After Mason (1996) *Biology of Freshwater Pollution*, Pearson Education Ltd.

Temperature

Organisms that have the ability to control temperature through processes such as sweating, vasodilation and vasoconstriction are called homeotherms; poikilotherms, on the other hand, are organisms whose temperature varies with the temperature of the surrounding environment. Plants are also affected by temperature. If the temperature falls too low then the water within plant cells can freeze, expand and cause irreversible damage. However, plants adapted to cold environments where temperatures often fall below freezing have been shown to use sucrose and sugars to stablilise cell membranes which prevents freezing damage. High temperatures can also cause direct damage to plants through scorching by direct sunlight (just try leaving a sensitive houseplant on a sunny windowsill).

Temperature also has indirect effects upon organisms due to its importance for affecting the water status of soils and plants. At high temperatures evapo-transpiration rates increase significantly, reducing the amount of water available for uptake by plants, but also increasing the evaporation of water from plant leaves. This will result in wilting of plants ill-adapted to these conditions (try this again with that plant on the sunny windowsill!). Plants adapted to regions with high temperatures have mechanisms to reduce evapo-transpiration (e.g. thick, waxy cuticles), storage cells that can store water and deep roots for accessing water (Figure 6.6).

Figure 6.6 A typical succulent showing characteristic adaptations to water-limiting environments.

Temperature can also affect oxygen concentrations. When the temperature of water is raised the solubility of oxygen in water decreases so warm waters hold less oxygen. Many aquatic organisms rely on a good supply of oxygen within waters as mentioned above, so if temperatures are too high, death of organisms may result.

Water

We have already seen what happens when water is limited as a result of high temperatures. However, when relatively static water is present in excess due to poor drainage, this can also cause problems. Too much water retained in soils can result in anaerobic conditions because oxygen diffuses about 10,000 times more slowly through water than through air. Organisms within the soil (particularly bacteria) consume oxygen and therefore its concentration gradually becomes depleted if is not replaced sufficiently rapidly by diffusion from the surrounding air. Therefore temperature, oxygen concentration and water status are all very closely linked.

Organisms that are adapted to saturated environments have evolved a variety of methods by which they can tolerate the low concentrations of oxygen. Wetland plants, for example, have evolved a specific structure called aerenchyma, that essentially consists of large tubes that enable plants to transport oxygen taken in from the air through their leaves, down into the roots (Figure 6.7).

Nutrients

All organisms need a variety of chemical elements in order to survive and reproduce. The presence of these essential nutrients within the environment is critical to whether an organism can exist or not; it is rare within the environment to encounter 'perfect' conditions and therefore one or more nutrients will be limiting; i.e. the lack of one or more nutrients will prevent optimum growth. To some extent mobile animals can move away from nutrient poor areas to ensure they get enough essential nutrients. For plants, however, the environment in which they exist is the only source of these elements.

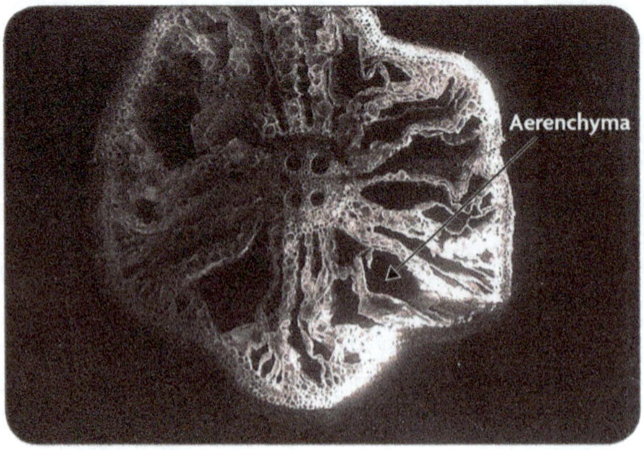

Figure 6.7 Cross-section through the root of *Phragmites australis* (Common reed) showing the development of aerenchyma.

The key elements required by organisms are divided into macronutrients (those that are needed in relatively large amounts) and micronutrients (those that are only needed in trace amounts). Macronutrients include nitrogen, phosphorus, potassium, calcium and magnesium; examples of micronutrients include copper, zinc, iron, molybdenum and boron. Organisms may acquire these nutrients through the consumption of food, or by taking them up directly from the environment. The latter is the only method by which primary producers are able to obtain their nutrients and therefore the environment controls the levels available.

There are many controls on the concentrations and chemical forms of elements within the environment which are covered in Chapters 5 and 7. Essentially what we need to consider are the underlying geology and the resulting soil type. For example, soils formed upon calcium carbonate-rich rocks (e.g. limestone) have a high pH (at least early in their evolution). Iron (Fe) is only soluble at low pH and therefore within these soils it is in a form that is unavailable for uptake by plants. Therefore plants that grow in such environments (known as calcicoles) have adapted to tolerate very low concentrations of iron (and some other elements). When organisms are not able to access sufficient quantities of either macro- or micro-nutrients then growth can be significantly affected and this is often characterised by visual symptoms (Figure 6.8).

Nutrients in excess may also affect the presence of organisms. Naturally occurring outcrops of mineral ores, for example, may produce soils that are rich in metals such as Zn and Cu (both essential nutrients), but also non-essential or potentially toxic elements such as Pb. Some plants have adapted to these soils and are able to tolerate them either by limiting uptake of the

Figure 6.8 *Urtica dioica* showing chlorosis (yellowing) of the leaves in response to a deficiency in nutrients. The inter-veinal chlorosis here is probably indicative of magnesium deficiency (see Chapter 7).

metals into the plant, or by developing biochemical mechanisms of coping with the high concentrations within tissues. In some cases, the plants actively take up extremely high concentrations of metals and are known as hyperaccumulators; an example of one of these is *Thlaspi caerulescens* (alpine pennycress) (Figure 6.9).

It is not simply the case that organisms react and adapt to an environment; they in turn affect the environment by inducing physical and chemical changes. This realisation is the key to understanding why, when we interfere in some way with an ecosystem, it can have such a dramatic effect upon it. For example, the introduction of *Rhododendron ponticum* in the UK has dramatically altered many areas largely through the presence of its dense canopy which prevents other plants from growing. There is also some evidence for the release of chemicals from its roots that inhibit the growth of other plants within the vicinity of the root system. This raises another question regarding the response of organisms to their environment: why do changes in one species dramatically affect other species? This is due to the interactions of organisms with each other, known as biotic factors.

Figure 6.9 *Thlaspi caerulescens* growing on remains of a lead mine in Wales, UK.

6.4.2 Biotic factors

There are a number of ways in which organisms interact with each other and such relationships are not always balanced. If one organism benefits to the detriment of another then this is known as parasitism/predation. For example, the flea is a typical parasite in that it removes blood from its host and gives nothing back. You might think that the occasional flea bite is just irritating but a heavy infestation on smaller animals such as birds can result in death. If neither organism benefits but does not suffer from the relationship it is commensalism, clear examples of which are difficult to find in nature. If both organisms benefit this is symbiosis and we have already seen examples of this in the early formation of eukaryotic cells. There are many classic examples of this relationship such as the sea anemone and clownfish where the clown fish benefits from the protection that the poisonous tentacles of the sea anemone offer, and the anemone benefits from the activity of the clownfish in keeping the area clean (eating left-over food and dead tentacles) and it is thought that the presence of the bright clownfish attracts other fish (a food source). The N_2 fixing rhizobia on roots of clover and legumes are another well-known example of symbiosis, as discussed in Chapter 5. Finally, if both organisms are negatively affected it is known as competition.

6.5 Organisms as environmental indicators

Recognition of how organisms adapt to certain environmental conditions can be used as an indicator of environmental conditions. For example, we have already seen that *Asellus* can tolerate low concentrations of oxygen within rivers and therefore the presence of large numbers of this species can be indicative of oxygen limiting conditions. Birds have often been put forward as good indicators of a particular habitat type and its condition. Thus the ovenbird (*Seiurus aurocapillus*) is thought to be an indicator of closed-canopy, mature forests with a sparse understory (Carignan and Villard, 2002). Indicator organisms have been used to indicate many different environmental conditions including salinity, temperature, oxygen, presence of particular pollutants and pH, as well as habitat types and their ecological condition.

The adaptation of particular organisms to the presence of pollution has been used to monitor environmental conditions within rivers. Macroinvertebrates are common organisms within aquatic environments and are heterogeneous in their response to the presence of organic pollution (as discussed earlier in the section on oxygen). This response has been used to generate a score called the Biological Monitoring Working Party Score which indicates the extent of organic pollution within rivers (Hawkes, 1998). Each family of macroinvertebrates is given a score from 1–10 reflecting its tolerance of organic pollution, with 1 being the most tolerant and 10 being the most sensitive (Table 6.2). A sample of macroinvertebrates is taken from the river and sorted to determine the presence of each family; where the presence is recorded then the score is added to the tally. This is a presence/absence score and therefore the number of individuals is unimportant; it is simply whether they are present in the sample or not.

Table 6.2 Scoring system for the Biological Monitoring Working Party Score.

	Macroinvertebrate family	Score
Mayflies	*Siphlonuridae, Heptageniidae, Leptophlebiidae, Ephemerellidae, Potamanthidae, Ephemeridae*	10
Stoneflies	*Taeniopterygidae, Leuctridae, Capniidae, Perlodidae, Perlidae, Chloroperlidae*	
River bug	*Aphelocheiridae*	
Caddisflies	*Phryganeidae, Molannidae, Beraeidae, Odontoceridae, Leptoceridae, Goeridae, Lepidostomatidae, Brachycentridae, Sericostomatidae*	
Crayfish	*Astacidae*	8
Dragonflies	*Lestidae, Agriidae, Gomphidae, Cordulegasteridae, Aeshnidae, Corduliidae, Libellulidae*	
Caddisflies	*Psychomyiidae, Philopotamidae*	
Mayflies	*Caenidae*	7
Stoneflies	*Nemouridae*	
Caddisflies	*Rhyacophilidae, Polycentropodidae, Limnephilidae*	
Snails	*Neritidae, Viviparidae, Ancylidae*	6
Caddisflies	*Hydroptilidae*	
Mussels	*Unionidae*	
Shrimps	*Corophiidae, Gammaridae*	
Dragonflies	*Platycnemididae, Coenagriidae*	

Table 6.2 (*continued*)

	Macroinvertebrate family	Score
Water bugs	*Mesovelidae, Hydrometridae, Gerridae, Nepidae, Naucoridae, Notonectidae, Pleidae, Corixidae*	
Water beetles	*Haliplidae, Hygrobiidae, Dytiscidae, Gyrinidae, Hydrophilidae, Clambidae, Helodidae, Dryopidae, Elminthidae, Chrysomelidae, Curculionidae*	
Caddisflies	*Hydropsychidae*	5
Craneflies	*Tipulidae*	
Blackflies	*Simuliidae*	
Flatworms	*Planariidae, Dendrocoelidae*	
Mayflies	*Baetidae*	
Alderflies	*Sialidae*	4
Leeches	*Piscicolidae*	
Snails	*Valvatidae, Hydrobiidae, Lymnaeidae, Physidae, Planorbidae*	
Cockles	*Sphaeriidae*	
Leeches	*Glossiphoniidae, Hirudidae, Erpobdellidae*	3
Hoglouse	*Asellidae*	
Midges	*Chironomidae*	2
Worms	*Oligochaeta**	1

* Oligochaeta are a class rather than a family.
Source: National Water Council (1981) *River Quality: the 1980 survey and future outlook*. National Water Council.

As a general indicator, the following scores reflect different levels of pollution:

Score	Pollution
0–10	Heavy or gross
11–40	High
41–70	Moderate
71–100	Low
>100	Clean

Source: National Water Council (1981) *River Quality: The 1980 survey and future outlook*, National Water Council.

The score can also be compared against a score that is expected for the type of river that was sampled, which gives a measure of whether it is poorer in biological quality or as expected. This can then give an indication of pollution. In addition to this, further analysis can be undertaken and the average score per taxon (ASPT) can be calculated which then eliminates the effect of sample size. With larger samples it is more likely to sample a greater variety of organisms, but by calculating the ASPT this factor is removed. It should be noted that this score is based upon organic pollution and therefore should not be applied to other types of pollution.

6.6 Populations

A population can be defined as a group of individuals of a single species that simultaneously occupy the same general area. It is notoriously difficult however to define a population precisely, largely due to the issues with defining a species, but also because it is difficult to define the 'area'. The characteristics of a population are shaped by the interactions between individuals and their environment, and natural selection can modify these characteristics.

The most important characteristics of a population are the density and spacing of individuals. The growth and decline of a population gives the demographics of the population. Additions to a population result from births and immigration, depletions from death and emigration.

The traditional view of population growth is a simple curve that reaches a maximum known as the 'carrying capacity' of the environment which is the number of individuals that the area and its resources can support (Figure 6.10).

Population sizes fluctuate through time as a result in changes in the balance between additions and depletions. These are dependent upon biotic and abiotic factors. Many ecologists have attempted to produce general rules to categorise population properties but in most cases this has proved to be very difficult. However, there is one rule that appears to be true in many cases and that is the species area relationship. This rule basically states that the greater the area, the greater the number of species (Figure 6.11).

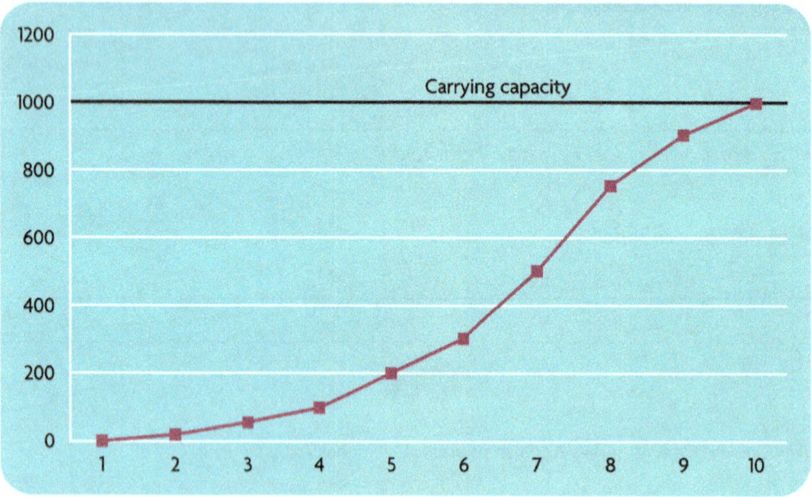

Figure 6.10 Plot showing the relationship between population growth and carrying capacity. The vertical axis represents the number of individuals in the population, and the horizontal axis an arbitrary timescale (e.g. days, months, years, decades).

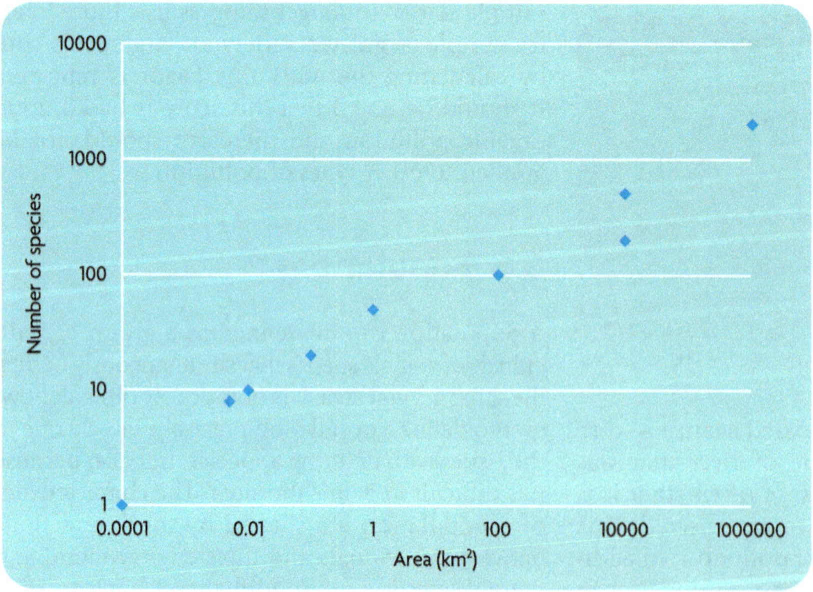

Figure 6.11 Species area relationship demonstrating that the number of species increases as the area of habitat increases. Note that both axes have logarithmic scales, so the plot is still approximately linear.

A number of theories have been suggested to explain the relationship illustrated in Figure 6.11. Firstly there is the sampling effect, where basically a greater number of samples is taken to fully represent the larger area and therefore there is an increased chance of sampling a greater number of species. Secondly, in larger areas there are a greater range and number of habitat types (environmental heterogeneity) which can therefore support a greater number of different species. Finally, in very large regions, there is a higher rate of speciation and a lower rate of extinction which results in a wide range of species. The species area relationship can be represented by the equation:

$$S = CA^z \qquad \text{or} \qquad \log S = \log C + z \log A$$

where S is the number of species, C is the intercept, A is the area and z is the slope for log transformed data. The value of z in the original model was considered to be related to the isolation of the environment but it is clear that many other factors can affect the slope, including species pool size, inclusion/exclusion of transient species and the balance between immigration and emigration.

It has also been observed that the species richness in a small area is dependent upon that of the larger area in which it is embedded. In other words local species richness tends to increase with regional species richness (e.g. Cornell and Lawton, 1992).

This theory was used as part of a more complicated theory proposed to explain the establishment and development of communities. The theory, known as the Island Biogeography Model, was proposed by MacArthur and Wilson in 1967 and has had a major impact upon the study of ecology in terms of understanding basic ecological processes; it also has been applied to nature reserve management and conservation. It centres upon community establishment and development upon islands.

The theory states that there is a balance of populations (i.e. equilibrium) on an island that depends upon the size of the island and its proximity to sources of organisms, reflecting the balance between immigration and extinction. Let us examine these ideas in more detail.

It was thought that there are two key factors that control the balance of the population, the proximity of the island to sources for colonisation and the size of the island itself. If we consider the proximity first, when an island is very close to a source of potentially colonising organisms (the mainland or another island) then immigration will be high initially. The rate of immigration then slows as niches become filled and species arriving are the same as those that have already colonised. In remote islands the rate of immigration is much lower as the species cannot disperse as easily from the source area.

In terms of extinction rate, this is also lower in the initial stages of colonisation. Initially there are less species and if each species is assumed to have an equal chance of extinction, then as the number of species increases the extinction rate also increases. However, in small islands there is less variety in habitats and therefore there is more competition between species; this means that the extinction rate is higher than on large islands. In addition the species population will be smaller and small populations are more prone to extinction, therefore increasing the extinction rate. By incorporating all these considerations we arrive at the final relationship illustrated in Figure 6.12.

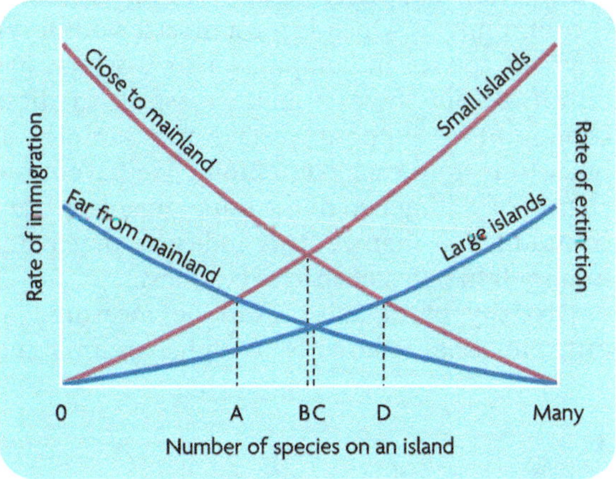

Figure 6.12 Graphical representation of the effects of island size and remoteness upon numbers of species. For example, the intersection at species number A represents the number of species on a small island far from the mainland, with a low immigration rate because of distance from the mainland and high extinction rate due to the small number of habitats on the small island (from MacArthur and Wilson, 1967).

Source: MacArthur and Wilson (1967).

There have been many challenges to this theory and amendments made, but its fundamental principles are still used within ecology today. In particular it has been used to make predictions of extinction rates if areas of habitat are reduced as a result of human activity. It has been applied also to the design of nature reserves which have been seen as 'islands' of habitat in a sea of human activity (see Chapter 25).

6.7 Communities

A community is an assemblage of populations in an area or habitat. Communities differ dramatically in their species richness (the number of species there are) and relative abundance of organisms (the species evenness), and these are fundamental measurements ecologists require. However, measuring these aspects of communities is challenging and there are problems associated with how representative these measures are of the real community.

Let's look at two communities of organisms (Figure 6.13). Firstly we ask ourselves which of these communities has the greatest species diversity?

If we look at species richness we realise that it is community A which is most diverse with three different species in comparison to B which has only two. But there is another aspect to these communities. Suppose we take two samples of organisms from each community. We have more chance of getting two of the same organism from community B than A. This indicates that community B has greater species evenness.

Diversity of organisms can be measured at different scales referred to as alpha, gamma and beta diversity. Local diversity (alpha) refers to the number of species in a community and if we are looking on a regional scale then we calculate the average number of species per local community. On larger geographic scales the number of species per continent or region is referred to as gamma diversity. A third scale, which is the change in species composition along an environmental or geographical gradient, is known as beta diversity. Table 6.3 illustrates how these different diversity measures reflect the species composition in any given area.

In region 1, where there are a greater number of different species within each individual woodland, alpha diversity is greatest with an average species number of 6. In contrast, region 2 has less diversity within each woodland but has greatest diversity of species within the region as a whole with 10 different species present. In region 1 many of the same species are found within all three woodlands whereas the woodlands within region 2 are very different in their species composition. However, in region 3 this difference in species composition between the woodlands is greatest, with no individual species found in more than one woodland. Therefore there is a distinct change in community composition across the region and beta diversity is high as a result.

Ecologists use these two components of diversity to establish an overall index of biodiversity and there are many different types of these *biological indices*. The Shannon diversity index is one that is commonly used to evaluate the diversity within a community; it uses the formula:

$$H = \sum_{i=1}^{s} (p_i)(\ln p_i)$$

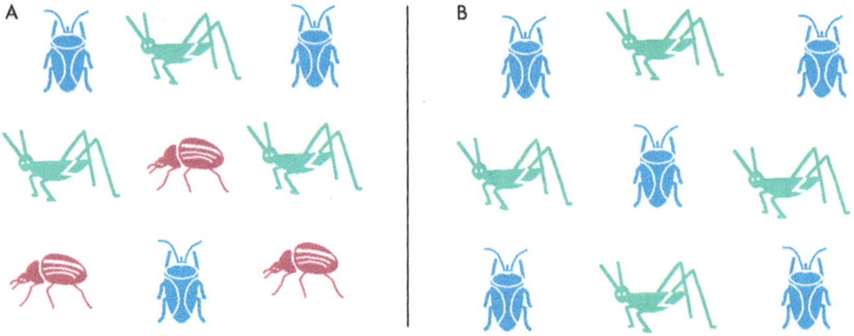

Figure 6.13 Two different communities illustrating the difference between species richness and species evenness.
Source: Based on Purvis and Hector (2000).

Table 6.3 Variation in species composition within woodlands and associated measures of diversity.

	Woodland 1	Woodland 2	Woodland 3	Alpha	Gamma	Beta
Region 1	• *Anemone nemorosa* • *Oxalis acetosella* • *Hyacinthoides non-scripta* • *Chrysospelnium oppositifolium* • *Allium ursinum* • *Galium odoratum*	• *Oxalis acetosella* • *Hyacinthoides non-scripta* • *Chrysospelnium oppositifolium* • *Allium ursinum* • *Galium odoratum*	• *Anemone nemorosa* • *Oxalis acetosella* • *Hyacinthoides non-scripta* • *Chrysospelnium oppositifolium* • *Allium ursinum* • *Galium odoratum* • *Phyllitus scolopendrium*	6	7	1.2
Region 2	• *Anemone nemorosa* • *Oxalis acetosella* • *Hyacinthoides non-scripta*	• *Chrysospelnium oppositifolium* • *Allium ursinum* • *Galium odoratum* • *Phyllitus scolopendrium*	• *Chrysospelnium oppositifolium* • *Phyllitus scolopendrium* • *Pteridium aquilinum* • *Arum maculutaum* • *Primula vulgaris*	4	10	2.5
Region 3	• *Anemone nemorosa* • *Oxalis acetosella* • *Hyacinthoides non-scripta*	• *Chrysospelnium oppositifolium* • *Allium ursinum* • *Galium odoratum*	• *Phyllitus scolopendrium* • *Pteridium aquilinum* • *Arum maculutaum*	3	9	3

Where:

H is the Shannon diversity index
p_i is the proportion of individuals in the total sample belonging to the ith species
$\ln p_i$ is the natural logarithm of p_i

Lead-in question

Why do communities change with time?

Organisms do not simply respond to their environment in terms of abiotic and biotic factors; they also affect their environment. If you think of grass growing in a soil, the roots affect the structure of the soil; they also take up nutrients and water from the soil, changing its chemistry; they release compounds into the soil, and the annual cycle of the grass returns organic matter to the soil, altering both chemistry and structure. Plant roots can also act as a focus for bacterial communities, especially in the root rhizosphere, which is vital for geochemical transformations in the soil. So all this activity leads to important changes in the environment and paradoxically can actually make the environment less well suited to the organism that has affected it. This is how succession occurs.

We can take a reed-bed succession as a typical example. Within a reed-bed environment there are areas of open water that are shallow towards the edge. Because the water is shallower at the edge it allows reeds (species such as *Phragmites australis*) to grow which can tolerate standing water to a certain depth. These reeds form dense stands with high rooting density and they slow water flow resulting in the deposition of silt around the plants. Over years this siltation builds up the land surface and the water becomes more shallow and eventually dries out to some extent. The reeds are less well adapted to dry soils and therefore other plant species are able to invade the area. Each of these plant communities has associated communities of other insects and animals that are specifically adapted to the plant species within them. Each of these stages of reed-bed succession will be evident within a reed-bed area. This is why conservationists whose aim is the preserve certain ecosystems need to actively manage areas to prevent natural succession from changing the area.

6.8 Ecosystems

The term ecosystem refers to a community and its abiotic factors together. There has been some attempt to classify ecosystems (as is the want of scientists) and this has led to the term biome. A biome is a major type of ecosystem defined largely according to climatic conditions but also includes some consideration of other abiotic factors such as soil type. These biomes are stratified both vertically and horizontally leading to a very broad-brush division of ecosystems. It should be remembered that individual organisms and populations may be able to move between ecosystems and biomes are therefore not self-contained systems.

Using the ecosystem approach to ecology (discussed further in Chapter 25) we can determine the key relationships between the different organisms and their surrounding environment. There are two ways in which we can do this: (1) by determining the flow of energy through the system and (2) by establishing the cycling of nutrients. In both cases, energy and nutrients are major *inputs* to the ecosystem; for energy this is usually in the form of sunlight (although there are systems such as hydrothermal vents where energy inputs are in other forms), but for nutrients this will depend upon the biogeochemical cycle in question (see Chapters 5 and 17). For example, the main input of nitrogen to natural (unpolluted/minimally managed) terrestrial ecosystems is in a gaseous form whereas for phosphorus it will be from weathering of rocks. Through improving our knowledge of the way in which ecosystems work we can understand how disturbances or changes to the ecosystem will impact upon each component. Thus if we increase carbon dioxide levels in the atmosphere we know that this will impact upon the carbonate chemistry within the marine ecosystem and its impacts upon the organisms within it.

6.8.1 Energy and ecosystems

The energy that enters an ecosystem is used by the 'primary producers' to transform inorganic compounds into organic compounds, primarily through the process of photosynthesis. Plants are obviously the key primary producers in most ecosystems. However in some systems, bacteria can use oxidation of inorganic compounds as an energy source. The organic compounds that result are essentially a store of this energy and the amount of energy that is assimilated by these primary producers is termed the *gross primary productivity*. Some energy is used by the organisms for other processes such as respiration and so the difference between the gross primary productivity and that used in cellular processes is termed the *net primary productivity*.

The productivity of any ecosystem is an important measure of the effectiveness of its energy capture and is often used as a key measure of ecosystem function (see later). There are diverse controls on primary productivity within ecosystems, many of which are related to climatic conditions (e.g. temperature, precipitation) but they also include availability of nutrients and the physical attributes of soil. These factors can limit the capture of energy but can also influence where the energy is stored. For example, where light is limited, plants can allocate more energy to the production of leaves in order to capture any light available.

Primary productivity in turn will affect the amount of *secondary production*. When primary producers are consumed (by heterotrophs) energy is transferred within the ecosystem. These consumers also utilise some of the energy for respiration and metabolism and any energy that remains is put into the growth of tissues and reproduction.

This remaining energy is the secondary production. However, not all consumers utilise the energy that they take in from consumption of primary producers to the same degree, so secondary production can vary significantly. The consumers can, in turn, be eaten by other consumers and therefore there is a further transfer of energy. This can happen repeatedly and this transfer of energy through the ecosystem is termed a *food chain*. Those organisms that obtain all the energy through the same number of these steps form a *trophic level*. Due to the inefficiencies of energy production at each level, the amount of energy that is contained within each level tends to decrease. If primary production within an ecosystem is limited by environmental factors then the energy at each trophic level will also be limited.

6.8.2 Nutrients and ecosystems

Within an ecosystem nutrients may be present in many different forms, both organic and inorganic. There are many inputs of nutrients into ecosystems including soil formation, precipitation, surface runoff and airborne particle (and for some elements, gaseous) deposition. The specific amounts of nutrient, the pathways of transport and the form or forms will vary greatly from element to element and from ecosystem to ecosystem. However, in all systems primary productivity is dependent upon the supply of essential nutrients that are required for normal cellular processes and to form important cell organelles. Thus the primary producers act as a store of these nutrients. On death the material decays and the nutrients can be returned to the soil via the process of mineralisation (usually mediated by the activity of fungi and bacteria) where they are once again available for uptake by the primary producers. When primary producers are consumed then there is a transfer of the nutrients up the food chain, just as in the case of energy.

In the case of nutrients certain activities of organisms can actually cause the loss of nutrients from the system. For example, microbial activity can convert nitrate (NO_3^-) to N_2O and N_2, both of which are gases and therefore can be released back into the atmosphere. It should be noted that some organisms can actually leave the ecosystem and therefore they can act as a loss of energy and/or nutrients. Figure 6.14 gives a general model of nutrient cycling within an ecosystem. The

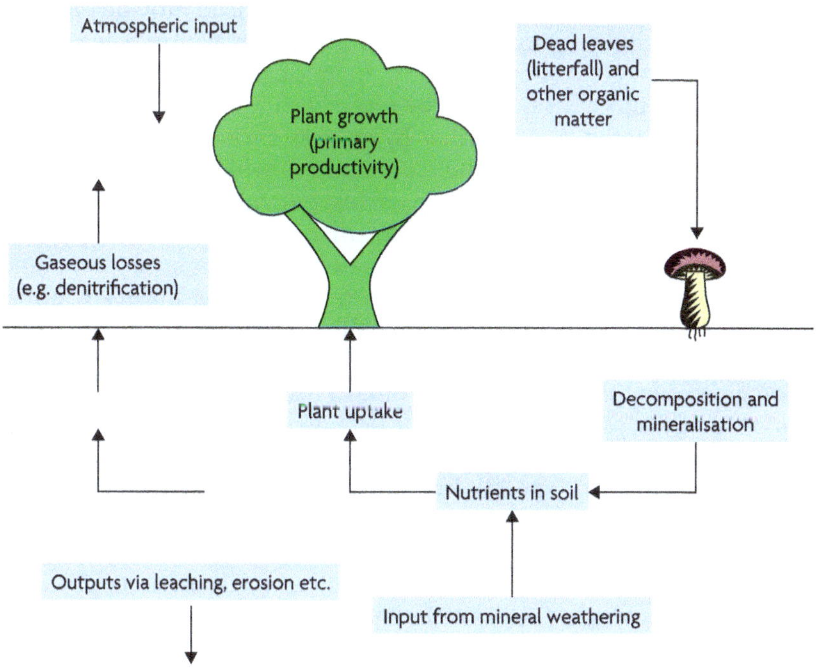

Figure 6.14 A generalised model of nutrient cycling within an ecosystem (see Chapters 4 and 5 for more specific examples).

specifics of each cycle will vary with the ecosystem and the nutrient (see Chapters 4, 5 and 17) but the key components of inputs, primary productivity, decomposition, cycling and outputs are always present.

6.8.3 Ecosystem function

Ecosystems may have a number of functions within a landscape which can be categorised as *ecosystem processes* or *ecosystem values* or *services*. The former represent vital transformations that contribute to the long-term stability of the landscape and can include such processes as oxygen consumption and production, carbon mineralisation, organic matter production and sedimentation. Any one ecosystem can have many different ecosystem processes.

Ecosystem services are the functions performed by an ecosystem that are deemed beneficial to humans. This is a relatively recent concept and has led to attempts by some environmental scientists to put an economic value on ecosystems (the study of environmental economics). Examples of these values include provision of food, timber or chemicals, protection of water quality through removal or transformation of pollutants, and regulation by forest growing upstream of discharge downstream in a river (reducing flood risk downstream). This aspect is discussed more fully in Chapter 25.

Remember that each ecosystem is made up of many different communities together with the abiotic components and each organism and/or community may be responsible for one or more of the processes occurring within the ecosystem. Ecologists have attempted to predict how ecosystem function would respond to changes in community structure and this has led to a number of hypotheses.

The first of these states that organisms within a system have no function whatsoever, so if you removed that organism the ecosystem would continue to function as normal (essentially the null hypothesis).

The second major theory is the diversity–stability hypothesis which states that as the number of species in a community decreases, the ability of the community to resist changes in environmental conditions (such as productivity or disturbance) will also decrease (Figure 6.15).

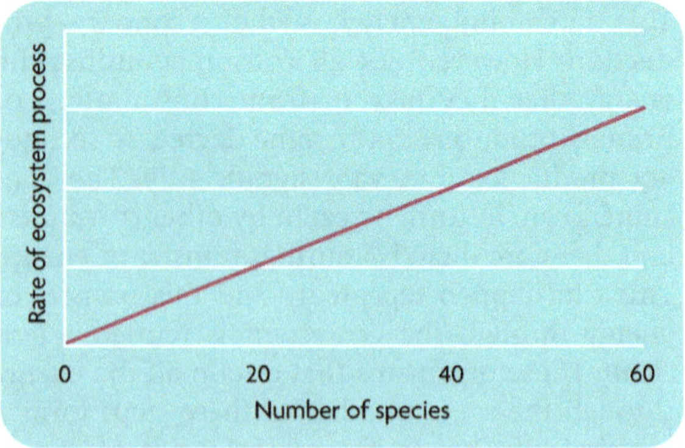

Figure 6.15 Plot representing one possible consequence of adherence to the diversity–stability hypothesis. As presented above the rate of the ecosystem process of interest declines linearly with the number of species present in the ecosystem, as an imposed stress causes a decline in the number of species.

Rather than simply considering the number of species in a community we can separate species into functional groups. The redundancy hypothesis uses this division and suggests that a community can withstand losses of individual species within functional groups as other species within the same group will continue to provide the function. However, if all species within a functional group are lost then clearly the ecosystem processes will be severely compromised.

A similar hypothesis also considers some species to be redundant, in which case the response will be stepped as individuals are lost from a system. It is referred to as the rivet hypothesis as it uses the analogy of rivets in an aeroplane. The loss of a few rivets may not make any difference but eventually the loss of a particular rivet may cause a rapid (or even catastrophic) failure; similarly, eventually loss of a key species may trigger a dramatic decline in ecosystem processes.

The rivet hypothesis is related to the keystone species hypothesis in which particular species within a community either determine the integrity of the community as a whole and/or have a disproportionate effect on ecosystem processes (Power *et al.*, 1996). Prairie dogs are an example of a keystone species. It has been suggested that prairie dogs have a number of roles within the prairie ecosystem including providing burrows that act as shelters for vertebrates and invertebrates, altering plant species composition and vegetation structure, creating open habitats and affecting nutrient cycling rates. It is clear that a number of other species rely on the presence of prairie dogs, including black-footed ferrets, mountain plovers, ferruginous hawk, golden eagle, swift fox, horned lark, grasshopper mouse and deer mouse. The activities of prairie dogs therefore directly or indirectly affect the biodiversity of the prairie ecosystem and thus they qualify as a keystone species (Kotliar *et al.*, 1999).

The final theory related to organism function is referred to as the *idiosyncratic hypothesis*. This suggests that ecosystem function and processes will change in response to changes in species diversity but the magnitude and direction change is unpredictable because of the complexity of the roles and interactions of species within communities.

These theories have been tested by many ecologists using both controlled laboratory and field experiments but they have failed to generate a definitive answer. It does appear that, in many systems, there is some evidence of complementarity. For example, in a key work by Hector *et al.* (1999) the effect of plant diversity upon ecosystem function, as measured by primary productivity, was investigated using experimental assemblages of grassland species. Productivity within the grasslands declined with the loss of plant diversity (Figure 6.16a) and with loss of functional group richness (Figure 6.16b and see also Figure 25.5). In addition it was found that there was a greater production of biomass in moderately diverse polycultures (more than one species) than in monocultures of the maximum yielding species of the polyculture, which suggested some positive interaction between species. This could be niche complimentarity, where different species access different resources, therefore allowing for more complete utilisation of the available resources (e.g. nutrients, soil at different rooting depths, light) and/or there are positive mutualistic interactions between species in the mixed species assemblages. The importance of interactions and complimentarity has also been found in other habitat types and for other groups including

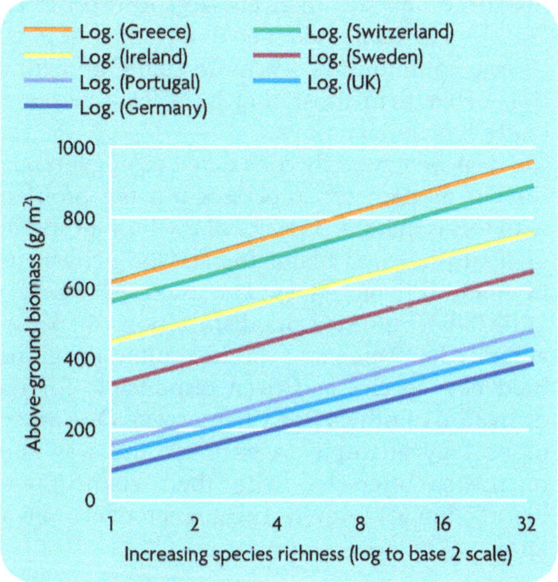

Figure 6.16a Relationship between above-ground biomass and species richness found for seven European countries. Scatter of points around individual lines is not shown because of excessive overlap, but was very considerable, which indicates the true complexity of the relationships.
Source: Based upon data in Hector *et al.* (1999).

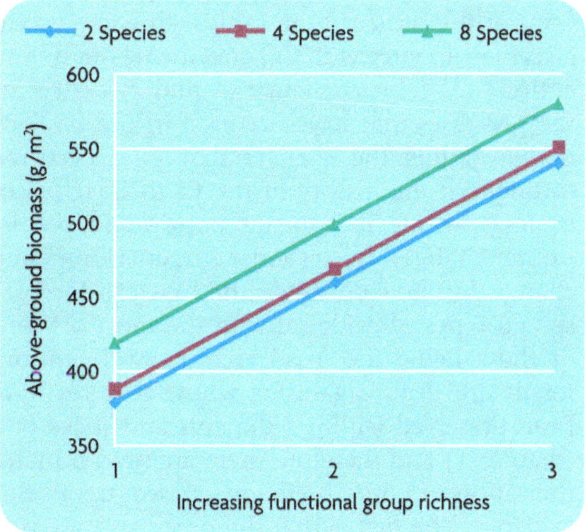

Figure 6.16b Graphs showing that biomass production increases with functional group richness.
Source: Based upon data in Hector *et al.* (1999).

macroinvertebrates in rivers (e.g. Jonsson *et al.*, 2002). In general, therefore, it appears that intact diverse communities are more stable and function better than those that have lost species or entire functional groups.

So far we have only considered ecology in relation to modern day processes but the problem with this is that the presence of the human species on Earth has had major impacts on ecosystems in many different ways. For example, we have artificially controlled the distribution of species, enhanced movement of species around the globe, bred new species and been responsible for the extinction of others (see Chapter 25). One aspect of ecology attempts to examine the way that organisms interacted with their environment before humans evolved. This branch of ecology is known as palaeoecology.

Figure 6.17 Sedimentary rock containing predominantly a large number of ammonites, an extinct group of marine invertebrate animals. Ammonites are often valuable index fossils for linking the rock layer to a specific geological date.
Source: Photograph courtesy of Lapworth Geological Museum/ Dr Jason Hilton.

Lead-in question

What evidence is there for ecological processes before the evolution of *Homo sapiens*?

6.9 Palaeoecology

It is relatively simple to make ecological observations at the present time (although some ecologists may disagree with this statement) but it is an entirely different challenge to find evidence for the processes that have occurred in the past. By understanding the changes that have occurred throughout the history of the Earth in terms of ecology we are in a better position to predict future changes in response to environmental change, but how do we do this? We can look for several types of evidence with the most obvious of these being the fossil record. By definition, fossils are dead organisms whose features have been preserved within sediments and rocks (see Chapter 1) and therefore there are several major limitations to interpretations based upon this evidence. The first is that the organisms are preserved in death, although on very rare occasions organisms may be killed so quickly that they are in the act of normal behaviour. For example, there are specimens of woolly mammoths with vegetation still in their mouths. Secondly the preservation of organisms requires very specific environmental conditions, notably a lack of oxygen (to reduce the amount of bacterial activity) and a rapid burial process. The marine environment is often a key area in which preservation occurs and therefore there is a bias towards marine life in the fossil records (Figure 6.17). As a result the fossil record is extremely patchy and it is likely that many organisms, particularly those without shells or other hard parts, have simply vanished without a trace. This introduces major bias in the fossil record in terms of the types of organisms and the environment in which they live within the fossil record. However, despite these drawbacks, we can make a lot of inferences about past life and changes on Earth using fossils.

The examination of life in the past has helped us to understand how organisms have evolved over time and how global biodiversity has changed. We have already seen that the natural processes of evolution cause some organisms to become extinct and that there is a continual turnover of species as evolution occurs (the background extinction rate). We have also seen that there can be periods of high rates of extinction termed mass extinctions. In addition to this we can also see significant changes to communities of organisms within the fossil record and we can use the study of palaeoecology to link this to significant changes in the abiotic conditions of an ecosystem.

In addition to the fossils themselves we also often can use information gathered from the surrounding rocks to determine the nature of the environment in which the organisms were living. In Chapter 1 we saw how different rocks form under different conditions. Characteristics of the rocks may sometimes provide us with information on parameters such as salinity, current strength, water depth, moisture status, water chemistry and temperature. This can then tell us something about the conditions that different organisms required for life. One final piece of evidence from the rocks is that of trace fossils. These provide evidence for the presence of different organisms without their actual remains, and can be burrows, footprints, tracks, root channels, imprints or faecal pellets. Rocks found in the north-east of England showed unusual footprints of a dinosaur with elongated toes (Figure 6.18), and this has been interpreted in conjunction with sedimentary evidence as being evidence of an animal swimming in shallow water. This type of track can therefore provide us with clues as to the behaviour of organisms within their environment while they were still alive. Dinosaur footprints are not always that easy to spot unless you look very carefully (Figure 6.18).

Figure 6.18 A fossilised trail left in mud from a dinosaur with elongated toes.

Source: Photograph courtesy of M.A. Whyte.

Chemical analyses of faecal pellets can provide information on the possible composition of the animal's diet, although unfortunately it is almost impossible to assign faecal pellets to a particular organism in the geological record as you can imagine! Although we can make many interpretations based upon the fossil record. Paleoecologists often use comparisons with modern organisms to facilitate their interpretation of what they can see.

There has been increased interest over recent decades in the effect of climate upon the distribution of organisms and we can look to the past to provide us with information on how species responded to dramatic shifts in environmental conditions. For example a study of global warming that occurred in the Palaeocene–Eocene boundary (around 55.8 million years ago) when temperatures increased by 5–10 °C over a period of between 10 and 20 thousand years showed significant changes to the floral communities (Wing *et al.*, 2005). Examination of fossil evidence showed an increase in the range of particular communities with movement of species north beyond their normal range and that this occurred within 10,000 years. There was also movement of species between continents, possibly facilitated by the presence of land bridges. This may provide an indication of the potential responses of communities to present day anthropogenic global warming.

POLICY IMPLICATIONS

Ecology

- Policy makers need to understand how ecology provides insight into how organisms interact with each other and with their environment.
- They should be aware of how organisms play a vital role in controlling their surrounding environment at a range of scales, through processes such as the production of oxygen and cycling of nutrients.
- They need to be aware of the inextricable links between the atmosphere, plants, soils, animals and fresh water resources.
- They need always to regard humans as part of an ecosystem and consider the impacts that humans have on the composition and functioning of ecosystems and therefore on biodiversity.
- They need to use an understanding of ecological interactions to determine how anthropogenic activity affects key processes and ecosystem services to improve management of the Earth's systems.
- Their appreciation of key ecological concepts should provide the tools needed to underpin long-term management decisions.

CHAPTER REVIEW EXERCISES

Exercise 6.1

Go to your nearest green space, for example your garden or a local park, and identify the different habitats that you can see. Count how many different faunal and floral species you can see within a small area.

What are the spatial and temporal challenges in quantifying biodiversity within your chosen area?

Suggest how you think that the species diversity will change over the next year and over the next 10 years, justifying your conclusions.

Discuss how the biodiversity compares to other habitats that you can find within your chosen area and to those that are thought to be biologically diverse (e.g. rainforests, natural wetlands).

Hint: remember not all biodiversity is above ground!

Exercise 6.2

Choose two organisms from contrasting environments. Describe how they are specifically adapted to their local environmental conditions. Discuss how each organism affects the surrounding environment and other organisms within the same area.

Exercise 6.3

Where there has been disturbance of the land (for example where building works have been carried out), rapid colonisation of the area by certain species (e.g. the common poppy) often occurs. Explain why these organisms are the first to colonise and discuss how the community will probably change over time if there is no further disturbance.

Exercise 6.4

In Figure 6.12, the significance of intersection point A was explained in the caption to the figure. Explain the significance of interception points B, C and D in the same figure.

REFERENCES

Carignan, V. and Villard, M.A. (2002) Selecting indicator species to monitor ecological integrity. *Environmental Monitoring and Assessment*, **78**, 45–61.

Childress, J.J. and Seibel, B.A. (1998) Life at stable low oxygen levels: Adaptations of animals to oceanic oxygen minimum layers. *Journal of Experimental Biology*, **20**, 1223–1232.

Cornell, H.V. and Lawton, J.H. (1992) Species interactions, local and regional processes and limits to the richness of ecological communities: A theoretical perspective: *Journal of Animal Ecology*, **61**, 1–12.

Grotzinger, J., Jordan, T.H., Press, F. and Siever, R. (2007) *Understanding Earth*, 5th edn. W.H. Freeman and Co., New York.

Hawkes, H.A. (1998) Origin and development of the biological monitoring working party score. *Water Research*, **32**, 964–968.

Hector, A., Schmid, B., Beierkuhnlein, C., Caldeira, M.C., Diemer, M., Dimitrakopoulos, P.G., Finn, J.A., Freitas, H., Giller, P., Good, J., Harris, R., Hogberg, P., Kuss-Danell, K., Joshi, J., Jumpponen, A., Korner, C., Leadley, P.W., Loreau, M., Minns, A., Mulder, C.P.H., O'Donovan, G., Otway, S.J., Pereira, J.S., Prinz, A., Read, D.J., Scherer-Lorenzen, M., Schulze, E.-D., Siamantziouras, A.S.D., Spehn, E.M., Terry, A.C., Troumbis, A.Y., Woodward, F.I., Yachi, S. and Lawton, J.H. (1999) Plant diversity and productivity experiments in European Grasslands. *Science*, **286**, 1123–1127.

Jonsson, M., Douglas, D., Malmqvist, B and Guerold, F. (2002) Simulating species loss following perturbation: assessing the effects on process rates. *Proceedings of the Royal Society. London* B269, 1047–1052.

Kotliar, N.B., Baker, B.W. and Whicker, A.D. (1999) A critical review of assumptions about the prairie dog as a keystone species. *Environmental Management*, **24**, 177–192.

MacArthur, R.H. and Wilson, E.O. (1967) *The Theory of Island Biogeography*. Princeton University Press, NJ.

Mason, C.F. (1996) *Biology of Freshwater Pollution*. Longman.

Mittermeier, R.A., Myers, N., da Fonseca, G.A.B. and Olivieri, S. (1998) Biodiversity hotspots and major tropical wilderness areas: Approaches to setting conservation priorities. *Conservation Biology*, **12**, 516–520.

Myers, N., Mittermeier, R.A., Mittermeier, C.G. da Forseca, G.A.B. and Kent, J. (2000) Biodiversity hotspots for conservation priorities. *Nature*, **403**, 853–858.

Power, M.E.D., Tilman, J.A., Estes, B., Menge, W.J., Bond, L.S., Mills, G., Daily, J.C., Castilla, J., Lubchenco, J. and Paine, R.T. (1996) Challenges in the quest for keystones. *Bioscience*, **466**, 9–20.

Purvis, A. and Hector, A. (2000) Getting the measure of biodiversity. *Nature*, **405**, 212–219.

Stork, N.E. (1997) Measuring global biodiversity and its decline. In: *Biodiversity II: Understanding and Protecting our Biological Resources*, edited by M.L. Reaka-Kudla, D.E. Wilson and E.O. Wilson, Joseph Henry Press, Washington, DC, 41–68.

Wilson, E.O. (1992) *The Diversity of Life*. Harvard University Press, Cambridge, Mass.

Wing, S.L., Harrington, G.J., Smith, F.A., Bloch, J.I., Boyer, D.M. and Freeman, K.H. (1995) Transient floral change and the rapid global warming at the Palaeocene–Eocene boundary. *Science*, **310**, 993–996.

CHAPTER 7

The evolution and functions of soils

Malcolm Cresser

Learning outcomes

By the end of this chapter you should:

- Know what soil is and why it is so important.
- Be aware of the key facets of soil as a medium for plant growth.
- Understand why soil pH is so important and so often measured.
- Recognise the importance of soil organic matter.
- Realise the importance of soil flora and fauna to soil fertility and element cycling and soil formation.
- Be more aware of the dynamic nature of soils.
- Be able to recognise some key soil types and the causes of differences between them.
- Appreciate the importance of soil physical properties to plant growth, drainage and erosion risk.
- Know how soil fertility is assessed.
- Understand the constraints to sustainable use of soils.

7.1 What is soil?

Soil is vital to crop and animal production, to controlling the chemical compositions of the atmosphere and surface waters, and even to climate. In pre-historic times it was crucial to the production of all our fossil fuels. Yet to the uninitiated, it may not seem to be particularly inspiring. If you wander through a forest like that in Figure 7.1, trees can sometimes impress by their sheer size if not their beauty. Ground flora or fungi may sometimes have a 'wow' factor and stop you in your tracks. But soil, in isolation, rarely stimulates much thought, save perhaps about muddy boots from a muddy track.

We saw in Chapter 4 that plants, including trees, acquire their carbon, hydrogen and oxygen for biomass production via photosynthesis from the atmosphere, and in Chapter 5 that their nitrogen too, though now predominantly coming from soil, originates initially from the atmosphere.

Figure 7.1 A forest outside Melbourne, Australia.

Figure 7.2 Soil exposed by an eroded river bank in northern England. The soil itself has been deposited here from higher slopes by erosion.

But all the other elements that plants require come from the soil.

So why does soil fail to impress? Partly it's because we usually only see it when it is cultivated or being eroded, for example beside a river as in Figure 7.2; partly it's because if you've never looked at soils carefully, you might mistakenly have the impression that they are uniform, drab coloured, inert, and, quite frankly, rather boring. In practice nothing could be further from the truth. Although surface layers of their mineral matrix may not impress when viewed with the naked eye, they are highly dynamic and teaming with life. They have to be to support the vegetation that they do. So soil is the upper layer of the lithosphere, most often with a mineral matrix derived from weathering of rocks either in situ or after movement by erosion by water, ice or wind as described in Chapter 1, mixed with organic matter produced as discussed in Chapters 4 and 5, and with biomass that includes fungi, bacteria, actinomycetes, archaea and a diverse range of invertebrates.

7.2 The soil profile

Even exposure of soil by erosion still fails generally to create an impressive image because the true colours of the underlying soil layers, also known as horizons, are concealed. They may

Figure 7.3 Scraping away the surface from a soil exposed by erosion often reveals visibly distinct soil layers or horizons, in this case revealing a podzol profile.

Figure 7.4 Collecting samples from a freshly dug soil pit. The student on the left is using a soil colour chart to classify the colour of each soil horizon.

often be exposed by scraping away the surface with a spade or trowel, as in Figure 7.3. The soil profile exposed here has several quite distinctive horizons, with a layer of partially decomposed plant litter overlaying a dark, organic matter rich horizon, which in turn lies over a parallel, ash-coloured horizon, and, below that, an orangey-brown horizon rich in iron (as Fe^{3+}) oxide. Such a soil profile is known as a podzol, and we can immediately tell a lot about the soil from its appearance. We will look at this in more detail at various points later in this chapter.

In the absence of any suitable eroded bank or roadside cutting, it is necessary to dig a soil pit to examine a soil profile so that soil horizons can be described, soil colours can be assessed, and representative samples taken from each horizon for analysis (Figure 7.4).

When digging soil pits in agricultural fields especially it is important to ensure that neither humans nor farm animals are put at risk of injury by tripping. Care should be taken not to trample upon the side of the pit on which the soil profile is to be described. Spoil from the pit is usually deposited carefully on a groundsheet to one side so that it can be returned, and turves removed repositioned (Figure 7.5). This is less of an issue in forests, especially if there is little or no vegetative ground cover, but again pit positions should be clearly marked to prevent accidents (Figure 7.6).

Figure 7.5 Putting turfs and spoil on a groundsheet or heavy duty plastic sheet close to the pit edge makes filling in after sampling much simpler and neater when digging soil pits. Small intact cores are also being collected here, but only to 30 cm depth. If the soil is to be examined to a depth of around 1 m, the area excavated is usually much larger than here, to facilitate safe working.

Figure 7.6 Exposing a forest soil profile. Tree roots may be a problem in forest soils when digging soil pits.

Sometimes exposing soil profiles provides evidence of anthropogenic disturbance that may not be apparent at the surface. Figure 7.7, for example, was a pit found to contain glass bottles and jars close to the surface. Further items were soon unearthed (Figure 7.8), and it was found eventually that the site had been a small lake used for landfill 50 years earlier, and then covered.

When an agricultural soil profile is exposed by digging a pit, sometimes only one or two layers may be apparent, and the soil may be quite uniform in appearance to 15 to 25 cm depth. This is partly due to the mixing effect of cultivation over many years. Often, however, such soils have high populations of earthworms that are very efficient soil mixers too. Thus when we see a podzol profile with its clearly defined horizons as in Figure 7.3, we can conclude that the degree of mixing is very small, and earthworms are almost certainly absent. Just like plants, all soil fauna have preferred ranges of environmental acidity and podzols are generally too acidic for earthworms. So a lot goes on in soils out of sight below the surface.

7.3 Why is soil so important?

Lead-in question

What can you name that does not originate from soil?

In Chapter 4 we discussed the carbon cycle at some length, and how, in the pre-historic era, early humans subconsciously set up and tested experimentally hypotheses such: 'Mixing shell sand or animal manure with soil may improve plant growth'. Early humans, once beyond the hunter-gatherer stage, would undoubtedly have recognised the importance of soil for food

Figure 7.7 and 7.8 Items unearthed from a soil pit at Hob Moor in North Yorkshire, providing very clear evidence of previous land use.

production. Being unaware of the nature of photosynthesis and the importance of the carbon and nitrogen cycles, it would naturally have been assumed that soil provided whatever it was that plants needed to grow. These early people would have soon found too that soil fertility declined if crops were harvested year after year. It is no wonder then that uses of manures and liming materials were soon tested. They came into regular use in some societies to replace lost soil nutrient elements thousands of years ago, long before the nature of the nutrients that were being replaced had been established. Early humans were very aware of the importance of soil for production of food and timber, and of the need to look after it. There can be little doubt too that they were soon aware of the need for water and of the risks of erosion when soils were mismanaged.

Ironically, if you stop someone in the street today and ask them why soil is important, the answer would still probably reflect these early views, namely for food production and (though probably less likely!) timber production. Those a little more environmentally aware might throw in 'sustainable' too, but few would say soil controls every physical thing that we need. Through its roles in the carbon cycle it controls the atmosphere, returning the carbon dioxide (CO_2) needed for photosynthesis and thus controlling the oxygen concentration in the atmosphere. Via its roles in water cycling it controls the purity and chemistry of surface waters in rivers and lakes. Biogeochemical weathering of soil minerals controls the chemistry of surface waters and their associated sediments and hence the rates at which chemical elements are returned to the oceans. Fossil fuels owe their origins to ancient vegetation largely grown in soils, having been modified over geological timescales. So even the squashed yellow plastic fish in the middle of Figure 7.8 owes much to the role of soil in prehistoric hydrocarbon formation, and the glass jars would have needed fossil fuel consumption.

In Chapter 4 we considered why shell sand would have been used as a liming material back in the stone age, as a consequence of seeing growth improvement around areas where residues from shell-fish meals had been dumped on acidic soils. There would have been no understanding of the neutralising process involved:

$$CaCO_3 + 2H^+ \rightarrow Ca^{2+} + H_2O + CO_2$$

The practice would have been adopted on a trial and error basis. Nowadays we can look at the mussel shell in Figure 7.8 that has survived more than 50-years burial and conclude that it must have been sitting in a near-neutral or slightly alkaline soil. Measurement showed that the soil had a pH of 7.5. It is appropriate at this point to look more closely at why soil pH is such an important soil parameter.

7.4 The importance of soil pH

7.4.1 The problem of soil acidity

Figure 7.9 shows the roots of a severely stunted wheat plant carefully dug out of a soil at a pH of 4.5. The plant was removed from one of a series of experimental plots set up at the Scottish Agricultural College Farm at Craibstone, near Aberdeen in Scotland. The plots were set up for farmers to demonstrate the importance of liming to the optimum soil pH. In a series of plots, the pH is adjusted to increase in steps of 0.5 pH unit from pH 4.5 to pH 7.5. It is not that easy to spot the wheat plants growing at pH 4.5 in

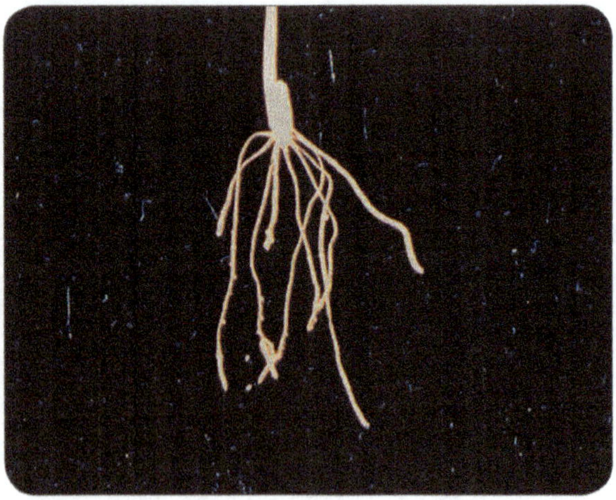

Figure 7.9 Roots of a wheat plant from a very acid soil. Note that roots are stubby; root branching starts but doesn't continue.

Figure 7.10 The poor growth of wheat plants at a soil pH of 4.5; note the more vigorous growth of the competing weeds.
Source: Photograph courtesy of Dr Tom Batey.

Figure 7.11 The quite good growth of potatoes at a soil pH of 4.5.
Source: Photograph courtesy of Dr Tom Batey.

Figure 7.10, because they are being outgrown by more acid-tolerant weed species.

The plant shown in Figure 7.9 is displaying classic symptoms of aluminium toxicity. Aluminium becomes much more soluble as soil pH falls. As a consequence, acidic mineral soils contain more potentially mobile hydrogen ions (H^+) and cationic aluminium species such as Al^{3+}, $Al(OH)^{2+}$, $Al(OH)_2^+$. Large proportions of these positively charged cations are associated with negatively charged cation exchange sites on the soil clay minerals and organic matter components (see section 4.6.1) to provide a balance of charge. The much smaller amounts in soil solution are still sufficient to adversely impact upon plant growth. The increasing solubility of aluminium as soil pH falls may be represented by the following equations:

$$Al(OH)_3 + H^+ \rightarrow Al(OH)_2^+ + H_2O$$

$$Al(OH)_2^+ + H^+ \rightarrow Al(OH)^{2+} + H_2O$$

$$Al(OH)^{2+} + H^+ \rightarrow Al^{3+} + H_2O$$

Not all plant species are equally sensitive to soil acidity. For example, oats and potatoes are much more acid tolerant. Figure 7.11, for example, shows potatoes growing quite adequately at the Craibstone experimental farm at pH 4.5.

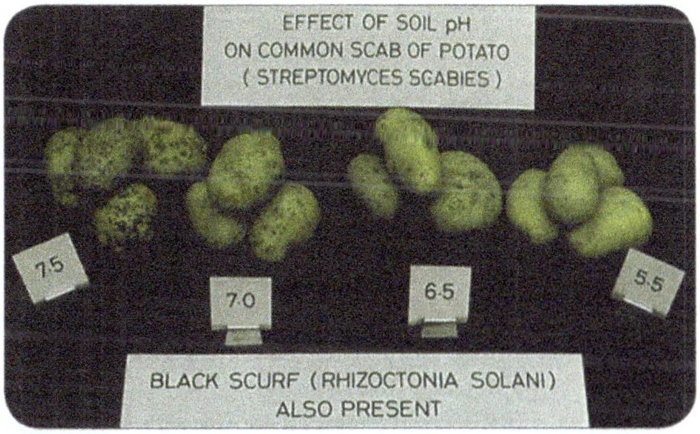

Figure 7.12 The adverse effect of increasing soil pH on the occurrence of common potato scab.
Source: Photograph courtesy of Dr Tom Batey.

Indeed, farmers normally would not lime before growing potatoes because of the increased likelihood of greater occurrence of common potato scab (*Streptomyces scabies*, Figure 7.12). This might mean a crop being rejected by over-fastidious members of the buying public or supermarkets.

It is rarely straightforward to demonstrate the effects of the concentration of any single elemental or ionic species of interest upon plant growth in soil. This is because if you add a simple ionic species to a soil its chemical form may change very rapidly as a consequence of absorption, precipitation, chemical complexation (for example with soil organic matter) or microbial uptake and/or immobilisation. For this reason it is necessary

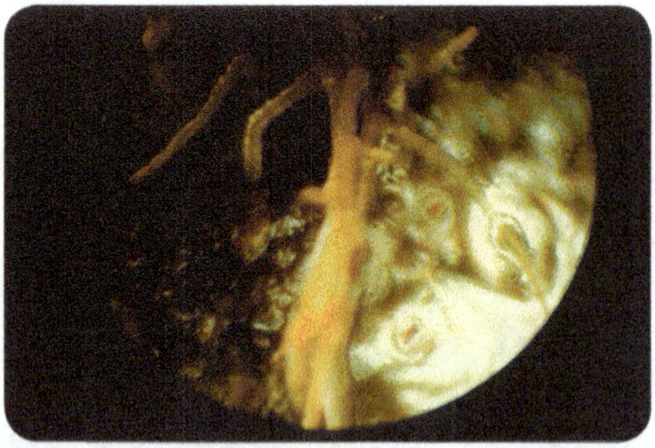

Figure 7.13 Boroscope photograph of the poor root growth of heather (*Calluna vulgaris*) in a soil that has become excessively acidified by acid rainfall. Note that the photographs are taken through the curved walls of an access tube (in this instance a test tube), so the images are not particularly sharp. However they are clear enough to demonstrate the stunting of root growth, which is all that was required on this occasion.

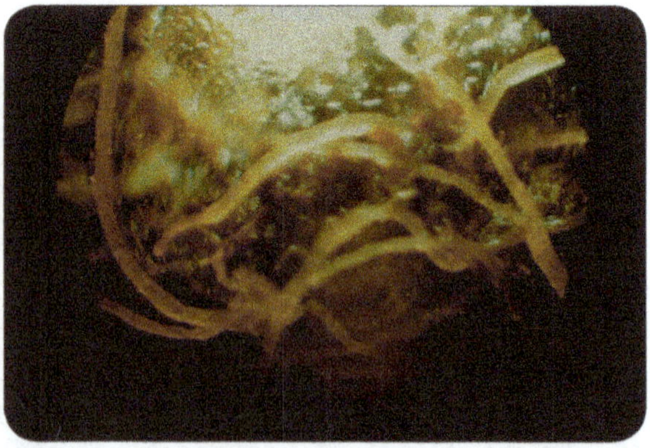

Figure 7.14 The beneficial effect of heather root growth of liming the soil in Figure 7.13 to slightly raise soil pH.

to investigate the effects of aluminium toxicity using solution culture (hydroponic) experiments, rather than directly using plants growing in soil.

Soil pH is not just important to agricultural crop species. Native wild plants and garden plants too have preferred pH ranges for good growth. Enthusiastic gardeners may often go to extreme lengths to acidify areas of their gardens so that they can grow acidophilic (acid loving) plants such as heathers, rhododendrons and azaleas in otherwise unsuitably alkaline soils. But soil can be too acid even for these acidophilic species. For example, Figure 7.13 shows the roots of a *Calluna vulgaris* (heather) plant growing in an organic soil that has been acidified by decades of acid rainfall. The stunting of the root system is clearly visible in the photograph. In this instance the roots have been photographed through a transplant glass access tube using a boroscope. The latter is effectively a high quality, narrow-bore periscope with a camera at the top end and a camera lens at the lower end surrounded by a fibre-optic light source to provide sufficient illumination to allow photographs to be taken. The reader might be more familiar with their use in the medical profession for internal examinations. They are useful in plant and soil science because they may be used in a time-lapse photographic mode to allow rates of root extension to be measured. Figure 7.14 shows the beneficial effect of treating the same soil as that in Figure 7.13 with a small amount of lime.

7.4.2 Trace element problems with slightly alkaline soils

So far in section 7.4, we have concentrated upon problems likely to be seen in plants grown in soils that have become too acidic. Problems also arise in soils that are too alkaline, however. Figure 7.15, for example, shows wheat plants growing at pH 7.5 and 7.0 at the Craibstone experimental farm. The plants on the left are clearly slightly

Figure 7.15 The chlorotic appearance of wheat growing at soil pH 7.5 (left) and pH 7.0 (right).

Figure 7.16 Closer view of the wheat crop in Figure 7.15 growing at soil pH 7.5.

Figure 7.17 Leaves of the wheat crop growing at soil pH 7.5.

shorter, and all the wheat plants appear to be a pale yellow-green colour from a distance.

A closer inspection (Figures 7.16 and 7.17) of the plants grown at pH 7.5 shows that the overall appearance is due to a general chlorosis but also necrotic stripes running along the cereal leaves. This, and the overall lime-green colouration, are typical symptoms of manganese deficiency.

Just as Al^{3+} becomes more soluble at low pH (more acidic conditions), and less soluble at higher pH as the reactions in section 7.4.1 are reversed, so manganese (Mn^{2+}) precipitates out at higher pH as MnO_2, though other forms also occur (Adams, 1984):

$$2Mn^{2+} + O_2 + 4OH^- \leftrightarrow 2MnO_2 + 2H_2O$$

Note that in the process, the manganese is oxidised from the Mn^{2+} state to the Mn^{4+} state too. We will see later that this oxidation of Mn^{2+} may be reversed under anaerobic conditions. Indeed, sometimes rolling of soils may be used to create anaerobic aggregates as a way of overcoming manganese deficiency induced by over liming. For now, suffice it to note that manganese is an essential plant nutrient that the soil has to supply via plant-root uptake.

Manganese is not the only element that becomes less available at higher pH values. The solubility of iron (Fe^{3+}), another essential plant nutrient element, falls in a similar way to that of aluminium. However, manganese deficiency is generally a more common problem than iron deficiency. Agricultural advisors therefore usually suggest only adding sufficient lime to maintain soils in a very slightly acidic state, typically pH 6.2 in Scotland or pH 6.5 in England, for example. The lower value in Scotland reflects the higher soil organic matter contents generally found at more northern latitudes, as organic matter affects manganese availability. Other trace elements essential to plants, such as zinc and copper, also become less available under alkaline conditions.

7.4.3 Soil pH effects on other cationic nutrient elements

The essential nutrient element base cations, calcium (Ca^{2+}), magnesium (Mg^{2+}) and potassium (K^+), attached to the negatively charged cation exchange sites of the soil solid phase are relatively much more available for plant uptake than those within the structures of the soil minerals discussed in Chapter 1. In acid soils, however, H^+ and cationic aluminium species dominate a higher proportion of the exchange complex, so bioavailability of base cation nutrient elements and sodium is generally lower.

The availability of calcium also tends to be lower if the soil has a pH value above 8.5. This is because calcium carbonate is sparingly soluble and starts to precipitate out as calcite at this pH. Thus the cation exchange sites of soils at pH values above 8.5 tend to become progressively more and more dominated by sodium (Na^+), making Ca^{2+} less available. Magnesium tends also to become less available for similar reasons.

7.4.4 Soil pH effects on anionic nutrient elements

Lead-in question

When OH^- (hydroxide ion) concentration increases at high pH, what effect do you think this would have on the mobility and availability of nutrient elements absorbed in soils in anionic forms?

Several important nutrient elements, namely sulphur, phosphorus, boron, molybdenum and nitrogen, occur in negatively charged anionic forms in soils. Anionic species with multiple negative charge, such as phosphorus species (HPO_4^{2-}) or sulphate (SO_4^{2-}) especially are held to a substantial extent by anion exchange reactions whereby they displace hydroxide (OH^-) ions from hydrous oxides of aluminium and iron or the surfaces of silicate minerals. The equations are typically written as:

$$Al(OH)_3 + SO_4^{2-} \leftrightarrow AlOHSO_4 + 2OH^-$$

or

$$Al(OH)_3 + H_2PO_4^- \leftrightarrow Al(OH)_2H_2PO_4 + OH^-$$

As the pH of a soil increases, the presence of more OH^- anions shifts the equilibrium back to the left, so phosphate and sulphate tend to become more plant-available in slightly alkaline soils. Conversely they are strongly absorbed (fixed) in acid soils. However, for phosphate in soils between pH 6.5 and 8.5, calcium phosphate starts to precipitate out; higher phosphate concentrations only tend to be found therefore above pH 8.5 when sodium cations start to dominate both the soil cation exchange complex and the soil solution chemistry.

Boron occurs as $B(OH)_3$ and also as $B(OH)_4^-$ anions, so it too tends to be relatively less available under acid conditions. However, like phosphate, it tends to become unavailable in alkaline soils below pH 8.5 which have with high calcium concentrations. Indeed boron deficiency is sometimes induced by over liming. Figure 7.18 shows small spruce trees in an experiment in Sweden where too much lime had accidentally been added to overcome effects of acid rain. The trees developed serious boron deficiency which slowed growth to such an extent that lichen growth was substantially enhanced.

Figure 7.18 Lichen growth established on spruce trees where growth has been almost totally stopped as a consequence of boron deficiency induced by over liming.

Molybdenum exists in soils as the molybdate anion, MoO_4^-. As soil pH, and thus OH^- concentration, are raised, the availability of molybdenum increases exponentially.

The high mobility of the nitrate anion (NO_3^-) compared with that of the cationic form of N as ammonium (NH_4^+) was discussed at length in Chapter 5 and need not be considered in the context of anion exchange, which is generally regarded as of little or no consequence for nitrate. Once produced in soil, nitrate is generally highly mobile.

7.4.5 Soil pH, microbial activity and nutrient supplies

When we were considering the nitrogen cycle in Chapter 5, we saw that typically 95 per cent of the nitrogen stored in soils is in the form of organic matter nitrogen. Because of this the effects of soil pH upon activities of microbial species in soils are very important to plant N supply. Cresser *et al.* (2008) have recently suggested that plant biodiversity in natural ecosystems depends upon a close match of the dynamics of soil mineral N supply to those of plant N uptake requirements. If this hypothesis is correct then the effects of soil pH on litter decomposition rate generally, and on ammonification and nitrification in particular, must be part of the evolutionary process. Mineralisation is much slower under acidic conditions, where fungi dominate over bacteria, so nitrogen availability is lower in more acidic soils such as the podzol in Figure 7.3. Although bacteria and actinomycetes flourish under alkaline conditions, nitrogen availability also falls at high pH; partly this is because of the effect of pH on nitrifying organisms and partly because ammonia (NH_3) may be lost at high pH by volatilisation.

It is important to realise that nitrogen is not the only element supplied by the decomposition of plant litter and other organic residues. Such processes are also important to the supplies of phosphorus and sulphur, and indeed of all other major and trace nutrient elements. But remember that microbes need nutrients too.

7.4.6 What naturally controls soil pH?

As soils evolve naturally from their parent minerals, their pH will change quite naturally over time. The resultant pH will depend on the five major factors of soil formation, namely:

- Parent material
- Climate
- Time
- Vegetation
- Topography.

To see how these factors interact to give soils of a particular pH at any point in a landscape, we need to consider how base cation elements are cycled naturally, as indicated in Figure 7.19.

We saw in Chapter 1 that the calcium contents of rocks and minerals vary over a wide range. Thus the rates of release of calcium and other base cations from a rock such as quartzite are very low compared with their rates of release from basic and ultra-basic rocks, which are in turn lower than rates of release from limestone or dolomite. Carbonic acid, H_2CO_3, is formed by dissolution of the elevated concentrations of carbon dioxide found in the soil atmosphere from decomposing organic matter. It is a major factor regulating losses of cations such as Ca^{2+} from soil systems, since it provides the charge-balancing counter anion (HCO_3^-). Thus $Ca(HCO_3)_2$ and $Mg(HCO_3)_2$ are potentially mobile products of weathering (Reid *et al.*, 1981).

Whether or not these weathering products leave the soil depends upon climate. In periods where the amount of precipitation (rain, snow, etc.) exceeds the amount of water returned to the atmosphere via evaporation and transpiration, drainage and runoff will occur and transport $Ca(HCO_3)_2$ and $Mg(HCO_3)_2$ out of the soil into surface waters or down to groundwater. Alternatively, on sloping sites, the weathering products may at least partially move down slope in laterally flowing water, functioning as base inputs to lower slopes. Thus over many years the base status of soils may fall, at a rate dependent upon climate, position on slope, topography and carbon cycling within the soil system.

The rate of loss of base cations from the ecosystem will, however, be reduced by retention on negatively charged cation exchange sites associated with the soil solids and by the uptake of base cations by vegetation. Nevertheless the long-term net result will be soil acidification over the long term wherever precipitation exceeds evapo-transpiration.

On the other hand, in regions where evapotranspiration exceeds precipitation on an annual basis, for most of the year biogeochemical weathering products cannot leave the system. They must be retained within the soil profile, where they will react at elevated pH to precipitate calcite:

$$Ca(HCO_3)_2 \rightarrow CaCO_3 + H_2O + CO_2$$

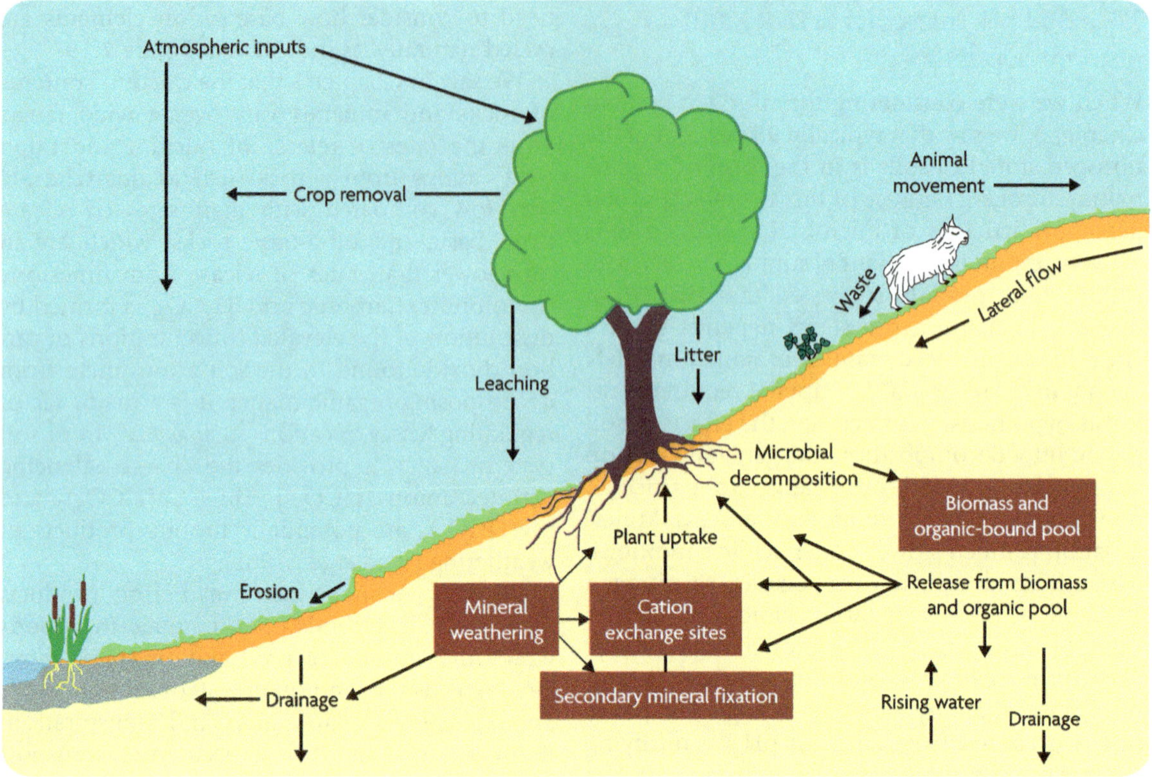

Figure 7.19 Schematic representation of the cycling of a base cation such as calcium (modified from Cresser *et al*., 1993).

Such soils will remain alkaline since they contain more than sufficient alkalinity to neutralise any acid inputs relating to, for example, base cation uptake by plants.

In such arid areas the input of soluble salts, especially sodium chloride, as wet and dry deposition from the atmosphere can lead to the development of saline soils (Fullen and Catt, 2004). High salinity from such salt accumulation may seriously restrict the growth of plants, as in the field outside Perth in Western Australia shown in Figure 7.20.

Figure 7.20 Effect of salinity build up on plant growth in a dry region of Western Australia. The accumulation of a white salt crust over the surface is clearly visible.

Source: Photograph courtesy of E.A. Fitzpatrick.

Salinity is a problem in many parts of the world in both natural and managed soil systems (Holden, 2005). The use of irrigation to allow sustained cultivation of saline soils is discussed later in Chapter 14, but it is worth mentioning here that, at a field scale, irrigation and drainage must be used in combination. It may seem curious on first thoughts to have to think about installing drains in dry, arid areas, but remember there has to be some way of taking the dissolved salts away from the site. If this is not done water rises back to the surface from the subsoil and salts are re-deposited back at the surface. It is important too that the water used for irrigation contains more calcium and magnesium than sodium. If it doesn't, then eventually the soil cation exchange sites will become sodium dominated, and the soil pH may then exceed 8.5 (the value buffered by the presence of calcite). The soil pH then might rise as high as 10.

7.5 The physical properties of soils

Lead-in question

What else do soils provide for plants apart from nutrient elements?

Soils contain three phases, the soil solids (a solid phase), the soil atmosphere (a gaseous phase), and soil solution (a solution phase). Nutrients may pass from the surface of soil solids via the solution phase to plant roots or microorganisms.

7.5.1 The soil atmosphere

The soil atmosphere is very important to both plants and microbes as it has to provide sufficient oxygen for respiration. We saw the importance of decomposition of organic residues in soil when discussing the carbon cycle in Chapter 4. The conversion of organic carbon to carbon dioxide (CO_2) obviously consumes oxygen. The oxygen consumed must be replaced by entry of air from the atmosphere above the soil surface for respiration to continue.

The entry of atmospheric oxygen by diffusion may be quite rapid if the soil has a well developed aggregate structure (or plenty of coarse-sized solid particles) and a network of pores interconnected in three dimensions. However, this is only true as long as the pores are not full of water. Diffusion of oxygen through water is much slower than its diffusion through the typical mixture of nitrogen and carbon dioxide gases in the soil atmosphere. Thus soils that are waterlogged for extended periods often become anaerobic. This will change the microbial population so that only facultative or obligate anaerobes will function, and plants may be damaged or eventually killed off. Nitrate is reduced to dinitrogen gas (N_2) or nitrous oxide (N_2O) by denitrifying microbes, a process known as denitrification. Organic matter starts to decompose to produce organic fatty acids and esters, and often the gas ethylene ($CH_2{=}CH_2$), which can seriously damage plants. Probably more house plants are killed by over watering than by under watering, especially as the ethylene induces epinastic curvature of leaves which to the inexperienced often is mistaken for drooping due to drought.

Figure 7.21 shows a clod of soil taken from an anaerobic soil profile. We saw in section 7.4.2 that manganese (IV) can be reduced to manganese (II) in anaerobic soils. Similarly iron (III) is reduced to iron (II):

$$Fe^{3+} + e^- \leftrightarrow Fe^{2+}$$

Figure 7.21 A clod from a poorly draining, gleyed soil showing the effect of reduction of Fe^{3+} on matrix colour and local orange-brown zones of re-oxidation.

Because there is so much $Fe(OH)_3$ in soil, which imparts its intense orange-brown colour to soil horizons, reduction of Fe^{3+} to Fe^{2+} under anaerobic conditions causes a dramatic colour change in soil. Hydrous oxides of Fe^{2+} are a dull olive-grey colour. This can be clearly seen in Figure 7.21. However, there are also bright orange patches around old root channels or worm holes, where oxygen re-entry was easier, and locally Fe^{2+} has been re-oxidised.

7.5.2 Soil texture

The properties of mineral soils vary markedly with their particle size distribution. In the UK, and indeed universally, any mineral particles with a diameter < 2 mm are classified as soil particles. Above 2 mm the particles become stones or gravel. If all the particles are between 0.6 and 2 mm in diameter, they would be described as coarse sand in the UK. The United States Department of Agriculture, USDA, however, further splits coarse sand into very coarse sand (1 to 2 mm) and coarse sand (0.5 to 1 mm). Note the slightly lower limit. Medium sand is from 0.2 to 0.6 mm in the UK, or 0.1 to 0.5 mm in the USDA system. Particles from 0.06 to 0.2 mm are medium sand in the UK, but this term is applied to particles from 0.1 to 0.5 mm by the USDA system. Fine sand particles are from 0.06 to 0.2 mm in the UK system, but 0.05 to 0.1 mm according to the USDA. Smaller particles down to 0.002 mm are known as being silt sized, while those even smaller are clay-sized particles in all systems. There is a third classification system, the international system (Cresser, Killham and Edwards, 1993), but we needn't consider that here.

If you take a large pinch of a soil that is strongly predominated by sand-sized particles in the palm of your hand, the particles do not stick to each other when dry. If you rub the soil between the thumb and first finger when wetted with water, it feels harsh and gritty. The moist soil can be shaped into a ball, just as you could make a sandcastle on the beach, but if you try to role the wet ball about in your hand, or even try to lift it from the sides, it disintegrates (i.e. the particles are not at all cohesive, Figure 7.22). Nor are the wet particles adhesive (they do not stick firmly to your hand), so if you rub a thumb across the palm it would leave a clean line (Figure 7.23). Such soils retain water very poorly so are prone to drought problems when cultivated. They dry out to individual grains rather than aggregates, so would be prone to erosion in strong winds when dry. When very wet, however, they would be susceptible to erosion by water because of their lack of cohesive properties.

Assessing the *dominant* particle size of a soil from how it feels and handles when adjusted to a suitable moist state is known as hand texturing. The resultant name is known as the soil's texture.

(a) (b)

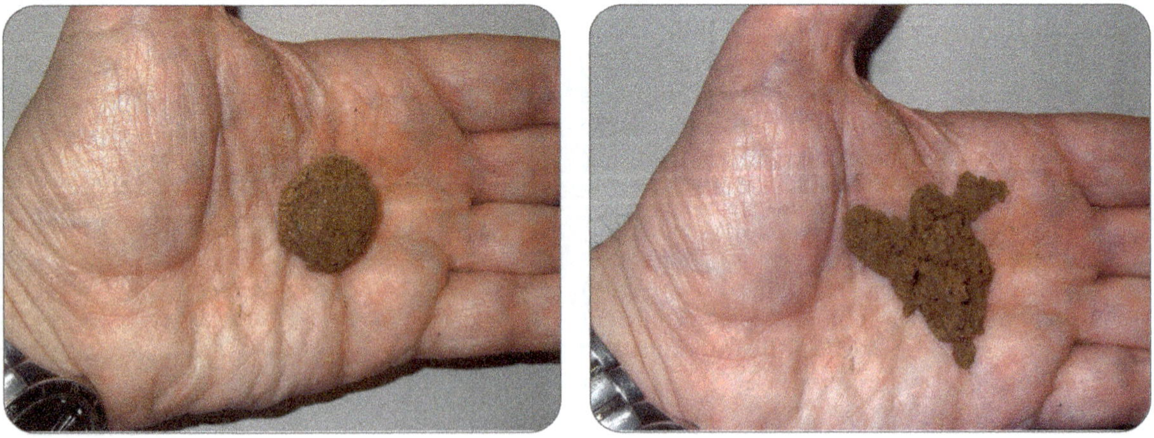

Figure 7.22 Appearance of a coarse sandy soil moulded into a ball (a) and after trying to pick the ball up during texture assessment (b). Some individual particles can be clearly seen because of lack of cohesion.

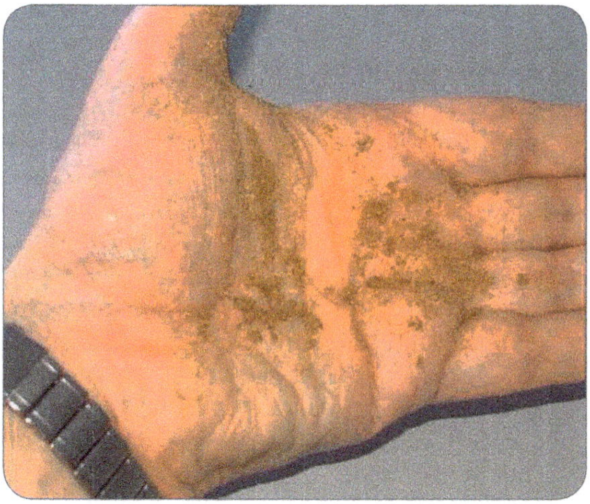

Figure 7.23 The same coarse sandy soil as in Figure 6.22 is not adhesive, so if you rub a finger across the soil residue on the hand it leaves a clean line (vertical here).

Figure 7.24 A selection of sand-sized particles fractionated by dry sieving. The single dish at the top contains grit and stones that would not pass through a 2-mm sieve. The sand grains then pass from coarse to fine, the latter having passed through a 0.068 mm sieve (lower right). These then are all less than 0.068 mm in diameter.

The main characteristics assessed during hand texturing are grittiness, smoothness, cohesiveness and adhesiveness. Texture names are universal, and give an immediate indication of risk of erosion by wind or water, potential drainage and drought susceptibility problems, and potential cultivation problems. We will therefore look at a few more examples so that this will become clearer.

Sand-sized particles, as we have seen, cover a quite wide range of particle sizes, going down from coarse sand (2 mm) to fine sand (0.05 mm or 0.06 mm). All feel gritty and abrasive when rubbed in a moist state, though coarse sand feels appreciably harsher and most potentially abrasive. In hand texturing we look for *dominant* characteristics, so a small number of coarse sand particles would not put 'coarse sand' or 'coarse sandy' in the texture name. Figure 7.24 shows the appearance of the full range of sand sizes. These were separated by dry sieving, but dry sieving has practical limitations for soils which often contain quite stable aggregates. Moreover, sieving could not be used for clay-sized particles, because they are simply too small.

The shapes of mineral particles are generally very irregular, so we use the term 'diameter' rather loosely. Often the term 'equivalent spherical diameter' is employed, because soil particle size distributions are often measured quantitatively by allowing the particles to sediment out in cylinders of water at constant temperature (White, 2006). In such methods the sedimentation rate calculations assume that all particles are spherical.

At the other end of the particle size range scale to sands, we have seen that the smallest particles are called clay sized, or just clay. Soils with a strongly predominant content of clay-sized particles are very cohesive (i.e. the particles stick to each other). Thus once water has been thoroughly mixed into the soil, which often takes several minutes, the ball of soil formed can be rolled into threads a mm across, which may be wound into a spiral without breaking, as in Figure 7.25. The pore size between particles is minute, so water drains through such soils very slowly. In Figure 7.25 water added to the small moulded bowl shape is not draining through the soil.

If a finger is rubbed across the surface of a suitably wetted clay soil it leaves a silky shine, as seen in Figure 7.26. Because the particles are much more cohesive than adhesive, the residue left on the hand after texturing is confined to lines in the skin (Figure 7.27).

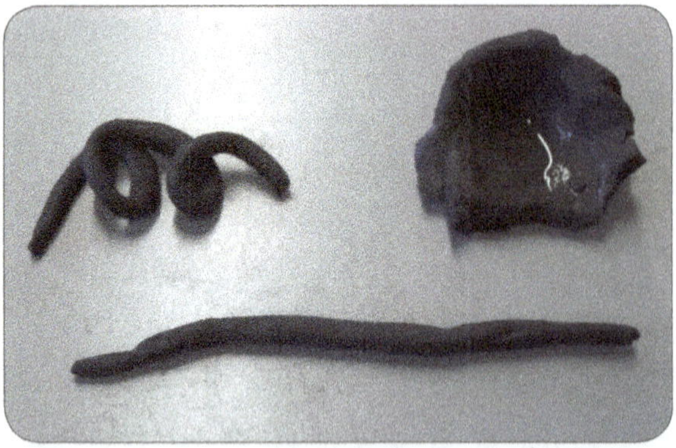

Figure 7.25 Clay soils are readily rolled into threads that can form rings, or indeed into other shapes.

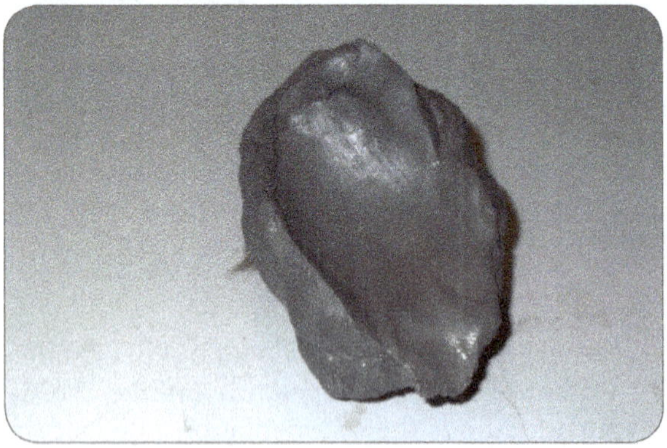

Figure 7.26 Clay soils are said to 'take a shine' when moist soil is rubbed with the finger.

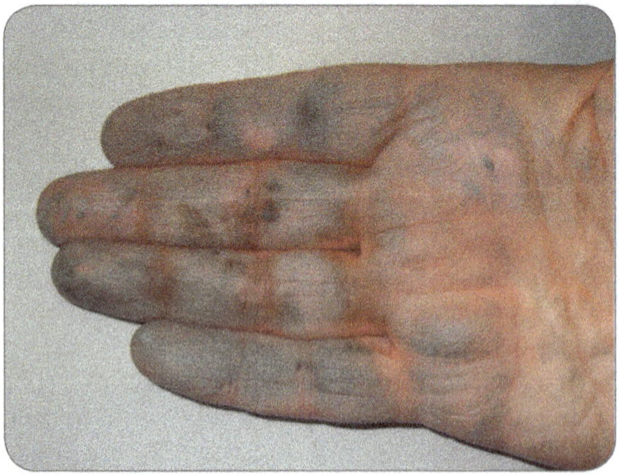

Figure 7.27 Clay residue only left in lines on the skin, as clay is much more cohesive than adhesive.

Because clays allow extremely slow water penetration when wet they are often waterlogged and may become anaerobic. This was the case in the clay soil in Figures 7.25 and 7.26, and the waterlogging has reduced the Fe^{3+} to Fe^{2+}, which is responsible for the soil colour in this sample. However, clay soils are not always anaerobic, so colour alone should never be used as a guide to texture. Rely on the 'feel' of the moist soil (Batey, 1984).

Clay soils can present a number of cultivation problems to the farmer because of their strong cohesive properties. They can form massive blocky structures as they dry out, making it difficult to create a good seed bed. Often cracks can be seen at the surface of soils with a high clay content, as in Figure 7.28. When wet, because they often restrict water infiltration and drainage to an excessive extent, fields with heavy textured clay soils often require installation of field drains. The feasibility of that depends upon local topography of course, as the water must be able to drain under gravity.

Silt-sized particles lie in size between the small diameter of very fine sand particles and the even smaller diameter of clay particles. When very dry, silt particles can have a very fine powdery feel and, as they are so fine, they are very prone to wind erosion. When slightly moist a very easily handled, friable structure develops, as shown in Figure 7.29. However, when more water is worked

Figure 7.28 Because of their cohesive properties, when clay soils dry out they often shrink to form large structural blocks, as here. In a profile these may form large, elongated prism-shaped structural units with flat vertical faces.

Figure 7.29 A very slightly moistened silt soil forming a very easily handled and friable structure.

into the soil it acquires a smooth, slippery, silky feel, not unlike that of wet soap. It also becomes incredibly sticky and cohesive when being rubbed between thumb and fingers (Figure 7.30).

It is easy to realise why silty soils have to be managed very carefully. When too dry they are prone to erosion by wind, especially in the absence of protective crop cover. When slightly moist they are fairly easy to cultivate, but cultivation, when too wet, can be a farmer's worst nightmare as farm machinery gets coated with the extremely sticky soil. Conditions need to be just right to form a good seedbed. Soils with high contents of both silt and clay are similarly hard to cultivate, but tend to form more stable aggregates when they dry out, so are less prone to wind erosion. One consolation for silt soils is that they retain water better than sandy soils because of their smaller pore sizes, so they are less prone to drought problems in climates that are only slightly moist.

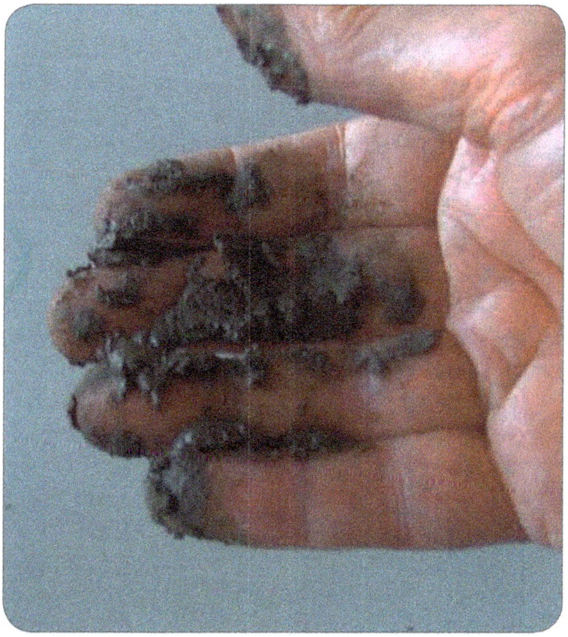

Figure 7.30 Silt soils have a silky feel when rubbed in a moist state, but are very sticky (adhesive) rather than having the more cohesive properties of clay. Hands tend to get very messy when texturing silts or silty clay soils!

For simplicity, so far we have discussed soil texture to a large degree as if soils only contained sand particles or silt particles or clay-sized particles. In the real world such soils would be exceptional, and most soils contain a mixture of sand-, silt- and clay-sized particles. When hand texturing we look for the dominant characteristics of the moist soil detected by feel. Thus, suppose a soil feels gritty and can be formed into a ball in the palm of the hand, but the ball can be rolled about and passed from hand to hand without breaking up; it is clearly a little more cohesive than a simple sand. However, if it can't be rolled into threads the amount of clay-sized particles is far from dominant. The same soil might leave the hands very slightly stained, even though it does not feel sticky or smooth. Such a soil would be a loamy sand.

Loam evolved as a term for a soil with an excellent balance of sand, silt and clay characteristics, with no one being dominant. It thus would be a near ideal soil for cultivating plants with good retention of moisture and nutrients (we'll discuss the reason for that later), a tendency to form nice small and quite stable aggregates, good drainage, low erosion risk, and ease of cultivation and seedbed preparation. Loam is the gardener's dream soil (and thus sold in bags at a premium in garden centres!).

You may have noticed that the term 'loamy' was used as an adjective above when we described a soil as a loamy sand. This means that the cohesiveness was better than that of a sand, but still too weak for the soil to be described as a sandy loam. The latter would, in turn, be slightly more gritty, but less cohesive, than a loam. 'Silt' or 'silty' and 'clay' are also used as adjectives in this way, so you might like to consider what the characteristics of a silty clay would be compared with those of a silty clay loam. Which would drain better? Would either feel very harsh? How cohesive and adhesive would each be? Which would be easier to cultivate? You can find an excellent detailed description of how to hand texture soils in a short monograph on soil husbandry by Batey (1988).

7.5.3 The importance of soil texture evaluation

Lead-in question

Does soil texture have any importance other than for assessing potential risks of erosion by wind and/or water, water retention, drought susceptibility, drainage problems and cultivation difficulties?

You should be convinced by now that although simple, inexpensive (or even free), rapid, and able to be done in the field, hand texturing is extremely informative for evaluating soil quality and potential soil management problems. It has one other important value however, not incorporated into the above question. Namely it can provide an indication of the ability of a soil to retain nutrient elements.

If ever you have tried to dissolve large, water soluble crystals in water, you may well have found that it took a long time and lots of vigorous shaking or stirring. This is because a given mass of material in large particles has a much smaller surface area than the same mass in much smaller particles. We can show this very simply by comparing the surface area of a cubic crystal 2 mm by 2 mm by 2 mm with the total surface area of 1000 cubic crystals each 0.2 mm by 0.2 mm by 0.2 mm. Both would have a volume of 8 mm^3 (1 cube with a volume of × 8 mm^3 in the first case or 1000 little cubes each with a volume of 0.008 mm^3 in the second case). Therefore if density was the same, the mass of the single cube would be identical to the mass of the 1000 little cubes. The surface area of the bigger cube would be 6 × 2 mm × 2 mm, or 24 mm^2, because a cube has 6 faces. The total surface area of the 1000 smaller cubes would be 1000 × 6 × 0.2 mm × 0.2 mm, which is 240 mm^2, i.e. 10 times larger. With a much larger surface area to react with water molecules, we would expect the smaller cubic crystals to dissolve much more readily.

When mineral crystals are broken down into smaller fragments, we saw in Chapter 1 that negatively charged sites are created at the surface of the crystal fragments. To maintain charge neutrality at the surface, positively charged cations are retained at the particle surfaces. However, these cations are exchangeable, so if we added an excess of potassium cations (K^+) or ammonium cations (NH_4^+), these cations could be absorbed, displacing the exchangeable cations initially present. It is intuitively obvious therefore that clay-sized particles will make a major contribution to the cation exchange (CEC) of a soil. Soils with higher clay contents generally retain added cation nutrients much more effectively than more sandy soils. Therefore the risk of losses of added cationic fertilisers during subsequent rain showers is much higher for sands and loamy sands than it is for loams or clay loams.

Newcomers to the science of soils sometimes get confused between the mineralogical term 'clay minerals', which was introduced in Chapter 1, and the term 'clay' as used in the context of particle size analysis. Clay minerals may be clay-sized particles, but they may be orders of magnitude larger too! Conversely, many clay-sized particles in soils will not be clay minerals. We saw in Chapter 1 that there are also cation exchange sites on some clay minerals associated with isomorphous substitution.

The particle size distribution of soils generally depends upon the conditions under which the soil was formed and/or the parent material from which it was formed. For example, the silt in Figures 7.29 and 7.30 came from a soil that evolved in the distant past from an old river estuary that was very broad. The slow moving water in the river had allowed silt-sized particles to deposit and accumulate in the river bed. Sand-size particles had deposited elsewhere, from slightly faster moving water up river, whereas clay-sized particles had tended to remain in suspension.

Particle size distribution is not just important to the evolution of CEC in soils. In Figure 7.19 we saw that mineral weathering is important to maintain the supply of elements in soils as they evolve over time. Mineral weathering too is favoured generally by smaller particle sizes and their associated larger surface area. However, the situation is not quite as straightforward as the simple dissolution of crystals in water, because as we saw in Chapter 4, carbonic acid production from dissolution of respired carbon dioxide in the soil solution has to provide the associated mobile anions (HCO_3^-). Thus mineral biogeochemical weathering will proceed in the films of moisture enveloping soil mineral particles in between rainfall events, making time slightly less critical.

7.6 Soils as suppliers of plant nutrients

In Chapter 4 we saw that plant biomass production in the carbon cycle depends upon solar energy providing the energy needed for photosynthesis to proceed and on the availability of water and of carbon dioxide present in the atmosphere. Production of the plant organic matter also requires a substantial amount of nitrogen, which is tightly recycled in natural ecosystems as described in Chapter 5, and supplemented via fertiliser applications in agricultural soils at rates sufficient to replenish the amounts removed in harvested crops. Other plant nutrient elements are supplied via the biogeochemical weathering of soil minerals or from decomposing organic residues as was indicated in Figure 7.19, for example, for base cation nutrients such as calcium, magnesium and potassium. These may be supplemented to a modest extent by atmospheric inputs of terrestrial dust or aerosol material of oceanic origins which provides useful inputs of magnesium, sodium, sulphate, chloride and boron.

In the days prior to the invention of synthetic chemical fertilisers, 5- or 7-year rotations would be used in fields to help to sustain crop production. These rotations would have 2 or 3 years where no crop would be harvested, but grass and clover would become established, and used for grazing animals. Growth of the clover would help restore the soil nitrogen status, and organic matter recycling in the fallow periods would allow at least partial restoration of the base cation status of the cation exchange sites as weathering would proceed without crop removal. Return of composted animal and/or human waste to the fields would also help maintain soil fertility. The biggest problem arises if crops and grazing animals are removed from the site and consumed elsewhere with no waste recycling locally. A fattened cow is not just a cow; it's a walking pile of calcium, magnesium, potassium, phosphorus, sulphur, nitrogen and trace elements. That's why for many years gardeners have been keen to use dried blood and bone meal as 'natural' slow release organic fertilisers, and why urbanisation may be regarded as the biggest problem in the nitrogen cycle (Cresser *et al.*, 2008).

7.7 The dynamic nature of soils

The point was made at the start of this chapter that soil is not regarded by most people as an exciting medium teaming with biological activity that is crucial for all life on the planet. Few would regard it as being at all dynamic. Most would never have questioned how it might have formed in the first place or how, for example, the 12 tonnes or thereabouts of organic nitrogen it contains per hectare ever got to be here. If the climate change problem has a silver lining it is probably because it has made many more of us aware of the importance of at least maintaining the storage of vast amounts of carbon in the world's soils.

Soils do not just change over timescales appropriate when considering soil evolution however. They also may be dynamic over much shorter timescales. The pH of soil solution, for example, may change significantly over a few hours as a soil dries out, thereby increasing mobile anion concentrations and the concentration of H^+ in solution in an acidic soil. A storm sweeping in from over the ocean may deposit high concentrations of sea salts in rainfall; this too may very rapidly (but temporarily) lower the pH of acid forest or moorland soils by increasing mobile anion concentrations. The extra H^+ ions in solution come from cation

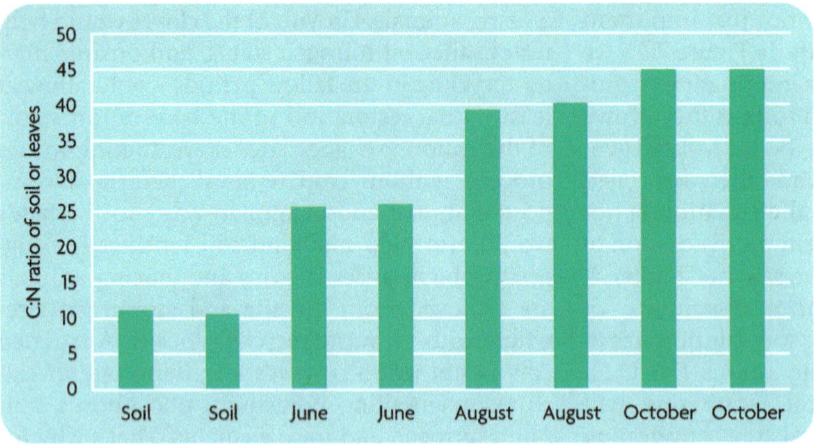

Figure 7.31 Soil and leaf C:N ratio for hazel tree foliage sampled in June, August and October, the last just prior to litter fall.

exchange sites in the soil as Na^+ and Mg^{2+} of the sea salts are absorbed. Cation exchange reactions are very rapid.

Plant/soil systems have evolved together in natural systems to ensure that essential nutrients such as nitrogen are provided as and when the plant needs them. Thus litter from deciduous trees has a high C:N ratio (about 40 to 50) when it falls in winter, compared with the much lower C:N ratio of leaves in summer (Figure 7.31).

As microbes start to decompose the litter, they tend initially to immobilise any organic N that is mineralised in the microbial biomass because of the very high C:N ratio of the litter substrate. This, plus the fact that litter decomposition is generally slow under colder, wintry conditions, helps conserve the soil nitrogen supply over winter and minimise nitrate leaching losses. As soils start to warm up in spring, the litter C:N ratio falls progressively until, just when the tree needs it in mid to late spring, mineral N species (ammonium and nitrate) start to be released in plant-available form.

7.7.1 Nutrient dynamics and soil fertility

We are now in a position to consider the nutrient dynamics for optimal soil fertility and sustainability. The soil must ideally be able to supply nutrient elements in plant-available chemical forms at the rate at which they are required by the plant for optimal growth and at a depth range that matches the root architecture in three dimensions. It is not useful, for example, if nitrate is being extensively produced either in winter when plant uptake is low or below the rooting zone at any time of the year. If it is, it is likely to be lost from the system by leaching.

In a similar way, it may be assumed that rates of release of nutrient elements by biogeochemical weathering or by decomposition of organic residues are likely to be quite well matched to plant uptake requirements in natural ecosystems. However, unlike the case for nitrate, cation exchange and anion exchange reactions are likely to help retain other nutrient ionic species liberated during colder winter months.

7.8 The importance of organic matter in soil revisited

We saw in Chapter 4 that not all of the carbon from organic residue decomposition is converted to CO_2 that escapes to the atmosphere or to carbonic acid to participate in soil mineral weathering. Some is incorporated into soil organic matter. In temperate parts of the world, agricultural soils typically contain organic matter to a level of 5 to 6 per cent. In hotter climates the figure may be lower, and in wetter and/or colder regions and in more acidic, semi-natural soils, where decomposition is slower, much higher.

In Chapter 4 we saw that organic matter content of soils is important because of its role in contributing to soil cation exchange capacity (CEC) and hence soil buffering capacity, and its importance to moisture retention and soil structural stability.

Its contribution to soil colour has a pronounced impact upon soil radiative properties. Its presence is crucial to sustaining soil microbial biomass and its associated contribution to the supply of plant nutrients.

These aspects need not be considered again here, though we are now in a better position to consider the importance of soil organic matter in the context of soil profile evolution.

7.8.1 Role of organic matter in soil formation

In the very first stages of soil evolution from physically weathering rocks organic matter is initially absent. The colonisation of bare rock by lichens is often the first stage of soil formation. Growth is slow, but the symbiotic relationship between algae and fungi allows fixation of di-nitrogen (N_2) from the atmosphere and photosynthetic production of organic matter along with slightly accelerated weathering of rock minerals. Additional organic matter may come in from insects, birds and small mammals and eventually the conventional carbon cycle comes into play, further facilitating biogeochemical weathering and allowing some organic matter to start to accumulate. The latter provides a source of mineralisable organic N and soil moisture retention, which in turn allow bryophytes, and eventually higher plants, to become established. Over century timescales a soil profile starts to evolve.

We can estimate roughly how long it might take for a soil to obtain the typical pool of nitrogen from the atmosphere that an uncultivated soil might contain today, possibly around 5000 kg ha^{-1}. If we assume that lichen and free living algae fixed only around 2 kg ha^{-1} per year we would only need 2500 years. If seeds of clover or other N_2 fixing plants were brought in by birds or animals, this period could easily be reduced. This build up of organic matter would be associated with generation of biogeochemical weathering rates typical of those observed at the present time. Of course, these calculations are over-simplified, in that they do not allow for any leaching losses of inorganic or organic nitrogen, or for any erosion losses, but they do suggest soil evolution is relatively rapid compared to geological timescales.

The soil type that evolves will depend upon the potential weathering rates of the minerals in the soil parent material. For rocks such as quartzite or granite, the capacity to release base cations such as calcium is limited and, as a consequence, the accumulation of exchangeable calcium on the cation exchange sites and the ability of weathering to replenish base cations lost to plant uptake or leached out with HCO_3^- (Figure 7.19) is also limited. If the climate is such that the amount of precipitation significantly exceeds the amount of water returned to the atmosphere via evapotranspiration, acidic conditions evolve, and a typical podzol profile will form, as shown in Figure 7.32.

The acid conditions slow the decomposition of organic matter as they are unfavourable for bacterial decomposition, although fungal decomposition

Figure 7.32 A podzol profile typically found in freely draining upland and lowland soils in moist climates where the parent material is a rock of low base status.

continues. As a consequence, organic matter starts to accumulate at the surface, producing a sequence of dark-coloured, organic-rich layers. The upper layer contains a substantial proportion of clearly identifiable litter fragments, and is thus known as a litter layer (or litter horizon). Below that decomposition is a little more advanced in what is traditionally known as an F, or fermentation, horizon. This in turn overlies a horizon in which the material is much more decomposed, with few identifiable litter fragments, known as the H, or humified, horizon.

The acid conditions in these upper organic horizons favour dissolution over many years of iron and aluminium hydrous oxides, leaving leached horizons containing predominantly white mineral residues generally rich in quartz. These can clearly be seen against a dark organic-rich horizon (the A horizon), and even more clearly where the leached mineral matter totally dominates, giving a pale, ash-coloured mineral E (eluvial) horizon.

Below the E horizon is a B horizon (an illuvial horizon) in which iron and aluminium hydrous oxides precipitate out, the Fe^{3+} imparting a bright orange-brown colouration which is clearly visible in Figure 7.32. Often organic matter is absorbed in this horizon too, though not in this example, the humus imparting a darker discolouration. At the bottom of the profile the apparently unaltered parent material is known as a C horizon.

7.9 Soil flora and fauna

The reason that the soil horizons are so distinctive in Figure 7.32 is because the acid conditions and the associated acidophilic vegetation residues are also unpalatable to earthworms. Earthworms play an important mixing role in soils at a higher pH, transporting plant residues to depth and subsoil material upwards as they move through the soil profile. They shred plant material and as they digest it the residues become very heavily colonised with soil bacteria, facilitating decomposition. Thus if a soil profile is developed from a more base-rich rock, mixing may be very thorough down to 25 or even 50 cm depth. Typically a Cambisol or Brown Earth profile develops, with only two horizons above the soil parent material C horizon. The upper A horizon is usually perceptibly, but only slightly, darker than the underlying B horizon. The vegetation colonising such a soil is generally appreciably different from that on the more acid pozols.

It is important to realise that soil profiles found are also very dependent upon local topography. In a hollow, for example, even from an acid parent rock, base cations may accumulate to a large extent, and wetness rather than acidity will slow down the decomposition of organic matter. Mineral subsoils may become waterlogged for much of the year and gleying is often seen. Initially a gley profile will establish, but eventually the wetness may slow decomposition to such an extent that a deep (> 50 cm) peat profile, known as a histosol, may build up. In the UK, at the time of writing, there is much interest in modifying the drainage characteristics of marshy peat bogs in an attempt to encourage peat accumulation as a way of increasing the amount of carbon stored in soils with a view to combating CO_2 build up in the atmosphere and the associated climate change. Care is needed, however, because another more potent greenhouse gas, methane (CH_4), is emitted from peat bogs under the prevailing anaerobic conditions.

7.10 Assessing soil fertility

Lead-in question

What is a fertile soil?

The concept of soil fertility is a little more complex than it seems on first thoughts. A soil might be regarded as 'fertile' if it is fit for purpose; agriculturally the purpose generally is growing healthy plants at a near optimal rate to give a near optimal yield. If you just think in chemical terms this means there must be a good match of the dynamics of nutrient availability and plant uptake. However, attaining such a match means that the chemical, physical and biological properties of the soil *all* have to be appropriate. Moreover, different plants have different soil requirements, for example different optimum pH, different nutrient needs, and different water demand. So while it seems easy to define soil fertility in terms of 'fitness for purpose', it is much harder to decide what has to be measured in practice if routine monitoring of soil fertility is required, for example as a check on the sustainability of management practices.

The 'purpose' in 'fitness for purpose' may not be crop or plant growth at all. It might be important, for example, to have a soil management target that encompasses both economic production of high quality crops while at the same time protecting surface and groundwaters within a soil's catchment. For soil fertility and the remainder of this chapter, however, we will assume that crop growth is the only target. At least then we can list the chemical, physical and biological requirements of a soil that can be regarded as fertile. This has been done in the following sections.

7.10.1 The ideal soil chemical properties

For a soil to be regarded as fertile, it should comply with the following chemical criteria:

- It should provide major nutrients (e.g. N, P, K, Ca, Mg, S) at the rates and in the forms required for optimal plant growth throughout growth.
- It should provide trace nutrients (e.g. Fe, Zn, Cu, B, Mn, Mo) at the rates and in the forms required for optimal plant growth throughout growth.
- Neither the soil solution nor the soil atmosphere will contain any chemical species at a concentration that adversely affects plant growth.
- The spatial and temporal distribution of plant-available nutrients in soil will be appropriate for the root architecture.
- It will be capable of retaining nutrients in a plant-available form.
- It will be capable of supplying all the nutrient needs of microorganisms as well as those of the target plant species.

It follows from the above list of chemical requirements that soil pH and base saturation will be appropriate for the plants being grown, that the soil will not contain high concentration(s) of soluble salts, that the soil will be sufficiently aerobic, and the soil will retain sufficient moisture. Note that a soil's physical and biological properties influence its chemical properties. Remember too that a soil's chemical properties will vary across three-dimensional space and time.

7.10.2 The ideal soil physical properties

The physical properties of a fertile soil should be as follows:

- The soil will provide firm anchorage for plant roots.
- There will be no significant impedance to root growth.
- The soil will provide sufficient water throughout growth to avoid water stress.
- The soil will remain aerobic throughout growth.
- It will be possible to plant and to cultivate the soil without causing physical damage.
- The soil will not be prone to erosion.

From this list it should now be obvious why a considerable amount of space was devoted earlier in the chapter to the discussion of assessment and importance of soil texture.

7.10.3 The ideal soil biological properties

A fertile soil should possess the following microbiological characteristics:

- Microbial activity will release nutrients by mineralisation at a rate and at depths appropriate for plant growth.
- Microbial activity will result in an appropriate rate of litter decomposition.
- Microbial activity will not produce any compound that reduces plant growth below the optimal rate.
- Soil microorganisms will not directly adversely affect plant growth (i.e. no plant pathogens).
- Soil microorganisms will not deplete nutrients to the extent that availability to plants is restricted.

With so many possible determinants it is not surprising to find that assessments of soil fertility often depend heavily on the farmer's first hand experience of how well his crops are doing compared with previous seasons, rather than on a full suite of chemical, physical and biological measurements. The latter would normally be prohibitively expensive.

7.10.4 The objectives of soil management

The primary aim of soil management has traditionally been to optimise soil fertility with respect to crop yield/quality. Commonly though, a compromise is made between crop yield or crop quality and optimisation of economic yield. Moreover it is increasingly being recognised that it may sometimes be desirable to compromise between optimisation of crop economic yield and minimisation of pollution of drainage waters (surface and/or groundwaters) or the atmosphere when formulating management policy. Other driving forces are also sometimes important, for example landscape or biodiversity conservation, or desirability of change to organic agriculture.

7.10.5 Quantifying chemical fertility

There is much interest in measuring the forms of metal ions in soil solution (both for metals we want and metals we don't). The total concentrations of elements present in soil are not useful for assessing their potential bioavailability, because a high proportion of the total may be locked up with soil mineral structures, and only released over centuries. The bioavailability of metal ions in soil solution depends upon the ionic species present. Metal elements may be present as simple cationic species such as Zn^{2+}, Cu^{2+} or Hg^{2+}, or as more complex cations and/or anions, or as neutral species, e.g. $HgCl^{+}$, $HgCl_2^{\circ}$, $HgCl_3^{-}$ or $HgCl_4^{2-}$ (in this example the Cl^{-} concentration would have a big impact on speciation). They may also be present as soluble organic complexes. Speciation may be modelled if matrix ion concentrations and pH are known. The distributions of species have a marked effect upon their toxicities to both microorganisms and to higher plants, and hence upon their eco-toxicity.

The relative abundances of different ionic species may vary considerably over time, depending upon both the absolute amount of the element present in the solution phase and also on how the concentrations and relative abundances of other ionic species vary. Therefore pot experiments and field experiments are often used to find chemical extractants which, if used under specified conditions, will extract an amount of the element of interest that correlates well with plant uptake of that element. Such procedures are called 'operationally defined', because the conditions (time, temperature, shaking conditions, etc.) must be followed exactly. They are widely used to assess the availability of major and trace nutrient elements in soils. Clearly it is desirable to know whether a deficiency or toxicity problem is likely to occur before planting the crop, so that remedial measures (e.g. fertiliser applications) may be made. This is obviously better than waiting for the crop to fail and then analysing the plant tissue to identify the problem just in time to be too late!

An example of an operationally defined procedure for assessment of soil phosphate status is the extraction of a fixed proportion of 'available' phosphate from soil by dilute ethanoic acid (more commonly still known as acetic acid). Typically 10 g of soil is extracted with 200 ml of 1M ethanoic acid by shaking intermittently for 2 h. The suspension is then filtered and the extracted phosphate determined by colorimetric analysis. The resultant phosphate concentration, after conversion to mg of P per g of dry soil, can be related to the available phosphorus status of the soil, and used as a basis for recommending fertiliser application rate.

Available zinc, for example, may be assessed by shaking a specified mass of soil with a specified volume of a solution of EDTA (ethylenediaminetetraacetic acid) of specified concentration for a specified time at a specified pH. DTPA (diethylenetriaminepentaacetic acid) is preferred by some researchers to test for available zinc in soils, but both procedures give useful information.

The selection of such procedures is purely empirical, though there may be tenuous links to theory (e.g. it could be suggested that the extracting powers of EDTA are similar to those of organic acids in root exudates).

It is important to remember that such extractants do not usually extract all the plant-available pool of an element; rather, they extract a constant fraction of the plant-available pool (constant enough for predictive purposes).

Other examples of operationally defined procedures for assessing element status include:

- Phosphate extracted with $NaHCO_3$ or an anion exchange resin.
- Boron extracted with boiling water.
- Copper extracted with 0.05M EDTA at pH 7 at 1:5 (m:v).
- Mg, P and K extracted with 2.5 per cent ethanoic acid (also referred to as acetic acid in earlier work).
- Cobalt extracted for 16 h with 0.43M ethanoic acid at 1:20 ratio (m:v); interpretation of the result is adjusted according to the drainage status of the soil sampled.
- Molybdenum extracted with 1M ammonium ethanoate at pH 7 at 20 °C for 16 h at a soil:solution ratio of 1:16.

To assess concentrations of potentially toxic elements, soil solution may be sampled directly by extraction using centrifugation techniques, or with a rhizon sampler by applying suction, as discussed in Chapter 17. These are inert teflon or ceramic porous tubes with a very small pore size to which a vacuum may be applied to suck through the soil solution. Alternatively it is possible to collect drainage water from a zero-tension lysimeter (water draining freely under gravity), again as described in Chapter 17. The reader might be forgiven for asking why soil solution doesn't provide the best indication of available nutrient element pools in soils. However, if sampled under growing plants, the result would depend upon plant uptake immediately prior to the sampling time, so available pool might not be assessed well at all.

Potentially toxic trace metals bound to exchange sites are generally regarded as being relatively bio-available, and may be assessed in a number of ways. Most commonly, shake-and-centrifuge techniques are used (typically three sequential steps). Magnesium solution (Mg^{2+}) is commonly used as the displacing trace element cation, as a 1M solution; it is effective at displacing divalent metals from the exchange sites.

The leaching methods commonly used for determining exchangeable base cations are less often used, as they are inconvenient for a sequential extraction procedure.

Simple experiments involving aqueous solutions in the absence of soil are sometimes used to help understand interactions between ions in solution. For example, Figure 7.33 shows how, in aqueous solution, pH affects the extent to which Zn co-precipitates with hydrous oxides of Fe^{3+} and Al^{3+}, and how it is even more effectively co-precipitated if phosphate is also present. It is difficult though to interpolate from what happens in soil solution to what happens in soils, because the latter are biologically active, and speciation of trace metal elements may change quite rapidly after they have been added to soils, to an extent that depends upon a whole range of soil chemical characteristics.

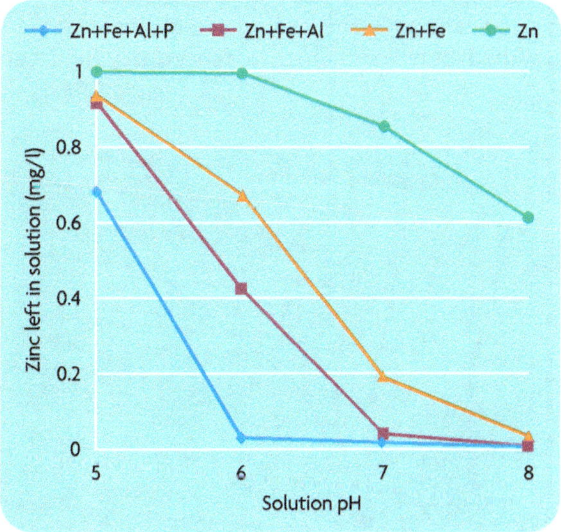

Figure 7.33 Changes in the solubility of zinc with increasing pH in the absence of other metals and phosphate, compared with its solubility in the presence of a small excess of iron, or excesses of iron and aluminium, or excesses of iron, aluminium and phosphate (from Cresser *et al.*, 1993).

7.11 The sustainable use of soil

Politicians keen to establish their 'green credentials' often aspire to have soil fertility monitored to show that soil fertility is not declining over the years. In the author's opinion such quests are ill advised, because of the complexity in quantifying soil fertility. They would be better advised to attempt to arrange monitoring to answer much more sharply focused questions.

An important fact to remember is that removal of a harvested food crop or farm animals or trees from a site automatically means removal of nutrient elements. The pools of elements thus removed must be replaced, if not from weathering or biological fixation of N_2 from the atmosphere or atmospheric deposition, then from fertiliser use. Whatever the source of nutrient replenishment, it is important in the long term to keep nutrient supplies appropriate to the current land use and also to maintain nutrient balances. We will see in the chapter on food production that this is often not easy, especially when foodstuffs are being moved all around the world. Soil sustainability is an issue that needs to be addressed at global scales as well as at local scales.

CASE STUDIES

A case study in Yorkshire – How does free range chicken production affect an acidic farmland soil?

Jennifer Beeston was interested in whether free range chicken production had beneficial effects on farm soils. She located a farm where chickens had been contained in fenced-off field sections for known, but different periods of time. She sampled soils in duplicate, from 0–10 cm, 10–20 cm and 20–30 cm, from areas where there had never been chickens and areas where chickens had been present for 2 years, 6 years, or over a much longer period. The soil samples were then analysed in triplicate, so each point in the plots below is the mean of six values.

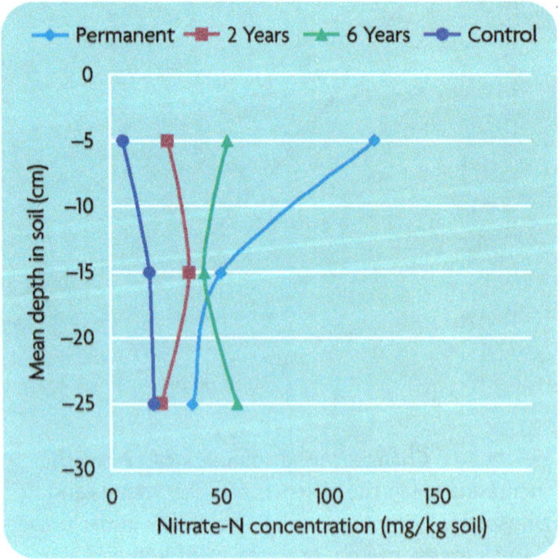

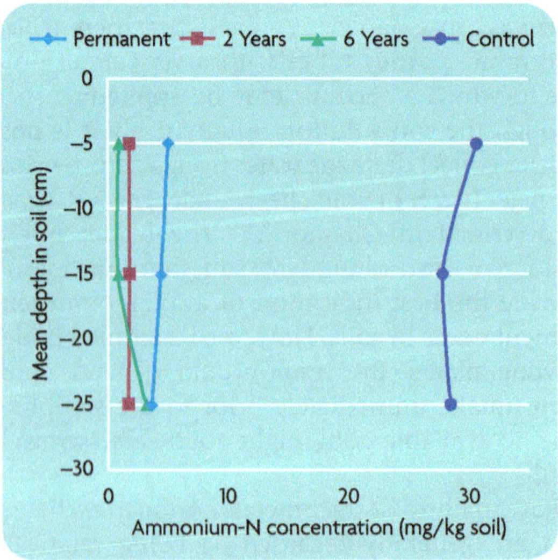

The plots show that compared with the chicken-free control soil, there was much less extractable ammonium-N at all three depths, even after only 2 or 6 years. However, the extractable nitrate-N concentrations increased significantly. When Jennifer measured the soil pH, she found that the areas under chickens were much less acidic, which had increased the nitrification rate. This was almost certainly due to the liming effect of calcareous grit fed to the chickens to ensure that they laid hard-shelled eggs.

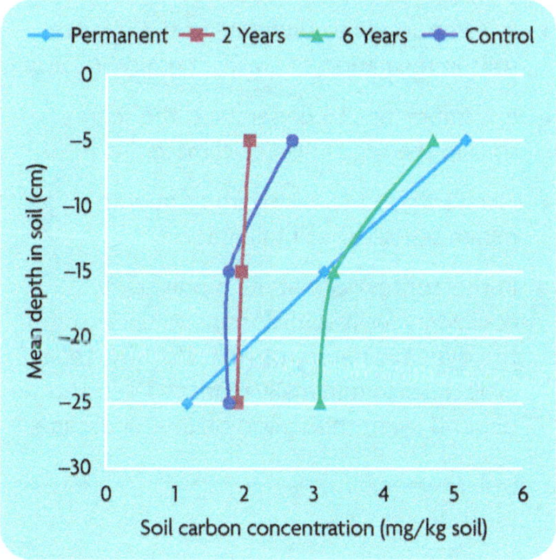

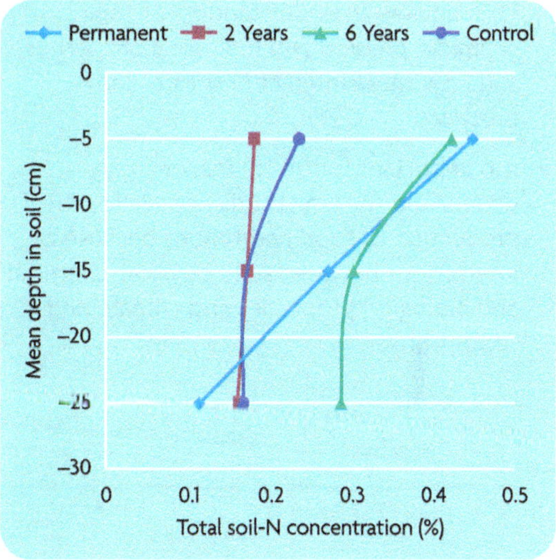

The second pair of plots above show that, over a period of 6 years or longer, the organic matter content of the soil increased substantially because of the presence of the free-range chickens, and the effect could be seen down to a depth of 30 cm. It was observed, when samples were being taken, that earthworm activity was higher in the 6-year and permanent housing sites, although lime effects from calcareous grit would, in any case, be expected to slowly penetrate to 30 cm depth over a period of 6 years. Statistical analysis confirmed the significance of the trends discussed. The results are interesting, because they suggest that careful consideration needs to be given to whether free-range chicken production in such soils could aggravate problems of nitrate leaching to surface waters and possibly to groundwaters too.

POLICY IMPLICATIONS

Soils

- Policy makers need to be aware of how the world's soils should be managed to optimise their sustainable uses for diverse purposes. This requires an understanding of their chemical, physical and biological properties and how these interact in soils as they function.
- It is important for policy makers to understand the limitations of soils for the production of food and timber because of interactions between soil's physical, chemical and biological properties and climate and topography.
- Policy makers need to be aware of the finite capacity of soils to function as pollution sinks over the long term.
- It is important to understand the roles of soils in regulation of water chemical quality.
- It is important to understand the roles of soils in the regulation of climate.
- Planners need to be aware of how to minimise risk of soil salinisation.
- In the context of natural biodiversity conservation, it is important to understand the dynamic nature of soils and how their properties in minimally managed ecosystems depend upon their position in a landscape.

CHAPTER REVIEW EXERCISES

Exercise 7.1

The volume (V) of a sphere of radius r is given by the equation:

$$V = 4\pi r^3/3$$

Its surface area (A) is given by the equation:

$$A = 4\pi r^2$$

Derive an equation for R, the ratio of surface area to volume, and calculate the value of R when r is 0.001 mm, 0.01 mm, 0.1 mm and 1 mm. Plot a graph of R against r on a log/log scale.

What does this graph tell you about the effect of particle size upon surface area?

Exercise 7.2

Soils were sampled at 10 cm intervals to a depth of 80 cm from a soil profile under an ancient acid grassland outside York in England. The pH values of the moist soil samples were measured as pastes both with deionised water and with 0.01M calcium chloride solution. The table on page 183 shows how soil pH varied with depth.

Depth (cm)	pH in water	pH in $CaCl_2$
0–10	3.92	3.52
10–20	3.99	3.60
20–30	4.15	3.75
30–40	4.28	3.87
40–50	4.42	4.12
50–60	4.53	4.28
60–70	5.01	4.84
70–80	5.45	5.30

Plot graphs of soil pH versus depth for the two methods of pH determination and explain the spatial trends observed and the difference between the two sets of results.

Exercise 7.3

Why is it very difficult to design experiments with soils to show the extent to which the concentration of H^+ ions in soil solution directly influences the growth of a particular plant species? Do you think we need such experiments in practice to optimise food production?

Exercise 7.4

The author encourages students to see cows in a field as large, walking piles of nitrogen, phosphorus, sulphur, calcium, magnesium, potassium, zinc, molybdenum, etc. Why does he suggest this?

REFERENCES

Adams, F. (ed.) (1984) *Soil Acidity and Liming*, 2nd edn. American Society of Agronomy, Crop Science Society of America, Inc., Soil Science Society of America, Inc., Madison, Wisconsin, USA, 380 + xi pp.

Batey, T. (1981) *Soil Husbandry*. Soil & Land Use Consultants Ltd, Aberdeen, Scotland.

Cresser, M.S., Killham, K.S. and Edwards, A.C. (1993) *Soil Chemistry and its Applications*. Cambridge University Press, Cambridge, 191 + xi pp.

Cresser, M.S., Aitkenhead, M.J. and Mian, I.A. (2008) A reappraisal of the terrestrial nitrogen cycle – What can we learn by extracting concepts from Gaia theory? *Science of the Total Environment*, **400**, 344–355.

Fullen, M.F. and Catt, J.A. (2004) *Soil Management: Problems and Solutions*. Arnold (Hodder Headline Group), London, 269 + xviii pp.

Holden, J. (ed.) (2005) *An Introduction to Physical Geography and the Environment*. Pearson/Prentice Hall, Harlow, England, 664 + xxvi pp.

Reid, J.M., MacLeod, D.A. and Cresser, M.S. (1981) Factors affecting the chemistry of precipitation and river water in an upland catchment. *Journal of Hydrology*, **50**, 129–145.

White, R.E. (2006) *Principles and Practice of Soil Science: The Soil as a Natural Resource*, 4th edn. Blackwell Publishing, Malden, Maryland, USA, 363 + xi pp.

CHAPTER 8

Climate change

Nicola Carslaw

Learning outcomes

By the end of this chapter you should:

- Understand the difference between weather and climate.
- Be able to explain and differentiate between climate variability and climate change.
- Understand the scientific concepts that explain climate change.
- Be able to explain the general features of atmospheric and oceanic circulation and their origins.
- Appreciate historical climate variability and the proxies used to study it.
- Understand the mechanisms that cause natural climate variability.
- Be able to put recent climatic warming into the context of the longer term temperature records.
- Be able to explain why anthropogenic causes of climate change are the most likely reason for the observed warming trends in recent decades.
- Understand what climate models predict for the future and be aware of the uncertainties associated with them.

8.1 Introduction

Our lives are defined by our climate and it is apparent that life becomes very difficult when the climate becomes extreme and falls outside our normal range of experience. Take, for example, the floods in the UK during the summer of 2007; they caused widespread chaos, with many homes and other buildings flooded, roads and railways brought to a standstill, and many people displaced. The sustained period of snow and cold temperatures during November–December 2010 in the UK also caused widespread disruption to travel and schools across much of the country.

We tend to think of really catastrophic events linked to climate as having occurred a long time ago, such as the onset of the Ice Ages. However, as recently as the seventeenth and eighteenth centuries, sustained bad weather led to poor harvests in parts of Europe; famine killed one-third of the population in Finland around the end of the seventeenth century. Also, in the 1930s, drought across the Great Plains of North America caused massive social disruptions (the 'Dust Bowl' years). We tend to take our climate for granted when most of the time it suits our lifestyle needs, but history has shown us that changes to the climate can cause catastrophic effects.

Most of the changes in climate throughout history have occurred over a long timescale and been largely due to natural causes. However, something different is now happening to our climate – humans are changing it, and at a rapid rate. We hear regular reports that summers are getting hotter, polar ice is shrinking, sea levels are rising and residents of some low-lying South Pacific islands have already been evacuated to higher ground. Indeed, climate change is the most serious environmental problem to emerge in recent decades, and one that requires our immediate and concerted attention.

Despite the overwhelming scientific evidence that humans have caused recent warming as presented in this chapter, there are still those who deny it. For instance, US Senator James Inhofe famously said 'With all of the hysteria, all of the fear, all of the phoney science, could it be that manmade global warming is the greatest hoax ever perpetrated on the American people?' No wonder many members of the public are confused when scientists and politicians seem to be at odds with each other. Clearly, as well as a scientific dimension, there is also a political dimension: our fossil fuel dependence and the interests associated with it often lead to opposing views.

Some climate-change sceptics argue that because there is great uncertainty with future predictions for climate change, it is foolish to rush into action that may put economies at risk. However, this argument is weak when one considers that the uncertainty inherent in the predictions also means that things could equally be much worse than expected. The time for action is now.

8.2 Climate and weather

While weather refers to what is happening at any one time, climate represents the probability that the weather will behave in a certain way, based on averages over long periods of time. Climate records could not tell you what the weather was likely to do at any particular place at a specific time, but they could help you predict the annual mean temperature for next year, or perhaps the average rainfall over a region. According to the IPCC:

> *Climate is usually defined as the 'average weather', or more rigorously, as the statistical description of the weather in terms of the mean and variability of relevant quantities over periods of several decades (typically three decades as defined by WMO). These quantities are most often surface variables such as temperature, precipitation, and wind, but in a wider sense the 'climate' is the description of the state of the climate system.*

The IPCC goes on to say that:

> *Climate change refers to a change in the state of the climate that can be identified (e.g. by using statistical tests) by changes in the mean and/or the variability of its properties, and that persists for an extended period, typically decades or longer.*

In other words, one hot summer does not constitute climate change, but year after year of hotter and drier weather may do. Although the change in climate may be focused on one parameter, it is more likely that there will be more general shifts in weather patterns leading to changes in several parameters (colder, wetter and windier for instance). The changes may not be the same all over the world either: some places may get hotter and drier while others get colder and wetter. Changes are, however, often part of a general global warming or cooling.

One further consideration is the effect of extreme weather events on overall climate. There can be some blurring of the distinction between climate and weather where a changing frequency of extreme events influences interpretation of changes in climate. For instance, an extreme summer flood may obscure the fact that summers are generally getting drier. A further complication is that climate change is predicted to increase the frequency of extreme events such as flooding and heat waves. Care must be taken when interpreting climate records that such extreme events do not mask the signs of climate change occurring.

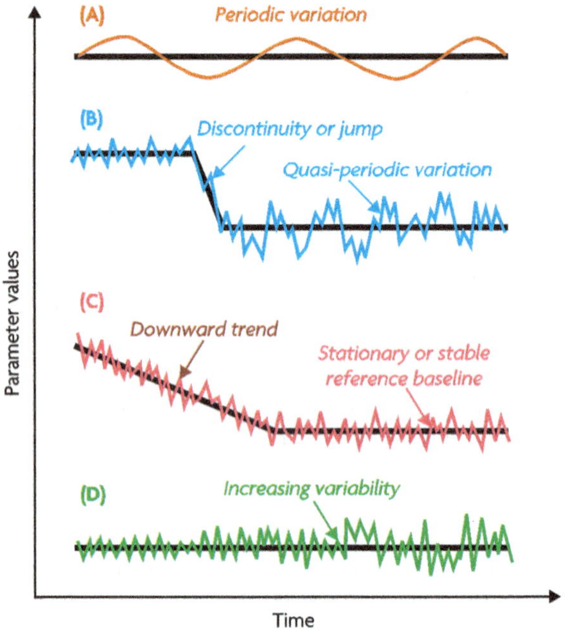

Figure 8.1 Changes in a meteorological parameter with time (e.g. temperature) demonstrating the difference between various instances of climate variability and climate change.

Source: http://www.asp.ucar.edu/colloquium/2000/Lectures/hurrell1.html

8.3 Climate change or climate variability?

We have already discussed the fact that weather is highly variable, but climate can also be variable. It is important that we can distinguish variability in climate from climate change. Making this distinction is confounded by the fact that human-induced changes are small compared with the variability associated with, for example, conditions at the Poles versus the Equator, in winter versus summer and even day versus night. Although daily meteorological parameters may depart from the climatic mean, the climate is considered to be stable if the long-term average does not significantly change. For instance, consider Figure 8.1.

Assuming that the parameter in question is temperature, curve (A) could perhaps represent either the daily (maximum during day, minimum during night) or annual (maximum in summer, minimum in winter) variation in temperature at a particular location. Although there is variability, the average is constant and hence there is no climate change. Curve (B) represents a sudden cooling, such as may be seen following a large volcanic eruption: the sulphate aerosol that fills the atmosphere following such an event can cool global temperatures for a few years afterwards. Variability also increases after the jump, but the average temperature is stable over this period. Curve (C) represents a gradual cooling, followed by a steadying off to a new average value. Such a pattern does constitute a change in climate, and something similar may have been observed as the Earth went into an Ice Age. Finally, curve (D), shows a scenario with increasing temperature variability, but no change in average value. The long-term average is stable in this case, so there is no climate change.

Both climatic averages and the probability of climate extremes are, by definition, statistical measurements based on probabilities, not certainties. This makes the absolute detection of climate trends difficult to predict and very difficult to measure, except by looking at long-term historical data.

8.4 Underlying scientific principles

8.4.1 What is radiation?

Any object at a temperature above absolute zero (0 K or −273.16 °C) transmits energy to its surroundings by radiation. This radiation is in the form of electromagnetic (EM) waves travelling at the speed of light between the sun and the Earth. Without this EM radiation, there would be no life on Earth. EM radiation covers a wide wavelength range from very long wavelengths of thousands of metres to wavelengths smaller than an angstrom ($1Å = 1 \times 10^{-10}$ m). However, the most important part of the EM spectrum in terms of climate change is the visible to near infrared region between 0.4–15 μm as we will see in subsequent sections.

8.4.2 Black-body radiation

A black body is any object that is a perfect emitter and a perfect absorber of radiation. The object does not have to appear 'black' and the sun's and the Earth's surfaces behave approximately as black bodies. This property is useful, as it allows us to use various physical laws to explore the implications for the climate on Earth.

The Stefan–Boltzmann Law for instance, relates the total amount of radiation emitted by an object to its temperature:

$$F = \sigma T^4$$

where, F = total amount of radiation emitted per square metre (W m^{-2}), σ = the Stefan–Boltzmann constant = 5.67×10^{-8} W m^{-2} K^{-4} and T = temperature of the object (K).

For instance, the surface of the sun has a temperature of 6000 K. How much radiation is emitted per square metre of its surface?

$$F = 5.67 \times 10^{-8}\ \mathrm{W\,m^{-2}\,K^{-4}}\ (6000\ \mathrm{K})^4$$
$$= 7.3 \times 10^7\ \mathrm{W\,m^{-2}}$$

Note that the units of K^{-4} and K^4 cancel in this calculation leaving the answer in units of W m^{-2}.

If we now consider the same calculation for the Earth with a surface temperature of 288 K:

$$F = 5.67 \times 10^{-8}\ \mathrm{W\,m^{-2}\,K^{-4}}\ (288\ \mathrm{K})^4 = 390\ \mathrm{W\,m^{-2}}$$

Not surprisingly, the surface of the sun emits much more energy per square metre than the surface of the Earth. Because of the temperature being raised to the fourth power in the Stefan–Boltzmann Law, hot bodies radiate energy far more efficiently than cooler ones. A doubling of temperature leads to an increase of 2^4 or 16 times more emitted energy.

The sun and the Earth emit radiation over many wavelengths, but there is one wavelength where an object emits the largest amount of radiation. This wavelength (λ_{max}) can be calculated using the Wien displacement law:

$$\lambda_{max} = 2897/T$$

where 2897 μm K is a constant. Note that the temperature input into this equation is in Kelvin so the answer is in units of microns ($1\ \mu m = 1 \times 10^{-6}$ m). Let's return to our examples of the sun and the Earth.

At what wavelength does the sun emit most of its radiation?

$$\lambda_{max} = 2897/6000 = \sim 0.5\ \mu m$$

At what wavelength does the Earth emit most of its radiation?

$$\lambda_{max} = 2897/288 = \sim 10\ \mu m$$

At room temperature, black bodies tend to emit at wavelengths in the IR region of the spectrum. However, as the temperature of a black body increases, the wavelength emitted decreases towards the higher energy end of the EM spectrum.

We have therefore found important differences between solar and terrestrial energy. Energy emitted from the sun is very powerful and is focused at around 0.5 μm, whereas that emitted from the Earth is at a much lower energy focused at around 10 μm. We can think, therefore, of solar energy as incoming shortwave radiation which *heats* the Earth. Outgoing terrestrial radiation on the other hand is referred to as longwave radiation (LWR), which acts to *cool* the planet. The balance between these two quantities is what determines changes in climate.

By the time the solar energy reaches the Earth's atmosphere, it is measured (from a satellite) to be roughly 1366 W m^{-2}. This value is known as the solar constant and is defined as the amount of incoming solar electromagnetic radiation per unit area, measured on the outer surface of Earth's atmosphere, in a plane perpendicular to the rays. The solar constant includes all types of solar radiation, not just the visible light. Ozone and oxygen high in the atmosphere absorb much of the high energy UV (see Chapter 3), so most of the solar radiation reaching the Earth's surface is concentrated in a band between 0.3 to 2 μm.

Energy emitted from the Earth is focused between 4–50 μm. The principal atmospheric gases (N_2 and O_2) do not absorb much of this outgoing radiation, but carbon dioxide (CO_2), ozone (O_3) and water vapour (H_2O) do. The region from 8–14 μm is often referred to as the *atmospheric window* as terrestrial radiation can escape to space over this region (Figure 8.2). However, if the concentrations of these three and other so-called greenhouse gases (GHGs) absorbing in this region increase, there are major implications for radiative balance.

The Greenhouse Effect

So how can the presence of CO_2, H_2O and O_3 in the atmosphere lead to warming at the surface? There is a detailed explanation of this process in Burroughs (2009), but the basic process is summarised here. The reason that we have a 'Greenhouse Effect' has much to do with where in the atmosphere these three key GHGs absorb and re-emit energy.

Different GHGs absorb and emit selectively at different wavelengths according to their individual characteristics (Figure 8.2). Most absorption of LWR takes place in the atmospheric window near the surface, owing to the large amount of water vapour there. The LWR is absorbed and then re-emitted, including towards the ground, leading to heating. Indeed, the presence of H_2O, CO_2 and O_3 in the atmosphere has led to a natural warming effect over history, with contributions of 21 K from H_2O, 7 K from CO_2 and 2K from O_3. The natural Greenhouse Effect has warmed our planet to a comfortable temperature. Without it, the planet would be too cold for us to live on (~255 K).

What has changed in recent history is the balance of these GHGs in the atmosphere, and in particular, the CO_2 concentration. Where CO_2 absorbs and emits most strongly, the radiation escaping to space comes from high in the atmosphere where the temperature is low (~220 K). As the CO_2 concentration increases, the band around 15 μm where it absorbs broadens. With more CO_2 molecules in the atmosphere, more of the outgoing LWR will be absorbed by these molecules rather than by H_2O at the surface and hence the net emission of LWR to space will be reduced (as the Stefan–Boltzmann Law tells us that cooler bodies radiate less effectively than warmer bodies). Therefore, the temperature of the lower atmosphere has to rise to compensate for reduced emission from the CO_2 absorption higher up in the atmosphere. This is the basic physical process underlying the Greenhouse Effect, and global warming refers to its enhancement by human activities.

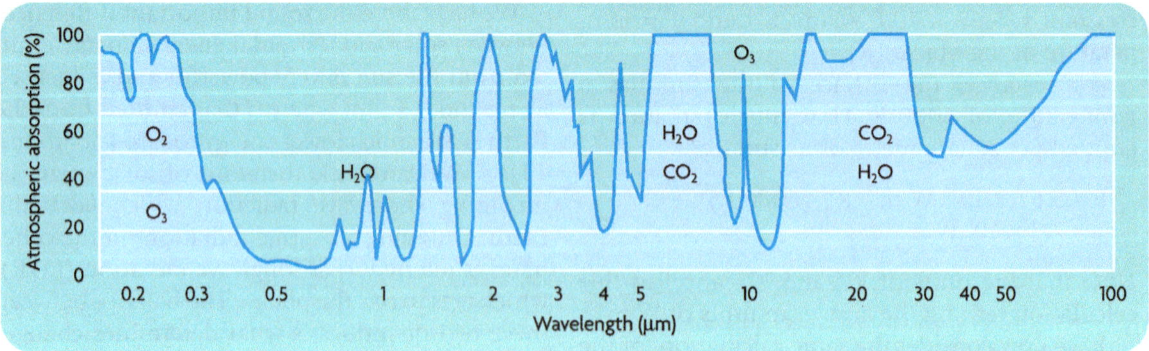

Figure 8.2 Atmospheric absorption by the major GHGs in the wavelength region between 0.2–100 mm.
Source: http://www.eduspace.esa.int/subtopic/images/07-atmosvindue.gif

Lead-in question

Can heat energy move around the planet's surface?

8.5 Atmosphere and ocean global circulation

The presence of an atmosphere around the Earth causes warming relative to the case with no atmosphere. However, there is another important element of the climate system to consider and that is heat redistribution around the planet. If you consider the Earth's energy balance in terms of incoming and outgoing radiation (Figure 8.3), there is an imbalance of radiation between the Equator and the Poles.

Solar energy is focused at the Equator, where the angle of inclination of the sun to the surface is 90°. At the Poles, the sunlight has to travel through more atmosphere to reach the surface, resulting in a sine wave appearance of the global distribution of solar energy (Figure 8.3).

The outgoing LWR on the other hand is reasonably evenly distributed over the surface of the globe, although there is a slightly higher emission at the Equator compared to the Poles owing to the Stefan–Boltzmann Law (more efficient emission at higher temperatures).

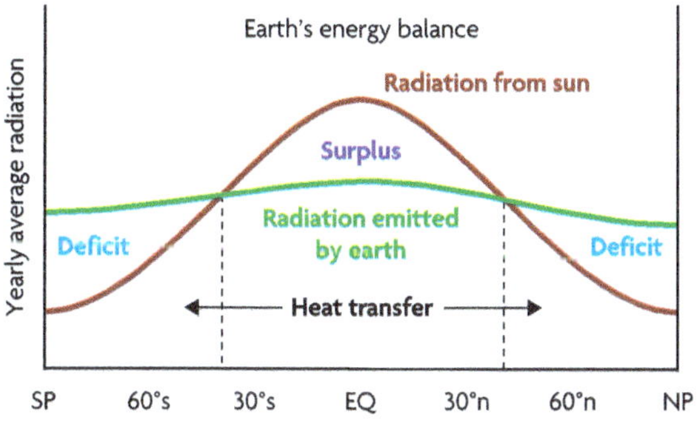

Figure 8.3 Radiation from the sun and the Earth across the globe and the resulting energy imbalance and heat transfer. Source: http://apollo.lsc.vsc.edu/classes/met130/notes/chapter2/ebal3.html

Effectively then, there is an energy imbalance between the Poles and the Equator with an excess of heat at the Equator and a deficit at the Poles. In order to redress this imbalance, heat transfer takes place from the Equator to the Poles, and this occurs by three primary mechanisms:

- the weather transports sensible heat;
- the weather generates and transports latent heat;
- ocean circulation.

8.5.1 Sensible heat

Sensible heat is potential energy in the form of thermal energy or heat: the body in question must have a higher temperature than its surroundings. The thermal energy can be transported via one (or more) of three mechanisms: conduction, which is the transfer of heat through molecular motions; convection, which is the transfer of heat through the mass movement of a substance; radiation, which is the transfer of heat through electromagnetic waves.

8.5.2 Latent heat

When water evaporates, heat must be supplied to convert water from the liquid to the gas phase, and this heat is known as latent heat. The water vapour can be transported elsewhere and condense as droplets and rain or even ice, at which point the latent heat is released. In this way, latent heat can be moved from warmer to cooler areas. The highest flows of latent heat occur near subtropical oceans, where high temperatures and a plentiful supply of water encourage evaporation of water.

8.5.3 How are these heat forms moved around the atmosphere?

The temperature difference between the Equator and the Poles leads to a pressure difference. Winds act to remove this pressure difference and move from areas of high to low pressure.

As the ground is heated, the air becomes less dense and rises owing to convection. This leaves an area of low pressure at the surface as the air rises, and higher pressure aloft. Above cooler surfaces, air becomes more dense than surrounding air over warmer surfaces and sinks, thus leading to a high pressure area at the surface and low pressure aloft. The pressure gradients thus created cause air to move horizontally from the high to low pressure area, hence forming winds. Such circulation cells form the basis of global atmospheric circulation and the transport of sensible and latent heat from the Equator to the Poles.

The tricellular model is often used to describe the general circulation of the atmosphere and contains three circulation cells known as the Hadley, Ferrel and Polar cells (Figure 8.4).

Near the Equator, hot, humid air rises and flows towards the Poles. At around 30° latitude in each hemisphere, the air starts to descend, with the returning branch flowing at the surface toward the Equator. The Polar cells are driven by cold dense air sinking at the Poles, with the resulting air moving towards the 60° latitudes, where it rises once more and returns to the Poles aloft. Finally, the Ferrel cell lies between the 30° and 60° latitudes. It is not directly thermally driven as for the other two cells, but represents an area of cyclonic disturbances that can transport heat between the equatorial and polar regions (Figure 8.4).

Rising air is associated with precipitation, such as over the Tropics and the 60° latitudes. Warm rising air takes with it moisture, which condenses into raindrops as it cools. Sinking air tends to be dry, and this can be seen when one considers the locations of the major deserts around the world, which all lie around the 30° latitudes. Note that the polar regions are also effectively deserts: there is very little precipitation in these regions.

However, this picture is complicated by the action of the Coriolis force upon the surface flow. The Coriolis force is caused by the rotation of the Earth and deflects the air to the right in the northern hemisphere and to the left in the southern hemisphere. The force varies from zero at the Equator to a maximum value at the Poles.

Between 0° and 30°, the resulting surface winds head in a north-westerly (from the south-east) direction in the southern hemisphere and south-westerly in the northern hemisphere. Between 30° and 60°, the surface winds are westerlies. You may know this if you have taken a flight from Europe to the US. The return flight from the US is always faster than the outgoing flight when the westerly wind is behind you. Finally, the polar regions are subject to easterly winds (Figure 8.5).

This view is somewhat idealised; Figure 8.5 shows that the presence of land masses distorts the circulation cells somewhat, such that a series of high and low pressure systems are likely to be present in specific locations at any one time.

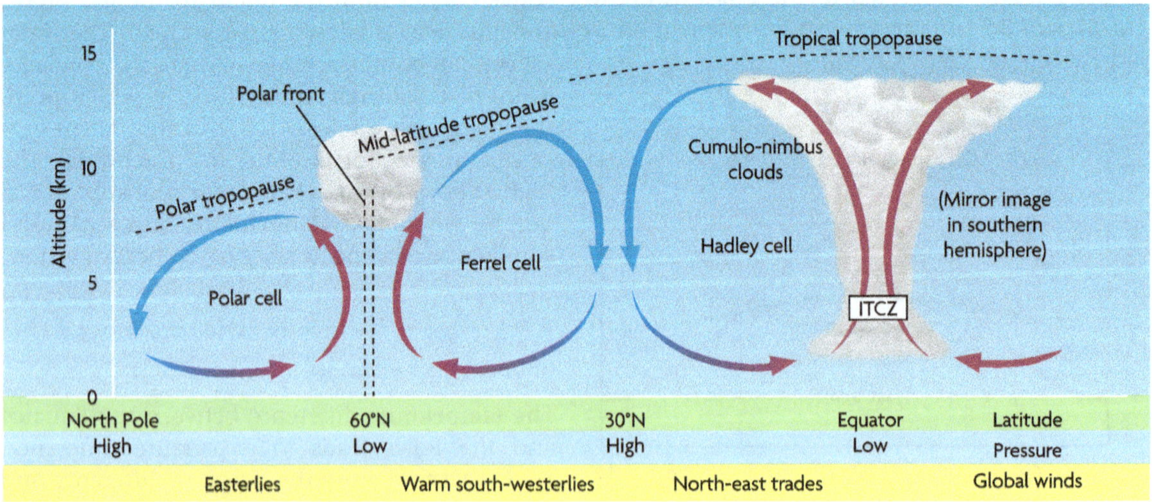

Figure 8.4 The tricellular model for atmospheric circulation containing the Hadley, Ferrel and Polar Cells in each hemisphere.
Source: Waugh (2009).

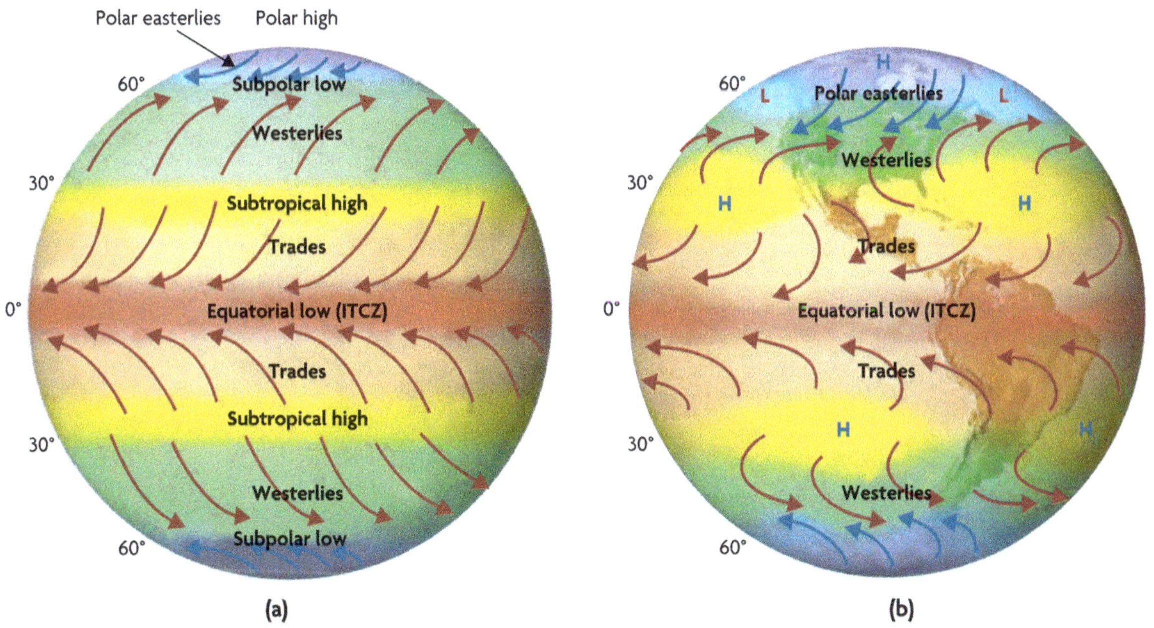

Figure 8.5 General circulation of the atmosphere where (a) shows the idealised picture without land masses and (b) includes a more realistic picture with the effect of land masses included.
Source: http://www.ux1.eiu.edu/-cfjps/1400/circulation.html

8.5.4 Ocean circulation

Ocean circulation can be split into surface circulation and deep water circulation for further consideration.

Surface water circulation

The surface layer is generally considered to be the top 100 m of the ocean and is well mixed by waves, tides, and the wind. It lies in a distinct layer above the colder and denser deep waters, separated by a boundary layer called the pycnocline.

As for the atmosphere, the circulation of surface waters is affected by the Coriolis force and the location of land masses. These interactions produce huge circular patterns called current gyres (Figure 8.6). The surface circulation moves warm water from the Equator to higher latitudes, so helping to transfer heat.

For those living in the UK, the Gulf Stream has a large effect on their climate. Edinburgh lies at a similar latitude to Moscow, Newfoundland and Labrador, yet the Gulf Stream modifies the UK climate so that whereas the average minimum temperature in January is minus 16 °C in Moscow, it is 1 °C in Edinburgh. This example shows the important modifying effect of the oceans on the climate.

Deep water circulation

The largest source of deep water is surface water that sinks in the North Atlantic Ocean. This water sinks slowly downward until it reaches a level of equal density and then spreads out. This so-called 'Great Ocean Conveyor Belt' then continues its journey south, past the Equator and into the Southern Hemisphere. The water continues past Antarctica and into the Pacific and Indian Oceans, where some of the deep waters are warmed and so rise again to the surface. The cycle is completed when this water returns to the North Atlantic, becomes cool and dense and sinks again (Figure 8.7).

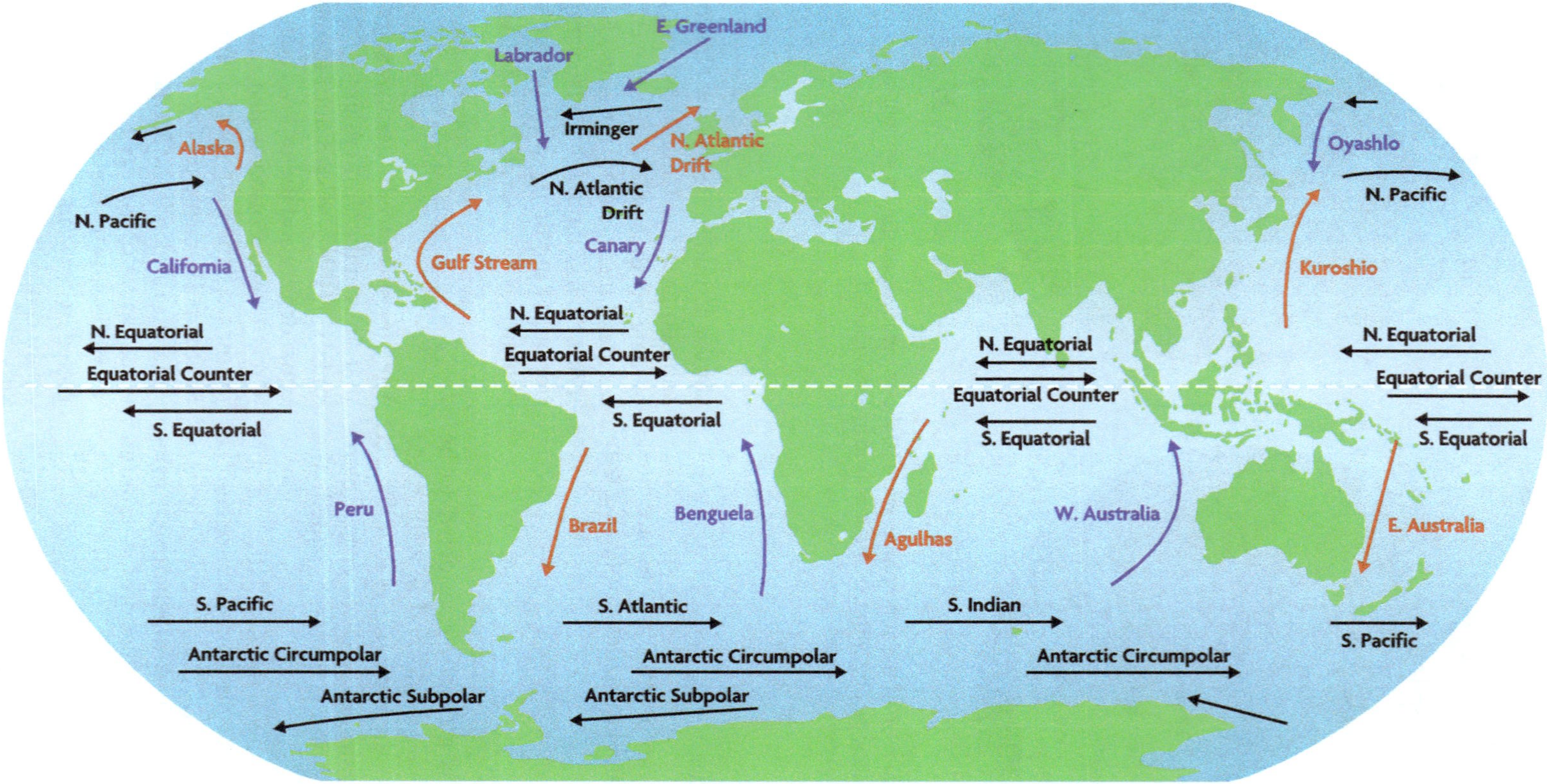

Figure 8.6 The major surface currents in the oceans. Red arrows denote warm currents and the blue arrows, cold currents.
Source: http://www.windows.ucar.edu/tour/link=/earth/Water/ocean_currents.html. Image courtesy of Windows to the Universe.

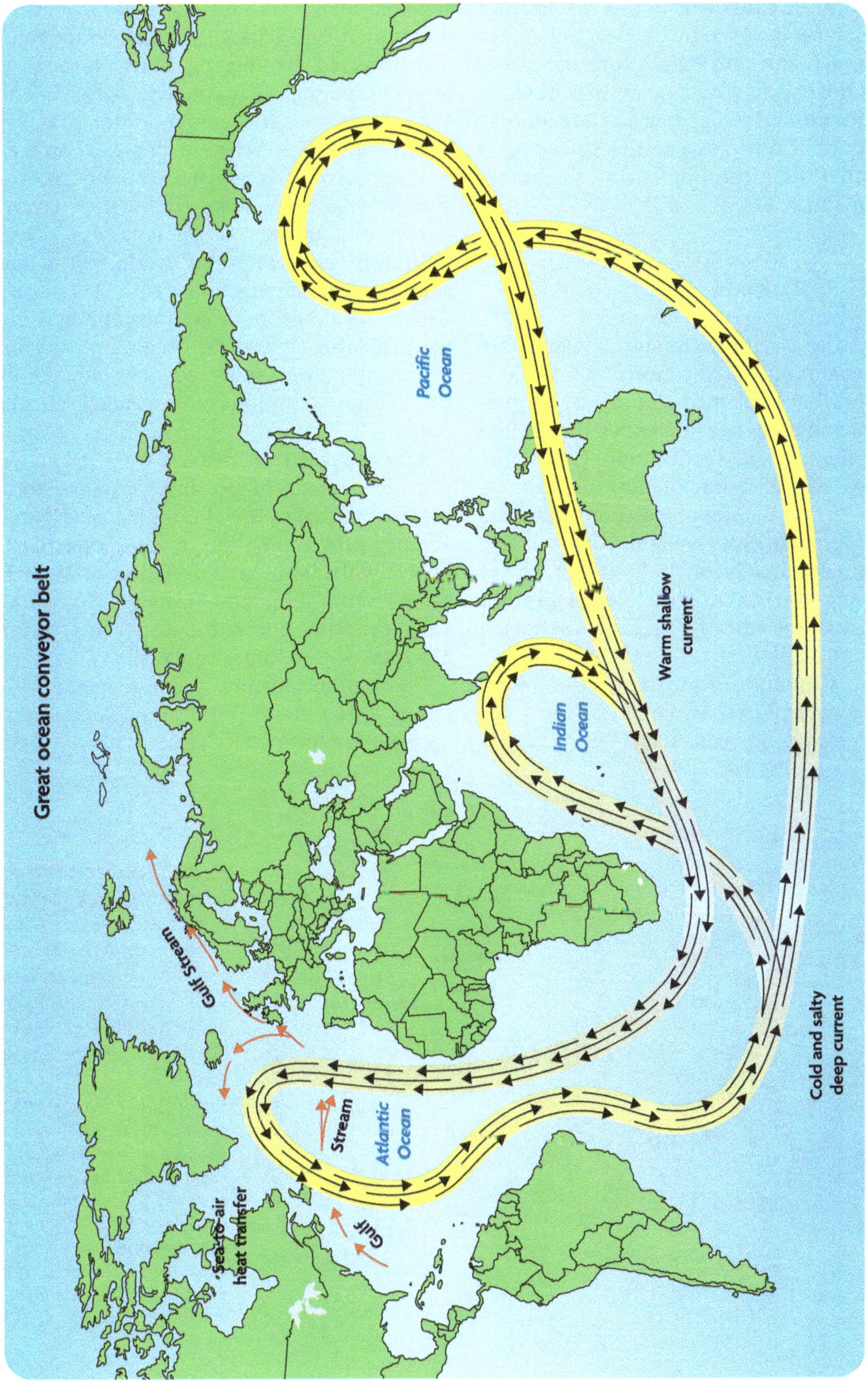

Figure 8.7 The Great Ocean Conveyor Belt moving warm water from Equator to Poles and cooler water in the opposite direction.
Source: http://www.grida.no/publications/vg/climate/page/3085.aspx

This section has shown that heat transfer around the planet is driven by the heat imbalance between the Equator and Poles. Both the atmosphere and the oceans play crucial roles in redistributing heat around the planet. The oceans play a more important role closer to the Equator and the atmosphere is more important at latitudes greater than about 20°.

8.6 Past climates

In order to work out if the climate change we are experiencing currently is of concern, it is important to study what happened in the past. In particular, knowing how the climate behaved before humans influenced it significantly (so before 1750), gives an indication of how the climate might have varied naturally. However, there are no direct measurements available before 1750 and so we have to rely on indirect (or proxy) measurements. Even then, there are gaps in our knowledge and the use of proxies is associated with major uncertainties.

There are a number of proxies that are commonly used to explore past climates. The most commonly used is ice core data, but it is also possible to use tree ring data, ice extent, fossils and pollen, and geological evidence. Ice core data are extracted by digging down through deep layers of ice, mainly in Antarctica and Greenland. We can go back to between 750–800,000 years ago (or 8 glacial cycles) in this way. An ice core is composed of layers representing subsequent summers and winters. The winter layers are darker and thicker, as there is typically more pollution (e.g. soot) and precipitation than in summer. By analysing an ice core, it is possible to determine atmospheric composition and the age of each layer, as well as obtaining information on precipitation levels, temperatures, the extent of volcanic eruptions and general circulation patterns.

Often, several proxies are used in tandem to provide a range of possible scenarios for how climate may have varied through time (Figure 8.8). Temperature anomalies are often used in such figures which show the deviation of a particular value from a long-term (usually 30 year) mean. Such a practice allows the reader to spot unusually high or low values very easily.

Throughout history, gross temperature variations often have physical explanations. The increase in temperature around 10,000 years ago

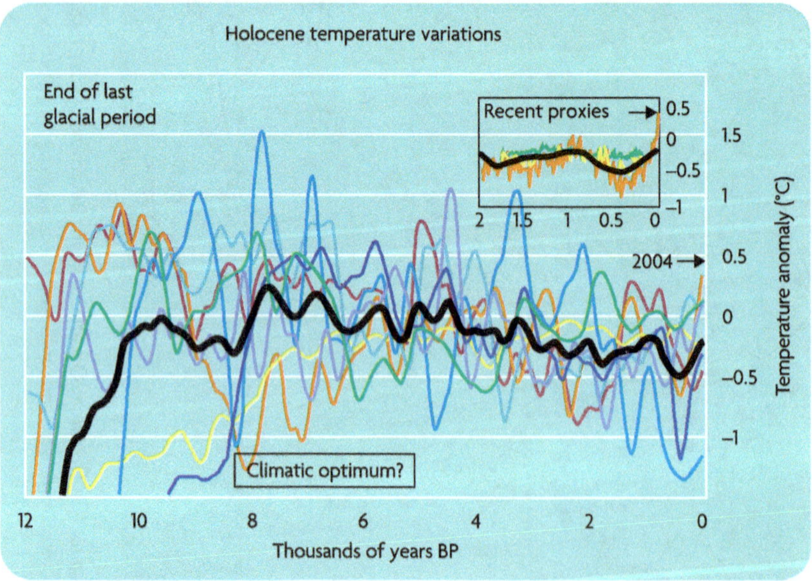

Figure 8.8 Reconstructed temperatures over the last 12,000 years using proxies (different coloured lines represent different proxies such as ice core, sediment core, pollen records, etc.). The solid black line shows the average for guidance.

Source: From Wikipedia: http://en.wikipedia.org/wiki/File:Holocene_Temperature_Variations.png). Copyright: This figure was prepared by Robert A. Rohde from publicly available data and is incorporated into the Global Warming Art project. Most, but not all, of the original data is available from http://www.ncdc.noaa.gov/paleo/data.html

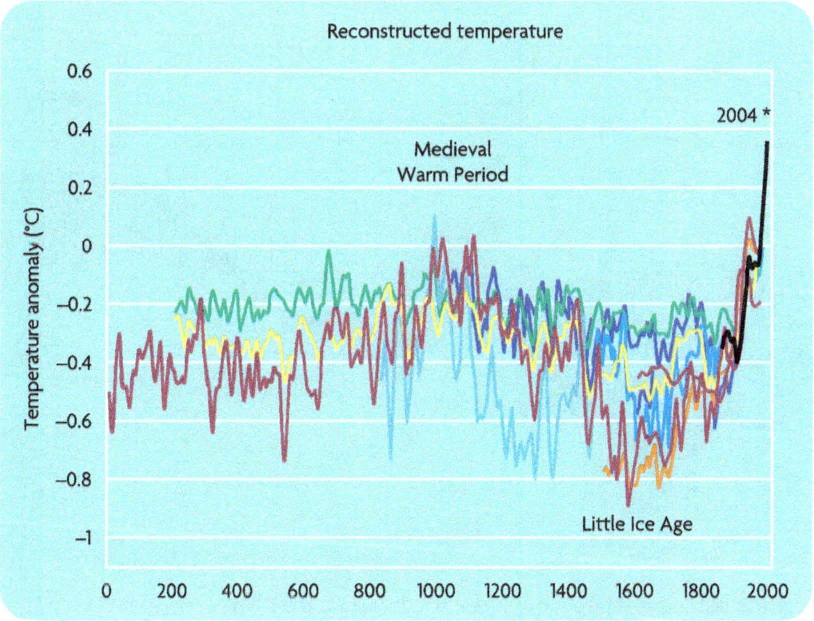

Figure 8.9 A comparison through the last 2000 years of 10 different published reconstructions of mean temperature changes. More recent reconstructions are plotted in redder colours, and older reconstructions in bluer colours. Observational data are shown in black. The single, unsmoothed annual value for 2004 is also shown for comparison.

Source: From Wikipedia: http://en.wikipedia.org/wiki/File:2000_Year_Temperature_Comparison.png). Copyright: The original version of this figure was prepared by Robert A. Rohde from publicly available data, and is incorporated into the Global Warming Art project.

for instance shows the Earth coming out of the last great Ice Age (Figure 8.8). However, it is often difficult to attribute specific features to any one effect. For instance, fossils of tropical plants and animals occur on the Antarctic continent, but these date from when the continent was closer to the Equator rather than being indicative of past climate change. Proxies are useful, but should be treated with caution!

Figure 8.9 shows temperature anomalies over the last two thousand years. This period includes the 'Little Ice Age', a period of sustained cold temperatures between the sixteenth and nineteenth centuries and the 'Medieval Warm Period' between 950–1100.

These two periods are the cause of much debate over their exact start and end dates, the size of the temperature anomaly as well as their global extent (particularly the Medieval Warm Period), as can be seen from the variation between proxy data. Nevertheless, they do demonstrate that temperature variation happened well before humans were influencing climate significantly. Perhaps the most notable aspect of this figure, however, is recent years where temperature rapidly increases according to all available proxies and measurements and is higher than at any other point during the last 2000 years.

Figure 8.10 focuses on the last 160 years since measurements became available, and shows again how dramatic warming has been in recent years, particularly the last 30 or so.

The year 2010 was the joint third (with 2003) warmest year on record, beaten only by 1998 and 2005. The period 2001–2010 (0.44 °C above 1961–90 mean) is 0.2 °C warmer than the 1991–2000 decade (0.24 °C above 1961–90 mean). After 1998, the next nine warmest years were all within the decade from 2001 to 2010.

In summary then, while there have been temperature fluctuations in the past, the rate of heating we are experiencing now is unprecedented since the end of the last Ice Age. For a more detailed review of past climates, the reader should refer to the excellent book by Burroughs (2005), *Climate Change in Prehistory*.

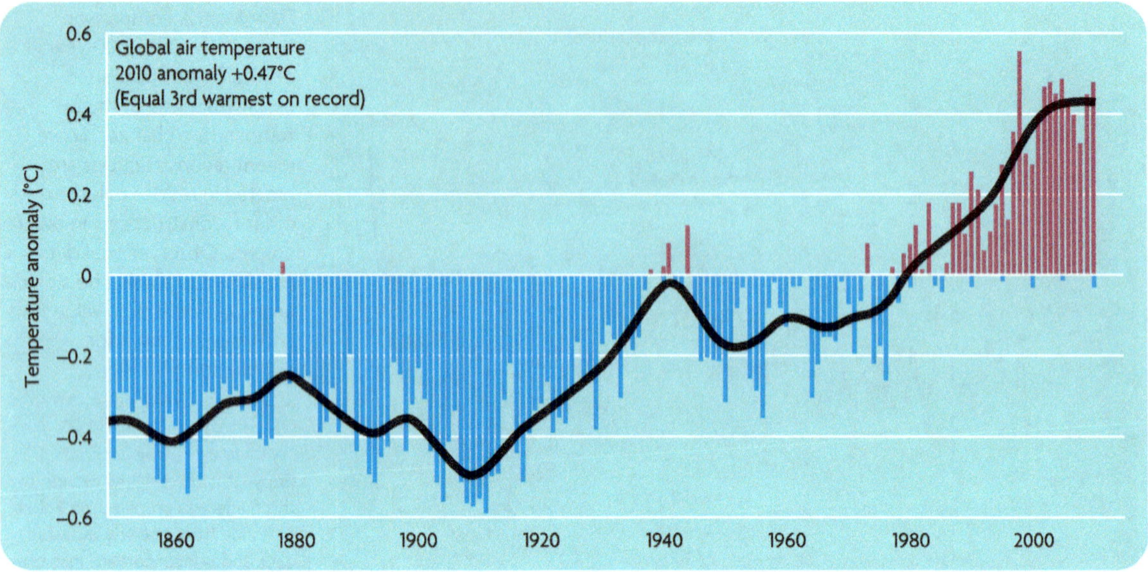

Figure 8.10 Temperature anomalies for the last 160 years from observational data from the Climatic Research Unit at the University of East Anglia and the UK Met. Office.
Source: http://www.cru.uea.ac.uk/cru/info/warming/

8.7 Natural causes of climate change

There are a number of natural processes that have affected climate in the past. The most important of these is change over time in the amount of solar radiation received at the surface of the Earth, owing to changes in the Earth's orbit. However, volcanic eruptions have also had an impact, as have changes in sunspot activity. Each of these processes will now be discussed in more detail.

8.7.1 Volcanic eruptions

Volcanic eruptions can affect the climate over both the short and long term and there is evidence of them having done so throughout history. For instance, hemispheric cooling of about 0.5 °C following the eruption of Tambora in 1815, caused crop failures and famine in North America and Europe (Oppenheimer, 2003).

During such eruptions, sulphur dioxide (SO_2) is thrown high into the atmosphere. The large amount of SO_2 emitted can form sulphate aerosol, with a residence time in the stratosphere of 1–3 years. The sulphate aerosol can form a layer around the Earth.

Sulphate aerosol particulates are small, of the order of 0.5 μm diameter, and effectively scatter incoming solar radiation (remember from section 8.4.2 that incoming solar radiation is centered at this wavelength). Consequently, increased amounts of sunlight are scattered back to space following a large volcanic eruption and the surface cools. Although ash is also emitted in large quantities during volcanic eruptions, the coarse size of the ash particles means that its residence time in the stratosphere is low. It is the sulphate aerosol that affects the climate, even though the ash is the most visible product of an eruption.

In summary, large explosive volcanic eruptions have typically been associated with net surface cooling in the 1–3 years following the eruption (Oppenheimer, 2003).

8.7.2 Solar flux variations

It has been known for some time now that solar flux varies, and that this variation is connected to the number of sunspots at any time. There is an 11-year solar magnetic cycle associated with the natural waxing and waning of solar activity (Figure 8.11): solar luminosity is higher during periods of high sunspot activity. Note that there are 23 peaks in the 250 years between about 1750 and 2005.

Also clear in Figure 8.11 are three distinct periods:

- the long *Maunder Minimum* when almost no sunspots were observed;
- the less severe *Dalton Minimum*;
- increased sunspot activity during the last fifty years, known as the *Modern Maximum*.

It is widely believed that the low solar activity during the Maunder Minimum and earlier periods may be among the main causes of the Little Ice Age. Similarly, the Modern Maximum is believed to be partly responsible for recent global warming, especially the temperature increases between 1900–1950. However, the change in resulting irradiance is too small (about 0.1 per cent) to produce the observed rise in temperature since the late nineteenth century.

8.7.3 Changes in the Earth's orbit

In 1920, Milutin Milanković proposed that orbital changes and the associated changes in solar radiation received by the Earth could affect the climate. The motion of the Earth can be simply described as involving a rotation about its axis through the poles with a period of one day and an orbit round the sun with a period of one year. However, this simplified picture is made more complete by considering other effects, more commonly known as Milankovitch cycles and summarised in Figure 8.12.

The orbit of the Earth around the sun is not perfectly circular owing to its proximity to Jupiter and Saturn and the shape of this orbit, or its *eccentricity*, changes cyclically with a period of approximately 96,000 years. When the orbit displays maximum eccentricity (mildly elliptical), there is more variation in the solar radiation received by the Earth over the period of a year compared to when it has low eccentricity (nearly circular).

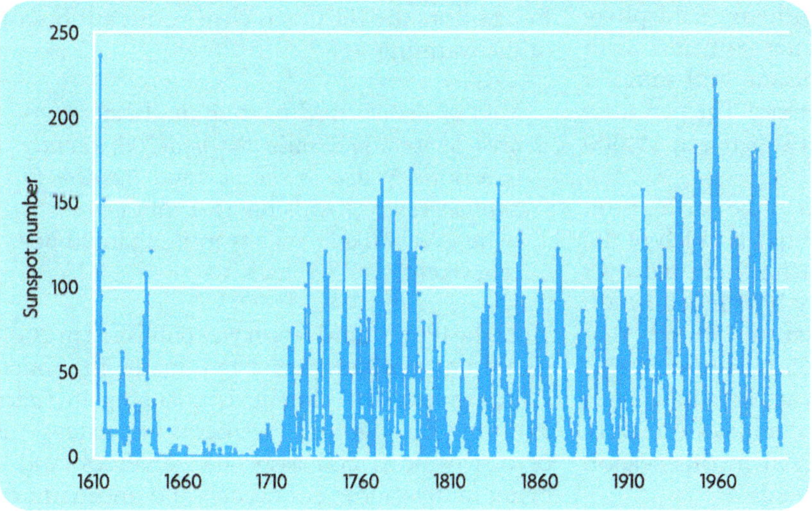

Figure 8.11 Sunspot number observations since approximately 1600. Early observations were sporadic (compiled by Hoyt Schatten 1998a, 1998b) but continuous monthly averages have been available since approximately 1749 based on an average of measurements from around the world (as reported by the Solar Influences Data Analysis Center, World Data Center for the Sunspot Index, at the Royal Observatory of Belgium).

Source: Replotted data from http://en.wikipedia.org/wiki/Sunspots

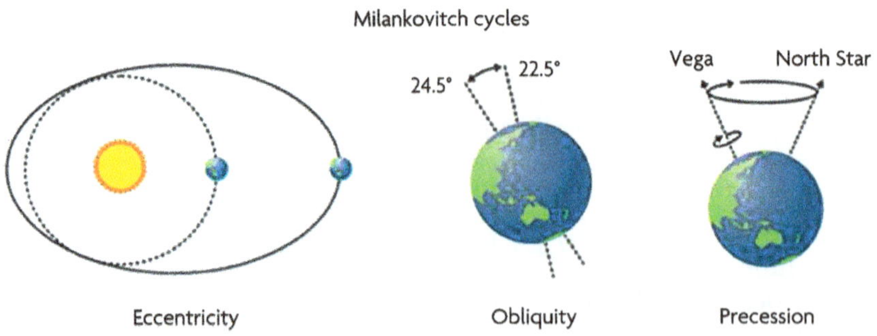

Figure 8.12 The three Milankovitch cycles. The Earth's orbit is elliptical, so eccentricity varies depending on the shape of the ellipse. Secondly, the obliquity is a measure of the varying tilt of the Earth. Finally, the precession of the equinoxes means that seasons can be hotter or colder than average.
Source: http://www.skepticalscience.com/co2-lags-temperature-intermediate.htm

The Earth's axis of rotation varies between about 21.5° and 24.5° within a period of about 41,000 years. Under conditions of high obliquity (or tilt), the seasons become exaggerated as the polar regions point more directly either towards or away from the sun: winters become colder and the summers warmer.

Finally, the Earth's orbit around the sun is elliptical, and the sun sits at one focus of the ellipse. The distance from the Earth to the sun is therefore different at opposite sides of the orbit (i.e. 6 months apart). Consequently, at certain times the largest distance between the sun and the Earth will occur during the northern hemisphere winter, making the northern hemisphere seasons more extreme: winter coincides with the greatest distance from the sun and summer coincides with the smallest distance. Conversely, if the furthest approach occurs during the northern hemisphere summer, the northern hemisphere seasons are less extreme: winter coincides with the smallest distance from the sun and summer coincides with the greatest distance. The period of precession of the Earth's orbit is between 19,000 and 21,000 years.

When these three variations in the Earth's orbit are considered, they can account for much of the variation in temperature that has occurred over Earth's history, though exactly how much is still cause for debate. In particular, the impact that these variations have on the 65°N latitude seems critical in understanding how ice ages have been triggered in the past. Crucially, none of the natural causes of warming discussed in this section can explain recent warming trends.

8.8 Human influences on climate

For the vast majority of scientists, there is no doubt that the climate is warming, that natural causes cannot be the reason and that manmade changes to the composition of the atmosphere are responsible. The latest Intergovernmental Panel on Climate Change (IPCC) report (the Fourth Assessment Report, or AR4) released in 2007 concluded that:

> *Warming of the climate system is unequivocal, as is now evident from observations of increases in global average air and ocean temperatures, widespread melting of snow and ice, and rising global mean sea level.*

In addition, the IPCC was clear about attribution of the warming:

> *Most of the observed increase in globally averaged temperatures since the mid-20th century is very likely due to the observed increase in anthropogenic greenhouse gas concentrations* [where 'very likely' means more than 90 per cent likely].

Since the IPCC's third report was published in 2001, significant progress has been made in understanding how the climate is changing in space and time, through improvements and extensions of observations and associated analyses, and a better understanding of models and uncertainties.

Further details on the AR4, which forms the basis of this section and other IPCC publications can be found at: http://www.ipcc.ch/publications_and_data/publications_and_data_reports.shtml. Note that the AR4 considers data available up to the end of 2005 ahead of the report's publication in 2007.

8.8.1 IPCC summary of GHG trends

One of the conclusions from the AR4 was that the global atmospheric concentrations of CO_2, methane (CH_4) and nitrous oxide (N_2O) have increased markedly as a result of human activities since 1750. In the case of CO_2, fossil fuel combustion and land-use change (e.g. deforestation) have been the main causes, whereas increases in N_2O and CH_4 are linked to intensification of agricultural activities, with some contribution from fossil fuel use for CH_4. The atmospheric concentrations of these species now far exceed pre-industrial values determined from ice cores spanning many thousands of years (Figure 8.13).

Figure 8.13 shows the particularly rapid increase in concentrations of all three gases recently. One of the most alarming conclusions from the AR4 was that the annual CO_2 growth-rate was larger during 1995–2005 (average: 1.9 ppm/yr), than since the beginning of continuous direct atmospheric measurements (1960–2005 average: 1.4 ppm/yr) although there is year-to-year variability in growth rates.

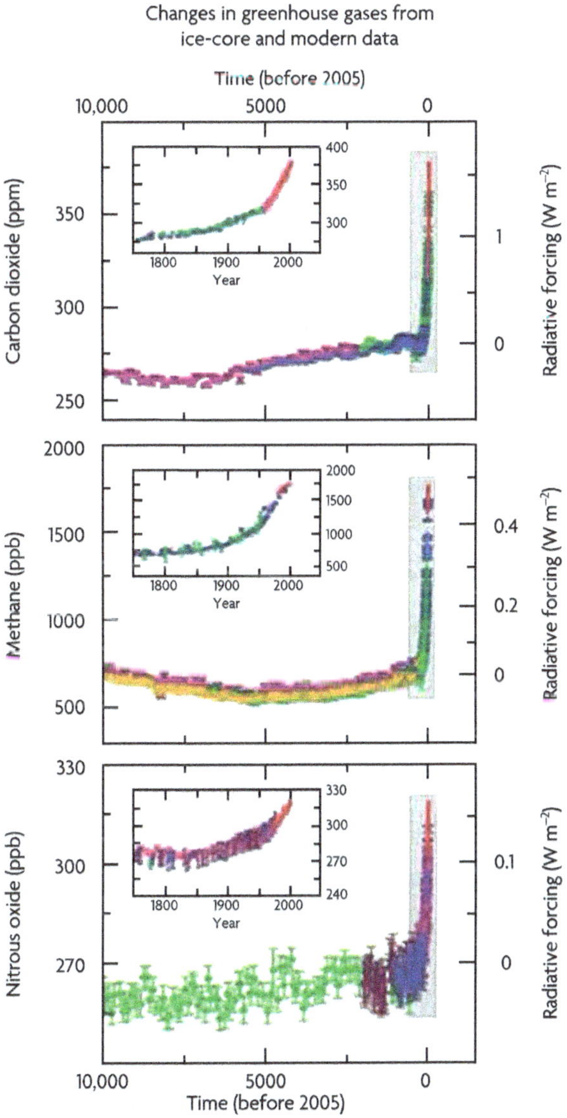

Figure 8.13 Atmospheric concentrations of CO_2, CH_4 and N_2O over the last 10,000 years (large panels) and since 1750 (inset panels). The measurements from ice cores are shown with different coloured symbols for different studies and the atmospheric samples are shown by red lines. In addition, the corresponding radiative forcings are shown on the right-hand axes of the large panels (from the IPCC AR4, 2007).

8.8.2 IPCC summary of GHG impacts on radiative forcing

So far in this section we have learned that GHG concentrations have increased significantly since 1750, and we also know that these gases cause warming by absorbing outgoing LWR in the atmospheric window (section 8.4.3). Scientists measure the impact of increasing concentrations of GHGs through investigating changes in the *radiative forcing* brought about by a change in the climate system. According to the IPCC:

> *Radiative forcing is the change in the net, downward minus upward, irradiance (expressed in W m^{-2}) at the tropopause due to a change in an external driver of climate change, such as, for example, a change in the concentration of carbon dioxide or the output of the Sun.*

A positive forcing causes heating, while a negative forcing causes cooling. Radiative forcing is thus a useful way of comparing the effects of different components of the climate system, although somewhat simplistic as these different components also interact with each other in reality. The IPCC AR4 reported *very high confidence* (confidence level that this finding is correct of at least 9 out of 10) that the globally averaged net effect of human activities since 1750 has been one of warming, with a radiative forcing of +1.6 [+0.6 to +2.4] W m^{-2}.

Figure 8.14 shows in detail how this figure has been derived. There are a number of components that have been considered to calculate this overall radiative forcing figure in addition to changes in the GHG concentrations. The figure shows that the combined radiative forcing due to increases in the concentrations of CO_2, CH_4, and N_2O since 1750 is +2.30 [+2.07 to +2.53] W m^{-2}. The IPCC AR4 concluded that such a rate of increase was *very likely* to have been unprecedented in more than 10,000 years. The concentrations of the other GHGs, halocarbons and tropospheric ozone have also contributed to further warming over this period, with contributions of 0.34 and 0.35 W m^{-2} respectively.

Note that a convenient way of comparing the impacts of different GHGs is through their global warming potentials or GWPs. The IPCC defines the GWP as:

> *An index, based upon radiative properties of well-mixed greenhouse gases, measuring the radiative forcing of a unit mass of a given well-mixed greenhouse gas in the present-day atmosphere*

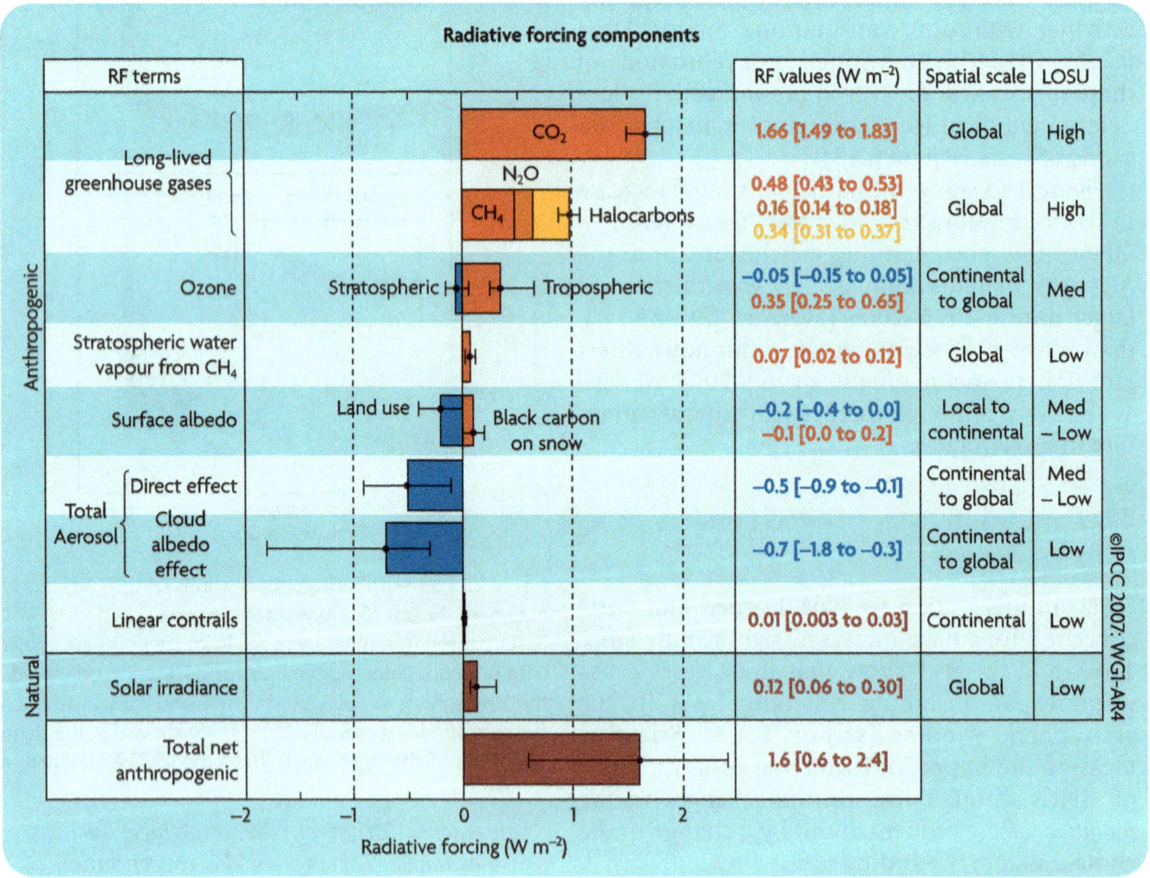

Figure 8.14 Radiative forcing components from the IPCC FAR (2007). The radiative forcing values of each component are shown with the possible range of values. Also shown are the spatial scale over which the effects occur, and the level of scientific uncertainty (LOSU) in the estimates.

integrated over a chosen time horizon, relative to that of carbon dioxide. The GWP represents the combined effect of the differing times these gases remain in the atmosphere and their relative effectiveness in absorbing outgoing thermal infrared radiation.

Some gases have a very long lifetime in the atmosphere (such as the CFCs) and although released in small quantities relative to CO_2, can cause damage over much longer periods. Consequently, over a 100-year time frame, the GWP of CFC-11 with a lifetime of 45 years has a GWP of 4750 when compared to the assigned value of unity for CO_2.

Stratospheric ozone has played a slightly different role over this period. The amount of ozone in the stratosphere has actually been depleted over this period owing to the widespread use of CFCs before they were banned in 1989 (see section 3.11.3). This depletion of ozone meant that less heat was generated through the absorption reactions of the Chapman mechanism described in section 3.7. Consequently, changes in stratospheric ozone concentrations since 1750 have led to a slight cooling of 0.05 W m^{-2}.

The stratospheric water vapour contribution is associated with more uncertainty. Oxidation of methane can increase the concentration of water vapour in this part of the atmosphere, but there may be contributions from other processes and these may vary at different altitudes within the stratosphere (see the AR4 for more details). Overall, changes in stratospheric water vapour concentration are thought to have led to a positive radiative forcing of 0.07 W m^{-2}.

The surface albedo is split into two separate components: land use and black carbon on snow. The *albedo* (A) of an object is the extent to which it reflects light, and can be defined as the ratio of reflected to incident EM radiation. Bright surfaces such as fresh snow and white clouds have high albedos (~0.7–0.85) whereas darker surfaces such as forests and dark soil have lower albedos (~0.1). The albedo of forested land tends to be lower than that of cleared land; increased leaf area within the canopy and multiple reflections around it result in incident radiation being efficiently absorbed. Therefore, forest clearance leads to an increase in the albedo, more incident radiation being reflected back to space and so a cooling at the surface. It has been estimated that, since the Industrial Revolution, land use changes such as these have been responsible for a cooling of 0.2 W m^{-2}.

Black carbon falling on snow causes warming. Aerosols are produced anthropogenically in the atmosphere through high temperature processes such as combustion. As aerosols settle on the snow, they lower the albedo, leading to more absorption of radiation at the surface and an overall warming since 1750 of 0.1 W m^{-2}.

Atmospheric aerosols can cause both a direct and an indirect effect. In the direct effect, aerosols can scatter and absorb both shortwave and longwave radiation, so affecting the overall energy balance of the Earth. For instance, as aerosols are often bright particles, they can reflect sunlight back to space. The magnitude of this effect depends on the size and composition of the aerosols and on the reflecting properties of the underlying surface and has been estimated to cause a cooling of 0.5 W m^{-2}.

In the indirect effect, or cloud albedo effect, aerosols can act as cloud condensation nuclei affecting both the formation and properties of clouds. As aerosol concentrations have increased, so the albedo of clouds has increased, causing more incoming shortwave radiation to be reflected back to space and consequently, a cooling at the surface. This effect is thought to be responsible for a cooling of 0.7 W m^{-2}.

Finally, linear contrails form high cirrus clouds in the atmosphere. These clouds reflect solar energy back to space, but also absorb outgoing LWR. The latter process dominates in this case, so these contrails have been responsible for a slight warming of 0.01 W m^{-2}.

Overall, the net effect of human activities has been to warm the atmosphere by 1.6 W m^{-2}. The GHG warming we have experienced to date would have been much stronger, had it not been for the simultaneous cooling effect caused by the increased concentrations of aerosols in the atmosphere over the same period. This process is known as *global dimming*. Owing to the small size of these aerosols, they can penetrate the human respiratory system very effectively and are thought to be harmful to health. Consequently, national governments have put into place measures to improve urban air quality through reducing aerosol concentrations, such as fitting particle traps to buses. However, it remains to be seen whether such policies may actually lead to an accelerated rate of warming as the 'protective' aerosols are removed from the atmosphere. This issue illustrates nicely how dealing with an environmental problem can often cause unexpected knock-on effects. This issue is discussed more fully in Chapter 21, which is about pollution swapping.

Note that the increase in radiative forcing owing to changes in the solar irradiance (natural causes) accounts for only ~7.5 per cent of that caused by human activities. This observation is further confirmation, should the reader still need it, that natural causes cannot be the reason for the warming we are currently experiencing.

8.8.3 IPCC summary on temperature and sea level rise

Radiative forcing is a somewhat abstract concept for many, but we can investigate what changes in this term have meant in terms of effects at the ground. The temperature increase over the 100 years from 1906–2005 was given as 0.74 [±0.18] °C in the AR4. In addition, the linear warming trend over the last 50 years (0.13 [±0.03] °C per decade) was found to be nearly twice that for the last 100 years. Global average sea level rose at an average rate of 1.8 [±0.5] mm/year over the period from 1961 to 2003, again with a faster rate over 1993 to 2003, of about 3.1 [±0.7] mm per year. The total twentieth century rise is estimated to be 0.17 [±0.05] m.

There have also been changes in Arctic ice extent, widespread changes in precipitation amounts, ocean salinity, wind patterns and aspects of extreme weather including droughts, heavy precipitation, heat waves and the intensity of tropical cyclones. These observations support the conclusion that it is *extremely unlikely* (less than 5 per cent) that global climate change of the past 50 years can be explained without external (human) forcing, and *very likely* (more than 90 per cent) it is not due to known natural causes alone.

Lead-in question

We've found out a lot about climate change in the past, but isn't what happens next more important? How do we know what will happen in the future?

8.9 Predicting the future

So we have a pretty good idea of the observed changes to date, but the big question is what will happen in the future? In order to try and ascertain future climate changes, it is necessary to use models. Such models need to account for a range of factors including atmospheric and oceanic circulation, the atmospheric radiation budget, the hydrological cycle, chemical reactions in the atmosphere and the ocean, the energy flow in rocks and soils and the role of the biosphere. These models can then be tested by attempting to reproduce observed features of current and past climate change. If there is good agreement, we can assume that the models represent the physical processes reasonably well and hence have confidence that future predictions are based on sound science. First, we need to consider what the future might look like.

8.9.1 Scenarios for the future

The IPCC have defined four 'storylines', each of which assumes a distinctly different outcome for the future. These storylines aim to incorporate a significant portion of the underlying uncertainties in the key drivers for climate change. The four storylines cover a wide range of parameters such as demographic change, economic development, and technological change, and all assume the absence of any climate change initiative. The storylines are then further divided to give a range of possible futures that can be tested within models (http://www.grida.no/publications/other/ipcc_sr/). While no one storyline is likely to provide an exact representation of the future, we should get an idea of the likely outcomes for climate if we follow a certain path.

Within the A1 storyline, there is rapid economic growth, a global population that peaks in mid-century and declines thereafter, and the rapid introduction of new and more efficient technologies. It is assumed that there is convergence among regions, capacity building, and increased cultural and social interactions. At the same time, a substantial reduction in regional differences in per capita income is assumed. The A1 family is further subdivided: A1FI assumes a fossil-intensive future; A1T a future with non-fossil fuel energy sources; A1B represents a balance across a range of sources.

The A2 storyline and related scenarios assume a much more heterogeneous world where self-reliance and preservation of local identities are dominant themes, with economic and technological development having a regional focus. Global population continues to increase.

The B1 storyline and related scenarios describe the same sort of convergent world and population structure as A1, but with service and information technologies becoming dominant and the introduction of cleaner technologies rather than rapid economic growth being a focus. There is also more emphasis on environmental sustainability and improved equity.

Finally, the B2 scenarios consider local solutions to economic, social and environmental sustainability, unlike the global emphasis of B1. Global population is assumed to grow at a lower rate than for A2, and there are intermediate levels of economic development, with less rapid but more diverse technological change than for A1 and B1.

In summary, A1 and A2 focus on economic growth while B1 and B2 have environmental protection as a focus. A1 and B1 consider global solutions, while A2 and B2 operate more on a regional or local level.

8.9.2 Climate models and uncertainties

Since the third IPCC report, we have a much wider range of models with improved representation of climate science. In addition, more observational data have become available with which to test the model results. Consequently, confidence in climate models has improved significantly; models can now reproduce the increased rate of warming observed between 1995 and 2005.

However, confidence in model estimates is higher for some climate variables (e.g. temperature) than for others (e.g. precipitation). There also remains considerable uncertainty with how clouds are represented in models and how some of the feedback systems work.

Clouds act like black bodies, absorbing almost all of the outgoing LWR. In addition, the high albedo of clouds, along with their global ubiquity, leads to a doubling of the albedo compared to a cloudless Earth. Cloud thickness and temperature at cloud top determine how much energy a cloud can radiate to space. An individual cloud may warm or cool the Earth, depending on its thickness and height above the surface. The transient nature of clouds and their variable properties are very difficult to parameterise accurately in climate models.

Another important issue is climate feedbacks. First, what do we mean by feedback? Imagine a system where quantity A changes which, in turn, affects quantity B. The change in quantity B then affects quantity A again. If quantity A subsequently increases, this is a positive feedback loop, whereas a consequent decrease in quantity A is described as negative feedback.

We can take this rather abstract definition and apply it to a real aspect of the climate system, namely water vapour. As the climate warms, the temperature increases and water evaporates more readily to form water vapour. As water vapour is a GHG, it can absorb and re-emit radiation in the lower atmosphere, contribute to further warming, with more consequent evaporation etc. This example illustrates a positive feedback in the climate system.

Similarly, there is also an ice-albedo feedback. As the atmosphere warms, more ice melts in the polar regions and there are larger areas of open water. As water has a much lower albedo than ice, less short-wave radiation is reflected back to space and more is absorbed at the surface, leading to further warming, further ice-melt etc.

Although these and many other feedback loops are understood in principle, parameterising them in models is tricky. Further, it is thought that some of these feedback loops may reach a tipping point where the climate moves to a completely new state. For instance, at some level of warming, the Greenland ice sheet will melt completely. The passage of a large volume of fresh water into the North Atlantic may 'shut down' the Gulf Stream and the climate for much of the northern hemisphere would cool. However, defining when such a tipping point may occur is very difficult.

In order to get around some of these problems, the different storylines described in section 8.9.1 have been used to initialise a range of climate models. As different models give different results, by using a range of them, we get a range of possible future values (e.g. estimates of future temperatures) for a range of future scenarios. Although we can't say precisely what the temperature will be in the future, we can predict a range of likely values with an average value, that represents the best of our scientific knowledge at the moment in the form of the different climate models, and the range of possible futures in terms of the four storylines. For more information on the models used to run these scenarios, the reader is referred to the relevant section in the IPCC Fourth Assessment Report on modelling at: http://www.ipcc.ch/pdf/assessment-report/ar4/wg1/ar4-wg1-chapter8.pdf

8.9.3 Model results: a brief summary

The range of possible surface temperatures for a selection of the storylines described in section 8.9.1 are shown in Figure 8.15. The figure shows that the B1 scenario is the best in terms of environmental sustainability, but even this scenario commits us to a 1.8 °C temperature increase relative to 1980 to 1999 values (with a range of 1.1–2.9 °C). The worst future scenario is the fossil fuel intensive future, A1F1. The most likely temperature rise by 2100 for this scenario is 4.0 °C (2.4–6.4 °C). If global warming exceeds 1.9–4.6 °C (cf. pre-industrial era), contraction of the Greenland ice sheet would outweigh precipitation and eventually (over millennia), the ice sheet would

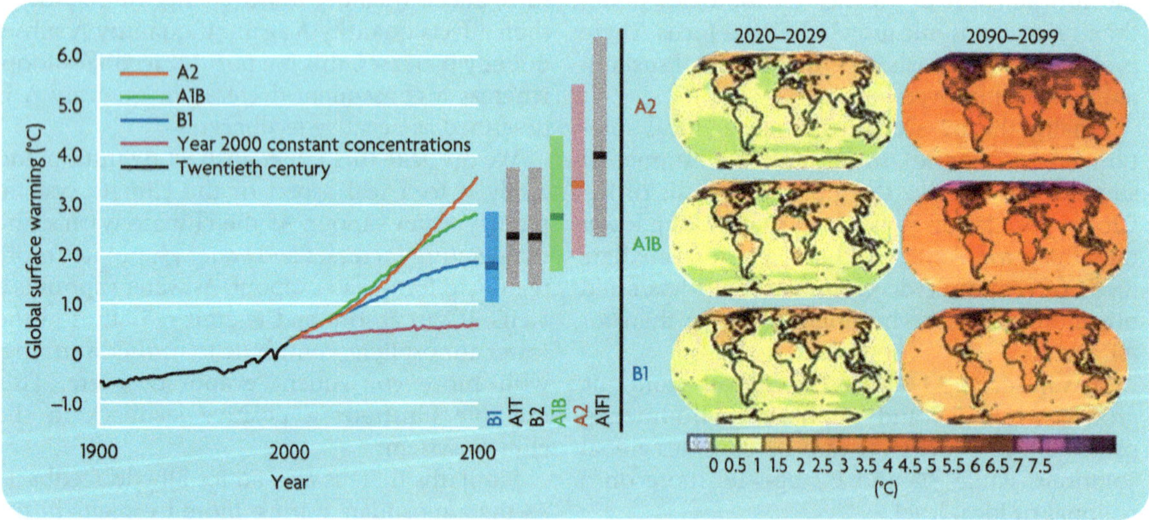

Figure 8.15 Projected surface temperature changes up to the end of the twenty-first century for different future scenarios relative to the period 1980–1999. In the left-hand panel, the solid lines show multi-model global averages of surface warming (relative to 1980–99) for the scenarios A2, A1B and B1, shown as continuations of the twentieth century simulations. The pink line is for the experiment where concentrations were held constant at year 2000 values. The coloured bars to the right indicate the best estimate (solid line within each bar) and the likely range assessed for six future scenarios (from a hierarchy of independent models and observational constraints). The central and right panels show the Atmosphere-Ocean General Circulation multi-Model average temperature projections for the A2 (top), A1B (middle) and B1 (bottom) scenarios averaged over decades 2020–2029 (centre) and 2090–2099 (right).
Source: http://www.ipcc.ch/graphics/syr/fig3-2.jpg

melt leading to extreme sea level rise. Even if emissions were halted so that concentrations remained at 2000 levels, we are still committed to a likely temperature increase of 0.6 °C (0.3–0.9 °C) by the end of the century relative to 1980–1999 levels. This is due to the timescales required for removal of CO_2 from the atmosphere, which has a very long lifetime.

Sea level rise is predicted to be between 0.18 and 0.38 m by the end of the century for the B1 scenario and between 0.26 and 0.59 m for the A1F1 scenario. Even the realisation of the B1 scenario would lead to major disruption for many coastal communities and the A1F1 outcome would be catastrophic.

Precipitation is much harder to predict with any confidence. Even the sign of the change is a topic of dispute over large parts of the globe. The models do tend to agree that the polar regions are likely to see increased precipitation while parts of North Africa, already very arid, are likely to become even drier.

8.10 Conclusions

Understanding radiative balance is crucial for examining any changes in Earth's climate. Anything that can influence this balance (globally or maybe one part of the planet) has the potential to affect the climate.

Continued emissions of GHGs at or above current rates will cause further warming and induce many changes in the global climate system that would very likely be larger than those we observed during the twentieth century. This would have serious impacts on our way of life as weather patterns become increasingly extreme and sea level rises. Further, owing to the past emissions of CO_2, there is already a large amount of warming we are already committed to, even if we halt emissions now. It is imperative, therefore, that as a society, we find ways to come to international agreements to cut our emissions drastically and soon to avoid some of the more calamitous events.

POLICY IMPLICATIONS

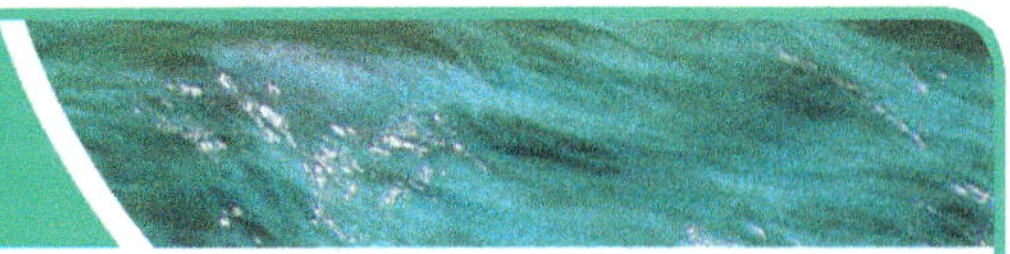

Climate change

- Policy makers need to be aware of the widely accepted causes of global warming and the nature and origins of the most important gases contributing to climate change effects with a view to formulating adequate pollution abatement strategies.
- Policy makers should be aware of what has caused rapid climate change in the past, and realise why what is happening now is very different.
- It is crucial that policy makers accept now that rates of climate change over recent decades have been too great to have arisen as a consequence of natural processes and that human activities are already starting to have a major effect upon climate.
- Policy makers need to be aware of national and global consequences of ignoring climate change, as most recently predicted by the IPCC.
- Policy makers need to be aware of the nature of uncertainties associated with climate change models and that these uncertainties must not be used as a basis for delaying immediate implementation of counteracting measures.
- Policy makers should be aware that the use of hind-casting to test the reliability of current climate change models suggests that they are now working reliably.

CHAPTER REVIEW EXERCISES

Exercise 8.1

What is the difference between climate variability and climate change, and how would you represent this difference graphically?

Exercise 8.2

How much radiation does your body emit, assuming a body temperature of 37 °C?

Exercise 8.3

Calculate the wavelength at which your body emits most radiation, assuming a body temperature of 37 °C.

Exercise 8.4

How do atmospheric circulation cells aid the redistribution of heat around the planet?

Exercise 8.5

Describe the three orbital changes in the sun that can affect climate.

REFERENCES

Burroughs, W.J. (2005) *Climate Change in Prehistory: The end of the reign of chaos*. Cambridge University Press, Cambridge.

Burroughs, W.J. (2009) *Climate Change: A Multi-disciplinary Approach*. Cambridge University Press, Cambridge.

Hoyt, D.V. and K.H. Schatten (1998a) Group sunspot numbers: A new solar activity reconstruction, Part 1. *Solar Physics*, **179**, 189–219.

Hoyt, D.V. and K.H. Schatten (1998b) Group sunspot numbers: A new solar activity reconstruction, Part 2. *Solar Physics*, **181**, 491–512.

IPCC Third Assessment Report (2001) *Climate Change 2001*, http://www.grida.no/publications/other/ipcc_tar/?src=/climate/ipcc_tar/wg1/518.htm.

IPCC Fourth Assessment Report (2007) *Climate Change 2007*, http://www.ipcc.ch/publications_and_data/ar4/wg1/en/annexessglossary-a-d.html.

Oppenheimer, C. (2003) Climatic, environmental and human consequences of the largest known historic eruption: Tambora volcano (Indonesia) 1815. *Progress in Physical Geography*, **27**, 230–259.

Solomon, S., D. Qin, M. Manning, Z. Chen, M. Marquis, K.B. Averyt, M. Tignor and H.L. Miller (eds) (2007) Contribution of Working Group I to the Fourth Assessment Report of the Intergovernmental Panel on Climate Change. Cambridge University Press, Cambridge, UK and New York, NY, USA.

Waugh, D. (2009) *Geography: An Integrated Approach*. Nelson Thornes Ltd., Cheltenham.

CHAPTER 9

Organic matter in rivers

Malcolm Cresser

Learning outcomes

By the end of this chapter you should:

- Understand the origins of natural organic matter in river water.
- Be more aware of the climatic and terrestrial factors that regulate the concentrations of dissolved organic matter in river water.
- Appreciate the importance of dissolved organic matter in the biogeochemical cycles of elements, especially carbon and nitrogen.
- Be able to explain the conflicting hypotheses currently being put forward to explain long-term changes in dissolved organic carbon concentrations in rivers and lakes.
- Be aware of why current changes in concentrations of dissolved organic matter are of concern to the water industry.
- Know approximately how much carbon and nitrogen may be transported each year by rivers to the oceans.

9.1 Introduction

In Chapter 2 we looked at the cycling of water from the oceans to the atmosphere and then back to the oceans, either as direct precipitation off shore or via drainage from terrestrial systems subjected to precipitation. We saw too that substantial fluxes of elements derived from biogeochemical weathering are transported back to the oceans by rivers, both as suspended sediment and as solute (dissolved) species. Both forms contribute to new rock formation and the rock cycle over geological timescales, as discussed in Chapter 1. Much of the carbon transported is as hydrogencarbonate anions (HCO_3^-), where these are the negatively charged counter anions balancing the positive charge on mobile base cations released from mineral weathering in soils and rocks. However, not all the organic matter derived from plants, animals and soil biotic components is converted all the way through to carbon dioxide as it is decomposed by soil microorganisms. As we saw in Chapter 7, some is converted to stable soil organic matter chemical components. Though only sparingly soluble, these too are important to the transport of elements in river systems. This is especially true for carbon and nitrogen.

We also saw in Chapter 2 (Figures 2.23 and 2.24) that water flowing through surface, organic matter-rich surface soil horizons during storms is likely to have a higher concentration of dissolved organic carbon (DOC) than water draining from lower, mineral-rich soil layers. This was represented by pathway B in Figure 2.23. Even when precipitation is so intense that precipitation rate exceeds infiltration rate so that overland flow (pathway A in Figure 2.23) occurs, there is still plenty of time, as the water winds its way down slope between vegetation stems and leaves, for DOC concentration to increase via contact with the soil surface. Sometimes too, on lower slopes, water initially flowing through surface soil horizons may issue from the surface as return flow (pathway D in Figure 2.23).

Often the presence of this elevated concentration of dissolved organic matter in rivers is clearly visible, as is the case in Figure 9.1 for example, even when the organic carbon concentration is only at around 10 mg l^{-1}. The river shown in Figure 9.1 flows on into a series of reservoirs used to supply potable water. Most consumers would be reluctant to bathe in such highly coloured water, let alone drink it. Therefore the discolouration of water by organic matter is a real cause for concern to the water industry because of the cost of its removal prior to feeding into supply networks.

Figure 9.1 Water in the River Etherow, which drains an upland basin between Barnsley and Manchester in England, displaying a brown discolouration due to the presence of dissolved organic matter.

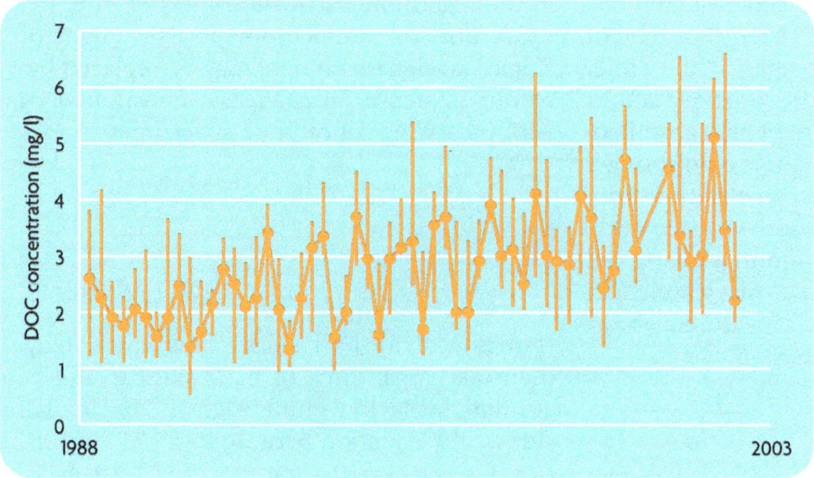

Figure 9.2 Changes over 15 years in the median values and associated interquartile ranges of DOC concentrations for 10 lakes across the UK (from Evans *et al.*, 2005). Four samples were collected each year.

Source: Draft figure provided by Dr Chris Evans. Evans *et al.* (2005).

Perhaps even more worrying to the water industry is the observation that DOC concentrations have been increasing for more than a decade in rivers and lakes in many catchments in the UK (e.g. Evans *et al.*, 2005; Figure 9.2) and elsewhere in the northern hemisphere. Therefore there is a real need to understand the processes that regulate how much DOC occurs in surface waters. Only when we know that will we be in a position to manage the problem in a cost-effective way. At the time of writing there are divergent opinions among the scientific community about the precise causes of this increasing DOC trend and how widespread a problem it may be for lakes and rivers.

Lead-in question

From where do you think that the DOC in rivers originates?

9.2 The origins of DOC and water discolouration

Readers who enjoy rambling across minimally managed countryside may well have noticed that rivers often appear to look more highly coloured in upland areas where the soils are highly organic or peaty (as indeed they are in much of the River Etherow catchment in Figure 9.1). It is tempting, therefore, especially as highly organic soils are generally darker in colour, to simply assume that the colour, and hence the DOC concentration, merely reflect the slow dissolution into drainage water of some of the organic matter in near-surface soils that are organic matter-rich. While there is a partial element of truth in this hypothesis, it is, however, a gross over-simplification. For a start the assumption is being made (albeit silently in the above statement!) that every organic carbon-containing species in drainage water contributes to the absorbance of visible light. However, we know that soils may contain some organic molecules that absorb ultra-violet light but not visible light, which means they would contribute to the DOC concentration, but not to colour. If it is also assumed that there is a direct linear relationship between the colour of water and its DOC concentration, this would imply that the amount of light absorbed for every microgram of dissolved organic carbon is constant. This would mean that the molecular composition of the organic matter is uniform over time, which is definitely not the case.

9.2.1 The link between DOC concentration and water colour

So far in this chapter we have used the term 'colour', which is descriptively useful but not quantitative. Natural waters that are 'coloured' must be absorbing some visible light. If a surface water looks orange-brown in colour, it must be transmitting light at the red/orange/yellow end of the spectrum, but absorbing light at the violet end of the visible spectrum. To quantify colour we need to measure the amount of light absorbed. In practice the absorption of light is even stronger in the near ultra-violet region of the spectrum, from 200 to 400 nm, and absorbance measurements are often made at 350 nm in a quartz cell (because quartz is transparent at this wavelength but glass is not). The parameter measured, absorbance, A, is defined by the equation:

$$A = -\log_{10}(I_t/I_0)$$

where I_t is the intensity of the transmitted light passing through the cell and its contents and I_0 is the intensity of the incident light before any absorption has occurred. The reason that we use absorbance calculated in this way is that for a single compound at low concentration in solution in a cell of constant optical path length, absorbance is directly related to concentration. We can express this as another equation, known as Beer's law:

$$A = \varepsilon c l$$

where ε is a constant that is characteristic of the absorbing molecule at any specified wavelength, c is the molecular concentration, and l is the optical path length.

Although Beer's law is empirical, it may be derived from a simple single assumption. Consider photons of light passing through absorbing molecules of a single substance in solution; we may assume that there is a fixed probability of each photon being absorbed by each molecule. Suppose, for simplicity of calculation, that for a concentration of c moles per litre, half of n incident photons would be absorbed over a 10 mm path length. The transmitted light intensity would then be n/2. What happens if we double c to 2c? The probability of absorption remains the same for the remaining n/2 photons, so the transmitted light intensity would fall to n/4. The corresponding intensities at concentrations of 3c, 4c, 5c etc., would be n/8, n/16, n/32 etc., so as concentration increased in linear steps from c to 5c, transmitted light intensity, I_t would fall off in the logarithmic sequence n/2, n/4, n/8, n/16, n/32. But for monochromatic light (light of a single wavelength) n here may be replaced by I_0, so our sequence for change in transmitted light with increasing concentration becomes:

$$I_t = I_0/2,\ I_0/4,\ I_0/8,\ I_0/16,\ I_0/32,\ I_0/64, \text{ etc.,}$$

or:

$$I_t = I_0/2,\ I_0/2^2,\ I_0/2^3,\ I_0/2^4,\ I_0/2^5,\ I_0/2^6, \text{ etc.}$$

If we now divide both sides of this equation by I_0, then take logarithms of both sides to the base 10, and bring the minus sign to the left-hand side of the equation because log ($^1/_2$) equates to (log 1 – log 2) = (0 – log 2) = –log 2, we get:

$$-\log_{10}(I_t/I_0) = \log 2,\ 2\log 2,\ 3\log 2,\ 4\log 2,\ 5\log 2,\ 6\log 2, \text{ etc.}$$

The term 'log 2' in this expression is effectively a constant that stems from our initial arbitrary assumption that half of the photons would be absorbed at concentration 'c', so we can replace log 2 by a constant, k, characteristic of the absorbing molecule. Thus as c increases to 2c, 3c, 4c, 5c, 6c, absorbance ($-\log_{10}(I_t/I_0)$) changes in the order k, 2k, 3k, 4k, 5k, 6k. In other words, absorbance increases linearly with concentration.

According to Beer's law, a graph of absorbance versus concentration for an absorbing substance in solution should be a straight line passing through the origin. This is true so long as the molecules don't start to interact with each other in such a way that the probability of a photon being absorbed is lowered, which unfortunately sometimes happens at high solute concentrations, causing curvature towards the concentration axis. Usually though, calibration graphs are linear up to absorbance values in the range 0.6 to 0.8 for single absorbing species.

Figure 9.3 shows the relationship between ultraviolet (u.v.) absorbance at 350 nm and TOC concentration for series of samples taken at regular intervals through two sequential storms for a catchment at Glen Dye in NE Scotland. There are two noteworthy points about this graph. The first is the curvature, which shows that the absorbance per unit mass of carbon, declines as TOC concentration increases. The second point is that the

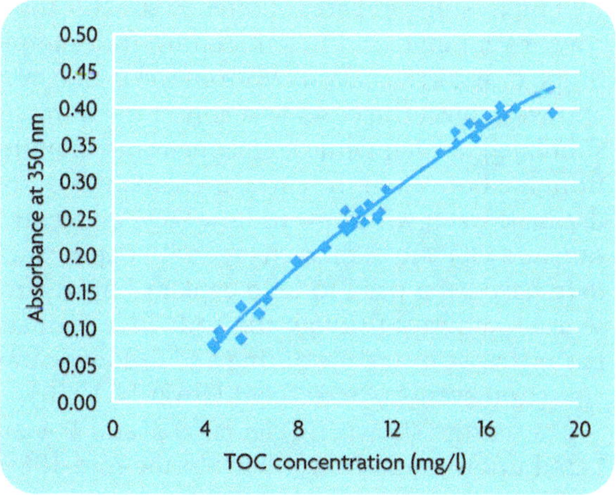

Figure 9.3 Change in absorbance with TOC concentration in samples collected at regular intervals from a river in NE Scotland throughout two consecutive rain storm events. Note that we have used Total Organic Carbon (TOC) concentration here rather than DOC. TOC often includes colloidal particulate organic carbon. This may be removed by filtration through a suitable membrane filter if DOC is to be measured.

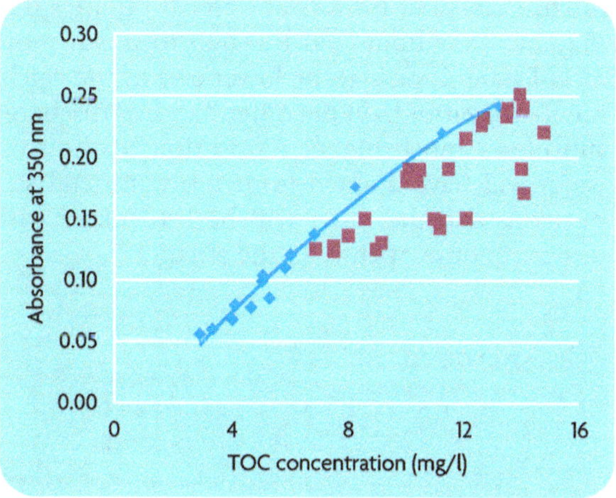

Figure 9.4 Change in absorbance with TOC concentration in samples collected at regular intervals throughout a single storm during periods of rising (blue points) and falling (red points) discharge from a stream at Peatfold in NE Scotland.

graph does not pass through the origin (0,0). This indicates that some of the DOC at low flows does not absorb at 350 nm or that some solute species other than DOC also absorbs. However, no other species present was found to absorb at this wavelength. Both points indicate that the nature of the DOC is not constant, but is varying as DOC concentration changes over time. We need to consider carefully, therefore, why this change in organic matter composition occurs.

The variation in the chemical nature of the DOC in river water throughout a single storm is even more clearly visible in Figure 9.4. On the rising limb of the storm flow, although the absorbance versus TOC concentration graph passes close to the origin, the plot is again a curve. However, the absorbance per unit mass of carbon is much more variable in this plot, and consistently lower, during the falling limb than during the rising limb.

Because of the variability in the absorption spectral characteristics of DOM, over recent decades standard and stable coloured reference solutions have been used to quantify water colour due to the presence of DOM. The American Public Health Association (APHA) has published a method for quantitatively describing yellow/orange colouration of water. It is based on optical comparison of water samples with platinum/cobalt reference solutions. Relatively inexpensive instruments are available that measure the platinum/cobalt colour (sometimes called the APHA color or Hazen color). The colour values are typically expressed in Hazen units or in mg/l Pt.

9.3 The importance of hydrological pathways

Let us now consider the differences in Dissolved Organic Carbon (DOC) concentration in two catchments in Norh East Scotland. The soils at the Peatfold catchment are less acidic than those at Glen Dye, so there are more areas of deep peat and peaty podzols at Glen Dye than at Peatfold. Therefore, even during periods of subsiding discharge after rainfall ceases at Glen Dye, a large proportion of the water in the River Dye is still likely to have drained slowly out of the highly organic and acidic soils. Moreover, the catchment of the River Dye is much larger than that at Peatfold; therefore at a sampling point at the lower end of the catchment the water will be a mixture of water from headwater tributaries that may be a few hours old through to much more recent water inputs at the lower end of the catchment. At Peatfold, on the other hand, as discharge subsides a much higher proportion of the river water will have drained from more mineral soil horizons and the water will be less mixed with respect to time of entry to the stream.

Lead-in question

How can we show that organic-rich surface horizons in soils can contribute more to DOC and water colour than lower, more mineral rich horizons?

Figure 9.5 illustrates the appearance of a typical podzol soil profile in the Glen Dye catchment. Just looking at the soil profile it appears that some organic matter has probably dissolved in water draining down the profile from the organic surface horizon and then is at least partially re-deposited lower in the profile, to give the darker brown staining at depth. Below that, the freely draining mineral soil material is very pale in colour and chemical analysis shows that it has a very low organic carbon concentration.

Figure 9.5 A typical podzol soil profile, widespread through upland areas of northern Europe and other regions of the northern hemisphere. Note how DOC in water that drains from upper, organic-rich soil horizons tends to be retained at depth. Therefore water draining from the paler mineral soil horizons towards the bottom of the soil profile tends to have a low DOC concentration.

If intact soil cores are collected in suitable tubes covering a range of different depths, for example 0–100 mm, 0–150 mm, 0–250 mm and 0–500 mm, and we subject the cores to a constant volume of simulated rainfall while collecting water draining from the bottom of the cores, then analysis of the drainage water will show us how DOC concentration and DOC-to-C ratio change with depth. Such data have been used to help explain how hydrological pathways through the soils in a drainage basin change over time during and following precipitation events (Cresser and Edwards, 1988).

The results shown in Figures 9.3 and 9.4 are based upon studies made back in the mid-1980s by Edwards and Cresser (1987). These researchers were interested in seeing how reliable estimates of DOC concentrations made from measuring u.v. absorbance were, because, if valid, such estimates are simple and rapid to make, and do not require an analytical instrument which is quite expensive to buy and to run (Figure 9.6). However, they found that the equations predicting DOC from u.v. absorbance varied significantly from catchment to catchment, so a separate calibration may be required for each catchment for precise estimations of DOC. Even earlier, because the

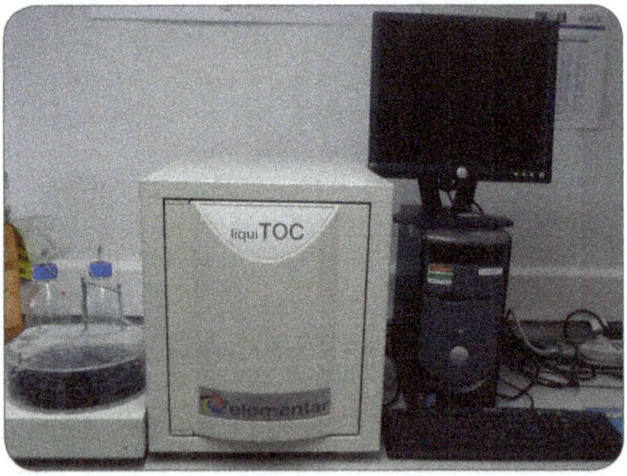

Figure 9.6 An automatic analyser designed to measure the total concentrations of organic and inorganic carbon species in water samples. For DOC determination the sample from an auto-sampler is acidified and purged of CO_2 from inorganic carbon prior to catalytic oxidation in a two-stage furnace, removal of halogens and determination by measurement of infra-red absorption by CO_2 gas produced.

ratio of absorbance at 350 nm to C changes with discharge, Reid *et al.* (1980) had shown that it is better to use separate predictive equations for low flow, high flow and medium flow conditions when estimating DOC concentration, even within a single catchment.

Collecting series of large intact soil cores to see how drainage water varies with depth in a soil profile is quite laborious. Sometimes an alternative approach is used, whereby soil solution is sucked out of the soil using a tube sealed at one end fabricated from a porous membrane material with very small pores. A vacuum is applied to the tube to suck water through the membrane, and the water inside the tube is then analysed. Such a sampling device is known as a tension lysimeter. Tension lysimeters were used to obtain the data illustrated in Figure 9.7 which shows how DOC concentration in soil solution varies with depth and seasonally throughout the year at a site intensively monitored as part of the UK NERC Environmental Change Network. The DOC concentrations are higher over the late summer (August) to early autumn (September to October) period. This temporal trend is much more pronounced for the soil closer to the surface.

It is not possible to explain the temporal trend without further research. It may well be that fresh litter inputs in early autumn are a contributing factor, as such material would be readily biodegradable at the late summer/early autumn temperatures. Slight increase in soil moisture content could also be important at this time of year. Some of the DOC could originate from foliar leaching, which can be considerable (Tukey, 1970). We could speculate that the decline in DOC concentration between November and February is due to impacts of lower temperatures in winter months, either on the solubility of soil organic matter or on microbial degradation of plant litter, but the DOC concentration remains lower through summer months too, so this cannot be the sole explanation. Later we will consider whether a similar seasonal trend to that seen in soil solution from surface soils is seen in river water too.

9.4 TOC concentration duration curves

We have seen that one of the important controls on DOC concentration in surface fresh waters

Figure 9.7 Seasonal variations in the concentration of DOC in soil solution samples sucked out of soil by tension lysimeters at Moor House in England. The upper (red) plot is for a deeper soil, and the lower (blue) plot for a more shallow soil. Data are taken from the UK Environmental Change Network data resource.

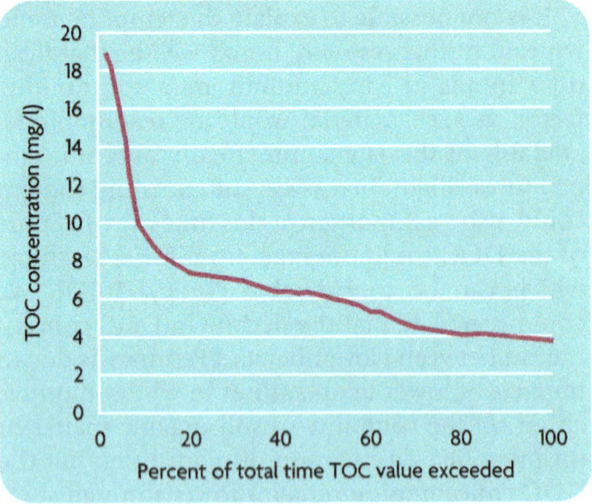

Figure 9.8 A typical TOC duration curve for a small upland catchment in the UK, showing the percentage of total time a specified TOC concentration is equalled or exceeded.

is hydrological pathway. However, some hydrological pathways may only become significant on relatively rare occasions, for example during the middle period of prolonged storm events. When a large data set is available for DOC concentration (e.g. > 50 values), it may be useful to plot a graph showing the percentage of total time that any specified TOC concentration value is equalled or exceeded. An example of such a plot is shown in Figure 9.8 for a relatively small upland catchment in Scotland. This plot shows, for example, that the DOC concentration only exceeds 10 mg of DOC per litre for 9 per cent of the year over which the data were collected.

Such plots, which are known as DOC duration curves, are very informative for the water industry and indeed to anyone who needs to design a sampling strategy to assess how much organic carbon is being transported in river systems. There is frequently a degree of similarity between the shapes of a river's flow duration curve (discussed in Chapter 2) and that of its DOC duration curve; however, this is not always the case because DOC concentration in river water is controlled by several other factors as well as discharge. If a duration curve is plotted for absorbance at 350 nm rather than DOC concentration, this too will often differ from the DOC duration curve, for reasons that should be clear from the previous section.

9.5 Other clues to hydrological routing from water chemistry

In section 9.3 we considered how analysis of leachates from intact soil cores of differing depths, or of soil solution sampled with tension lysimeters, can sometimes help us elucidate the origins of DOC in drainage waters. However, analytical data from river water analysis also help us to understand how the origins of the river water change during the course of storm events. For example, generally soils naturally acidify from the top down. Therefore the organic-matter rich surface horizons are generally more acidic than underlying mineral soils; as a consequence, generally, at least in minimally managed catchments where no lime has been applied, there is an inverse correlation between water DOC concentration and pH throughout a storm (remember that more acidic waters have a lower pH value). Thus high DOC concentration is usually related to lower alkalinity (Figures 9.9 and 9.10).

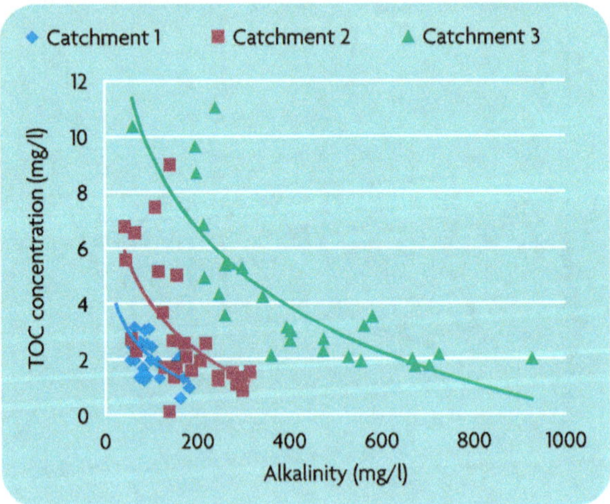

Figure 9.9 Relationships between TOC concentration and alkalinity in river water samples collected every 14 days in three upland catchments in NE Scotland. Catchment 3 (green line) had 23.8 per cent basin peat, catchment 2 (red line) had 15 per cent and catchment 1 (blue line) none. The catchments were predominantly heather moorland and rough grazing or montane.

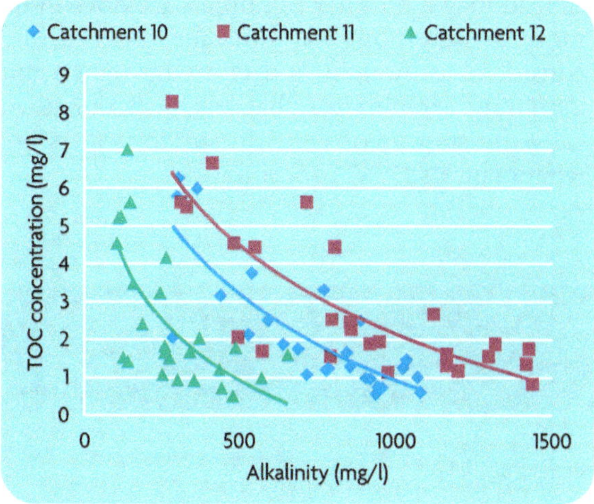

Figure 9.10 Relationships between TOC concentration and alkalinity in river water samples collected every 14 days in three more upland catchments in NE Scotland. Catchment 11 (red line) had 7.3 per cent basin peat, less than catchment 12 (green line) which had 10.1 per cent but more than catchment 10 (blue line) which had no basin peat. The catchments were quite small (0.95–4.02 km^2) and at higher altitude than those in Figure 9.1, so were all dominated by heather moorland and montane vegetation.

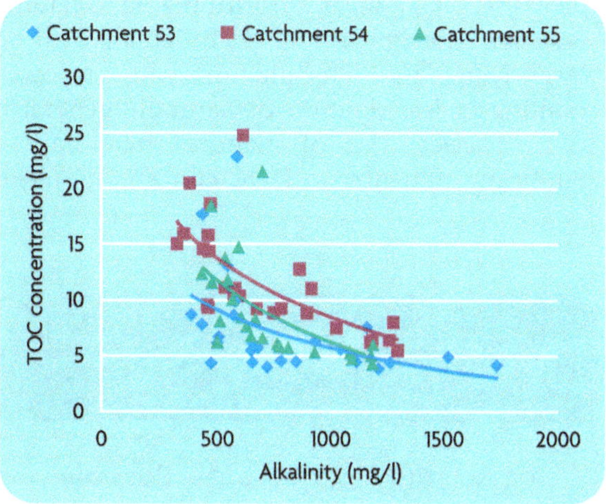

Figure 9.11 Relationships between TOC concentration and alkalinity in river water samples collected every 14 days in three lower-lying catchments in NE Scotland. All were below 263 m above sea level. Catchment 54 (red line) had 10.2 per cent basin peat, catchment 55 (green line) had 3.7 per cent and catchment 53 (blue line) had 2 per cent. The catchments were predominantly agricultural, with 34–37 per cent arable land and 3.5–8 per cent grassland, but all three contained some coniferous woodland too (11.8–13.4 per cent).

Sources: Figures 9.9–9.13 were generated from data collected by Richard Smart and Andrew Wade when working with the author and colleagues on a NERC research programme on large-scale processes in Ecology and Hydrology.

Lead-in question

Will there be a simple single equation relating TOC concentration to alkalinity for a series of different upland catchments?

From the earlier sections of this chapter, and from the contents of Chapters 1 on geology and Chapter 7 on soils, it should be clear that alkalinity generation in soils will vary from soil to soil and catchment to catchment. Thus mineral soils of different depth and/or different mineral composition may be found under superficially similar acidic organic horizons. Moreover, large expanses of boggy peatland are especially good sources of DOC. Therefore, in Figure 9.9, at any chosen alkalinity there is a higher DOC concentration in a catchment with higher basin peat content. However, to simply try to use the percentage of basin peat to make predictions of DOC concentration would be over optimistic. Figure 9.10 shows corresponding data for three further catchments, all at higher altitude than those for Figure 9.9. The link to the percentage of basin peat differs here, almost certainly because of differences in topography, climate and the spatial positioning of the peat compared with the river channel.

If other catchments in the same area contained more agricultural land, it might be expected that DOC concentrations in river water would be lower as the consequence of liming over many years should be much lower soil organic matter contents. However, Figure 9.11 demonstrates that this is not the case in practice. The increased alkalinity of the waters is apparent from comparison of the three sets of plots, but DOC concentrations go to higher values and never fall below 3 mg l^{-1}. We cannot conclude, though, that the higher DOC values are attributable to natural processes. The agricultural catchments are all on lower lying land. Precipitation will be lower, resulting in less dilution of naturally generated solute species. Use of manures may have a significant influence on DOC concentrations, as indeed may a number of human activities in the more densely populated lowland areas. Clearly considerable thought and effort has to go into designing experiments that establish unequivocally the sources of DOC in surface waters and what controls DOC concentration!

CASE STUDIES

A case study in Scotland – Possible evidence for reactions between DOC and alkalinity generated in soil

Stutter, Smart and Cresser (2002) compared the annual fluxes of weathering-derived alkalinity and base cations in samples collected every two weeks over a year from almost 60 sites spread throughout the tributary network of the River Dee in North East Scotland. To do this, the concentrations were expressed in terms of moles of charge (mol_c) of each cationic or anionic species per litre, so that it is possible to add up the base cation concentrations of a natural mixture of calcium, magnesium and sodium positively charged cations for each individual sample in a meaningful way. Potassium was not included, as its flux was negligible in relative terms. The $mol_c\ l^{-1}$ concentrations were then multiplied by the flow rate at each sampling site at the time of sampling, to provide values of cation and anion fluxes. By assuming the flux at time of sampling may be applied to each two-week period, and summation of the 26 individual fluxes, annual fluxes may be estimated.

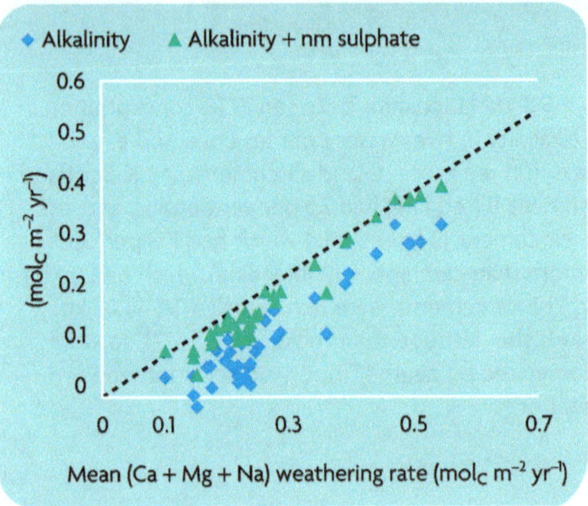

The researchers compared weathering-derived alkalinity and base cation fluxes (the blue diamonds in the above figure); they found that all points were below the 1:1 line (the dashed line in the plot). This shows that there is an excess of base cations over alkalinity. It was thought that sulphuric acid from acid rain might have been neutralising some of the alkalinity, but when the estimated non-marine derived sulphate flux was added to the alkalinity flux, the points (green triangles) still fell below the 1:1 line, many appreciably so.

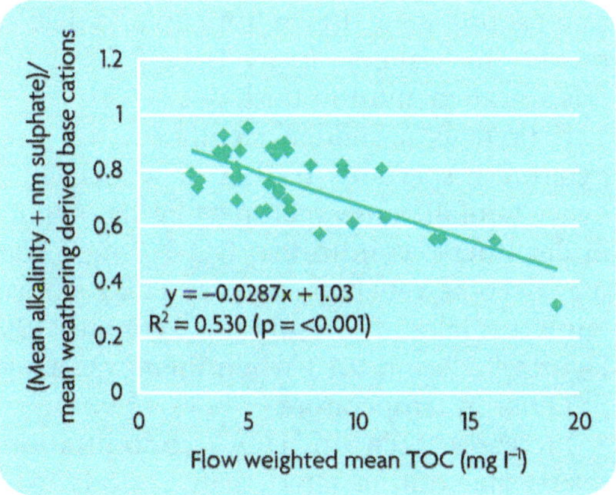

It was hypothesised that organic acids associated with the TOC might explain the apparent anion deficit. To test this idea, they plotted the ratio of mean alkalinity plus non-marine sulphate flux to weathering-derived base cation flux against flow-weighted mean TOC concentration for each catchment. They found, as shown in the plot above, that the ratio fell significantly to increasingly lower values (i.e. further below 1) when the mean TOC concentration was higher. This strongly supports their hypothesis and shows an important role of DOC in the transport of the base cations in some catchments.

Other anions such as nitrate could contribute to the anionic charge deficit too, but for the upland catchments that dominated the study area, nitrate concentrations are low for most of the year, so the contribution of nitrate could not explain the results. Therefore it seems highly likely that the phenolic and carboxylic acid groups associated with the DOC are contributing to the transport of base cations such as Ca^{2+} and Mg^{2+}.

9.6 What happens during freezing conditions?

Lead-in question

What happens to hydrological pathways when it snows?

So far we have considered only how precipitation that falls as rain is likely to move through a catchment. In many regions of the northern and southern hemisphere, however, precipitation in winter months may fall as snow rather than rain, especially at high altitude. We therefore need to consider what changes are likely to occur to hydrological pathways during snowfall events compared to precipitation events. Figure 9.12 shows how TOC concentrations in samples collected every two weeks vary over a 12-month period for three rivers in north-eastern Scotland. Figure 9.13 shows how air temperature varied over the same period.

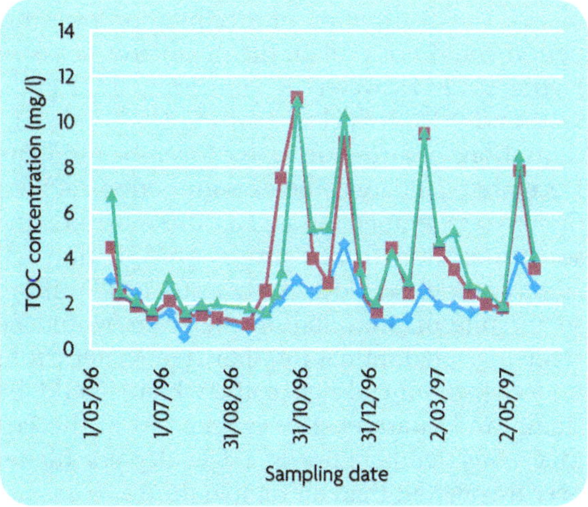

Figure 9.12 Seasonal trends in TOC concentration in water in three Scottish rivers, all sampled on the same day twice in each month from June 1996 through to June 1997.

It is clear that the peak values of TOC concentrations in Figure 9.12 occur at the same time of year in all three catchments. This suggests that the variation in TOC concentration is being largely

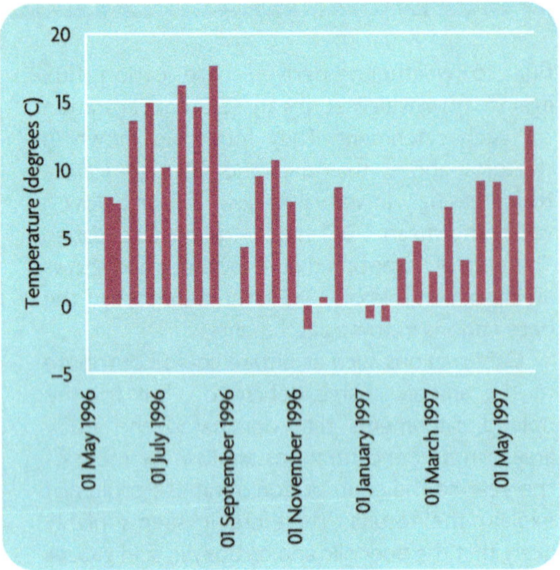

Figure 9.13 Seasonal trends in air temperature on the days the Scottish rivers were sampled to produce the graphs in Figure 9.12.

controlled by the same factor in all three catchments. The distribution of precipitation over this period could not explain the distinctive peaks in Figure 9.12. However, if you compare Figures 9.12 and 9.13, it is clear that the low TOC concentrations occurring in early November and late December–January coincide with periods at zero or sub-zero temperatures. Over these periods the soil surface was frozen and/or precipitation fell as snow. Both would severely restrict infiltration, so in spite of precipitation the rivers would be receiving predominantly drainage water from underlying mineral-rich soils, which would be significantly warmer. This is reflected in the fact that water temperature is a few degrees higher over this period than air temperature.

It is unwise to try to make definitive statements based upon a small number of catchments all within an area of just over 100 km^2, so at this point it is worth reconsidering the time series data shown for lakes in Figure 9.2. It can clearly be seen that the minima in the annual cycles occur in mid-winter each year. Peaks in DOC concentrations tend to be associated with heavy rainfall events in autumn months when soils are often already wet and near surface through-flow or overland flow are favoured. Where potable water supplies are drawn from upland river catchments, the water industry is very familiar with this so-called autumn flush.

If there is a sudden significant rise in temperature, especially if this is associated with heavy rainfall, snow melt may be very rapid, and under these conditions DOC concentrations may become very high. Such an effect almost certainly explains the results for February 1997 in Figures 9.12 and 9.13 when there was a sudden 5 °C rise in temperature.

Increases seen in DOC concentrations in river water during rapid thaw periods may not just be attributable to changes in hydrological pathways. It has been known for many years that when soils are subjected to freeze/thaw cycles, the rupture of microbial and plant component cells may release substantial fluxes of soluble organic matter (Edwards *et al.*, 1986). In the field it is difficult to differentiate experimentally between the two potential contributing mechanisms, so there is a tendency to make semi-quantitative estimations based upon laboratory simulation experiments. Moreover, in many catchments the rate of snow melt is spatially very variable, depending, for example, upon both aspect and altitude.

9.7 How important is land use?

We have already seen in section 9.5 that agricultural land use can significantly influence DOC concentrations, but that it is difficult to factor out the relative importance of the processes occurring. We can at least list some of the possible effects of land use and soil management here though, if only to indicate the complexity of the issue.

- Liming of naturally acid soils increases the decomposition rate of organic matter, which would have both short and long-term impacts.
- Ploughing in crop residues could contribute to autumn flushes of DOC.

- Bare soil will have different temperature and moisture retention compared with a cropped soil (see Chapter 8).
- Forest trees change hydrological pathways through interception losses, which differ between deciduous and non-deciduous species and with forest age and tree architecture.
- Organic C fluxes from foliar leaching and root exudation will vary with crop type and seasonally.
- Anthropogenic drainage schemes can influence C turnover rate in soil and fluxes of water and organic C transported to streams or rivers.
- Animal management, and management of associated manures, may markedly influence fluxes of C transported to adjacent rivers.
- Wastewater treatment discharges and permitted industrial discharges may have significant effects on TOC concentrations, especially in more urban areas.
- Water abstraction, for whatever purpose, may impact upon TOC concentrations downstream.
- Fish farming activities may influence DOC.
- Heather burning on grouse moors in the UK may influence hydrological flow pathways, soil temperature, precipitation interception, and organic matter decomposition rates.
- Litter inputs from bank-side vegetation may have direct impacts and impacts upon in-stream processing of organic matter.

The above list is by no means intended to be exhaustive, but rather to indicate why it is so difficult to attribute causes to the sort of increases apparently occurring in Figure 9.2. Remember too that these potential impacts need to be considered alongside the impacts of changes in climate, which may be oscillatory or uni-directional trends in temperature and/or precipitation amounts and intensities.

9.8 The possible causes of long-term DOC increase

Attempts have been made to attribute the 65 per cent increase in DOC in British lakes and streams over the 1990s to the observation that mean air temperature was 0.66 °C higher then than during the three preceding decades. Experiments with peat suggested that this rise would have stimulated DOC mobility (Freeman *et al.*, 2001). The proposed mechanism is interesting, and not a simple temperature effect upon DOC solubility. It invokes an enzymic latch mechanism; phenolic compounds accumulate in peat bogs because of very low activity of phenol oxidase under anaerobic conditions. The phenolic compounds inhibit the activity of hydrolase enzymes which are important to decomposition. Any lowering of water table could therefore facilitate decomposition of organic matter, increasing DOC concentrations. Tranvik and Jansson (2002) questioned this hypothesis, suggesting hydrological variations might well be more important, stressing the importance of proportion of wetland in catchments to DOC concentrations. However, we have already seen in this chapter that the latter idea alone is far from adequate to explain TOC spatial trends, let alone temporal trends, without also considering spatial distributions of soil types within drainage basins. Freeman *et al.*'s interesting response to the criticism was to suggest that the crucial effect of temperature was on organic matter decomposition rates and possibly an effect of greater drying out of organic soils in summer (Evans *et al.*, 2002). Certainly the latter effect could be important. So also could changes over time in acidifying pollutant deposition (which are considered in Chapter 16), since recovery from acidification effects known to have lowered DOC concentrations in the past might well be perceived as a more feasible hypothesis to explain why these concentrations are now increasing. Roulet and Moore (2006) suggested that the mechanism by which decreases in sulphur deposition could increase DOC concentration would be primarily a chemical one rather than biological. The author believes that this misses the crucial point that surface soil pH would increase as sulphur deposition fell, facilitating the decomposition of organic matter (i.e. increasing microbial activity). Litter decomposition is known to have slowed down in response to acidification from pollution deposition.

Because of the complexity of the issue it is perhaps hardly surprising that a more recent comprehensive study has shown decreases in DOC concentrations in many rivers as well as increases in many others across the UK (Worrall and Burt, 2007). A later assessment by Worrall and Burt (2010), which considered long-term changes in DOC/colour ratios, favoured the role of the reversal of acidification concept, but alongside other factors such as temperature change and rising atmospheric carbon dioxide effects. At the time of writing the debate looks set to continue for a few years yet, despite the importance of the issue to the water industry and the global carbon balance. More research is needed too on leaching of DOC from terrestrial vegetation (Tukey, 1970) and from aquatic plants in rivers and estuaries (Gallagher *et al.*, 1976). Although the latter topics were extensively researched in the 1960s and 70s, in the author's opinion they are not adequately reconsidered in the context of DOC concentrations in rivers.

9.9 How big are DOC fluxes to the oceans?

So far we have been concentrating upon what regulates the concentrations of TOC or DOC in river waters. We have seen that, in upland catchments of temperate areas, TOC concentrations generally increase with discharge, although discharge alone is not sufficient to explain all of the seasonal or long-term variations in DOC concentration. At the time of writing, there is also considerable interest in knowing how much carbon is transported by river systems to the oceans, because of growing concern about possible changes in the global carbon cycle. One of the earliest attempts to estimate these fluxes at national scales and beyond was that of Hope *et al.* (1997). These researchers collated available data for rivers in Europe draining temperate forest, boreal forest or temperate moorland/grassland; they then did a similar survey for rivers across New Zealand, the former USSR and North America, but including temperate grassland and wetland as dominant land cover types. Their results are summarised in Figure 9.14.

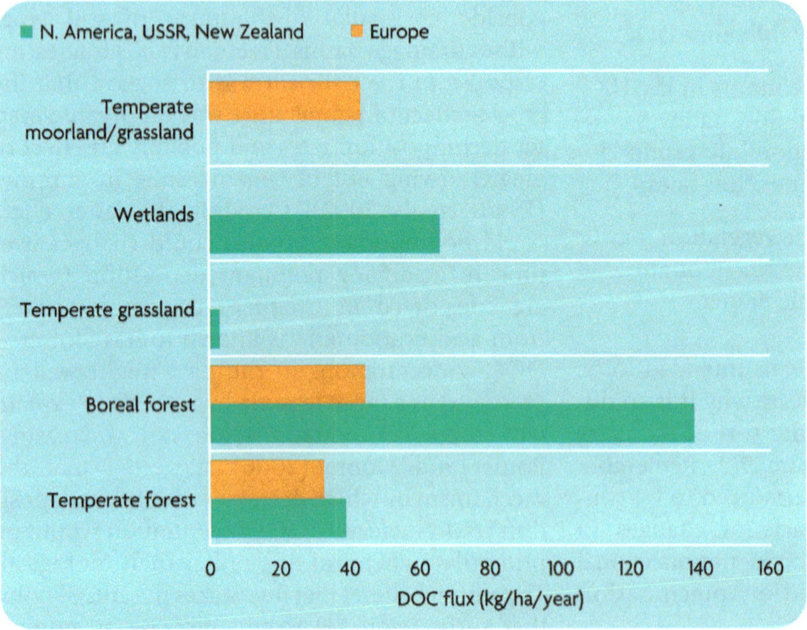

Figure 9.14 Mean values of fluxes of carbon transported from catchments with differing land use as dissolved organic carbon, estimated for groups of rivers in North America, the former USSR and New Zealand (green bars) and Europe (orange bars).

Source: Data from Hope *et al.* (1987).

The ranges of values found (not shown here) were quite high, as should be expected from consideration of the many factors that would influence DOC concentrations; as well as the factors considered in this chapter already that may influence DOC concentrations, clearly spatial climatic variations and topographic variations at the continental scale would be major contributory factors to the variation observed.

In the same review, mean particulate organic carbon (POC) fluxes were also estimated (Hope *et al.*, 1997), though less data were available from particulate organic carbon monitoring. Generally these were somewhat smaller than the DOC fluxes, but the POC fluxes were still considerable.

Upland soils in much of the northern hemisphere are often quite acidic, and as a consequence alkalinity (HCO_3^-) generation associated with biogeochemical weathering is generally low. Moreover, some of the alkalinity generated reacts with organic acid groups that constitute part of the DOC, with the release of CO_2 back to the atmosphere. This process may occur in both soil and river water itself. When Dawson *et al.* (2002) compared the transport of organic and inorganic carbon in rivers in north-eastern Scotland and mid-Wales in the UK, they obtained the results shown in Table 9.1.

From this data, 88 per cent of the C loss in Scotland and 69 per cent of that in Wales was as DOC, which emphasises the importance of reliable DOC measurement in this type of upland ecosystem. Obviously for catchments with more base-rich parent materials, or in limed lowland catchments, carbon loss as alkalinity is likely to be relatively much more important. If you look at Figure 9.10, for example, and convert mg HCO_3^- – C l^{-1} to mg C l^{-1}, it is clear that for all three catchments HCO_3^- – C concentration would exceed TOC concentration over the year.

Table 9.1 Comparison of losses of carbon in organic and inorganic forms, all expressed in kg ha^{-1} y^{-1}, from upland catchments in Scotland and Wales in the UK; taken from Dawson *et al.* (2002).

	Brocky Burn, Scotland	River Hafren, Wales
DOC – C	169	83.5
POC – C	18.5	27.4
HCO_3^- – C	1.12	1.28
CO_2 – C	2.62	8.75

9.10 Is DON important too?

So far this chapter has concentrated almost exclusively on carbon transport, especially in organic forms. However, although the nitrogen content of dissolved organic matter is much lower than its carbon content, much of what has been written about controls on DOC mobility applies equally to dissolved organic nitrogen (DON) mobility. Until relatively recently concern about nitrogen losses focused very strongly on nitrate because of concerns over impacts of nitrate from intensive agriculture on human health and eutrophication. Losses as ammonium-N or DON have been largely ignored until the start of the present millennium. Yet as van Breemen (2002) pointed out, in regions such as South America, 70 per cent of the loss of nitrogen in rivers is as DON. Parakis and Hedin (2002) also emphasised the dominant importance of DON losses in N cycling in relatively unpolluted South American forests.

The N cycle, and the role of DON to some extent, were considered in detail in Chapter 5. It is worth mentioning here though that organic N losses via drainage to streams must always have been important in upland areas in the pre-pollution era. We saw in Chapter 5 that there is a finite limit to the amount of N that can be stored in a soil profile of somewhere around 12 tonnes of N per hectare. If a soil is 10,000 years old and the natural input from unpolluted air is about 2 kg ha^{-1} y^{-1}, then if all the N input was stored the profile would contain 20,000 kg or 20 tonnes per hectare. This exceeds the N storage capacity, indicating that some N was always being lost. This probably would have been predominantly in the form of organic nitrogen in the upland areas, losses being both as DON and via eroded particulates, although denitrification could also contribute significantly too in some soils.

POLICY IMPLICATIONS

Organic matter in rivers and lakes

- Policy makers need to know how national- and global-scale carbon budgets are changing in the context of assessing risk of climate change. DOC fluxes transferred from land to the oceans are an important part of these C budgets.
- Understanding the processes regulating DOC concentrations in rivers and lakes/reservoirs is a crucial prerequisite to catchment management in ways beneficial to the water industry to minimise water treatment costs.
- Long-term monitoring of DOC and DON concentrations is a potentially useful indicator of environmental change.
- Monitoring changes in the near u.v. and visible region absorption spectral characteristics of DOM could be considered by monitoring agencies as a method to provide early warning of changes in soil ecological processes at relatively low cost.
- Comparisons of DOC concentrations in comparison with expected concentrations for a site during routine monitoring could provide a useful indication of pollution incidents.
- It is important for environmental managers to understand the links between alkalinity and the neutralisation of organic acidity and acid deposition from the atmosphere if adequate alkalinity is to be maintained in surface fresh waters in acidification-sensitive areas.
- Policy makers need to be aware of how attempting to increase soil carbon store to ameliorate effects of rising concentrations of the greenhouse gas carbon dioxide in the atmosphere could influence DOM concentrations in surface waters.

CHAPTER REVIEW EXERCISES

Exercise 9.1

Intact soil cores were collected from a freely draining acid brown earth soil profile at a site in the British uplands in heavy duty, 150 mm diameter, PVC pipes. They were collected to total soil depths of 120 mm, 210 mm and 500 mm. A plate with a small central drainage hole was attached to the bottom of each core so that drainage water could be collected, and each was subjected to simulated rainfall for three hours. The DOC concentrations in the drainage water were 18.5, 9.30 and 4.2 μg ml^{-1} for the 120 mm, 210 mm and 500 mm deep cores respectively.

Explain the most probable reasons for the differences between the values found. How are such data useful in interpreting changes in DOC with time in upland streams that occur during and following storm events?

Exercise 9.2

A set of 24 river water samples was collected over a year at approximately two-weekly intervals from a stream at the outlet of a small upland catchment in northern Europe, and their TOC concentrations and u.v. absorbance values at 350 nm were measured. Seventeen of the samples were taken under base flow conditions and seven were thought to be at moderate to high-flow conditions. Plot a graph of absorbance (on the vertical axis) against TOC concentration, and give possible reasons for the shape of the graph and the cause or possible causes of variation about the best fit curve.

Month	TOC mg/l	Absorbance at 350 nm
January	2.9	0.056
January	4.6	0.078
February	3.3	0.06
February	3.95	0.068
March	4.1	0.08
March	12.1	0.205
April	5	0.1
April	5.3	0.085
May	5	0.105
May	3.9	0.07
June	2.9	0.055
June	2.7	0.045
July	2.8	0.05
July	3.1	0.06
August	5.25	0.11
August	10.1	0.184
September	2.8	0.05
September	2.95	0.055
October	11	0.15
October	14	0.25
November	8.6	0.15
November	12.1	0.149
December	5.05	0.108
December	11.5	0.195

Also plot a bar chart showing how TOC concentration varied over the year and give possible explanations for any temporal trend or trends observed.

Exercise 9.3

List 10 soil, climatic or catchment topographic features that could influence the DOC concentration found at any particular time of the year in a river draining an upland moorland catchment in northern Europe.

Exercise 9.4

In section 9.9, the point was made that large variations in DOC fluxes, expressed on a kg/ha/y basis, should be expected for a single type of land use when rivers are considered on a continental scale. How and why would climate and topography influence DOC fluxes at this scale?

REFERENCES

Cresser, M. and Edwards, A. (1988) Identification of changes in hydrological pathways during storms by analysis of leachate from intact soil-core lysimeters. *Scottish Geographical Magazine*, **104**, 84–90.

Dawson, J.C., Billett, M.F., Neal, C. and Hill, S. (2002) A comparison of particulate, dissolved and gaseous carbon in two contrasting upland streams in the UK. *Journal of Hydrology*, **257**, 226–246.

Edwards, A.C. Creasey, J. and Cresser, M.S. (1986) Soil freezing effects on upland stream solute chemistry. *Water Research*, **20**, 831–834.

Edwards, A.C. and Cresser, M.S. (1987) Relationships between ultraviolet absorbance and total organic carbon in two upland catchments. *Water Research*, **21**, 49–56.

Evans, C.D., Freeman, C., Monteith, D.T. Reynolds, B. and Fenner, N. (2002) Reply to: Terrestrial export of organic carbon. *Nature*, **415**, 862.

Evans, C.D., Monteith, D.T. and Cooper, D.M. (2005) Long-term increases in surface water dissolved organic carbon: Observations, possible causes and environmental impacts. *Environmental Pollution*, **137**, 55–71.

Freeman, C., Ostle., N. and Kang, H. (2001) An enzymic 'latch' on a global carbon store – A shortage of oxygen locks up carbon in peatlands by restraining a single enzyme. *Nature*, **409**, 149.

Gallagher, J.L., Pfeiffer, W.J. and Pomeroy, L.B. (1976) Leaching and microbial utilization of dissolved organic carbon from leaves of *Spartina Alterniflora*. *Estuarine and Coastal Marine Science*, **4**, 467–471.

Hope, D., Billett, M.F. and Cresser, M.S. (1997) A review of the export of carbon in river water: fluxes and processes. *Environmental Pollution*, **84**, 301–324.

Parakis, S.S. and Hedin, L.O. (2002) Nitrogen loss from unpolluted South American forests mainly via dissolved organic compounds. *Nature*, **415**, 416–419.

Reid, M., Cresser, M.S. and MacLeod, D.A. (1980) Observations on the estimation of dissolved organic carbon from u.v. absorbance for an unpolluted stream. *Water Research*, **14**, 525–529.

Roulet, N. and Moore, T.R. (2006) Browning the waters. *Nature*, **444**, 283–284.

Stutter, M., Smart, R. and Cresser, M.S. (2002) Calibration of the sodium base cation dominance index of weathering for the River Dee Catchment in North-east Scotland. *Applied Geochemistry*, **17**, 11–19.

Tranvik, L.J. and Jansson, M. (2002) Terrestrial export of organic carbon. *Nature*, **415**, 861–862.

Tukey, H.B. (1970) The leaching of substances from plants. *Annual Review of Plant Physiology*, **21**, 305–324.

Van Breemen, N. (2002) Natural organic tendency. *Nature*, **415**, 381–382.

Worrall, F. and Burt, T.P. (2007) Trends in DOC concentration in Great Britain. *Journal of Hydrology*, **346**, 81–92.

Worrall, F. and Burt, T.P. (2010) Has the composition of fluvial DOC changed? Spatio-temporal patterns in the DOC-color relationship. *Global Biogeochemical Cycles*, **24**, GB1010,doi:10.1029/2008GB003445.

CHAPTER 10

James Lovelock, Gaia and beyond

Malcolm Cresser

Learning outcomes

By the end of this chapter you should:

- Be aware of who James Lovelock is and how he changed the way most scientists now think our planet functions.
- Understand what the Gaia hypothesis states and why its concepts are so important.
- Understand why we need to think of everything that man does in the context of how the planet functions and go beyond a narrow, individual national perspective when managing the environment.
- Realise that the Gaia hypothesis embraces the earlier evolutionary concepts advocated by Darwin and others.
- Be able to use the Daisyworld model to explain how plants can regulate a planet's temperature to suit their own requirements.
- Be capable of viewing evolutionary concepts from a Gaian and planetary system perspective.
- Be able to discuss where the material in other chapters in this book fits within Gaia.

10.1 Who is James Lovelock?

James Lovelock (b. 1919) is usually described as 'an independent scientist, inventor and author' (frontispiece of Lovelock, 2000). A quick perusal of his c.v. shows he has been a Fellow of the Royal Society since 1974, has won the Amsterdam Prize for the Environment, the Nonino Prize, the Volvo Environment Prize, the Blue Planet Prize from Japan and was awarded a CBE (Commander of the British Empire) in 1990. If any one scientist deserves to have his photograph in the body of the present text, it is undoubtedly James Lovelock (Figure 10.1). But why are he and his ideas so important?

Figure 10.1 James Lovelock.
Source: © Bruno Comby – EFN – Environmentalists For Nuclear Energy – http://www.ecolo.org – with permission. Detail from this image is additionally included on the chapter opening page.

10.2 A brief introduction to the Gaia hypothesis

Early in his career, Lovelock came up with one of his most important *tangible* inventions, the electron capture detector, a highly sensitive detector for gas chromatography that facilitated the determination of chlorofluorocarbons (CFCs) in the atmosphere. Later this in its own right became a major contribution to environmental science because of its role in development of our understanding of the causes of the so-called hole in the ozone layer (as discussed in Chapter 3). Lovelock worked at NASA at the time when space exploration was about to become an exciting reality; he was therefore faced with the fascinating, but demanding, challenge of deciding what could feasibly be measured by spacecraft if we wanted to look for signs of life on other planets. To Lovelock it suddenly became clear that our own planet's atmosphere had the composition that it did, dramatically different from the atmospheres of Venus and Mars (see Chapters 1 and 3), because of the way life functioned. The facts needed to reach this inescapable conclusion were not new; what was radically new was the idea that biota regulated the planetary atmosphere and temperature to maintain conditions that facilitated their continued survival.

Thus James Lovelock's Gaia hypothesis was born in the 1960s (Lovelock, 1972). In the Gaia hypothesis our planet was presented as a single, giant, self-regulating organism; the innovative underpinning idea was that Earth had evolved naturally in such a way that it was sustaining a stable environment suitable to support all the interacting forms of life living upon it (Lovelock, 1979). In Lovelock's own eloquent words: 'The Gaia hypothesis was first described in terms of life shaping the environment, rather than the other way round' (Lovelock, 2005). Thus the planet controlled both the composition of its atmosphere and its temperature to sustain life. The Gaia hypothesis encompassed Darwin's conceptually more limited theories of evolution (see Chapter 25), since evolution was just one facet of Gaia's operational process (Lovelock, 1979).

There were several unfavourable criticisms of the whole concept of Earth as a single, unified living organism, and these have been reviewed ably by Turney (2003). Critics apparently chose to think that Earth was being presented as some kind of thinking super-organism, rather than as a planet where everything interacted more or less symbiotically with everything else. This was, of course, massively widening the scope of symbiotic relationships compared to the concept presented for clover and rhizobia in Chapter 5. To the present author it seems as if the reservations expressed about Gaia were perhaps more to do with the critics' blinkered interpretation of the language used to put complex concepts within the hypothesis across to the public at large, rather than to problems with the Gaia hypothesis *per se*. Sadly, many career scientists think science has to be presented in a rather dull way to be taken seriously. As Lovelock himself put it when talking about presentational style in his first book on Gaia: 'My disclaimer was about as much use as the health warning on a packet of cigarettes to a nicotine addict' (Lovelock, 2000).

As they evolved, Gaian concepts were considered over a range of timescales, from fractions of a second through years and decades to evolutionary and geological timescales. Illustrative examples over these scales are included in Lovelock's excellent series of books on Gaia (Lovelock, 1979, 1988, 1991, 2000, 2006). In the later books especially, Lovelock pays considerable attention to how Earth's natural systems struggle to combat pollution impacts associated with man's attempts to run the planet to best meet his own perceived needs, disrupting its natural modus operandi. Titles such as *The Revenge of Gaia* (Lovelock, 2006) did nothing to quash the criticisms of presentational style. Nor were they intended to, because their intended audience was the much wider population that needed to wake up to the reality of how mankind was (and still is) abusing the planet.

10.3 Planet Daisyworld

Watson and Lovelock (1983) published a model in the early 1980s designed to illustrate, in a simplified way, how plant communities could evolve to regulate temperature on a planet's surface as conditions changed on the star that the planet was orbiting. The idea involved a planet called Daisyworld with two possible plant species, black daisies and white daisies. At the simplest level, when the star was quite dim, only the equatorial region might be warm enough to support plant life. Black daisies would have a distinct competitive edge because they would absorb more solar radiation, and thereby help warm the planet; so they would flourish, whereas white daisies would reflect light and not compete favourably. As the

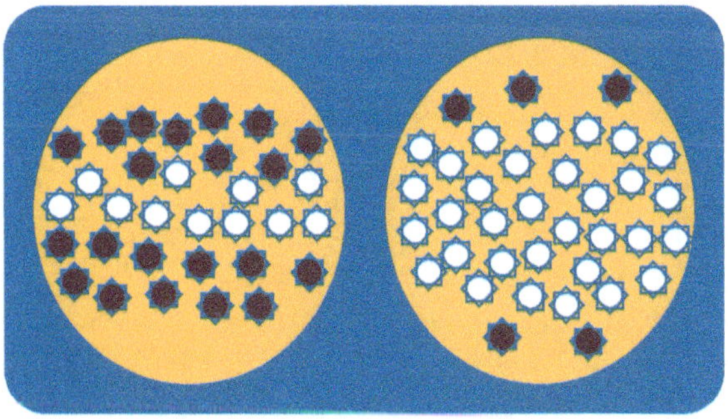

Figure 10.2 Author's simplified representation of Planet Daisyworld. Once it has become too hot for the dark daisies around the equator, so more reflective white daisies become established, replacing the earlier dark daisies (left). If solar energy continues to increase the area dominated by white daisies would continue to increase (right).

planet warmed, the black daisies would become established progressively further north and south of the equator.

Suppose then the sun started to emit more radiation. It could become too hot for the dark daisies in equatorial regions, and the white daisies would gain a competitive edge there, as indicated in Figure 10.2 (left). By reflecting the solar energy they would reduce the temperature of the planet surface. If white daisies migrated north or south too quickly the temperature would become too cold for them, so their spread would be checked, and dark daisies favoured. Eventually though, if the star became hotter, white daisies would predominate except in polar regions that receive less solar energy per unit area because of the curvature of the planet's surface (Figure 10.2, right).

It is important to realise that in the absence of the temperature buffering effect of plant colour then as the solar radiation increased the planet temperature would have displayed a corresponding steady increase, as indicated in Figure 10.3. Later the model was modified by predicting impacts of adding herbivores such as rabbits and carnivores such as foxes. A full and very readable account of the outcome can be found in Chapter 3 of Lovelock (1991). Suffice it to say here that their impact on the temperature buffering was minor.

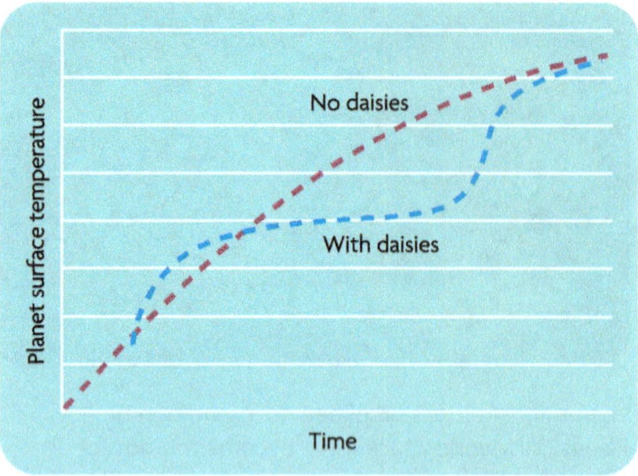

Figure 10.3 Typical output for Daisyworld model, showing how growth of dark and white daisies can stabilise the planet surface temperature for a long period of time in spite of changes in solar energy output. Initially black daisies increase the temperature but as solar energy input increases, white daisies start to cool the planet.
Source: Adapted from Lovelock (1991).

Cynical readers may be starting to think: 'Gosh, this is all a bit toy town!', but if you do you have missed the point of the model, which was to demonstrate in a very simple way the concept that biota can influence climate to maintain a temperature best suited to their own needs. Several other potential and more realistic feedback mechanisms might be postulated for the real world. For example, we know that increasing carbon dioxide (a greenhouse gas) concentration in the atmosphere can contribute to global warming. But from consideration of the carbon cycle in Chapters 4 and 16, rising concentrations of carbon dioxide in the atmosphere can also increase the growth of plants, thereby removing carbon dioxide from the atmosphere. This would be a potential cooling feedback mechanism, at least partially offsetting the global warming effect. Rising carbon dioxide concentration also can accelerate the physiological growth stages in plants (Woodin *et al.*, 1992), so senescence in annual plants will occur earlier. This could have a beneficial effect upon temperature by reflecting more incoming radiation from the senescent vegetation. Thus we need to consider a complex set of interactive positive and negative feedback mechanisms.

Note here that, as in all aspects of the Gaia hypothesis, we are talking about natural processes and evolutionary processes and it's important to make this clear. So never write things like: 'Plants like the extra carbon dioxide and decide to grow faster, because they know that if they do they'll help stop global warming.' If you do you may find some very unflattering comments on your work!

10.4 The nitrogen cycle from a Gaian perspective

In Chapter 5 we considered the diverse components of the nitrogen cycle as typically represented and taught; it's now worth revisiting the topic, but this time to see what else we can conclude if we adopt more of a Gaian perspective (Cresser *et al.*, 2008). Lovelock considers the nitrogen cycle at some length, stressing its importance to the planet through the contribution it makes to atmospheric pressure, its importance to all life forms as an essential nutrient, and its diluent effect on atmospheric concentration of oxygen (Lovelock, 1991). Oxygen concentration in the atmosphere would not have to be much higher than it currently is for fire to become an uncontrollable problem. Lovelock points out that without microbial conversion of oxidised nitrogen, especially nitrate, back to nitrogen gas (N_2) via denitrification, over geological timescales lightning would long ago have converted atmospheric N_2 to nitrate, which would be transferred to the oceans by the hydrological cycle (Chapter 2) and stored there as a stable end product. Having depleted the oxygen, it would then react with carbon dioxide using the same energy source. Currently life as we know it on the planet depends upon a delicate balance between oxygen production via photosynthesis and consumption by respiration and by oxidation of N_2 in storms.

We have to assume that natural ecosytems have evolved over many tens of millennia or longer to sustainably provide environmental niches to support the enormous diversity of organisms that exist under favourable conditions. In Chapter 25 the advantages of high species diversity are explained, but because of the complexity of the interactions between life forms with each other and with their environment, we cannot yet explain in specific and quantitative detail how the chemical composition of the atmosphere is buffered as well as it seems to be. Sadly the situation is not nearly as black and white on Earth as on planet Daisyworld.

One of the many questions that James Lovelock asks is: 'Why do animals pass urine rather than exhaling N_2?'. Is this evidence for Gaia in operation? The author would like to supplement this question with another, namely: why have some plants evolved to be deciduous or to die back totally every winter? Prior to our relatively modern day tendency towards urbanisation, human and animal excrement and urine, and plant litter, recycled nutrients, especially nitrogen, quite efficiently. Clearly plants need a steady dynamic supply of nutrients during periods of active growth. For nitrogen this meant that soils had to evolve to store nitrogen which they could release as and when needed; they have done this by accumulating pools of organic matter. We saw in Chapter 7 that soil organic matter contains typically 90–95 per cent of the soil nitrogen; it also stores cationic nutrients on its cation exchange sites and helps with water retention and the establishment of stable soil structures. Thus a plant that evolved to help build up soil organic matter content in a very freely draining soil in an arid area has a better chance of survival than one that does not. Deciduous trees remove far less carbon dioxide from the atmosphere in winter months which could help maintain the concentration of the greenhouse gas at the very time of year when it is most useful.

It was mentioned in Chapter 7 that leaf litter from a deciduous tree contains a high C:N ratio compared with the underlying soil by the time leaves fall from the trees in autumn. To remind you, Figure 7.31 is reproduced again (as Figure 10.4); note how the C:N ratio of the leaves increases markedly from June to October. This has a distinct advantage for nitrogen cycling under deciduous plants (Cresser *et al.*, 2008). Because the litter has a high C:N ratio when it first falls, any ammonium and nitrate produced by mineralisation of the fresh litter is likely to be retained by the microorganisms colonising the litter. As winter approaches and temperature falls, litter decomposition slows down, so nitrogen is retained in the litter/soil system over winter. Thus it appears that deciduous litter has evolved naturally, with a high C:N ratio, to help retain mineral N species produced during winter months when tree N uptake is low. As litter decomposes progressively, its C:N ratio falls until microbial N immobilisation is less favoured. By the time it warms up in spring, plant-available N increases just when it is needed by the tree.

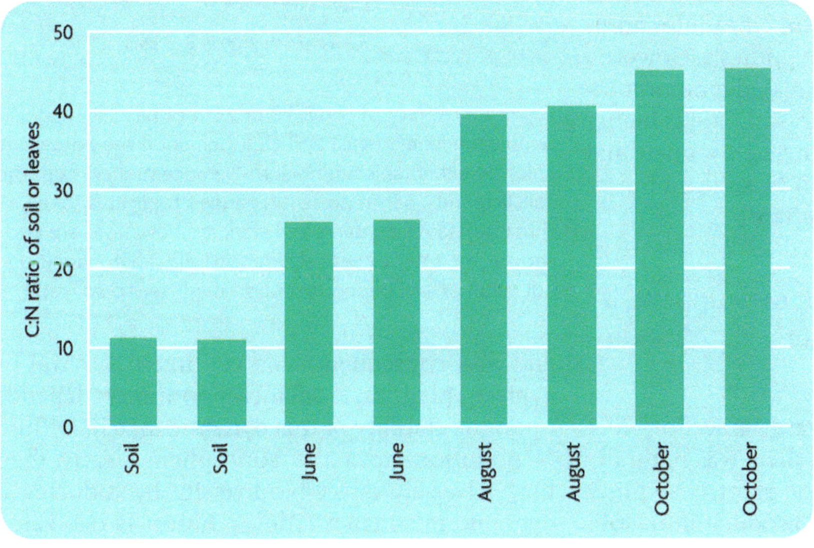

Figure 10.4 Comparison of the C:N ratio for duplicate foliage samples from hazel trees (*Corylus avellana*) sampled in June, August and October with the C:N ratio of the underlying soil.

Our hypothesis then is that, as a consequence of natural evolutionary processes, there is naturally a dynamic match between plant N requirement and mineral N production in soil. Plant continued occurrence is favoured when this dynamic match is good. Later in Chapter 16 we will consider how mineral-N pollution in the atmosphere can disrupt this natural balance.

10.4.1 A simple experiment to demonstrate that fresh litter retains N

The data in Figure 10.4 were obtained by Claire Stephens and Ishaq Mian, two students working with the author at York. Leaves from hazel (*Corylus avellana*) were sampled in June, August and October to get a range of C:N ratios. Sub-samples were dried and ball-milled for C and N analysis. An adjacent woodland mineral soil was sampled, sieved and thoroughly mixed. Either 0, 0.5, 1.0 or 2.0 g of litter (dried and homogenised by pushing through a 4 mm sieve) and 4 ml de-ionised water were added in triplicate to 15 g sub-samples of field-moist soil in loosely capped glass jars. The concentrations of KCl-extractable ammonium-N and nitrate-N were measured after seven days.

The results of the experiment are shown in Figure 10.5 and Figure 10.6 for extractable nitrate-nitrogen and ammonium-nitrogen respectively. It is very clear that much less nitrate or ammonium could be extracted from the soils which had litter added compared to the soil with zero litter addition, strongly supporting the hypothesis. However, the results also suggest that increased ammonium-N production when the litter-to-soil ratio is higher may not be completely countered by microbial immobilisation of ammonium-N, so the amount of litter deposited too is important.

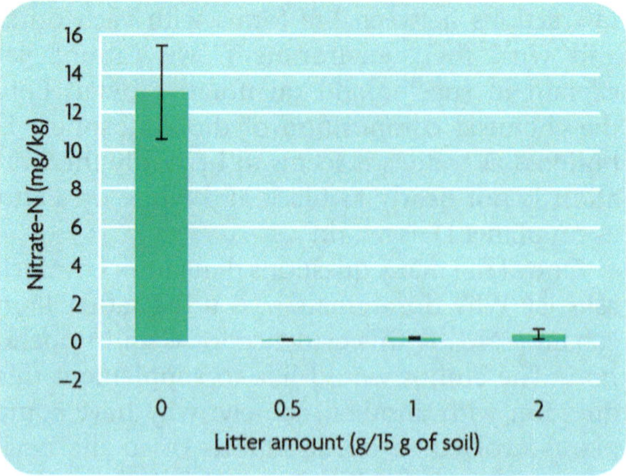

Figure 10.5 Effect of 0, 0.5, 1.0 or 2.0 g of hazel leaves, sampled in June, August and October, upon the potassium chloride-extractable nitrate-N concentration from the soil/litter mix. Bars represent means of triplicates for all three leaf-sampling dates and standard deviations are also shown. The litter very effectively immobilised nitrate-N.

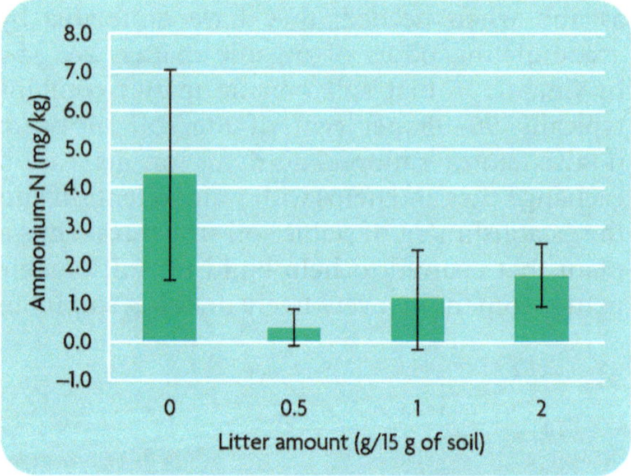

Figure 10.6 Effect of 0, 0.5, 1.0 or 2.0 g of hazel leaves, sampled in June, August and October, upon the potassium chloride-extractable ammonium-N concentration from the soil/litter mix. Bars represent means of triplicates for all three leaf-sampling dates and standard deviations are also shown. The litter effectively immobilised ammonium-N, but to an extent dependent upon the litter-to-soil ratio.

10.4.2 Can a Gaian/evolutionary approach help understand atmospheric N pollution impacts?

The hypothesis was put forward in section 10.4 that in pristine environments there was naturally a match between dynamics of plant N requirement and those of mineral N production in soil, and as a consequence the continued survival of a given plant in a plant community depends upon the stability of this dynamic match. Diffuse N pollution from the atmosphere clearly then may adversely affect biodiversity by inducing a dynamic mis-match. The pollution is deposited throughout the year. Therefore its deposition may clearly provide a competitive edge to a plant species that had evolved naturally to use N inputs earlier in the year. That may be under-storey vegetation or another tree species. The environmental impacts of atmospheric pollution will be discussed more fully in Chapter 16, but historically little thought has been given to the need for a supply and demand balance in bio-available N. This is unfortunate, because the author and several of his colleagues now believe that this is what is causing biodiversity change in heavily N-impacted areas (Cresser *et al.*, 2008). The dynamic balances of ecosystem processes assume far greater importance when a Gaian approach is adopted.

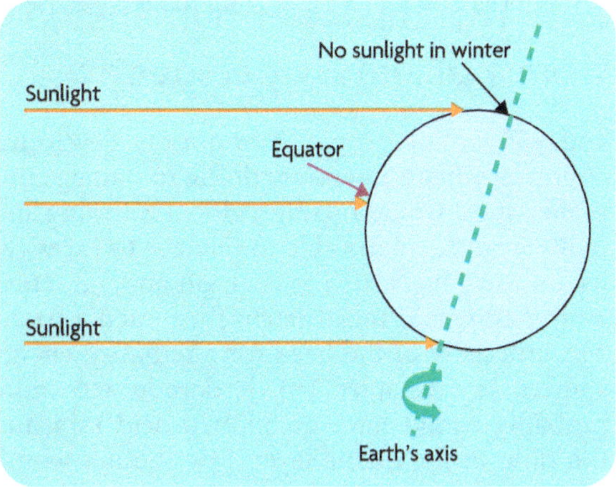

Figure 10.7 Simplified schematic diagram showing how the tilting of the Earth's axis leads to seasonal variations in climate.

Lead-in question

What do we assume when using *annual* N deposition fluxes to assess N pollutant impacts?

Nitrogen pollution deposition impacts are invariably assessed and quantified in terms of annual fluxes, typically in tens of kg per hectare per year. By ignoring the potential of a Gaian/evolutionary approach we immediately are assuming that seasonality effects are of little or no consequence to the impacts of pollutant deposition. But we saw in Chapter 1 that some 4.5 billion years ago the Earth suffered a major collision from a body thought to be about the size of Mars; at that point in time the moon was formed and the Earth became tilted upon its axis as it rotated around the sun (Figure 10.7). This cataclysmic event led to the major seasonal differences that still occur throughout each year and had major consequences for climate and for the longer-term evolution of biodiversity on the planet. To the author it does not seem sensible to assume that pollution impacts will not vary with season. Seasonal differences in climate must have impacted upon what evolved where.

There are several other reasons why we need a fresh approach to assessing pollutant impacts upon biodiversity. The author has previously questioned several aspects of the use of annual fluxes for assessing N pollution impacts. For example, using annual fluxes assumes that bioavailability doesn't depend on N species concentration in precipitation or in soil solution or on soil water content (Cresser, 2007). He and colleagues suggested that the widespread acceptance of a flux-based approach was simply the result of an illogical jump from units that had been commonplace in the past in agriculture (Cresser *et al.*, 2008). Certainly interactions of N species with foliage following interception are not independent of their concentration, so may differ between wetter and drier areas. Mobile anion concentrations (see Chapter 7) in soil solution will often be lower in wetter areas, so their impact too may differ with regional precipitation. Nitrogen species can move laterally down slope, and rapidly so over bare rock outcrops. We assume tacitly that when we quote N species deposition fluxes in kg ha^{-1} the soil surface is horizontal and the stone content of the soil doesn't matter, which is bordering on naïve in many upland areas. These issues have all been raised in the literature before (Cresser, 2000; Cresser, 2007), but are still largely being ignored. To the author it does seem much more appropriate to use the concept that plants have evolved in such a way that their environment is suited to their continued growth, a fundamental underpinning concept of Gaia, as a starting point for assessing causes of biodiversity change in natural and semi-natural ecosystems.

10.5 The impact of humankind from a planetary perspective

Today if you ask people at random in developed countries what humans are doing to damage our planet, most would mention global warming and many would even be able to relate it to excessive use of fossil fuels such as coal, oil and gas. Their awareness would mostly stem from media coverage, especially of the need for alternative energy sources (see Chapter 18). A decade ago more probably would have mentioned acid rain and the hole in the ozone layer. Few though would have made the point that atmospheric pollution was a global issue in the sense that what any of us do anywhere on Earth can impact upon plant and animal life thousands of km away. 'Global issue' to most would primarily have meant people all over the planet were starting to become concerned about impacts of pollution. Pollution impacts by too many are still perceived from local regional or national perspectives rather than the broader planetary perspective demanded by the Gaia hypothesis.

Lead-in question

How should we consider human activities from a planetary perspective and what would a Gaian response be?

Clearly the obvious pollution issues mentioned briefly above are very relevant in the context of Gaia and form the focus of much research. For example, ecosystems will respond naturally to rising carbon dioxide concentrations by enhanced plant growth (provided no plant nutrient becomes limiting), a potential feedback mechanism that simultaneously helps slow global warming. But what about intensive agriculture? Food production on a massive scale absorbs massive amounts of carbon dioxide from the atmosphere. The food is often not consumed locally, but transported over large distances to meet the needs of city dwellers. In the past food would be consumed close to where it was produced and much human excrement disposed of nearby. So carbon would be re-cycled locally, as were other nutrient elements. Little damage would be done to the atmosphere from this process.

It could be argued that the carbon can still be recycled 100s of km away from the sources of food production with no planetary ill effects, and to a large extent this is probably true. However, what about past practices of using rivers to dispose of human waste by dumping it in the oceans, and what about the other plant nutrients? Clearly for elements where global cycling is not totally dominated by gaseous components the planet's compensatory feedback mechanisms will be seriously over-stretched and may no longer be able to cope. Indeed, it has been suggested that urbanisation and disposal of sewage sludge are far more serious global problems from a planetary perspective than acid deposition from the atmosphere ever was (Cresser *et al.*, 2008). This topic will be discussed further when the environmental problems associated with food production and global food webs are discussed in Chapter 12.

10.6 Conclusions

When James Lovelock introduced the Gaia Hypothesis he was certainly *not* suggesting that the Earth is a single, thinking living organism. Rather, he was advocating that as life on Earth has evolved, and is evolving, the feedback effects of plants and animals upon the environment evolve in such a way that conditions remain suitable for the organism or group of organisms concerned. If evolutionary process departed from this concept, species could effectively become self-destructing. Lovelock believed, and the present author agrees, that we should always consider how anthropogenic activities may disrupt these natural feedbacks when searching for sustainable strategies to manage the planet. We must consider environmental management from a planetary perspective, not just from a blinkered local, regional or national viewpoint.

POLICY IMPLICATIONS

The Gaia Hypothesis

- It is crucial that policy makers appreciate how biota influence their environmental conditions and vice versa, and all possible feedback mechanisms need to be very carefully considered when formulating environmental policies.
- Policy makers need to be aware of the need to develop policies for sustainable environmental and resource management that encompass a global perspective, and not just local regional and national self-interests.
- Policy makers need to be aware of the need to consider every aspect of human activity in the context of the risks of over-exploitation of natural resources and ecosystem services on the planet.
- Policy makers need to realise that human activities may cause excessively rapid changes in ecosystem conditions, so that there is a high risk of natural feedback mechanisms being incapable of being able to respond sufficiently rapidly to buffer against environmental change.
- Policy makers need to be aware that the scale of urbanisation and food redistribution between point of production and point of consumption means that many natural feedback mechanisms that contribute to the stability of the planetary system are being seriously over stretched. Ultimately, these issues will become as important as the impacts of excessively rapid consumption of fossil fuels.

CHAPTER REVIEW EXERCISES

Exercise 10.1

At the time of writing this chapter it was rapidly approaching the 25th anniversary of the Chernobyl nuclear power plant disaster in the Ukraine, the world's worst nuclear accident. Still no one in Ukraine is permitted to live within 30 km of the wrecked nuclear plant, although a few elderly individuals have insisted on returning. A recent report on BBC television filmed abandoned villages slowly being swallowed up by newly generating forest.

Discuss critically the extent to which the forest development could be regarded as Gaia in operation, with natural feedback mechanisms restoring conditions to those that were needed to buffer the planetary system prior to urbanisation.

Exercise 10.2

To what extent, and by what mechanisms, do you think that the advances in medical science that have contributed significantly to the global population explosion are contributing to the over-exploitation of the Earth's natural feedback mechanisms for stabilising environmental conditions on the planet?

Exercise 10.3

Using the information from the present chapter and from Chapters 2 and 14 as a starting point, critically appraise the extent to which current exploitation of natural fresh water resources, including both surface waters and groundwaters, poses a risk to the long-term buffering of the planetary system.

REFERENCES

Cresser, M.S. (2000) The critical loads concept: Milestone or millstone for the new millennium? *The Science of the Total Environment*, **249**, 51–62.

Cresser, M.S. (2007) Why critical loads should be based on pollutant effective concentrations, not on deposition fluxes. *Water, Air and Soil Pollution: Focus*, **7**, 407–412.

Cresser, M.S., Aitkenhead, M.J. and Mian, I.A. (2008) A reappraisal of the terrestrial nitrogen cycle – What can we learn by extracting concepts from Gaia theory? *Science of the Total Environment*, **400**, 344–355.

Lovelock, J. (1972) Gaia as seen through the atmosphere. *Atmospheric Environment*, **6**, 579.

Lovelock, J. (1979) *A New Look at Life on Earth*. Oxford University Press, Oxford. Reissued with new preface and corrections in 2002, pp. xix plus 148.

Lovelock, J. (1988) *The Ages of Gaia*. W.W. Norton, New York, 1988. New edition, Oxford University Press, Oxford, 1995.

Lovelock, J. (1991) *Gaia: The Practical Science of Planetary Medicine*. Reprinted as *Gaia: Medicine for an Ailing Planet*, Gaia Books, London, 2005.

Lovelock, J. (2000) *Homage to Gaia*. Oxford University Press, Oxford.

Lovelock, J. (2002) *A New Look at Life on Earth*. Oxford University Press, Oxford. Originally published 1979. Reissued with new preface and corrections in 2002, pp. xix plus 148.

Lovelock, J. (2005) *Gaia: Medicine for an Ailing Planet*. Gaia Books, London.

Lovelock, J. (2006) *The Revenge of Gaia*. Allen Lane/Penguin, London.

Turney, J. (2003) *Lovelock and Gaia: Signs of Life*. Icon Books, Cambridge, UK.

Watson, A.J. and Lovelock, J.E. (1983) Biological homeostasis of the global environment: The parable of Daisyworld. *Tellus B*, **35**, 286–289.

Woodin, S., Graham, B., Killick, A., Skiba, U. and Cresser, M.S. (1992) Nutrient limitations on the long-term response of heather [*Calluna Vulgaris* (L) Hull] to CO_2 enrichment. *New Phytologist*, **122**, 635–642.

CHAPTER 11

Manmade chemicals and the environment

Alistair Boxall

Learning outcomes

By the end of this chapter you should:

- Understand the main pathways of emission of manmade chemicals to the environment.
- Know what happens to a chemical once it is released to soil or water.
- Understand the different effects that a chemical can have on ecosystems and how to assess these.
- Understand the concepts of hazard and risk assessment.
- Be aware of the different approaches that can be used to minimise the risks of chemicals in the environment.
- Be aware of the implications of future environmental change in terms of impacts of chemicals on the environment and human health.

11.1 The plethora of substances and materials we use in our everyday life

Until the early 1800s, society relied on natural resources to provide the chemicals that were needed for everyday life. This began to change though when Friedrich Wöhler made the organic chemical urea from the inorganic compound ammonium cyanate and the discipline of synthetic chemistry was born. Many of the early chemical synthesis discoveries, however, were serendipitous, and it was not until the late 1800s that chemists began to design and make substances in a systematic way.

Today we make and use a large number of synthetic chemicals in our everyday lives. These include components of personal care products, pharmaceuticals, paints and coatings, detergents and cleaning products, plastics, pesticides and flame retardants (Figure 11.1). It is estimated that there are now over 100,000 chemicals in use in Europe in amounts exceeding 1 tonne per year. Each of us will use 100s of chemicals in our day-to-day life; it has been estimated, for example, that in some countries a woman may typically apply treatments containing up to 200 chemicals to her face each day!

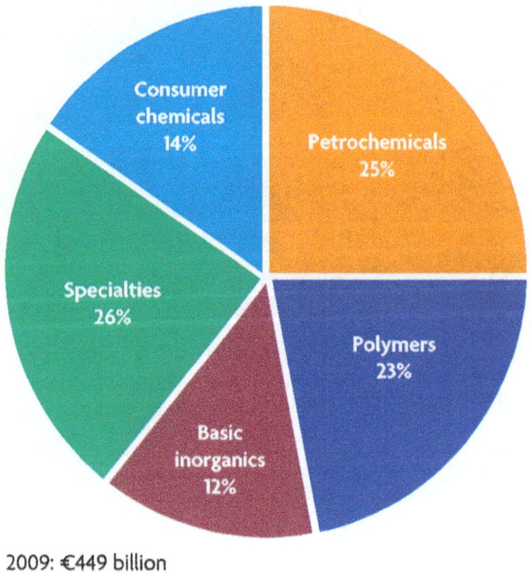

Figure 11.1 Breakdown of chemical production in different sectors in Europe in 2009 (from Cefic, 2011).

We are also getting much more sophisticated in terms of the chemicals that we develop. An example of this is the recent emergence of nanotechnology, which involves the development of materials with one dimension or more in the range of 1–100 nm (20,000–2,000,000 times smaller than a pin head). Nanotechnology provides us with access to substances with very special properties. These materials are now being used in almost every product sector, including in tennis rackets, socks and underpants, paints, cosmetics and medicines and it looks as if use of these materials will grow massively in the future. The nanoparticles we are using at the moment are relatively unsophisticated and include metal and metal oxide particles, fullerenes ('Bucky balls') and carbon nanotubes. However, it is expected that much more sophisticated particles will be developed over the next decade. We will talk more about nanoparticles later.

It is inevitable that, during use, these manmade chemicals will be released to the natural environment. Chemicals are also produced as a result of our day-to-day activities. Thus automobiles emit a complex mixture of hydrocarbons as a result of fuel combustion and waste incineration can result in the production of many persistent organic chemicals, including dioxins. This chapter therefore discusses how manmade chemicals can be released to the environment, what happens to them once they are in the environment and the potential effects of these substances on ecosystem and human health. Towards the end of the chapter we will consider how the impacts of manmade chemicals on the environment can be assessed and controlled and how the impacts of manmade chemicals might change in the future.

11.2 How do manmade chemicals get into the environment?

Chemicals may be emitted to the environment at all stages of a product lifecycle, including during the manufacturing process, during use, and following disposal of the used product. Emissions to the environment can be from a point source (e.g. a pipe or a chimney) or may be diffuse. Point-source emissions are relatively easy to control, whereas the control of diffuse emissions is much more challenging. The next few paragraphs describe how chemicals with different uses may be released to the environment.

Figure 11.2 Source of emissions of pharmaceutical effluent from a processing site in Hyderabad, India, 2006. Many tankers arrive each day with wastewater to be treated from bulk pharmaceutical manufacturers in the area, but products in treated water have found their way into local rivers and groundwater supplies of some local villages.

Lead-in question

Think of some examples of chemicals used in different sectors. How do you think these are released into the environment?

There are a number of examples of accidental releases of chemicals to the environment from manufacturing processes; for example, you may remember a recent alumina plant accident in Hungary in 2010 which released red mud to surrounding rivers. A number of similar incidents have occurred further into the past. In the Seveso disaster in Italy in 1976 dioxins were released to the atmosphere, and in the Minamata incident in Japan (which is discussed in more detail later) mercury was released to the marine environment. However, in general, in developed countries, releases of chemicals from manufacturing processes are very well controlled as companies employ good practice in order to reduce releases of chemicals and chemical by products and usually have treatment systems in place designed to prevent chemicals that are released from the manufacturing process from entering the environment. The situation can, however, be very different in less well developed countries (Figure 11.2). Recent monitoring at pharmaceutical manufacturing sites in India has shown that up to 45 kg of one antibiotic is emitted from manufacturing sites in the Hyderabad area each day – this is equivalent to 45,000 daily doses of the drug! Emissions of this type are unforgiveable as in many cases the control options are installed at the sites but are not used in order to save money and avoid hassle.

A wide range of substances is released to the air, soils and waters from urban environments, including household chemicals such as detergents, personal care products and pharmaceuticals, building and construction materials, hydrocarbons and heavy metals from roads and pesticides used to control weeds and other pests in gardens, golf courses and on highways and pavements. Household chemicals will typically be released to the sewer network and subsequently enter wastewater treatment plants where they may be removed by degradation processes or through adsorption to sewage sludge or be released in the sewage effluent to surface waters. Hydrocarbons, heavy metals, chemicals used in buildings and construction and pesticides will run-off from streets, car parks, buildings and park areas and will either be transported to adjacent water bodies (see Chapters 14 and 19) or collected by the sewer system. Substances released to the atmosphere, e.g. by road traffic, may also be transported in the air and then deposit onto soils and waters.

Chemicals used in agriculture may be released directly or indirectly to the environment. Fertilisers and pesticides will be released to the soil environment when they are sprayed onto fields. During the application process, a small proportion may also enter nearby water bodies due to spray drift. Veterinary medicines and growth promoting compounds may be excreted to soils by animals on pasture or, if used to treat housed animals, may be released to the soil environment when the faeces and urine from these animals are applied to land as a fertiliser. Many of these chemicals will also be used in the farmyard where they may be spilt onto farmyard hard surfaces and, during periods of rainfall, transported to surface waters.

Following use, products containing chemicals may be recycled or be disposed of. Disposed chemicals will typically be sent for landfill disposal, incineration or for some specialist treatment such as chemical oxidation or reduction. If sent to landfill, the chemical may leach to surrounding rivers and aquifers, although in developed countries most modern landfill sites are well engineered and monitored to prevent such emissions. If operated under appropriate conditions, incineration will break down most organic chemicals so emissions to the wider environment will be limited.

11.3 What happens to chemicals once they are in the environment and what factors determine this?

Once a chemical is released to the natural environment, it may associate with particulate material such as soils or sediment, be degraded, be transported to other environmental compartments or be taken up into biota (Figures 11.3 and 11.4). The behaviour of a chemical in the environment will be determined by the underlying physical properties (including water solubility, lipophilicity, volatility and sorption potential) of the chemical and the characteristics of the receiving environment. In the following sections information on the fate and transport of chemicals in the environment is reviewed.

The interaction of a chemical with particulate matter such as in soils, sediments and sludges will affect the degree of transport of the chemical around the environment; it may also affect the availability of the chemical to organisms and the potential for the chemical to be degraded.

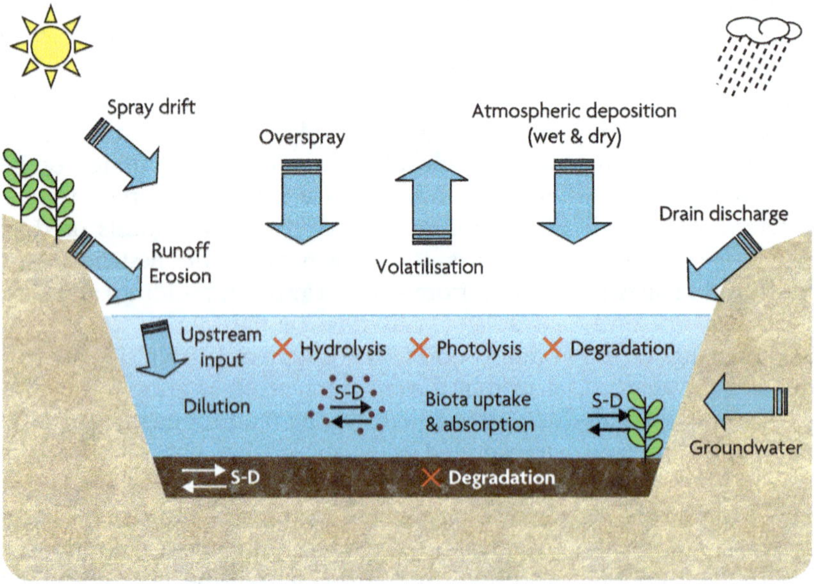

Figure 11.3 Inputs and fate and transport processes that occur in surface waters. S = sorption; D = desorption.
Source: Figure kindly provided by Dr Igor Dubus.

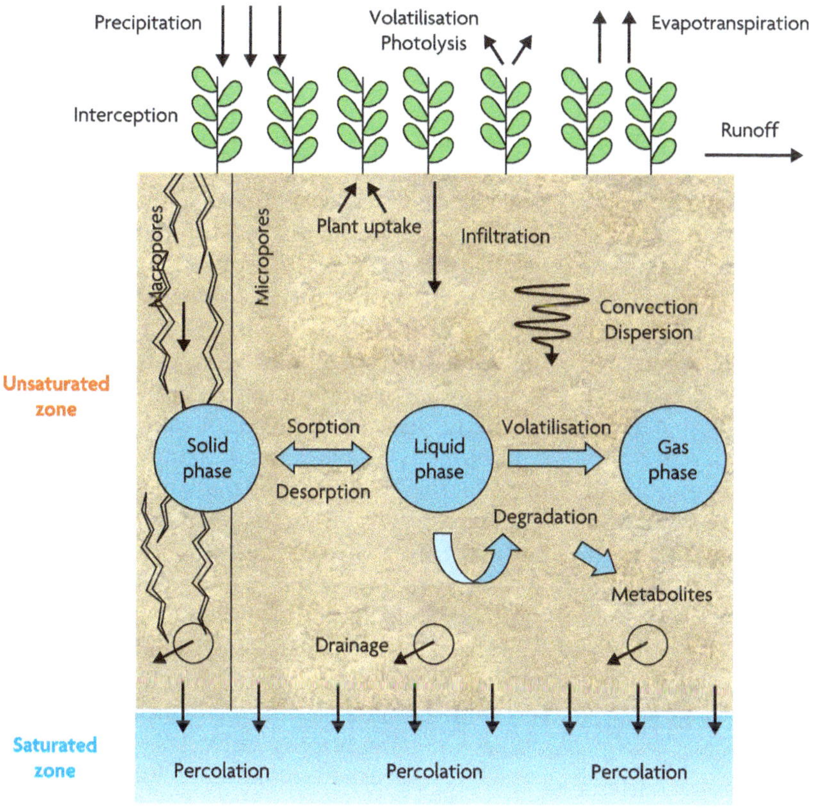

Figure 11.4 Fate and transport processes that occur in the soil compartment.
Source: Figure kindly provided by Dr Igor Dubus.

Chemicals can associate with particulate matter through either adsorption or absorption. Adsorption occurs when the chemical associates with the surface of the particle whereas absorption occurs when the chemical moves inside the three dimensional structure of the particle. The association with a particle is due to a range of bonding mechanisms, including Van der Waals bonding (the weakest), ionic bonding and covalent binding (the strongest). When measuring the interaction of a chemical with particulate material, we usually don't know whether the interaction is due to adsorption or absorption so you will generally see the behaviour referred to as the 'sorption' behaviour of a chemical. It is also possible for a chemical to be dissociated from the particle and we call this process desorption. The sorption behaviour of a chemical is strongly influenced by the properties of the chemical, such as its hydrophobicity, charge and the presence of ionisable functional groups. The sorption behaviour of chemicals is also strongly influenced by the characteristics of the solid matrix including pH, cation exchange capacity, organic carbon content, the nature of the organic carbon, metal oxide content and ionic strength; as a result the same chemical may behave very differently in different matrices.

The persistence of a chemical in the environment will also be very important in determining its impact. Chemicals can be degraded abiotically by photolysis, hydrolysis and chemical oxidation or reduction. Photolytic degradation is caused by energy from sunlight. Photolysis can be direct, where energy from the sunlight is absorbed into the chemical molecule resulting in degradation, or indirect, where energy is taken up by another chemical molecule (such as a plant pigment or a humic acid) and the energy is then transferred from the other chemical molecule to the chemical of interest, resulting in degradation. Hydrolysis is caused by the reaction of the chemical with water molecules. Oxidation or reduction occurs when the chemical reacts with an oxidising agent or a reducing agent. These mechanisms of degradation are often applied in drinking water treatment processes.

Biotic degradation is performed by biological organisms, in particular bacteria and fungi. Two types of biodegradation processes occur, aerobic degradation and anaerobic degradation. In aerobic degradation, an organic chemical is degraded by a microorganism in the presence of oxygen. The ultimate products of this reaction are carbon dioxide and water (equation 11.1). As the microbes are essentially using the chemical as a food resource another product is new biomass. Anaerobic degradation is performed by a totally different set of microbes (oxygen gas is actually toxic to obligate anaerobe organisms). In this case, the microbes use oxygen bound up in other molecules to degrade the chemical to carbon dioxide, methane and ammonia. Again more bacteria are formed in the reaction (equation 11.2).

$$\text{Organic} + O_2 \rightarrow CO_2 + H_2O + \text{aerobic bacteria} \qquad 11.1$$

$$\text{Organic} \rightarrow CO_2 + CH_4 + NH_3 + \text{anaerobic bacteria} \qquad 11.2$$

The above equations are extremely simplistic and in reality an organic chemical will be degraded via a number of pathways and a number of steps (e.g. Figure 11.5). The intermediate compounds in the degradation pathways are referred to as transformation products. In some instances, a chemical may be degraded to a transformation product that is persistent. There is therefore increasing recognition when we are considering the impacts of chemicals on the environment that we not only consider the parent compound but also consider potential impacts of the transformation products (Boxall *et al.*, 2004).

Like sorption, the persistence of a chemical is very much determined by its properties and the characteristics of the environment that it is in. We know that certain functional groups are particularly sensitive to different abiotic and biotic degradation mechanisms. Photolysis will be affected by turbidity of the system and degree of tree cover; hydrolysis is very dependent on the pH of a system; and biodegradation is dependent on the diversity of the microbes present in a system, the biomass of these microbes, oxygen concentration and the presence of other chemicals. Previous exposure to a chemical can also be very important in determining the rate at which a chemical degrades. For example, studies looking at the persistence of pesticides in soils indicate that pre-exposure to a pesticide can stimulate the growth of microbes that are well suited to degrading the compound. Consequently, when pesticides are applied to these systems in the future, degradation is faster than would be expected in a previously unexposed but otherwise similar system.

Chemicals can be transported around the environment via a number of pathways. If applied to soils, they may run off the soil surface, leach to groundwaters or volatilise into the air compartment. In rivers and streams, chemicals may be transported to estuaries and oceans, they may infiltrate into groundwater and they may be volatilised to the atmosphere. Chemicals that are particularly persistent may be transported over very long distances. Global distillation (sometimes called the grasshopper effect) is the process by which persistent organic pollutants (POPs) such as DDT and polychlorinated biphenyls are transported from warmer to colder regions of the Earth such as the Poles. If POPs are released to the environment at lower latitudes, a proportion will evaporate when ambient temperatures are warm; the evaporated chemicals may then be transported by winds until temperatures become cooler. At this stage the chemical will condense out. When temperatures warm again, the chemical will re-evaporate and be further transported. By 'hopping' in this way, the chemical will be transported from low to high latitudes (Wania and Mackay, 1996; Figure 11.6).

The grasshopper effect is a slow process that relies on successive evaporation/condensation cycles. It is only effective for semi-volatile chemicals that break down very slowly in the environment. It explains why many POPs have been found in the Arctic environment, including in the bodies of animals and humans, even though most of the chemicals have not been used in the region in appreciable amounts.

The fate and transport of engineered nanoparticles are even more complicated than those of

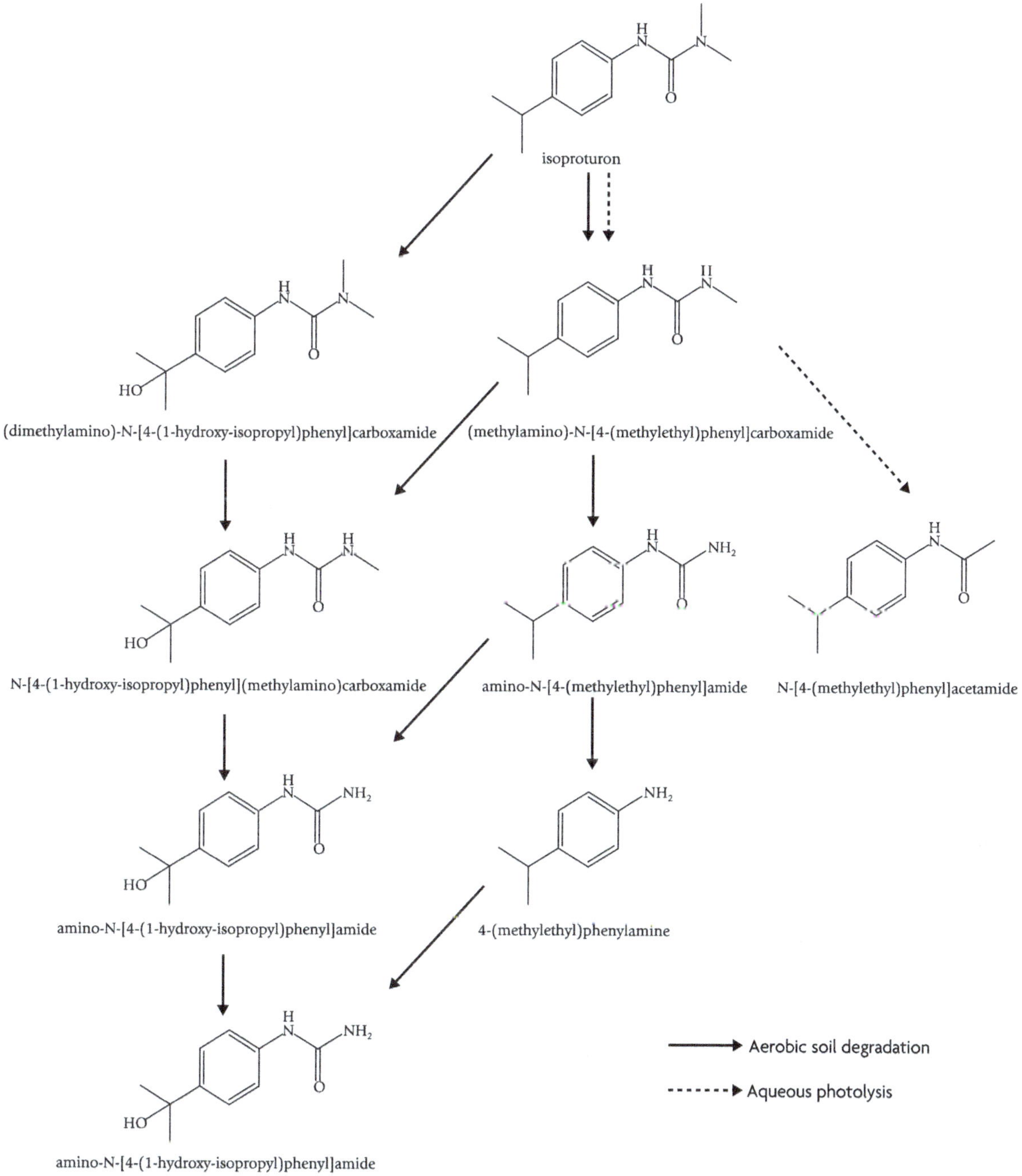

Figure 11.5 Degradation pathway for the herbicide isoproturon in the environment. The figure illustrates how the compound is degraded via a number of steps to produce a range of transformation products. The precise route of degradation can be different in different systems.

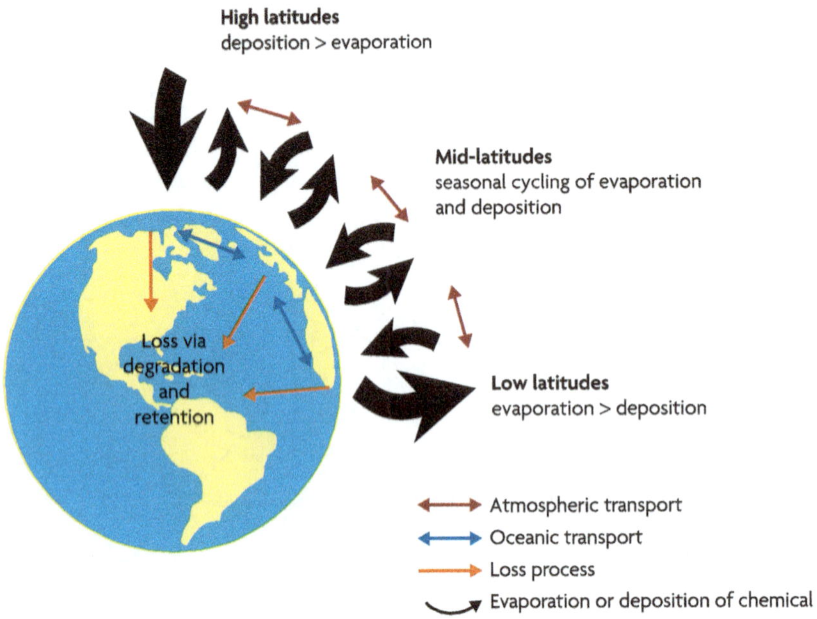

Figure 11.6 The grasshopper effect: pathways and processes involved in the long-range transport of semi-volatile persistent pollutants (adapted from CCEC, 1997).

traditional manmade chemicals that we are used to working with. A nanoparticle will be made of a core material (e.g. a metal, metal oxide or carbon) and may have a chemical capping to give the particle special properties. Following release to the water environment a range of things can therefore happen to the particle (Figure 11.7). The particles may 'bump' into each other and then join together to form aggregates or agglomerates. The aggregated particles may then settle out. The core of some nanoparticles may dissolve over time, meaning that the particle will gradually get smaller and smaller. Finally, the surface capping may be degraded or transformed by physical, chemical or biological processes.

The importance of these individual fate processes and the rate at which they occur will depend on the chemical characteristics of the core material, the chemistry of the capping and also the distribution density (concentration) of the particles in solution; in some instances all of these processes will be happening at the same time. Environmental properties such as pH and the presence of dissolved organic matter will also affect the fate of a nanoparticle. Taking this all together, if a nanoparticle is released to an aquatic system, it may be altered to a highly complex

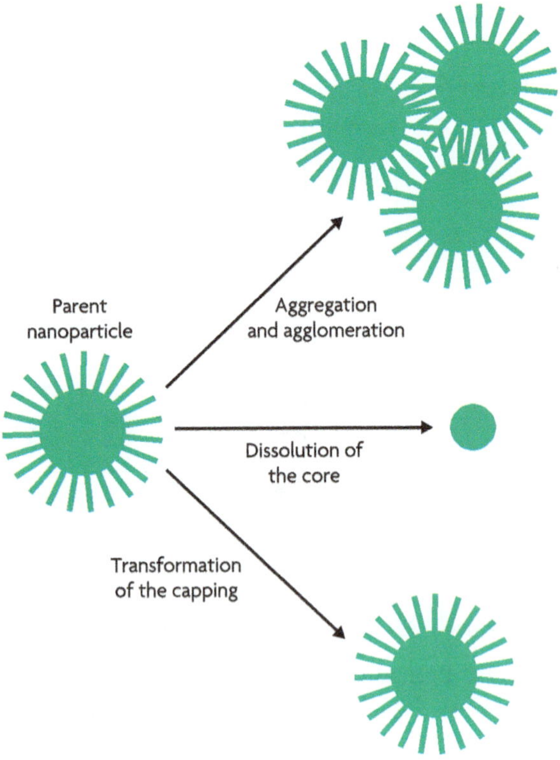

Figure 11.7 Different processes that can occur to determine the fate for an engineered nanoparticle in an environmental system.

mixture of different sized particles which differ in terms of their surface properties. This causes a real headache for environmental chemists who are trying to assess the behaviour of nanoparticles in the environment and the whole area is currently a 'hot topic' for research. It also makes the assessment of the effects of a nanoparticle very challenging as different sized particles and different surface functionalities are likely to behave very differently in terms of toxicity.

11.4 How do chemicals interact with organisms?

For a chemical to affect an organism, it must usually be taken up into the organism, although occasionally substances may interact externally with the organism causing toxicity. For example, some polymers may adsorb to gills and will then effectively suffocate a fish or invertebrate. Once in the organism, the chemical may then be metabolised or be transported to a site of action. The different processes determining how chemicals interact with an organism are illustrated in Figure 11.8 and discussed in Escher and Hermens (2004).

Uptake of chemicals into an organism can occur via a number of routes, passive or facilitated diffusion, active transport or endocytosis. Passive diffusion involves the random movement of molecules or particles from the environment into the organism or vice versa. Assuming the concentration of the chemical in the environment of interest is constant, the chemical will continue diffusing into the organism until it has reached a steady state; i.e. the concentration increases over time until it levels off. We refer to the ratio of the concentration in the water to the concentration in the organism at steady state as the bioconcentration factor (BCF). Different chemicals will have different bioconcentration factors. In general, lipophilic (fat loving) compounds will have high BCFs whereas hydrophilic (water loving) substances will have low BCFs.

Facilitated diffusion is similar to passive diffusion but in this case substances are transported across a biological membrane by means of a carrier molecule. Both passive and facilitated diffusion do not require energy. In active transport a substance is transported against a concentration gradient by a protein called a carrier protein, and this process requires expenditure of energy.

Active transport is used by organisms to transfer substances that are essential for an organism to function from the outside of a cell to the inside. Active transport will therefore generally be important for the uptake of manmade chemicals that

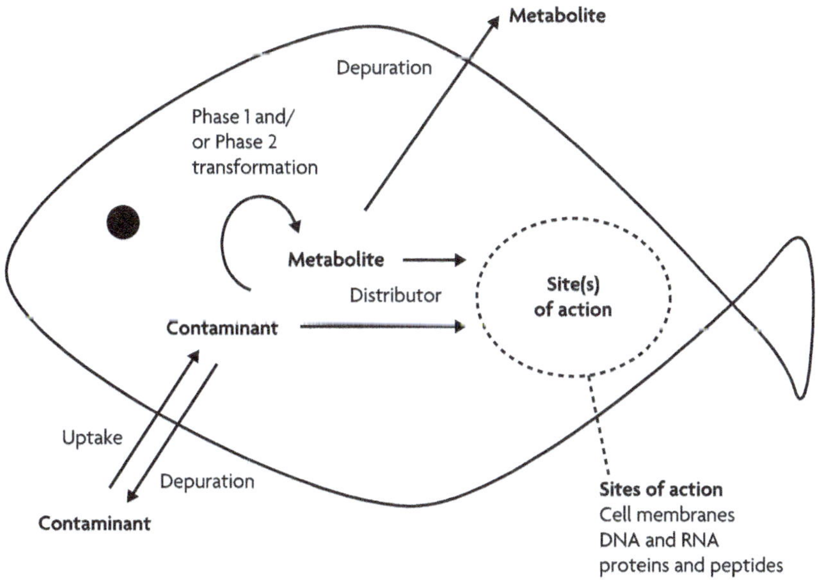

Figure 11.8 Conceptual diagram illustrating the different uptake, metabolism and distribution processes that can occur in organisms in the natural environment.

are similar to these endogenous chemicals. In endocytosis, cells in an organism will engulf the chemical or particle and then incorporate it inside the cell. There is increasing evidence suggesting the endocytosis is very important for the uptake of engineered nanoparticles.

Once in the organism, the chemical may be transformed by the organism into other chemicals. We call this process metabolism and the products of this process are referred to as metabolites. Metabolism is a mechanism that has evolved in organisms to detoxify chemicals. Generally, for organic chemicals, metabolism will alter a substance so that it is more soluble and therefore can be more easily excreted. A number of general metabolic reactions can occur and these are classed into two classes, phase 1 and phase 2 (Table 11.1). The purpose of phase 1 metabolism is to alter the structure of a molecule so that it can then undergo phase 2 metabolism. In phase 2, a new functional group is attached to the phase 1 product to produce a substance that is easily excreted.

In some cases, metabolism will make a substance more toxic. A good illustration of this is the metabolism of paracetamol by humans (Figure 11.9). Humans will metabolise paracetamol by three routes: glucuronidation, sulfation and N-hydroxylation followed by glutathione conjugation. The latter pathway results in the formation of an intermediate compound called

Table 11.1 Major phase 1 and phase 2 metabolism reactions that can occur in organisms. Typically, phase 1 reactions will alter the functionality of a chemical so that it is amenable to phase 2 metabolism. The metabolism process generally produces substances that are more soluble and therefore more readily excreted.

Phase 1	Phase 2
Oxidation	Sulphation
Reduction	Glucuronidation
Hydrolysis	Glutathione conjugation
Dehalogenation	Acetylation
Hydration	Amino acid conjugation

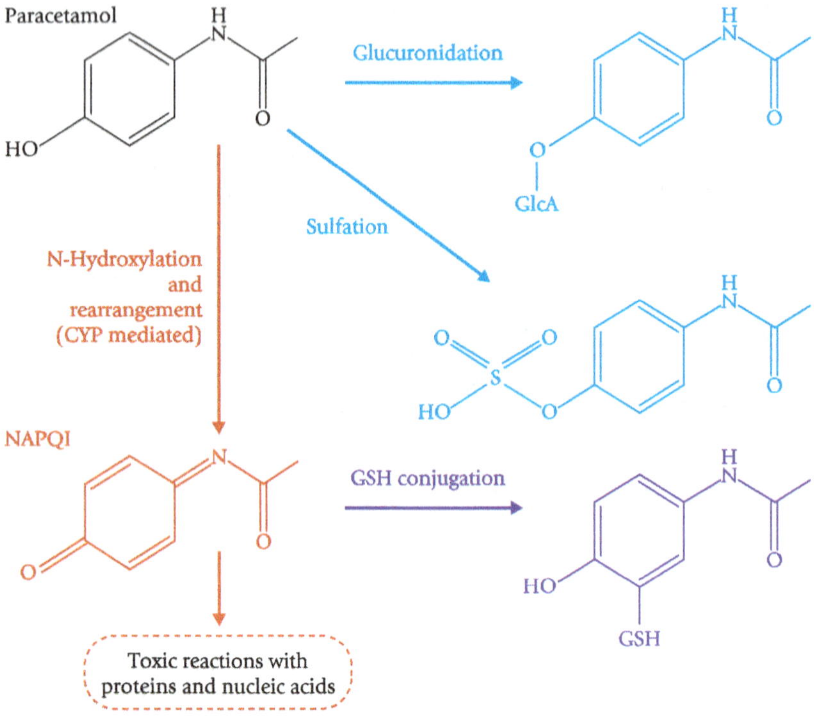

Figure 11.9 Metabolic pathways for paracetamol in humans.
Source: http://en.www.wikipedia.org/wik/iFile:Paracetamol_metabolism.svg. This material is in the public domain.

N-acetyl-p-benzo-quinone imine (NAPQI) which is a very reactive and toxic molecule. Under normal use, glucoronidation and sulfation are the major transformation pathways. However, if a human takes an overdose of paracetamol, these pathways become overloaded and the N-hydroxylation pathway becomes more important meaning NAPQI is produced. This causes the horrific effects associated with a paracetamol overdose. Similar toxification reactions probably occur in other organisms although our understanding of this area is much less developed than for humans.

Once in an organism, chemicals (and their metabolites) can reach the site of action where they elicit an effect. Chemicals can elicit an effect in a number of ways that we call mechanisms of action: they may interact with the membranes of cells and cell organelles by either damaging lipids and proteins, blocking membrane transport channels, or affecting the chemistry of the membrane which inhibits some of the important membrane functions; they may damage DNA and RNA; and they may bind to enzymes or receptor sites. The mechanism of action of a chemical will be determined by the type and position of functional groups on the molecule, its size and shape. For a compound to bind to a receptor site, it will need to be of the right shape and size to fit into the receptor. The same chemical may act by a number of mechanisms of action and act on different species in different ways.

In the real world, an organism will be exposed to a number of chemicals at any one time. It is possible that exposure to one chemical will affect how another chemical is taken up, metabolised or interacts with the site of action. Sometimes these interactions will result in a lowering of the effect of a chemical (we call this antagonism) but in some instances it may increase the toxicity of a chemical (we call this synergism). The area of mixture interactions is a very active topic for environmental research and scientists are trying to understand the mechanisms by which these interactions occur so that, in the future, we can better predict those mixture combinations that could cause problems in the environment.

If effects occur at the biochemical level then these can lead on to a range of effects on the organism. A wide range of effects can occur on single organisms, including destruction of cells and cell organelles, effects on feeding, reductions or increases in respiration rate, inhibition of photosynthesis, impairment of reproduction, behavioural effects and mortality. These effects can then lead on to effects on the population of an affected species or to effects on communities. To illustrate the different effects that chemicals have on organisms and how these then translate to effects at the ecosystem level, it is worthwhile looking at a group of substances as a case study. One group that has received significant attention over the past two decades is the endocrine disrupting chemicals.

An endocrine disruptor chemical (EDC) is an exogenous substance or mixture that alters the function(s) of the endocrine system and consequently causes adverse health effects in an intact organism, or its progeny, or (sub)populations. Over the past twenty years or so there has been increasing interest in the impacts of EDCs on organisms in the natural environment. Many chemicals are known to be capable of causing endocrine disruption. They can have different types of activity (e.g. some are estrogenic whereas others are androgenic or anti-androgenic) and can elicit their effects via different mechanisms. Most work to date has focused on the estrogenic EDCs and some of this work is summarised below. A more thorough review of the topic can be found in Sumpter (2005).

The issue of exposure to estrogenic chemicals was first recognised when scientists spotted that wildlife such as alligators and fish were becoming feminised. For example, field studies done in the UK in the early 1990s found that fish downstream of sewage treatment plants were being feminised. The effects observed included an increase in vitellogenin (an egg yolk pre-cursor) concentrations in fish and intersexuality (e.g. feminisation of male fish). Studies were done to identify the causes of these effects and for domestic effluents it is believed that the effects were probably due to combined exposure to natural (e.g. 17β-estradiol and estrone) and synthetic (e.g. ethinylestradiol – the active ingredient in the contraceptive pill) estrogens that are excreted by humans and which are biologically active even at the low concentrations seen in the environment. Numerous laboratory studies have also been performed to further characterise the effects of estrogenic compounds on fish and a wide range of effects have been reported.

There is significant debate as to whether the effects seen in the types of studies described above really matter in terms of the functioning of an ecosystem. In an attempt to answer this question, Canadian scientists performed a seven-year study where they dosed a whole lake with ethinylestradiol at concentrations that could be expected in surface waters and observed the effects on the fathead minnow population. The data supported the results of the previous monitoring studies and showed that exposure led to feminisation of the male fish through the production of vitellogenin mRNA and protein, impacts on gonad development and altered egg production in females. The exposure also led to the near extinction of the fish species. Based on these findings, the researchers concluded that estrogens that are released to the environment can indeed have an impact on the sustainability of wild fish populations. However, this conclusion is still debated as some scientists argue that the concentration of EE2 used in the study (5–6 ng l^{-1}) is higher than concentrations seen across the broader environment.

11.5 Chemicals in the environment and human health

As well as affecting aquatic and terrestrial organisms, chemicals in the environment may adversely affect human health. Consumers might be exposed to these chemicals in several ways, such as via consumption of crops that have accumulated the chemical from contaminated soils, from livestock that have accumulated chemicals through the food chain, from fish exposed to contaminated waters, or from abstracted ground waters and/or surface waters contaminated with chemicals but then used for potable supplies. Chemicals in the environment may also affect human health indirectly. Below we will explore two case studies to illustrate both direct and indirect effects of chemicals on human health.

CASE STUDIES

Mercury in Minamata City, Japan

Our first case study in this chapter is not strictly based upon a manmade chemical but the impact did result from the use of the substance in a manufacturing process. Minamata city is located on the Western coast of Kyushu, Japan's southernmost island. The local bay (Minamata Bay) was the main source of fish and shellfish for the local population which was an important component of people's diet. In 1932 the Chisso Corporation, which had operated in the Minamata area since 1907, began to manufacture acetaldehyde which was used as a precursor in plastic production. Soon after the Second World War, the production of acetaldehyde in Minamata boomed. Around this time, fish were seen to float to the surface of the bay and the cats in the area began to exhibit strange behaviour – they would stagger around, go into convulsions or start whirling in circles. Often the cats could not control the direction they were heading and would end up falling into the sea where they died – the residents referred to this as 'cat suicides'.

In 1956 the first case of humans having the disease was recognised in a little girl who showed signs of neurological damage and eventually started having convulsions. Soon after, other children started to demonstrate the symptoms, babies started being born with the disease and eventually adults started to develop symptoms.

In 1992, the number of people officially diagnosed with Minamata Disease was 2,252 people; 1,043 of them had died. Another 12,127 people were waiting to be tested.

By the end of 1956, epidemiological data had shown that the disease was probably caused by heavy-metal poisoning arising from consumption of fish and shellfish of Minamata Bay. Eventually, the source of the poisoning was tracked down to the release of mercury from the Chisso plant that was using the metal as a catalyst in the acetaldehyde production process. Following the release to the bay, the mercury was sorbing to the sediment in the bay and then being transformed by microorganisms into a substance called methyl mercury. Methyl mercury is a highly lipophilic and bioaccumulative compound. It was therefore readily taken up by fish and shellfish. When the contaminated fish were eaten by the residents of Minamata, the methyl mercury was found to be transported to the human brain where it elicited its toxicological effect.

CASE STUDIES

Decline in vulture populations

Our second case study in this chapter occurred much more recently. Over a 15-year period in areas of India and Pakistan, populations of a number of vulture species were observed to be declining. The populations of three species had declined to levels where extinction of the species was possible. In response to the observed population decline, scientists worked to find the cause of the effect. The death of the vultures was found to be associated with renal failure and visceral gout. The symptoms that were seen suggested that these effects were due to exposure to a toxic chemical.

The scientists then used detective work to try to find out the cause of the population declines. Carcasses of affected birds were initially tested for known chemical causes of renal disease (including cadmium and mercury) and non-chemical causes (e.g. infection with avian influenza, bronchitis and West Nile viruses) – all of the results were negative. As the primary food source for the vultures was domestic livestock, they then hypothesised that veterinary pharmaceuticals might be the cause. A survey of vets and veterinary retailers was therefore performed to identify drugs that are known to be toxic to kidneys and which are absorbed orally.

Only one drug met these criteria and this was the non-steroidal anti-inflammatory drug, diclofenac. Analysis of vulture carcasses for this compound showed that birds that had died of renal failure had indeed been exposed to diclofenac whereas as birds that died from other causes (including being hit by a car and lead poisoning) did not contain diclofenac. A study was also done where captive vultures were treated with diclofenac and the results showed that exposure to the chemical did indeed cause renal failure. It was therefore concluded that the population declines were due to diclofenac exposure and that the source of the diclofenac was the consumption of livestock treated with the drug which was widely sold by vets and retailers in India and Pakistan. Diclofenac has now been banned for veterinary use in these regions and work is ongoing to revive the vulture populations.

While the compound had a severe impact on the vulture populations, it is thought that the use of the compound has also had an indirect negative effect on human health. Vultures are natural scavengers, cleaning up carrion of wildlife and domestic livestock, and also of cattle, sacred to Hindus, which cannot be consumed by people. Moreover, to avoid contamination of the earth, water and fire, people of the Parsi faith, descendants of the Zoroastrians of the Persian empires, have traditionally left their dead to vultures for disposal. The vulture population decline is therefore thought to have had a range of socio-economic, cultural and human health impacts. For example, Markandya *et al.* (2008) reviewed the economic implications of the human health impacts of the decline in vulture populations. Livestock carcasses are the main food source for vultures but are also eaten by dogs. As the vulture populations have declined, the dog populations have increased. As dogs are the main source of rabies in humans in India, it is probable that the incidence of rabies in humans has increased and hence mortality has increased. Markandya *et al.* (2008) estimated that the vulture decline had likely caused many thousands of extra deaths in the human population as illustrated here.

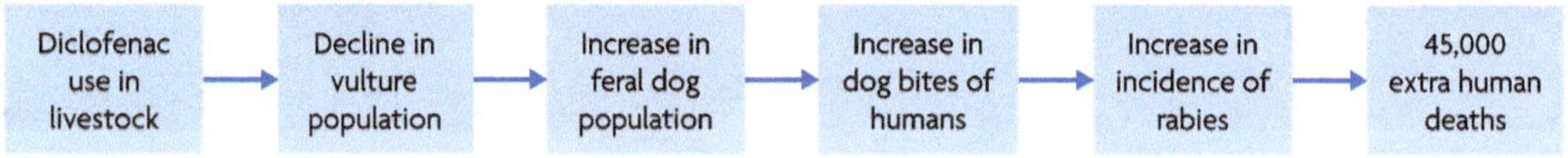

The cascade of effects that linked diclofenac use in livestock to increased mortality in areas in India and Parkistan.

Source: Based on Markandya *et al.* (2008).

11.6 Assessing risks of chemicals

In order to ensure protection of the environment from manmade chemicals and to prevent the types of incident described in the two case studies above, manufacturers of synthetic chemicals in many regions of the world are now required to assess the potential impacts of any particular chemical before it is put on the market. The process by which potential impacts of chemical products are assessed is called *environmental risk assessment* (ERA). In Europe, ERAs are required on all new pesticides, biocides, human medicines and veterinary medicines.

Policy makers also recognise now that a large number of substances have been manufactured and placed on the market for many years, sometimes in very high amounts, yet we actually know very little about the risks of these chemicals to human health and the environment. In Europe, one exciting development is the REACH (Registration, Evaluation, Authorisation and Restriction of Chemicals) Regulations that aspire to address the significant data gaps that exist for many chemicals that we use every day. The REACH Regulations place greater responsibility on industry to manage the risks from chemicals and to provide safety information on the substances. Manufacturers and importers are required to gather information on the properties of their chemical substances, which will allow their safe handling, and to register the information in a central database run by the European Chemicals Agency (ECHA) in Helsinki. The Regulation also calls for the progressive substitution for the most dangerous chemicals when suitable alternatives have been identified.

In order to understand what is meant by environmental risk assessment it is important to be familiar with the concepts of hazard and risk. These terms have different meanings; *hazard* is the inherent potential for something to cause harm

Table 11.2 Examples of chemical sources, pathways and receptors for contaminated water.

Example source	Pathways	Example receptors
Contaminated surface water	Water	Pelagic fish, invertebrates, algae
	Sediment	Benthic invertebrates and fish, sediment-dwelling worms, molluscs
	Food chain	Mammals and birds
	Drinking water	Humans
	Irrigation	Crops and humans

whereas *risk* is the likelihood that harm will actually be done by the realisation of the hazard for a particular situation. Risk of a chemical to the environment is therefore determined by the hazard of the chemical, the likelihood of the chemical being released to the environment and the likely level of exposure of the chemical once it is in the environment.

A concept frequently used in environmental risk assessment is that of the source – pathway – receptor (e.g. Table 11.2). In this model the pathway between a hazard source (for example a chemical emission) and a receptor (for example a particular ecosystem) is investigated. The pathway is the linkage by which the receptor could come into contact with the source (a number of pathways often need to be considered). If no pathway exists then no risk exists. If a pathway exists linking the source to the receptor then the consequences of this need to be determined.

Lead-in question

Table 11.2 describes the source pathway relationship for contaminated surface waters. Develop a similar table for other systems, e.g. contaminated soil, contaminated groundwater.

Numerous guidelines, that are a bit like scientific recipe books, exist that describe how to assess the environmental risk of a particular product type. In the next few sections, we will work through the different stages of the risk assessment process and then illustrate the approach using a hypothetical example.

When considering the risks of chemicals in the environment, there are a number of hazards that we typically consider, including toxicity to aquatic and terrestrial organisms, persistence, bioaccumulation, and toxicity to humans. Information on the potential hazards is usually obtained from laboratory tests. It is clearly impossible to assess the toxicity of a chemical to all species. Therefore the general approach to assess hazard to organisms in the environment or to humans is to run tests using surrogate species chosen to represent the wider environment or to mimic the human body.

For aquatic and terrestrial risk assessment schemes, these surrogate species are selected to cover the different taxonomic groups that can occur in the environment of interest. Therefore to assess the toxic hazard of a chemical to the aquatic environment, tests will usually be done on algae, invertebrates and fish whereas terrestrial studies will measure effects on soil microbes, invertebrates (e.g. earthworms or springtails) and plants. Most of these studies involve concentration response studies. The general approach used involves the preparation of a range of concentrations of the chemical in the environmental medium of interest (e.g. soil or water). Organisms are then introduced to the system and exposed for anything from a few days to many months. During the exposure, different endpoints can be monitored including the death of the organisms, growth, number of offspring produced and behaviour. If the test is short term, the test is called an 'acute' study whereas longer-term studies are referred to as 'chronic' studies.

The endpoint results from the studies are then plotted against concentration to give the concentration response relationship (e.g. Figure 11.10). From such relationships, we can extract a number of summary statistics that are then used in the risk assessment, including, for example, the median lethal concentration (LC_{50}) or the median effect concentration (EC_{50}), the no-observed effect concentration (NOEC – the highest concentration tested that results in an effect that is not statistically different from the control) and the lowest observed effect concentration (LOEC – the lowest concentration tested that results in an effect that is statistically different from the control). Median lethal concentration is defined as the concentration of a harmful chemical, most commonly in aqueous solution, that would be expected to cause the death of 50 per cent of a specified population of organisms under a specified set of experimental conditions, so to evaluate this the vertical axis would show the percentage mortality. If a substance has an EC_{50} or LC_{50} to aquatic organisms of less than 1 mg/l, then it would be regarded as very toxic.

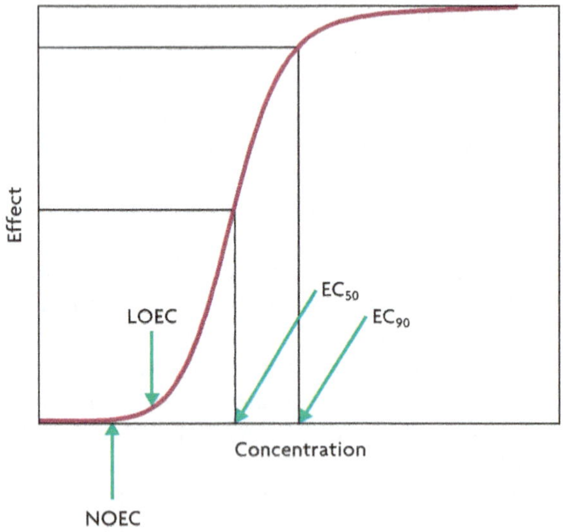

Figure 11.10 A typical concentration response relationship that would be obtained from an ecotoxicity study. NOEC = no observed effect concentration; LOEC = lowest observed effect concentration; EC_{50} = concentration causing 50 per cent effect; EC_{90} = concentration causing 90 per cent effect.

However, just because a substance is very toxic, it does not mean that it will have an adverse impact on the environment. The way that the chemical is used could mean that it may never be released to the natural environment so organisms will never be exposed to the chemical. Therefore, when assessing the risks of a substance, in addition to assessing the hazard of the chemical, we also need to consider whether it will be released to the environment, how much will be released to the environment and what the resulting concentration will be in the environment – this process is called exposure assessment.

While we could go out and measure concentrations of a chemical in the environment and use this data as a measure of exposure, for chemicals that are not yet on the market (which is usually the case for new chemicals needing a risk assessment), this will not be possible as they are not yet being released to the environment. To overcome this problem, mathematical models are often used to estimate the concentrations that will occur in a given environmental system. The output of these models is a predicted environmental concentration (PEC) for the chemical in the system of interest. A plethora of these models is available for different emission types such as emissions to wastewater treatment plants or the application of a pesticide to a crop and for a range of scales.

Once we have our measures of the hazard of a substance and the prediction of exposure, we are then in a position to begin to establish the risk of the substance to the environment. In its simplest form, risk is characterised by comparing the predicted exposure concentration with a concentration where we think no effects are likely to occur on the system (often referred to as the predicted no effect concentration (PNEC)) to give a risk characterisation ratio (equation 11.3):

$$RCR = PEC/PNEC \qquad 11.3$$

If the RCR is greater than 1, then the risk to the environment is regarded as unacceptable. If it is less than 1, then the risk to the environment is regarded as acceptable and the use of the substance would be considered safe.

The obvious way to derive a PNEC is to look at the concentration response data for the chemical and the species tested and identify a concentration where effects are not seen. However, there are considerable uncertainties in extrapolating from laboratory toxicity studies to the real environment. For example, we do not really know how sensitive our laboratory test organisms are compared to other species that will occur in a natural system; we do not know whether our chemical will interact with other pressures in natural systems which perhaps do not occur in the laboratory; and we don't know whether the timescale used in the laboratory study is predictable of the exposure conditions that occur in the real world. To cope with these unknowns, uncertainty factors are used along with the concentration response data to derive the PNEC using equation 11.4.

$$PNEC = EC_{50}/UF \qquad 11.4$$

The uncertainty factor used will depend on the type of hazard data that are available. If the tests are short term (i.e. a few days) and mortality is used as a test endpoint, then an uncertainty factor of 100 to 1000 might be used whereas if the tests are longer term (weeks – months) with endpoints such as growth and reproduction, then an uncertainty factor of 10 might be used as these tests are probably closer to reality.

Lead-in question

Can you think of any other factors that mean that extrapolations from laboratory studies to effects in the natural environment are uncertain?

We have now seen the general process of environmental risk assessment, but to illustrate how the approach works in practice it will probably help to consider a worked example.

WORKED EXAMPLE: a simple risk assessment of a pharmaceutical

A pharmaceutical company has just developed a new heart treatment. The treatment will be taken daily by adults at a dose of 0.02 mg kg^{-1} body weight d^{-1}. The company have forecast that the drug will be taken by 0.1 per cent of the population. Tests have also been done to look at the metabolism of the drug in patients and these indicate that around 80 per cent of the drug is metabolised. A series of ecotoxicity studies have been performed on the drug and the results are shown below. Studies have also looked at the removal of the drug in sewage treatment and these indicate that only 10 per cent of the drug will be removed in typical treatment processes. Using this information, the company now want to establish whether the use of the drug will pose a risk to the environment.

Ecotoxicity data for the heart treatment:

Algae 96 h NOEC for growth = 100 mg l^{-1}

Water flea 21 d NOEC for reproduction = 33 mg l^{-1}

Fish early life stage 32 d NOEC = 0.1 mg l^{-1}

Step 1: problem formulation

When faced with a problem like this, the first thing to do is to consider the source–pathway–receptor relationship for the chemical. In this example, we know that the drug will be used

➤

and excreted by humans in urine or faeces. The main pathway by which it can reach a receptor is therefore via the sewage system and then to surface waters. A proportion of the substance is also removed in the sewage system. If this removal is a result of sorption to sewage sludge particles, it is also possible that the compound will be released to agricultural fields when sewage sludge is applied to land as a fertiliser – to keep things simple, we won't consider this pathway any further here. Based on this we can conclude that the environment could well be exposed to the chemical and that the pathway we need to consider is human → sewage system → river → aquatic organisms.

Step 2: exposure assessment

We now need to assess the potential level of exposure of the aquatic environment to the drug. In the particular situation we can use fairly simple equations and assumptions to estimate the exposure concentration. In Europe, on average, a human will emit around 200 l of wastewater per day (this is a combination of bath and shower water, water used for washing and cooking). A typical sewage treatment effluent will be diluted around 10 times by the receiving river.

$$PEC = \frac{A \times P \times F_{exc} \times F_{treat} \times \text{population size}}{WW \times D \times \text{population size}} \qquad \text{Equation X (note population cancels)}$$

Where: PEC is the predicted environmental concentration in river waters; A is the amount used by a patient per day (determined from the daily dose and the average body weight of a human (70 kg) = 1.4 mg); P is the proportion of the population using the drug (0.001); F_{exc} is the fraction of drug excreted unmetabolised by the patient (0.2); F_{treat} is the fraction of the drug released from the wastewater treatment works (0.9); WW is the amount of wastewater produced per person per day (200); and D is the dilution factor of the sewage effluent by the river (10). When we substitute all of the numbers into equation 1, we arrive at a predicted concentration in river waters of 0.00014 $\mu g\ l^{-1}$. While this calculation seems incredibly simple, recent comparisons of predictions of concentrations for commonly used pharmaceuticals with measured concentrations in rivers indicate that the calculation actually works quite well.

Step 3: hazard assessment

We now need to determine a concentration in water where we wouldn't expect effects to occur. Ecotoxicity data are available for three trophic levels in aquatic systems and based on these data, it appears that fish are the most sensitive with a NOEC of 0.1 $mg\ l^{-1}$. As fish are the most sensitive, we use this data in the risk assessment. The guidelines for pharmaceutical environmental risk assessment state that an uncertainty factor of 10 should be used to derive a predicted no effect concentration from a chronic study. Applying this factor, we arrive at a PNEC of 0.01 $mg\ l^{-1}$.

Step 4: risk characterisation

We have our predicted exposure calculation and a predicted no effect concentration for the drug so it is possible to calculate a risk characterisation ratio. The RCR is calculated by dividing the PEC by the PNEC. This gives an RCR for our drug of 0.00001. As this value is lower than 1, we can conclude that the use of the drug poses an acceptable risk to the aquatic environment.

The worked example is an extremely simple one and quite often, when we assess chemicals using simple models, a risk characterisation ratio of greater than 1 is obtained. If this happens, then it is necessary to refine the risk assessment using more sophisticated modelling approaches and ecotoxicity experiments. Predictions from more sophisticated exposure modelling approaches are often lower than predictions obtained using simple exposure algorithms. By using more sophisticated ecotoxicity experiments that better mimic the real environment, it is possible to lower the uncertainty in the test result which means that a lower uncertainty factor is justified. There are numerous higher tier approaches that can be used to assess ecotoxicity; these include laboratory studies on a wider range of species or the use of microcosms or mesocosms (Boxall *et al.*, 2004). Microcosms are artificial, simplified ecosystems that are used to simulate and predict the behaviour of natural ecosystems under controlled conditions whereas mesocosms are designed to mimic natural conditions, in which environmental factors can be realistically manipulated (Figure 11.11a and b).

Figure 11.11a Image of laboratory microcosms used for ecotoxicological testing of chemicals.

Figure 11.11b Sampling of a mesocosm; this particular facility is designed to 'mimic' communities that we found in a dyke in the Netherlands.

Source: Photographs for Figures 11.11(a) and (b) kindly provided by Professor Paul van den Brink, Alterra, Wageningen, Netherlands.

The environmental risk assessment described above is also very simple in that it uses only one measure of hazard and one measure of exposure. In the real world, concentrations of a chemical are likely to vary over both space and time. For example, if we take the heart treatment in our worked example, concentrations in surface waters will probably be greater in regions with high incidents of heart disease compared to regions with low rates of heart disease. It is possible to begin to account for these variations in exposure by using distributions of exposure in the risk assessment rather than one single value. For our heart treatment we could collate information on heart disease for different regions in a country as well as information on population densities, sewage flows and river flows in each region. The exposure algorithm would then be run numerous times to derive region-specific estimates of exposure and from this a frequency distribution of exposure concentrations would be developed (Figure 11.12a). The degree of overlap of the exposure distribution with the predicted no effect concentration would then tell us how widely the risk will be distributed across the country.

The risk assessment can be further refined by generating data on distributions of effects for species within the environment and comparing these with exposure distributions (Figure 11.12b). Based on the degree of overlap of the two distributions it is not only possible to determine how widely the risk is distributed but also the proportion of a biological community that would be at risk. The interpretation of assessments of this type is much

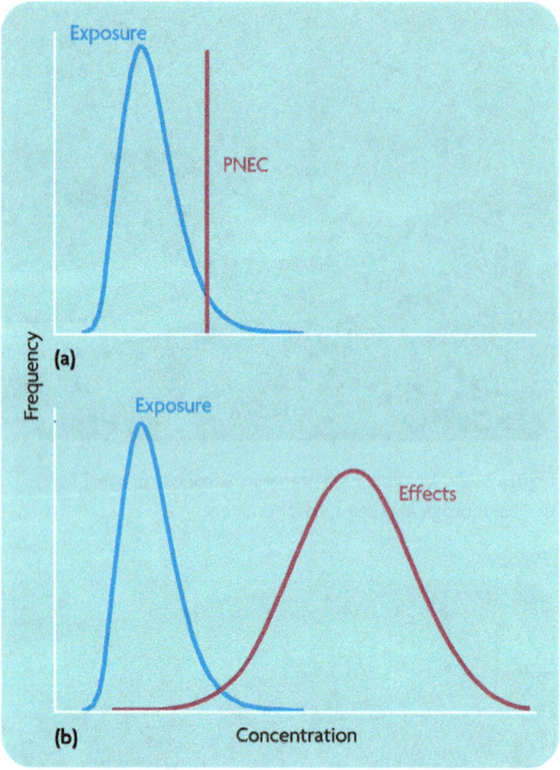

Figure 11.12 Characterising risk using (a) distributions of exposure of a chemical; and (b) distributions of effects and exposure for a chemical.

more complicated than the situation where one exposure and one effect value are compared.

11.7 Could existing risk assessment schemes have predicted some of the impacts described above?

While environmental risk assessment schemes have been in place for some groups of chemicals for some time, often it is not until after a chemical is widely in use that problems resulting from environmental exposure are identified. This begs the question: 'Could risk assessment schemes have identified some of the adverse impacts that we have discussed earlier in the chapter?'.

If we take endocrine disruptors, past environmental risk assessment schemes would not have identified the risks of these chemicals as endocrine disruption wasn't an endpoint that was investigated in the ecotoxicity tests and it is only in the past couple of decades that we have recognised the importance of endocrine disrupting activity as an environmental pressure. As a result of the observations of endocrine disruption in the field, improved test methods are being developed for assessing whether a substance has endocrine disrupting properties or not and these tests now need to be applied in the risk assessment for some groups of chemicals (e.g. pesticides).

Predicting the impact of diclofenac on vultures and the subsequent impacts on the ecosystem in India and Pakistan would however be much more difficult to predict. The pathway by which vultures were exposed to the diclofenac was very unusual and the vulture species seemed to be much more sensitive to the drug than data for laboratory test organisms suggested. With hindsight, if the exposure pathway had been considered alongside some of the reported side effects of the drug in humans, we may have been alerted to the problem. This example demonstrates that when assessing the risks of a chemical, we should think more broadly than we do at the moment and consider all potential exposure pathways that can occur in different regions as well as all other data that might be available on the activity and side effects of chemical exposure. Advances in molecular biology may also help as these are beginning to provide us with much more information on how chemicals interact with organisms and on the sensitivity of different organisms to chemical stressors.

The diclofenac example also highlights the need for ongoing monitoring studies on populations around the globe so that we can identify when things start to go wrong as early as possible and take measures to control chemical pressures.

11.8 Managing risks

In the event that a chemical is identified as posing an unacceptable risk to the environment, there are a number of options that exist for managing or mitigating the risks. One approach to remove the risk is to ban a chemical or not put the chemical on the market in the first place. For example, in India and Pakistan, diclofenac is now banned for use in the treatment of livestock and it is hoped this will allow the vulture populations to re-establish.

Over recent years there has been a steadily increasing drive within some sectors of the chemical industry towards the synthesis of 'greener' (sometimes called 'benign-by-design') chemicals. These chemicals are usually designed to be degraded rapidly in the environment and not to contain functional groups that are toxic to organisms. However, this management strategy is not always possible as some substances need to be persistent and have certain functionality in order to be effective in their use. A good example of such a class of compounds is the cytotoxic drugs that are used in the treatment of cancers. These drugs need to be highly toxic in order to treat the cancer and need to persist long enough to deliver their therapeutic effect. Banning of these substances is not an option as society is so reliant on them – imagine the uproar if an effective cancer treatment was banned based on environmental concerns. Green chemistry methods and technologies can also be applied to minimise the use of hazardous chemicals in industrial processes and the release of these to the environment (see Chapter 23).

For many classes of chemical, changes in the way that the chemical is used can help to reduce or eliminate environmental impacts. Thus for chemicals used in agriculture a number of approaches can be used, including changes in treatment timings and intensities for pesticides, changes in application rates and timings of fertilisers, development of recommendations on when not to apply chemicals and fertilisers (e.g. where slopes or hydrological conditions are such that they will promote transport of agrichemicals to water bodies), and specification of buffer zones to protect water bodies. Numerous schemes exist to encourage farmers to adopt good practices; a good example of this is the Voluntary Initiative (VI) in the UK. The VI was developed by farming unions and pesticide manufacturers in the UK in response to a threat from the UK government to introduce a pesticide tax because of a perceived high level of pesticide contamination of water bodies in the UK. The VI uses educational approaches (see Figure 11.13) to encourage farmers to use pesticides in such a way that emissions to the environment are kept to an absolute minimum. The scheme also encourages regular maintenance and cleaning of spray equipment to avoid over-application of pesticides.

Classification and labelling approaches may also help to minimise risks. An interesting instance of a scheme of this type is a system running in Sweden which is a voluntary scheme that targets active pharmaceutical substances where information on their environmental impacts is made publicly available on websites and in information booklets. The idea is that doctors and patients consider the information when discussing which drugs should be prescribed for treatment of a medical condition. The scheme also encourages doctors to use a number of strategies to ensure that only the amount of drug that is necessary is prescribed in order to reduce the amount of unused medicines being thrown down the sink or being disposed of in the trash.

It is, however, inevitable that we will need to dispose of unused chemicals at some point. There are many approaches that can be used to ensure that chemical disposal does not cause negative impacts on the environment. Sticking with the example of pharmaceuticals, in Europe, drug take back schemes of unused/expired medication are an obligatory post-pharmacy stewardship approach that reduces the discharge of pharmaceuticals into environmental waters and minimises the amounts of pharmaceuticals entering landfill sites. Although the contribution of improper disposal of pharmaceuticals to the overall environmental burden is generally believed to be minor, drug take back schemes are still considered to be important in reducing the impacts of pharmaceuticals on the environment. The problem is that the general public are not aware (just ask your friends and family if they are aware of these systems) that these schemes exist so higher levels of public awareness and education on the environmental consequences of the disposal of unused/expired drugs are needed for such schemes to succeed.

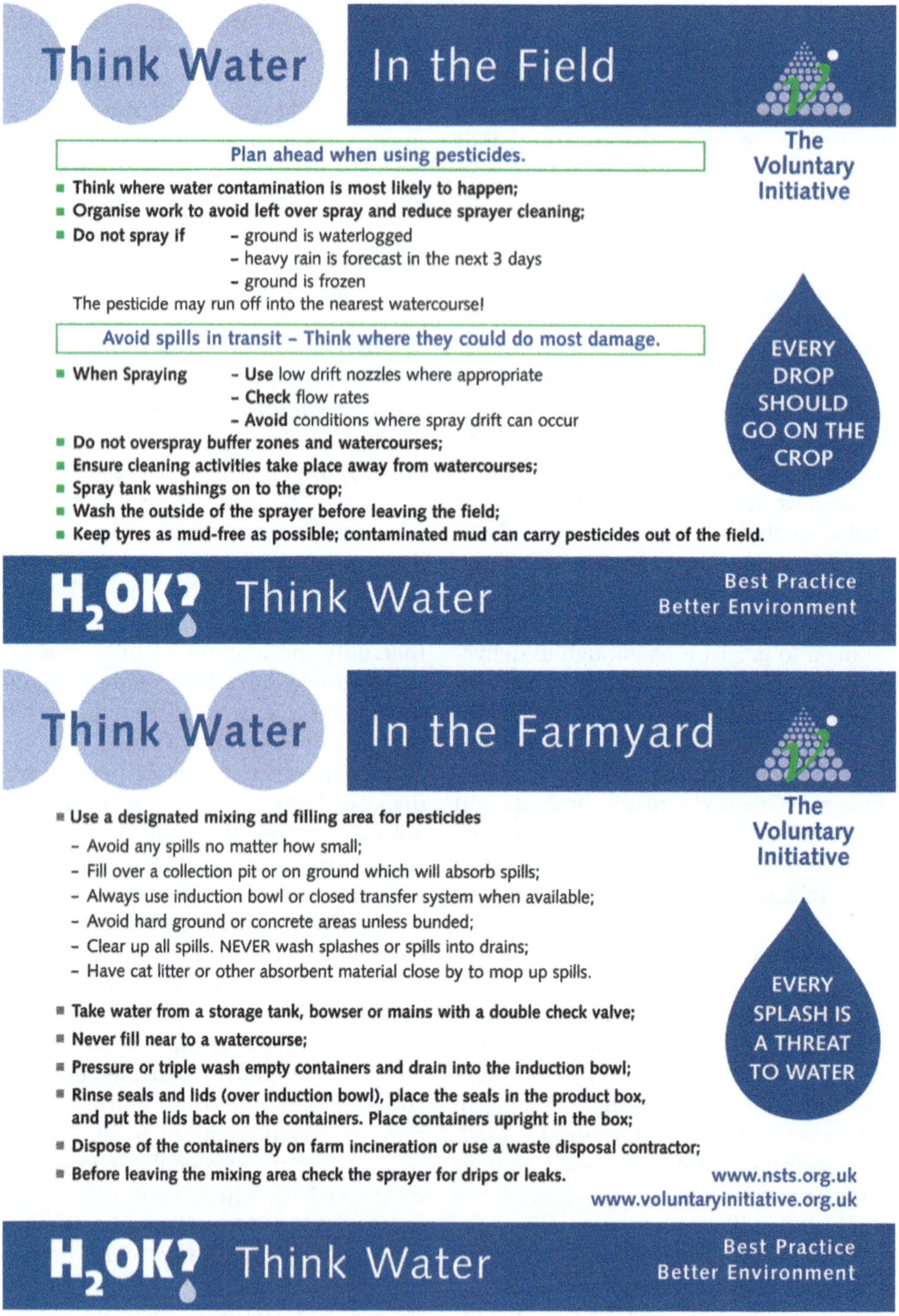

Think Water
In the Field
The Voluntary Initiative
Plan ahead when using pesticides.
Think where water contamination is most likely to happen;
Organise work to avoid left over spray and reduce sprayer cleaning;
Do not spray if - ground is waterlogged
- heavy rain is forecast in the next 3 days
- ground is frozen
The pesticide may run off into the nearest watercourse!
Avoid spills in transit – Think where they could do most damage.
When Spraying - Use low drift nozzles where appropriate
- Check flow rates
- Avoid conditions where spray drift can occur
Do not overspray buffer zones and watercourses;
Ensure cleaning activities take place away from watercourses;
Spray tank washings on to the crop;
Wash the outside of the sprayer before leaving the field;
Keep tyres as mud-free as possible; contaminated mud can carry pesticides out of the field.
EVERY DROP SHOULD GO ON THE CROP
H2OK? Think Water
Best Practice Better Environment
Think Water
In the Farmyard
The Voluntary Initiative
Use a designated mixing and filling area for pesticides
- Avoid any spills no matter how small;
- Fill over a collection pit or on ground which will absorb spills;
- Always use induction bowl or closed transfer system when available;
- Avoid hard ground or concrete areas unless bunded;
- Clear up all spills. NEVER wash splashes or spills into drains;
- Have cat litter or other absorbent material close by to mop up spills.
Take water from a storage tank, bowser or mains with a double check valve;
Never fill near to a watercourse;
Pressure or triple wash empty containers and drain into the induction bowl;
Rinse seals and lids (over induction bowl), place the seals in the product box, and put the lids back on the containers. Place containers upright in the box;
Dispose of the containers by on farm incineration or use a waste disposal contractor;
Before leaving the mixing area check the sprayer for drips or leaks.
EVERY SPLASH IS A THREAT TO WATER
www.nsts.org.uk
www.voluntaryinitiative.org.uk
H2OK? Think Water
Best Practice Better Environment

Figure 11.13 Voluntary Initiative information cards aimed at farmers in the UK to encourage them to use pesticides in an environmentally sustainable way. The cards encourage good practice in the farmyard in order to reduce spillages and splashes of pesticides, as well as good practice in the field, such as avoiding use of pesticides at certain times where entry to water bodies is most likely, and the use of clean and well-maintained equipment.

Source: Crop Protection Association UK Ltd.

Even if green chemistry methods are employed in the design and production of chemicals and stewardship approaches are encouraged to promote good practice in the use of chemicals, chemicals will continue to be released to the sewer system, surface waters and soils. The impact of these chemicals can be managed through the use of wastewater and drinking water treatment systems. One of the main purposes of wastewater treatment is to reduce the concentration of organic matter in the wastewater. If this was released to surface waters, microbes would use the material as a food resource and this would de-oxygenate the water, killing fish and other sensitive organisms.

Wastewater treatment will also remove many manmade chemicals. In developed countries, a typical treatment system will include four phases: preliminary treatment, primary treatment, secondary treatment and tertiary treatment (Figure 11.14). Preliminary treatment uses physical treatments such as screens and grit settling to remove sticks, rags and sand. Primary treatment uses settling to remove organic solids. In secondary treatment, microorganisms are used to break down organic material either aerobically (primarily) or anoxically/anaerobically (in biological nutrient removal processes). In tertiary treatment, additional methods such as chemical oxidation or filtration are used to remove contaminants that cannot be treated out at the primary or secondary stages. Depending on their structure and properties, manmade chemicals can be removed at the secondary treatment phase by microbial degradation or through sorption to the biological sludge. In tertiary treatment, the chemicals are removed through abiotic degradation processes or through sorption to a solid material such as activated carbon. Some chemicals, however, do seem to be resistant to many of the wastewater treatment technologies that are commonly used.

Drinking water treatment not only removes chemical contaminants from water but also microbes. Like wastewater treatment, it involves a number of stages. Typically water is abstracted and screened. Coagulants are then added in order to remove dissolved organic compounds. Towards the end of the process, methods such as chemical oxidation and sorption onto activated carbon are used to remove chemical contaminants. Finally, disinfection agents, such as chlorine, are added to control levels of pathogenic bacteria. While these treatment methods will remove many chemical contaminants, the processes can convert some substances, including endocrine disrupting compounds, from a benign form to a more toxic form. For example, chlorination of the phenolic compound triclosan which is an antimicrobial agent widely used in soaps and toothpastes, leads to chlorination of the aromatic ring which is then cleaved to release chloroform which is a known carcinogen. A good discussion of this topic is provided by Sedlak and von Gunten (2011).

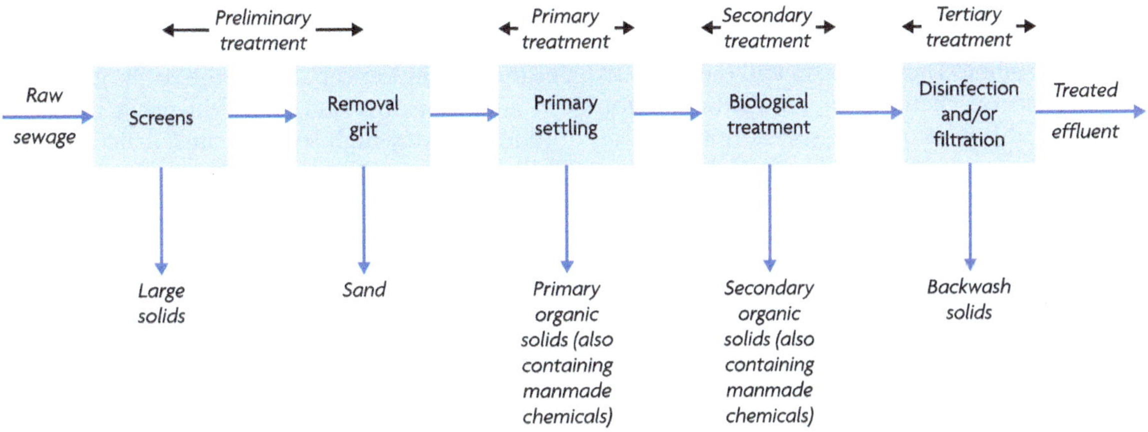

Figure 11.14 Traditional operations in a municipal sewage treatment plant.
Source: Craig Adams.

11.9 Chemical impacts in the next 100 years

So far we have discussed examples of previous impacts of chemicals on ecological and human health. However, our environment is rapidly changing so impacts of chemicals on human health could become more severe in the future, due to changes in exposure caused by alterations in land use, demographics, climate, physico-chemical properties of the environment (e.g. acidification), water availability and increased urbanisation (including a move towards mega-cities). Organisms, including humans, may become more sensitive to chemical pressures and the effects of multiple stressors on ecosystems and human health are likely to become increasingly important. Some changes may also bring benefits to the ecosystem and human health. In the following section we will consider how some of these anticipated changes will affect the risk of chemicals; this discussion focuses on implications of climatic changes. More detailed reviews of the implications can be found in Bloomfield *et al.* (2006), Boxall *et al.* (2009) and Noyes *et al.* (2009).

The climate around the globe is forecast to change. For example, in the UK, climate change models predict that temperatures will increase, that winters will be wetter, summers drier and that intense rain events will be more frequent (see Chapter 8). Frequent periods of intense rainfall, and possible breach of coastal defences due to storm surge, are expected to lead to increased risks of flooding. These changes will have important impacts on chemical inputs to the environment and the subsequent dispersion around the environment, resulting in changes in human, animal and plant exposures. Broadly speaking, climate change will affect risk by: (a) affecting the types and quantities of chemicals and pathogens that are released to or formed in the environment; (b) affecting the transport and fate of pathogens and chemicals in the environment; and (c) affecting the sensitivity of receptors (including ecosystems, livestock and humans) to a particular contaminant. Examples of some of these changes are discussed below.

The inputs of chemicals to the environment are likely to change in the future. Climate change will affect the abundance and seasonal activity of agricultural pests and diseases. The use of pesticides and other biocides could therefore increase in the future and more effective pesticides will be required in some instances. Climate change will also cause changes in agricultural practices. Higher temperatures will facilitate the introduction of new pathogens, vectors or hosts leading to increased use of biocides, and also of human and veterinary medicines.

Changes in contaminant transport routes will also affect contaminant inputs. Flood events can transport pathogens, dioxins, heavy metals, cyanide and hydrocarbons from a contaminated area to a non-contaminated one. Climate change is predicted to increase the frequency of heavy precipitation events, and therefore transport of historical contaminants from previously undisturbed sediments could occur. As irrigation demands increase, due to warmer and drier summers, water of poorer quality may well be applied to crops resulting in additional contaminant loadings. Changes in temperature and precipitation will also increase aerial inputs of volatile and dust-associated contaminants into a system.

Both transport pathways and fate processes for chemicals and pathogens will be affected by changes in climate conditions and this will affect exposure levels. The significance of a particular pathway or process depends on the underlying properties and form of the contaminant (Figure 11.15). Climate change is likely to lead to a greater frequency of macropore and overland flow events as the infiltration capacity of the soil is exceeded. In addition, drier summers will result in longer periods of very high soil moisture deficits, which will increase soil surface hydrophobicity and increase runoff during more intense summer rainfall. A secondary effect of dry summers will be an increase in soil shrinkage cracks, which will result in more extensive and better-connected macropore systems. Matrix flow will also be influenced under climate change, but to a lesser extent than the pathways described above. Matrix flow may increase with wetter winters resulting in higher soil moisture contents and hydraulic conductivities. In the summer, soil moisture will be

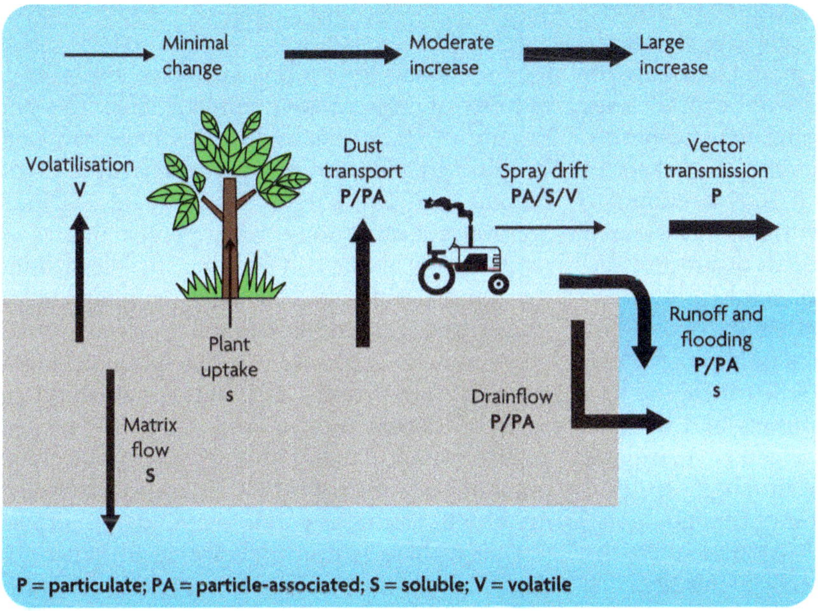

Figure 11.15 Predicted impacts of climate change on major environmental pathways for exposure to pathogens and chemicals. Letters indicate which contaminant classes are likely to be transported via an individual pathway. Large letter = large extent; small letter = small extent; P = particulate contaminants; PA = particle-associated contaminants; S = soluble contaminants; V = volatile contaminants (from Boxall *et al.*, 2009).

lower and hydraulic conductivities reduced. These lower summer soil moisture contents often will be offset by increased irrigation thus restoring the matrix flow route temporarily, especially if the irrigation is poorly targeted.

Flood events have already been demonstrated to enhance the contamination of water bodies by pesticides. Flooding will increase, aiding the dispersion of chemicals following immersion in floodwater. Changes in precipitation levels and patterns will also affect river flows, changing both annual totals and seasonal patterns of flow. For rivers draining impermeable catchments, flow may decline rapidly when runoff decreases and effluent discharges will make up a higher proportion of river flow. Low flows will threaten effluent dilution, resulting in greater pathogen loading, although the effects may be offset by the increase in the amount of time between discharge and abstraction, allowing more time for pathogen and chemical decay to occur.

Volatile organic and inorganic contaminants can be transported via a combination of volatilisation and dispersion. The extent of the transport is dependent on the surface temperature, air temperature and wind speed, all of which are predicted to change as a result of climate change. Dust can be released into the atmosphere during soil tilling and crop harvesting and is an important transport pathway for particulate and particle-associated contaminants such as bacteria, fungal and bacterial spores, steroids, pesticides and polycyclic aromatic hydrocarbons. Soil dust has already been linked to human health impacts. The predicted hotter drier summers will lead to increased drying of soils and an increase in surface dust and hence increased transfer into the environment.

As well as affecting contaminant transport, climate change will also affect fate processes that determine the persistence and form of a contaminant in an environmental compartment. Biodegradation, transformation and volatilisation are expected to increase, while sequestration of sorptive contaminants will decrease. The significance of these changes on exposure will vary. Flood immersion will result in anaerobic conditions in soils affecting the speciation, degradation and transport of selected contaminant types.

The impact of climate change on the viability and fate of pathogens in the environment, and the stability and mobility of genes that encode attributes of public health significance is much more difficult to assess due to a lack of knowledge. The various pathways and processes should not be considered in isolation as different climate-sensitive factors may have conflicting effects on human exposure. One of the main consequences of warmer water temperatures will be an increase in the metabolic rates of freshwater ectotherms. This means that contaminants that are taken up via ingestion or respiration will bioaccumulate faster within an organism, resulting in increased toxicity and a possible reduction in organism tolerance. Faster bioaccumulation will result in enhanced biomagnification up the food chain, potentially having disastrous consequences for top predators. This increase in metabolism will contribute to a decrease in dissolved oxygen levels and the two effects together may cause the organism to expend more energy, exacerbating the situation further.

As well as affecting contaminant inputs, transport and fate, climate change will also affect the structure and functioning of ecosystems and this in turn will affect the sensitivity of an ecosystem to a particular contaminant. For example, in aquatic systems, an increase in primary productivity as a result of a longer growing season, increased temperature and higher levels of nutrient availability are expected, as well as a shift in the timing and magnitude of the spring bloom of phytoplankton which can have adverse effects on the macroinvertebrate community in terms of food availability and quality. Zooplankton abundance will increase, but a decline in the average species body size is predicted. Therefore, if a community has an absence of fish and is dominated by large zooplankton, it is likely to be much more severely affected by climate change. There will be a reduction in available habitat for cool water species and species with a high oxygen demand. The competitive balance within a community may change and there will be an increased risk from invasive species. Finally, changes in flow will severely impact fish reproductive strategies. However, it is not clear yet what these changes will mean in terms of the sensitivity of an ecosystem to chemicals.

As you can see from the above, the assessment of the implications of future changes on chemical risks is not an easy task. Some changes will increase impacts whereas others will decrease impacts. To deal with this highly complex issue, experts from diverse disciplines will need to work together more closely.

11.10 Is natural good?

In the previous sections we have explored a number of cases where manmade chemicals and the use of chemicals by humans have had catastrophic impacts on the environment and on human health. As a result of such incidents, there is a perception among many people that manmade chemicals are bad and that societies around the world should be using more natural products and using fewer synthetic chemicals. If we were to move towards a society without synthetic chemicals, human health and wellbeing would be severely affected and we wouldn't be able to produce the amount of food needed to feed the global population (see Chapter 12). It is also a complete myth that 'natural is good; manmade is bad'.

Firstly, there is no difference at all between a chemical that is obtained from a natural source and the same chemical that is synthesised by a chemist. There are many examples of natural chemicals that are extremely toxic to humans and the environment. Dioxin is the most toxic manmade compound but it is still a million times less toxic than botulinum! Sarin is a chemical warfare agent that is even less toxic than the dioxins. One teaspoon of botulinum could kill a quarter of the world's population, even though some people choose to inject it in the form of Botox® to reduce or eliminate wrinkles around the face. The toxin blocks nerve impulses and temporarily paralyses the muscles that cause wrinkles. This gives the skin a smoother appearance. In fact, five of the seven most deadly known compounds occur in nature (Table 11.3).

Some people also argue that humans and ecosystems have been exposed for 1000s of years to natural chemicals so that they have become resistant to those chemicals whereas exposure to manmade chemicals has been much more recent so resistance has not developed. Again this is a misconception; botulism has been around for millennia and we are still not resistant to it. Moreover, organisms are constantly evolving and making new natural chemicals. There are also numerous examples where organisms in the environment have developed resistance to manmade chemicals, for example antibiotics, anti-parasitic compounds and some insecticides. Finally, you will sometimes

Table 11.3 The seven most deadly chemical compounds to man. LD_{50} is the dose that causes 50 per cent mortality in laboratory animals such as cats, rats and mice. (Table taken from Royal Society of Chemistry presentation on natural vs manmade chemicals.)

Toxin	LD50 mg/kg
Botulinum toxin A	0.00000003
Tetanus toxin A	0.000005
Diptheria toxin	0.0004
Dioxins*	0.03
Muscarine	0.2
Bufotoxin	0.4
Sarin*	0.1

* Manmade chemicals

Source: http://www.rsc.org/learn-chemistry/resource/res00000140/ready-made-careers-presentations-natural-or-man-made-chemicals

hear that synthetic chemicals bioaccumulate in our bodies more than natural chemicals. While many synthetic chemicals do indeed bioaccumulate, there are many examples of natural chemicals (e.g. vitamins A and D) that are lipid soluble and which accumulate in fatty tissues.

In light of the above, and given that many natural chemicals have not been tested and assessed as rigorously as many of their synthetic counterparts, we should be much more considered when making decisions on which chemicals should and shouldn't be used.

11.11 Concluding remarks

This chapter has provided a taste of our knowledge of how chemicals enter the environment and effect ecosystems, as well as illustrating approaches for assessing and managing impacts of chemicals in the environment. Manmade chemicals are essential for everyday living. However, over the past few decades, there have been numerous cases where manmade chemicals that are released to the environment have had adverse impacts on ecological and human health. By using all available knowledge to assess the environmental risks of chemicals, it is possible to use chemicals in an environmentally sustainable way. There will, however, always be surprises and risks that could well change in the future, so it is critical that we begin to better monitor ecosystems and human health to identify problems at as early a stage as possible.

POLICY IMPLICATIONS

Manmade chemicals in the environment

- It is crucial that policy makers appreciate the very diverse pathways by which synthetic chemicals may find their way into the environment.
- Possible environmental impacts of synthetic chemicals need to be very carefully considered when formulating environmental policies.
- Policy makers need to be aware of the nature of toxicity tests and the limitations arising when extrapolating from results of experiments with single potentially toxic compounds tested under well defined laboratory conditions to possible effects in the 'real world'.
- Policy makers need to be aware of the fact that predicted changes in climate may result in changes in the exposure levels of biota, including humans, making it necessary to continually assess risk and, if necessary, modify management policies accordingly.
- Policy makers need to realise that there is often a potential conflict of interest between environmental protection and other policy areas such as maintaining food security.

CHAPTER REVIEW EXERCISES

Exercise 11.1

A thoughtless postgraduate student decides to dispose of 100 g of a sparingly soluble and potentially toxic organic chemical by flushing it down a toilet at an inner-city University. Produce a sketch to illustrate the potential pathways by which the chemical could subsequently be distributed throughout the environment.

Exercise 11.2

Measurement of the light output of a bioluminescent microorganism provides an indication of its overall activity. The effect of the concentration of an organic compound on light output is shown in the table below.

Concentration (mg/l)	Light intensity (arbitrary units)
0.0	100
0.1	100
0.2	98
0.3	94
0.4	87
0.5	77
0.6	59
0.7	31
0.8	15
0.9	5
1.0	0.5

Plot a graph of light intensity against concentration and use the plot to quantify NOEC, LOEC and EC_{50} for the compound.

Exercise 11.3

The effect of the concentration of an organic compound on light output from the microorganism in the previous question was also studied in the presence of 20 mg/l of dissolved organic carbon added as fulvic acid, and the results are shown in the table below.

Concentration (mg/l)	Light intensity (arbitrary units)
0.0	100
0.1	100
0.2	100
0.3	99
0.4	97
0.5	93
0.6	87
0.7	79
0.8	68
0.9	52
1.0	32

Plot a graph of light intensity against concentration. How have NOEC, LOEC and EC_{50} values for the compound changed? Briefly discuss possible causes of the change.

REFERENCES

Bloomfield, J.P., Williams, R.J., Gooddy, D.C., Cape, J.N. and Guha, P. (2006) Impacts of climate change on the fate and behaviour of pesticides in surface and groundwater – a UK perspective. *Science of the Total Environment*, 369, 163–177.

Boxall, A.B.A., Brown, C.D. and Barrett, K. (2002) Higher tier aquatic toxicity testing for pesticides. *Pest Management Science*, **58**, 637–648.

Boxall, A.B.A., Sinclair, C.J., Fenner, K., Kolpin, D.W. and Maund, S. (2004) When synthetic chemicals degrade in the environment. *Environmental Science and Technology*, **38(19)**, 369A–375A.

Boxall, A.B.A., Tiede, K. and Chaudhry, M.Q. (2007) Engineered nanomaterials in soils and water: How do they behave and could they pose a risk to human health? *Nanomedicine*, **2**, 919–927.

Boxall, A.B.A., Hardy, A., Beulke, S., Boucard, T., Burgin, L., Falloon, P.D., Haygarth, P.M., Hutchinson, T., Kovats, R.S., Leonardi, G., Levy, L.S., Nichols, G., Parsons, S.A., Potts, L., Stone, D., Topp, E, Turley, D.B., Walsh, K., Wellington, E.M.H. and Williams, R.J. (2009) Impacts of climate change on indirect human exposure to pathogens and chemicals from agriculture. *Environmental Health Perspectives*, **117**, 508–514.

Cefic (2011) Chemicals industry profile. Downloaded from http://www.cefic.org/Global/Facts-and-figures-images/Graphs%202011/FF2011-chapters-PDF/Cefic-FF%20Rapport%202011_11_ChemIndProfile.pdf

Escher, B.I. and Hermens, J.L.M. (2004) Internal exposure: Linking bioavailability to effects. *Environmental Science and Technology*, **38**, 455A–462A.

Kidd, K.A., Blanchfield, P.J., Mills, K.H., Palace, V.P., Evans, R.E., Lazorchak, J.M. and Flick, R.W. (2007) Collapse of a fish population after exposure to a synthetic estrogen. *PNAS*, **104**, 8897–8901.

Markandya, A., Taylor, T., Longo, A., Murty, M.N., Murty and S. Dhavala, K. (2008) Counting the cost of vulture decline – An appraisal of the human health and other benefits of vultures in India. *Ecological Economics*, **67**, 194–204.

Noyes, P.D., McElwee, M.K., Miller, H.D., Clark, B.W., Van Tiem, L.A., Walcott, K.C., Erwin, K.N. and Levin, E.D. (2009) The toxicology of climate change: Environmental contaminants in a warming world. *Environment International*, **35**, 971–986.

Oaks, J.L., Gilbert, M., Virani, M.Z., Watson, R.T., Meteyer, C.U., Rideout, B.A., Shivaprasad, H.L., Ahmed, S., Chaudhry, M.J.A., Arshad, M., Mahmood, S., Ali, A. and Khan, A.A. (2004) Diclofenac residues as the cause of vulture population decline in Pakistan. *Nature*, **427**, 630–633.

Sedlak D.L. and von Gunten, U. (2011) The chlorine dilemma. *Science*, **331**, 42–43.

Sumpter, J.P. (2005) Endocrine disrupters in the aquatic environment: an overview. *Acta Hydrochimica et Hydrobiologica*, **33**, 9–16.

Wania, F. and Mackay, D. (1996) Tracking the distribution of persistent organic pollutants. *Environmental Science and Technology*, **30**, 390A–396A.

CHAPTER 12

The production of food and its environmental impacts

Malcolm Cresser and Craig Adams

Learning outcomes

By the end of this chapter you should:

- Recognise the growing demands that food production makes upon the Earth's natural resources.
- Understand the disruption of natural element cycling caused by food production for global markets.
- Be aware of how fresh water availability may limit the production of food to sustain a growing world population.
- Be more aware of some key problems associated with global food webs and excessive urbanisation.
- Appreciate the importance of human waste as a potential input to sustainable agriculture.
- Have a better understanding of the pros and cons of organic agriculture.
- Be able to discuss the potential risks and benefits of genetically modified foods to human health and the environment.
- Start to understand some key aspects of food processing to meet global demands.

12.1 The nature of the problem

12.1.1 Globalisation of food supplies

In Chapter 2 we saw how the global population had grown, increasing from around 1 billion in 1800 to more than 6 billion by 2000. Forward projection suggests 9 billion mouths to feed by 2050. Although these raw numbers present a frightening challenge in their own right, they are really just the tip of an iceberg. In the developed world people have come to expect to be able to eat whatever they fancy whenever they fancy it. Thus while the author's recent shopping trolley just before Xmas, and in the middle of an exceptionally cold winter, contained locally produced potatoes, brussels sprouts, carrots, parsnips, a cauliflower and meat, it also contained a variety of citrus fruits from as far away as South Africa, green beans from Kenya, asparagus tips from Peru, baby corn from Thailand, scallops from Canadian waters, tomatoes from Holland, peppers from Israel, blackberries from Mexico, walnuts from the USA and strawberries from Morocco (see examples in Figure 12.1).

12.1.2 Globalisation and element cycling

Most of us looking at Figure 12.1 would simply see a potentially appetising assortment of fruit and vegetables, rather than packets of plant nutrient element resources mined from around the world, undoubtedly topped up by addition of fertilisers as and when required. However, in the context of global cycling of nutrient elements, that is also what they are. Worse still, their transport around the planet, and their production, require inputs of energy, often via fossil fuel combustion. The argument is often put forward that less energy may be needed for transnational transport than that required for heating and lighting if produced at the point of consumption. This may be true, but avoids the issue of whether it would be better for the environment if we all still mainly ate seasonal locally produced foods.

12.1.3 Preservation of food supplies: refrigeration

Another problem with the global distribution of food is the need to provide conditions that minimise loss of quality during distribution. For example, the strawberries in Figure 12.2 are protected from bruising to some extent by packing in rigid plastic cartons. However, they are starting to decay only three days after purchase. Protection involves careful protective packaging, often after disinfection and/or in inert atmospheres, as well as storage at low temperatures, usually only a few degrees above freezing. For some food products though, short storage life makes transport by air freight the only viable possibility.

Figure 12.1 A few typical examples from the multi-national contents of a shopping trolley in the UK in mid-winter.

Figure 12.2 Fungal attack on food during transport and storage is one of the problems faced even when food is consumed close to the point of production, but it becomes a potentially even bigger problem in the context of global food webs.

Generally packets are transparent and carefully designed to make the produce that they contain look as attractive as possible to tempt potential customers to buy an item they hadn't planned to buy. To appease the conscience of the environmentally aware consumers though, the recyclable nature of the packaging is often emphasised on the labelling. The overall strategy clearly works, as most supermarket customers charging along the isles at the weekend seem happy to pay an often significant premium for pre-packaged goods; many do not compare pre-packaged and bag-your-own prices, so probably do not even realise that they are paying more.

Even in the past, storage of local produce to meet food needs in winter was a problem. Grain, of course, could be stored in barns or perhaps, at the household level, in lead-lined chests to reduce loss to rodent pests. Fish and meat products could be salted, dried and or smoked to aid conservation. Various devices were used to preserve root vegetables in acceptable condition. However, food choice was far more restricted, especially for the poor.

Once reliable electricity supplies became available, the invention of refrigeration facilities played a major part in the 'developed' world on the evolution of the 'we can eat what we like when we like' mentality. Keeping temperatures low to preserve food in summer was also a problem, however, long before electrical refrigeration was available. For those that could afford it, insulated ice stores, often built partially below ground, could store ice blocks gathered in winter to be used in summer months for cooling. These might be local winter ice, or ice brought down from mountainous areas not too far away, as was the practice in Tehran, for example, or imported ice. Figure 12.3 shows an old ice store that can still be seen (though not in use) in York in England (Smith *et al.*, 2010).

Well before electrical refrigeration was employed, use was made of the cooling effects of evaporation as a cooling mechanism. The latent heat of vaporisation of water is 80 calories per gram. If a porous ceramic container is created and kept moist, water will evaporate from the container surface and its temperature will fall as a consequence. So too will the temperature of the container contents, perhaps a litre of milk.

In remote rural areas where air temperature is high but with no electricity supply, huts incorporating the water cooling system described above may be used to slow down the deterioration until refrigerated trucks arrive to collect the harvested crop. Thus it can arrive at a supermarket a thousand km away still looking freshly gathered (Figure 12.4). Figure 12.5 provides an indication

Figure 12.3 An old ice store that still exists close to the city walls in York. The wall cavity is well insulated and part of the storage is below the surface.

Source: Photograph courtesy of Elizabeth Smith.

Figure 12.4 A typical supermarket pack of fresh beans, imported to the UK from Africa.

(a)

(b)

(c)

Figure 12.5 (a–c) Effect on beans of storage at 4 °C (six beans on left in each picture) or 25 °C (six beans on right in each picture) for 12 h (a), 35 h (b) and 120 h (c). Even after 12 h some of the beans are starting to curl, and have become less crisp and more flexible.

of the visual effect of green bean deterioration at 20 °C over a few days. The deterioration would be much faster at higher temperatures.

Although the fact that fresh food doesn't stay fresh for very long, especially if grown and/or stored in warm climates, is a problem, warm climatic zones are beneficial in some respects because that may allow more than one crop to be grown per year at a single site; as long as never subjected to water or nutrient stress the period between sowing and crop harvest appears to be attractively short. However, between crop gathering and processing, the crop starts to deteriorate.

12.1.4 Preservation of food supplies: canning, bottling, pickling and drying

In the days before a domestic refrigerator was the norm in almost every household in developed countries, alternative methods of foodstuff preservation had to be found. Sterilisation by heating was commonplace to preserve fruits, often in syrup and sometimes as jams. Pickling could provide a protective environment that was sufficiently acidic to inhibit the growth of degenerative microorganisms. Drying was not just used for cereal grains; it could also be used for many vegetables, and especially those like peas and beans. Rehydration was achieved relatively rapidly in boiling water. Such practices are still in use today of course for 'instant' soups and similar instant meals. Dehydrated foodstuffs were popular because of ease of transport when many did not have a car to carry their shopping and because of their long shelf-life for those with no fridge at home. Canning had the benefits of shelf-life, but was less attractive if you had to carry shopping home on foot.

12.1.5 Food disinfection and sterilisation

It was mentioned earlier that disinfection is often used to protect produce from degeneration prior to packaging for sale. On first thoughts it would seem that all that is required is a sterilisation method that controls fungi that could reduce shelf-life; subsequently the method should have no potential adverse impact upon human health when the foodstuff is later consumed. In that respect sterilisation by irradiation seems close to ideal as long as it is carefully performed. However, disinfectants such as ozone, hydrogen peroxide or chlorine may also be employed to eradicate pathogenic organisms (bacteria, viruses or protozoa) that might otherwise enter the food chain (Demirkol *et al.*, 2008).

Unfortunately there is another potential problem associated with disinfection techniques, even when the reagent used is not itself potentially a direct health risk to consumers. It is possible that potentially beneficial antioxidants contained in fruits and vegetables may be decomposed upon exposure to the disinfecting agents. For example, as much as 70 per cent of the biothiols were lost from spinach treated with 5 per cent hydrogen peroxide for 30 minutes (Qiang *et al.*, 2005). In a later study, free chlorine or hydrogen peroxide exposure was found to adversely affect the concentrations of biothiol compounds, cystein and γ-glutamylcysteinylglycine, in strawberries, whereas aqueous ozone treatment did not (Demirkol *et al.*, 2008). From a nutritional and health perspective, therefore, careful thought needs to be given to the optimisation of disinfection methodology. At the time of writing however, the ultimate consumers are generally blissfully unaware of what has happened to the food they buy, except perhaps when food is sourced locally.

Lead-in question

What's food shelf-life got to do with environmental science?

12.1.6 The 'Display until' and 'Use by' problem

You may well be wondering by this stage in the chapter why the author keeps writing about shelf-life and food conservation. The answer of course is that waste of food is potentially a major disaster in terms of long-term food security and feeding a growing population. The energy and raw material consumption for food production and transport is the same, whether we eat the food or throw it in a bin. So effectively, when food is wasted, environmental problems are being created for no useful purpose. Yet in developed countries wastage figures may be up to around 30 per cent of the food purchased, especially for the more affluent members of society.

There is a curious irony in the fact that availability of refrigeration facilities at home increases the risk of high wastage. This is because the 'weekly big shop' is rarely planned as carefully as it should be. Shoppers yield to temptation just because a product looks 'nice' on the day and are tempted too by BOGOF (buy one get one free) offers in their local supermarkets. The latter often involve products with short use-by periods; by making apparently generous offers, supermarkets can pass part of their potential waste disposal problems associated with surplus stock on to the unsuspecting customers, who later decide they don't really want a similar meal twice within a few days.

Health risk-aware consumers are often over-precautionary in their interpretation of the meaning of 'display until' or 'best-before' information of perishable products. Some tacitly seem to assume that something dreadful must happen at midnight on the day concerned, so many unopened packages join the trash. Yet if fruit and vegetables still look in good condition, they are almost certainly perfectly edible.

12.1.7 Does waste have to be wasted?

In the recent wars when foodstuff was in short supply, any food waste in several countries was often collected to be fed locally to pigs. Nowadays more innovative approaches are starting to be used, such as using food waste for biogas production, and policy makers are again starting to give more thought to how food waste can be put to good use.

Waste is a particularly problematic issue when it goes to landfill, which basically is itself a finite resource. Recently in the UK a green energy company, GWE Biogas Limited, received government funding towards the costs of construction of an £8 million+ Anaerobic Digestion (AD) facility on the outskirts of Driffield (Driffield Today, 2011; Figure 12.6).

Figure 12.6 Anaerobic digestion facility for efficient conversion of food waste to electricity via biogas production at Driffield in Yorkshire, UK.
Source: Photograph by courtesy of Tom Megginson of GWE Biogas Ltd.

The project involved building a facility to convert food waste to green electricity from some 50,000 tonnes of food waste received on site each year from local authorities, supermarkets, food producers, and manufacturers. The facility has the potential to produce enough electricity to power more than 2000 homes or light 15,000 houses locally, and either heat around 400 homes or provide heat to businesses on a nearby industrial estate. The great benefit of the system is that the by-product will be a safe, odourless bio-fertiliser to displace manufactured chemical fertilisers currently used on the company's associated farmland. State-of-the-art equipment employed by GWE Biogas de-packages food products and then recycles the packaging (Figure 12.7).

Figure 12.7 Facility for automated removal of packaging from food waste at the waste food disposal plant at Driffield in Yorkshire, UK.
Source: Photograph by courtesy of Tom Megginson of GWE Biogas Ltd.

Not all food waste occurs in shops or domestic premises. Supermarkets are highly competitive and have high purchasing power. In the process of making sure that they can make profits and meet customer needs at competitive prices they often subject producers to challenging demands with respect to quality, quantity, price and delivery guarantees. Sometimes, either deliberately or inadvertently, this means that some farmers are, on occasions, left with a crop with no market, and hence potential waste. Crops may sometimes be ploughed back in or composted, so the problem becomes more related to the producer's cash flow rather than a mainly environmental issue.

12.1.8 Does it have to 'look nice'?

Another issue that leads to food waste is the actual or assumed desire for food items that look 'perfect'. Fruits and vegetables must be blemish-free, the 'right' size and the 'right' shape. The problem here is what wholesalers and retailers are prepared to accept from primary producers and what they will reject. Imperfections tend to be more acceptable, however, to purchasers of organic produce.

It should be remembered that supermarkets especially, but well-run smaller shops too and even market stalls, do everything that they can to tempt customers to succumb to temptation to buy items that they don't really need. It is sound commercial sense for them, but every time you do you are contributing in your own small way to a potential environmental problem.

As argued earlier, if a farmer composts, or ploughs back in, an unmarketable crop it is at least environmentally friendly to a degree in terms of localised nutrient recycling. However, we still need to think too of the farmer's livelihood when we talk about sustainable agriculture.

12.2 More food needs more land and water

In Chapter 7 we saw some of the limitations to using soils for crop production. One of these was the development of saline soils in arid and semi-arid areas. If such soils are carefully managed, by irrigation with water containing adequate concentration of calcium and relatively less sodium, and drained so that salts do not accumulate close to the surface, these soils can be productive. Soil salinity is a global problem, but unfortunately management for food production is not always a viable prospect.

Figure 12.8 shows a young orchard plantation in Iran. Here water from precipitation over the distant mountains is being used as a supply for a trickle irrigation system to sustain the trees so that their roots are not subjected to salt accumulation. Great care is still needed to prevent contamination of groundwater however. Remember, as discussed in Chapter 7, irrigation under these circumstances has to have an associated drainage system so that dissolved salt is removed from the soil system.

Often exploiting marginal land involves substantial engineering effort, for example to provide adequate water for irrigation whenever required or to cultivate soils on slopes in a way that minimises soil loss by erosion. Figure 12.9, for example, shows terraces on the slopes of a long-extinct volcano crater in Tenerife. With added phosphate the volcanic soils become fertile and productive, and water may be obtained from up slope.

Figure 12.8 A trickle irrigation being used in Iran to allow crop growth in an arid climatic zone. Each tree has its own water supply. Salt does tend to accumulate, but outside the rooting zone. Surface salt deposits between rows may be removed by hand, though the process is highly labour intensive.

The low-lying coastal areas of Tenerife are extremely dry, as we saw in Chapter 2. The island therefore makes extensive use of treated wastewater for irrigation purposes (Figure 12.10). The chemical quality requirements for irrigation water supplies are less stringent than those for potable water supplies.

Lead-in question

How significantly do global food networks influence global nutrient cycling?

Figure 12.9 Terraced slopes inside an extinct volcano crater in Tenerife being used for food production.

Figure 12.10 If used for irrigation, wastewater can provide useful quantities of nitrogen and phosphorus. Careful assessment needs to be made of risks from pathogenic organisms, however.

12.3 Food and global nutrient cycling

12.3.1 Evolution of a potential cycling problem

The obvious, if superficial, answer to the above lead-in question has to be substantially. If we think back to pre-historic times when food supplies depended upon hunting and gathering, and the human population was small, then nutrient elements would have been recycled very locally. Nutrient elements in food (surplus to the proportion temporarily locked up in human growth!) were transferred back to nearby soils in locally deposited faeces and urine. In effect, humans recycled nutrients in much the same way as any other animal depicted in Figure 7.19.

Once villages and towns evolved centuries later, human waste was initially collected and disposed of locally. Over time, however, sanitation improved and disposal became progressively less localised. Most of the population would have been too delighted with less smelly streets to think about whether what they were doing was sustainable. Rivers became highly contaminated, passing 'waste' and the nutrients it contained to coastal waters. Health benefits would have been considerable so long as drinking water remained uncontaminated. As a consequence of the 'waste is waste' approach, often sewage management would mix industrial and domestic effluents with storm water runoff. As we will see later, subsequently this constrained the potential use of recycling of the nutrients in wastewater, because it could also contain contamination that was potentially harmful to humans, plants or both.

As trade developed, food started to be moved over longer and longer distances to meet market requirements. There would have been little thought of the plant nutrients being transported in the food, aside perhaps from the farmers' awareness of the need to use fertilisers imported from elsewhere as and when yields started to decline. When yields of crops were modest and the population remained relatively low because of far shorter life expectancy than now, little thought would have been given to the finite nature of supplies of mineral fertilisers.

Faith in continuity of inorganic fertiliser supplies would almost certainly have been encouraged by the development of the Haber process for converting nitrogen (N_2) gas in the atmosphere to ammonia, as N was often the plant nutrient limiting crop yields. Rock phosphate too seemed more than plentiful. It seemed that developments in science and technology would provide answers to any problems that might emerge, even if populations grew and more land was claimed for agriculture. It is easy to see how farmers could have been led to believe that the need to maintain soil fertility by crop rotations incorporating symbiotic nitrogen fixing plants such as clover to maintain soil nutrient nitrogen supply to plants would become a thing of the past.

In parallel with this came the increasing emergence of world trade, with larger ships, and then refrigerated shipping and aircraft, transporting new foods across continents. Improved methods of food production and food processing further increased the global exchanges of nutrients. At the time of writing, however, attention still focuses almost exclusively on how to improve crop yields and quality and the profit margins of producers and the retail trade over the decadal timescale. Much less thought generally has given to long-term sustainability until very recently.

12.3.2 Maintaining soil fertility prior to development of chemical fertilisers

We wouldn't have to travel too far back in time to be in a world where all food production depended upon the use of natural, and predominantly local, resources. In the 1700s in the north of Europe a pineapple might have graced only the table of the very wealthy, but as a curiosity and conversation piece rather than the dessert. At that time, crop yields would generally have been lower than those achieved nowadays as plant breeding and selection was still in its infancy. This was not a problem when population density was relatively much lower. Indeed, in terms of sustainable agriculture, it could be regarded as positively beneficial as lower yields resulted in lower rates of nutrient element removal with harvested crops. Nevertheless, if for example a field yielded 3 tonnes (3000 kg) of wheat per hectare, and the wheat contained 4 per cent nitrogen, then 120 kg of nitrogen would be taken away with the harvested crop.

Clearly that is not sustainable for more than a few years unless nitrogen is returned in some way. It could be, and was, replenished by adding animal manures or human excrement. However, common practice was to use crop rotation too to help maintain N status. For example, if grass/clover was grown for two years out of each five-year cycle, the clover could fix around 40 kg of dinitrogen (N_2) from the atmosphere. The grass/clover over the two years would also help partially replenish base cation nutrients (Ca^{2+}, Mg^{2+}, K^+) and P and trace element status, as these elements were released by biogeochemical weathering. At the time these methods were being developed, management decisions would have been trial and error based, rather than being based on simple back-of-envelope calculations to assess sustainability of course. Farmers soon became aware that crop rotation had other benefits. In particular, it reduced the risk of build up of plant pathogens that sometimes could occur if the same susceptible crop was grown year after year in the same soil.

Intercropping, especially with N-fixing crops, was clearly highly desirable. In such systems replenishment of soil nutrient status can occur alongside depletion by a crop that is to be harvested. If, for example, peas or beans are used, a second harvestable crop is achieved as well as useful green manure as the crop residues are ploughed back into the soil.

CASE STUDIES

A case study in China – Sustainable food production

If the human aspirations over much of the developed world to eat whatever-we-like whenever-we-want-it were more modest than at present, longer-term sustainability might appear to be a much brighter prospect. Even then, though, the constraints on long-term use of sewage sludge to recycle nutrients for agricultural production would remain a problem. This issue is discussed in more detail later in this chapter. Meanwhile it is worth considering a case study in a part of China where every effort has been made to develop a sustainable system. This has been done by optimising the production of appropriate amounts of components of a diverse diet for a specified number of people, while making sure all waste from the production and consumption is recycled as efficiently as possible.

The following two figures illustrate the type of landscape that evolves as a consequence. Areas of individual crop types are carefully planned and planted to meet the needs of local human and animal populations, optimising the growth conditions for each crop type. The system includes areas of water that are used for production of both rice and fish, and algae contribute to maintaining the N supply of the system.

Diverse crops planted in easily managed areas to meet human and animal food needs, with an area used for fish and paddy rice production behind.

Source: Photograph from Huang Guo-Qin.

A view of a shallow lake area used for fish and rice production. The system is also attractive to ducks, another food resource.

Source: Photograph from Huang Guo-Qin.

By careful planning and optimisation of growth conditions and planting times, up to three crops may be harvested per year from a given area of soil. The approach is known as ecological cropping. Crops are planted according to their ecological characteristics and their requirements for optimal growth are matched to spatial and temporal variations in local soils as well as spatial and temporal variations in precipitation, temperature, sunlight, and so on (Huang *et al.*, 2003). Therefore crop species can make optimal use of natural resources and produce high yields and high product quality. Thus the agricultural efficiency of the system is very high. The ecological cropping incorporates both intercropping and relay cropping.

Field experiments at Jiangxi Agricultural University, Nanchang, China between 1986 to 2002 on local upland red soils in Jianxi Province indicated that intercropping and relay cropping reduced invasions by pathogens and pests (Huang *et al.*, 2003). For example, *Rhizoctonia solani* and *Bipolaris maydis* attack in a late maize-mung bean intercropping system was lower than that in mono-cultured maize systems. Maize pests in the middle and late growing periods, including *Rhopalosiphum maidis*, *Ostrinia farnalis* and *Haptonchus luteotus*, injure the ears of maize, often greatly reducing yield. There were fewer species and smaller quantities of maize pests in a canola intercropped with Chinese milk vetch relay cropped with maize relay cropped with maize intercropped with mung bean than in another simpler cropping system and the injury percentage by major pests was low. Intercropping and relay cropping can reduce diseases and pests but also can increase the quantity and quality of crop yields, which is of benefit to the production of green food and organic food (Ye and Huang, 2002).

Several diverse types of intercropping and relay cropping are now being used in the South of China. For example, rapeseed is relay cropped with maize intercropped with soybean and then sesame intercropped with mungbean. Similarly, broad bean is relay cropped with sweet potato intercropped with maize. Many other combinations may be found (Huang and Mao, 2000). Nitrogen fixing legumes are often found as intercrops in rotations, as shown in the top-right figure.

The system, however, goes well beyond simply using ecological agriculture. Human and animal waste, coupled with waste vegetable products not useful for animal bedding or feed, and waste sediment periodically removed from the fish/rice production, are used to generate biogas for energy. If necessary this may be supplemented as an energy source by burning coppiced wood taken from higher ground. Ash residues and residues from the biogas production are returned to the land as fertiliser. Thus the system is as close as possible to a closed system, but provides a diverse diet including rice, maize, beans, soya, fish, pork, chickens, ducks and even fruit (as seen in the figure below), as well as sustainable energy.

An example of intercropping between maize.
Source: Photograph from Huang Guo-Qin.

Fruit crops may also be incorporated within the sustainable cropping system where climate is appropriate.
Source: Photograph from Huang Guo-Qin.

12.3.3 The problems with recycling sewage sludge

Sewage sludge is the residual matter of municipal wastewater treatment works. It is derived from domestic wastewater and human faeces and urine, domestic products and contaminants, and industrial products and contaminants. It is predominantly organic because the nutrients contained are primarily derived from the waste from human organisms, but it is far from 'organic' in the context of being a potential fertiliser useable for the production of certified organic produce. This is unfortunate, because sewage sludge is rich in valuable nutrients that really need to be recycled, and especially phosphorus. Moreover, disposal to agricultural land is potentially a cheap method of disposal compared to alternatives such as land fill or incineration.

Sewage sludge could improve a number of physical, biological and chemical properties of soil, as well as adding nutrient elements. For example, it increases humus content (and thereby moisture retention and soil structure), it modifies soil pH, and it potentially increases microbial biomass and nitrogen mineralisation and enzyme activities.

Unfortunately, contamination with heavy metals, synthetic chemicals and pathogens can seriously adversely impact upon the safe recycling of nutrients in sewage sludge. Some heavy metals are required at trace levels in the human body, but problems occur with bioaccumulation and/or higher quantities being present. Although many heavy metals will not be absorbed by plants to a dangerous extent, some, such as cadmium, zinc and mercury, may be. The chemical and structural composition of soil affects the likelihood of heavy metal contamination of food crops. The US EPA regulations only take nine heavy metals into account in risk assessment of sewage sludge use.

In the 1988 US National Sewage Sludge Survey, 100 different synthetic organic compounds were found, and the mean number of synthetics detected in a sample was 9. The link to food production and processing was obvious, as 42 pesticides were found with at least one being present in nearly every sample. These could potentially leach into the soil, though this would present more of a problem to organic agriculture. At the time of writing, the effects on soil/plant/water ecosystems of many of these synthetic chemicals, especially in multi-component combinations, are unknown due to a limited amount of appropriate scientific research. Synthetics of concern also include endocrine disrupters from birth pills and a range of pharmaceutical compounds. In terms of potential sludge use in organic agriculture, unacceptable genetically modified organisms eaten and digested could end up on agricultural land. Some of these issues were discussed in Chapter 11.

Most scientists would agree that the precautionary principle should be adopted when making environmental management decisions, particularly with regards to protecting human health. However, largely due to high analytical costs, US EPA regulations do not include routine testing for toxic pollutants that are banned in the USA or are detected in less than 5 per cent of sewage sludge samples.

The pathogen numbers contained in soil can be reduced via treatment techniques. If the lifetimes have not been correctly calculated pathogens may be retained in the sludge in significant numbers, suggesting a potential route of entry to the human food chain.

In conclusion, sewage sludge may be perceived intuitively as an eco-friendly organic fertiliser because of its recycling of valuable essential nutrients and potential economic advantages, with the potential to play a major role in sustainable agriculture. Unfortunately it often contains a complex mix of organic and inorganic contaminants detrimental to agricultural productivity, ecological health and human and animal health.

Typically in the UK for example, only about half of sewage sludge is returned to agricultural land (Figure 12.11). To a large extent this is because of the constraints imposed upon the concentrations of potentially toxic heavy metals that may be allowed to accumulate according to current policies and risk assessment. The Soil Association does not allow sewage sludge, effluents or sludge-based composts to be used in the production of

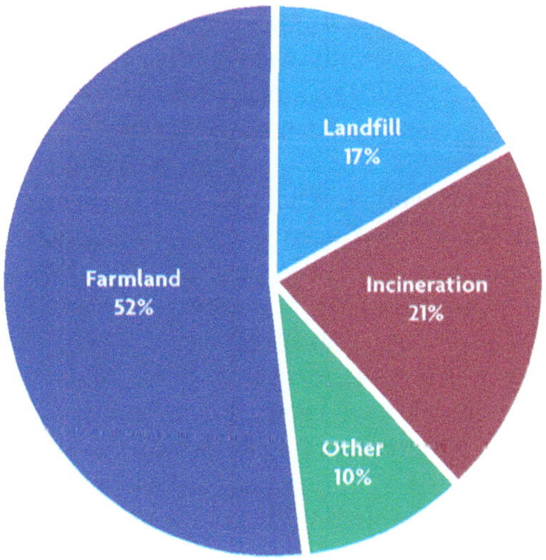

Figure 12.11 Pathways for disposal of sewage sludge in the UK 1999/2000.

Source: Redrawn based on data from www.defra.gov.uk/publications/files/pb6655-uk-sewage-treatment-020424.pdf

organically certified produce. This is, and for the foreseeable future will remain, a constraint of the sustainability of large scale organic agriculture.

12.3.4 Pollutant and natural food contamination from soil

The ever increasing demand for land for food production is not always without risk. In Taiwan, for example, the author noted that small patches of derelict land between buildings in cities were often pressed into use for production of rice and other crops, without always ensuring that the soils are free from heavy metal and other types of pollution resulting from previous land use. For more than 30 years cadmium in particular has been of concern because of its known adverse impacts upon human health and its relatively high bioavailability to crops (Street *et al.*, 1978). Ironically, especially in the past, sometimes potentially hazardous concentrations of cadmium have been inadvertently added to agricultural soils with phosphate fertilisers (Swaine, 1962).

One of the most worrying problems associated with production of food contaminated with a potentially toxic element over recent times, however, was not a consequence of direct contamination from industrial or urban waste or agrochemicals. It was due instead to naturally occurring arsenic found at unacceptably high concentrations (around 2 mg l^{-1}) in groundwater used for potable water supplies or for irrigation. The arsenic was present as a consequence of natural geochemical processes in certain types of groundwater-bearing rocks. In Bangladesh and west Bengal in particular this has resulted in high concentrations of this potentially carcinogenic element in both drinking water and food, and especially rice grown in paddy fields (Abedin *et al.*, 2002). However, the problem occurs elsewhere, including parts of the USA. Irrigated soils have been shown to contain up to around twenty times more arsenic than nearby non-irrigated soils in Bangladesh. The rice straw, which is often used as animal feed, displays disturbingly high arsenic concentrations, so elevated arsenic concentrations may also be found in meat in the area. There is a cruel irony in the fact that the problem only surfaced after locals were encouraged to construct deep wells to provide safer drinking water, but the 'safety' was only with respect to low concentrations of pathogenic organisms.

12.4 Food contamination from pesticide residues

Over the past three decades there has been growing concern over the use of synthetic agrochemicals for food production. This was triggered at least in part much earlier by Rachel Carson's now famous book *Silent Spring* (Carson, 1962). The book drew the attention of a (generally) previously ignorant public to the damaging potential and actual impacts of pesticides on the environment. Carson especially was concerned about impacts on birds, noting that DDT (dichlorodiphenyltrichloroethane) caused thinner egg shells and could lead to subsequent reproductive problems or even death. She questioned the information being distributed by the chemical industry and criticised officials responsible for protecting the environment and health for their generally uncritical acceptance of industry's claims and products.

It was 10 years later, however, that use of DDT was banned in the USA. In fairness to the chemical industry, though, it must be pointed out here too that many human deaths from malaria were prevented by the use of DDT. New chemicals are now extensively tested prior to being licensed, though it is impossible to test for every conceivable long-term effect. So most people would be happier if they could eat food which had not been treated with synthetic organic chemicals.

12.5 Organic agriculture

12.5.1 What is organic agriculture?

An internet search for 'Organic Agriculture' quickly shows just how much interest there is worldwide in food produced without the use of synthetic chemicals. In the UK, the Soil Association site provides a useful concise explanation of what organic farming means http://www.soilassociation.org/Whyorganic/Whatisorganic/tabid/206/Default.aspx. Organic food production originates because of a perceived direct link between human health and the 'purity' of food consumed. The emphasis is largely on food quality from the perspective of chemical contamination, rather than quantity or dietary choice. Strict regulations are in place in different countries around the world to regulate what organic farmers may and may not do. Often, however, much emphasis is placed upon the protection of wildlife and the natural environment, not just upon protection of human health. Organic farming is governed by five main principles if the produce in the UK is to be certified as being 'organic' at point of sale:

1 Synthetic pesticides are not used and use of all pesticides is minimised. Attempts are made instead to manage farms to encourage wildlife to help control pests and to use crop rotations and to produce healthy plants to reduce damage from disease.
2 Synthetic chemical fertilisers are prohibited, but crushed natural mineral fertilisers may be used. Attempts are made to develop and maintain fertile soils by growing mixtures of crops and crop rotation. Clover and legumes are used to fix dinitrogen (N_2) from the atmosphere as intercrops or as green manure crops which are cultivated in. The latter have the benefit of also replenishing (by biogeochemical weathering) nutrients other than N that have been removed in harvested crops at other stages in a rotation.
3 A free-range lifestyle for farm animals is mandatory.
4 The *routine* uses of drugs, antibiotics and worming products are not permitted.
5 Genetically modified (GM) crops and foodstuffs are banned.

12.5.2 To what extent does organic farming reduce resource depletion?

A key objective of organic farming, apart from allaying consumer fears about food contamination with synthetic chemicals, is to minimise depletion of soil nutrient resources. It purports to rely on using 'closed' systems to maintain fertility on farms. Therefore composted farmyard waste and manure are returned to the land. These are supplemented by green manures incorporating plants that fix nitrogen in rhizobia associated with their roots to help build and maintain soil fertility as discussed under point 2 above and earlier in Chapter 5. This agro-ecological approach does indeed contrast to the approaches of more conventional intensive agricultural production, in which yields depend on additions of large quantities of needs-matched manufactured fertilisers and pesticides. The problem with the intensive approach, as often pointed out by advocates of more organic agriculture, is that the finite inputs eventually will run out. The Soil Association especially highlights the potential phosphorus supply problem. However, nutrients are removed off site even in organic agriculture when animals or crops are sold.

12.5.3 The phosphate problem

A review published in 2004 suggested that the world's most economically exploitable phosphate reserves would last more than a century (Cisse and Mrabet, 2004). At the start of the millennium 70–75 per cent of those reserves were in the USA, Morocco, China and Russia; significant reserves could also be found in Tunisia, Jordan, Brazil and Israel and smaller amounts in a few other countries (Cisse and Mrabet, 2004). However, a more recent report by the Soil Association (2010) indicated that supplies of phosphate rock suitable for fertiliser usage are being depleted more quickly than thought. It suggested, correctly in the author's view, that shortage of supply and potentially dramatic price increases for phosphate pose a major new threat to global food security. The report highlights the urgent need for both conventional and organic agriculture to become much less reliant on rock phosphate. It points out that, currently, 158 million tonnes of phosphate rock is mined globally each year but that we may hit a peak in potential phosphate supply as early as 2033. Subsequently phosphate supplies will become increasingly scarce and therefore much more expensive.

It is encouraging that the report recognises that this problem has to be confronted by both conventional intensive farmers and organic farmers, although it does suggest that organic farms should be appreciably more resilient to the problem. It emphasises that we need, as a matter of urgency, to change how we deal with human excreta, pointing out that only about 10 per cent of human excreta is returned to agricultural soils and that urine contains more than 50 per cent of the phosphorus excreted by humans. It is important to realise, however, that somehow we need to achieve a better geographical match between where phosphorus is being removed in harvested crops and where it is being returned to soil. This goes back to having to deal with impacts of the global food web on biogeochemical cycling at the local scale.

A further important point raised in the Soil Association (2010) report is that we really need to change what we eat. This is especially true for much of the so-called developed world. Consuming less meat (or consuming meat less often) would reduce the demands on the remaining accessible phosphate resource, because vegetable-based production is more efficient in its use of phosphorus than livestock production.

Figure 12.12 Two cans of clearly labelled tomato puree made from genetically modified tomatoes, as sold in supermarkets in the UK in 1996. At that time the supermarkets were, very sensibly, keen to embrace the new technology.

Source: Photograph courtesy of Nigel Poole.

12.6 Potential roles and risks of genetic modification

12.6.1 Benefits and potential health risks of GM crops

In section 12.5.1, it was mentioned that growing genetically modified crops is banned in organic agriculture, but so far the nature and justification for producing genetically modified crop species has not been described. Traditional plant breeders have, for many years, selected and propagated specimens that have occurred naturally or via cross-pollination to achieve greater resistance to disease, soil salinity or climate extremes, greater shelf-life, or better flavour or crop growth characteristics.

Back in the early 1990s, genetic modification apparently offered similar potential benefits, which apparently could be achieved much more efficiently. Genes with a desired trait could be transferred artificially from one species to another. That process can occur naturally if, for some reason, exogenous DNA penetrates cell membranes in a species. To genetically modify species artificially, the desired genes are transferred deliberately using a micro-syringe or coated on gold nano-particles fired at selected target cells, or in association with an attenuated virus genome. So when, back in 1996, one author was given a lecture slide (Figure 12.12) by a colleague from the biotechnology industry showing two tins of genetically modified tomato puree, he enthusiastically used it for a few years as a pointer to how food yields could be improved in decades to come.

There was a marked divergence over the next few years in public willingness to consume GM foods between the USA and the UK and Europe. In North America the technology was widely embraced, and GM crops are grown now on a massive scale. In the USA, more than 90 per cent of the soybeans, cotton, rapeseed and sugar beet grown by 2010 was genetically modified, as was more than 85 per cent of the maize. Data from the USDA National Agricultural Statistics Service (NASS, 2010) provide an indication of the proportions of soy, corn and cotton crops grown in various states that have been genetically modified to achieve insect or herbicide resistance or both. GM crops there are well and truly entrenched in the human food chain. Their use has spread from there to developing countries too, so, for example, GM cotton is widely grown in India. The takeover was stimulated by the yield benefits associated with herbicide- and pest-resistance. But in the UK progress has gone down a far more precautionary route, and this is the picture for most of Europe.

The public in Europe, including numerous members of the scientific community, rapidly became concerned about potential health risks and environmental risks. To the biotechnology industry the enormous commercial potential of GM crops meant that most of the research conducted on health effects on animals, mainly on mice and rats, was not in the public domain. Ethically the products could not have been tested rigorously on humans and, in any case, concern was about possible longer-term effects that might take years or even decades to appear in humans. To the public it appeared that GM crops were being rushed through for profit without adequate safety testing. It was a common public perception, though probably a misconception, that GM foods had been accepted as safe just because they didn't look or taste particularly different – a GM tomato was still a tomato so, if it looked the same and tasted the same as any other tomato, then it must be safe. The public was unaware of the extent of the extensive testing on small mammals.

But while GM crops were not apparently being grown or openly sold in Europe, by 2007, 58.6 per cent of the world's 216 million tonne soybean crop was genetically modified; European Union Member States were importing about 40 million tonnes of soy, primarily for use in feedstuffs for cattle, pigs and chickens. Without the protein this soy provided, Europe would not have been able to maintain its current level of livestock productivity (GMO Compass, 2008). In addition, soybeans are used to produce numerous food ingredients and additives, for example in chocolate, ice cream, margarine, and baked goods (GMO Compass, 2008). Undoubtedly therefore, many members of the EU unwittingly directly or indirectly consume produce which contains at least some GM materials. The EU approves certain GM feedstuffs for animals, but sometimes the delay in approval means that farmers may have to pay a premium to acquire non-modified equivalent materials in the interim (Defra and FSA, 2009).

Much of the research on genetically modified crops is targeted towards production of higher yields, for example by building-in resistance to insect pests or resistance to herbicides, thereby allowing herbicide use, post-crop emergence, to eradicate competitive weeds better. The yield increases are often far greater in developing countries than in regions of the world where agricultural practices have already been optimised over many years using conventional crops. Genetic modification may also be used to produce new, more attractive looking or better-tasting varieties, or to improve crop growth under difficult conditions. Examples of the latter include reducing aluminium toxicity in acid soils or salinity effects in saline soils. It is easy to see why these concepts seem very attractive to both commercial companies, who can sell the crop seeds on global markets, and also to farmers, especially in developing countries. For sound commercial reasons it may be deemed desirable to ensure that seed cannot be saved from year to year.

When occasional papers were published suggesting genetically modified potatoes or soy had significant adverse effects on the health of small mammals, they received a disproportionate amount of attention from newspapers. Headlines flagged the potential toxicity of what were to be dubbed 'Frankenstein foods' with little serious attempt to evaluate all the available evidence robustly and public confidence plummeted.

Lead-in question

How can you show whether or not a GM food is safe to small mammals and humans?

Hopefully, readers of this chapter will appreciate the need for great care in the design of experiments to evaluate toxicity of a novel crop species. Matched replicate animal populations are needed and the only difference between the treatments the matched groups subsequently receive must be that one set has GM food and the other set has non-GM food which is *otherwise identical*. To be confident that any adverse health or mortality effect observed is due to the genetic manipulation, the feedstuffs should have been grown under identical conditions. Thus if, for example, only the GM crop had been treated with herbicide post-emergence because it was herbicide resistant, health impacts could be due to residues of herbicide or its metabolites rather than the genetic manipulation *per se*. This still, of course, could indicate a potential health problem, but one associated with the late herbicide use rather than genetic manipulation. Sadly the media rarely concerns itself with such points of detail when a good headline is at stake. So if sheep die after grazing on GM crop residues, the problem simply becomes killer GM foods.

Even when such toxicity testing experiments have been correctly designed some problems remain in extrapolating from effects on small mammals observed within weeks or months to potential effects on human health that might occur over much longer timescales and with a very mixed diet. It is perhaps little wonder then that in many countries precautionary approaches still prevail and drive legislation.

Sometimes those concerned with assessing potential health risks associated with GM foods resort to looking for differences in general human health trends over the two decades since GM food started to be widely used; they compare the trends between countries where GM foods are widespread and those where they are banned. Even with that approach, however, it is difficult unequivocally to assign significant differences detected to the presence or absence of GM foods in the diet. There may be several other changes occurring over two decades at different rates in two different countries.

It would be comforting perhaps to be able to conclude this chapter with a statement that GM foods pose no health risk of any kind, but inappropriate to do so. The best that can be said is that the risk appears to be very low; most would regard it as an acceptable risk when somehow we will eventually have to feed 9 billion, but meanwhile we may all be willing or unwilling participants in a long-term global experiment.

12.6.2 What are the potential environmental risks of GM crops?

So far in this section we have only considered the potential health risks of GM crops to humans and small mammals. The reluctance to see growth of GM crops in Europe, however, also stems in part from environmental concerns. In particular there is concern that herbicide resistance could be transferred from the crop to which it had been introduced to other weed species. For example, scientists from the University of Arkansas found populations of wild plants with genes from genetically modified canola in the United States. They noted that canola can interbreed with 40 diverse weed species globally, and more than half of those occurred in the USA. This clearly raised questions about the production of herbicide resistant weeds which could compete very effectively with other plant species.

This risk of horizontal gene transfer is of particular concern to organic farmers who believe, quite sensibly, that there is a high risk of cross-pollination from a GM crop to an organic crop of the same species growing in the vicinity.

Supporters of GM biotechnology also suggest complementary benefits however. Reduced tillage for weed control may reduce soil erosion problems for example, and reduced use of pesticides should be possible, hence reducing groundwater and surface water contamination. The debate on the pros and cons will undoubtedly continue long after this book is published.

12.7 Should we eat less meat?

12.7.1 Meat production, health and the environment

The sustainability problems associated with eating fish are considered later in Chapter 15, so fish stocks have not been discussed in this chapter. Meat though should be considered here before leaving the topic of food. We should at least briefly consider the desirability of continuing to eat red meat as often as many of the developed world's population currently does, and of enjoying such large portions. Over the past decade obesity and the associated human health problems have become a growing cause for concern in numerous countries. Consumption of large quantities of saturated fats from meat and animal products is a major contributory cause, especially when accompanied by over-sized portions of carbohydrate. Clearly generally eating less would benefit both human health and the environment, with less food per capita needed and less waste to dispose of. There can be no doubt that a well balanced vegetarian diet can be perfectly healthy. Indeed, often significant health benefits are claimed (Appleby *et al.*, 1998; Mattson, 2002). Care is needed though to ensure that the population groups being compared are identical in all other respects, which is not a trivial task. Perhaps curiously, most of the author's vegetarian friends and colleagues are motivated in their choice by personal convictions related to not eating animals rather than by health issues.

There can be no doubt that, compared to crop protein production, animal rearing for red meat as food consumes far more by way of resources used per unit mass of protein produced. Back in 2001, the USDA found that 80 per cent of the agricultural land in the USA was used to produce crops to feed animals and a major proportion of the US water supply was used for irrigation (Vesterby and Krupa, 2001). Moreover, animal rearing makes a major contribution to the emission of greenhouse gases such as methane and nitrous oxide, and grazing animals can make a major contribution to nitrate leaching to fresh waters and to ammonia and ammonium concentrations in atmospheric deposition down wind. They may also contribute to pathogenic organisms in surface waters, which is a problem in some parts of the world.

It should be realised, of course, that a significant percentage of the animals reared are to produce dairy products, but the latter are potentially, if consumed in moderation, far better for human health as part of a well-balanced diet. There is clearly though a real need for policy makers to consider more carefully what can be done to influence dietary choices as the population continues to expand.

POLICY IMPLICATIONS

Food production

- Policy makers need to understand the implications of the global food web to nutrient element transfer between regions and continents in the context of the finite nature of nutrient resources.
- Policy makers need to be able to assess critically the potential human health and environmental risks associated with the expanded use of GM crops.
- Policy makers need to understand the potential ways in which the use of GM crops may help increase global food supplies to help match the world's growing population.
- Policy makers need to understand the potential finite nature of fertiliser element supplies, especially for phosphorus, and to recognise the longer-term need for localised recycling of human waste.
- Policy makers need to understand the need to minimise food waste and to develop integrated policies for waste management that embrace environmental and resource protection.
- Policy makers need to consider how food retailing could be better managed to minimise non-essential resource use and waste generation.
- Policy makers globally need to more fully integrate policies for sustainable use of water and other resources to meet consumer needs in the long term.
- More consideration needs to be given to the balance between vegetable protein and animal protein in the human diet in the context of population growth and environmental protection.

CHAPTER REVIEW EXERCISES

Exercise 12.1

Why is the agricultural production system described for Jiangxi Province in China more sustainable than organic agricultural systems as currently practised in Great Britain?

Exercise 12.2

In an organic farm, the average crop yield is currently 5 tonnes per hectare per year, with an average nitrogen concentration of 5 per cent. How much nitrogen, in kg, is removed per hectare per year with the harvested crop?

The farmer applies 2 tonnes of organic manure and compost per year to his fields on average, with a nitrogen concentration of 6 per cent. How much nitrogen is applied per hectare per year in this way?

If an additional 30 kg of nitrogen per hectare per year is input via biological nitrogen fixation during crop rotations, and 15 kg of nitrogen per hectare per year is deposited as nitrogen pollution from the atmosphere, how sustainable is the farm soil nitrogen supply?

Exercise 12.3

Do you believe that the precautionary principle should continue to drive policy on the growth and sale of GM crops in Europe? If you believe it should continue, how long should this be for? Justify your conclusion against an anticipated world population growth to 9 billion.

Exercise 12.4

Use the internet to find out what you can about the use of land for the production of crops for biogas production. Do you believe that the conversion of land use from forestry or food production to the production of crops for biofuel production is a sustainable answer to the energy crisis? Justify your decision.

REFERENCES

Abedin, J., Cresser, M.S., Meharg, A.A., Feldmann, J. and Cotter-Howells, J. (2002) Arsenic accumulation and metabolism in rice (*Oryza sativa* L.). *Environmental Science and Technology*, **36**, 962–968.

Appleby, P., Thorogood, M., Mann, J. and Key, T. (1998) Low body-mass index in non-meat eaters: the possible roles of animal fat, dietary fibre and alcohol. *International Journal of Obesity and Related Disorders: Journal of the International Association for the Study of Obesity*, **22**, 454–460.

Carson, R. (1962) *Silent Spring*, Houghton Mifflin, Boston.

Cisse, L. and Mrabet, T. (2004) World phosphate production: Overview and prospects. *Phosphorus Research Bulletin*, **15**, 21–25.

Cresser, M., Killham, K. and Edwards, T. (1993) *Soil Chemistry and its Applications*. Cambridge University press, Cambridge, 192 pp.

Defra and FSA (2009) GM Crops and Foods: Follow-up to the *Food Matters* Report by Defra and the FSA, downloadable from: http://archive.defra.gov.uk/environment/quality/gm/crops/index.htm, accessed June 2011.

Demirkol, O., Cagri-Mehmetoglu, A., Qiang, Z., Ercal, N. and Adams, C. (2008) Impact of food disinfection on beneficial biothiol contents in strawberry. *Journal of Agricultural and Food Chemistry*, **56**, 10414–10421.

Driffield Today (2011) http://www.driffieldtoday.co.uk/news/local/government_funding_for_new_anaerobic_

digestion_facility_in_driffield_east_yorkshire_1_831180, accessed June 2011.

GMO Compass (2008) http://www.gmo-compass.org/eng/grocery_shopping/crops/19.genetically_modified_soybean.html, accessed June 2011.

Huang, G.Q., Cresser, M. and Andrews, M. (2003) Increasing the Quality of Agricultural Products by Developing Ecological Agriculture in China. *Aspects of Applied Biology: Special Issue on Crop Quality: Its role in sustainable livestock production*, **70**, 63–70.

Huang, G.Q. and Mao, X.D. (2000) On the eco-agriculture in Jiangxi Province. *Acta Agriculturae Universitatis Jiangxiensis*, **22**, 178–184.

Mattson, M.P. (2002) *Diet-Brain Connection: Impact on Memory, Mood, Aging and Disease*. Springer, the Netherlands, 280 pp.

NASS (2010) Acreage report of the National Agricultural Statistics Service, published by the USDA on 30 June 2010.

Qiang, Z., Demirkol, O., Ercal, N. and Adams, C. (2005) Impact of food disinfection on beneficial biothiol contents in vegetables. *Journal of Agricultural and Food Chemistry*, **53**, 9830–9840.

Smith, E.A., Reed, D. and Ramsbottom, A. (2010) *Discovering Dringhouses: Aspects of a Village History*. Dringhouses Local History Group, York Publishing Services Ltd, York, UK, 100 pp.

Soil Association (2010) *A Rock and a Hard Place: Peak Phosphorus and the Threat to Our Food Supply*. Report published by the Soil Association, Bristol, November 2010, 13 pp., available online at: http://www.soilassociation.org/LinkClick.aspx?fileticket=eeGPQJORrkw%3d&tabid=1259.

Street, J.J., Sabey, B.R. and Lindsay, W.L. (1978) Influence of pH, phosphorus, cadmium, sewage sludge, and incubation time on the solubility and plant uptake of cadmium. *Journal of Environmental Quality*, **7**, 287–290.

Swaine, D.J. (1962) *The Trace-Element Content of Fertilizers. Technical Communication No. 52*. Commonwealth Bureau of Science, Rothamsted Experimental Station, Harpenden.

University of Arkansas (2010) First wild canola plants with modified genes found in United States, from: http://newswire.uark.edu/article.aspx?id=14453, accessed June 2011.

Vesterby, M. and Krupa, K. (2001) *Major Uses of Land in the United States, 1997*. USDA Economic Research Service Bulletin No. 973, Washington DC.

Ye, F. and Huang, G.Q. (2002) Maize diseases and pests of field ecosystems under different structures on dry-land red soil. *Chinese Journal of Eco-Agriculture*, **10(1)**, 50–51.

CHAPTER 13

Wildlife disease: an emerging problem

Piran White, Monika Böhm and Michael Hutchings

Learning outcomes

By the end of this chapter you should:

- Understand the nature of the threat of wildlife disease to humans and livelihoods.
- Understand why disease may be a threat to species survival.
- Distinguish between types of disease: endemic and epidemic.
- Understand how and how fast disease may be transmitted between animals.
- Start to understand how disease in wildlife and livestock may be managed.
- Understand how we attempt to predict risk of disease.
- Start to appreciate risks from new and emerging diseases.

13.1 What is wildlife disease?

Infection and disease are natural occurrences in plant and animal populations, including populations of humans. Disease, which we can take to mean ill-health or some reduction in the normal functioning of an organism, is the response of an organism to being invaded by another organism. In common language, the term disease is also used to refer to non-infectious illnesses such as most cancers. However, in this chapter we will restrict our definition to ill-health caused by 'infectious agents', or parasites.

Typically, disease will cause ill-health among infected individuals, and may contribute to their death, either directly or indirectly, through making them more susceptible to other direct causes of death such as predation and starvation. The effect of disease at the population level is therefore to reduce the population size of infected hosts. One of the key facts about infectious disease is that it can be spread between individuals. Thus, once disease establishes in a single individual, it can spread quite rapidly to other individuals in that population. A few diseases are restricted to one or two host species. However, most diseases are generalist in nature, and able to infect many different species. Thus, disease in one host can have repercussions for infection or population declines in other species. These effects of disease are a major cause for concern, since they can have considerable adverse impacts on wildlife conservation and, where livestock populations are involved, livestock health and economic productivity.

13.2 The significance of wildlife disease

Although diseases are an integral part of natural systems, there are occasions when it is desirable to control disease in wildlife. Because of their generalist nature, many diseases can affect wildlife, livestock and humans alike. A disease which is transmissible from animals – either wild or domestic – to humans is referred to as a 'zoonotic' disease (or zoonosis) (see Box 13.1 for some useful definitions of disease-related terms). These diseases can enter the human population

BOX 13.1 Terms used to describe disease and its effects

Term	Definition
Infectious agent	An organism which can live in or on another organism
Infection	The presence of infectious agents within or on a host
Host	An organism, in or on which an infectious agent lives and multiplies
Dead-end host	A host which is not able to transmit infection to another organism
Maintenance or 'reservoir' host	A host in which infection can persist by intra-species transmission alone, with no need for any infection from other sources
Spillover host	A host which is unable to maintain infection indefinitely in the absence of re-infection from another source
Zoonosis	An infection that can be transmitted from wildlife and/or domestic animals to humans

via the ingestion of contaminated meat, milk or water as well as occupational or accidental contact with infected animals. Emerging diseases in particular have been associated with the presence of wildlife populations harbouring the disease. Due to the immediate health risk to humans, the control of zoonotic diseases in wildlife or the prevention of disease spread from animals to humans is of paramount importance. Similarly, generalist wildlife disease can spread to livestock populations, from which zoonotic infection may become more likely. Infection of livestock from wildlife populations also causes suffering in livestock as well as financial losses to farmers and a reduction in economic productivity of the agricultural sector (e.g. in the case of bovine tuberculosis, as considered later).

Wildlife disease is also of particular concern where species of conservation importance are implicated. Disease epidemics have the power to drive already small or fragmented populations to extinction, either as a direct effect of the infection or by weakening the population to such an extent that other factors can subsequently cause population extinction. Infection with chytrid fungi species has caused dramatic declines in populations of frogs and toads worldwide, many of which were already of conservation concern, due to high rates of mortality associated with the infection. Sometimes, diseases are carried into native animal populations by invasive species, which act as carriers of the disease but are not themselves directly affected by the infection. In the UK, for example, the non-native grey squirrel, originally from North America, is a carrier for squirrel poxvirus, which is fatal to the native red squirrel and threatened the UK red squirrel population.

As a result of the importance of wildlife disease to human health, agricultural production, and biodiversity conservation, disease eradication or management programmes for wildlife have become more widespread throughout the world. Any strategies aimed at disease control have to be tailored to the wildlife species in question, as well as to the particular pathogen involved. An understanding of the biology of infectious agents and their effect on infected individuals is therefore vital in order to develop an appropriate strategy to combat the disease.

13.3 Agents of disease: microparasites and macroparasites

Diseases are caused by small organisms, known as 'infectious agents', which invade the body of a susceptible organism. This animal becomes a 'host' for the disease, and for some time at least will provide the disease with the essential requirements for its survival. Because these infectious agents are living within a host and feeding off the host, they are also known as 'parasites'. While the infectious agent or parasite is living in the host, the host itself becomes a vehicle to promote the survival of the parasite. As well as providing it with somewhere to live, the host may also enable its reproduction and potentially the spread of new infectious agents to other potential hosts or into the environment. There they may develop further or be picked up by other potential hosts. In this way, populations of parasites are sustained over time.

13.3.1 Microparasites and macroparasites

Parasites and infectious agents have many varied lifestyles, and by far the most useful way to think about their different lifestyles is based on their size. In the early 1970s, Roy Anderson and Bob May proposed that infectious agents could be categorised into microparasites and macroparasites. 'Microparasites' are of small size, have short generation times, high rates of reproduction, and multiply directly within the cells of the host. For such parasites, the duration of infection is normally short-lived in relation to the lifespan of the host. 'Macroparasites' are larger, have longer generation times, and live between cells or in body cavities such as the gut. They grow in the host but multiply by producing infective stages which are released from the host to infect new hosts. Macroparasites may infect a host continually over a long period of time.

The major microparasites that infect animals are bacteria, such as the bacterium that causes tuberculosis, viruses such as the rabies virus, and protozoa such as trypanosomes that cause sleeping sickness and *Plasmodium* species that cause malaria. Many are transmitted directly between hosts. For example, the flu virus in humans can be transmitted directly through sneezing, and the bacteria which cause paratuberculosis or Johne's disease in cattle and other ruminants can be transmitted directly by oral ingestion of infected faeces. Other microparasites are transmitted indirectly via other organisms, which are known as 'vectors'. Infections with those are termed vector-transmitted diseases. The tsetse fly (*Glossina*) is a vector responsible for transferring trypanosomes from wild and domestic mammals to man, and various species of *Anopheles* mosquito are responsible for transferring *Plasmodium* spp. from wild mammals to man. However, in these cases, the fly does not only act as a vector; the parasites actually multiply within the fly, so that it also serves as a host. Some diseases may need to pass through several host organisms to complete their lifecycle.

Macroparasites include parasitic helminth worms such as platyhelminth worms or flatworms (tapeworms, schistosomes) and nematodes or roundworms. Macroparasites also include lice, fleas, ticks and mites (see, e.g., Figure 13.1). Directly-transmitted nematodes are one of the most important human parasites, in terms of the number of people infected and their potential for causing ill-health.

Since the time of Anderson and May's work, some new types of disease have emerged. The most notable from the perspective of wildlife disease are infectious cancers, such as the facial tumour disease that affects Tasmanian devils (Figure 13.2), and the rogue proteins (prions) that are implicated in Transmissible Spongiform Encephalopathies (TSEs). These infections are best considered within the microparasite category, since they share most of the same characteristics as the other, longer-recognised, microparasites.

(a)

(b)

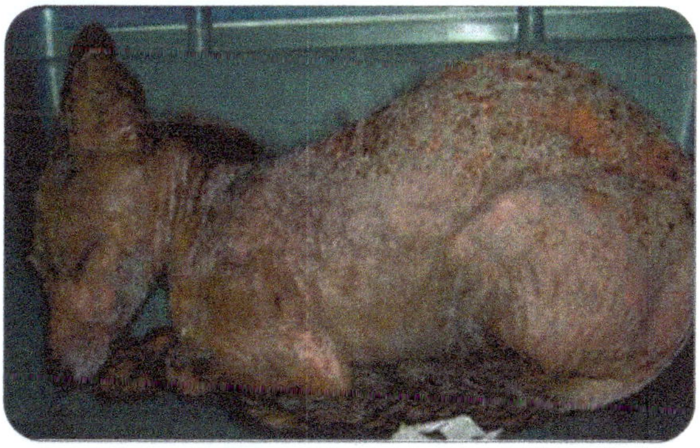

Figure 13.1 Early (13.1A) and late (13.1B) stages of infection for foxes with sarcoptic mange. The infection is caused by a mite that burrows into the fox's skin. Macroparasitic infections such as mange are treatable, but once they get beyond a certain stage, it is unlikely that the animal will survive.

Source: Images from the National Fox Welfare Society.

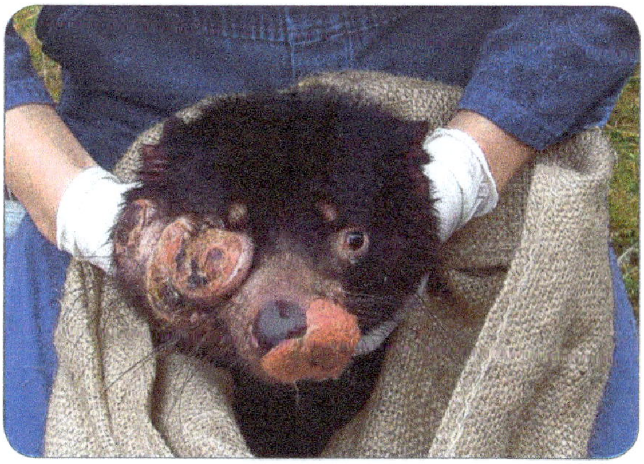

Figure 13.2 Tasmanian devil infected with facial tumour disease.

Source: Image reproduced by kind permission of the Save the Tasmanian Devil Program.

13.4 Immune responses of hosts

Vertebrate hosts are able to show an immune response to some microparasites, whereby specific antibodies can be released in response to certain invading organisms. These antibodies can attack the parasite in a variety of ways, and may enable the host to recover from the infection. In addition, the 'memory' of these antibodies may persist, allowing the host to become immune to re-infection. Viral and bacterial microparasites tend to elicit strong immune responses, which can result in long-term immunity. In contrast, protozoan microparasites and macroparasites tend to elicit weaker ones. Thus, these infections tend to be more persistent and immunity does not develop.

13.5 Disease transmission

One of the key factors affecting the population dynamics of infectious diseases is the rate of transmission. This determines the rate at which the disease will spread through a host population and is therefore of great importance for the control of a disease. The factors that affect the rate at which parasites are transmitted between hosts vary according to type of parasite.

For directly-transmitted microparasites, in which transmission occurs by direct physical contact, the net rate of transmission is directly proportional to the frequency of encounters between infected hosts and susceptible hosts. The transmission rate of directly-transmitted infections is normally greater in dense populations than in sparse ones. The exception to this is for sexually-transmitted infections, since mating activities are maintained irrespective of population density, at least until population densities become so low that the process of finding mates becomes more difficult. Where infection is caused by a free-living infectious agent (indirect transmission), transmission is usually proportional to the frequency of contacts between hosts and infectious agents. For vector-transmitted microparasites, the rate of transmission from vectors to hosts and vice versa is proportional to the frequency of contacts between the vector and the host. This is known as the 'host biting rate'.

The study of infectious diseases (epidemiology) uses various terms to describe the distribution and abundance of parasites or hosts in an area. In microparasite infections, hosts are either infected or not. The most widely used descriptive statistic for microparasites is the prevalence of infection. This is the proportion or percentage of a host population that is infected with the parasite. However, for macroparasites, the severity of infection of a host varies according to the number of parasites it carries. Thus, the number of parasites in, or on, a particular host is referred to as the intensity of infection, or the 'worm burden'. The mean intensity is the mean number of parasites per host and may include those not infected. Macroparasites tend to be unevenly distributed among individual hosts, so that within a single host population, many individuals have relatively light parasite burdens, while a few individuals are heavily infected (Figure 13.3).

The population dynamics of infectious diseases have much in common with predator–prey interactions, and much emphasis in the literature is on the use of mathematical and computer models to predict patterns of prevalence and requirements for control. The originators of this approach were Roy Anderson and Robert May,

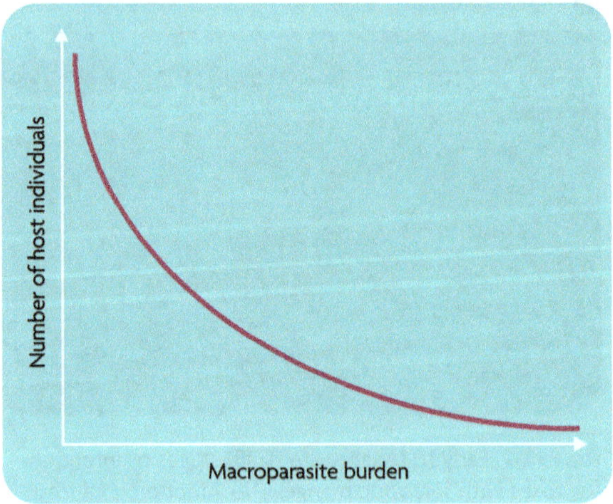

Figure 13.3 Typical distribution of a macroparasite within a host population. High levels of infection are concentrated in relatively few individuals and the distribution often takes the form of a negative binomial.

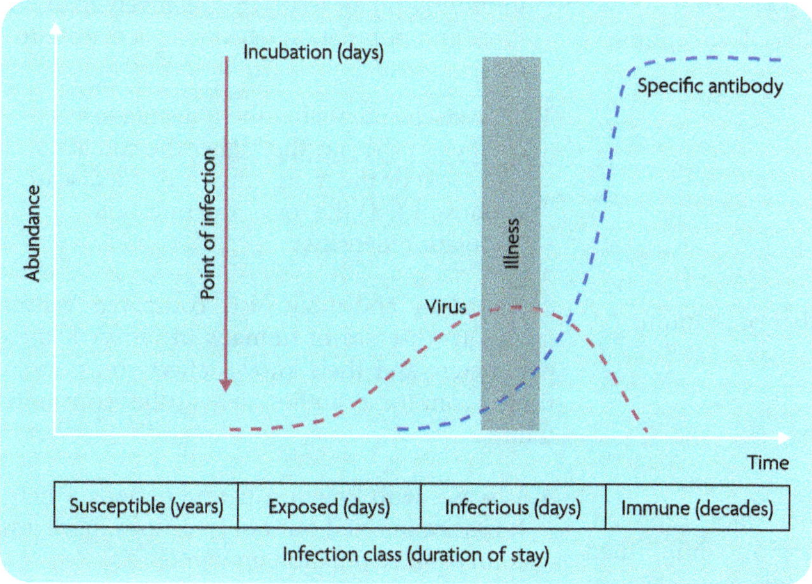

Figure 13.4 Course of a typical acute viral or bacterial infection in a host individual showing the corresponding progression through four categories of infection.

Source: Adapted from Nokes and Anderson (1988).

who outlined their ideas in two seminal papers published in *Nature* in 1979. The basic concepts are not complicated, and these models serve to demonstrate how relatively simple assumptions and relationships can be used to build powerful models with wide-ranging implications (see, for example, Figure 13.4). Although the current literature abounds with many more complex models, they can still be traced back to the original concepts developed by Anderson and May.

The models developed under this basic approach are called SEIR or just SIR models, which classify the host population into different disease categories, depending on the disease in question, into susceptible (S), exposed (E; in diseases with a latent stage, i.e. individuals have been exposed to infection but are not yet infectious themselves), infectious (I) and recovered or immune individuals (R). Similarly, for a disease without recovery, which is always fatal, they may be termed SI models. Various rates then describe the transition of individuals from one category to the next.

13.6 A simple SIR model for a microparasite infection

The total population of a host, N, can be described as consisting of three categories of individual: uninfected or susceptible, X, infected, Y, and immune, Z. New animals are born into the population at a birth rate, a, and animals die from the population at a death rate, b.

For a direct infection, the rate at which the infection will be acquired by susceptible animals will be proportional to the number of encounters between susceptible and infected animals, and will be βXY where β is the transmission coefficient (i.e. the proportion of contacts that result in infection).

The mortality rate for infected animals is $b + \alpha$, where α is the disease-induced mortality which operates in addition to the natural mortality, b. Some animals may recover from infection, and they do so at the recovery rate, v. Recovered animals are initially immune, but this immunity can be lost again at a rate γ (for permanent immunity $\gamma = 0$).

These simple assumptions give the following equations that describe the dynamics of the disease–host interaction:

$$dY/dt = \beta XY - bY - \alpha Y - vY$$

(rate of change in the number of infected hosts)

$$dN/dt = (a - b)N - \alpha Y$$

(rate of change in the total host population)

$dX/dt = a(X + Y + Z) - bX - \beta XY + \gamma Z$
(rate of change in the number of susceptible hosts)

which simplifies to:

$$dY/dt = \beta XY - (b + \alpha + v)Y$$

and

$dZ/dt = vY - \gamma Z - bZ$
(rate of change in the number of immune hosts)

which simplifies to:

$$dZ/dt = vY - (b + \gamma)Z$$

From these equations, further, more applied, equations can be derived using more complicated mathematical methods. The two most relevant are the following. We may substitute r for $a - b$ (i.e. r is the intrinsic population growth rate); then, if

$$\alpha > r\,[1 + v/(b + \gamma)]$$

the disease has the ability to regulate the host population. This is a particularly important property for a disease. If not, the population will continue to grow in the presence of the disease.

Another important concept in disease epidemiology is the threshold population density. This is the density of hosts that is required for the disease to persist in a population. This is given by:

$$N_T = (\alpha + b + v)/\beta$$

From these equations, certain generalisations about disease–host systems can be made:

1 For a disease to regulate a host population, the disease-induced mortality rate α must be high relative to the intrinsic growth rate r of the population.

2 The ability of a disease to regulate the host population will be decreased by lasting immunity (small γ) and high rates of recovery from infection (large v).

3 High levels of natural mortality (b), disease-induced mortality (α) and recovery rate to immunity (v) all result in a relatively high threshold population density for a disease to persist.

4 Conversely, the higher the transmission coefficient (β) (i.e. the higher the contact rate *per se* or the more efficiently the disease is transmitted) the lower the threshold population density.

Anderson and May also compared certain infectious diseases of humans in terms of these parameters and their known effects on populations. From these studies, they further concluded that:

5 Diseases with long incubation periods, where hosts are infected but not infectious, have less impact on population growth.

6 Vertical transmission, whereby the disease is passed from mother directly to offspring, lowers the threshold population density required for persistence of a disease. This can also be the case for pseudo-vertical transmission, whereby the rate of disease transmission between mother and offspring is considerably enhanced purely by their spending a large amount of time in close physical proximity.

13.7 Patterns of disease

Diseases may show different patterns in populations (Figure 13.5). Some diseases are characterised by epidemics, or rapid changes in the prevalence of infection, whereby the disease has a sudden and dramatic effect on the host population. Other diseases persist for long periods of time, showing only small changes in prevalence and a relatively constant effect on population size. These are known as endemic infections. However, these terms are descriptive of the pattern of disease rather than the diseases themselves. Indeed, many diseases such as measles in humans can persist either as an endemic infection or as recurrent epidemics.

So how do we know which diseases should produce which patterns? Diseases in which the

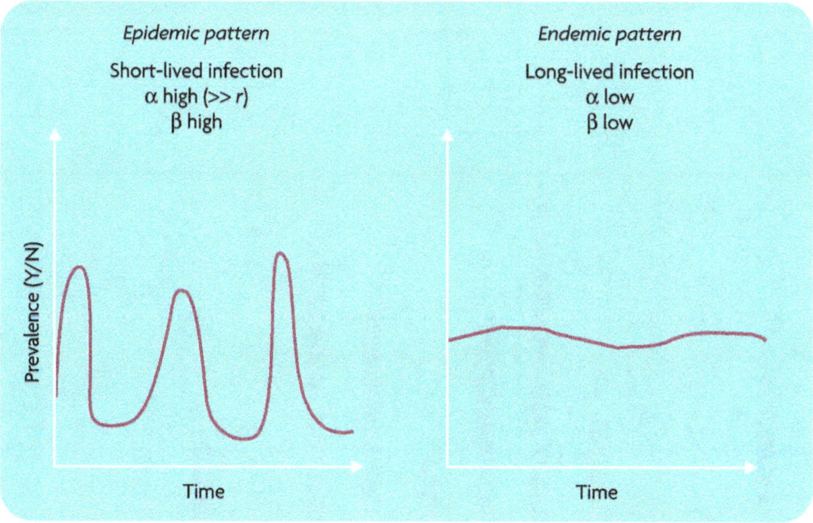

Figure 13.5 Typical patterns of epidemic and endemic infections over time. Epidemic patterns of disease are typified by marked fluctuations in abundance and may be cyclic.

duration of infection is short, or those in which disease-induced mortality far exceeds the intrinsic growth rate of the population, will tend to produce epidemic patterns. Thus, if the natural rate of increase of the host population is much less than the disease-induced mortality, and if the host population is at a level greater than the threshold population required for the disease to persist, the introduction of an infection will have a huge effect on the population size, which will be rapidly decreased to below the threshold population once more. As a consequence, the disease will fall to such a low prevalence that it is likely to go extinct through stochastic processes. In contrast, diseases in which disease-induced mortality is relatively low and the infection is long-lasting are usually endemic in character.

One phenomenon often associated with disease is cyclicity. Time-dependent cyclicity, such as seasonal variations, can be brought about by seasonality in the efficiency of transmission, which may reflect environmental conditions or the behaviour of the host. For example, the seasonal peak of rabies in striped skunks in Illinois in the spring reflects behavioural patterns in the host population because of increased mating activity during the winter. Other infections, such as the gut nematode *Trichostongylus tenuis* in red grouse (*Lagopus lagopus scoticus*) can exhibit cyclicity over longer time periods (Figure 13.6), although the causes for this cyclicity are sometimes not well understood.

13.8 Diseases of multi-host communities

Many diseases are generalist in nature and have the ability to infect not just one, but also multiple host species. This may have positive effects on biodiversity since the presence of a shared disease can lead to coexistence between host species via indirect competition where they would otherwise exclude each other. Disease dynamics in multi-host communities may be complicated by variation in the resistance and tolerance to the disease between the different host species. Similarly, disease transmission from one species to another may be facilitated in some cases (for example, in species sharing common feeding grounds) and inhibited in others, and this in turn will have an effect on multi-host disease dynamics.

In order for a disease to persist within the community, the concept of a host population density threshold is replaced by a similar concept of a threshold community configuration. If inter-species transmission occurs freely, host species may

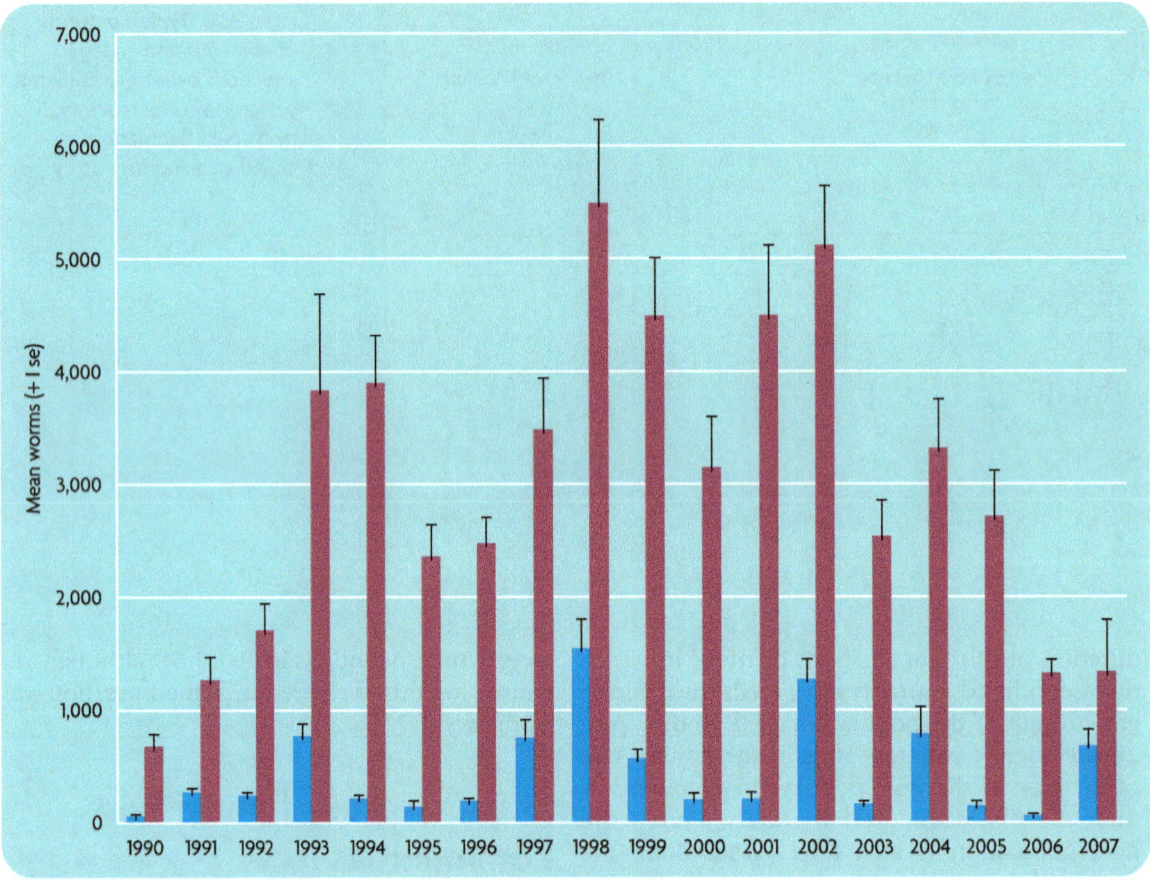

Figure 13.6 Abundance of *Trichostrongylus tenuis* in adult (red) and young (blue) red grouse on eight moors in the North of England, showing evidence of cyclicity over 5–7 year periods.

Source: Graph supplied by kind permission of the Game & Wildlife Conservation Trust.

combine to provide a single reservoir for the pathogen and hence generate a common 'community threshold'. On the other hand, if there is no disease transmission between the hosts, the existence of the alternative host is irrelevant to the establishment of the pathogen in either of the available hosts. Where disease transmission is weak or uni-directional between hosts, spillover from a reservoir population to non-reservoir species and dead-end hosts may occur. In strongly territorial species inter-species disease transmission may be much stronger than intra-species transmission, so that in these systems an alternative host is required to achieve pathogen establishment and persistence within the population.

13.8.1 Leishmaniasis

One of the most widespread diseases carried by vertebrate hosts and which causes much suffering in humans is leishmaniasis, caused by a protozoan parasite *Leishmania*. Among the protozoal diseases of humans, leishmaniasis probably ranks second only to malaria in terms of its medical and economic importance. One estimate put the number of new cases worldwide at about 400,000 every year. Phlebotomine sandflies are the only known vectors of leishmaniasis and they can pick up the parasite through blood meals from a vertebrate host. This can then be passed on to man through the bite of the sandfly. The principal hosts of the disease in many areas are dogs. This is especially true on the edges of large cities or in rural communities in South American countries, where foxes may also be involved in the disease cycle. Infection is also common in rodents, canids and marsupials.

Infection in the vertebrate host is normally benign and not visibly apparent. However, in humans, it produces severe disease in the form of lesions. Two forms of the disease have been recorded: cutaneous leishmaniasis, in which the lesions affect the skin, and visceral leishmaniasis, in which they affect the intestines.

One of the problems of leishmaniasis is that it may be present in a large number of different vertebrate host species in any one area. For example, one study in north Brazil found *Leishmania* in 13 different wild mammal species. The reservoir of infection may also be shared between wild and domestic hosts such as donkeys and horses. This makes it a particularly difficult disease to control because it is not only contacts between humans and domestic animals or humans and wild animals that are important but also those between domestic and wild animals.

13.8.2 Trypanosomiasis

Trypanosomiasis, or sleeping sickness, is another disease that is the cause of much suffering to humans and domestic animals in various parts of the world. Trypanosomes are protozoan parasites which are found in the blood and tissues of many vertebrate hosts, and are transmitted by the bite of an infective tsetse fly of the genus *Glossina*. Favoured host species are wild game animals such as buffalo, kudu and bushbuck. Wild game animals seem to have a greater tolerance to infection with trypanosomes than domesticated ones. One of the problems of controlling trypanosomiasis, like leishmaniasis, is that many animals can act as hosts for the disease. Thus, early attempts to control the disease in Natal centred on the destruction of impala and bushbuck. However, they did not control warthog or bushpig, which also acted as reservoirs of infection. As a result, control was unsuccessful.

Most trypanosome infections arise directly through bites from tsetse flies. However, it has also been found that carnivores have a particularly high rate of infection. It seems that carnivores are only rarely directly infected through being bitten by tsetse flies, and that they acquire the infection mainly through eating infected prey. Lions and hyenas were the carnivores examined. Both species live in social groups and mutual grooming of cuts and sores between members of the group will present further opportunities for the disease to spread. Rodents may also be important reservoirs of the disease in some areas, and infection rates in rodents may be as high as 50 per cent.

13.8.3 Rinderpest

One of the most famous examples of a disease of domestic animals with reservoirs in wild mammals is rinderpest. Until the late 1880s, the Sahara desert had acted as an efficient barrier to the spread of rinderpest into the southern half of the African continent, primarily because the ungulate species that live in the desert exist at densities that are too low to support a continual infection of rinderpest. However, in 1889, a major pandemic was initiated by the accidental introduction of the disease into Somalia, probably by cattle imported from India for the Italian armies. It subsequently spread along trade routes throughout the continent, spreading south through Zimbabwe to reach South Africa in 1897, and also spreading west to reach the west coast in 1890–1892.

As a result of this first epizootic, 5.3 million cattle died, and the mortality in wild ruminants was also very heavy, up to 95 per cent in some species of wildlife. Some estimates indicate that at least 90 per cent of Kenya's buffalo were eradicated. Buffalo, eland, warthog and bushpig suffered most badly to start with, although giraffe, kudu, roan antelope, bushbuck and wildebeest were also badly hit later on. This order reflects the differences in susceptibility between the species.

There was also evidence of knock-on effects of rinderpest for tsetse abundance. Mortality in wild ruminants was so high that the tsetse fly was severely reduced or even died out in some areas, for example in the Kruger National Park, because it had so few suitable hosts on which to feed.

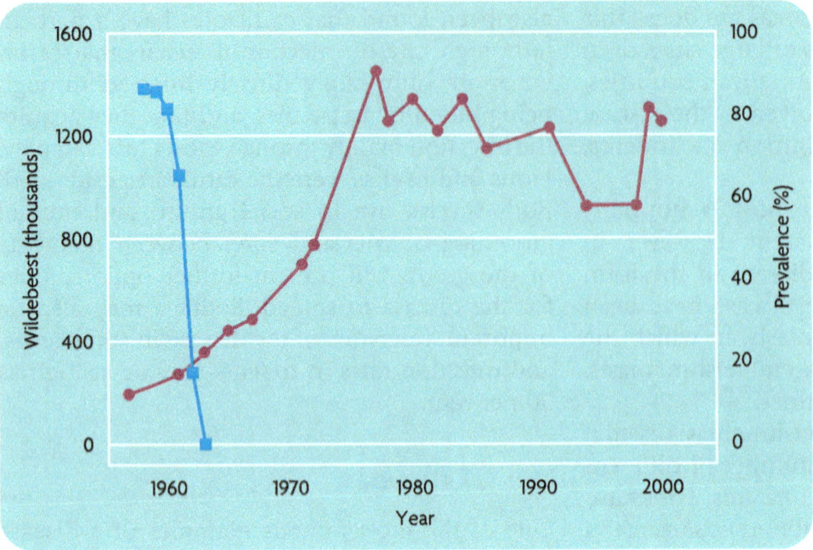

Figure 13.7 Changes in rinderpest (blue squares) and wildebeest (red circles) populations in the Serengeti ecosystem.

Source: From: Holdo *et al.* (2009) *PLoS Biol*. 7(9): e1000210. doi: 10.1371/Journal.pbio.1000210.

Losses of wild mammal hosts were so substantial that the first rinderpest epidemic effectively burned itself out through most of the continent. However, it remained enzootic and occasionally epizootic in parts of north-east and north-west Africa. Since then there have been several incidences of epizootics reaching down towards central Africa, most notably in the early 1960s, but also more recently in two separate areas of the Serengeti in 1982, when a large number of buffalo died. Overall though, rinderpest has declined substantially. In the Serengeti, the original eradication of rinderpest in the early 1960s led to a 700 per cent increase in the wildebeest population, indicating that in some situations, disease can play a role in ecosystem regulation (Figure 13.7). Rinderpest is now largely under control through coordinated cross-border vaccination campaigns, and scientists in the Food and Agriculture Organisation (FAO) are confident that it will be eradicated in the near future. If so, this would be the first time that humans have succeeded in eradicating an animal disease.

13.8.4 Rabies

One of the diseases in which wildlife play a major part and which strikes the most fear into people is rabies. Rabies has been known to occur for over 2000 years. It was recorded by the Greek physician Hippocrates and even at that time it was understood that the clinical symptoms in man were related to previous attacks by mad dogs. It is the symptoms of the disease and the fact that they are followed by virtually inevitable death that have made it a disease to be feared. The disease is caused by the rabies virus invading and multiplying in the central nervous system of the victim.

An infected animal excretes virus in the saliva and frequently in the urine. The virus cannot penetrate unbroken skin, although it can enter the body across intact mucous membranes, such as the lining of the nose or mouth, or the conjunctiva of the eye. Some incidences of the disease being transmitted from person to person through corneal implants have been reported. The commonest way of transmitting the virus is by a bite from a diseased animal depositing the virus in the tissues of a susceptible one. However, transmission can also occur through social grooming or as an aerosol in enclosed spaces, since virus particles can be excreted in water droplets in the breath. This is extremely rare in humans, but can occur in animals which show a high degree of aggregation at various times. Thus, bats which may roost together in large numbers, and those species of canids which live in social groups, such as foxes, jackals, wild dogs and wolves, are particularly good hosts for the disease. They actively defend a territory, within the group they mutually groom, and they often share an underground burrow system.

The time between exposure to the virus and the onset of symptoms may vary widely, ranging between four days to many years in humans. However, it is generally between 20 and 90 days. In foxes and dogs the incubation period is shorter, commonly around 14 to 28 days. There is still no cure once the symptoms have started, and death ensues in between three and seven days. This period is the time when the animal will be able to transmit the infection. At this stage there are two patterns of symptoms which the disease may take.

In the *furious* form the victim displays excitement, aggression, may suffer from hallucinations, and also develops hydrophobia, a fear of water. In the case of animals, the victim will attack other animals, man and even inanimate objects without provocation. After a time, the animal becomes paralysed and eventually dies. In the *dumb* form the furious stage is bypassed and the victim develops increasing paralysis. The furious form is more common in humans but less common in animals. In either case, eventual death is thought to be due to encephalitis (inflammation of the brain), which leads to respiratory problems and cardiac arrest. Once the symptoms start, death is virtually inevitable, although not all infected animals necessarily develop symptoms.

Rabies can be maintained in two cycles. *Canine* or *urban rabies* is the cycle most important in developing countries, at least from the perspective of causing human deaths (90 per cent of all human rabies deaths are thought to be due to canine rabies). In this cycle, rabies is maintained in populations of feral dogs in and around human settlements. This is the form of rabies most common in Africa and South America. This was also the form of rabies which was present in Britain before it was eradicated at the start of this century and which predominated in western Europe up until that time. The other cycle is the *sylvatic rabies* cycle. In this, rabies is maintained within a wildlife reservoir. It is thought that although these two cycles can be distinguished, they are to some extent linked. Domestic dogs and cats provide the link to man in both cycles.

All warm-blooded animals are susceptible to rabies to varying degrees. Bats, rodents and even deer may be important hosts for the virus in some areas, but some species may constitute spillover hosts rather than true reservoirs for infection (Figure 13.8). In general, herbivores are considered to be dead-end hosts because they are unlikely to pass the disease on to other species. However, they may pass the virus readily to

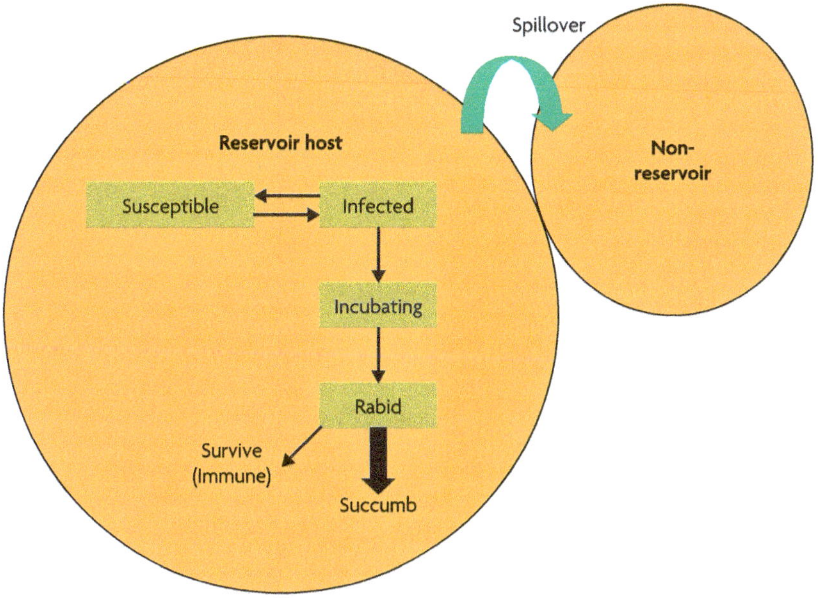

Figure 13.8 Schematic diagram showing compartments in rabies transmission within reservoir hosts and between reservoir and spillover hosts. Spillover hosts are not able to maintain the disease in the absence of external sources of infection, and some spillover hosts, especially herbivores, may constitute dead-end hosts.
Source: Sterner and Smith (2006), *Biological Conservation* 131, 163–179, Elsevier.

members of the same species. An outbreak of rabies in deer in Richmond Park in London at the beginning of the twentieth century is an example of this. Also, an outbreak occurred among kudu in Namibia in the late 1970s which resulted in 20,000 deaths. Carnivores are normally the main vectors in the sylvatic cycle. Different species are the most important vectors in different parts of the world. In Western Europe, it is the red fox; in the United States, the fox, the striped skunk and the racoon are all important in different areas. In fact, one of the problems of managing rabies in many areas of the world is that there are many species that can be involved in maintaining the disease. The most common domestic victims are cattle. In areas of endemic rabies, cattle commonly account for >20 per cent of diagnosed cases and they make up around 9 per cent of rabies cases in Europe.

13.9 Wild rodents as hosts of disease

Rodents play a large part as animal reservoirs for diseases of importance to man. Although many wild rodents have relatively little direct contact with man, they may nevertheless serve to maintain infectious agents of particular diseases in particular places for long periods. The classic example of this is the flea which causes plague or Black Death. The causative agent of the disease is the bacterium *Yersinia pestis*, which is transmitted from one rodent host to another by the flea. The most important reservoirs worldwide are the black and brown rats. However, other *Rattus* species may be important in some areas, and the multimammate rat is an important host in Africa. The number of cases of plague has fallen considerably in recent years due to increases in hygiene, earlier diagnosis of the disease, better treatments and improved rodenticides. However, even in 1990, human plague was reported from 12 countries with a total of 1250 cases and 137 deaths. A third of these cases were in Vietnam. More than 200 different species of rodent and lagomorph have been implicated in the epidemiology of plague in different locations, so it is another example of a multi-species system.

In the United States, cats are frequently found to suffer from/contract plague due to their contact with infected rodents and other small mammals while hunting. There were 117 reported cases of plague in cats in New Mexico between 1977 and 1988. Other diseases can be passed from rats and mice to cats as well, and in these circumstances, cats may form a link between the rodent and human populations. In addition, while there is a reservoir of infection in wild rodents, disease may also be transmitted to rodents such as house mice which live in close proximity to humans and domestic animals. This can often lead to outbreaks of disease in man or domestic animals.

In addition to plague, there is a wide range of other diseases associated with rodents. Of these, leptospirosis is probably now the most widespread. Leptospirosis is a bacterial disease, and the type that affects humans is *Leptospirosis interrogans serovar icterohaemorrhagiae*. Although the distribution of the disease is worldwide, it is especially common in the tropics in areas with heavy rainfall. The disease in man is often called Weil's disease or rice-field worker's fever.

Transmission from an animal host to man or from one animal to another is usually due to exposure to water containing urine of infected animals and infection is thus particularly frequent among people whose work brings them into regular contact with infected water, such as workers in rice fields, sugar cane plantations, sewage systems and abattoirs. Higher-than-average levels may also occur in people pursuing recreational water sports.

The most common rodent reservoir is the brown rat *Rattus norvegicus*, although other species, including the bank vole, may be quite frequently infected. Epidemics can occur in cattle, pigs, horses, sheep and goats, although these are normally caused by *Leptospirosis hardjo*. In cattle, there may be a sharp drop in milk production and milk may be yellowish, viscous and stained with blood. High proportions of brown rats may be positive for the infection in any one area. Prevalences recorded include 77 per cent in Detroit, 21 per cent in Israel and 39 per cent in Rio de Janeiro.

Commensal rodents (a form of symbiosis in which one organism derives a benefit while the other is unaffected), such as both brown rats and house mice, may also serve as hosts for other diseases of importance to livestock including salmonellosis (*Salmonella typhimurium*), cryptosporidiosis (*Cryptosporidium parvum*), a gastrointestinal infection, and toxoplasmosis (*Toxoplasma gondii*), a reproductive disease. Salmonellosis is currently on the increase in both cattle and pigs. Levels of rodent infestation have also shown general increases in Britain over recent years, especially in rural areas. However, the reasons for these increases in infestation levels are not understood. Also debatable is the extent to which rodents are in fact to blame or whether poor animal husbandry/hygiene is more important.

13.10 Brucellosis

Wildlife is also thought to be involved as a reservoir for brucellosis in many areas of the world. Brucellosis is a reproductive disease of ungulates including domestic cattle caused by the bacteria *Brucella abortus*. It results in offspring being aborted. Transmission can occur by horizontal (sexual), vertical (trans-placental) and pseudo-vertical (suckling, proximity) or diagonal routes (see Figure 13.9). The diagonal route is the only means by which the disease can be passed from bison to cattle, although abortion is quite rare in bison.

In Africa, buffalo are again believed to be the main host, although it is debatable whether the levels of infection in buffalo merely reflect a self-perpetuating problem in cattle rather than the buffalo actually continually reseeding infection into the cattle.

Wildlife hosts for brucellosis may be a major problem in some developed countries too. In the United States, brucellosis is mainly controlled through the detection and slaughter of infected cattle. However, in some areas, it continues to persist. One such example is around Yellowstone National Park in Wyoming. The disease has been recorded in the National Park since 1916, and represents a potential threat to the cattle industry in areas to the north of the park. The two principal wildlife hosts for the disease are the bison (*Bison bison*) and elk (*Cervus elaphus*). The prevalence in bison is probably around 10–20 per cent. There are also indications, as with most wildlife diseases, that a threshold population density for the continued maintenance of the pathogen exists. In this case, it appears to be a herd of about 200 animals.

There is currently much conflict between the cattle farmers who want these wild reservoirs controlled, and the conservationists who want to conserve them. Conservationists and farmers are also arguing over the extent to which bison and elk are acting as a reservoir for reseeding

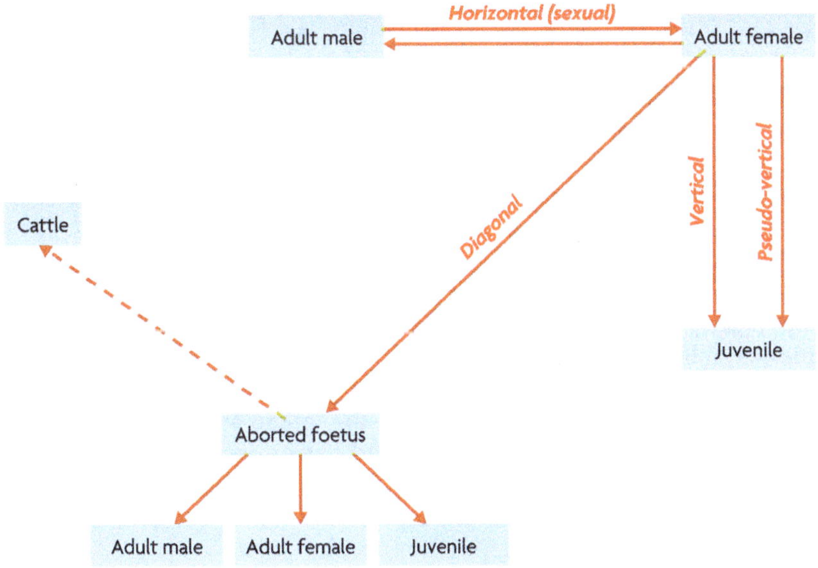

Figure 13.9 Transmission pathways for brucellosis in bison. Transmission to cattle can occur only via the investigation of aborted bison foetuses by cattle.

infection to the cattle, or simply as a spillover species, which acquire infections from cattle, but which cannot readily transmit the infection back to cattle. It is known that brucellosis only causes abortion relatively rarely in bison, which supports the conservationists' argument that bison are acting as a spillover rather than a true reservoir. However, the threat is considered particularly great because nationwide campaigns of testing and slaughter have been successful in eradicating it from most cattle herds, yet it persists here.

13.11 Bovine tuberculosis

In developed countries, many of the diseases of farm livestock that are a threat to man have been largely eliminated. This has mainly been due to advances in medical and veterinary health services. A good example of this is tuberculosis (TB). TB was a significant cause of death in the early part of the century and in the 1930s and about 30,000 people in England and Wales died from TB each year, 7 per cent of these being due to *Mycobacterium bovis*, the bacterium responsible for bovine tuberculosis. The introduction of skin testing to detect infected cattle, the subsequent slaughter of infected animals and the pasteurisation of milk all contributed to driving levels of TB in both humans and cattle to very low levels by the mid-1970s. However, the disease has never been completely eradicated from many parts of the developed world.

Bovine tuberculosis appears to be similar to brucellosis in that a wildlife reservoir may often maintain the disease and therefore constitute a continued risk to livestock. Badgers (*Meles meles*) act as a reservoir for bovine tuberculosis in Britain and Ireland, possums (*Vulpecula trichosurus*) act as a reservoir for the disease in New Zealand, and there are fears that bison (*Bison bison*) may be serving a similar purpose for cattle near Yellowstone National Park in the United States.

In Britain, cattle herds with TB are concentrated in the south-west region of the country. This region is only about 10 per cent of the total land area of Britain but it contains 25 per cent of the badger population (Figure 13.10). It was the discovery of a dead tuberculous-infected badger on a farm in Gloucestershire in 1971 that first suggested that badgers might be acting as a reservoir host. Subsequent investigations showed that in areas where TB in cattle occurs, up to 20 per cent of badgers may be infected with TB. A characteristic of the disease in cattle is that affected herds tend to be repeatedly found in the same limited areas. This patchiness may arise by chance alone but it is more likely to be due to the fact that some feature of the environment favours transmission of the disease to cattle in these areas. Patchiness in infection exists at both the macro-scale and also the micro-scale among badger social groups, and patchiness at the local scale is increasingly understood in terms of behavioural decisions made by individual animals when foraging (see Box 13.2).

Figure 13.10 Bovine tuberculosis in cattle in Britain is highest in areas where high densities of badgers are found, suggesting that badgers play a significant role in maintaining the infection in the ecosystem.
Source: White *et al.* (2008) *Trends in Microbiology*, 16(9).

BOX 13.2 Spatial heterogeneity of disease risk at the local scale

Just like distributions of other organisms, diseases are often found in spatial and temporal clusters, and may arise due to the aggregations in the distribution of the wildlife host, or their excretory products. As a result, targeting disease control at foci of infection within populations may combine the benefits of cost-effectiveness with a minimum impact on the wildlife host population.

Social organisation and habitat use of the wildlife host determines the distribution of disease risk to livestock populations. In a territorial system, for example as seen in the European badger, limited overlap between neighbouring territories may lead to localised disease hotspots at the territory level. Similarly, badger latrines and setts, which can harbour serious numbers of viable *Mycobacterium bovis* populations (the causative agent of bovine tuberculosis), were found to be clustered across space, causing distinct hotspots of high environmental disease risk. This suggests that reduction of disease risk to cattle could be enhanced by controlling access around these hotspots.

The level of disease risk posed to livestock by host excretions depends greatly on the grazing and avoidance behaviour of the livestock species in question. Most herbivores generally avoid grazing swards contaminated with faeces or urine. Cattle, however, do not avoid the faeces of the European rabbit *Oryctolagus cuniculus*.

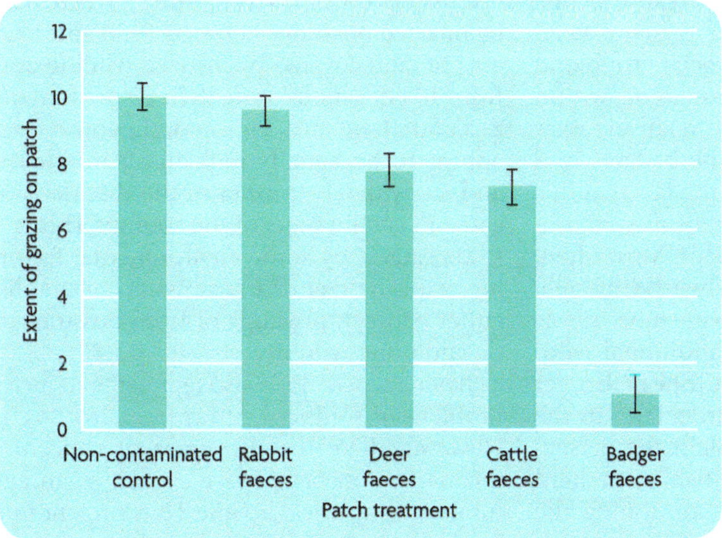

Grazing of contaminated sward patches by cattle. Swards contaminated by rabbit, deer, cattle and badger faeces. Cattle show no avoidance of swards contaminated by rabbit faeces relative to non-contaminated controls. Some avoidance of swards contaminated by deer and cattle faeces is evident, but avoidance of swards contaminated by badger faeces is greatest.

Source: Adapted from Smith *et al.* (2008) *Behavioral Ecology*, 20(2), 426–432. doi:10.1093/beheco/arn143.

This, combined with recent research findings that highly dispersed defecation patterns, such as those of rabbits, invite more contacts by grazing livestock than defecation patterns using a few well-defined latrine sites, may considerably heighten the risk of transmission of paratuberculosis from rabbits (the wildlife host for the disease in the UK) to cattle.

In wildlife–livestock disease systems, assessment of livestock disease risk across space is of major importance in order to focus disease control strategies at those areas where disease transmission risk is highest. Thereby, efficacy of control can be maximised and costs minimised.

It is likely that badgers obtained the infection early this century when bovine tuberculosis in cattle was widespread. The infection in badgers is very long-lasting. Some badgers may never pass to an infectious stage, while others may become infectious after a few months. The infection may cause lesions in the kidneys and lymph nodes. It is thought that kidney lesions are most important from the point of view of the transmission of the disease to cattle. Heavily infected badgers may excrete bacteria in their urine, faeces and sputum. Cattle can contact badger excretory products on pasture where badgers have been foraging for earthworms, which form the main component of the badger diet in Britain.

Transmission between badgers probably occurs through bite wounds, social grooming of cuts or grazes, or possibly through animals investigating scent marks at the edges or interior of their territories. Bovine tuberculosis is a disease which tends to persist in specific locations rather than spread rapidly through the population. Because of the social behaviour of badgers, with as many as 25 animals belonging to the same social group and sharing the same main set, transmission within the group should theoretically be relatively easy. Indeed, on some occasions all the badgers in a group may be infected. However, the fact that this is relatively rare implies that the transmission process itself is not particularly efficient. Most often, just one or two animals in a group may be infected.

In the UK, recent studies on bovine tuberculosis have tried to elucidate whether additional wildlife reservoirs other than the European badger exist for the disease in cattle, particularly in the deer population. This could explain the persistence of bTB infection in the British cattle herd despite past measures aimed at controlling the disease in both badgers and cattle. Current evidence suggests it's unlikely that additional wildlife reservoirs can explain the previous failure of bTB control operations in the UK. However, with deer populations increasing in number and expanding in range throughout the UK, deer may become a significant part of the host community in the future. The changing nature of the wildlife host community may also have implications for the distribution of TB across Britain in the future (see section 13.13).

In New Zealand, the persistence of a reservoir of TB in possums poses considerable problems. The persistence of TB in livestock threatens the NZ$3.6 billion beef, dairy and deer exports. The brushtail possum was first implicated in the disease in the early 1970s. More recently, bovine TB has also been found in feral pigs, cats, ferrets and deer. In deer, the prevalence may reach 25 per cent. Prevalence in possums only reaches levels of 1–5 per cent. However, there are now around 70 million possums in New Zealand, and their high density poses a real threat to cattle. Also, it is known that individual possums carry high loads of bacteria and readily excrete the organism. Possum-to-cattle transmission is thought to occur when cattle investigate disorientated, diseased possums wandering in pastures close to their brush habitat. There are also correlations between reactor rates (i.e. cattle reacting positively to the tuberculin test) and TB prevalence in possums.

The threat to cattle is currently on the increase due to an ever-increasing area of land covered by infected possums. Between 1980 and 1990 the area of land covered by infected wildlife doubled from 10 per cent to more than 20 per cent of New Zealand's land area. However, the threat does not appear to be equally high in all infected areas, and, as with TB in Britain, the disease is aggregated in certain parts of the country. This is likely to be caused by some environmental factor that increases the risk of transmission of the disease, either between possums or from possums to cattle and other wildlife.

Bovine tuberculosis also represents an increasing problem in southern Africa (see also Box 13.3). However, there the disease is of concern from a conservation perspective as well as an agricultural one. In Africa, bovine TB is present in cattle in the majority of countries, although there are strong regional differences in the number of outbreaks, cases and deaths. Only seven nations in Africa apply disease control measures and consider bovine TB as a notifiable disease. Although measures to control bovine TB in domestic stock are becoming established, the infection has relatively recently infected certain populations of native wild bovids, most notably the African buffalo. This species is considered the main reservoir throughout Africa and is thought to

BOX 13.3 Management of multi-host disease: bovine tuberculosis in Southern Africa

Bovine tuberculosis (TB) is found through southern Africa. The presence of multiple wildlife hosts for TB poses a particular problem with regard to management and eradication programmes of disease in cattle. Any control programmes solely focusing on the main buffalo host are doomed to failure due to the presence of infection in other African wildlife, such as kudu, lechwe and warthog.

Extending single-host disease models to account for disease in multi-host communities is a vital step to gain insights into how community structure affects the establishment and maintenance of an infection in general and what role different host species play in maintaining real infections in particular.

These models can then also elucidate the effects of management strategies, directed at specific hosts, on the persistence of these infections within the community. For example, TB infection is passed on to large carnivores via predation. Conceptual TB models including these inter-species pathways suggest that proactive control of herbivore host populations (such as buffalo and kudu) may help in reducing the overall level of infection in the host community and therefore benefit large carnivore populations. These in turn are a vital part in Southern African conservation programmes and are paramount to tourism revenue. Such proactive control has already been implemented in some parks focusing on TB eradication from buffalo herds by large-scale test and slaughter techniques which involve mass-corralling of buffalo.

While undoubtedly useful for assessing multi-host disease dynamics and resulting management strategies, parameterisation of multi-host disease models relies on large quantities of good-quality data on disease-induced mortality, infectious periods of different hosts and intra- and inter-specific contact and transmission rates. At present,

Occurrence of bovine TB in cattle and/or other wildlife in southern African countries between 1998–2004.

Country/Territory	1998	1999	2000	2001	2002	2003	2004
Angola	+	...	+	...	+	+	+
Botswana	–	–	–	–	–	–	–
Lesotho	–	...	...	–	–	–	...
Malawi	+	+	+	+	+	+	+
Mozambique	+	+	...	+	(2001)	+	+
Namibia	(1995)	(1995)	(1995)	(1995)	(1995)	(1995)	(1995)
South Africa	+	+	+	+	(+)	+	+
Swaziland	+	...	+	+	+	(2002)	+
Zambia	+	...	+	...	+	+	...
Zimbabwe	(1990)	(1990)	(1996)	(1996)	(1996)	(1996)	(1996)

Key: +, infection reported or known to be present; –, infection not reported and date of last outbreak unknown; . . . , no information available; (year), year of the last reported occurrence of disease; (+), disease limited to specific zones.
Source: Renwick *et al.* (2006), *Epidemiology and Infection* 135, 529–540.

➤

such detailed data are either based on small sample sizes, or in most cases non-existent, particularly with respect to inter-specific contact rates. Future research should therefore focus on obtaining the necessary data for reliable disease models.

In the long run, vaccination of humans, domestic animals and wildlife appears to be the only viable option for TB control over much of Southern Africa. Finding an effective vaccine and delivery strategy to suit a large range of wildlife host species is the main stumbling block of any wildlife vaccination strategy against multi-host disease. For example, oral baits can only be used for non-living vaccines which are often not as effective as live vaccines, while the process of oral baiting may be an ineffective approach for species that are not used to artificial feeding, such as the African buffalo. Alternative delivery strategies have been suggested, such as aerosol vaccines distributed by helicopter, or self-replicating recombinant vaccines containing important mycobacterial antigens.

be responsible for infection of other sympatric wildlife and the possible re-infection of cattle. However, the more recent detection of other potential maintenance hosts indicates that bovine TB in Africa exists as a multi-host pathogen within a multi-species system. In addition, because the number of interacting large mammal species in certain savannah regions of Africa is perhaps higher than that in any other geographic area of similar size, bovine TB has spread rapidly through these ecosystems. In the Kruger National Park (KNP) in South Africa, a survey in 1998 revealed an average bovine TB prevalence of 38.2 per cent in the southern region of the Park, 16.0 per cent in the central region and 1.5 per cent in the northern region, reflecting the spread of the disease since its introduction in the south. The high prevalence seen in some buffalo herds is due almost entirely to intra-specific transmission (buffalo in the immediate social group of those infected receive high and possibly multiple exposures to the disease, although this may vary depending on the severity of lesions in the individual). Other species in the KNP commonly infected are the greater kudu and various carnivores. TB was first diagnosed in kudu in the Park in 1996. The common 'buffalo strain' of *M. bovis* has been isolated from some kudu. However, a different genotype has also been found in a group of kudu in the KNP, which indicates that, as well as being susceptible to the dominant strain of infection, kudu may also be able to maintain a separate infection cycle, which may have implications for their potential role as maintenance hosts.

The first reported carnivore infections in the KNP occurred in 1995 when lions, and then cheetahs, were diagnosed with bovine TB. Following this, leopards were diagnosed in 1998 and a further 50 cases of lion infection have since been confirmed. Most of the confirmed cases have occurred in the southern and central region of the KNP, corresponding positively with the region of high prevalence of bovine TB in buffalo. The KNP has a lion population of approximately 1700 of which around a third live in areas where there is high tuberculosis prevalence in buffalo. Most carnivores become infected with bovine TB from eating infected prey animals, in which the most infectious material is present in infected organs such as lungs and lymph nodes. The risk of lions becoming infected with bovine TB may be increased because of simultaneous infection with feline immuno-deficiency virus (FIV), potentially making them more susceptible to bovine TB. Lions are social cats, and when compromised by infection, they retain the support of other members of the pride; therefore infected individuals have a better opportunity of surviving for longer than solitary species without this support system. The time from infection to death has been estimated to be between two to five years. Leopards are solitary, and do not have a social support system like lions. Therefore, once infected, the disease is likely to be fatal to leopards within a shorter period of time.

The existence of bovine TB in free-ranging mammals in southern Africa poses significant threats to conservation and tourism. Bovine TB is virulent in lions and other top carnivores, and these species exist at relatively low densities in most of the infected areas. If these were the only potential host species, the infection probably would not be able to persist. However, gregarious large herbivores such as the buffalo and the kudu occur at higher densities, and the pathogen may persist in these host species, and be maintained in the absence of any additional disease source. Consequently, if inter-specific transmission occurs between buffalo or kudu and large carnivores, or if the infection passes from these species to carnivores indirectly via another prey species, the impacts of the infection in reducing populations of large carnivores will be exacerbated, with consequent adverse impacts on tourism revenues.

13.12 Controlling wildlife disease

13.12.1 Culling

The traditional approach to controlling wildlife disease is to reduce populations of the host species by culling. According to the theoretical models, if the host population or host community can be reduced below the threshold required for disease persistence (N_T), then the disease will die out. This is effectively what happened with rinderpest following the first epidemic in southern Africa. Thus, strategies for the management of disease have been based around the control of the principal hosts. Such population control is an obvious solution, and may, at least in the short term, prove effective.

However, there is increasing evidence that culling of wildlife populations can cause disruption to the established patterns of social behaviour and movement, such that the remaining animals will range more widely and may therefore make more contacts with other animals in the population. The increased frequency of these contacts can then lead to increases in disease transmission. In undisturbed populations, especially of group-living carnivores such as foxes, badgers and lions, contacts are frequent within a social group, but can be quite rare between different social groups. For these animals, contacts can be risky and may result in injury or even death. Therefore, contact behaviour with neighbouring groups is minimised and behaviour between familiar neighbours is quite ritualised. However, when populations are reduced by culling, these established social structures break down, and as well as increased movements and contacts, the nature of the contacts changes too. Animals will be more aggressive towards strangers than familiar neighbours, and this will increase opportunities for disease to spread further.

The best evidence for the impact of perturbation on disease dynamics comes from the Randomised Badger Culling Trial (RBCT), which was set up in Britain in 1999 to determine the extent of badger involvement in TB in cattle and the effectiveness of culling badgers in reducing TB infection in cattle. The field trial showed that badger culling over 100-km^2 areas could result in a decrease in cattle TB inside these areas, with effectiveness increasing with distance from the boundary, but that TB in cattle increased in areas immediately surrounding the culled area. However, culling does not always result in perturbation. For example, there is no evidence for any such perturbation effects in Ireland following intensive badger control. Levels of TB prevalence in badger populations are fairly consistent between Britain and Ireland, but there are substantial differences in the areas over which badgers have been controlled in the two countries and the efficiency with which this has been done. In Ireland, badger culling is currently being carried out proactively over areas ranging in size between 188 and 305 km^2, with the overall policy aim of maintaining badger populations at <20 per cent of their original densities over 30 per cent of the area of agricultural land. In Britain during the RBCT, badgers were culled using cage traps in ten separate 100-km^2 areas, and trapping efficiency varied between 35 and 85 per cent in three areas. The use of restraints (snares) in Ireland allows for a high efficiency of badger culling, which is likely to minimise problems due to perturbation, but animal welfare-based legislation prevents the use of restraints to catch badgers in Britain. Models of TB in badgers have shown that the reduction of TB in badger populations requires a badger culling efficiency of at least 70–80 per cent. The use of restraints by experienced field staff in Ireland can approach these efficiencies, whereas to achieve such high levels of badger removal over large areas based on cage trapping alone would require a huge investment in time and resources and, even if it were achievable, it would be extremely unlikely to yield any net economic benefits.

Figure 13.11 Feral goats on the Great Orme in North Wales. This feral goat population has been the subject of research to investigate the effectiveness of fertility control using Gonacon™. Originally introduced to the Orme peninsular in around 1900, the herd had reached about 200 individuals by 2002, at which point they were starting to have a negative impact on the flora of the peninsular. The goats are a local tourist attraction, and culling is not considered as an acceptable form of population control.
Source: Photograph by Piran White.

13.12.2 Fertility control

There is an increasing realisation that the most effective form of population control is likely to consist of several different techniques used together in an integrated pest management strategy. This realisation has also renewed interest in methods such as fertility control.

In small populations of wild mares and white-tailed deer in the United States, pregnancies have been prevented by injection with preparations of the glycoprotein membranes that surround the eggs (*zonae pellucidae*) from porcine (pig) ovaries (PZP). These sterilants are useful for limited populations: they are strictly reversible because the effects wear off, but they may remain effective for several years. However, because of the fact that animals have to be individually vaccinated, PZP injections are impractical over large areas or for large populations. For example, the feral horse population of Australia exceeds 300,000. Treating these animals in this way would be impractically expensive.

Similar methods, but using a different compound, have been developed for fertility control in kangaroos. In this instance, pregnancies do occur, but lactation ceases when the young in the pouch are at an early stage of development and they die. Animal welfare groups support this because they consider it is preferable to shooting adult kangaroos. However, it suffers from the same problem as that for controlling feral horses in that over a large area the expense makes it impractical. This sort of temporary sterilisation results in temporary population reduction rather than permanent solutions, although the effects may remain for some time. For some wildlife populations, permanent reductions in population density or even complete eradication may be required.

A more permanent solution to population reduction via fertility control may lie with immunocontraceptives. Recent advances in molecular biology and reproductive immunology have led to the ability to develop immunogens that induce a specific immune response in the target animal against its own reproductive proteins, thereby preventing successful fertilisation (Figure 13.11). This immune response is the same as that expressed towards infectious agents such as viruses. Individuals that have these immunogens will be incapable of reproducing successfully since fertilisation will be prevented from taking place.

Immunogen delivery must be humane, species-specific and safe. To date, most success has been achieved using injectable immunocontraceptives. One example is Gonacon™, a compound that has been developed by the United States Department of Agriculture. Gonacon™ works by inducing an animal's body to make antibodies against its own gonadotropin-releasing hormone, one of the key hormones associated with the reproductive cycle.

Gonacon™ is not yet available commercially, but in experimental trials, it has been effective in reducing fertility in a number of species, including white-tailed deer, California ground squirrels, feral pigs and wild horses. In these species, success rates are around 80 per cent in the first year, declining to around 50 per cent in the second year. Fertility can be reduced in individual white-tailed year for up to five years. In Britain, fertility control using Gonacon™ is being tested in populations of wild boar and feral goats.

Even if a fertility control agent can be shown to work, it still requires a high time investment to actually deliver it to a population, since all individuals to be treated must first be caught. This is clearly impractical for very large or dispersed populations. As a consequence, researchers in Australia have been trying to develop ways in which the immunogen could spread itself instead using viral vectors. The benefit is that the viruses themselves are infectious diseases that can be transferred between host organisms. When the viral vectors that transmit these immunogens can be introduced into a wildlife population, they will be transmitted between individuals, and the immunogens will be transmitted with them. By the use of species-specific contagious viruses, the immunogens can be spread throughout the target population only.

In the 1990s, the Australians set up a Co-operative Research Centre to tackle this problem and investigate the potential role of immunocontraceptives in the control of rabbits, foxes and mice. They identified potential candidate viruses for rabbits and mice as myxoma virus and cytomegalovirus respectively. However, trials were disappointing and they have now stopped the research programme. In addition to the technological problems of immunocontraception, it is also considered by some people as interfering with organisms, and the current backlash against processes such as genetic modification means that it would be extremely unlikely to get the go-ahead in Australia or New Zealand at the present time. Even if made to work, the expense would mean that it would only be feasible for a few high profile systems. If it ever does come into practical use, it would probably be used as a part of integrated control, rather than as a control method on its own.

13.12.3 Vaccination

In terms of the control of wildlife diseases themselves, the biggest advances have come with vaccines. For some diseases which are transmitted from wildlife to domestic animals, vaccination of domestic stock has proved successful in reducing or eradicating the disease in the wildlife reservoir as well as in domestic stock. An example of this is the control of rinderpest in Africa. Although rinderpest is still not completely eradicated it has been reduced to very low levels through an intensive cattle vaccination campaign.

Besides vaccinating the domestic stock against a disease, wildlife managers can alternatively try to vaccinate the wildlife hosts themselves. An example of a disease for which this approach has been most fruitful is rabies. Rabies has been relatively effectively controlled in developed countries using either vaccination of wildlife alone, or a combination of depopulation by culling and vaccination.

Oral vaccination campaigns across western Europe have had a great impact in reducing the numbers of wildlife rabies cases in recent years. It has been estimated that in Europe around 60 per cent of the fox population needs to be vaccinated to ensure successful eradication of the disease. Overall in France and Germany, rabies cases among both domesticated and wild animals dropped by 90 per cent between 1989 and 1993. However, although the overall trends are favourable, cases have been reported from previously unaffected areas in Italy. Also, the number of cases recorded in the first part of 1994 in Germany actually exceeded those recorded in the same period in 1993. Currently in Germany there is political pressure to stop the vaccination programme, primarily to save money, but scientists are urging against this, as fox populations are now recovering.

In North America, both depopulation and vaccination strategies have been employed to control rabies. Mathematical models can also be used to predict rabies spread and the requirements for control (see Box 13.4). Oral vaccines for red foxes have been found to be effective in reducing the incidence of the disease in areas where it is enzootic, such as in Ontario in Canada. In these situations, depopulation campaigns alone have

BOX 13.4 Spatial models of rabies spread and control

A large number of mathematical models have aimed at describing and predicting the spread of the rabies in different parts of the world. Early rabies models used a standard *SI* (susceptible and infected individuals only; no recovery) or *SIR* model with an added spatial diffusion coefficient *D* in order to account for the dispersal of infected wildlife hosts across the landscape. These models allowed the estimation of the wave front velocity at which the infection was advancing across a landscape. Landscape heterogeneity may greatly affect rabies spread by inhibiting or facilitating the wave's progression; in order to predict not just the velocity of rabies spread but its most likely direction and the areas which may be at an increased risk, landscape heterogeneity has been incorporated into some more recent models. Spatial stochastic simulation models have been used to predict the likely spread of the disease and the efficacy of different control regimes in wildlife in various countries.

In the United States, various wildlife species including foxes, racoons and coyotes can act as wildlife hosts for rabies. In the south-eastern seaboard of the USA, racoon rabies expanded rapidly during the 1980s and 1990s. An intensive programme of oral vaccination appeared to be successful at controlling the disease, but then a new outbreak was discovered further west in 2004. Spatial stochastic simulation models are being used to predict the potential rate of rabies spread and the efficacy of different control strategies. The results from such modelling exercises can be extremely valuable for helping to develop the most effective and efficient intervention strategies to contain and eradicate disease.

Vaccination is also being looked at increasingly to solve the TB problem in Britain and Ireland. Vaccines for badgers have been developed and are currently being tested in Ireland. Vaccines pose considerable technological challenges, not only in terms of their manufacture, but also because they need to be disseminated efficiently to the host population. It is too early to tell whether this approach will be any more successful than culling. However, one significant advantage of vaccination is that it should not disrupt the social structure of the population and therefore should not entail a perturbation effect.

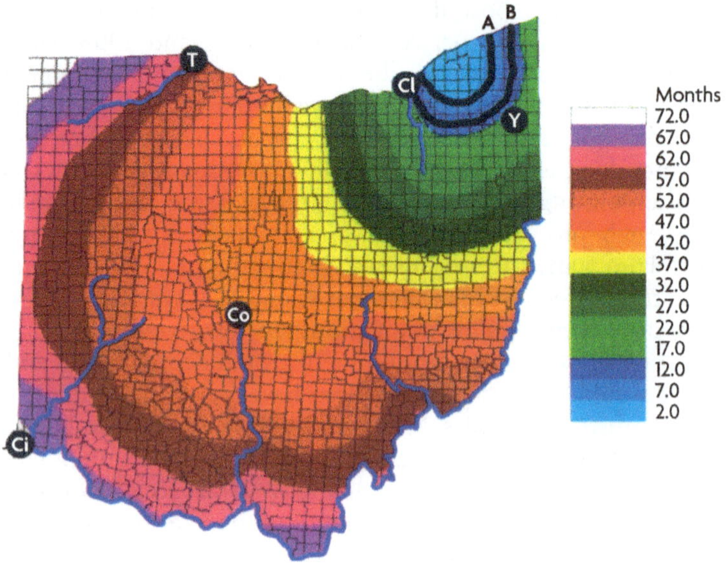

Prediction of racoon rabies spread across Ohio from Chardon Township in the north-east in the absence of any oral vaccination campaign. Each colour band indicates a given time interval to arrival at a township. The width of the bands corresponds to velocity of spread, with wider bands associated with more rapid spread. Major cities are Cleveland (Cl), Youngstown (Y), Toledo (T), Columbus (Co), and Cincinnati (Ci). The two black lines labelled A and B correspond to the area where cases have been detected (A) and the expected position of the wave front given a long-tailed distribution of incubation periods.

Source: From: Russell *et al.* (2005) *PLoS Biology* 3(3): e88. doi:10.1371/journal.pbio.0030088.

been unsuccessful. One of the main reasons for the lack of success of depopulation strategies is thought to be the high reproductive capacity of the main vectors, red foxes, combined with their high dispersal ability. Thus, in areas that are depopulated, foxes will rapidly move in from adjoining areas and breed successfully, quickly bringing the population up to above the threshold density once more. Another contributory factor is the presence of a substantial secondary reservoir of infection in striped skunks.

13.13 Systems-based approaches

Although we have been successful in eradicating many diseases in different parts of the world, some diseases persist as intractable problems. Past attempts to manage these diseases which have focused on one particular method have not been successful, and it is recognised increasingly that we need a multi-faceted approach to these problems. In other words, we need to consider questions of disease control at a broader scale, and recognise the disease as just one component of a much more complex system. Using this holistic, systems-based approach, we can start to see many and more varied, potential solutions to disease control. For example, surveillance work has shown that one of the ways in which cattle and badgers make contact with one another is when badgers enter farm buildings to feed on cattle cake and other food left out for the cattle. Thus, simple measures such as preventing badger access to cattle sheds could have dramatic impacts on the transmission and maintenance of TB.

Identifying risk factors at the farm level can therefore help in reducing disease risks at the local scale. Similarly, consideration of the problem at broader landscape scales will help us to understand the overall risks better and use the resulting information to adjust disease management accordingly. One technique that can be very useful in this respect is risk analysis. By considering the different risks posed by different situations in different localities, we can identify areas of significantly higher risk, sometimes known as 'hotspot' areas. Plans can then be drawn up in order to avoid, or significantly reduce, the identified risks.

At the systems level, one of the things we can do to analyse risk is to quantify the potential host community for a disease. Risk maps based on the distribution of known or suspected hosts and vectors can be a valuable tool for identifying potential high-risk areas where introduced disease may become established in the host community. Although TB in British wildlife is concentrated in badgers, other animals such as deer can also act as hosts for the disease. Deer are known to act as an important host for TB in parts of New Zealand, so they have the potential to assume a greater importance in TB host community in the UK as well. Since the British deer population is currently expanding both in distribution and numbers, this means that the size and diversity of the potential TB host community is also expanding in certain areas.

In standard approaches to modelling, as we have seen above, the spread of contact-transmitted infections is based on the principle of mass action or frequency-dependent transmission, whereby the number of new cases arising per unit time is proportional to the density of susceptible and infectious hosts. Although the assumptions of mass-action are far removed from reality for many wildlife disease hosts, the general overviews produced by these types of models are still useful in a strategic sense, for example when making predictions of disease spread over large spatial areas. For the bTB disease system at the landscape level, we can assume that single or multiple species guilds of potential hosts at higher densities are more likely to support bTB infection. Based on this assumption, we can therefore use a simple multiplicative relationship between the relative abundance scores of each of our potential bTB host groups (cattle, badgers and deer) to indicate the relative capacity of the host community in different areas to maintain TB and hence constitute a persistent risk to cattle (Figure 13.12).

This synthesis of existing datasets on host distributions highlights the existence of substantial risk of TB persistence in livestock–wildlife host communities in existing TB hotspots, but it also highlights other areas, hitherto unaffected by TB, where the infection has a relatively higher chance of gaining a persistent foothold in the host community if introduced. Hypothetical disease risk posed by cattle and badgers (Figure 13.12a) is

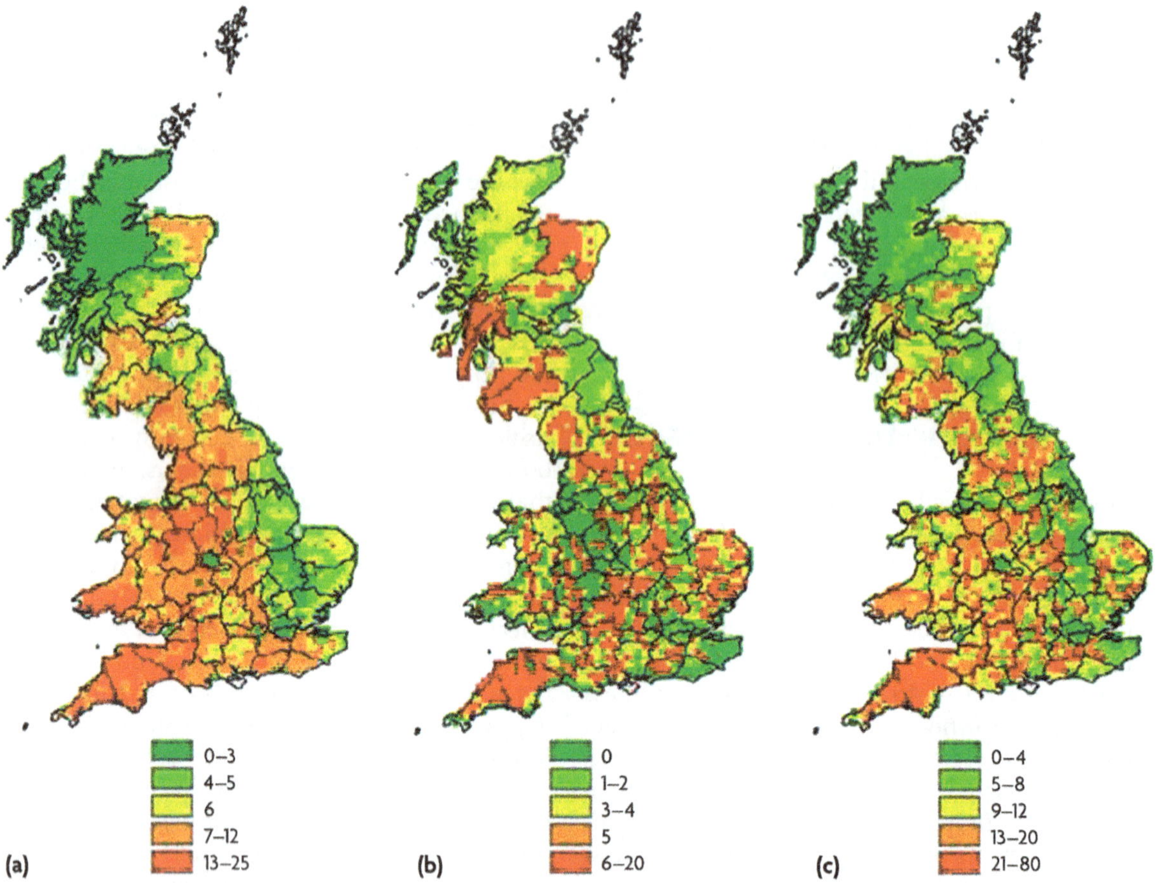

Figure 13.12 Maps of Britain showing hypothetical risk scores for TB persistence in the host community per 10 km square based on host communities consisting of (a) cattle and badgers, (b) cattle and deer, and (c) cattle, badgers and deer combined. Lower risk areas are identified by green shading and higher risk areas are identified by red shading. County boundaries are also shown.

Source: White, P.C.L. *et al.* (2008) *Trends in Microbiology*, 16(9).

highest in the traditional TB 'hotspot' areas such as the south-west of Britain, south-west Wales and Cheshire, with high risk values also obtained in Cumbria, south-west and north-east Scotland and central North Yorkshire. The disease risk posed by cattle and deer combined (Figure 13.12b) is highest in south-west Britain, the southern Midlands (including the Cotswolds), central North Yorkshire, Cumbria and south-west and north-east Scotland. Some of these areas coincide with the badger hotspots, although there are additional high-risk areas identified in parts of Scotland. The disease risk posed by the full potential host community (cattle, badgers and deer) reflects both these distributions, with the highest overall risk areas in south-west Britain, the southern Midlands, Cumbria, North Yorkshire, south-west and north-east Scotland (Figure 13.12c). For a disease such as bTB, which has proven so difficult to eradicate once it becomes established in the wider ecosystem, it is imperative that there is greater surveillance in areas where the host community presents a higher risk of disease persistence, so any infection can be tackled before it becomes widespread in the ecosystem. Risk maps such as those in Figure 13.12 can be used to direct surveillance effort, which should be a priority for disease control and are essential for developing a strategy with a more proactive, rather than reactive, approach to disease control. An increased appreciation of the importance of understanding the whole disease system has also led to more research into the ecology of disease transmission, and this has undergone significant advances in recent years with the advent of new technology.

Until the late 1990s, the only means of monitoring free-living wild vertebrates was by means of radio-tracking. This technique is still used widely, and can provide quite detailed information on the movements of animals, as well as their likely patterns of interaction. For example, in a badger population in North Yorkshire, it has been shown that interactions between groups are normally rare, but that individuals within the same group interact more often than would be expected by chance. However, there are also differences in social behaviour between individuals of the same social group, and these differences will affect the way in which diseases are transmitted through a social network.

In the late 1990s, the first proximity data loggers were developed, which could be fitted to animals and monitor their interactions with other animals fitted with the same type of device. These devices mean that it is now possible to monitor all the close interactions of wild animals for periods of up to a year or more, providing an unrivalled record of social behaviour and potential disease transmission events. This provides us with the potential ability to map the details of both intra-specific and inter-specific networks of potential disease spread. The resulting data could be used potentially to identify high-risk individuals within wildlife and livestock populations (see Box 13.5). This type of research therefore provides a means for developing effective, targeted control management based around certain high risk subgroups within the population, which is likely to represent a more effective option for reducing disease than untargeted mass culling, which has dominated disease control strategies in the past.

BOX 13.5 Use of proximity loggers to quantify the potential for disease transmission

In epidemiological models of disease transmission, the basic reproductive rate of the disease, R_0, is a function of the contact rate between infected and susceptible individuals, and model fit relies greatly on the accurate estimation of these underlying contact rates. Contacts, however, are difficult to study under field conditions, so that model parameters often rely on estimated or derived values.

Recent advances in biotelemetry have brought about the development of proximity data-logging devices, which have opened up new opportunities for the study of animal interactions in the field. The data loggers transmit unique identification codes via a continuously pulsing UHF signal, while simultaneously receiving these signals from other loggers within a specified detection range. The resulting information in terms of the ID of the data logger contacted, the date, start time and length of the contact are then stored in memory.

These devices have helped to reveal highly heterogeneous contact networks within populations of wildlife and domestic animals, including possums, racoons and cattle. For example, data logger research on possums in New Zealand showed that relationships between contact rates and population density can be asymptotic rather than linear, which is likely to have contributed to the failure of one-off possum culling operations to control bTB in New Zealand. Since many disease models and resulting management considerations have previously been based on the assumption of homogeneously mixing populations, these newest findings highlight the need to view contact networks within populations as complex heterogeneous structures with equally complex management implications.

➤

The data loggers were recently also used to monitor interactions between badgers and cattle in a first attempt to measure inter-specific interaction rates in the pasture environment.

The study highlighted differences in the connectedness of individuals within the network: inter-specific contacts involved those individual cows, which were highly connected within the cattle herd – and which may be of dominant status. This helped to identify 'high-risk' individuals for both intra- and inter-specific. Findings such as these emphasise the need for a more targeted approach to disease management strategies, by reducing disease transmission from these high-risk individuals.

A badger fitted with a proximity logger mounted on a leather collar.

Source: Photograph by Piran White.

(a)

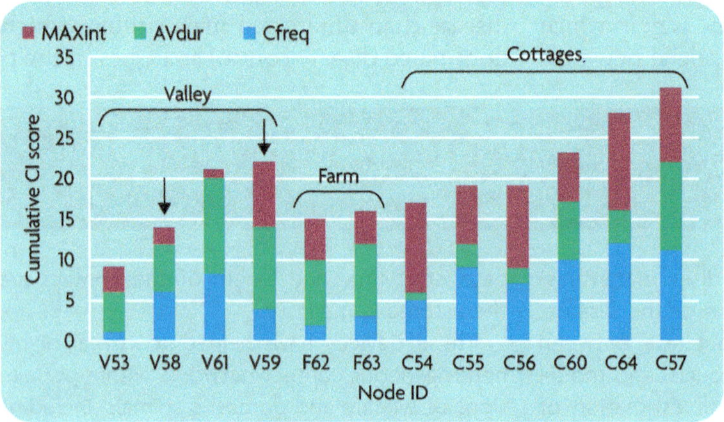

(b)

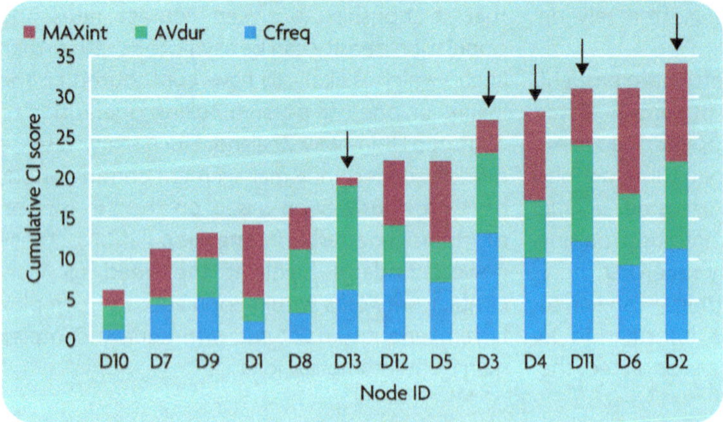

Level of interactions (Connectedness Index scores) detected by proximity data loggers. Node ID represents different individuals. Levels of connectedness are shown for (a) badgers only (letters signify different social groups) and (b) cattle from a study in North Yorkshire. Arrows mark individuals implicated in inter-species interactions. Contact patterns in both badger and cattle populations vary widely, both between individuals and over time. Interactions between badger social groups were infrequent, although all badgers fitted with data loggers were involved in these inter-group contacts. Contacts between badgers and cattle occurred more frequently than contacts between different badger groups. Moreover, these inter-specific contacts involved those individual cows, which were highly connected within the cattle herd.

Source: Böhm *et al.* (2009) *PLoS ONE* 4(4): e5016. doi:10.1371/journal.pone.0005016.

13.14 Disease and climate change

Long-term environmental changes, such as those invoked by climate change, may be particularly significant in influencing the range and spread of vector-transmitted diseases that require an intermediate invertebrate host, since climate strongly influences the distribution of key vectors or intermediate hosts by affecting host susceptibility and survival. Furthermore, warmer winters are expected to aid over-winter survival of pathogens through increases in precipitation and humidity in some areas.

Tick-borne diseases, such as Lyme disease and tick-borne encephalitis (TBE), are predicted to expand their range with climatic warming. Tick-borne encephalitis is a flavivirus infection of the central nervous system, which can cause serious illness and fatalities in humans. The virus now occurs at higher altitudes than previously recorded and is expected to shift northward in distribution under current climatic forecasts. Incidence of the disease in Sweden has already increased substantially since the 1980s, and has been attributed to a change toward milder and shorter winters.

Some helminth parasites are also predicted to increase in abundance in northern temperate latitudes since the development of the free-living stage of their lifecycle is closely linked to environmental conditions, in particular temperature and rainfall. In the UK, climate change is expected to result in an overall warmer and wetter climate, which will extend the duration of suitable environmental conditions for development of free-living stages. Examples of helminth parasites which are already showing changes in distribution as a result of climate change include: *Nematodirus battus*, which has increased in Scotland, Wales and northern England; *Haemonchus contortus*, previously confined to southern regions of UK, now increasing in Scotland; and *Fasciola hepatica*, which has shown 12-fold increases in some EU countries.

Some other types of disease may also pose an increased risk as a result of climate change. One example that is posing an increasing threat to British livestock is bluetongue. Bluetongue disease is a viral disease of ruminants (domestic and wild), which is transmitted by midges of the genus *Culicoides*. There are differences in the virus serotypes and/or *Culicoides* species that act as principal hosts in different parts of the world, but both vector insects and ruminant animals are essential for maintaining the infection. In Europe, *C. imicola* is the principal insect vector. Changing climatic conditions are likely to lead to increased abundance of *C. imicola* and other Culicoides species, and hence a greater potential capacity of the host community to maintain and transmit bluetongue virus (see Box 13.6).

BOX 13.6 Climate change and bluetongue virus in Europe

Bluetongue disease has expanded its European range northward over recent years, due to increased over-winter survival of the virus and a northward expansion of *C. imicola* linked to milder climate. Six serotypes (1, 2, 4, 8, 9 and 16) have entered Europe since 1998 via four different routes: (i) from the east via Turkey/Cyprus; (ii) from the eastern part of north Africa (Algeria, Tunisia) into Italy and the western Mediterranean Islands; (iii) from Morocco into southern Spain and Portugal and (iv) via an unknown route into North Western Europe. Serotype 8 was first detected in northern Europe in 2006, when an epidemic affected Belgium, the Netherlands, Germany, France and Luxembourg. The first case of bluetongue in the UK occurred in 2007. Although no positive cases have been found since 2009, all of England, Wales and Scotland is now designated as a Lower Risk Zone. With future predictions of climatic warming, risk zones

➤

for the disease are expected to expand in size, reaching into higher latitudes. Climate-aided northward expansion of *C. imicola* may bring the virus into contact with northern *Culicoides* spp., providing a pool of alternative disease vectors and therefore increasing the disease risk and accelerating its northward spread.

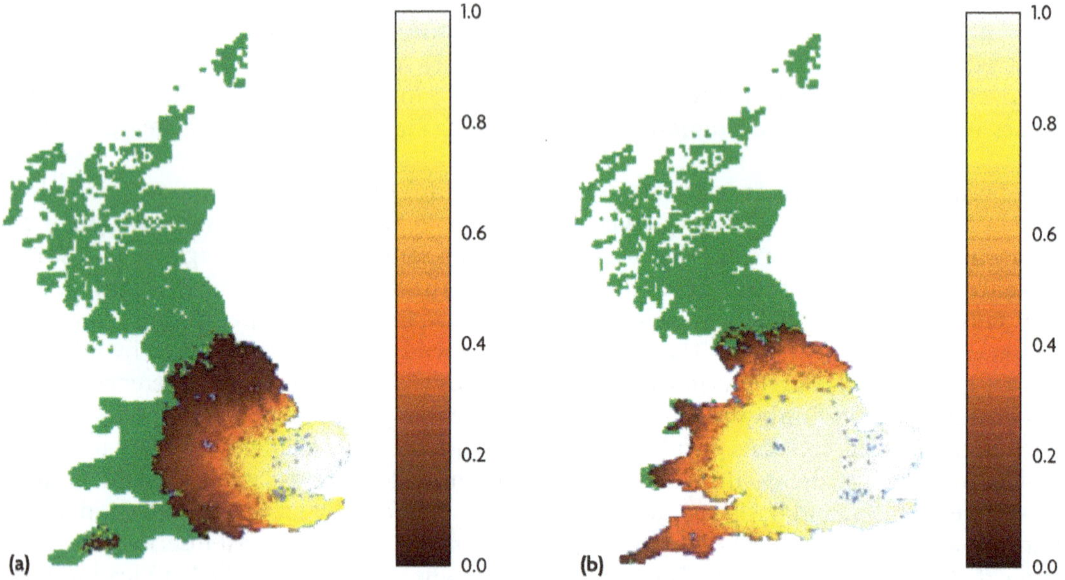

Potential spread of bluetongue virus serotype 8 in Britain in 2008 based on (a) 2007 and (b) 2006 temperatures. The maps show the cumulative risk (see colour bars) expressed as the proportion of simulated outbreaks (out of 30 which took off) for which at least one farm was affected by BTV within each 5-km grid square.

Source: Taken from: Szmaragd *et al.* (2010) *PLoS ONE* 5(2): e9353. doi:10.1371/journal.pone.0009353.

13.15 Conclusions

Diseases occur naturally in wildlife populations, as in populations of humans. Disease can cause ill-health in an infected individual, and infected individuals can pass disease on to other individuals in a population. Disease that affects large numbers of individuals in a population may cause the population to decline, which can be a problem for species of conservation concern.

Disease in wildlife populations may also be a problem where it is transmissible to humans or their livestock. For these diseases, wildlife species can act as 'host' populations, and sometimes, more than one species of wildlife may be involved in a 'host community'. Diseases with multiple hosts are sometimes very difficult to control, and some of these diseases have significant economic and welfare costs for humans and their livestock.

Infectious agents which give rise to disease can be categorised into microparasites and macroparasites. Microparasites include bacteria, viruses and some protozoa. Macroparasites include helminth worms, nematodes, fleas and lice. Newly emerging diseases, in particular those caused by prions and infectious cancers, are causing increasing problems in wildlife conservation and management.

The transmission of infectious agents between individual hosts sometimes occurs directly through close contact, but may also occur indirectly via contact of susceptible individuals with free-living infectious agents, infected excretory products or via other species, such as some biting flies, which can act as vectors for an infection.

Diseases can show epidemic or endemic patterns in populations, and frequently exhibit cyclicity that may be related to behavioural patterns of hosts or to conditions in the environment. The same disease can sometimes show an endemic or epidemic pattern according to local conditions.

Efforts to control disease in wildlife populations have been focused traditionally on reducing the population by culling, with the aim of reducing the host population below a threshold level which the disease requires to persist. However, culling can cause social perturbation in the host population and potentially increase the risk of transmission. For some species, culling may not be socially acceptable and in situations where disease occurs in multiple hosts, it may not be very effective. Fertility control offers a potential alternative means of control for the future, but the methods currently being tested require individual animals to be captured and injected, which is time-consuming and expensive. Vaccination of hosts against disease is being increasingly used to complement culling-based approaches, and for some diseases, it may be more effective than culling.

Disease is being considered increasingly as part of the broader ecological or livestock system. The emphasis of control is shifting from culling towards strategies which incorporate an understanding of risk, whether in terms of overlapping distributions of wildlife and livestock on a large scale or behavioural interactions on a more local scale. These approaches can help to support the development of more preventive, efficient and sensitive strategies to combat the risk of disease.

The transmission of some diseases is linked to environmental conditions, whether directly or via the changing distribution of vectors. Climate change is already having an impact on the geographical range of these diseases, and warmer and wetter conditions are likely to increase the risk of disease in northern temperate latitudes in the future. The establishment of effective mechanisms for disease surveillance will play a critical role in combating the growing threat of existing and emerging diseases to wildlife, humans and livestock.

POLICY IMPLICATIONS

Wildlife disease

- Policy makers need to understand the biology of infectious agents and how they affect animals and humans in order to develop an appropriate strategy to combat any particular disease.
- Policy makers should understand the value of risk maps to focus surveillance effort on areas that should be given priority for disease control.
- Policy makers should realise the importance of risk maps for developing a strategy with a more proactive, rather than reactive, approach to disease control.
- Policy makers need to understand the potential of predictive models of transmission in formulation of policy for effective disease control.
- Policy makers should realise that wildlife disease is also of particular concern where species of conservation importance are implicated, as disease epidemics may drive already small or fragmented populations to extinction.
- Policy makers need to realise that climate change is already having an impact on the geographical range of some diseases, and warmer and wetter conditions are likely to increase the risk of disease in northern temperate latitudes.

CHAPTER REVIEW EXERCISES

Exercise 13.1

Why is it important to control disease in wildlife?

Exercise 13.2

What are the main processes by which disease can be passed from wildlife to humans?

Exercise 13.3

What are risk maps and why are they particularly valuable when designing disease management strategies at a national scale?

Exercise 13.4

Explain why the social behaviour of animals is important in disease transmission, including specific examples in your answer.

Exercise 13.5

How do insects contribute to the transmission of disease between animals and humans?

REFERENCES

Anderson, R.M. and May, R.M. (1979) Population biology of infectious diseases: Part I. *Nature*, **280**, 361–367, doi:10.1038/280361a0.

Böhm, M., Hutchings, M.R. and White, P.C.L. (2009) Contact networks in a wildlife-livestock host community: Identifying high-risk individuals in the transmission of bovine TB among badgers and cattle. PLoS ONE 4(4): e5016. doi:10.1371/journal.pone.0005016.

Holdo, R.M., Sinclair, A.R.E., Dobson, A.P., Metzger, K.L., Bolker, B.M., *et al.* (2009) A disease-mediated trophic cascade in the Serengeti and its implications for cosystem C. PLoS Biol 7(9): e1000210. doi:10.1371/journal.pbio.1000210.

Nokes, D.J. and Anderson, R.M. (1988) The use of mathematical models in the epidemiological study of infectious diseases and in the design of mass immunization programmes. *Epidemiology and Infection*, **101**, 1–20. doi:10.1017/S0950268800029186.

Renwick, A.R., White, P.C.L. and Bengis, R.G. (2007) Bovine tuberculosis in southern African wildlife: A multi-species host-pathogen system. *Epidemiology and Infection*, **135**, 529–540.

Russell, C.A., Smith, D.L., Childs, J.E. and Real, L.A. (2005) Predictive spatial dynamics and strategic planning for raccoon rabies emergence in Ohio. PLoS Biol 3(3): e88. doi:10.1371/journal.pbio.0030088.

Smith, L., White, P.C.L., Marion, G. and Hutchins, M.R. (2008) Livestock grazing behavior and inter- versus intra-specific disease risk via the fecal-oral route. *Behavioural Ecology*, **20**, 426–432.

Sterner, R.T. and Smith, G.C. (2006) Modelling wildlife rabies: Transmission, economics, and conservation. *Biological Conservation*, **131**, 163–179.

Szmaragd, C., Wilson, A.J., Carpenter, S., Wood, J.L.N., Mellor, P.S. and Gubbins, S. (2010) The spread of bluetongue virus serotype 8 in Great Britain and its control by vaccination. PloS ONE 5(2): e9353.

White, P.C.L., Böhm, M., Marion, G. and Hutchins, M.R. (2008) Control of bovine tuberculosis in British livestock – there is no 'silver bullet'. *Trends in Microbiology*, **16(9)**, 420–427.

CHAPTER 14

The use and abuse of water cycling

Malcolm Cresser and Cumhur Aydinalp

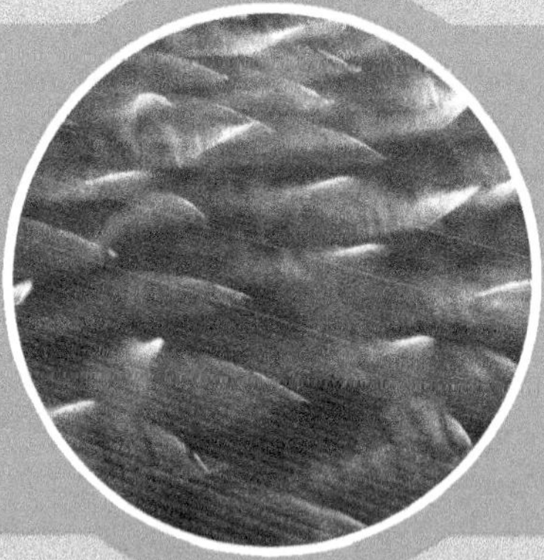

Learning outcomes

By the end of this chapter you should:

- Be more aware of the many roles that water fulfils in the modern world.
- Appreciate the difference between sensible, sustainable uses of rivers as a management tool and abuse of rivers when their short- or long-term capacity to deal with pollutant loads is exceeded.
- Appreciate how pollution should be managed carefully to protect potable water supplies from surface water and groundwater.
- Better appreciate the risks of salinisation of groundwater and surface waters from poor irrigation practices.
- Have an improved awareness and understanding of the inextricable links between soils and surface waters and groundwaters.
- Have some insight into how problems of saline water problems may be dealt with.
- Be better able to start to assess the water-related problems that will be associated with climate change.

14.1 Introduction

When you look down from above at a river like the Dee in north-east Scotland, winding its way gently through an attractive rural landscape – like the one in Figure 14.1, it's very easy to overlook the many roles that the river is fulfilling. But you don't have to think for too long to produce quite a long list of functions.

Lead-in question

What functions is the river in Figure 14.1 fulfilling?

Having read Chapter 2, if asked to list the roles that a river fulfils you might well start with its role in the general water cycle. Much of the water deposited as precipitation over high and low ground drains to the river channel, and passes on to the oceans; there it contributes water vapour back to the atmosphere as part of the global cycling of water. On first thoughts you might think that, so far, there is no scope for abuse of water cycling. But what happens if, for example, hill peats are drained or upland forests cleared? Clearly both of these processes could lead to faster delivery of drainage water to the river channel. In countries such as Bangladesh forest clearance can dramatically worsen the risk of flooding downstream. Therefore land management in headwater parts of drainage basins requires careful consideration of potential consequences of land use change.

The flow of water from the headwater parts of a drainage basin is not just important in the context of impact upon flood risk down-river. For many catchments we rely heavily on the delivery of large volumes of relatively clean and unpolluted water from the upland areas to dilute pollution entering the river downstream. For example, if some water is abstracted from the river for potable supplies then we need to ensure that the nitrate concentration remains below target threshold concentration values, typically 50 mg of nitrate per litre. Thus the cleaner water may be needed to dilute inputs of drainage water from agricultural areas lower in the catchment. We will return to this issue in section 14.2.

As we saw also in Chapter 2, rivers transport the products of biogeochemical weathering to estuaries and the ocean, and not just water. Thus over geological timescales, they play a major role in the rock cycle discussed in Chapter 1. Much of the material they transport is as solute species, for example $Ca(HCO_3)_2$ or $Mg(HCO_3)_2$, but some is transported in particulate forms from erosion.

Again on first thoughts, it might be concluded for an upland catchment area like that in Figure 14.2 that if there has been no deforestation

Figure 14.1 A view of the River Dee in north-east Scotland, looking inland from east to west, with the Grampian mountains in the far distance.
Source: Photograph courtesy of Dr John Creasey.

Figure 14.2 A section of a typical upland catchment area in northern Britain; this is the River Etherow to the east of Manchester.

and no improved drainage the system is free of human pressures. However, to the top left of the picture is a main road that is salted and gritted every winter, a topic discussed in Chapter 19. So the river has to dispose of a salt load, taking it back to the oceans; we will return to this issue later. What doesn't leave with the river will pass to groundwater. But there is another important anthropogenic effect on this river. Look at the water colour in Figure 14.3.

The water in Figure 14.3 is very brown in colour because of its high dissolved organic matter (DOM) concentration. The DOM at this site has been increasing consistently for more than a decade (Evans *et al.*, 2006). The authors believe this is primarily due to the reversal of the slow down of plant litter decomposition in the surface organic-rich soils in the catchment that was caused by decades of acidic pollution deposition. The reversal reflects a major decrease in the pollutant sulphur deposition in the area. Other possible reasons have also been advanced, however, and increasing DOM concentrations remain a topic for debate (Evans *et al.*, 2006). Waters were naively expected to deal with acid deposition in their catchments with no adverse effect on water chemical quality or aquatic biota that they support. Moreover the aquatic biota, from invertebrates to fish, may constitute part of the food chains of higher species. Supporting biota is another function of surface freshwaters. In rural areas where game fishing and recreational uses of water are important, so too is supporting rural economies.

There are other activities going on in the catchment in Figures 14.2 and 14.3 as part of the catchment management. There are small areas of improved grazing for example, and if these have

Figure 14.3 A small waterfall in the River Etherow, just a few 100 metres upstream of the section shown in Figure 14.2.

been limed then the turnover rate of organic matter will be increased, so more mineral N, especially nitrate, may be mobilised to the river (see Chapter 5 for a discussion of N cycling). The liming may raise the pH of the river water slightly, but increase the river's nitrate load. Heather in the catchment is burned periodically to encourage young growth beneficial to the grouse that are important to the rural economy. Some people are happy to pay to shoot the grouse. Unless carefully managed, heather burning too can change the pollution load that the river has to handle; for example it has been shown to increase nitrate leaching by reducing biotic uptake (Cresser *et al.*, 2004).

Rivers are also often used as conduits for pollution removal. For example, effluent from sewage treatment works is often discharged to rivers, where it is progressively diluted as it moves downstream. Discharge consents are generally based upon the maximum pollution load which calculations show will not excessively reduce the dissolved oxygen concentration in the river water. Such calculations should take into account that there is no photosynthetic oxygen production from algae and aquatic plants during periods of darkness, but oxygen is still consumed via respiration. This process is especially important in warm and shallow lakes where dissolved oxygen concentration may vary dramatically over each 24-hour period (Ansa-Ansare *et al.*, 2000).

From the above introduction it should be clear that we expect rivers and lakes to fulfil many useful functions, providing many ecosystem services; however the scope for humans to disrupt their normal functioning is enormous. Until very recently we have got things wrong in many countries around the world. In some we still are getting things wrong and indulging in temporarily convenient, but unsustainable, practices.

14.2 Nitrate from agriculture

If we consider what agricultural priorities were about two decades ago, over much of the world maximising crop yields would have been high on the list. As fertiliser and energy prices started to increase however, this priority rapidly changed to optimising economic yields. It was a few years still, though, before 'while minimising environmental damage' would have been added to this. This was because it was hoped that soils would retain any surplus fertiliser that had been added in excess of crop requirements, even after crops had been harvested in late summer and autumn months. If the soils could not retain any excess nitrate, it would leach and contaminate either surface waters or groundwater or possibly both. In pristine, minimally managed ecosystems the cycling of nitrogen was very tight, with minimal leaching of inorganic N species such as nitrate or organic N, even in winter (Chapters 4 and 19).

It has long been known that the nitrate concentration in a river for catchments with mixed land use that includes arable agriculture increases downstream with the increasing amount of arable agriculture in more lowland areas (Edwards *et al.*, 1990). Figure 14.4, for example, shows, for two major river catchments in Scotland, how strongly nitrate concentration correlates with increasing amount of agricultural land. The data fit quite well onto a single line, though this is partly only because precipitation distribution is

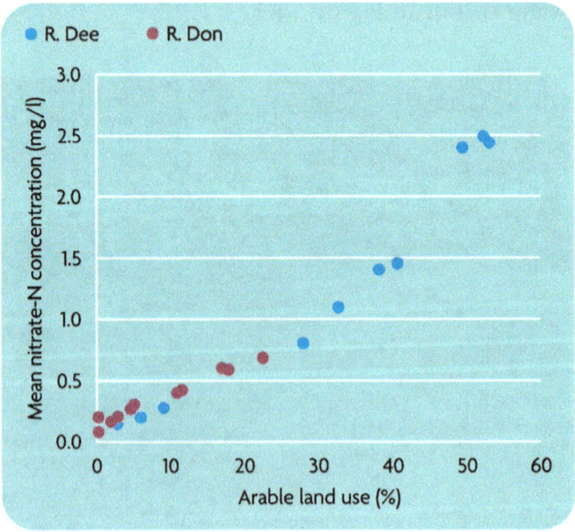

Figure 14.4 Correlation between mean nitrate concentration between 1980 and 1986 and percentage cover of arable land for two adjacent river catchments in north-east Scotland, the River Dee and the River Don.
Source: Based upon data presented by Edwards *et al.* (1990).

similar for the two catchments. In wetter areas nitrate would be expected to be more dilute at a given percentage of arable land use. From plots such as this one it is inferred generally that the nitrate predominantly originates from use of manures and other N fertilisers on arable land. In the context of this chapter we simply expect the river to get rid safely of any excess nitrate from agriculture.

Leaching of nitrate from agricultural soils is commonplace and significant, but became a cause for concern when nitrate–nitrogen concentrations exceeded around 11 mg l^{-1} (which corresponds to around 50 mg of nitrate per litre, a possible health risk in potable water supplies). In Europe, every effort is made now to ensure that nitrate concentrations remain below this critical threshold, by imposing constraints in areas known as Nitrate Vulnerable Zones – NVZs) and on how, how much, where and when farmers can apply and store manure and apply nitrogen fertilisers. In effect this is to reduce the excessive abuse of surface waters, and ultimately ocean waters, which is where most of the nitrate in river water eventually ends up. When the nitrate concentration is unacceptably high, for whatever reason, the river system is being abused rather than just sensibly used.

There is often a consistent seasonal trend in nitrate concentration, with winter peaks, even in minimally managed upland catchments in areas impacted by heavy atmospheric N pollution (Cresser *et al.*, 2004). Maxima occurring in mid- to late-winter in such N-impacted areas in the UK are much clearer and more pronounced than they would be in pristine catchments; minima occur in summer months, but in such areas nitrate leaching often still occurs even in mid-summer. Figure 14.5, for example, shows the spatial variation in nitrate concentrations assessed by intensive spatial sampling in a dry week in mid-summer for the heavily N-polluted River Etherow catchment mentioned earlier. The summer minima invariably are attributed by scientists to higher uptake of nitrate by plants in warmer, summer months; the leaching of nitrate that still occurs in summer, and can be seen in Figure 14.5, is attributed to soil nitrogen saturation, as discussed later in Chapter 19.

Smart *et al.* (2005) examined the causes of differences between nitrate concentrations in a series of adjacent upland streams in the Netherbeck drainage basin in Cumbria in north-western England. They looked at correlations between nitrate concentrations and a wide range of catchment characteristics such as maximum altitude,

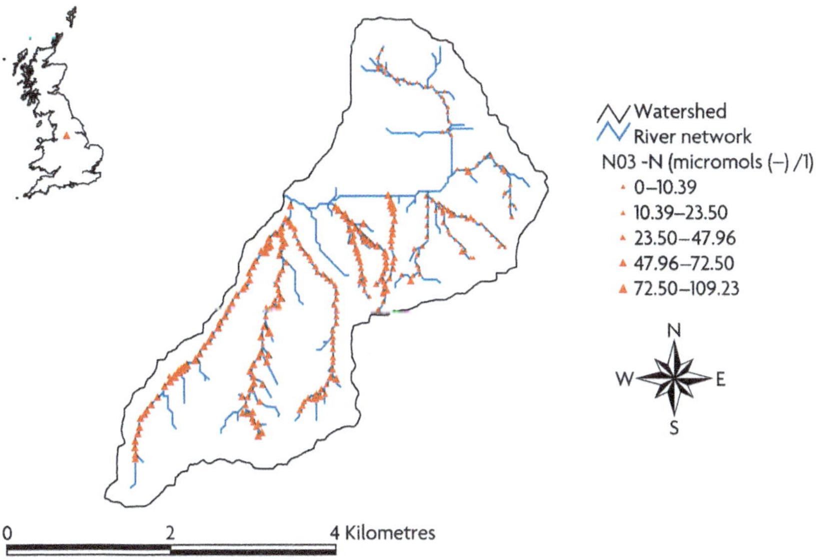

Figure 14.5 Spatial variation in nitrate concentration, here expressed in units of micromoles of nitrate per litre, across the River Etherow catchment. A micromole of nitrate is 62 micrograms, so the highest summer concentrations exceed 6 mg nitrate l^{-1}.

altitude range above sampling sites, slope distribution, bare rock percentage cover, vegetation percentage cover and time of year samples were taken. Having found which parameters had most effect, these parameters were then put into a simple empirical model to explain seasonal and spatial variations in nitrate concentrations. Slope and percentage of bare rock were the most important catchment characteristics, and seasonality was incorporated by fitting part of a cosine wave function to the data. Figure 14.6 shows what we mean by using a partial cosine function. Parts of sine waves and cosine waves are often used in empirical models which have distinct seasonal peaks and troughs. The fit of their model is shown in Figure 14.7.

You will see, nevertheless, that the fit of the earlier model is good for most sub-catchments in the local area, but less good for others, such as site 10. This suggests that another factor should probably have been incorporated into this model; for example catchment 10 might have a substantial area of very poor drainage. Models are often very useful for showing us that there is something that we did not include in our original perceptions of how the systems we are studying really work. Sometimes it's grappling with 'what's gone wrong' that leads to progress being made!

Recently Mian *et al.* (2010), when examining long-term data kindly provided by the Environment Agency, reported that at six of nine sampling sites along the length of the River Derwent in North Yorkshire seasonality was pronounced from 1988 to 1997, but subsequently was hard or impossible to detect. Figure 14.8 clearly shows this for one of the sampling sites. At the three remaining sites there was no seasonal trend at all in the moderately N-polluted area. Peaks in nitrate concentration occurred every winter (in Figure 14.8 months 1, 13, 25 etc. indicate January and months 12, 24, 36 etc. indicate December) in long-term time series plots of nitrate-N concentration in catchments with significant areas of arable land. It was suggested that where seasonality had markedly declined it was probably due to the impacts of higher energy and fertiliser costs, increasing farmer awareness of environmental issues, the impact of the Foot and Mouth outbreak in England in 2001 on subsequent animal numbers, and increased leaching of N from subsoils in summer as a consequence of atmospheric N pollution penetrating soil profiles over winter months. However, they did not consider the fact that the water in the Derwent is also used to dispose of treated sewage effluent from several treatment works (STWs) beside the river channel.

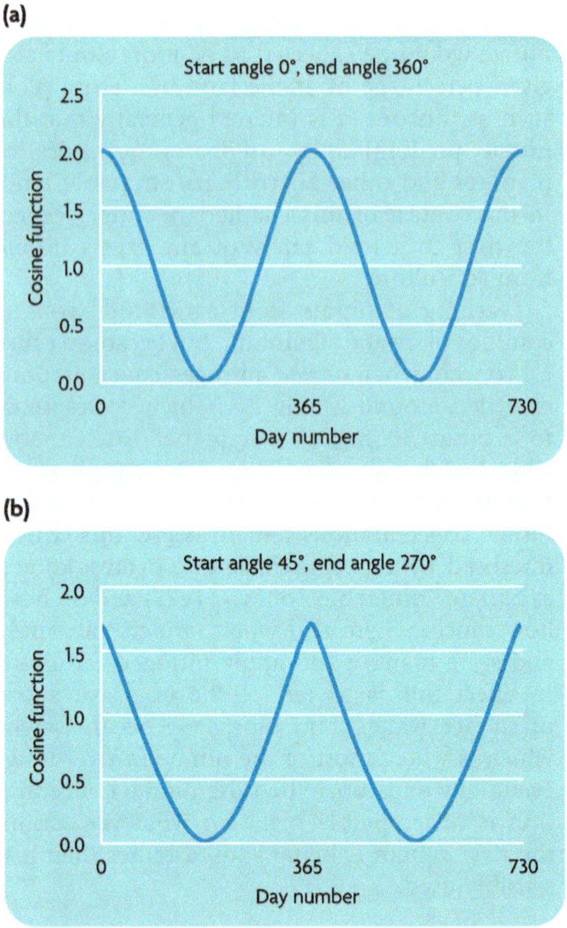

Figure 14.6 Sketch to show how a cosine wave can be fitted to experimental data. A cos function varies between +1 at 0° and 360° and –1 at 180°, passing through 0 at 90° and 270°. If we add 1 to this it oscillates between +2 and 0, as shown in (a) above. We can represent a day number of each year as one 365th of 360°, so the figures above represent two years of a cos function versus day number. If we want the curve to have sharper peaks at the start and end of each year, we can chose to use, for example, 45–270° (b) instead of 0–360° of the cosine function.

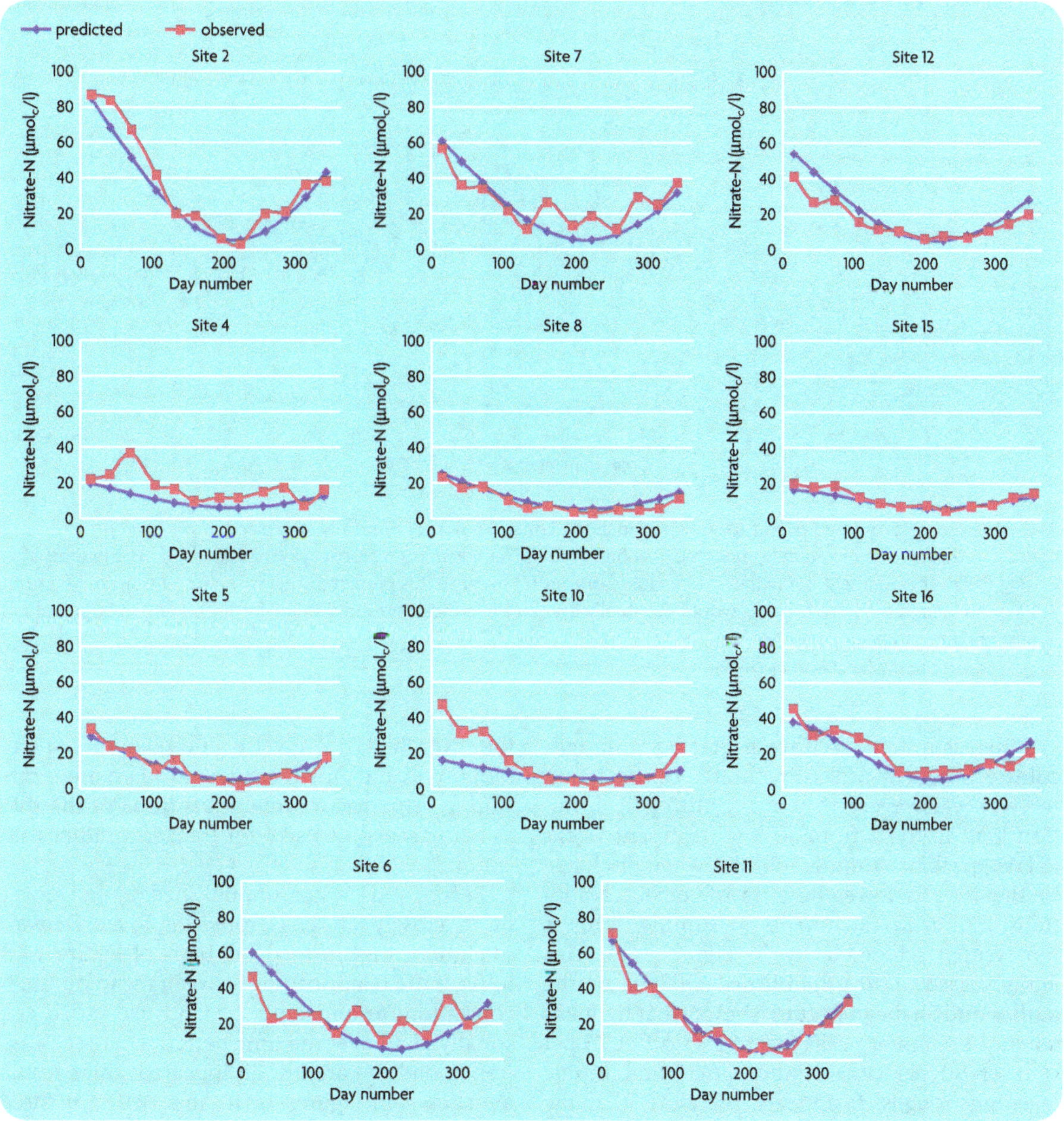

Figure 14.7 Fit of an empirical model developed to explain the spatial and seasonal variations in nitrate-N concentrations in 11 sub-catchments of the Netherbeck drainage basin in Cumbria, UK. Source: Smart *et al.* (2005). Day numbers run from 1 on 1 January to 365 on 31 December, so the minima occur in summer when growth is most active and plant uptake highest. We shall see later that low input of readily decomposable fresh litter in summer months also reduces nitrate leaching in summer. In this upland area the nitrate originates from nitrate pollution in the atmosphere, primarily coming from fossil fuel consumption in cars. The pollution is more than sufficient to N-saturate the soils, so we depend upon the rivers to get rid of the surplus. A later version of the empirical model included atmospheric N pollutant deposition as an input term, so that the model was transferable across the UK.

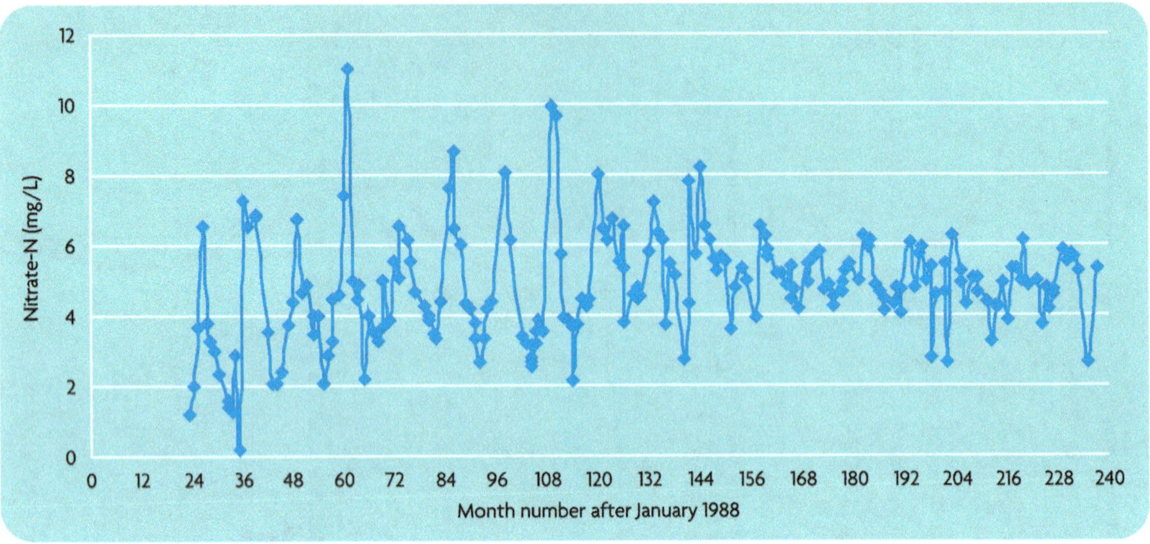

Figure 14.8 Long-term trend over a 20-year period in the concentration of nitrate-N, assessed by the Environment Agency, in the lower, agriculturally impacted, reaches of the River Derwent in North Yorkshire, England. At the time of writing the river catchment is a Nitrate Vulnerable Zone (NVZ) because it is predicted that if the trend from the first ten years continued, the nitrate-N peaks would eventually exceed the critical threshold value corresponding to 50 mg of nitrate per litre of water.

Source: Adapted from Mian, I.A. *et al.* (2009).

Whereas nitrate leaching associated with agriculture produces peaks in nitrate concentration in river water every winter, the nitrogen pollution load to rivers from sewage treatment works (STWs) peaks in summer months, when discharge in the river is lower because of greater evapotranspiration in catchments in summer. This is represented schematically in Figure 14.9. This figure, based upon unpublished analysis by the author and Shaheen Begum, shows what happens when the summer peaks from STWs offset 75 per cent or 90 per cent of the agricultural nitrate leaching troughs. It appears, therefore, that the loss of distinctive seasonality peaks in Figure 14.8 may well be due to increasing effects of better N fertiliser use on farms bringing down the agricultural winter peaks while the STW peaks may have increased marginally due to property development and modest population growth in the region. Note though that the river is still being 'abused' as a dumping ground for N pollution, but from fertiliser and manure use, STW effluent disposal and, for the headwater areas, atmospheric N deposition too. If the N load in the river had no adverse effects upon aquatic biota, and did not increase nitrous oxide emission in estuaries, we might just think of the river as being sensibly 'used' for transport rather than 'abused'.

Begum and Cresser (unpublished results) looked more closely at data from the River Derwent catchment that formed the basis of Figure 14.8, to see if the relationships between mean nitrate-N concentration and types of land use varied seasonally at all over the time period where seasonal trends had apparently disappeared; their results are shown in Figures 14.10 and 14.11 for three-monthly means over the years 1998 to 2006.

Two interesting facts emerge from Figures 14.10 and 14.11. The first is that the slopes of the regression equations and the intercepts in Figure 14.10 hardly vary at all from season to season, which confirms the lack of any clear seasonal trend over the assessment period. The second, and more interesting, trend noted by Cresser *et al.* (2011) is that the negative correlations of nitrate-N concentration to percentage of moorland rough grazing land use for the seasonally divided data are

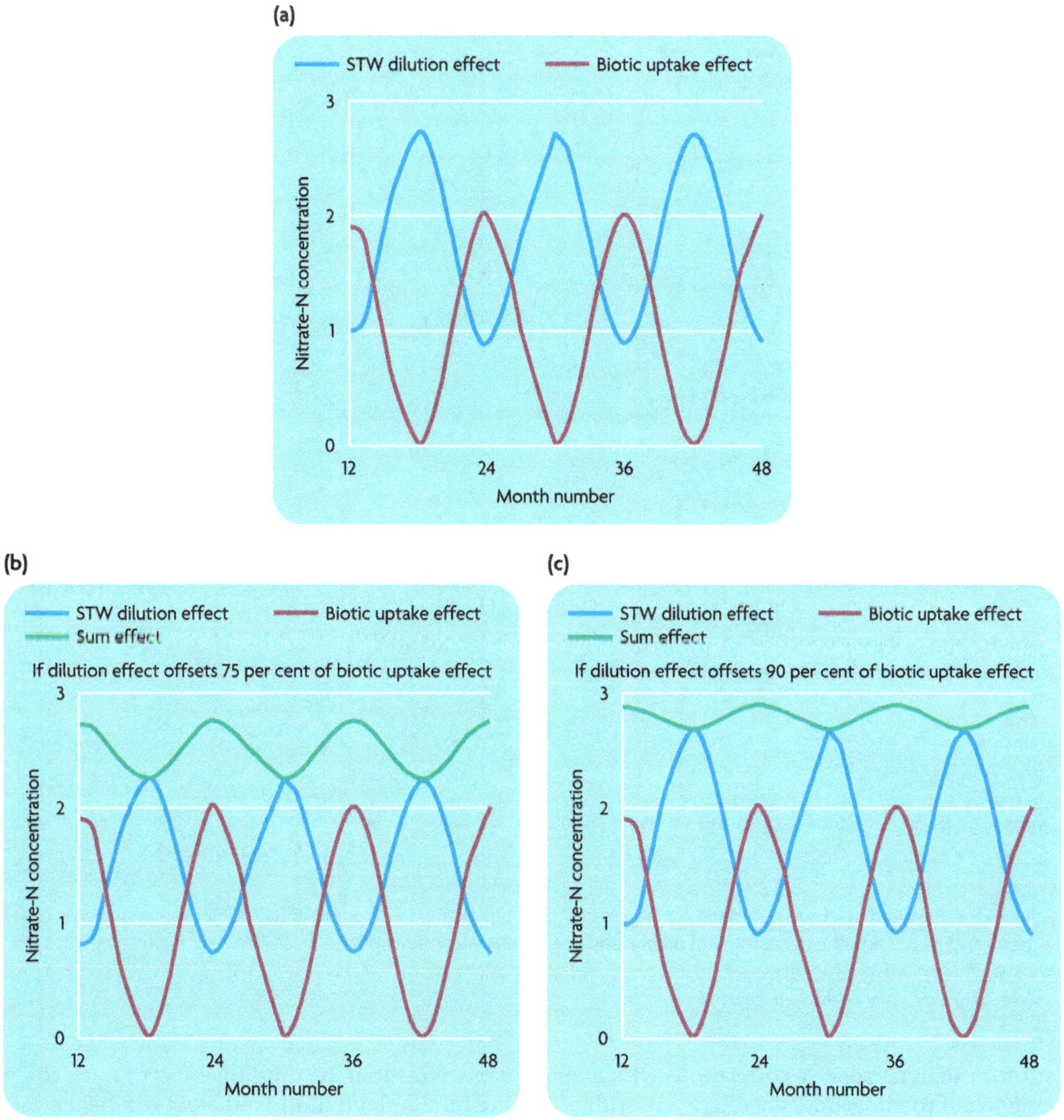

Figure 14.9 Simplified schematic representation of seasonal variations in the nitrate from agriculture and from STWs (a) and how they can partially cancel out apparent seasonal trends in a river ((b) 75 per cent balance and (c) 90 per cent balance). Note though that the nitrate load has not disappeared and we still expect the river system to get rid of it at its estuary or in the ocean. The trouble is that in anaerobic estuarine sediments it can partially turn to nitrous oxide (N_2O), a greenhouse gas.

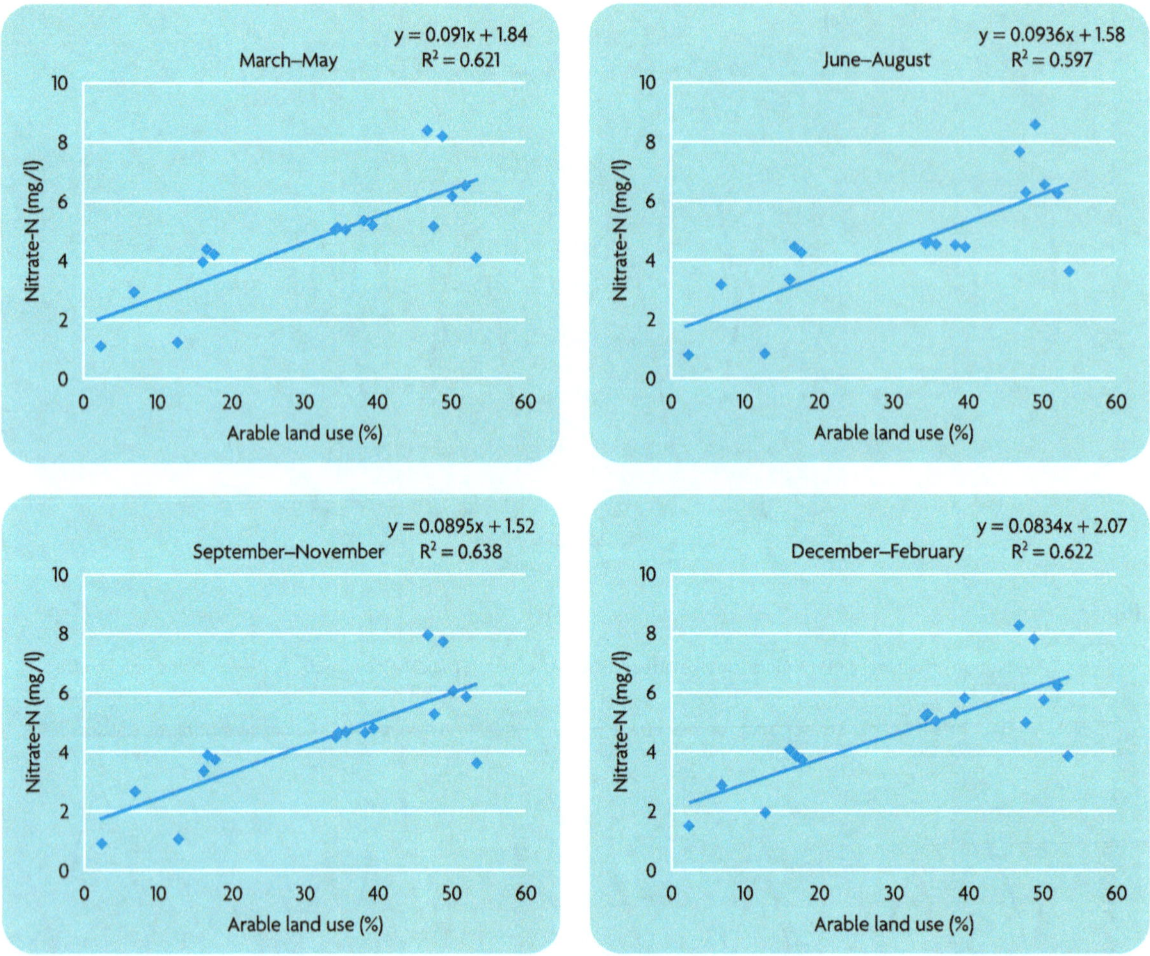

Figure 14.10 Relationships between mean nitrate concentration in river water from 1998 to 2006 in spring, summer, autumn and winter and the percentage of arable land use in sub-catchments of the River Derwent. Eight sampling sites were on the main river channel.

Source: Graphs courtesy of Shaheen Begum.

stronger than its positive correlations with arable land use. If you imagine a flat catchment with uniform arable land use and uniform distribution of precipitation and runoff to a river, then clearly nitrate-N concentration might be expected to be constant along the length of the river. In practice, however, in many typical catchments relatively less nitrate-polluted water from headwaters in uplands serves to dilute the more polluted drainage water downstream; on moving down a stream therefore, nitrate concentration increases because the *relative* contribution from headwaters decreases compared to that from drainage from agricultural areas. Note that the dilution effect of water draining from upland moorland areas will beneficially dilute the N pollution loads from both arable land drainage and STWs. Hence the greater strength of the correlations in Figure 14.11 compared with those in Figure 14.10.

Very recently another mechanism within soils under grassland has been suggested that may also contribute to the lack of pronounced autumn/winter maxima in nitrate-N leaching losses to streams from less polluted upland areas around headwaters. Cresser, Begum and Bhatti (2011) have suggested that plants have evolved to

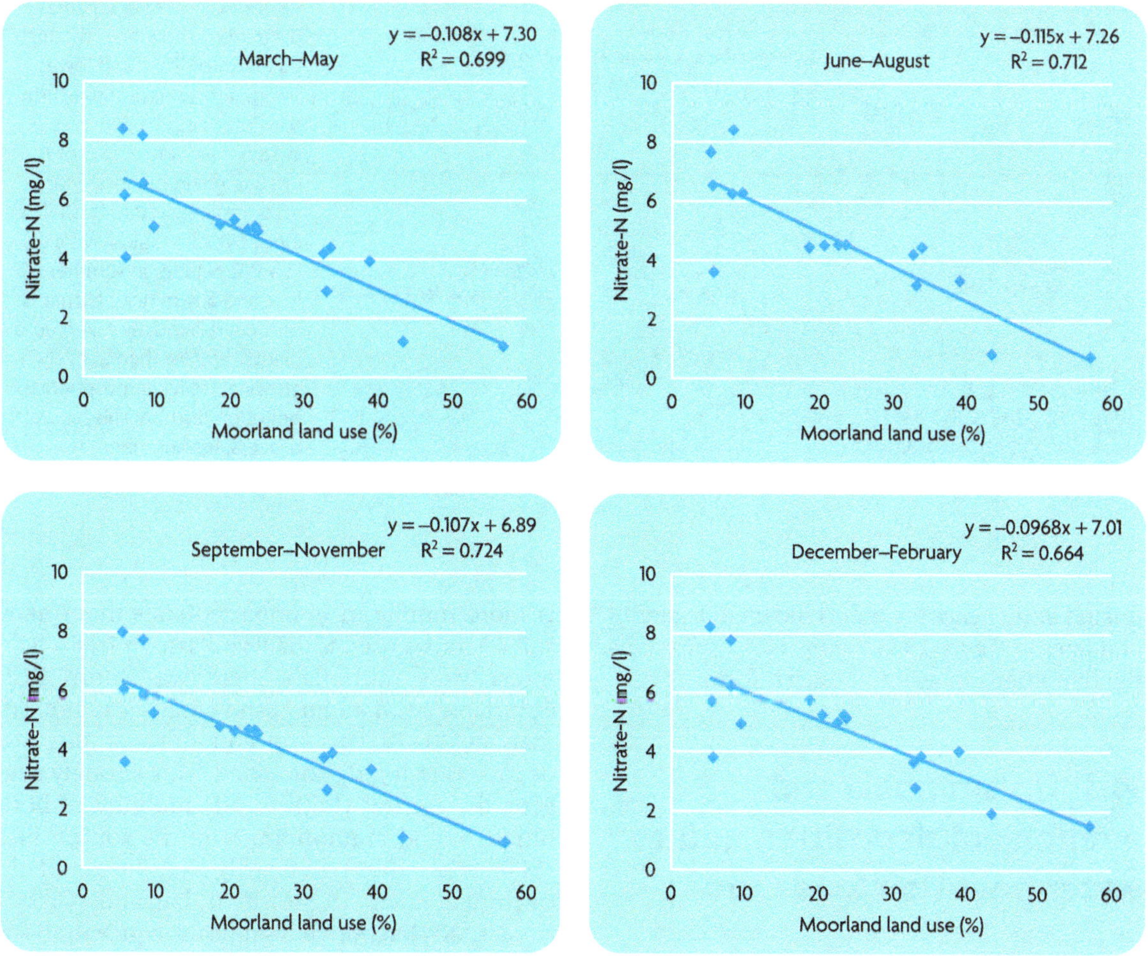

Figure 14.11 Relationships between mean nitrate concentration in river water from 1998 to 2006 in spring, summer, autumn and winter and the percentage of moorland and rough grazing land use in 17 sub-catchments of the River Derwent. Eight sampling sites were on the main river channel.

Source: Graphs courtesy of Shaheen Begum.

deal with seasonality so that litter is deposited (or roots senesce) with a high C:N ratio in autumn. Decomposition is initially slow in winter, and most mineral N produced as the litter is decomposed is retained by microbial biomass surrounding the litter because of the high litter C:N ratio. Eventually C:N ratio falls to a value at which mineral N becomes progressively more available in the soil. This occurs at a time when the plant is starting to grow actively however, and requires the mineral N. Thus plants have evolved to achieve a seasonal dynamic match of litter N release and plant N uptake requirements. In a soil under permanent grassland the fresh inputs to the readily mineralisable pool of degradable litter are probably minimal by July, restricting *new* mineral N production. Atmospheric pollution from mineral N disrupts this natural N cycle that has taken so long to evolve. This may modify biodiversity by initiating a dynamic mismatch, giving a competitive edge to plants that can benefit from N use earlier in the year. The authors believe that this is what changes biodiversity in minimally-managed grasslands. In the present context it also explains why rivers are called upon to deal with nitrate leaching even in mid-summer in heavily N-polluted

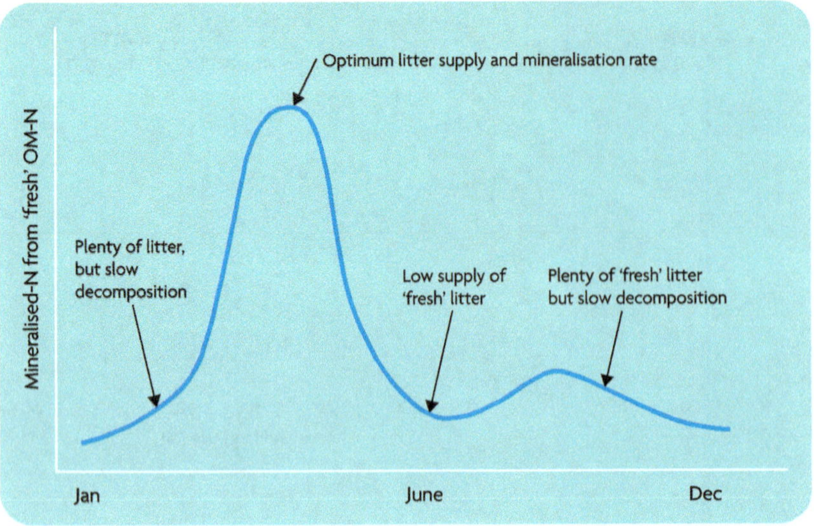

Figure 14.12 Simple schematic diagram to show how the high C:N ratio of litter and lower temperatures as autumn turns to winter help to conserve N in soils over winter in more pristine environments. More readily mineralisable fresh litter input (which would have a lower C:N ratio) in summer is low, and any mineral N that is produced is mostly taken up by plants that are then actively growing. Heavy atmospheric N pollution disrupts this natural process, leading to nitrate leaching even in summer (Based on Cresser *et al.*, 2011).

upland areas; hence the results shown in Figure 14.5 and later in Figure 19.7. Figure 14.12 summarises the hypothesis in a simple sketch form.

14.3 Do ammonia and ammonium deposition add to water pollution loads too?

Intensive animal rearing has resulted in substantial increases in the amounts of ammonia gas transferred to the atmosphere in many countries around the world, especially down-wind of poultry farms and intensive pig rearing units. Much of this ammonia is deposited directly (or as ammonium aerosols) relatively locally compared to oxidised N pollutant species, typically within a few kilometres of the emission source. Because ammonia deposited on acid soils rapidly forms ammonium cations (NH_4^+) it has long been widely assumed that it is immobile in soil, being strongly retained on cation exchange sites (see Chapter 7). When deposition levels were low, they would only add to the N pollution load of nearby rivers after nitrification in the soil had converted the ammonium-N to nitrate-N.

However, more recently, absorption isotherm experiments have indicated that ammonium-N is more mobile in N-impacted soils than many soil scientists used to believe. Also, in fresh, field moist soils, although there was more KCl-extractable ammonium-N and nitrate-N at 0–10 cm than at 10–20 cm, concentrations at the two depths were highly correlated. This strongly suggests the vertical (both ways!) mobility of both nitrate-N and ammonium-N in the soil profiles (Mian *et al.*, 2009). In heavily N-polluted upland catchment areas in northern England, Cresser *et al.* (2004) reported significant concentrations of ammonium in river water samples *throughout the year*. They found that ammonium was especially mobile during acid episodes that occurred in river water during storm events. It was suggested that, as the dissolved organic matter concentration increased during storms, the high correlation of ammonium with concentrations of H^+ and organic matter reflected the transport of ammonium cations balancing part of the negative charge associated with soluble organic acids from upper soil horizons. The authors concluded that the mechanism for ammonium mobilisation was as indicated in Figure 14.13.

In heavily ammonium-polluted soils in uplands that have been acidified by decades of acid deposition, ammonium is not transformed as quickly to organic N or nitrate as it would have been in less polluted soils. It therefore starts to accumulate on cation exchange sites in upper soil horizons,

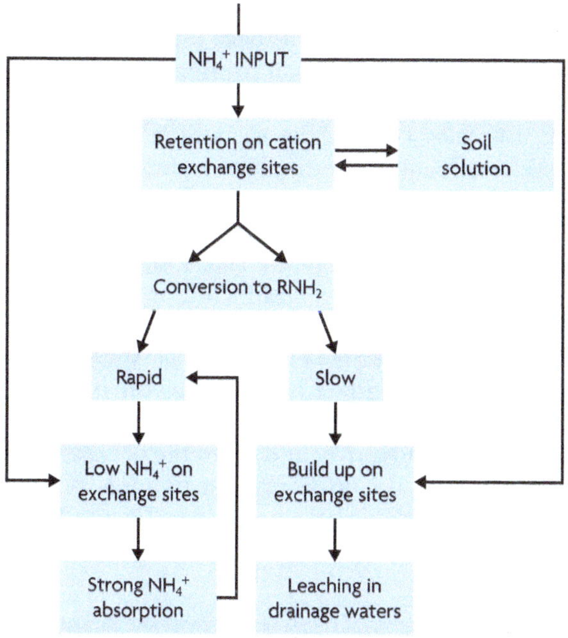

Figure 14.13 Suggested mechanism for the accumulation of ammonium in soils and its mobilisation to surface waters.

until eventually it starts to leak out of the soil into streams. Remember that cation exchange is an equilibrium process. This process must also provide higher input fluxes of ammonium ions to lower slopes.

14.4 Irrigation and salinity problems

We saw in Chapter 2 that, under natural, unpolluted conditions, the natural water cycle provides naturally clean and unpolluted water that is well suited to use for irrigation or indeed for drinking. Rainwater and snow contain soluble salts, but at very low concentrations, and plants have evolved to grow perfectly well when exposed to those concentrations. So water harvested from distant mountains and channelled to where it is required for irrigation, as in Figure 14.14, should not pose any particular problems, and indeed it doesn't so long as the water can drain away harmlessly. Difficulties do arise, however, if the irrigation water cannot remove any accumulating salts harmlessly via drainage or if waters from lakes or groundwater are used at rates faster than their *natural* recharge rate (see Chapter 2).

Lead-in question

What happens to the soluble salt in irrigation water when water evaporates from the soil surface in arid areas?

If attempts are made to establish crops on soils in arid areas simply by adding water when the plants need it, for a while everything will proceed as hoped. Initially the irrigation water will wash soluble salts down the soil profile. However, as the soil dries, water rises back to the surface and the salts become more concentrated; moreover salt from the irrigation water will add to the surface salt load after every watering event. Eventually the soil will become so saline that attempts to grow most crops will fail (Figure 14.15); the irrigation water will have been totally wasted.

Soils become 'saline soils' when they contain a concentration of soluble neutral salts sufficient to interfere seriously with growth of most plants. This happens when electrical conductivity of a saturated soil extract exceeds 4 decisiemens per metre (dS m^{-1}). This equates to 4000 $\mu S\ cm^{-1}$

Figure 14.14 An experimental site outside Tehran in Iran, being used to test trickle irrigation under polythene sheet mulching to reduce loss of water by evaporation from soil between high-value horticultural plants. The water is high quality, as it originates in the mountains that can be seen in the distance.

Figure 14.15 Small citrus trees failing in an arid area because of build up of salinity in the soil and soil solution, in spite of an attempt to use a straw mulch to reduce water loss from the surface.
Source: Photograph courtesy of Ken Killham.

(the units that you may well meet in older literature). Saline soils have a pH below 8.5, and less than 15 per cent of the cation exchange capacity is occupied by Na^+ (ESP < 15 per cent; ESP stands for exchangeable sodium percentage). If salt crusts form at the surface, they are white in colour.

Management of saline soils requires irrigation with water containing sufficient calcium and magnesium, *and drainage* as discussed in Chapter 7. When the soil ESP eventually exceeds 15 per cent, the soils are said to be 'saline-sodic', as long as the pH says below around 8.5. Leaching of saline soils can cause the soils to become 'sodic', unless the irrigation water contains significant concentrations of Ca^{2+} and Mg^{2+}. The exchangeable Na percentage rises well above 15 per cent, and because soluble Ca^{2+} and Mg^{2+} salts have been lost, conductivity falls below 4 dS m^{-1}. Na^+ from exchange sites can by hydrolysed, leading to pH values increasing to as much as 10. At this pH organic matter dissolves much more readily, and surface salt encrustations may be very dark or even black in colour. Gypsum ($CaSO_4$) addition and irrigation with good quality water, plus drainage, leads to recovery of sodic soils (Na_2SO_4 is less damaging than Na_2CO_3).

It is easy to see how abusive use of water for inappropriate irrigation can lead to serious soil salinity problems but sometimes also to water pollution problems. Suppose, for example, water is taken from a lake in an arid area and used to irrigate crops. If the water drains back into the lake from the cultivated area, its salt concentration will have increased slightly, adding to the salt concentration in the lake. For decades everything may be fine; eventually, however, the lake water will become sufficiently salty to be unsuitable for crop irrigation. It may also become sufficiently salty to modify the aquatic biodiversity in the lake. Once lakes reach this condition it is very difficult, or prohibitively expensive, to reverse the salinity problem.

So what is the answer? Ideally lake water should be channelled down from a reservoir on higher ground, and the drainage water from irrigated crops should pass to a river and on to the oceans where the salt can do no harm. Let us look at other examples of trickle irrigation in Iran (Figure 14.16)

Figure 14.16 Trickle irrigation in Iran being used in an attempt to establish trees for shade.

Figure 14.17 Another example of trickle irrigation, here in the United Arab Emirates. The site was being used to find the optimal amounts of water for the sustainable growth of tomatoes.

Figure 14.18 Liquid fertiliser being added to irrigation water.

or the United Arab Emirates (Figure 14.17). In Figure 14.16 the trickle irrigation is being used to supply water to a row of pine trees being grown to provide shade. It is important that soluble salts are moved away from the rooting zone so timing of the water supply is crucial. Thus drainage is needed as well as irrigation, which seems strange at first sight. However, there must be somewhere for the salty water to drain to. Here water from higher ground (note the mountains in the background in Figure 14.17) after use eventually drains away from around the plants. Salt is sometimes allowed to accumulate at the surface a few metres away from trees or crops and then scooped up and removed periodically by hand.

Salinisation of groundwater has become a serious problem in the United Arab Emirates because of over-abstraction of groundwater for crop production. After the discovery of the oil fields it was decided to invest heavily in crop production in an attempt to make the country self-sufficient in food. Much of the irrigation water drained back to groundwater, but each time the water goes through an irrigation/drainage cycle, the water becomes a bit more saline. This is clearly not sustainable over the long term. It was common practice when groundwater was used for irrigation to tap deeper groundwater supplies as soon as water started to become saline. However, eventually this can lead to sea water incursion (see Chapter 2). It is no wonder then that the UAE now invests oil revenues in desalination to provide bottled drinking water.

Irrigation is still widely used for food production in areas where without it nothing could be grown economically. Liquid fertilisers can be added to trickle irrigation water to optimise fertiliser application rate and minimise groundwater contamination, as in Figure 14.18. Ideally plant nutrient additions should be exactly matched over time to optimal plant nutrient uptake requirements. If this is achieved then the pollution burden in drainage water is minimised and fertiliser is used very efficiently. At least policy makers are starting to become aware of the constraints, sometimes sadly learning from experience when, after a few decades, things have all started to go horribly wrong. In fairness they face a real challenge if we are ever going to feed 9 billion people. If the fertiliser/irrigation combination is not used carefully then groundwater pollution problems may become significantly worse.

14.5 Contamination of surface waters from industrial waste disposal

Right from prehistoric times there were good reasons for people who wanted to settle to do so in parts of the landscape where they found a good and plentiful supply of clean freshwater.

Often they would find fertile valley floor soils beside a river that could easily be cultivated to produce food and timber too. Subsequently there would be a tendency for towns to develop in such areas and the population that accumulated would naturally exploit the water resource. In semi-arid climates that might well mean water would be abstracted from the water source for irrigation purposes as well as to meet the needs of the people, including those involved in local industries. Initially that apparently would work well, but in many cases the river would also be exploited for waste disposal too, including industrial wastewater that had to go somewhere. Thus rivers could become conduits for waste disposal, and as such could acquire both inorganic chemical and organic pollutants and significant populations of pathogenic organisms. It generally was tacitly assumed that these would all flow into the oceans where dispersal and dilution would render them harmless.

If you only had one town or human settlement on a river and the community was well organised so that pollution entered the river only down-river of any water abstraction point, this might have appeared to be a sustainable system, the water being effectively 'used' rather than 'abused'. The problem though became one of size. If the population grew too large there would be overlap of abstraction and effluent discharge points. Down-river of the town the organic loading might become so high that the water could become anaerobic, and thus potentially toxic to sensitive aquatic organisms. Heavy organic waste loadings in estuaries could lead to problems of anaerobic sediments being formed. In the presence of sulphate from sea water in tidal reaches this could lead to the production of hydrogen sulphide (H_2S), with the associated smell of rotten eggs pervading the atmosphere. As recently as the 1970s in Scotland one author can well remember the impact of organic waste disposal into the River Dee from the paper industry in Aberdeen leading to this problem. In an area known as Bridge of Don it was referred to semi-affectionately by the locals as 'the Don Pong'.

Although in more recent times more care has been taken to reduce such obvious abuse of fresh water resources, problems still occur. Consider Figure 14.19, for example, which shows how

(a)

(b)

Figure 14.19 Tractor adapted by a Turkish farmer to pump canal water for irrigation (a) of crops growing on vertisol soils (b).

enterprising farmers in the Bursa Plain in Turkey were adapting their tractors to make them function as powerful water pumps less than a decade ago. The water is abstracted from the local river and canal system to irrigate crops (Figure 14.19a) growing on the naturally heavily cracked vertisol soils of the valley bottom (Figure 14.19b). The pumps are so effective that they spread fine suspended sediment that is present in the water too.

Unfortunately the deeply cracked vertisol soils of the Bursa Plain are being irrigated with poor quality water containing pollution from industrial effluents. Although the soils are alkaline and contain free $CaCO_3$, the irrigation water becomes slightly more acidic on going downstream through the city (McClean *et al.*, 2003). This can be seen from measurement of water pH from site 1, which is above the city, to site 6 which is below the city (Figure 14.20).

Figure 14.21 shows the spatial variations in heavy metal concentrations in the water being used for irrigation at six points along the river. Concentrations of each were reasonably consistent over the four sampling weeks. In the context of salinisation, the loss of calcium carbonate would be a problem only if the irrigation water had a high sodium concentration. If it does, once the $CaCO_3$ has all dissolved, Na^+ will accumulate on cation exchange sites, and soil pH will rise. At this site, salinisation should not be a problem, however, as the stream system is fed from precipitation in mountainous areas to the east. In Turkey, the use of lake water for irrigation may pose much more of a salinity problem, especially in areas where soil drainage water passes back to the lake.

What is more of an imminent problem at this site is the combined effects of volume and acidity of the irrigation water as the soil calcium carbonate is being dissolved. Its concentration is an order of magnitude lower in the irrigated soils than in adjacent soils that are not being irrigated. Once the carbonate has all dissolved the soil will tend to acidify, increasing the mobility, and potentially the crop uptake, of most of the potentially toxic elements that have been accumulating in the soils. While using the river to 'dump' potentially toxic elements in the ocean might debatably be argued as use rather than abuse of the river, polluting the water and then using it for irrigation is undoubtedly abuse.

The irrigation water contains suspended solids, so irrigation is lowering the soil clay content (see Figure 14.22) by adding silt-sized and fine-sand-sized particles to the soil via the deep cracks seen in Figure 14.19b. Therefore the soil texture is slowly changing, so CEC is also falling as a consequence of irrigation.

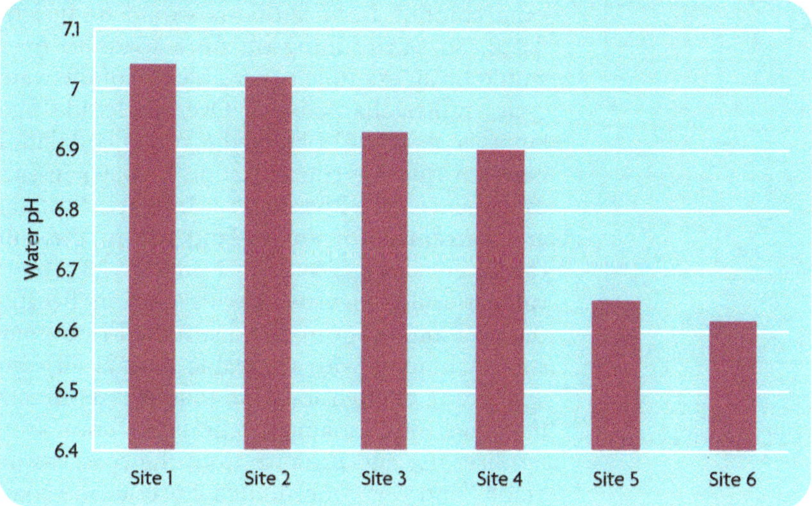

Figure 14.20 Effect of industrial discharges into the Nilufer River upon river water pH between site 1 (upstream the urban area) and site 6 (downstream).
Source: Based upon data from Aydinalp *et al.* (2005).

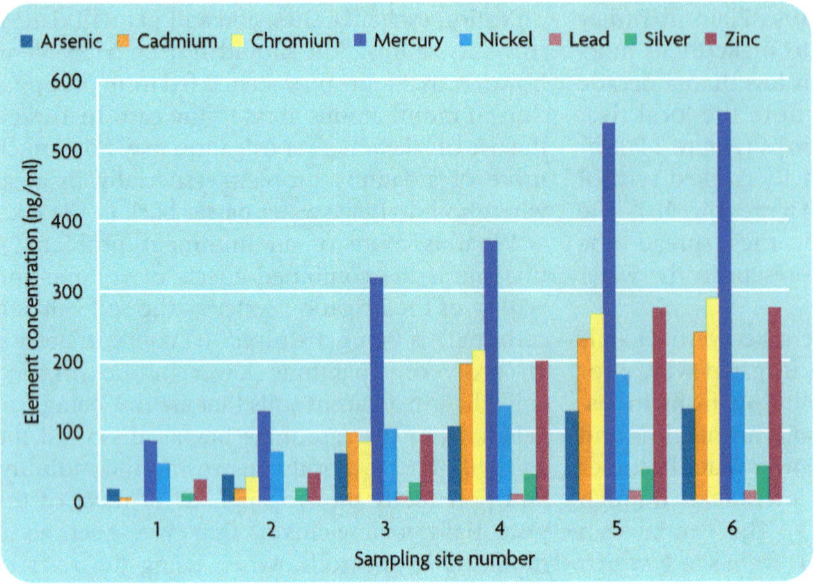

Figure 14.21 Effect of industrial discharges into the Nilufer River upon river water on concentrations of potentially toxic elements between site 1 (upstream of the urban area) and site 6 (downstream). The units are ng/ml, so the highest zinc concentration, for example, is approaching 0.6 $\mu g\ ml^{-1}$ or 0.6 ppm.

Source: Based upon data from Aydinalp *et al.* (2005).

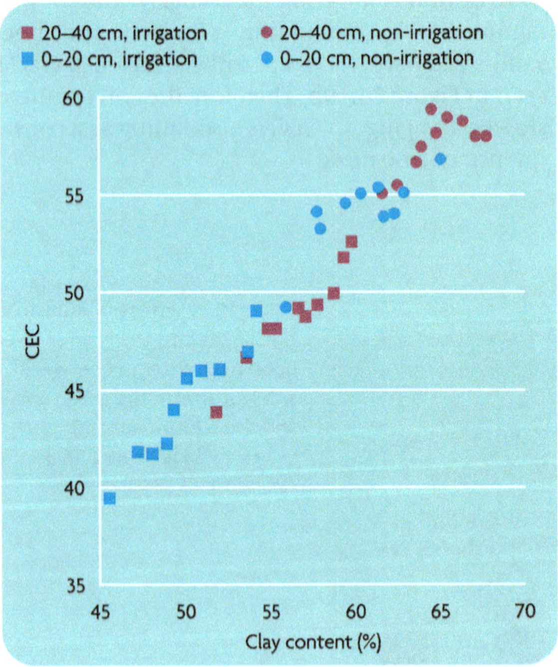

Figure 14.22 Change in soil clay content brought about by irrigation with water containing silt and fine sand from the Nilufer River, Turkey.

Source: From McClean *et al.* (2003).

14.6 The problem with road salt

Chloride ions are cycled naturally from sea spray to the atmosphere and then via precipitation and dry deposition back to terrestrial and fresh water surfaces and then, via streams and rivers back to the oceans. This has happened for billions of years, so it might be thought that the effects of a bit of extra salt from road salting and a bit of extra chloride from fertilisers would be insignificant. So salt running off from roads to rivers would be an example of sensible use of the water cycle, rather than abuse. Over a decade ago, Smart *et al.* (2000) observed a strong correlation between chloride concentration in water in sub-catchments of the River Dee catchment in Scotland and percentage of arable land use in the sub-catchments, as indicated in Figure 14.23. The R^2 value should be viewed with caution because the data are not normally distributed (too many sub-catchments with no arable land at all), but was strong enough to make them think initially that they were looking at effects of runoff from fertiliser use and distance from the coast, as the arable land was concentrated closer to the coast,

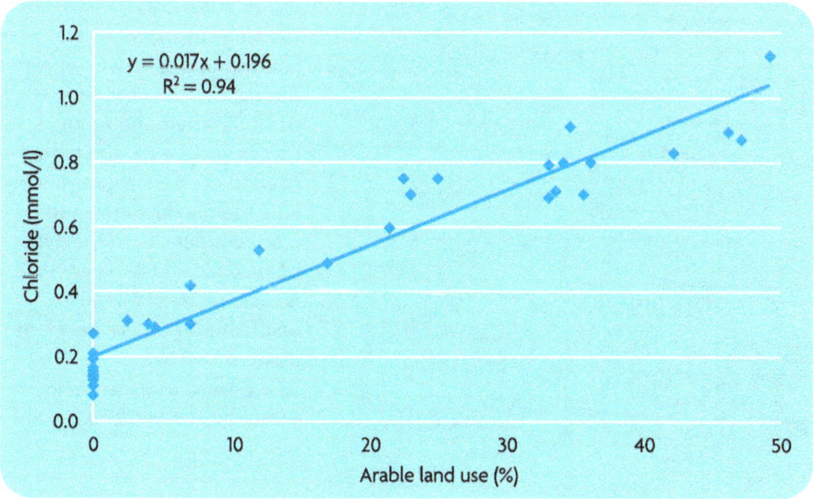

Figure 14.23 Relationship between mean chloride concentration in river water in sub-catchments within the River Dee drainage basin and the percentage of arable land use in the sub-catchments.
Source: Based upon data in Smart *et al.* (2000).

and non-arable land was in uplands inland to the west of Aberdeen.

Smart *et al.* (2000) proceeded to develop an empirical model for predicting chloride concentration in rivers at any point within the Dee catchment, with percentage of arable land use (A), percentage improved grassland (IG) and distance from the coast (d) as input variables. Thirty independent sites were used for calibration, and 24 other sites for validation. The predictive equation was:

$$[Cl^-] = 0.0131A + 0.00512IG - 0.0022d + 0.306$$

Calibration was excellent ($R^2 = 0.97$), as was the validation using data from independent sub-catchments. However, when later they looked at the enhancement of the *total amounts* of chloride in the river system (as opposed to concentrations) compared with sea salt inputs from atmospheric deposition, the enhancement was far too great to be explained by fertiliser chloride alone. It may be concluded that the additional chloride comes from road salting in winter months. Road density in northern Britain tends to be much higher where road networks link farms, villages and towns than in remote upland areas. Therefore the percentage of arable land use in catchments is strongly correlated in such areas with the total road length in the catchments.

At this stage we still have no evidence to suggest that the extra chloride being transported back to the ocean in rivers is an example of abuse of the hydrological cycle rather than sensible use. However, we shall see later (Chapter 19) that road runoff can quite dramatically increase the pH of naturally acidic soils down slope of roads in upland areas. This leads to increases in the mineralisation rate of soil organic matter and associated enhanced mobilisation of nitrate from soils to rivers. This is discussed further in Chapter 19, and illustrated here in Figure 14.24.

It might be argued that if the road salting increases the nitrate concentrations in some rivers then this is an abuse of the hydrological cycle; however, the increase in nitrate-N concentration is quite small, so it may still be concluded that this is simply an example of safe exploitation of rivers as a conduit for minimising salt accumulation. The same cannot be said from the perspective of increasing the salinity of groundwaters however. These may receive enhanced chloride inputs from both road salt runoff and fertiliser use, and the long-term sustainability of the practices therefore requires careful evaluation. Elevated chloride concentrations in potable water supplies are highly undesirable for health reasons.

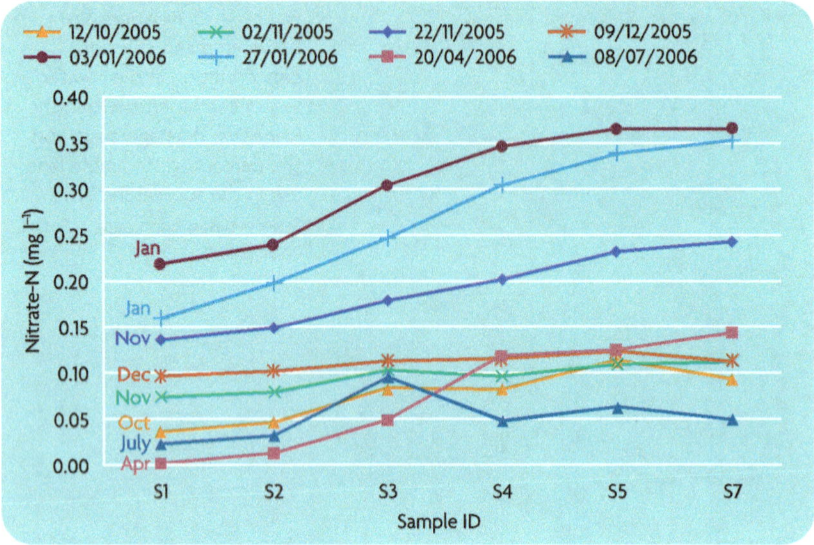

Figure 14.24 Data showing enhanced nitrate concentrations in river water, throughout the year, from point S1, where the A6 road first comes in close proximity to the river, to point S7 approximately 1000 metres down-river; the road runs parallel to the river between S1 and S7. Point S6 is excluded as it was an unaffected tributary between points S5 and S7.
Source: Based upon data from Green and Cresser (2008).

14.7 Is acid rain abuse of the hydrological cycle?

We will see in Chapter 16 that acidification of surface fresh waters and shallow groundwaters are widely recognised consequences of acid deposition. This is clearly an example of abuse for rivers because both the acid flushes that occur during storm events and the acidification of water under base flow conditions may lead to marked changes in aquatic biodiversity within the river. Policy makers have attempted to deal with the issue by using the critical loads approach discussed in Chapter 16, although it has been argued that this is not totally robust conceptually, partly at least because it deals with deposition fluxes of pollutants rather than their concentrations (Cresser, 2007). Concentrations are what biological receptors respond to in the short term, not fluxes. Concentrations better take into account regional differences in the hydrological cycle. Clearly such changes as acid flushes are unacceptable if we regard sustaining biodiversity as an important facet of river systems and one that should be protected. For shallow groundwaters acidification may result in attack on metal water pipes used for abstraction, so this too is a cause for concern. Although these aspects are very important, we will leave further discussion of them until Chapter 16.

14.8 Dealing with human excrement

We mentioned earlier in this chapter that it is important not to add so much organic effluent to rivers that it changes the dissolved oxygen status excessively, by creating anaerobic conditions. It must be accepted though that there is nothing 'natural' about flushing a toilet and using copious volumes of water to transport the waste to a septic tank or a sewage treatment works. How abusive this procedure is depends upon the source of the water used and where it would have ended up anyway if not intercepted. If a reservoir on high ground simply controls the rate of release of water to a river that it would eventually have passed to anyway, the level of abuse could be regarded as very slight. Similarly, localised harvesting of rainwater to flush toilets might be regarded as causing minimal adverse effects.

Perhaps the biggest problem with STWs (Figure 14.25) is the effect of urbanisation, which concentrates very large quantities of organic waste, usually at single, highly localised points. This means that even if the balance of water abstraction and wastewater disposal is acceptable in the context of the hydrological cycle, problems may still occur because of the high organic loading of rivers and the risk of generating anaerobic conditions fatal

Figure 14.25 Part of a wastewater treatment works in Tenerife. In the low lying, coastal areas of the island water is in very short supply, so much of the wastewater is purified and recycled.

to some aquatic organisms. Discharge consents are usually granted to avoid this situation, though under severe storm conditions both organic pollutant load and load of pathogenic organisms may reach undesirable levels. While we have come to live with this temporary abuse of rivers, swimmers and surfers sometimes protest vociferously when concentrations of faecal coliforms exceed critical threshold limits on their favourite beaches.

Discharge of steady streams of effluent into a river at controlled rates relies upon dilution and dispersion to reduce the risk of generation of anaerobic conditions. In some instances there may be advantage in splitting the effluent stream and having two or more well-spaced discharge points, rather than just one. With sufficient spacing and a well aerated river flow, this may result in smaller, lower impact, dissolved oxygen sags at several points rather than a single, potentially much larger and more serious decrease.

POLICY IMPLICATIONS

Use and abuse of facets of the hydrological cycle

- Policy makers need to understand how the hydrological cycle works and the diverse contributions it makes to many ecosystem services.
- In formulating pollution management policies, policy makers need to differentiate between the safe and sustainable *use* of water resources and their *abuse* due to over exploitation of any facet of water use.
- Policy makers need to adopt an integrated approach to catchment management, giving balanced consideration to sustainable use of soils and of surface and groundwater resources.
- Policy makers need to analyse plans for the use of water resources over a range of timescales so that they do not compromise long-term use by implementation of policies that are only viable in the short term.
- Policy makers need to give careful consideration to industrial uses of water, and particularly to limitations on the role of rivers for effluent disposal.
- More careful consideration should be given to the role of water for transport of domestic wastewater, faeces and urine.
- More careful consideration should be given to the possible adverse effects of irrigation water on groundwater resources.
- Nationally and internationally, appropriate use of the many ecosystem services provided by the hydrological cycle will need even more careful consideration in future as a consequence of population growth and climate change.

CHAPTER REVIEW EXERCISES

Exercise 14.1

Select a large river and a small stream close to the area where you live, and collect information about their catchments. For each surface water and for the groundwaters in the catchments, list the roles that the water is fulfilling and critically assess whether the water is being used sensibly and sustainably or is being abused.

Exercise 14.2

For both surface water resources and groundwater resources, prepare lists of chemical, physical and biological measurements that could be made over a range of timescales to assess whether or not a water resource is being used sustainably.

Exercise 14.3

A plant producing palm oil in Malaysia produces a very organic matter-rich treated effluent stream, which they discharge into a nearby river. Why might they decide to discharge the effluent at three separate sites down-river that are approximately one km apart, rather than at a single point?

Exercise 14.4

Since people have become more aware of the finite nature of fossil fuel resources, some have now started to describe water as 'the new oil' and some have predicted that 'water wars' will replace 'oil wars'. Write a critical evaluation of the potential for the global water cycle to provide the water needs of the planet when the population reaches 9 billion people.

REFERENCES

Ansa-Ansare, O.D., Marr, I.L. and Cresser, M.S. (2000) Evaluation of modelled and measured patterns of dissolved oxygen in a freshwater lake as an indicator of the presence of biodegradable organic pollution. *Water Research*, **34**, 1079–1088.

Aydinalp, C., Fitzpatrick, E.A. and Cresser, M.S. (2005) Heavy metal pollution in some soil and water resources of Bursa Province, Turkey, *Communications in Soil Science and Plant Analysis*, **36**, 1691–1716.

Cresser, M.S. (2007) Why critical loads should be based on pollutant effective concentrations, not on deposition fluxes. *Water, Air and Soil Pollution: Focus*, **7**, 407–412.

Cresser, M.S., Begum, S. and Bhatti, A. (2011) Causes of losses seasonal trends in nitrate concentration in some UK rivers. *IWA Specialist Group on Diffuse Pollution Newsletter*, No. 32, February 2011, 20.

Cresser, M., Smart, R.P., Clark, M., Crowe, A., Holden, D., Chapman, P.J. and Edwards, A.C. (2004) Controls on N species leaching in upland moorland catchments. *Water, Air & Soil Pollution, Focus*, **4(6)**, 85–95.

Edwards, A.C., Pugh, K., Wright, G., Sinclair, A.H. and Reaves, G.A. (1990) Nitrate status of two major rivers in N.E. Scotland with respect to land use and fertiliser additions. *Chemistry and Ecology*, **4**, 97–107.

Evans, C.D., Chapman, P.J., Clark, J.M., Monteith, D.T. and Cresser, M.S. (2006) Alternative explanations for rising dissolved organic carbon export from organic soils. *Global Change Biology*, **12**, 2044–2053.

Green, S.M. and Cresser, M.S. (2008) Nitrogen cycle disruption through the application of de-icing salts on upland highways. *Water, Air and Soil Pollution*, **188**, 139–153.

McClean, C.J., Cresser, M.S., Smart, R.P., Aydinalp, C. and Katkat, A.V. Unsustainable irrigation practices in the Bursa Plain, Turkey, *Proceedings of IWA DipCon 2003, 7th International Specialised Conference on Diffuse*

Pollution and Basin Management, University College Dublin, 17th–22nd August, 2003, 14-60–14-65.

Mian, I.A., Riaz, M., Begum, S., Ridealgh, M., McClean, C.J. and Cresser, M.S. (2010) Spatial and long-term temporal trends in nitrate concentrations in the River Derwent, North Yorkshire, and its need for NVZ status. *Science of the Total Environment*, **408**, 702–712.

Mian, I.A., Riaz, M. and Cresser, M.S. (2009) The importance of ammonium mobility in nitrogen-impacted unfertilized grasslands: A critical reassessment. *Environmental Pollution*, **157**, 1287–1293.

Smart, R.P., Cresser, M., Calver, L.J., Chapman, P.J. and Clark, J.M. (2005) A novel modelling approach for spatial and temporal variations in nitrate concentrations in an N-impacted UK small upland river basin. *Environmental Pollution*, **136**, 63–70.

Smart, R., White, C.C., Townend, J. and Cresser, M.S. (2000) A model for predicting chloride concentrations in river water in a relatively unpolluted catchment in the north-east of Scotland. *The Science of the Total Environment*, **265**, 131–141.

White, C.C., Smart, R. and Cresser, M.S. (2000) Spatial and temporal variations in critical loads for rivers in N.E. Scotland: A validation of approaches. *Water Research*, **34**, 1912–1918.

CHAPTER 15

Exploiting the sea for fish

Julie Hawkins and Callum Roberts

Learning outcomes

By the end of this chapter you should:

- Be more aware of how fishing techniques have evolved over historic times.
- Know approximately how much fish is currently caught, but understand why estimates of current fish yields are uncertain.
- Have a better idea of how much fish can be caught sustainably.
- Appreciate why fish stocks are declining, despite implementation of management strategies designed to make them more sustainable.
- Be aware of how models are used to estimate potential sustainable fish stocks.
- Understand the relationship between maximum sustainable yield and maximum economic yield.
- Be aware of the main tools at a manager's disposal for regulating catches and of their strengths and limitations.

15.1 Introduction

Fishing is as old as humanity itself. Deposits left by early humans in Blombos Cave, South Africa, show that people exploited shellfish 140,000 years ago and had learned how to catch sea fish by 100,000 years ago. Bone remains left in Gibraltar sea caves by Neanderthals show that they scavenged or hunted monk seals and dolphins 20,000 years ago. Commercial fishing by specialist fishers dates back over 3000 years in the Mediterranean, as evidenced by images on frescoes, mosaics and vases. In northern Europe, archaeological remains show that commercial sea fishing began in earnest around 1000 years ago as growing demand outstripped falling supplies from fresh waters.

The recent history of sea fishing has been fraught with problems. Fishing power climbed steeply over the last two centuries due to technological innovation and use of larger boats and fleets. For example, sail gave way to steam power in the late nineteenth century, which yielded to diesel engines in the mid-twentieth century. Monofilament line was invented in the 1950s and used to make larger, lighter nets that were harder for fish to see. In the 1960s echo-sounders heralded the onset of a high technology electronics revolution that now enables fishers to see fish and underwater seascapes in unprecedented detail. But for all this invention and innovation, catches are in decline. In this chapter, we examine the underlying principles of fisheries management and investigate why regulation has so far failed, in most cases, to deliver sustainable catches.

15.2 How much fish do we catch and where does it come from?

Since 1950, the Food and Agriculture Organisation of the United Nations (FAO) has collated figures on the size of global fish catches (Figure 15.1). In this chapter, the word 'fish' is used to refer to anything that is caught and eaten, whether finfish, shellfish, jellyfish or other. The FAO estimate that the total catch of wild fish from the sea currently falls in the range of approximately 80 to 90 million tonnes per year. Catch levels fluctuate from year to year, particularly because of the large contribution of Peruvian anchoveta to the total (roughly 1 to 15 per cent of total legal landings). This species lives in an area of upwelling currents in the Pacific off South America and the population rises and falls in phase with the strength of upwelling, which itself is controlled by the El Niño Southern Oscillation climate cycle.

The exponential increase in fish landings that had lasted since records began in the nineteenth century came to an end in the last decade of the twentieth century. When anchoveta catches are

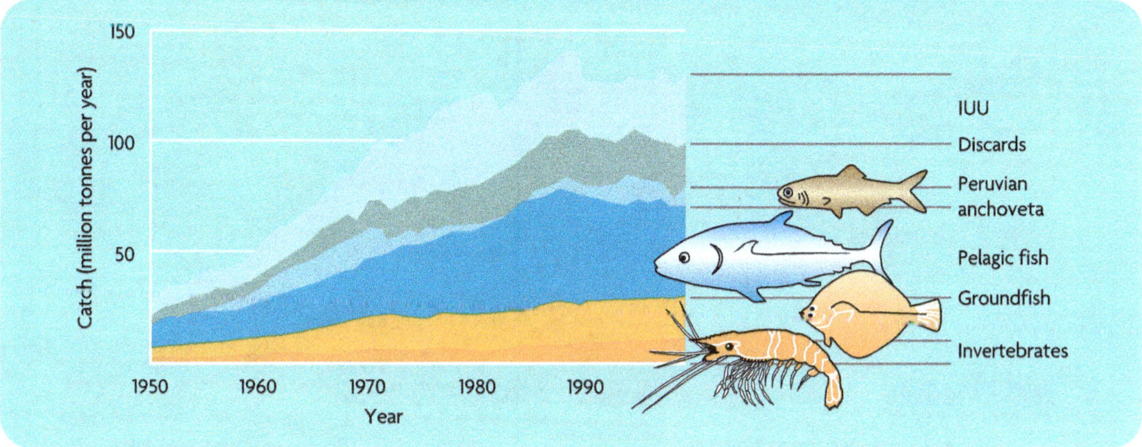

Figure 15.1 Trends in world fish production from wild stocks since 1950. IUU = illegal, unregulated and unreported catch (estimated); discards refers to fish caught but not landed (estimated).
Source: Pauly *et al.*, 2002.

excluded, and the figures are corrected for over-reporting by China (where officials were expected to meet production targets whether or not the fish were there), overall catches began a slow decline from 1988.

The FAO figures that are quoted above exclude subsistence catches in developing countries because they are not traded. Nonetheless small-scale fisheries provide important food sources to poor people. For example, coral reef fisheries probably exceed 7 per cent of the total reported world catch, although this is only an informed guess based on the area of reefs worldwide and typical fish yields from this habitat. Another estimate puts the annual global subsistence catch at around 20 million metric tonnes. A second reason why FAO figures do not properly reflect how much fish is really caught is that they exclude fish that are thrown away as bycatch (non-target species) and discards (target species that are either undersized, over-quota or caught by a vessel which does not have a quota). Estimates vary as to the amount of fish thrown away, but it is probably at least a quarter of the global landings total. A third problem is that around ten million tonnes of fish, at least, are caught illegally worldwide.

About 95 per cent of the world's fish catch is taken from shallow continental shelves (less than ~200 m deep) and upwelling zones. This is not simply because these areas are closest to coasts and ports, although of course they are. In general there is only enough light for photosynthesis in the sea to depths of about fifty to a hundred metres. Once nutrients sink deeper they are lost to ocean primary productivity unless returned to the surface from the bottom by storms and turbulence or upwelling currents (hence the high productivity of Peruvian anchoveta). Continental shelves and upwellings have plentiful supplies of nutrients (Figure 15.2), but over the vast

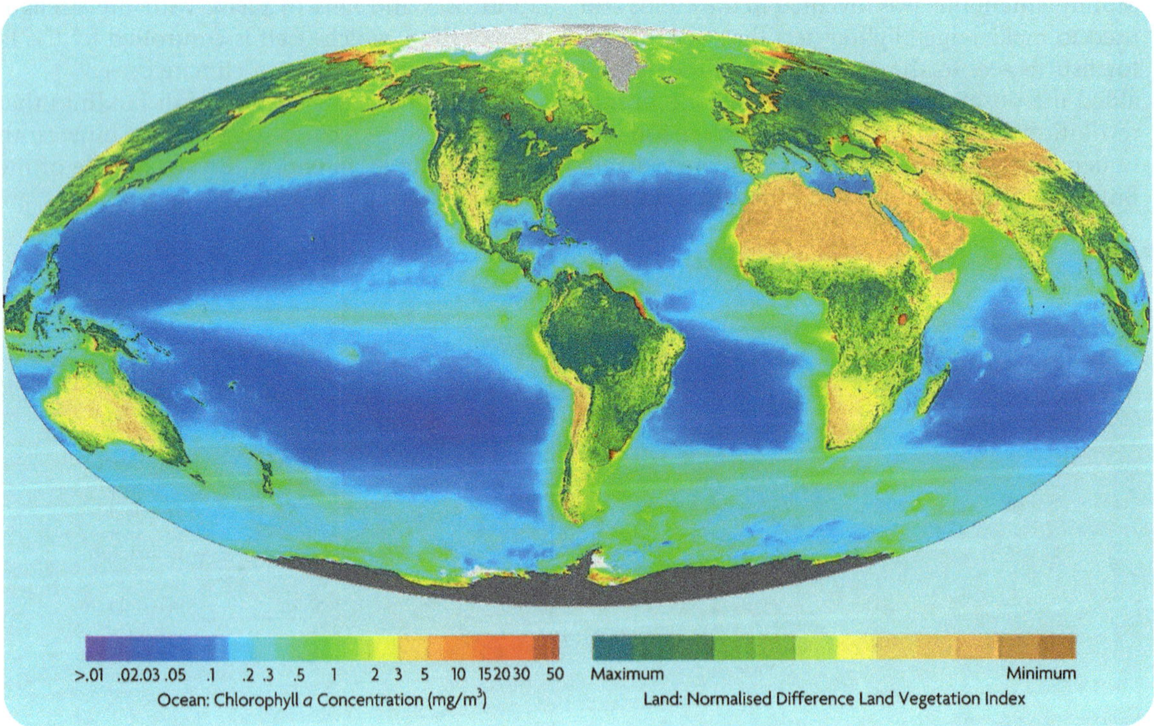

Figure 15.2 Global distribution of primary production in the sea and on land.
Source: http://oceancolor.gsfc.nasa.gov/SeaWiFS/TEACHERS/sanctuary_7.html. GeoEye Satellite image.

reaches of the deep ocean, marine life is limited by lack of nutrients, making the sea overall only about one sixth as productive per unit area as the land.

15.3 How much fish can we catch?

One of the aims of fisheries science is to determine how many fish can be taken from the sea without driving a population to collapse. The underlying principles of fisheries management are founded on mathematical models that simplify reality. At their most straightforward, they consider only one species at a time and ignore all extrinsic factors. Consider first what is known as the logistic growth curve. This shows the growth of a population over time from two individuals (for a sexually reproducing species) to a limit imposed by the availability of resources, known as the carrying capacity, or K (Figure 15.3a). The logistic growth curve is sigmoidal, i.e. s-shaped. In the first phase, growth is exponential because population size is below the environment's carrying capacity and resources such as food or space are not limiting. As the population expands it becomes limited by resource availability so density-dependent factors (competition, mortality) slow growth, eventually to zero, as the carrying capacity is reached. A life process is known as density-dependent when it varies in relation to population density.

The population growth rate at any point on the logistic curve is equivalent to the slope of a line that is a tangent to the curve. Expressing a logistic growth curve in terms of the population growth rate versus population size reveals that growth rates initially increase as the population expands, then peak and finally return to zero when the carrying capacity is reached (Figure 15.3b).

According to fisheries theory, unexploited fish populations (often called *stocks* by fisheries scientists) are at their carrying capacity, which means they are: limited by resources, population growth has ceased, and fish experience high levels of mortality, especially the young. Of course, populations can be limited in size by predation, or environmental shocks and disturbances as well as limited resources. But let's stay with this cartoon reality for now and see where its logic takes us.

One of the cornerstones of fisheries science is the surplus production model. This is based on the idea that, according to logistic growth, if you reduce a population in size below the carrying capacity, mortality will fall and growth rate will increase. By fishing down a population you can take advantage of this effect and benefit from being able to exploit the *surplus yield* (Figure 15.3c). In essence, as fishing rate increases, population size decreases, which boosts population growth.

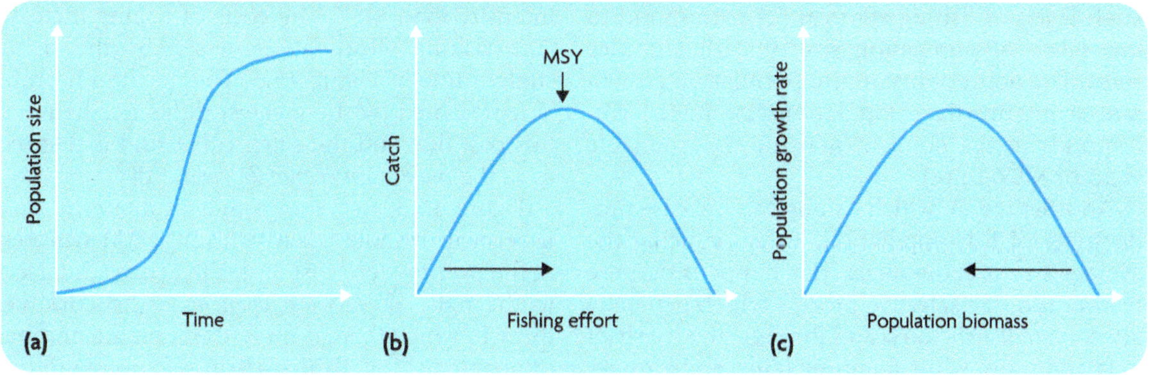

Figure 15.3 (a) Logistic growth curve for a population. (b) Logistic growth expressed in terms of population growth rate. (c) Surplus yield curve. Figures (b) and (c) can be considered mirror images of one another in terms of fish population size. As fishing increases, see arrow on (b), population size of fish decreases, see arrow on (c).

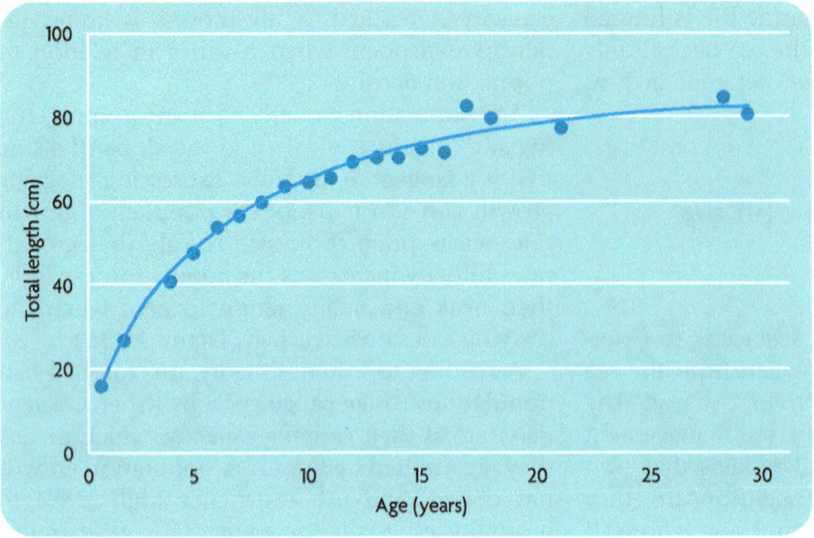

Figure 15.4 Growth in body size of the Caribbean Nassau grouper (*Epinephelus striatus*). Source: Redrawn from Sadovy and Eklund (1999). [Sadovy, Y. and A.M. Eklund, 1999. Synopsis of biological data on the Nassau grouper, Epinephelus striatus (Bloch, 1792), and the jewfish, *E. Itajara* (Lichtenstein, 1822). NOAA Technical Report NMFS 147.]

The surplus yield curve in Figure 15.3(c) shows that as fishing intensity rises, catches initially increase, then peak, then fall away. At low levels of fishing effort, before the peak in catch, the fishery can be considered 'underexploited' (a term that comes from the FAO and seems to imply that fish are only there for our use!). Around the optimum of maximum catch, or *maximum sustainable yield* (MSY), the fishery is 'fully exploited' with no scope for further increase in landings. At higher fishing intensities, catches drop and the species is considered 'overexploited'.

In terms of fish population size, the horizontal (x) axes of the surplus yield curve (Figure 15.3c) and that of Figure 15.3(b) are mirror images. High levels of fishing effort in (c) correspond to low levels of population size on (b), and vice versa. The point of maximum population growth rate at intermediate population size in (b) corresponds to the point of maximum sustainable yield or MSY, in (c).

Another way in which fishing can increase productivity of fish populations is by reducing the average size of individuals. This happens because fishing is a size-selective process which generally allows small fish to escape through the mesh of nets and traps, while retaining larger individuals. Hooks are also size-selective since small fish can't swallow big hooks. As time goes on, selective removal of large bodied individuals decreases average size and age of animals in targeted populations. Growth rates are higher in young fish (which expend energy in body growth rather than reproduction (Figure 15.4)). This boosts the rate of *biomass* increase of the exploited population (biomass is the total body weight present).

We mentioned above that maximum sustainable yield represents an optimum level of fishing, and indeed attaining MSY has often been the target for fishery managers. However, it is perhaps better seen as a 'social optimum' rather than an economic one because lower levels of fishing effort could deliver higher profits. If the fishery was owned by a company with a monopoly, that company would probably want to fish at a lower intensity than MSY. The costs of fishing increase with fishing effort, as expressed by the rising straight line in Figure 15.5. The maximum profit is represented by the greatest difference between this cost line and the catch curve, and is referred to as the *maximum economic yield*, MEY.

Figure 15.5 also shows that profitability of a fishery will dwindle to zero as fishing effort increases, at the point where cost and catch curves intersect to the right of MSY. In *open access* fisheries, i.e. those in which there is no restraint on the number of people fishing, fishing effort is expected to rise as long as there is profit to be made. The point of zero profitability is therefore also the point at which fishing effort should stabilise under open

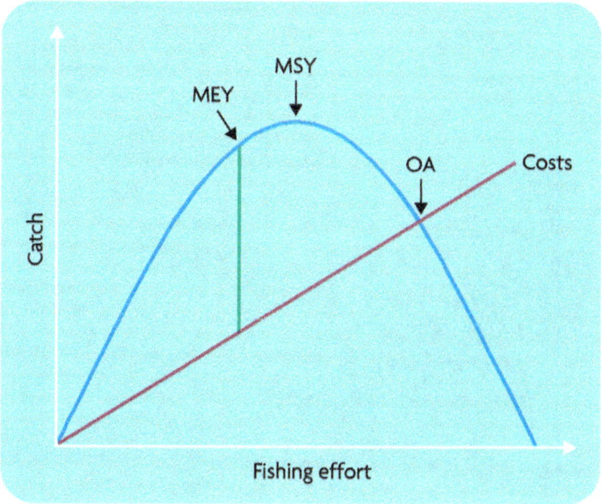

Figure 15.5 The relationship between maximum sustainable yield (MSY) and maximum economic yield (MEY). The vertical green line shows the point of maximum profitability of the fishery (the largest difference between the costs of fishing and the yields obtained). Note that MEY lies to the left of MSY, indicating that profitability is greatest at lower fishing intensities than typical management targets. The open access equilibrium of zero profitability is shown by the point marked OA.

access. As this also corresponds to an overfished state, open access is usually considered undesirable. Hence some form of regulation is needed to restrain fishing.

15.4 Effects of fishing on reproduction by fish populations

As explained above, in theory exploiting to the right level of fishing intensity should increase the productivity of a fish population. This is achieved by reducing mortality rates of young fish and by maintaining populations with a young age structure to maximise individual growth rates. However, in terms of maintaining fish stocks for the future, there is one major problem with this approach: fishing can have a large impact on egg production.

Fish tend to grow asymptotically throughout life. In the early stages of life they put most of their energy into body growth because bigger bodied fish have fewer enemies and are better competitors. Later in life they put more of their energy into reproduction. Egg production scales as roughly a cubic power of body length (Figure 15.6).

$$\text{Egg production} = \text{species-specific constant} \times \text{body length}^3$$

In other words, big fish produce many times more eggs than small ones. For example, it has been estimated that it takes 212 red snappers weighing 1 kg to produce as many eggs as one

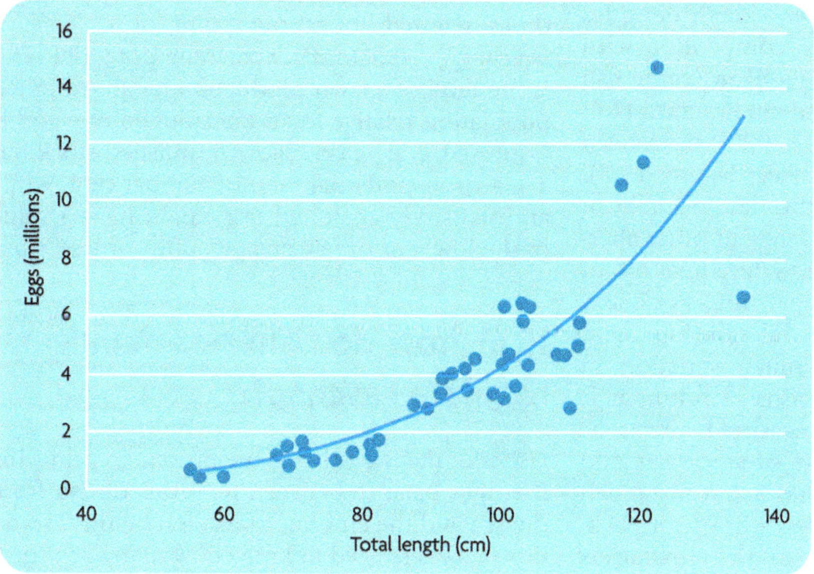

Figure 15.6 Egg output versus body length for northern Arctic cod (*Gadus morhua*).

Source: Redrawn from Kesbu *et al.*, 1998. [Kjesbu, O.S., P.R. Witthames, P. Solemdal, M. Greer Walker (1998). Temporal variations in the fecundity of Arcto-Norwegian cod (*Gadus morhua*) in response to natural changes in food and temperature. *Journal of Sea Research* 40(3–4): 303–321.]

10 kg individual, or 93 groupers of 0.5 kg to produce the same as a grouper of 10 kg (Roberts and Hawkins, 2000). Hence managing a fish stock for maximum productivity will significantly reduce the total egg production by a population.

15.5 The relationship between spawning stock size and recruitment

An elusive goal of fisheries science is to determine the relationship between the size of the reproductively active part of a fish population (called the *spawning stock*) and the number of offspring that survive to the size/age of first capture (*recruitment*). This is made difficult by the very high levels of variability in recruitment from year to year that marine animals experience. The size of a *year class*, or *cohort* as it is sometimes called, can vary by several orders of magnitude from year to year as a result of environmental factors that affect the survival of young fish. Such variability makes it difficult to determine any underlying relationship between stock size and recruitment.

A null model predicts that recruitment would be linearly dependent on the size of the stock. The more spawning fish there are, the more offspring they produce. Other models that incorporate density-dependent effects suggest that survival of young fish is lower at high densities than at low densities, due to competition with adults for resources, or predation on young fish by them. The density-dependent Beverton-Holt and Ricker models (Roberts, 2000) have been widely accepted and used (see Figure 15.7), despite the fact that very little data actually fit them any better than a wide range of other alternatives. The former predicts a levelling off of recruitment above a certain spawning stock size, while the latter has a humped relationship.

One reason there is so much variation in graphs of spawning stock versus recruitment is because neither can be measured precisely. Combined errors in estimates of stock size and recruitment may obscure what are really much closer relationships. Nonetheless, one use of stock-recruitment relationships is to help managers determine the minimum size of spawning populations necessary to avoid *recruitment overfishing*. Recruitment overfishing happens when exploitation is so intense that it impairs the ability of a population to replace itself. On the stock-recruitment graph, one point can be measured with certainty: if there are no fish there can be no recruitment. As spawning stocks increase above zero, recruitment rises, either to an asymptote or to a peak. It makes good management sense to avoid fishing a stock to below the level that delivers maximal recruitment. This level is sometimes referred to as the point of *replacement spawning potential ratio*. Spawning potential ratio is the level of spawn production by an exploited population relative to an unexploited one and is expressed as a percentage. An unfished stock has a spawning potential ratio of 100 per cent, while an intensively exploited one might have a value in the low tens of percentage points or less.

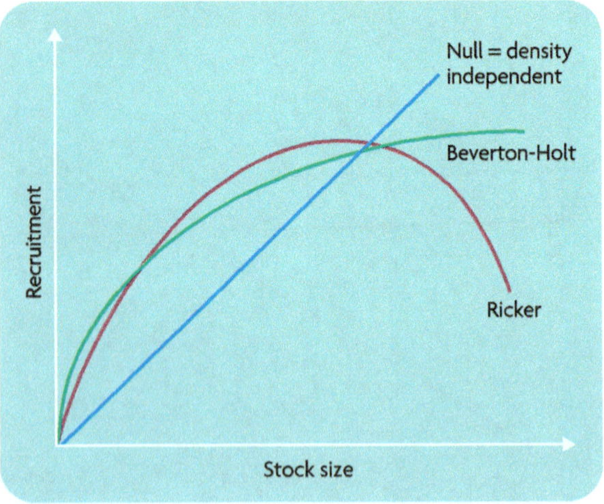

Figure 15.7 Three possible relationships between the size of a fish spawning stock and recruitment to the stock.

15.6 How do fisheries managers regulate catches?

Fisheries management is the process of achieving the right balance between removals of fish from the sea and the amount of fish left behind. To do this managers need to know how many fish there are in the sea (i.e. spawning stock size), and the productivity of the stock. There are many methods by which to estimate stock size (these are described in Box 15.1). Managers also need to decide what their goal is, whether it be management for MSY, avoidance of recruitment overfishing or some other goal. To estimate the appropriate removal rate for a given goal, managers generally use single-species population models. These models require information on life

BOX 15.1 How are fish counted?

Fisheries managers need to know how many fish there are in the sea to determine how many can safely be removed. Assessing the state of wild fish stocks is not as straightforward as counting sheep on a hillside, or birds on an estuary. For a start, you can't see them directly. Scientists resort to a variety of methods to estimate fish abundance and most involve catching fish using hook, net, trap, trawl, dredge or other suitable device. Fish caught are then identified, weighed, measured and counted. To arrive at a *stock assessment*, as these estimates are called, you must also estimate the volume that is sampled by the device to derive a measure of fish density. Average densities across many samples can then be multiplied by the volume of suitable habitat within the fishing grounds to determine how many fish are present. Usually stock assessments concentrate only on the fraction of fish that are catchable (the rest remain unsampled). The figures are also often expressed as *spawning stock biomass*, that is, the weight of mature fish present.

All sampling methods have limitations. Perhaps the most controversial, when it comes to fisheries assessment, is the fact that fishery scientists often use old-fashioned methods to sample fish. At first glance, their penchant for traditionalism seems peculiar. Why not adopt the latest technology to do the job in the most efficient way? The reason for conservatism in sampling is to enable comparability between old samples and new. Every time you change your method of data collection, you have to cross-calibrate old and new methods to benchmark new methods against old. Often scientists continue to use old methods to avoid this inconvenience. The problem is that fishers complain that the use of outmoded fishing methods by scientists underestimates the real numbers of fish in the sea and they challenge the figures produced (often finding a sympathetic ear from a fishery minister who doesn't appreciate the nuances of sampling any better).

A second reason that scientific surveys are criticised by fishers is that scientists are fishing in the wrong places. If only they would go to where the fish are, they would find that stocks are far healthier than the surveys indicate, say the fishers. Scientists generally sample the sea using a stratified random design. This means that they visit randomly selected sampling points each year, stratified according to the depth or habitat present. In reality, this sampling method gives a better picture of the state of fish stocks than is seen by fishers, who 'sample' in a very non-random way. Fishers are interested in beating 'average returns' and seek out concentrations of fish, so their catch rates tend to be higher than those obtained by scientists. But if you were to average densities from sites of fish concentration across the whole fishing grounds, you would end up with a misleading overestimate of the amount of fish present.

Fishery-independent surveys are expensive and provide only a part of the stock assessment picture built up by fishery scientists. Often, data from the fishery itself are used, because it is cheaper and there are far more vessels 'sampling' the sea than research boats. These data take the form of measures of *catch per unit of fishing effort* (CPUE), which is an index of the catchability, and by implication abundance, of fish in the sea. The problem, as we have just mentioned, is that fishers do not 'sample' stocks at random so their data may overestimate the real abundance of fish. For some species, especially those that form schools, catchability remains high even as abundance falls because the remaining fish aggregate together. This was a major contributing factor in the collapse of Canadian cod stocks in the early 1990s (Hutchings and Myers, 1994). Data from the fishing fleet suggested there were far more cod in the sea than scientific surveys suggested. Unfortunately, managers and politicians believed the fishery data, with disastrous consequences. The cod fishery has been closed since 1992.

CPUE measures derived from commercial fisheries suffer from another source of bias that complicates estimation of change in the size of fish populations over time, technological creep (described in section 15.6.3). Fishing power tends to increase with time. A trend of steady CPUE over time therefore probably conceals a decline in stocks. Time series of catch-per-unit-effort can only be properly interpreted after technological creep is factored out.

history characteristics of the species in question, such as growth rate, natural mortality, age at maturity and the relationship between spawning stock size and recruitment. They are used to estimate the target stock size that corresponds to the management goal set, and the appropriate amount of catch that can be safely removed in a given year. So far, so quantitative. However, delivering these levels of catch while maintaining enough fish in the sea is where fisheries management becomes something of an art – a blend of science, experience, intuition and guesswork. In the following sections we describe the main tools at a manager's disposal for regulating catches and discuss their strengths and limitations. There are five main categories of fisheries' regulation: limits on (1) landings, (2) gear, (3) fishing effort, (4) when you can fish, and (5) where you can fish.

15.6.1 Limits on landings

Imposed limits on landings are perhaps the most familiar form of regulation. They are the tool of choice within the European Union, for example, where they take the form of a Total Allowable Catch (TAC) that is set for each of the main species annually. This TAC is set separately for different parts of a species' range (e.g. the Irish versus North Seas), and is sub-divided into quotas that are given to different EU member states to catch in these areas. Total allowable catches have been much criticised, largely because they only limit what is landed, not what is caught (and usually either dies when returned, or is already dead). Perhaps their only advantage is bureaucratic convenience. They are easy to understand and can be monitored and measured at ports rather than requiring surveillance at sea.

Total Allowable Catches can work well in single-species fisheries where fish of one species can be caught in isolation from others. For example, herring form mono-specific shoals that can be targeted separately from other species by purse-seine boats (purse-seines are like large bags that can be set around fish schools). But most fisheries are multi-species, which means that gears catch more than one kind of fish at a time, often many. Fishers using traps on coral reefs may catch tens or even hundreds of different species for example; bottom trawlers in Europe and North America can catch dozens of different commercially valuable fish. Unfortunately, Total Allowable Catches ignore this complexity and are set species by species. Worse still, boats in multi-species fisheries are often only licensed to land just one or a few of the species they catch. Species for which they have no quota, or that are over quota, have to go over the side, usually dead or dying. Fishers also dump *high grade* catches, discarding good fish of less valuable sizes or species to fill their holds with higher value fish. Limits on landings can therefore lead to a huge waste of fish that are discarded at sea. In some cases, such as for North Atlantic redfish and haddock, more fish are caught and killed than are landed (Pauly and Maclean, 2003).

There are other problems with Total Allowable Catches. For example, to set them at the right level requires high levels of information on the size and composition of fish stocks. Total Allowable Catches also fail to protect the population or genetic structure of stocks, and they do not address habitat damage caused by fishing, or impacts on other wildlife.

There is another way of allocating quotas that is much in favour at the time of writing: *individual fishing quotas* (IFQs). These quotas are designed to give fishers a guaranteed share of a fish stock. They are intended to reduce problems of fishery regulation by reducing the incentive for people to catch more than is sustainable. Economists often argue that giving people ownership of resources, which are otherwise 'common property', will help prevent individuals from taking more than is sustainable. Without such property rights individuals stand to obtain private gain at a cost to the rest of society by acting selfishly. By contrast, if everybody exercises restraint by only taking a sustainable share of a resource, then everyone will be better off. The payoff to fishers from IFQs is that the more healthy the fish stock, the more can be taken sustainably, so each share will be worth more. Sometimes these catch shares can be traded, so fishers have the option to increase or decrease their stakes in the industry. In reality IFQs provide only a very partial solution to the problem of overfishing, because while IFQs give fishers a share in an annual total stock, fishery scientists still have to determine how large the Total Allowable Catch will be. Hence IFQs operate on the basis of single-species management with all its uncertainty in determining how much of a stock it is safe to catch. From a conservation perspective, one possible benefit of IFQs is that they could be bought by environmental organisations as a means of reducing overall fishing effort.

Other kinds of limits on landings are commonly applied in recreational fisheries. Typically, they include bag limits (how many fish or shellfish you can take in a day), or size limits. Where a size limit is specified, it is usually a minimum size limit to help prevent *growth overfishing* (i.e. the premature removal of fish from the water), and to protect spawners. However, since large fish can produce many times more offspring than small (Figure 15.6), sometimes a maximum size limit is set where removal of larger animals is prohibited. A combination of lower and upper size limits is known as a 'slot limit'.

These kinds of regulations can be effective in giving some protection to fish stocks, but like TACs they also have high information requirements. To set a bag limit, for example, you need to know how many fish there are in the sea, how many people catch them and how often they go fishing. Bag limits are also prone to problems of high grading and discarding, and since not all fish are landed through ports, they require quite intensive enforcement to work.

While not strictly a limit on landings, quotas on bycatch can be very effective. These are usually used to protect highly vulnerable bycatch species such as threatened marine mammals or birds, but may also be applied to commercially important species, such that if the bycatch quota is exceeded, the fishery is shut down. Examples include the longline fishery for swordfish in the North Pacific where there is a quota on the number of loggerhead and leatherback turtles that may be caught (46 and 17, respectively, in 2010), and the southern arrow squid fishery in New Zealand which limits the number of Hooker's sea lions that may be killed. However, for bycatch quotas to work, all fishing boats would require observers, whose job would be to record and report the bycatch. These people are required because few fishers can be trusted to self-report on things which lead to fishery closure.

15.6.2 Limits on gear

Limits on the kind of fishing gear permitted and its design have a very long history in fisheries management. Bottom trawling – the dragging of a net across the seabed – for example, has been banned at various times and places since the method was first mentioned in writing in the fourteenth century. These bans were introduced for reasons which included damage to fish habitats, interference with static gears like nets, traps or hook and line, and ability to catch too many juvenile fish.

Catching too many young fish is a problem that has vexed fishers for centuries. In a nineteenth century British Government enquiry, for instance, one fisherman complained about bottom trawl catches, saying, 'I have seen hundreds of baskets of skate [landed] not bigger than the loof [palm] of your hand. I would say there would not have been fewer than 150 skate in each basket that would not have been fit for any human food at all' (quoted in Roberts, 2007). The simple answer to this problem is to increase the mesh size on nets. Bigger holes let more small fish escape. Ideally, managers set the mesh to a size that allows fish to reach maturity and reproduce before they can be caught. This works best in single-species fisheries, but creates problems for fisheries in which more than one species is caught.

Fish come in many different shapes and sizes. They also encompass a broad spectrum of life histories. Some fish possess a package of characteristics that make them resilient to high fishing intensities. They include fast growth, early reproduction, small body size and short lifespan, among others. At the other end of the spectrum are fish that are more easily overexploited. They typically grow slowly, mature late, are large bodied and long-lived. In more familiar terms, resilient species live fast and die young. They are the rats and mice of the sea. By contrast, vulnerable species are the tigers and elephants of the sea, and their populations are easily depleted by fishing.

The manager's dilemma in multi-species fisheries is where to set the mesh size of nets or traps. Should they set it large enough to allow the biggest, most vulnerable species to reach maturity and reproduce? If they do, they won't catch many of the smaller species present and will lose out on some of the potential fish production. Alternatively, if they set meshes small enough to catch the smaller-bodied, more resilient species present, they risk overfishing the larger species, even to the point of their complete disappearance. Mesh sizes are always a compromise in multi-species fisheries. In practice, the history of fishing shows that we have generally set mesh sizes too small for the most vulnerable species. For example, bottom trawls would need to have enormous meshes to let mature bottom-living skates and angel sharks through. This is why trawling has eliminated several large species of skates and the angel shark from huge areas of their former ranges in the North Atlantic (Dulvy *et al.*, 2003).

Fishing gears can also be modified to help deal with some of the problems of unwanted bycatch. Shrimps and prawns have small bodies and so small meshed trawl nets are used to capture them. Not surprisingly, they catch enormous quantities of bycatch, typically 5 to 15 times the weight of target catch. When first emptied on deck, a prawn trawl is a study in carnage. Most striking is the huge number of juvenile fish, many from species that would be commercially valuable if they had been allowed to live longer. Larger skates and bottom sharks flap and writhe, while the occasional drowned turtle completes the scene of pathos. One remedy to fish bycatch is to introduce just ahead of the bag end of the trawl, a panel of large, square, stiffened mesh that won't close up when under tow. At this point back pressure from water leads fish to this 'exclusion panel' where those who are small enough to pass through the mesh are able to swim out (Figure 15.8). Turtle Excluder Devices can also be fitted for use in tropical waters. These pop turtles out through a flap in the net when they hit an angled grid that blocks their passage to the back of the trawl.

Most kinds of fishing gears can be modified in some way to improve their selectivity or reduce damage done to the environment, such as occurs when seabed habitats are hit by bottom trawls. But there are limits to this. In most fisheries, bycatch can never be completely eliminated and there will always be some habitat damage from gears set or towed across the seabed. Indeed, some gears are considered to be so damaging that they have been banned in most countries. Dynamite fishing has been popular for hundreds of years, and gained a boost in Asia and the Pacific from the ready availability of abandoned munitions. Likewise, poisons such as rotenone (a plant-based fish toxin), cyanide and bleach are frequently used on coral reefs. Most countries ban fishing with poisons and explosives. Bermuda and Florida have banned unselective fish traps

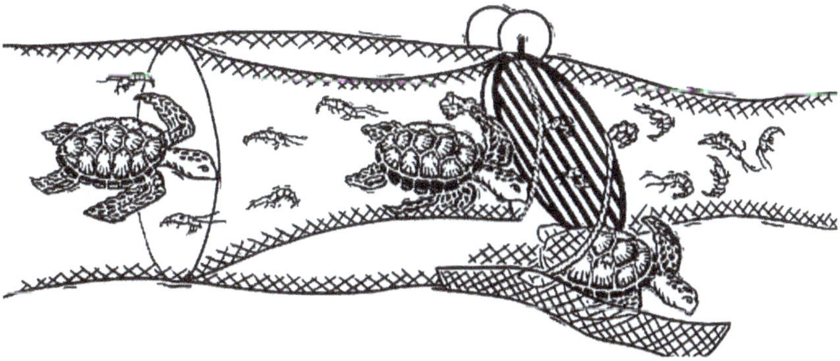

Figure 15.8 Turtle excluder device fitted to a shrimp trawl. Turtles that enter the mouth of the net are stopped from reaching the bag by a grid that diverts them out through a flap in the net. Shrimps pass through the grid.

Source: Image from http://seawifs.gsfc.nasa.gov/OCEAN_PLANET/IMAGES/I-71.gif. GeoEye Satellite image.

from their coral reefs (Butler *et al.*, 1993). The United Nations has banned the use of drift nets (dubbed 'walls of death') greater than three miles long from the high seas.

15.6.3 Limits on fishing effort

Limits on fishing effort represent a marked improvement over limits on landings because they prevent fish from being killed or habitats from being damaged in the first place. There can be no bycatch or other harm from fishing that does not take place. There are many ways to limit fishing effort. You can restrict the amount of fishing gear a person is allowed to set, such as the number of traps, the length of longlines (long fishing lines studded with thousands of hooks) or the size of nets, for example. You can also limit the time people spend fishing. In the European Union, for instance, some fisheries have limits placed on the number of days people can spend at sea. Another way to restrict fishing effort is to issue only a limited number of licences to fish. You can also reduce capacity by removing licences as fishers retire. This process is known as '*grandfathering out*'.

Overcapacity – an excess of fishing power in relation to the amount of fish available – is a widespread problem in world fisheries. In open access fisheries, as Figure 15.5 showed, more and more fishers tend to enter a fishery until there is no profit left to be made. This point lies to the right of the maximum sustainable yield, well into the zone of overfishing. In many fisheries, fleets were developed over long periods with little regulation and when management was eventually introduced, there were too many fishers. Hence, even regulated fisheries may have to deal with the legacy of an excess of fishing capacity. However, open access or lack of regulation is not the only reason for excess capacity: *technological creep* is a major cause too.

Fishers constantly try to improve their catching power by adopting improved fishing gears and adapting the way they fish. In part this is to compensate for declines in fish stocks, in part to improve their own profitability. Technological creep, as it is called, has in many fisheries increased catching power by approximately 2 to 5 per cent per year. Over time, fishing capacity builds up and managers must make periodic purges to get rid of excess capacity. One method popular in Europe is to decommission vessels. In such a scheme, governments buy out vessels to reduce fleet size. However, sometimes these schemes do not have the effects intended. If you offer a group of fishers money for their boats, those who are not very good at fishing or have outdated vessels will be the first to take the money. In addition, there has been much consolidation in modern fleets, so companies often own multiple vessels. Hence they may simply retire their oldest boats and use the revenue to upgrade the remainder with the latest gear. The result is a disappointment for government officials who had hoped decommissioning would solve the problem of excess fishing power.

15.6.4 Limits on when you can fish

Measures to reduce fishing effort can also lead to restrictions on when you can fish. For example, the Canadian halibut fishery became more and more overfished over the course of the twentieth century leading to increasingly stringent limits on the amount of time available to fishers to catch these species. By the time that the fishery switched to limitation by licensing, the fishing season had been reduced to just six days. Such short fishing seasons encourage *derby fisheries* in which large numbers of fishers compete for the best fishing spots. They are deeply unpopular with fishers because of the dangers involved, especially if they are forced to go to sea in bad weather. They don't make economic sense either, since the prices fetched for fish are depressed by glutting the market.

Fisheries managers have long used a second kind of temporal restriction to limit fishing mortality, namely *closed seasons*. These are often introduced to protect animals when they are spawning; for example seasonal closures are regularly used in crab and lobster fisheries.

15.6.5 Limits on where you can fish

Temporal restrictions are also sometimes combined with spatial protection. Many exploited species use different places at different times in their lifecycles. If, for example, a species is known to aggregate to spawn at a predictable place and time then fishers can target such sites to obtain enormous catches. Such practice can cause an immense threat to stocks. For example, the Nassau grouper, a chunky and tasty fish once common on Caribbean coral reefs, aggregates to spawn at traditional sites for a few weeks each year. When fishers in the US Virgin Islands discovered a spawning site in the 1970s, they reduced it from 50,000 fish to 15 individuals (yes, 15!) in just five years (Olsen and LaPlace, 1978). The aggregation has never recovered. In other parts of the Caribbean, spawning aggregations are protected seasonally to safeguard what remains of spawning populations of vulnerable fish like groupers and snappers.

Spatial protection – limits on where you can fish – has been around for almost as long as there has been commercial fishing. In the nineteenth century, for example, Scotland closed certain bays to bottom trawling to protect juvenile fish which used the bays as nursery grounds. In the southern North Sea, a large area of nursery grounds for plaice is protected from large trawl boats in what is known as the 'plaice box'. In the Bering Sea, the Red King Crab Savings Area (a nice name) protects juvenile crabs (http://www.fakr.noaa.gov/ref/00fig11.pdf).

Spatial protection can also be combined with gear restriction by limiting the use of particular gears to certain areas. This approach is usually taken to separate conflicting gears, such as to keep static gears like gillnets and traps away from mobile gears like trawls and dredges. More recently, gear closures have been used to help rebuild depleted stocks. On Georges Bank, which brackets the south side of the Gulf of Maine off America's east coast, several large areas have been closed to bottom trawls and scallop dredging since the mid-1990s, with spectacular success. These closures cover some 20,000 km^2 of habitat and have helped recover the severely overfished scallop and haddock stocks, revitalising these fisheries (Murawski *et al.*, 2000).

Experiments with protecting areas of the sea from fishing altogether suggest that such *marine reserves* can play an important role in achieving sustainable fisheries. The idea behind reserves is very simple: if fish are protected from fishing they live longer, grow larger and produce an exponentially increasing number of eggs. A plethora of studies have convincingly demonstrated that creation of reserves allows the rapid biomass build up of fish spawning stock. Stocks of exploited species frequently build up by two or three, even five or ten fold, within a decade of protection. For example, total biomass of commercially valuable species increased five times within a network of small protected areas on coral reefs of St Lucia (Figure 15.9).

The effects of protection on egg production in fish and other marine organisms such as lobsters can be calculated by using figures on differences in biomass and body size between stocks in exploited and unexploited areas. For example, the snapper, *Pagrus auratus*, produced 18 times more eggs in New Zealand's Leigh Marine Reserve compared to unprotected populations nearby (Willis *et al.*, 2003), and a tasty Chilean limpet produced 1000 times more eggs in protected compared to

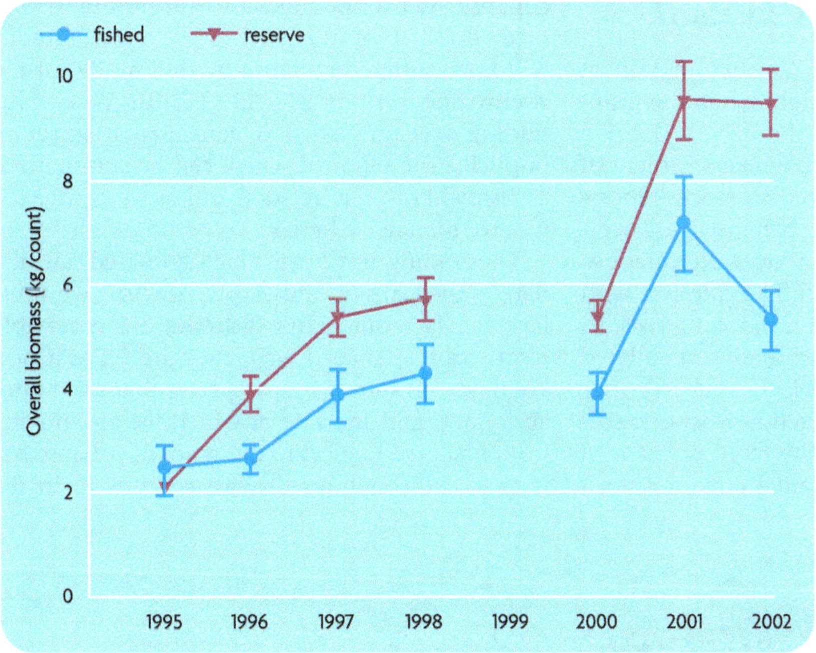

Figure 15.9 Increase in biomass of commercially important reef fish species in reserve and fished zones of the Soufriere Marine Management Area in St Lucia. Reserve zones were protected from all fishing in 1995, after the first survey had been completed. The points show means ± standard error.

Source: Hawkins *et al.* (2006). (Hawkins, J.P., C.M. Roberts, C. Dytham, C. Schelten and M. Nugues. (2006) Effects of habitat and sedimentation characteristics on performance of marine reserves in St Lucia. *Biological Conservation* 127(4): 487–499).

unprotected areas (Manriquez and Castilla, 2001). These spectacular real-life increases in reproductive potential derive from increased abundance of larger, older individuals in the population. Recall from section 15.4 that big fish are the engines of reproduction in fish populations, but they are the ones that are preferentially removed by fishing. For vulnerable species, marine reserves provide a critical means of recovering large-bodied animals and extended population age structures. This effect is thought to be particularly important to increasing the resilience of populations to environmental fluctuations and shocks, preventing a collapse in recruitments when times get tough.

As well as the transfer of eggs and larvae from reserves to fishing grounds via ocean currents, marine reserves may also export a *spillover* of juveniles and adults to surrounding fishing grounds. As stocks build up in reserves, so conditions become more crowded and there is expected to be density-dependent movement of animals from more crowded reserves to less crowded fishing grounds. Marine reserves also benefit fish and fisheries by promoting habitat recovery from the damage done by fishing gears, and they protect species caught as bycatch. People often argue that marine reserves in specific locations will not provide much benefit to fish that move around a lot or migrate. However, as noted above, such creatures can benefit from protection when they occur in areas of vulnerability during certain phases of their lives, such as when forming spawning aggregations, or occurring in nursery grounds or migration bottlenecks where fishers often target them intensively. They can also benefit from improvements in prey availability and habitat within reserves. Fisheries models suggest that establishing some 20–40 per cent of the sea as marine reserves could help maximise benefits to fisheries (Gell and Roberts, 2003).

Despite these benefits, marine reserves remain highly controversial in the fishing industry. The major reason for this is that fishers tend to resist anything that places greater restrictions on what they can do. Despite fishers' scepticism about reserves there is, however, little doubt that marine reserves provide one of the most powerful tools in the fishery manager's armoury to rebuild stocks and recover habitats.

15.7 Will we run out of fish?

While total global wild fish catches remain at levels close to their historic high, when the numbers are broken down fishery by fishery, the picture is less sanguine. The 2008 FAO assessment of the status of the world's major fish stocks reports that 28 per cent of stocks have collapsed, are over-exploited or recovering from over-exploitation, with a further 52 per cent fully exploited with no room for further expansion. Looking back to 1950, when fish catch statistics were first collated by FAO, two-thirds of the stocks monitored since then (65 per cent) have suffered at least one collapse, defined as a fall to less than 10 per cent of peak landings. If the cumulative curve of stock collapses is extrapolated into the future, it predicts that all of the species we exploit today will have suffered collapse by the middle of the twenty-first century (Figure 15.10). When this finding was published it generated a storm of publicity, the tenor of which can be summarised as 'fish will run out by 2048' unless we change the way we manage fisheries.

The finding has been much criticised and the date argued about endlessly. However, we must face the fact that while fisheries are renewable resources if managed well, they are finite if not. We have run out of many species that were once abundant, and the rate at which we are moving from first exploitation to the point of collapse has increased throughout the last century. If, in the

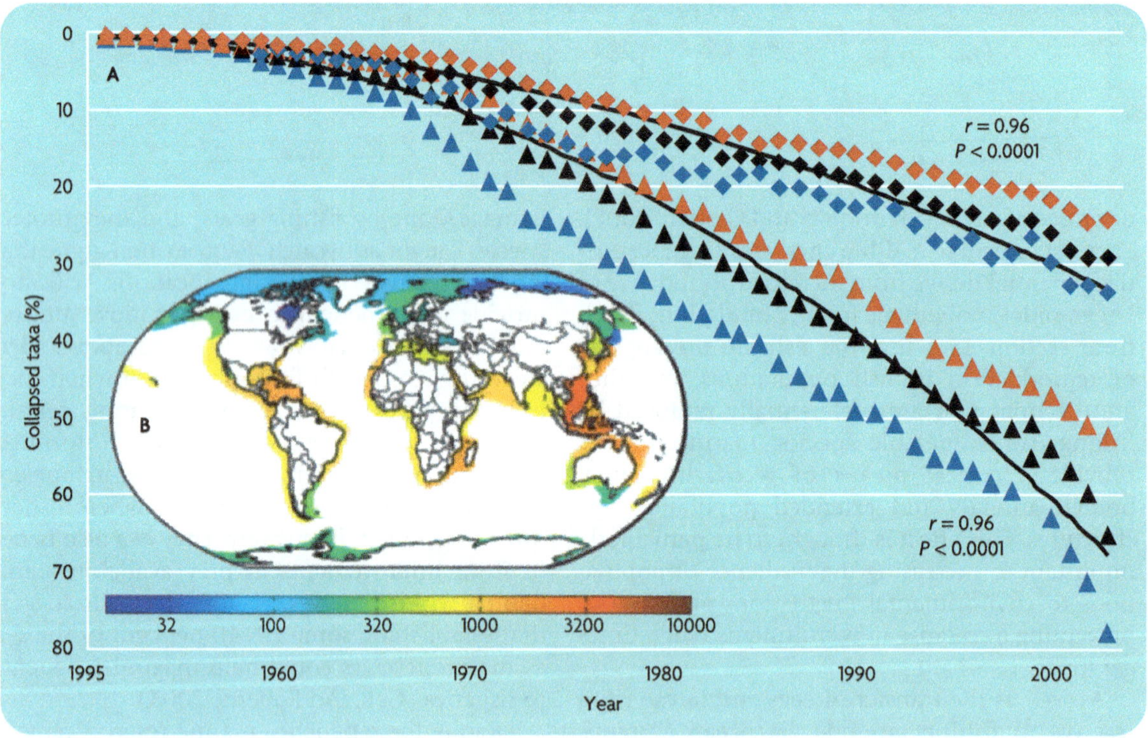

Figure 15.10 Collapses of fish stocks monitored by the Food and Agriculture Organisation since 1950. The upper black line shows the average percentage of stocks that are collapsed, while the bottom line illustrates the cumulative percentage of stocks that have experienced collapse at some point in the time series. Black symbols are averages across all stocks, while red symbols show values for stocks present in high diversity marine ecosystems, and blue symbols show values from low diversity ecosystems. The authors of the study caused controversy when they noted that if they extrapolated the lower line, it predicts that 100 per cent of monitored fish stocks will have experienced collapse by 2048.

Source: From Worm B., Barbier, E.B., Beaumont, N., Duffy, J.E., Folke, C. *et al.* (2006) Impacts of biodiversity loss on ocean ecosystem services. *Science* 314(5800): 787–790. Reprinted with permission from AAAS.

future, we want to continue to be able to eat wild caught fish, we will have to reform fisheries management.

15.8 Reforming fisheries management

Fisheries management is rarely covered in glory. There have been few genuine successes to contrast with the many high profile disasters, such as the collapse of cod on Canada's Grand Banks in the 1990s, or the abject failure of the International Commission for the Conservation of Atlantic Tunas to rein in gross overfishing of bluefin tuna. There are many roots of failure, among the most prominent of which are (1) oversimplified models, (2) lack of data on stocks, (3) lack of selectivity in targeting fish, (4) difficulties with applying the right tools to control fishing mortality, (5) habitat alteration and damage by fishing, (6) lack of precaution.

Against this background, many scientists now argue that we must shift from a species-centred approach to ecosystem-based management. We have to remember that fish are living beings that interact with complex environments and behave in complex ways. Experience has shown that it is not possible to successfully manage such complexity with conventional management tools and approaches alone. Instead, a more precautionary approach is required to reduce human imprint on the sea.

To be honest, much of what has been written about ecosystem-based management is rather vague and there are no commonly accepted principles. In simple terms it means reducing fishing pressure on target and non-target species and curtailing habitat damage. There are many tools that we can use to accomplish these ends. Below we outline a package of reforms which, if implemented, would go a long way towards achieving sustainability.

1 Reduce fishing capacity. Most fishing fleets are too large. They have got that way by lack of control, technological creep and sometimes through government subsidies. The lesson from history and from fisheries theory is that we can catch more by fishing less, and the larger stocks we would have at lower fishing effort would make fish easier to catch and therefore cheaper to obtain.

2 Abandon catch quotas and switch to measures that limit fishing effort. Quotas give a semblance of management whereas, in reality, there is little value in a measure that leads to excessive and largely unquantified mortality in target and bycatch species. By contrast, effort controls can give real protection to fish and habitats. Alongside the use of this approach we need to put in place mechanisms to prevent effort control being undermined by technological creep.

3 Ban discarding. So much good fish is wasted by discarding that the practice is abhorrent. With a ban on discards, hold space in a boat would be at a premium and fishers would have an incentive to catch fish more selectively, either by using modified gear or fishing in ways and places that minimise low-value bycatch. Norway bans discarding and the measure seems to be working well there. The government buys what bycatch is produced at a low price to make into meal for fish farms.

4 Require use of the best-available technologies to fish more selectively and reduce habitat damage. Many good bycatch reduction devices have been invented but never adopted. Partly this is because most also reduce the take of target species (although the reduction may only be slight), and partly because it costs fishers money to modify or exchange their gear. The only way to ensure that fishers take up better conservation-minded gear is to require them to do so by law. To help them swallow the pill, payments could be provided to help cover the cost of the new gear.

5 Ban or restrict the most damaging fishing gears. Some gears are more trouble than they are worth because they cause so much collateral damage. Fisheries would be better off without them. For example, the footprint of bottom trawling and scallop dredging on

the world's seas could be shrunk if these gears were only permitted in designated areas.

6 Implement large-scale networks of marine reserves that are off limits to fishing. Reserves have a fundamental role to play in rebuilding depleted fish stocks by creating more natural and resilient, extended population age structures, and by providing the opportunity for damaged habitats to recover. Reserves inject precaution into fishery management by helping to insure against future management failure.

7 Eliminate risk-prone decision making. This is one of the fundamental reasons for today's failures in fisheries management and is discussed at more length below.

In many countries, decisions on how much fish should be caught are taken by politicians. Although they usually have scientific advice to hand, they often do not pay much heed to it. In Europe, for example, member states collect data on the state of fish stocks which are analysed centrally by the International Council for the Exploration of the Sea. This body then provides advice to the European Commission on how much fish it is safe to catch. Hence ultimately it is Europe's Fisheries Ministers who make this important scientific decision and they do so at an annual bun-fest which is no better than a competitive bargaining forum. Not surprisingly, ministers seek the best deal they can get for their countrymen and vote providers, the net result being that larger quotas get set than were advised. On average, since the onset of the Common Fisheries Policy in 1983, Ministers have set TACs that are 25–35 per cent higher than advised.

Taking more fish than are produced, which is what such decisions permit, is a guarantee for unsustainable fisheries. Far from helping the industry, such decisions condemn it to death. If a farmer produces 200 lambs each year but takes 220 sheep to market, he will soon run out of sheep. Young children can grasp this logic yet it seems repeatedly lost or ignored by so many Fisheries Ministers.

Political decision making like this plagues international fisheries bodies, like the International Commission for the Conservation of Atlantic Tunas (or the International Conspiracy to Catch all Tunas, as many call it). In 2007, their scientists advised a quota of approximately 15,000 tonnes. The Commission set a quota of 29,500 tonnes, although under pressure later revised it down to 22,000. However, that year some 61,000 tonnes were removed from the sea (http://www.msnbc.msn.com/id/32975077/ns/world_news-world_environment/).

Fishing industry representatives often argue that they should be much more closely involved in the decision making on how much fish they should be allowed to catch, whereas in reality they already wield considerable influence over fishery ministers. In most walks of life, we are rightly wary of giving people the power to make decisions that will bring them direct benefits – think of putting builders in charge of granting planning permission, for example. Likewise giving fishers control over fishery management is like putting a fox in charge of the henhouse. In the USA, fisheries are managed by councils, where the majority of members are from the industry. These councils have presided over some disastrous fishery declines and only recently have had their power curbed by a law which now requires them to eliminate overfishing – in other words to follow scientific advice. Scientific advice is not always right, but the process of science is self-correcting. Advice is continuously revised on the basis of new information. We should use that advice. It makes no sense to pay for it and then ignore it.

POLICY IMPLICATIONS

Exploiting the sea for fish

- Policy makers need to understand the importance of sustainable management of fish resources in the oceans and what the threats are to sustainability.
- Policy makers need to make management decisions based upon sound scientific advice rather than short-term political gain.
- It is crucial that policy makers formulate policies based upon international, rather than narrow national, perspectives.
- Policy makers need to be aware of the nature of uncertainties associated with models used to estimate current fish stocks and possible future changes, but should not use uncertainty as a basis for avoiding implementation of essential policies that might be unpopular to the fishing industry.
- Policy makers should be aware of what is available to minimise bycatch wastage and make more effort to use what is available in management strategies.

CHAPTER REVIEW EXERCISES

Exercise 15.1

What effects do you think mining of coastal waters during the two world wars would have had on fish stocks?

Exercise 15.2

Why has so little effort been made to minimise the waste of fish resources associated with bycatch and what is the consequence of this poor fisheries management?

Exercise 15.3

What changes do you think need to be implemented at a global scale if fish stocks are to be managed sustainably? Justify your recommendations.

Exercise 15.4

To what extent have developments in technology over the past 100 years increased the risk to the sustainability of fish supplies and is there any way that they could be used by fisheries managers in future to improve sustainability?

REFERENCES

Butler, J.N., Burnett-Herkes, J., Barnes, J.A. and Ward, J. (1993) The Bermuda fisheries: A tragedy of the commons averted? *Environment*, **35**, 6–33.

Dulvy, N., Sadovy, Y. and Reynolds, J.D. (2003) Extinction vulnerability in marine populations. *Fish and Fisheries*, **4**, 25–64.

Gell, F.R. and Roberts, C.M. (2003) Benefits beyond boundaries: the fishery effects of marine reserves and fishery closures. *Trends in Ecology and Evolution*, **18**, 448–455.

Hawkins, J.P., Roberts, C.M., Dytham, C., Schelten, C. and Nugues, M. (2006) Effects of habitat and sedimentation characteristics on performance of marine reserves in St Lucia. *Biological Conservation*, **127**, 487–499.

Hutchings, J.A. and Myers, R.A. (1994) What can be learned from the collapse of a renewable resource? Atlantic cod, *Gadus morhua*, of Newfoundland and Labrador. *Canadian Journal of Fisheries and Aquatic Sciences*, **51**, 2126–2146.

Kjesbu, O.S., Witthames, P.R., Solemdal, P. and Greer Walker, M. (1998) Temporal variations in the fecundity of Arcto-Norwegian cod (*Gadus morhua*) in response to natural changes in food and temperature. *Journal of Sea Research*, **40**, 303–321.

Manriquez, P. and Castilla, J. (2001) Significance of marine protected areas in central Chile as seeding grounds for the gastropod *Concholepas concholepas*. *Marine Ecology Progress Series*, **215**, 201–211.

Murawski, S.A., Brown, R., Lai, H.L., Rago, P.J. and Hendrickson, L. (2000) Large-scale closed areas as a fishery-management tool in temperate marine systems: the Georges Bank experience. *Bulletin of Marine Science*, **66**, 775–798.

Olsen, D.A. and LaPlace, J.A. (1978) A study of a Virgin Islands grouper fishery based on a breeding aggregation. *Proceedings of the Gulf and Fisheries Institute*, **31**, 130–144.

Pauly, D., Christensen, V., Guénette, S., Pitcher, T.J., Sumaila, U.R. *et al.* (2002) Towards sustainability in world fisheries. *Nature*, **418**, 689–695.

Pauly, D. and MacLean, J. (2003) *In a Perfect Ocean. The State of Fisheries and Ecosystems in the North Atlantic Ocean*. Island Press, Washington.

Roberts, C.M. (2000) Why does fishery management so often fail? pp. 170–192 in Huxham, M. and Sumner, D. (eds) *Science and Environmental Decision Making*. Harlow: Addison Wesley Longman.

Roberts, C.M. (2007) *The Unnatural History of the Sea*. Island Press, Washington, and Gaia Thinking, London.

Roberts, C.M. and Hawkins, J.P. (2000) *Fully-Protected Marine Reserves: A Guide.* WWF Endangered Seas Campaign, 1250 24th Street, NW, Washington, DC 20037, USA and Environment Department, University of York, UK, 131 pp.

Sadovy, Y. and Eklund, A.M. (1999) Synopsis of biological data on the Nassau grouper, Epinephelus striatus (Bloch, 1792), and the jewfish, E. Itajara (Lichtenstein, 1822). *NOAA Technical Report*, NMFS 147.

Willis, T.J., Millar, R.B. and Babcock, R.C. (2003) Protection of exploited fish in temperate regions: high density and biomass of snapper *Pagrus auratus* (Sparidae) in northern New Zealand marine reserves. *Journal of Applied Ecology*, **40**, 214–227.

Worm, B., Barbier, E.B., Beaumont, N., Duffy, J.E., Folke, C. *et al.* (2006) Impacts of biodiversity loss on ocean ecosystem services. *Science*, **314**, 787–790.

SOME SUGGESTED FURTHER READING

Myers, R.A. and Worm, B. (2003) Rapid worldwide depletion of predatory fish communities. *Nature*, **423**, 280–283. A classic, although controversial, study on how fishing has depleted the seas of top predators like tuna and swordfish. It also illustrates why biodiversity is important to ecosystem function and services, and documents how overfishing is eroding that biodiversity.

Partnership for Interdisciplinary Studies of Coastal Oceans (2008). The Science of Marine Reserves, 2nd edn. www.piscoweb.org. A clear and downloadable introduction to the science of marine protection.

Pauly, D., Christensen, V., Dalsgaard, J., Froese, R. and Torres, F. (1998) Fishing down marine foodwebs. *Science*, **279**, 860–863. A highly influential study showing how removal of predatory fish is restructuring food webs in the sea.

Pauly, D., Christensen, V., Guénette, S., Pitcher, T.J., Sumaila, U.R. *et al.* (2002) Towards sustainability in world fisheries. *Nature*, **418**, 689–695. A well written primer on some of the ills of world fisheries and the solutions to their problems.

Roberts, C.M., Hawkins, J.P. and Gell, F.R. (2005) The role of marine reserves in achieving sustainable fisheries. *Philosophical Transactions of the Royal Society of London*, **B 360**, 123–132. Answers the paradox of how closing areas to fishing can help fisheries.

Roberts, C.M. (2000) Why does fishery management so often fail? pp. 170–192 in Huxham, M. and Sumner, D. (eds) *Science and Environmental Decision Making*. Harlow: Addison Wesley Longman. Illustrates some of the pitfalls of fisheries management as it has been practised.

Roberts, C.M. (2007) *The Unnatural History of the Sea*. Island Press, Washington, and Gaia Thinking, London. Documents the effects of 1000 years of fishing and hunting on marine life. The last few chapters provide an overview of fisheries management, what has gone wrong, and how it could be reformed to achieve sustainability.

Worm, B., Barbier, E.B., Beaumont, N., Duffy, J.E., Folke, C. *et al.* (2006) Impacts of biodiversity loss on ocean ecosystem services. *Science*, **314**, 787–790.

CHAPTER 16

Atmospheric pollution: deposition and impacts

Malcolm Cresser

Learning outcomes

By the end of this chapter you should:

- Be aware of the nature of the major chemical pollutants in the atmosphere.
- Be aware that these pollutants may be transferred over large distances and across international boundaries.
- Understand the nature and environmental impacts of acid rain, and why it became such an issue from the 1970s and still is an issue in parts of the world today.
- Understand the use of the critical loads concept as a strategy for formulation of pollution abatement policies.
- Be aware of the diverse timescales over which pollution impacts on environmental receptors occur, and why this is a problem for environmental managers.
- Be able to discuss the limitations of the critical loads approach as currently practised.
- Be aware of why ozone in the atmosphere is both essential and a problem.
- Have an improved awareness of the dynamic nature of ecosystems.
- Be able to discuss the relevance of atmospheric pollution in the context of biogeochemical cycles discussed elsewhere in this book.

16.1 Atmospheric pollution from a historical perspective

16.1.1 Introduction

Lead-in question

How would you have been aware of atmospheric pollution before the development of analytical chemistry?

Atmospheric pollution is obviously not a new phenomenon. Go far enough back in time and it is easy to imagine the early occupants of caves or primitive huts complaining about fumes from the fire causing them to cough and splutter. If they had been coughing and spluttering from birth, however, they might perhaps just have accepted it as a normal facet of life. It would not have been long though before the realisation dawned that they felt a bit better outdoors, and a hole in the roof to let the smoke out would be born, followed soon by a chimney, perhaps after the fire had been put out by heavy rain a few times. No wonder then that more affluent Romans two thousand years ago enjoyed good hot air under-floor heating.

As towns and cities developed, accumulation of atmospheric pollution from fires and inefficiently disposed of excrement would have become progressively more noticeable outdoors, but still accepted as an inescapable fact of life. The human sense of smell would have remained the primary detector of atmospheric pollution, and for a long while response was simply to hold something smelling a bit less unpleasant under the nose in a vinaigrette sponge. At this stage there would have been no knowledge of the chemical nature of the smells being encountered in the air.

By the middle of the nineteenth century it was recognised by Robert Smith that acidity in rainfall had become sufficient in Manchester in England to significantly damage plants (Franks, 1983). So plants too then had become biosensors, as well as humans. Damage was also being done to buildings, especially those fabricated from readily weatherable rocks such as limestone and dolomite (see, for example, Figure 1.48 in Chapter 1).

In the city of York in England, the acidic atmospheric pollution from the large number of chimneys of the laundries in the immediate vicinity of the cathedral was causing damage to the stonework of York Minster as long ago as the sixteenth century (Sir Ron Cooke, personal communication). The exact cause though, predominantly sulphuric acid from dissolution of sulphur dioxide emanating from chimneys and subsequent oxidation of sulphurous acid (as discussed in section 3.11.1), was not recognised at the time. Figures 16.1(a) and (b) show examples of more recent damage.

16.1.2 What suddenly changed our attitude?

Few things stimulate politicians into activity more than deaths that might (justifiably or otherwise)

(a)

(b)

Figure 16.1 (a) Example of damage to masonry at York Minster. The presence of white magnesium sulphate crystals on surfaces in (b) clearly points towards damage by sulphuric acid from the atmosphere.
Source: Photographs courtesy of Sir Ron Cooke.

be blamed on their incompetence as policy makers. The greatest incentive comes from inexplicably dead humans, and as we saw in Chapter 3 the Clean Air Act of 1956 in the UK was triggered mainly by excess human deaths associated with smogs caused by atmospheric pollution. By the twentieth century, though, analytical methodology was readily available that allowed identification and quantification of the pollutants present in the atmosphere and in precipitation. Figure 16.2, for example, shows typical maps being produced in the 1990s showing the concentrations of ammonium and nitrate in precipitation across Great Britain. The maps not only reflect the distribution of N pollutant emission sources; they also reflect the distribution of rainfall across the Britain, which is much drier on the east side of the country because of the rain shadow effect (as discussed in section 2.3 of Chapter 2). Less rain means lower dilution, contributing to the higher concentrations to the east. Similar maps were produced by the Centre for Ecology and Hydrology in the UK for wet deposition of pollution-derived and marine-derived sulphate and for dry deposition of atmospheric pollutants such as reduced N species (NH_3 and NH_4^+) and oxidised N species (NO_x).

Once pollutant deposition had been mapped in this way, scientists were able to start to look for correlations between measurable effects upon sensitive receptors and pollutant deposition concentrations and/or loads along pollution gradients. It soon became clear in the 1960s that there were strong correlations of visible damage to forests in the most polluted regions (Figure 16.3) or the acidification of surface waters that appeared to be causing the death of highly prized fish species to atmospheric pollutant loads. While such correlations alone do not unequivocally prove causal links, coupled with knowledge of what might be predicted on a theoretical basis they were more than sufficient to activate strong

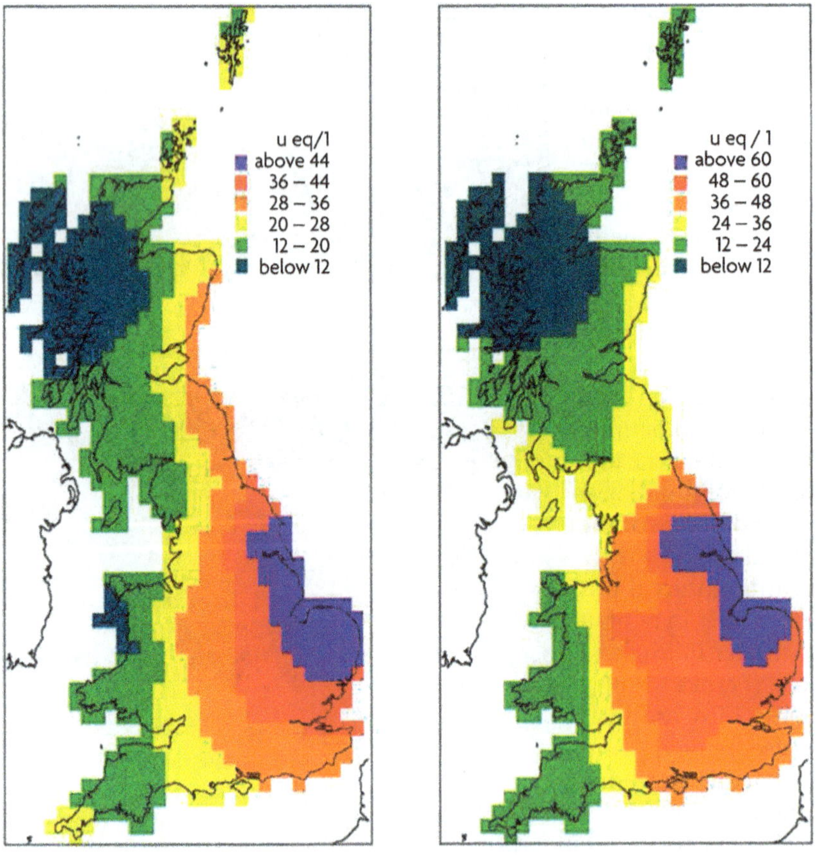

Figure 16.2 Maps showing the precipitation-weighted annual mean nitrate (left) and ammonium (right) concentrations in µmol l^{-1} for Great Britain between 1989 and 1992.

Source: Centre for Ecology & Hydrology, www.ceh.ac.uk. Note that the key shows the units as µequivalents l^{-1} which equates to µmol l^{-1} for the monovalent nitrate and ammonium ions. (May not reflect most up to date data/information.)

Figure 16.3 Forest in Czechoslovakia, close to the former border with East Germany, damaged by acid rain.

environmental lobbying. Something had to be done about acid rain.

16.1.3 The origins of acid rain – the dilemma for policy makers

Mining and exploitation of coal had underpinned the Industrial Revolution in Europe and North America, and from the middle of the twentieth century was being supplemented more and more by use of oil and natural gas. However, the combustion of fossil fuels, on which economic growth and national prosperity depended, was now clearly being identified as the cause of acid rain and major environmental problems. Policy makers needed to quickly identify the relative contributions of national emissions coming from industry, domestic power consumption and transport so that they could focus attention on key pollution sources. Thus national maps were soon also produced to show the spatial distribution of pollutant emission fluxes on a kg ha^{-1} yr^{-1} basis. These could then be compared with the deposition maps such as those that are shown in Figure 16.2 after they too had been converted to a kg ha^{-1} yr^{-1} basis.

It was tempting to look to nuclear power as a possible partial quick-fix solution, and this has happened in some countries. At that point in time much less consideration was being given to renewable energy (see Chapter 18) than is the case today. What the policy makers really needed to know first though was how much damage was being done, what was being damaged, and where the damage was significant. Only when they could answer these questions reliably could they formulate cost effective pollution abatement policies.

16.1.4 Atmospheric pollution does not respect national boundaries

When national maps of pollutant emissions and pollutant deposition fluxes were compared, it indicated clearly that reduced N species such as ammonium or ammonia largely tended to be deposited within a few km of their point of origin. Indeed, it was soon realised that deposition of fluxes of these species could be very high immediately down-wind of intensive pig and poultry rearing units. However, emitted oxidised sulphur and nitrogen species travelled for much larger distances.

For the UK the prevailing winds are predominantly from the west, sweeping in from over the Atlantic Ocean. This meant that the UK tended to receive lower loads of pollution than it emitted; as a consequence it started to be referred to as 'The dirty old man of Europe' by the southern Scandinavian countries that were recipients of much of its pollution. Policy makers were already familiar with trans-boundary pollutant transport in the context of management of water quality in some major European river systems. However, acid rain highlighted the problem for the first time in the context of atmospheric pollution.

At the time of writing we are all much more aware of the problems of global redistribution of pollution. Partly this is because of incidents such as the radioactive contamination emanating from the Chernobyl power plant nuclear accident or the impact of fine dust from the Icelandic Eyjafjallajokull volcanic emissions on air transport in 2010, at distances of many hundreds of km from the emission sources. Partly though it is because of the higher awareness of the greenhouse effect problem, the solution to which requires carbon dioxide and methane emissions to be considered on a global scale (see Chapter 8).

But back in the 1970s, agreeing to pollution abatement strategies at the continental scale was still a relatively novel challenge for policy makers. The UK government was sometimes accused of deliberately being slow to act and complacent, because it knew that most of its acidifying pollutant emissions had greater impacts elsewhere than 'at home'. In reality the slow pace of policy generation reflected the complexity of the problem and the need for a better understanding of the deposition effects.

Lead-in question

How can you tell unequivocally that damage is due to acid deposition rather than something else?

Figure 16.4 Part of an ornamental structure in York, allegedly damaged by acid rain. But how much of the deterioration is a normal weathering effect and how much is due to acid pollution in the atmosphere?

16.2 How can sensitivity to acid deposition be assessed?

If you look at a seriously damaged forest such as that in Figure 16.3, or a damaged building such as in Figure 16.4, and someone in authority says they were destroyed by acid rain, it is tempting to simply agree in a moment of panic. But can you be sure how much of the damage to the limestone in Figure 16.4 is attributable to acid deposition rather than just a few hundred years of rain or freeze/thaw effects on a fairly porous stone? Indeed, the lower level of visible damage on the more sheltered arch behind might suggest the latter are relatively more important. To be sure what we need is an experiment with controls and samples where the *only* difference is rainfall acidity. Without that the evidence is just circumstantial.

On first thoughts it might seem straightforward to investigate the impacts of acid deposition upon building materials, but this is not the case. Some building materials are more porous than others, and acid deposition occurs in both wet and dry forms. Rainfall acidity is very variable, not just between precipitation events but also within events as the first precipitation tends to scrub both terrestrial dust and acidifying pollutants from the air below cloud level (Edwards and Cresser, 1985). Moreover, impacts may change with time as the building material ages. They will also change with prevailing temperature and sometimes with rates of diurnal temperature change. It is no wonder then that often inferences are made from trends along pollution gradients, even though these may be associated with climatic differences along the same gradient and the conclusions are questionable.

So how about the damaged forest depicted in Figure 16.3? It is down-wind of a site that emits a diverse range of pollutants, and clearly the trees have been damaged. But can we say the damage is due to acid rain? Even if we could, are we seeing an effect of sulphuric or nitric acid in rain or an effect of ammonium deposition on the foliage or an effect on the soil? Or are we seeing an effect of sulphur dioxide or nitrogen dioxide on the foliage or on element cycling in the forest litter layer? Or are we seeing an effect of some totally different emission, perhaps of one or more toxic heavy metals or volatile organic compound? Or are we seeing a combined effect of many pollutants? Undoubtedly Figure 16.3 gives cause for concern, but without appropriate additional experimental

Figure 16.5 Effect of 3-y exposure to ambient plus supplementary rainfall with effective pH values 3.94 (left), 3.45, 3.23, 2.65 and 2.15 (right) on the growth of heather (*Calluna vulgaris*). Only one of the four replicates for each treatment has been selected to simplify the photograph, but appearance within each group of four replicates was very similar.

investigations we can't say were looking at an acid rain effect.

Now look at Figure 16.5. It shows results from a pot experiment in which heather plants were grown outdoors in homogenised peat soil from a relatively unpolluted area. The peat used was sampled in April 1997 at Shieldaig in NW Scotland, where the effective rain pH was approximately 4.1 (effective pH is calculated from the total wet plus dry H^+ deposition load per hectare divided by the volume of runoff, or precipitation minus evapo-transpiration per hectare). Hand-sorted, partially dried peat in sets of four replicate 10-litre pots was used to grow 3-y plants of a single *Calluna vulgaris* cultivar outdoors in Aberdeen. Twice each week, plants and pots were watered with synthetic rainwater, supplementing ambient Aberdeen rain, giving in total around 2000 mm of precipitation per year and the effective pH values 3.94, 3.45, 3.23, 2.65 and 2.15. From six months onwards, soil solutions were sampled monthly from the upper 8 cm of peat, and analysed. The plant shoots were harvested and analysed after three years, but simulated polluted rain treatments were continued.

It can be seen that after three years of acid treatment the plant growth for the middle acidity treatment is greatest, and visible damage is only obvious at the highest treatment level (the plant on the right in Figure 16.5). The question here is whether damage is being delayed by the buffering capacity of the peat soil. It is therefore informative to look at how the acidity of the soil solution was changing over time. Figure 16.6 shows this effect, and clearly there is a long delay in the peat acidification. Note that the peat soil solution pH is approaching the effective rainfall pH after three years, but is still not a perfect match.

Figure 16.6 highlights one of the problems in assessing sensitivity of plant/soil systems to acid deposition effects; there can be a long delay, perhaps a decade or more, before the ultimate effect is seen, even when quite extreme acid treatments are being applied. For this reason it was decided to continue treatment applications after harvesting the shoots of the heather plants to assess impacts upon potential re-growth of vegetation. Figure 16.7 shows the result.

Several bryophytes re-established for the H3 treatment post-harvest of the *Calluna*. They included *Hypnum*, *Sphagnum*, *Liverwort* and *Campylopus*. In one of the four replicates, *Mnium hornum* was dominant. *Erica tetralix* was also found in H3 treatment and control treatment peats. The Shieldaig peat used for this experiment, however, had a mineral content of almost 10 per cent, so for the H3 treatment, soil solution pH was still around 3.6, and had not dropped to 3.2.

There have been very few *long-term* experiments of this type, which embrace effects of pollution upon soils as well as possible direct effects upon plants, to underpin the assessment of longer-term atmospheric pollution impacts upon plants. You will see in Chapter 17, however, that far more studies have been made of the direct impacts of gaseous pollutants upon plants, often using open-topped growth chambers or enclosed atmospheres in solar domes. Unfortunately many of the plants used in such studies are grown in unrealistically fertile soils; results might well have been different if naturally less fertile soils were used as a growth medium instead.

There is another important factor to consider when attempting to assess the sensitivity of plants to acid deposition impacts if soil change is to be embraced too, and that is position in the landscape. Acid deposition is known to cause base cation leaching from soils, which is especially important for calcium in naturally acid upland

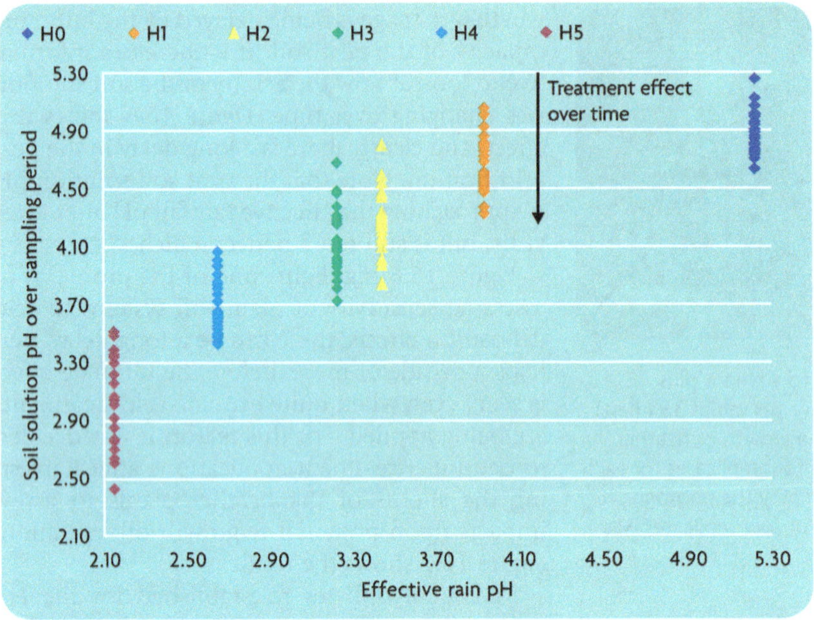

Figure 16.6 Effect on the soil solution pH of peat under heather (*Calluna vulgaris*) of exposure to ambient plus supplementary rainfall with effective pH values of 3.94 (H1), 3.45 (H2), 3.23 (H3), 2.65 (H4) and 2.15 (H5). Measurements were made monthly from month 6 to month 36.

Source: Based upon unpublished data of Calver, Cresser, White and Smart.

Figure 16.7 Re-growth of vegetation after harvesting *Calluna vulgaris* plants after exposure to ambient plus supplementary rainfall with effective pH values 3.94 (right), 3.45, 3.23, 2.65 and 2.15 (left); note the absence of new plant growth on the soil from the most acid treatment (left), but excellent recovery at an effective rain pH value of 2.65.

ecosystems. However, often the leached base cations flow laterally down slope, and therefore soils and plants on lower slopes initially may receive enhanced base cation inputs. Therefore, if we want to look for potential plant damage to plants intuitively we should look at ecosystems at the tops of slopes. Another problem then, however, is that climatic conditions at the tops of slopes may be more extreme, and therefore plant growth poorer anyway, so how can any poorer growth be attributed unequivocally to an acid rain effect? A further problem is that soil chemical, physical and biological properties may vary markedly over space and time (see Chapter 7). Clearly then if soil ameliorates the effects on a plant species of any specific load of atmospheric acidifying pollution, any observed impact in the field may vary dramatically over space and time, as well as with position in the landscape. Not only that, but plants in natural ecosystems rarely grow in monoculture, so we also need to consider how plants compete with each other and how plant biodiversity will be influenced.

An interesting aspect of research into the effects of acid deposition upon soils is that the vast majority of studies have been concerned with changes in soil chemistry and very few with changes in soil biology. Soil physical properties have received virtually no attention at all. However, it has been shown that the micromorphology of peat is apparently strongly influenced by atmospheric pollution, as can be seen in Figure 16.8.

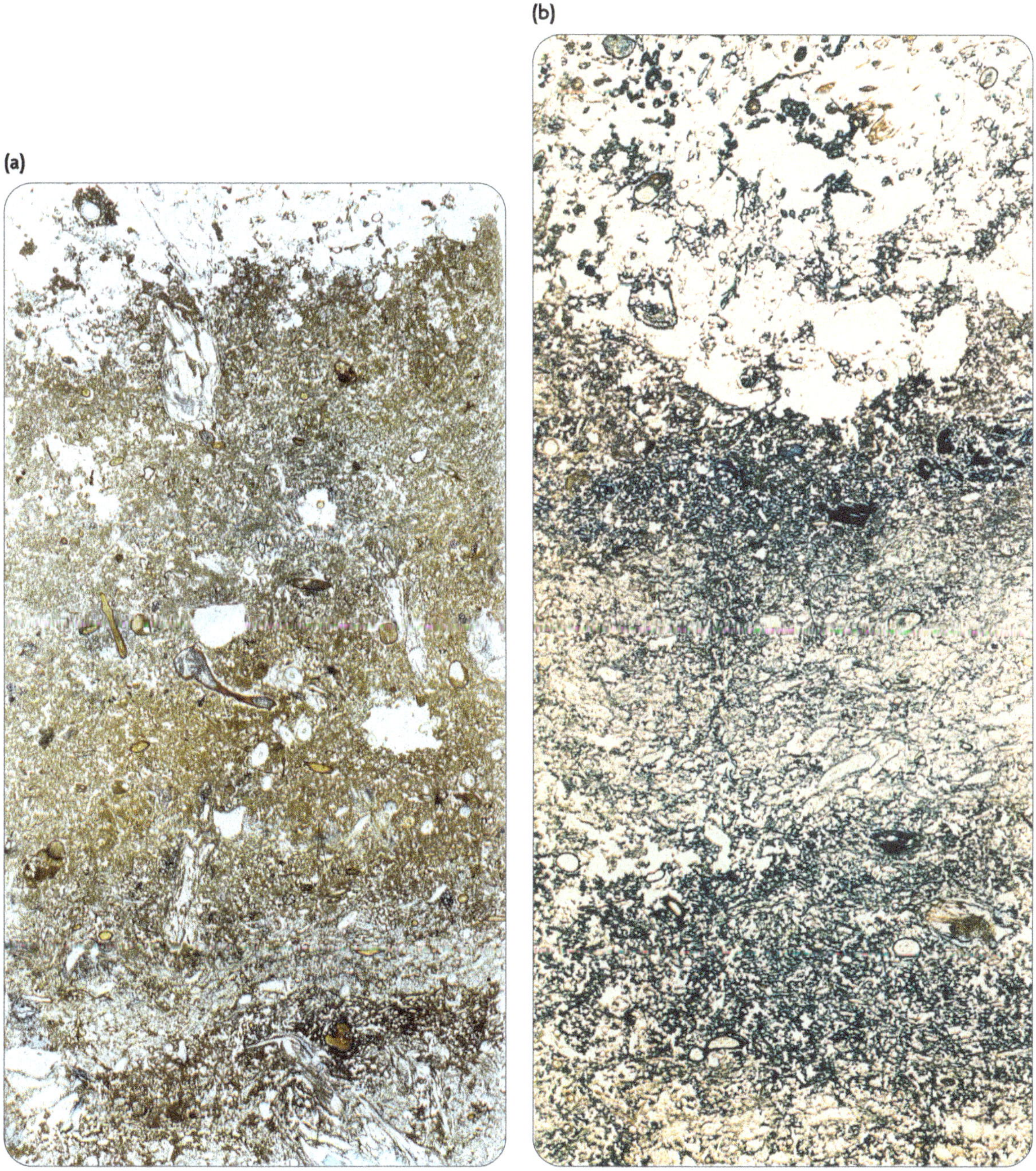

Figure 16.8 Thin sections, prepared by E.A. Fitzpatrick, of upper layers of intact peat soils collected from along a pollution gradient in Great Britain, between a relatively unpolluted site in the north-west of Scotland (a) through to a very heavily polluted site in the South Pennines (d). Each section is about 10 cm deep. Note that the depth of relatively poorly decomposed plant material increases along the pollution gradient.

Source: Photographs taken for the author by E.A. Fitzpatrick and Laila Yesmin.

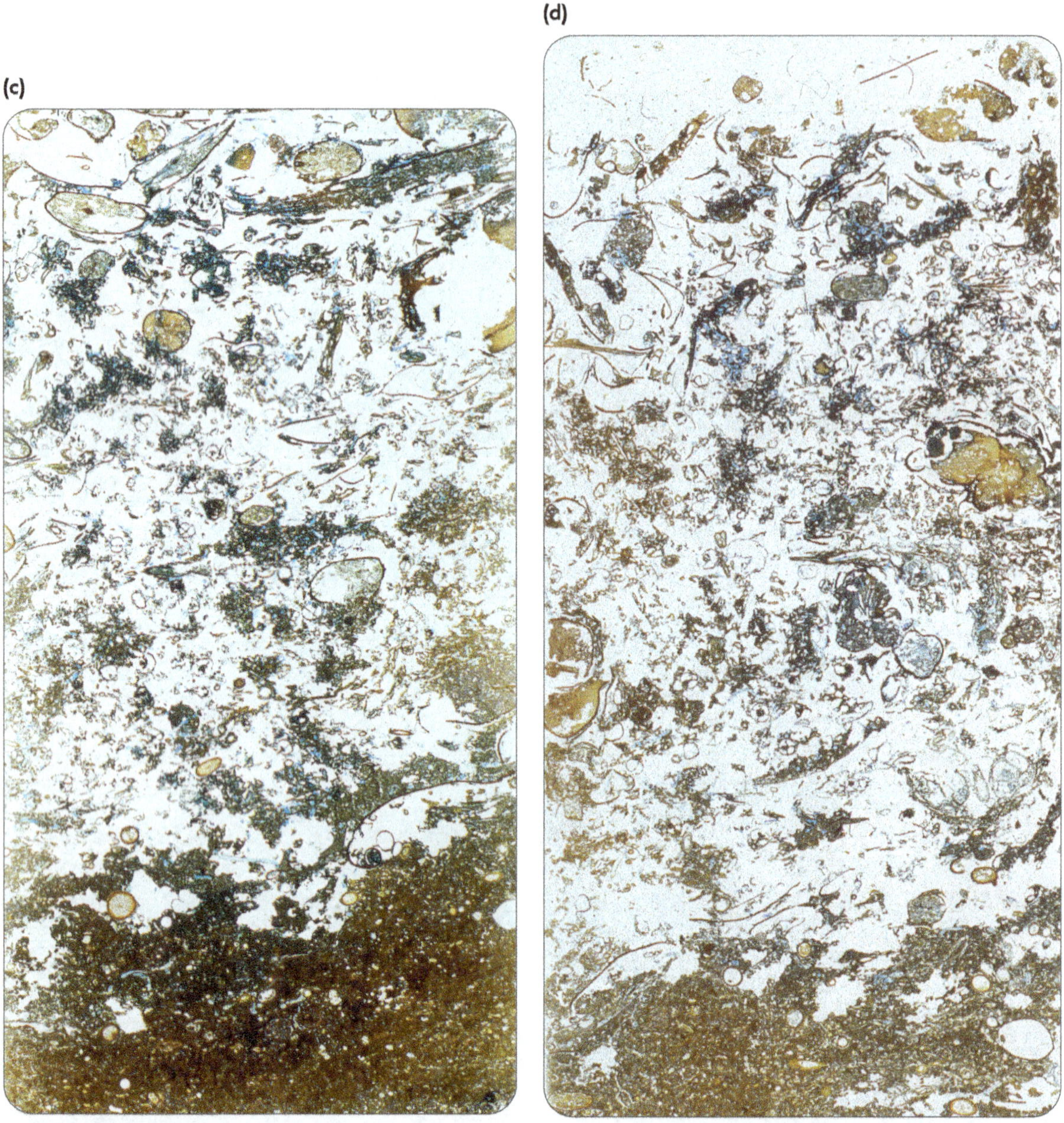

Figure 16.8 Continued

Such changes may have substantial influences upon the hydraulic conductivity of the peat and therefore upon the chemistry of adjacent surface waters (Cresser *et al.*, 1997).

Lead-in question

Is it easier to assess the sensitivity of surface waters?

In the UK, Kinniburgh and Edmunds (1986) were the first to produce a map of the sensitivity of groundwaters to acidification. They allocated the mapped units of the GB 1:650 000 bedrock geology map to one of four classes based upon their informed estimates of each unit's buffering capacity. Their estimates were based upon the mineralogy and associated anticipated geochemistry of the dominant rock type in each mapped unit. The resultant map was susceptible to uncertainty because of variations with individual mapped units, but represented an excellent starting point. Some of this variation was compensated for in a later variant of the map (Hornung *et al.*, 1995a).

On first thoughts you might expect that it would be much easier to assess the sensitivity of surface waters to damage from acidifying pollution deposition than the sensitivity of plant/soil systems. Sadly that is not the case. Nevertheless a general sensitivity map has been produced for surface waters in Great Britain, by overlaying soil sensitivity data and rock data using a GIS (Hornung *et al.*, 1995a). Thus highly sensitive soil over a highly sensitive rock type would give a highly sensitive surface water, whereas a highly sensitive soil over a moderately sensitive rock type would be prone to acid flushes only at high discharge; a low sensitivity soil would tend to give insensitive waters regardless of the sensitivity of the underlying geology.

We saw in Chapter 2 (section 2.8.2) that, in upland areas especially, river discharge may increase rapidly during storm events and subside when the rain stops; at the same time water chemistry may change dramatically, with a flush of acidity and often too a flush of dissolved organic matter, and a fall in concentrations of calcium and alkalinity. Because acid flushes are associated with changes in hydrological pathways, and more water draining through or over surface organic-rich soil horizons, acid flushes are perfectly natural phenomena; they would still occur in the complete absence of acid rain. However, two mechanisms influence the *severity* of the flush. In the longer term it will be greater in areas where the soil has been acidified by acid rain. This means that the acidification of rivers by acid deposition may take decades to attain the worst possible situation. The severity of acid flushes will be influenced by enhanced mobile anion concentrations too, as rapid cation exchange reactions regulate water acidity and when there is a higher concentration of SO_4^{2-} and NO_3^- in drainage water, there will be more H^+ ions too, so the water will be more acidic, both in the short term and in the longer term. When acid rain first became a major issue back in the 1960s, some researchers failed to grasp the fact that the ultimate extent of surface water acidification and soil acidification are inextricably interlinked.

There has always been a general consensus that surface waters with a high alkalinity (HCO_3^-) concentration are not at serious risk of damage from acid deposition, so long as the alkalinity reserves significantly exceed the acid inputs. This is a fairly safe assumption, but only as long as appropriate consideration is given to how alkalinity may decline sharply in river water in periods of high discharge. It is important therefore to know how aquatic biota will respond to temporary acid flushes. Unfortunately we know much less about that than about how they behave at different constant levels of acidity.

Recently Begum *et al.* (2010) have suggested that if river water sediment composition is known throughout river networks it may be used to predict the spatial distribution of mean and lowest alkalinity values of the river water. Figure 16.9 shows the locations of 851 sediment sampling points in the drainage basin of the River Derwent in North Yorkshire, England, for example. The sediments were all analysed by the British Geological Survey who provided elemental composition data. Using GIS, a small portion of the catchment area upstream of each sediment sampling point was assigned to each sediment sample, based upon hydrological flow paths estimated from topography using a digitised elevation map. This allowed the flow pathway-weighted mean sediment calcium percentage to be calculated.

The relationships were then studied between mean maximum and minimum alkalinity values, determined by the Environment Agency, and the flow path-weighted sediment calcium concentrations. The results for mean alkalinities, shown in Figure 16.10, display a strong correlation, suggesting that where such data sets are available

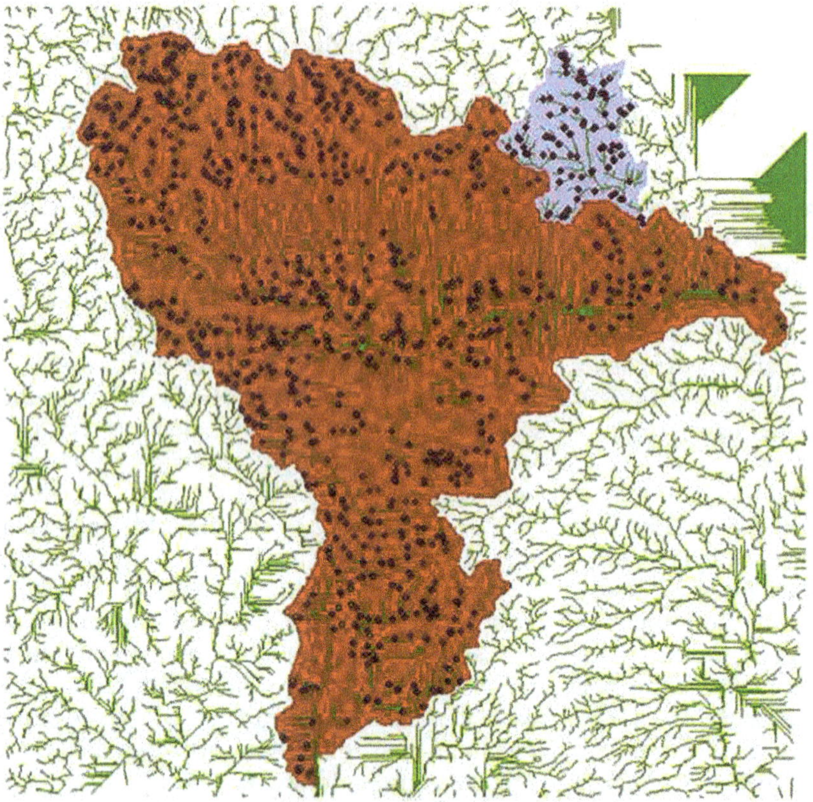

Figure 16.9 The distribution of 851 BGS sediment sampling points throughout the River Derwent drainage basin in North Yorkshire, England. The part of the catchment above Forge Valley is shaded in lilac, as this part of the basin is isolated by a Sea Cut draining to the North Sea as part of a flood protection scheme for the area downstream. Map regenerated from BGS data as described in Begum *et al.*, 2010.

Source: Courtesy of Shaheen Begum.

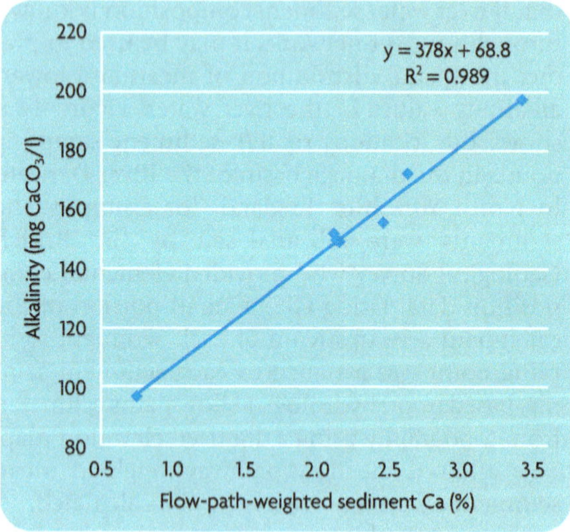

Figure 16.10 The correlation between mean alkalinity from monthly samples over a three-year period and flow path-weighted sediment calcium concentration for the River Derwent catchment in North Yorkshire.

Source: Based on the approach described by Begum *et al.* (2010).

they provide a useful starting point for identifying acidification-sensitive river waters. The correlation was significant for minimum alkalinity values but not quite as strong.

16.3 The critical loads concept

It should be clear from the content of this chapter so far that in the 1970s there was real concern internationally about the damage that acid rain apparently was doing to architectural heritage, to aquatic life in streams and lakes, to forests and to plants in minimally managed ecosystems. There was also speculation about possible damage higher up in food chains. But it should also be clear that obtaining robust quantitative evidence for the incidence and extent of such damage in terms of dose response relationships was far from simple. Even in the simple experiment described earlier for *Calluna* growing in a peat soil, for example, the trends shown in plots of plant yield (an example of a response) versus pollutant dose might be quite different after 1, 2, 3, 4 and 5 or more years, and they might have been different on different peats. Yet policy makers needed to act, and act quickly, if deterioration of buildings and ecosystems was not to continue. Somehow policy makers needed to agree internationally upon a strategy that would allow them to decide where acidifying pollutant emissions should be reduced and how much they should be reduced by. The approach that they adopted was the 'critical loads' approach, an idea first advanced in Canada in the 1970s.

Critical load is defined as 'a quantitative estimate of an exposure to one or more pollutants below which significant harmful effects on specified elements of the environment do not occur according to present knowledge' (Nilsson and Grennfelt, 1988; Hornung *et al.*, 1995b). The definition is interesting because it was put together in this form by an international group of scientists assembled at Skokloster in Sweden in 1988 to decide on a strategy for producing critical loads maps for soils; they added 'according to present knowledge' to reflect their concerns about the robustness of the approach, bearing in mind the complexity of the problem. To policy makers though the concept was very attractive; it was so simple that it would be understandable by the less well informed members of the general public; conceptually if critical load was not being exceeded, no damage was being done. Critical load maps would allow the policy makers to overlay deposition maps and critical load maps to show exactly where critical load was being exceeded and by how much.

16.3.1 Soil critical loads

Mineral soils

The first step at Skokloster was to assume that, provided the alkalinity generated by biogeochemical weathering (see section 7.4.6) per square metre in a soil was less than the acid load per square metre, then the soil would not be acidified. The current author reluctantly agreed to this at Skokloster in the interests of 'progress' but we will see later that this starting premise is highly questionable. The assumption was simply:

$$H^+ + HCO_3^- \rightarrow H_2O + CO_2$$

The second step was to make informed estimates of weathering rates of key soil minerals, which could be equated to critical load ranges, based upon the combined knowledge of the assembled scientists of research into weathering rate estimates for particular rock and soil types. Values ranged from < 0.2 $kmol_c\ ha^{-1}\ yr^{-1}$ for quartz and potassium feldspar through 0.2–0.5 $kmol_c\ ha^{-1}\ yr^{-1}$ for plagioclase, muscovite and < 5 per cent biotite, 0.5–1.0 $kmol_c\ ha^{-1}\ yr^{-1}$ for biotiote and amphibole, 1.0–2.0 $kmol_c\ ha^{-1}\ yr^{-1}$ for pyroxene, epidote, olivine and > 2.0 $kmol_c\ ha^{-1}\ yr^{-1}$ for carbonates (Hornung *et al.*, 1995b). You should be familiar with the properties of most of these minerals from Chapter 1.

The third step was to tweak the values according to environmental properties; for example critical load could be lowered in wetter areas, for coniferous forests, at high elevations, in very freely draining and/or thin soils, for soils with a low sulphate absorption capacity (see section 19.3), or where base cation deposition was low. Conversely they would be raised in drier areas, for deciduous forests, at low elevations, in poorly draining and/or deep soils, for soils with a high sulphate absorption capacity, or where base cation deposition was high. Moreover, in arable agricultural areas, clearly critical loads should be higher.

Finally at Skokloster it was pointed out by the policy makers that pollutant concentration *ranges* were no use for map production if the maps were to be used subsequently for planning pollution abatement strategies. So even though the ranges were there partly to reflect uncertainty, they were eventually altered to single values so that maps could be produced showing how much critical loads were being exceeded by across a country. A precautionary approach was reluctantly accepted and lower single values used in place of deposition load ranges.

It should be starting to become clear that this first approach to setting critical loads, now known as the level zero approach, was hardly an exact science. To be fair, attempts were made though to improve mineralogical classification early on (Sverdrup and Warfvinge, 1988). Hornung *et al.* (1995b) have published a concise but informative account of how the Skokloster classification was modified in Great Britain to provide reasonable critical load maps for acidity. An example of an early critical loads of acidity map for the UK can be seen in Chapter 19 as Figure 19.9.

Lead-in question

How would organic, peat soils differ from mineral soils when setting critical loads of acidity?

Peat soils

From the above discussion of mineral soil critical loads it is clear that, initially, informed guesses of soil mineral weathering rates were used to set critical loads for mineral soils. However, deep peat soils, unless receiving inputs of mineral material from erosion upslope, generally contain negligible quantities of mineral material. Ion exchange equilibria (see sections 7.4.6 and 7.5.3) therefore regulate surface soil chemistry and pH of peats rather than soil mineral biogeochemical weathering. Moreover, peat equilibrates with both precipitation inputs and drainage water. Thus effective input cation concentrations (including H^+ concentration) should be very similar to drainage water cation concentrations (again including H^+ ions). Note that we have used the descriptor 'effective' for concentration here. 'Effective' concentration of a cationic species is its total input flux divided by runoff; thus effective concentration takes into account both wet and dry deposition of acidifying pollutants and the fact the inputs in precipitation may be concentrated up by evapo-transpiration (Smith *et al.*, 1992).

Skiba and Cresser (1989) conducted a set of simulation experiments in their laboratory in which simulated rainfall solutions covering a range of pH from 3.5 to 5.0 were allowed to equilibrate with peat from a single site. At intervals they measured the pH of the peat as a paste with dilute calcium chloride and also the pH of the associated drainage water. Their peat soil pH results are represented graphically in Figure 16.11. They converted the volumes of equilibrating simulated rainfall to equivalent amounts of precipitation (in m) that the peat would experience in the field. The results suggested that 10 cm or so of peat could reach a new equilibrium pH value within 2–3 years in response to changes in rainfall pH. This relatively rapid change reflects the lack of buffering by weathering minerals in the peat. It's bad news in terms of rate at which damage might be seen, but at least gives hopes of relatively rapid recovery if atmospheric pollution levels are reduced.

The results in Figure 16.11 were obtained using variable hydrogen ion concentrations at a single rainfall base cation formulation. Because the equilibrium pH of the peat depends upon competition between the effective concentrations of all cation species in the precipitation, it was necessary to repeat the experiment using a range of different base cation concentrations. This was done by Smith *et al.* (1992). It was then possible

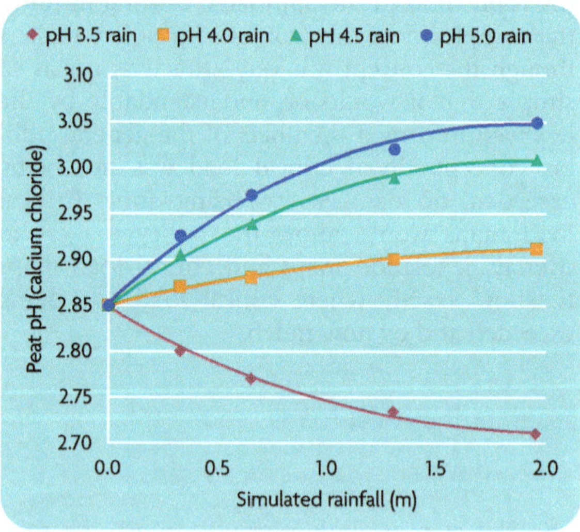

Figure 16.11 Graphs showing how peat pH, measured as a paste with calcium chloride, changes in response to effective rain pH for simulated precipitation at pH values of 3.5, 4.0, 4.5 and 5.0.

Source: Redrawn by the author based upon the results of Skiba and Cresser (1989).

to estimate what peat pH would have been anywhere in Britain under pristine, acid-rain free conditions, and how much acid deposition at any point would cause a pH fall of 0.2 pH units. This pH shift was deemed acceptable on the basis that there was no statistically significant evidence for biological damage being caused if pH fell by 0.2 pH units (Hornung *et al.*, 1995b). This concept was therefore used for setting peat critical loads in the UK for a number of years. Also 0.2 pH units reflected uncertainty in the pH assessment, so government was not having to agree to 'acceptable damage' (something generally they are very reluctant to do).

With hindsight, basing critical loads on measurement of peat pH values found as pastes with dilute calcium chloride, though a widely accepted method for routine determination of soil pH, was not a particularly good idea in the critical loads context. The more appropriate values are those of moist peat which embrace the mobile anion concentrations experienced under field conditions. These include sulphate and nitrate from atmospheric pollutant deposition; they are the acidity values that the plant roots and microbes in the peat will experience *and* also reflect drainage water acidity if this is also a potential issue. Figure 16.12 revisits the experiment of Skiba and Cresser (1989), this time looking at changes in drainage water pH. It is immediately apparent that the drainage water pH approaches the rainfall pH at pH 3.5 and 4.0; however, organic acid buffering cuts in at around pH 4.2 to 4.4.

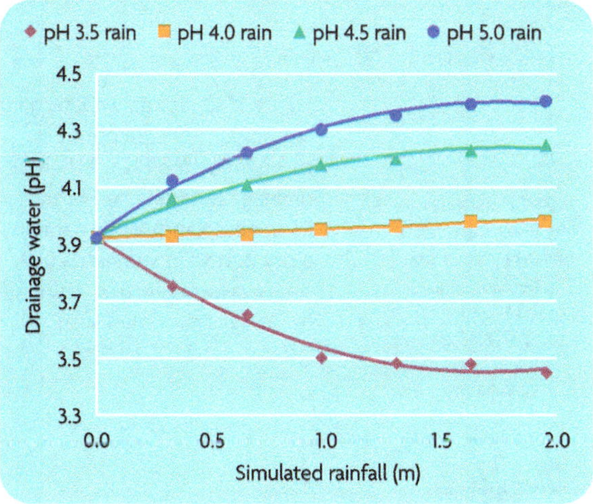

Figure 16.12 Graphs showing how the pH of drainage water from peat equilibrated with simulated precipitation at pH values of 3.5, 4.0, 4.5 and 5.0 changes in response to the effective rain pH.

Source: Based upon the results of Skiba and Cresser (1989).

The conclusion for such a reassessment was that it was not sensible to try to reduce acid loads below those that would cause an equilibrium drainage water pH of 4.4, because that is the pH that the plant communities and microbes in peats would experience anyway. Subsequently therefore this critical threshold value has served as the basis for setting critical loads of acidity in the UK. This raised the critical loads of acidity for many peat soils in the UK, but they still remain among the most sensitive soils in many areas because of the lack of any buffering from biogeochemical weathering of soil minerals.

The Level 1 (mass balance) approach for mineral soils

As we have seen, acidifying pollutants in the atmosphere move freely across national boundaries. Very early in the 1990s a coordination centre for effects (CCE) was set up in the Netherlands at RIVM, the National Institute of Public Health and Environmental Protection. They have done an excellent job in coordination of national efforts and producing maps at the European scale. They make best use of available data, accepting that different countries may have more or less input data on which to base their mapping. The RIVM website is an excellent source of information on maps and mapping techniques. Recognising the limitations of the Level Zero approach, a number of countries have attempted to use a base cation mass balance approach, the Level 1 approach. This considers base cation inputs from weathering and non-marine base cation inputs as offering buffering capacity against acidification. However, it takes into account that base cations (and associated alkalinity) may be removed from ecosystems by plant uptake, for example in maturing forest, or leached. Sometimes discussions of what values should be assigned to these parameters may become quite heated at international meetings.

The ecosystem type-based approach

Over the past decade more and more attention has been paid to trying to define critical chemical parameters for specific biological receptors of particular national interest. For example it might be decided that the ratio of calcium to free aluminium in soil solution in the rooting zone is crucial to tree species; however, different tree species have different critical threshold critical values. Therefore it might be deemed desirable to prepare separate critical loads maps for coniferous and deciduous woodlands. However, space does not allow us to consider this approach in any detail here.

The N mass balance approach

So far in this chapter we have concentrated upon assessment of acidification risk to natural or minimally managed ecosystems from the deposition of nitrogen and sulphur species from the atmosphere. It has been known for more than a century, however, that nitrate may be leached from agricultural soils, especially in autumn and winter months, and contribute to eutrophication problems in surface waters (see section 14.2). It has long been recognised also that the capacity of soils to accumulate nitrogen pollution is finite and that eventually N saturation may result in the leaching of nitrate to surface waters and/or groundwaters (Ågren and Bosatta, 1988; Aber *et al.*, 1989). Some of the field evidence for N accumulation in soils will be presented and discussed in section 19.2.1, so it will not be considered here in any detail. Suffice it to say here that the capacity of soils to retain pollutant N deposition is finite.

In Chapter 14 we emphasised the problem of nitrate leaching from remote upland areas of the UK that have been susceptible to high loads of N deposition (see, for example, Figure 14.5). We mainly tend to think of N breakthrough from soils in terms of finding nitrate in rivers at unexpectedly high concentrations during periods of active growth, e.g. in rivers draining peat soils in NE Scotland even in summer (e.g. Black *et al.*, 1993). We also mentioned in section 14.3 that ammonium may be mobile too in such areas. Figure 16.13 shows the spatial variation in ammonium-N concentration across the River Etherow catchment between Barnsley and Manchester in the UK, based upon results for 314 samples from an intensive sampling carried out over three dry days in summer at this heavily N and S impacted drainage basin.

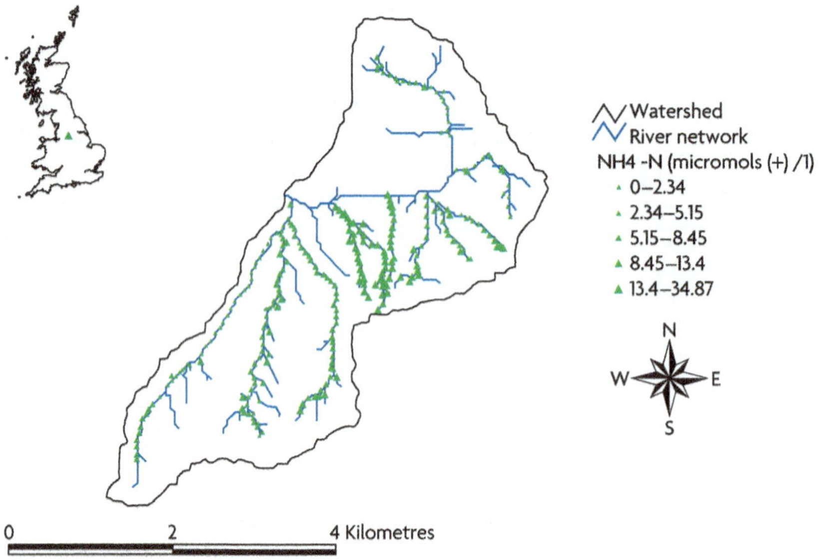

Figure 16.13 Spatial variation in ammonium-N concentration throughout the River Etherow drainage basin in the South Pennine uplands of England in a dry period in summer, 2002.

Source: Based upon data collected by Cresser, Smart and Clarke.

	pH	H^+	Alkalinity	Na dominance	DON
NH_4^++–N	−0.61***	0.71***	−0.67***	0.50***	0.78***
NH_3^-–N	−0.06	0.23***	−0.22***	−0.01	−0.08
DON	−0.70***	0.81***	−0.76***	0.39***	1.00

Figure 16.14 Correlations (*r* values) between N species for 314 samples from the Etherow in June 2002 and water determinants indicative of high contribution of water originating in acidic and/or organic soils to total discharge. (*** = $p \le 0.001$, ** = $p \le 0.01$, * = $p \le 0.05$).

Figure 16.14 is a correlation matrix showing the relationships between nitrogen species in the water samples and other determinants. Ammonium-N is strongly correlated with dissolved organic nitrogen (DON), suggesting that much of the ammonium is transported as a counter cation to organic acid anions in solution. The areas giving high DON concentrations are organic matter-rich and acidic, so the correlation with pH is negative; so too is the correlation with alkalinity, both because the soils yielding high ammonium are acidic and because organic acids neutralise alkalinity. Ammonium also correlates strongly with sodium dominance. This is defined as the ratio of [Na] to [Na] + [Ca] + [Mg], expressed as a percentage, where the square brackets denote concentrations in $mmol_c$ per litre (White *et al.*, 1999; Stutter *et al.*, 2002). High sodium dominance is an indicator of low inputs of calcium and magnesium from biogeochemical weathering; therefore the sodium dominance index is a useful method for assessing susceptibility of surface fresh waters (and the soils they drain from) to acidification, as we shall see later.

From the above discussion it is clear that the deposition of inorganic nitrogen species from the atmosphere may contribute to eutrophication problems associated with nitrate and, to a lesser extent, ammonium leaching, as well as to acidification problems. Consequently attempts have also been made to produce critical loads maps for nitrogen deposition to show where eutrophication problems are likely to arise. Such maps generally are based upon N mass balance calculations which take into account the possible sustainable sinks for N deposition. Thus the critical load for N to woodland would be based upon the mean annual accumulation of N by growing trees over a forest rotation plus the amount of N lost be denitrification plus the amount of N thought to correspond to acceptable N leaching losses (for example as organic N), plus the amount of N allowed to accumulate annually in soil. Conceptually this seems a simple and reliable approach, though once again heated discussions often have occurred at international meetings about the precise values that should be assigned to some of these fluxes. Many argue, including the author, that it is not sensible to allow any N accumulation in soil over the long term when calculating the steady state mass balance.

16.3.2 Fresh water critical loads

Because the pH and alkalinity of river water may vary so greatly over time during storm events, an averaging approach is generally opted for. The so-called steady-state water chemistry model (SSWC) suggested by Henriksen *et al.* (1986) has been widely adopted for lakes, especially by the Nordic countries. It has also been applied to rivers (White *et al.*, 2000). In essence the method is based upon assuming steady-state conditions with annual outputs and inputs in balance. This allows mean concentration/flux values to be used in calculating critical loads. The underpinning assumption is that a sustainable supply of acid neutralising capacity (ANC) may be reliably estimated. The pre-industrial concentration of weathering-derived base cations is estimated and taken as the long-term critical load on the basis that base cations from weathering correspond to the bicarbonate or ANC produced.

A minimum acceptable value for ANC for the water is selected in the critical load calculation to allow protection of a sensitive organism or organism group (receptors) of particular interest; the fresh water critical load is then the sum of acid anions (usually taken as nitrate plus non-marine sulphate) input which, when subtracted from the pre-industrial flux of base cations, gives the desired ANC (Henriksen *et al.*, 1992). Thus:

$$\text{Critical load} = Q([BC]_{nm-0} - ANC_{crit})$$

where the critical load is in $mmol_c\ ha^{-1}\ y^{-1}$, Q is the runoff from the site in $l\ ha^{-1}\ yr^{-1}$, $[BC]_{nm-0}$ is the estimated sum of mean non-marine base cation concentrations for pristine conditions in $mmol_c\ l^{-1}$ and ANC_{crit} is the selected critical acid neutralising capacity. The value of $[BC]_0$ is calculated using an equation which takes into account the exchangeability of base cations in the catchment soils, the estimated pre-acidification concentration of naturally occurring nitrate and non-marine sulphate (based on values for near pristine lakes) and the rate of leaching non-marine base cations.

16.4 How can critical loads be validated?

The conventional approach to full development of any model is to proceed from a conceptual stage via the development of predictive equations and then collection of data for calibration of the model through to validation. Validation either confirms that the model works or tells the modeller that he/she needs to think again. The problem with critical loads models is that all the early approaches to generating critical loads were steady state models; unfortunately, without using dynamic modelling, we rarely know how far down the route to steady-state ecosystems have progressed. Thus even where critical load is predicted to be exceeded it is possible that the 'damage' may still be years or decades away. Moreover, for mineral soils, the progression towards steady state will depend upon the soils' position in the landscape and local topographic and associated climate variations. Since soil acidification may be a precursor to fresh water acidification, the same problem arises for surface water critical loads of acidity or of pollutant N deposition. This makes validation difficult for many ecosystems.

An exception to some extent is peat soils, as we saw in section 16.3.1 that the re-equilibration of such soils with acidifying deposition is relatively very rapid compared with that for mineral soils because of the cation exchange nature of the equilibration process. Therefore Skiba *et al.* (1989), using analytical data collected by the Scottish Soil Survey, were able to demonstrate very clear links across Scotland between the acidity of peats and atmospheric deposition levels. Later White *et al.* (1995) clearly demonstrated in the Scottish uplands that the pH of acidic mineral soils under *Calluna* vegetation and derived from quartzite and acid sandstones, like that of nearby peat soils, was also driven by cation exchange reactions and atmospheric deposition, and not by biogeochemical weathering. Thus they found strong correlations between soil pH and the ratio of $[H^+]_{dep} / \sqrt{[Ca]_{dep}}$, where square brackets and dep denote appropriate deposition fluxes, as would be predicted from cation exchange theory. Across the pollution gradient the soils had acidified by up to a whole pH unit, which is a ten-fold increase in H^+ concentration.

The most convincing evidence for acidification of lakes primarily came from paleo-ecological studies of Battarbee *et al.* (1985). The diatom species found in surface waters vary substantially with water pH. Diatoms also leave identifiable siliceous residues in lake sediment cores which allow the water pH at the time they were deposited to be estimated quite precisely. If segments of core can also be dated, for example by using radiochemical dating techniques, then change in lake water pH over hundreds of years may be demonstrated. This, of course, is not direct evidence that the critical loads approach is being accurately used, but at least showed that the acidification problem was real. Long-term studies of recovery, as reflected in increasing alkalinity in response to decreases in acid deposition loads over the recent decade or so, should provide more convincing validation of methodology. Such long-term data may be found on the website of the UK acid waters monitoring network (UKAWMN).

16.5 Limitations of current critical loads approaches

16.5.1 Limitations to the critical loads concept for soils

With the wisdom of hindsight, and bearing in mind that critical loads initially were established in a bit of a rush internationally, it is worth revisiting the concepts underpinning our approach to critical loads of acidity for soils at Skokloster in 1988 (Cresser, 2007). First it was assumed that all the alkalinity (a) generated in soils or present in fresh waters is available for neutralising acidifying pollutant inputs (p), and that provided (a) > (p), no 'damage' to the ecosystem will occur. But what assumptions are being made (albeit tacitly) in this approach? First it assumes that in uplands, alkalinity isn't transferred down slope naturally with lateral water flow (or if it is it's of no consequence to soil chemistry). If you think about it, this is true only when biogeochemical weathering is negligible and/or CL is exceeded. Lateral alkalinity flow keeps soil pH and BS percentage higher on lower slopes. These lateral fluxes cannot sensibly be ignored when CLs are quantified using the (a) > (p) concept. Only when lateral flow of alkalinity approaches zero can the effect of acid deposition start to become significant on lower slopes, but this corresponds to a system where ion exchange regulates surface soil chemistry rather than weathering rate. Under these conditions cation relative concentrations and total mobile anion concentrations together regulate soil solution chemistry. Critical load calculations assume that only chemical species coming in vertically matter, and all water drains vertically down through soil, with no alkalinity inputs in water flowing down slopes. The counter argument of course says that critical loads represent steady-state conditions, which may take hundreds of years to be attained in a large catchment; that's why, they say, validation is so difficult.

The second tacit assumption is that pre-pollution, alkalinity production played no role in regulation of 'natural' soil pH, so all the alkalinity being generated is 'spare'. But pre-pollution, soil type, and hence soil horizon pH and BS percentage, still depended on weathering rate and topography and the other factors of soil formation discussed in Chapter 7. Much HCO_3^- was 'used up' in neutralisation of naturally produced acids (even pre-pollution) in soil. Even where weathering is still significant in surface horizons, we shouldn't assume the weathering products are 'all available' to neutralise incoming acidity. This is especially true if our estimates of weathering rates at Skokloster (and since) are based on measured BC annual fluxes in rivers. These fluxes depend upon the cation exchange chemistry of riparian zone soils at the time of measurement, and these soils may have been receiving beneficial base cation inputs leached from soils up-slope for hundreds of years. Therefore actual weathering rates, assessed near hill summits, may be much lower than those being used to calculate critical loads (White and Cresser, 1995; White *et al.*, 1995).

It is easy to see why the concept of critical loads was so attractive to policy makers early in the 1980s. For example, they could say if you had less than 10 kg of H^+ deposition per hectare for a particular soil it would not be damaged, but if you had more than 10 it would be damaged, and so on. The pollution-effects scientific community as a result became 'hooked' on a CL concept based on kg of S or N deposition per hectare. But is this sensible? Consider, for example, N deposition. There are many issues that really should be considered:

- What happens under rock outcrops in upland ecosystem soils? Clearly the amount of pollutant N received per square metre under a large rock outcrop may far exceed the vertical deposition per square metre.

- What happens with stony soils? If we base critical load on the mass of soil to a chosen depth under a horizontal square metre, but the soil is 50 per cent stones by volume, should we halve the critical load?
- What happens on slopes in upland ecosystems? Clearly the mass of soil on a slope but under a horizontal square metre will be higher than the mass of soil in a flat area under a horizontal square metre. Should we take this into account?
- What about N draining laterally rather than vertically? We know that N species, especially nitrate but also organic N and, to a lesser extent, ammonium, may move down slope in laterally flowing water. This means that the down-slope soils receive more pollutant N per square metre than the vertically deposited N per square metre.
- Do plants respond to annual loads of N or variation in concentration of N supply throughout the year? Soils evolved to conserve N in winter months, so litter has a high C:N ratio so that N produced by mineralisation is immobilised in microbial biomass (as we saw in Chapter 10).
- Atmospheric N deposition disrupts the delicate natural dynamic balance between plant N needs and soil mineral N production, as discussed in Chapter 10. This may be the cause of biodiversity change.

This list of potential limitations to the N mass balance approach is not exhaustive, but hopefully has raised enough points to make you realise that we still fall well short of a fully robust critical loads methodology.

16.5.2 Limitations to critical loads for surface waters

Concerns over fresh water acidification were triggered initially by observations of fish kills or skeletal deformities in fish such as salmon, especially in Scandinavian lakes, and this instigated much research on the toxicity of aluminium to aquatic organisms, especially in waters with low calcium concentrations. The biodiversity of snails, mussels and crustaceans also declined markedly as waters became progressively more and more acidic (Cresser and Edwards, 1987). The biggest problem though with critical loads models for fresh waters is the inextricable link between soil acidification and surface water acidification. This link makes it very difficult to predict the rates of change of surface water chemistry, not only during the acidification phase but also during a recovery phase when acid deposition loads are reduced.

16.5.3 Is there an alternative approach?

Cresser (2007), prompted by concerns such as those expressed in the previous section, has suggested that critical loads for soils should be based on pollutant effective concentrations rather than upon deposition fluxes. As we have seen, this alternative has already been successfully used for peats in the UK, and is suitable for the most sensitive mineral soils derived from quartzite and acid sandstones (White *et al.*, 1995). The acidity experienced by plant roots in natural ecosystems depends highly upon the mobile anion concentrations, not on mean mobile anion annual fluxes. Thus higher effective concentrations of nitrate and sulphate, and hence H^+ too, other factors being equal, will occur in drier areas than in wetter areas; remember that the calculation of effective concentration takes into account the concentrating influence of evapo-transpiration, so the difference between drier and wetter areas will be even more extreme. Plants respond to concentrations of species in soil solution, not to annual fluxes. True, if a pollutant flux at a given site increases, its effective concentration will vary too, but remember that critical load maps are there to predict probability of damage across regional to national and international scales.

The sodium dominance index

When discussing the correlation matrix in Figure 16.14, brief mention was made of the sodium dominance index. This index is based upon the observation that when inputs of calcium and magnesium from biogeochemical

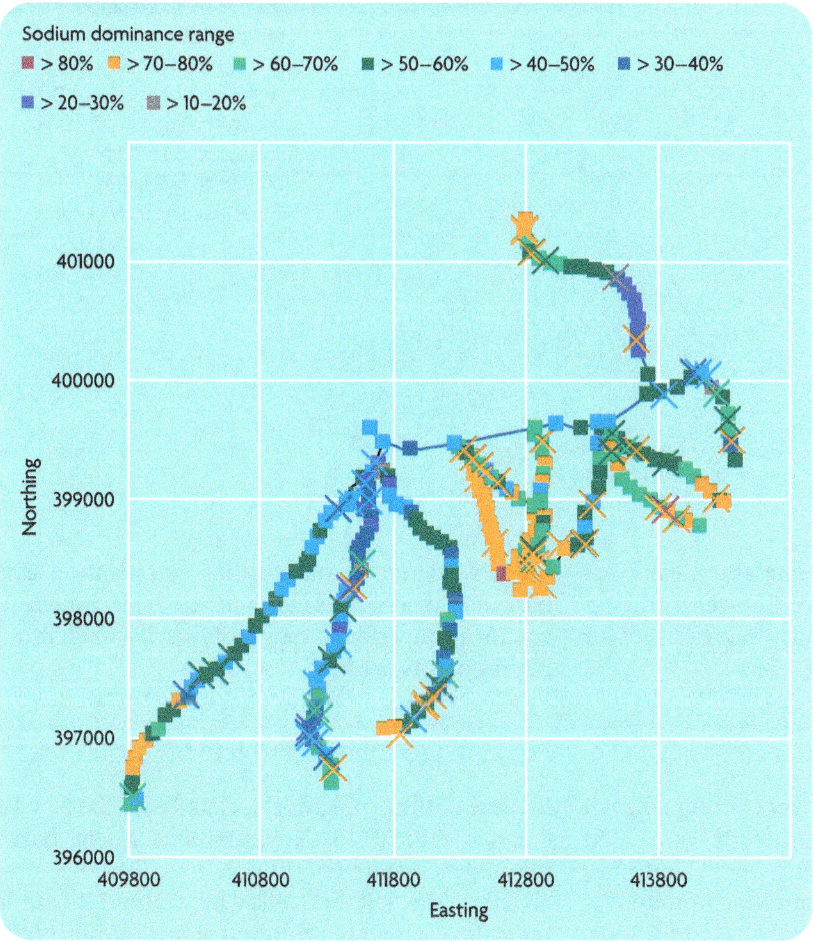

Figure 16.15 The spatial distribution of sodium dominance (100{[Na]/∑[Na] + [Ca] + [Mg]}, where square brackets denote concentrations of the ion species shown on a mol_c basis, across the River Etherow catchment in the South Pennines, UK.

mineral weathering in soils are low, the water draining from the soil, and hence solute composition of river water in the vicinity, is dominated by sodium of maritime origins (White *et al.*, 1999; Stutter *et al.*, 2002). Figure 16.15 shows the spatial distribution of sodium dominance, expressed as a percentage on a mol_c basis, in the River Etherow catchment. The link between Figures 16.13 and 16.15 is immediately obvious, as ammonium is being mobilised as a positively charged counter cation with dissolved organic acid anions that form most of the DOM. The high ammonium and high DOM concentrations are associated with the more organic-rich soils, which naturally occur in the more acidic parts of the catchment and have low weathering rates and high sodium dominance.

The results shown in Figure 16.16, taken from White *et al.* (1998), show that sodium dominance measured under base flow conditions is a good indicator of catchment soils with low weathering rates. Therefore there is a highly significant (R^2 = 0.791 for a simple quadratic equation) correlation between lowest recorded water pH (which is indicative of the severity of acid flush events) throughout the basin of the River Dee in North East Scotland and sodium dominance.

Results such as those presented in Figure 16.16 prompted White *et al.* (2000) to investigate whether sodium dominance could be used alone or with other parameters to predict critical loads of river water under diverse discharge conditions. They found that catchment maximum altitude

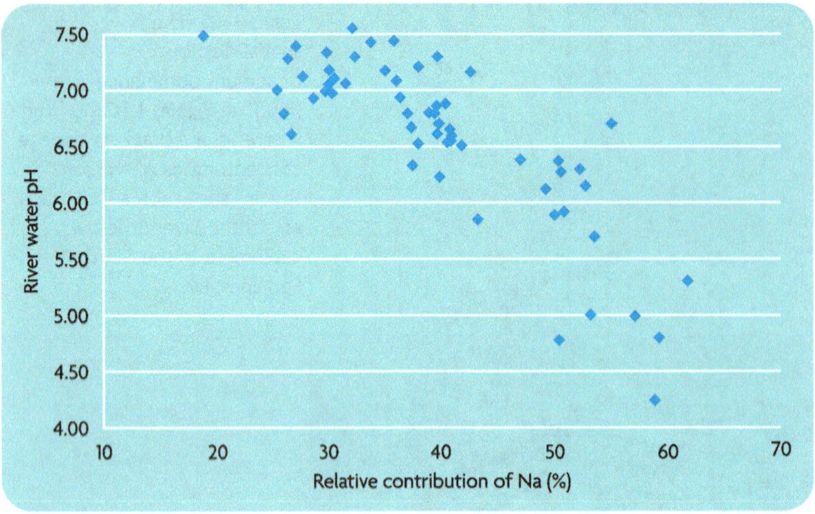

Figure 16.16 Correlation between lowest river water pH measured for 61 sub-catchments of the River Dee drainage basin in north-east Scotland and sodium dominance (per cent).
Source: Based upon White *et al.* (1999).

was a useful additional parameter as critical loads of acidity for stream waters were lower in higher catchments. The equation that they derived for mean annual data was:

$$CL = 57.9 - 3.46\ (\ln \text{max. alt.}) - 8.12\ (\ln \text{Na dom.})$$

Figure 16.17 illustrates the very strong correlation between predicted and directly calculated critical load. Strong correlations were also found for the most acidic conditions encountered at high flow and for the lowest base flow conditions. In the author's opinion the sodium dominance index warrants serious further investigation as a basis for assessing sensitivity of river waters to acidification effects, especially as its data input requirements are low.

Possible indicators of soil N pollution status

Because of the problems associated with the N mass balance critical loads assessment being based upon steady-state conditions, much thought was given in the UK and elsewhere about possible indicators of soil N pollution status of minimally managed ecosystems. For agricultural soils, in the

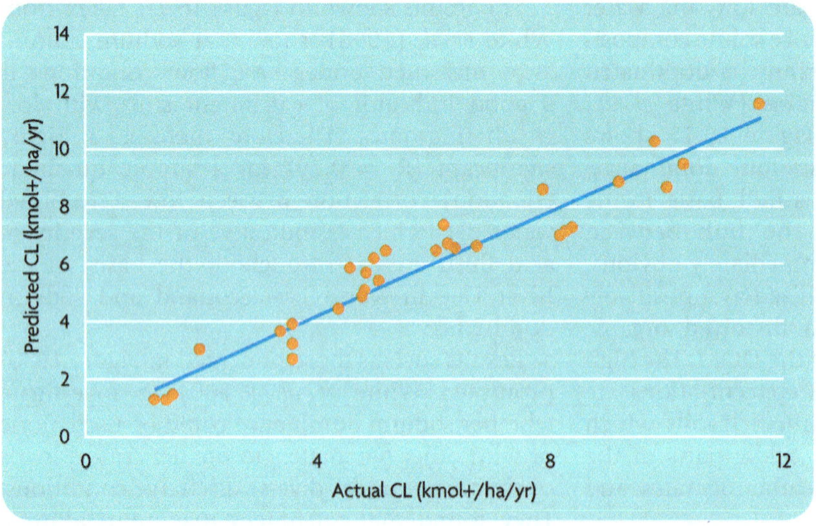

Figure 16.17 Calibration graph for predicting freshwater critical load (CL) from ln Na dominance and ln max. altitude, using annual mean data for River Dee catchment.
Source: Redrawn from White *et al.* (2000).

1970s and 1980s KCl-extractable ammonium-N and nitrate-N were extensively used in attempts to predict crop yields and N uptakes, but time and depth of sampling were (interactively) critical, which was a problem. Moreover, results varied between species and between years, as well as between soils. However, in experiments with *Calluna*/peat and with grass/peat microcosms, KCl-ammonium-N and nitrate-N concentrations increased significantly after 12 months of ammonium and nitrate treatments (Yesmin *et al.*, 1996). For upland ecosystems there are still several problems however. The results will depend on all N inputs/outputs and inputs/outputs of other elements as biogeochemical cycles interact. How they interact will depend upon climate, soil parent material, topography, soil age, soil management and land use, pollution inputs, soil depth, stoniness, up-slope characteristics, vegetation, drainage status, and even distance from the coast. In an unpublished PhD thesis from 1996, White looked at the ammonium saturation link to N deposition for L/F/H horizons of 36 *Calluna* moorland podzols derived from quartzites or sandstones in Scotland. She reported that:

- The Spearman's rank correlation coefficients between total N deposition and extractable ammonium or ammonium saturation were 0.520 and 0.574, respectively (both >99 per cent significance).
- Using ammonium saturation eliminates the effect of variation in CEC.

Ammonium saturation (i.e. percentage of the CEC occupied by ammonium) provides a better index than exchangeable ammonium alone of the extent to which the exchangeable ammonium is 'labile'.

Many prefer to look for N breakthrough, assessed by measuring N species concentrations in soil solution or drainage water. However, while we have seen earlier in this chapter that this is a useful indicator of damage already done, and possibly even of the extent of that damage, it is less well developed as an early warning indicator of ecosystem damage.

Many scientists have confidence in soil C:N ratio as an indicator of accumulating damage. In Chapter 19, Figure 19.6, it can be seen that the C:N ratio of peat falls markedly along an N deposition gradient in GB. However, a study by Billett *et al.* (1990) of Scots pine (*Pinus sylvestris*) mature forest soils highlighted the importance of taking into account what is happening in the carbon cycle, and not just considering the nitrogen cycle in isolation. When they re-examined the C and N concentrations in the surface horizons of the forest soils after an interval of almost 40 years they found that often the C:N ratio had increased rather than decreased (Figure 16.18). There was a large accumulation of nitrogen in the soils, but the carbon had accumulated so much faster than nitrogen in the acidified accumulating litter that C:N ratio had not declined. The fact that the N had accumulated therefore suggests that soil profile nitrogen storage (SPNS) might be a better indicator, once properly calibrated. This is discussed further in Chapter 19, section 19.2.1.

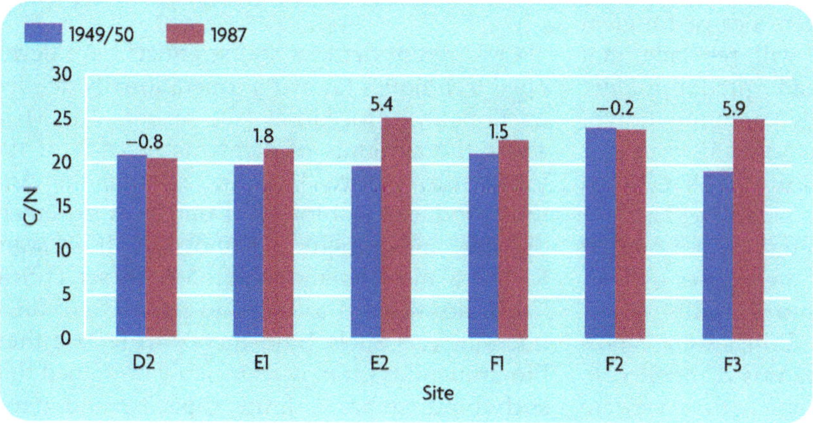

Figure 16.18 Changes in the C:N ratio of surface organic horizon soils at six mature Scots pine forest sites in a relatively remote region of north-east Scotland.

Source: Redrawn from Billett *et al.* (1990).

In upland areas that are heavily impacted by nitrogen pollution from the atmosphere, the rate of transformation of deposited ammonium to organic N or nitrate-N may slow dramatically compared with what happens on less polluted sites (see section 14.3 of Chapter 14). It is possible that a slow rate of change in ammonium concentration in ammonium-spiked soil might provide a useful test to indicate soil N saturation, though no such test has yet been developed adequately. Similarly, in anaerobic soils, high denitrification rate might provide a useful indicator. It is improbable though that soil N concentration *alone* will ever provide a useful indicator for soil N saturation, because it is so inextricably interlinked with soil organic matter content; the latter varies with many of the factors of soil formation discussed in Chapter 7, including parent material, climate, time, vegetation and topography.

16.6 Critical loads approaches for metal pollutants

So far in this chapter we have been concerned mainly with the impacts of gaseous molecular N and S species as pollutants in the atmosphere, although some of the N and S movement is as aerosol such as ammonium sulphate particulates. Metals too can move through the atmosphere. Alkyl lead compounds were very extensively used as an anti-knock agent in petrol a few decades ago, and these volatile organometallic compounds polluted not just urban air but also remote upland areas at appreciable distances from roads. In congested urban areas their potential neurotoxic effects were a cause for great concern (Smith, 1992). We will see later that accumulation in soil of lead, predominantly from the atmosphere, resulted in accumulation of up to 25 kg of the element ha^{-1} over 38 years in a Scots Pine forest at a remote Scottish Glen (Billett *et al.*, 1991 and Figure 19.3 in Chapter 19).

In the UK in 2005 the government announced rules for crematoria to limit the masses of mercury pollution discharged into the atmosphere when dental amalgam tooth fillings were vaporised in furnaces. The industry was compelled to fit mercury filtering equipment with a view to halving mercury emissions. Details of the abatement policy may be found on the Defra website. Other metal emissions come from vehicles and incinerators, as well as a range of industrial processes. In Europe it seemed a logical step to extend the critical loads approach to metallic pollutants and therefore we should at least briefly consider how this might be done and what the constraints are likely to be.

To use the critical loads approach usually we start by selecting one or more selective biological receptors of interest so that we can then use a dose-response relationship (as discussed in section 16.3) to relate an adverse biological response to a measurable chemical determinant. The problem for metal pollutants in soils and sediments is that the total amount of a metal pollutant in a soil does not normally provide a useful indication of the bio-available amount. The bio-available amount will depend upon the relative and total amounts of the pollutant metal of concern in each of the various soil components and other soil factors such as pH and redox conditions. Thus the bio-availability of a metal cation in exchangeable forms may be very different from that of the same metal bound up in amorphous or crystalline iron and aluminium hydrous oxide forms, or within carbonates or to organic matter or within inert mineral crystalline lattices.

Lead-in question

If you wanted to attempt to use a mass balance approach to set a critical load for a potentially toxic heavy metal, how would you set about it?

Over recent decades soil scientists have developed techniques known as operationally defined fractionation procedures (see section 7.10.5) to assess the amounts of elements in each of the forms listed above. However, relating the data thus produced to biological uptake is extremely difficult, and sometimes impossible. If we want to use a mass balance approach to set critical loads, do we need to know to what extent these fractions can safely bind metals to reduce their bio-availability? It may reasonably be argued that such binding has a finite capacity, and never provides a long term, sustainable solution to a metal pollution problem. It may allow soils to be abused for many decades before the steady-state situation is reached, but it is not a long-term solution. Eventually bio-availability will reach, and then exceed, a critical level. Ultimately therefore we can only consider inputs as acceptable if they are offset by outputs from harvested crops (such as a harvested forest in a forest cycle), an acceptable sustainable level of leaching to surface waters, and permanent losses by erosion as suspended sediment. Thus critical loads of metals should normally be very low for natural or minimally managed ecosystems. For agricultural systems the safe removal with harvest crops may push critical loads to higher values.

16.7 Are target loads an alternative?

It is tempting, because of the difficulties explained throughout this chapter in setting robust critical loads, to give up and set target loads instead. These may be set based upon national and international observations of what's happening in the field. For example, if there is no detectable change in Scots Pine woodlands anywhere where the deposition load is less than 15 kg N ha^{-1} yr^{-1}, it could be argued that it would be sensible to set that value as a universal target load for maximum acceptable N deposition. Some critical loads for N deposition have indeed been set on such a basis for semi-natural ecosystems. Of course it could be argued that it is too soon to set target loads as we're possibly not at steady state yet. If the damage occurs in a couple of decades time then the target load has to be reduced; the problem then becomes one of reversibility of damage, and that is another potentially complex issue.

16.8 Are critical levels a better alternative?

The conscientious and diligent reader may well have realised by now that the present author has serious reservations about the critical loads approach generally. Working with peat soils made him aware that effective concentrations are much more important to pollutant impacts than annual fluxes are for the reasons outlined. Even for mineral soils he believes that effective concentrations, or critical levels, are much more important than annual pollutant fluxes. Chapter 17 will show various methods for assessing the direct effects on vegetation of concentrations of pollutant gases in the atmosphere, and for gases critical levels are widely accepted as being the way to protect the environment. When setting such levels, if necessary, seasonality may be taken into consideration.

16.9 Why was the critical loads approach not applied to CO_2?

The critical loads concept is based upon tolerable loads of pollutant deemed sustainable in the long term that will not result in adverse deposition impacts on a selected biological receptor. For carbon dioxide the situation at first appears to be very different as the atmosphere is perceived as the sensitive receptor and the thrust is towards reducing climate change. Thus in section 8.8 of Chapter 8, when the dramatic increase in atmospheric carbon dioxide concentration was considered, all the emphasis was upon human impacts on global warming. The rate of increase was so fast and the potential threat of global warming so great that politicians generally decided that steps towards arresting the rate of increase had to be taken sooner rather than later. Wind farms and solar panels proliferated (Chapter 18) and green buildings such as that shown in Figure 16.19a–c shouted 'We're doing our bit'. Similarly, improved transport efficiency has become 'a thing to aspire to'.

Setting aside the potential global warming impacts, agronomists perceived increasing CO_2 concentration almost as a free fertiliser, as it would increase crop yields provided water and nutrient element supplies were maintained. But we saw in Chapter 4 that we need to know whether natural and minimally managed systems are sustainable as CO_2 concentrations in the atmosphere increase; in such systems plants often are growing on nutrient-poor soils, and may have

(a)

Figure 16.19 An example of a building (a) designed to be 'environmentally friendly', CEH Bangor. The cladding is with long-lived local slate and oak wood. The building has low emissivity glass windows that also maximise use of natural light (c). Skylights have solar panels built in for shading (b). The roof area collects rainwater for flushing etc. The gas fired combined heat and power unit is reputedly 90 per cent efficient and supplemented by ground source heat pumps and photovoltaic panels that double to provide some shading. The building employs extensive metering and natural/passive ventilation.

(b)

(c)

trouble meeting nutrient demands if the extra CO_2 increases photosynthetic rate; an added complication is that water use efficiency is changing because of changes in stomatal distributions on foliage (Woodin *et al.*, 1992).

Woodin *et al.* (1992) performed experiments with heather plants (*Calluna vulgaris*) growing in natural, nutrient-deficient soils at ambient and ambient plus 100 or plus 200 ppm CO_2; they observed strong suppression of growth in the first year, but enhancement in the second year. As was pointed out in Chapter 4, this highlights the problems with many experiments in this area, which have been run over relatively short periods and with annual crops; most, for reasons of economy, have involved unrealistically large and sudden increases in CO_2 concentration rather than a gradual increase over several years. The author believes that a real case could be made for setting critical levels for CO_2 in the atmosphere for such ecosystems, but as yet we do not have nearly enough information to do this.

We can be fairly sure from horticultural practices that physiological plant growth stages will be accelerated and flowering will occur earlier, as reported by Woodin *et al.* (1992). However, we know less about the consequences of this for insect populations and higher species on food chains. This is a lamentably under-researched area.

Ironically, although 'rising CO_2' and 'critical load' would rarely be seen in the same sentence, often a steady-state mass balance approach for carbon is used in models to make predictions about how best CO_2 and methane emissions can be brought under closer control. Much of this though is done with a view to seeing what ecosystem types and components should be targeted as part of an abatement strategy.

16.10 Ground level ozone

You will have seen in Chapter 3 that a layer of ozone in the upper atmosphere is recognised as having been important to the evolution of life on Earth because of its capacity to absorb harmful ultra-violet radiation from the sun (section 3.4); the depletion of this ozone layer was a cause of real concern (section 3.11.3) because of enhanced risk of skin cancer and malignant melanomas. Later in the same chapter it was explained how the emission of nitrogen oxides and hydrocarbons, primarily from vehicles in urban areas, may lead to the generation of ozone at ground level, generally building to peak concentrations in mid to late afternoon because of the importance of energy from sunlight in the dissociation of nitrogen dioxide ($N_2O \rightarrow NO + O$, see section 3.11.2). We saw too that methane can convert NO to NO_2, increasing the capacity for generating O and hence O_3. Ozone concentrations typically are around 30–40 ppb in the background atmosphere, but can exceed 10 times this concentration in peak periods in heavily polluted areas.

When the author was young (sadly more than five decades ago) railway posters occasionally urged visits to the seaside to breathe in the ozone in the healthy sea air. We know now that ozone is not good for human health (section 3.11.2) and, ironically, that the supposed healthy ozone smell was in practice more probably from volatile organics from decomposing seaweed. More recently there has been growing concern about possible adverse impacts of accumulated concentration hours of O_3 exposure above critical threshold levels upon the growth of crops and natural vegetation species.

Critical levels have been set for ozone based upon the concept of accumulated ozone time (AOT) because of the accumulative nature of damage due to exposure. Suppose, for example, that it has been decided that 40 ppbv (ppb by volume) is a critical threshold value, and continuous monitoring of ozone concentration shows that concentrations reach 60 ppbv for 5 hours a day every day over a 70-day period. This would mean the concentration would be 20 ppbv above the threshold for 5×70 hours, or 350 hours. The AOT40 would therefore be 20×350 or 7000 ppbv-hours. This is a simplified example to show how the concept works, and in practice of course concentrations might vary appreciably over the period when the critical threshold concentration is exceeded. This would be taken into account in the calculation of exposure. A critical level of 3000 ppbv-hours over three months is used in Europe for exposure of agricultural crops or vegetation of natural ecosystems to ozone.

Numerous studies have been made of the adverse effects of ozone upon agricultural crops and upon natural vegetation. For example, Felicity Hayes, Harry Harmens, Gina Mills and Phil Williams at the Centre for Ecology and Hydrology at Bangor collected 33 plant species from Snowdonia in the uplands of North Wales and exposed them to ozone under simulated episodic conditions over 10 weeks in solar domes and made weekly assessments of the incidence and extent of ozone damage. Figure 16.20 illustrates the ozone fumigation regime in control (domes 5

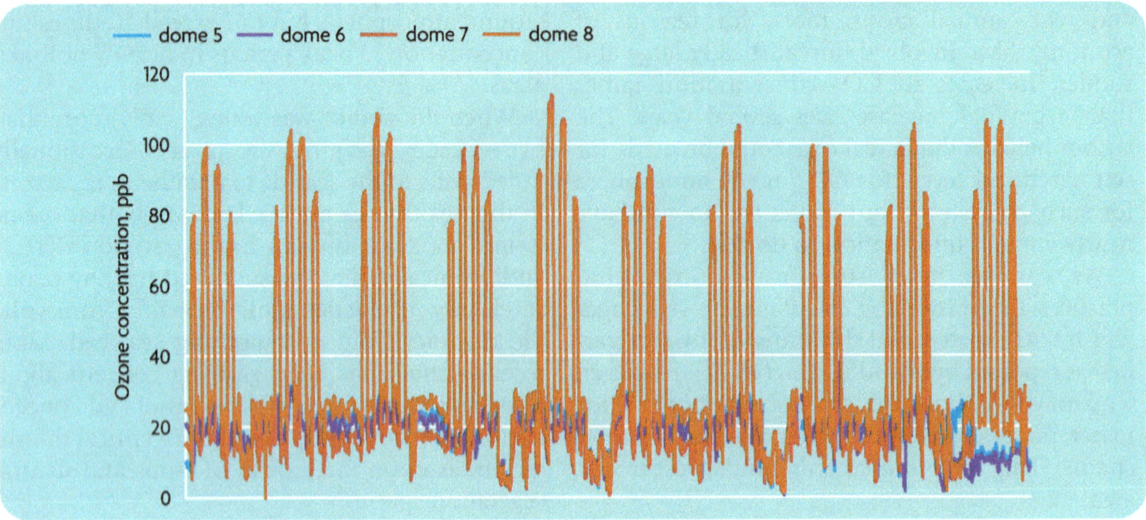

Figure 16.20 Changes in ozone concentrations (ppb) in solar domes at CEH Bangor during simulated exposure to ozone episodes over a ten-week period for control domes (5 and 6) and treatment domes (7 and 8).
Source: Chart kindly provided by Felicity Hayes.

and 6) and ozone-exposed (domes 7 and 8) solar domes (typical solar domes can be seen in Chapter 17, Figure 17.33).

Visible injury was noted after two-weeks exposure at the test levels used in *Potentilla erecta, Dryas octopetala, Oxalis acetosella, Eriophorum angustifolium, Carex panacea and Carex echinata* and after 10 weeks in *Nardus stricta*. Examples of damage to two of these species are shown in Figure 16.21.

Figure 16.21 Ozone specific injury symptoms on (a) *Carex echinata* and (b) *Dryas octopetala*.
Source: Photographs kindly provided by Felicity Hayes.

16.11 Conclusions

In many ways the content of this chapter differs deliberately from that of most earlier chapters. Much of the discussion centres around difficulties in quantifying tolerable pollutant burdens when attempting to establish pollution abatement policies for our highly dynamic planet. The author felt that is was important to convey to readers that establishing robust criteria for setting pollution loads is by no means a trivial task, and this poses a problem for both scientists and policy makers. Often the scientists disagree among themselves upon the best way forward, so compromise is usually necessary. At the same time the chapter highlights the fact that we do not always get everything right first time, and that environmental science is still to some extent in its infancy, with much scope for innovative and exciting developments.

In this chapter we have concentrated upon environmental effects of selected key atmospheric pollutants, rather than the requirements arising from implementation of abatement strategies; however, the thoughtful reader should be able to suggest numerous obvious options that would go towards reducing impact problems. These include, for example, improved public transport and other measures to reduce consumption of fossil fuels, improved combustion engine efficiency, enforced speed limits for road vehicles, and the removal of sulphur emissions from power stations and industrial emissions. You should by now be able to evaluate the impact such policies would have upon deposition impacts for all the atmospheric pollutants considered in this chapter.

POLICY IMPLICATIONS

Atmospheric pollution

- Policy makers need to be able to assess the *sustainable* concentrations of atmospheric pollutants over a range of timescales.
- Policy makers need to be aware of the need to assess actual and potential impacts of atmospheric pollution upon sensitive receptors including humans, animals, plants, soils, surface waters and groundwaters.
- Policy makers need to consider more carefully the robustness of the critical loads approach and the consequences of exceeding critical loads.
- Policy makers need to consider carefully the acceptability of using target loads in policy formulation as an interim measure when ensuring critical loads are not exceeded would be prohibitively expensive.
- Careful consideration needs to be given to integration of transport and energy policies and pollution abatement strategies when formulating policies.
- Policy makers need to be fully aware of the trans-boundary nature of air pollution problems and accept international as well as national responsibility when formulating abatement strategies.

CHAPTER REVIEW EXERCISES

Exercise 16.1

Why might you expect the reversal of acidification of water in an upland river in northern European in response to a rapid halving of sulphur deposition to be slow? Why would the rate of pH recovery vary with the size of the catchment?

Exercise 16.2

A scientist, convinced that acid flushes during sudden snowmelt were natural and nothing to do with acid rain, calculated the total amount of acid per hectare in the snow cover in his region and arranged for his students to spray sufficient dilute sodium hydroxide over the snow surface to neutralise all the acid in the snow pack to pH 7. He then showed that an acid flush still occurred during snowmelt. What could he conclude from this result?

Exercise 16.3

How might the interactions between the carbon and nitrogen cycles influence critical load of nitrogen estimations based upon the N mass balance?

REFERENCES

Aber, J.D., Nadelhoffer, K.J., Steudler, P. and Melillo, J.M. (1989) Nitrogen saturation in northern forest ecosystems. *BioScience*, **39**, 378–386.

Ågren, G.I. and Bosatta, E. (1988) Nitrogen saturation of terrestrial ecosystems. *Environmental Pollution*, **54**, 185–197.

Battarbee, R.W., Flower, R.J., Stevenson, A.C. and Rippey, B. (1985) Lake acidification in Galloway: A palaeoecological test of competing hypotheses. *Nature (London)*, **314**, 350–352.

Begum, S., McClean, C.J., Cresser, M.S. and Breward, N. (2010) Can available sediment data be used to predict alkalinity and base cation status of surface waters? *Science of the Total Environment*, **409**, 404–411.

Billett, M.F., Fitzpatrick, E.A. and Cresser, M.S. (1990) Changes in carbon and nitrogen status of forest soils organic horizons between 1949/50 and 1987. *Environmental Pollution*, **66**, 67–79.

Billett, M.F., Fitzpatrick, E.A. and Cresser, M.S. (1991) Long-term changes in the Cu, Pb and Zn contents of forest soil organic horizons from northeast Scotland. *Water, Air and Soil Pollution*, **59**, 179–191.

Black, K.E., Lowe, J.A.H., Billett, M.F. and Cresser, M.S. (1993) Observations on changes in nitrate concentrations along streams in upland moorland catchments. *Water Research*, **27**, 1195–1200.

Cresser, M.S. (2007) Why critical loads should be based on pollutant effective concentrations, not on deposition fluxes. *Water, Air and Soil Pollution: Focus*, **7**, 407–412.

Cresser, M. and Edwards, A. (1987) *Acidification of Freshwaters.* Cambridge Environmental Chemistry Series, Cambridge 136 pp. Republished as a paperback in 2010.

Cresser, M.S., Yesmin, L., Gammack, S.M., Dawod, A. and Billett, M.F. (1997) The physical and chemical 'stability' of ombrogenous mires in response to changes in precipitation chemistry. *Blanket Mire Degradation: Causes, Consequences and Challenges, British Ecological Society Mires Discussion Group Special Publication*, 153–159, MLURI, Aberdeen.

Edwards, A.C. and Cresser, M.S. (1985) Design and preliminary evaluation of a simple fractionating precipitation collector. *Water, Air, and Soil Pollution*, **26**, 275–280.

Franks, J. (1983) Acid rain. *Chemistry in Britain*, June, 504–509.

Henriksen, A., Dickson, W. and Brakke, D.F. (1986) Estimates of critical loads to surface waters. In: Critical loads for sulphur and nitrogen (eds J. Nilsson and P. Grennfelt). Nordic Council of Ministers, Copenhagen, pp. 87–120.

Henriksen, A., Kämäri, J., Posch, M. and Wilander, A. (1992) Critical loads of acidity: Nordic surface waters. *Ambio*, 21(5), 356–363.

Hornung, M., Bull, K., Cresser, M., Hall, J., Langan, S., Loveland, P. and Smith, C. (1995a) An empirical map of critical loads for soils in Great Britain. *Environmental Pollution*, **90**, 301–310.

Hornung, M., Bull, K.R., Cresser, M., Ullyet, J., Hall, J.R., Langan, S., Loveland, P.J. and Wilson, M.J. (1995b) The sensitivity of surface waters of Great Britain to acidification predicted from catchment characteristics. *Environmental Pollution*, **87**, 207–214.

Kinniburgh, D.G. and Edmunds, W.M. (1986) *The Susceptibility of UK Groundwaters to Acid Deposition*. Hydrogeolgical report 86/3, British Geological Survey, Wallingford, UK.

Nilsson, J. and Grennfelt, P. (eds) (1988) *Critical Loads for Sulphur and Nitrogen: Report 1988:15*. Nordic Council of Ministers, Copenhagen, Denmark.

Skiba, U. and Cresser, M.S. (1989) Prediction of long-term effects of rainwater acidity on peat and associated drainage water chemistry in upland areas. *Water Research*, **23**, 1477–1482.

Skiba, U., Cresser, M.S., Derwent, R.G. and Futty, D.W. (1989) Peat acidification in Scotland. *Nature*, **337**, 68–69.

Smith, S. (1992) Ecological and health effects of chemical pollution; Chapter 8 in *Understanding Our Environment: An introduction to Environmental; Chemistry and Pollution*, 2nd edn, Harrison, R.M. (ed.) Royal Society of Chemistry, Cambridge, pp. 245–295.

Smith, C.M.S., Cresser, M.S. and Mitchell, R.D.J. (1992) Sensitivity to acid deposition of dystrophic peat in Great Britain. *Ambio*, **22**, 22–26.

Stutter, M., Smart, R. and Cresser, M.S. (2002) Calibration of the sodium base cation dominance index of weathering for the River Dee Catchment in North-east Scotland. *Applied Geochemistry*, **17**, 11–19.

Sverdrup, H. and Warfvinge, P. (1988) Weathering of primary minerals in the natural soil environment in relation to a chemical weathering model. *Water, Air and Soil Pollution*, **38**, 387–408.

White, C.C. and Cresser, M.S. (1995) A critical appraisal of field evidence from a regional survey for acid deposition effects on Scottish moorland podzols. *Chemistry and Ecology*, **11**, 117–129.

White, C., Dawod, A., Cruickshank, K., Gammack, S. and Cresser, M. (1995) Evidence for acidification of sensitive Scottish soils by atmospheric deposition. *Water, Air and Soil Pollution*, **85**, 1203–1208.

White, C.C., Smart, R. and Cresser, M.S. (1998) Effects of atmospheric sea salt deposition on soils and freshwaters in northeast Scotland. *Water, Air and Soil Pollution*, **105**, 83–94.

White, C.C., Smart, R., Stutter, M., Cresser, M.S., Billett, M.F., Elias, E.A., Soulsby, C., Langan, S., Edwards, A.C., Wade, A., Ferrier, R., Neal, C., Jarvie, H. and Owen, R. (1999) A novel index of susceptibility of rivers and their catchments to acidification in regions subject to a maritime influence. *Applied Geochemistry*, **14**, 1093–1099.

White, C.C., Smart, R. and Cresser, M.S. (2000) Spatial and temporal variations in critical loads for rivers in N.E. Scotland: A validation of approaches. *Water Research*, **34**, 1912–1918.

Woodin, S., Graham, B., Killick, A., Skiba, and Cresser, M. (1992) Nutrient limitation of the long-term response of heather [*Calluna vulgaris* (L.) Hull] to CO_2 enrichment. *New Phytologist*, **122**, 635–642.

Yesmin, L., Gammack, S.M. and Cresser, M.S. (1996) Changes in N concentrations of peat and its associated vegetation over twelve months in response to increased deposition of ammonium sulfate or nitric acid. *Science of the Total Environment*, **177**, 281–290.

CHAPTER 17

How do we quantify biogeochemical cycles?

Malcolm Cresser

Learning outcomes

By the end of this chapter you should:

- Understand how inputs and outputs of nutrients to catchments are measured in practice.
- Be more aware of how to deal with changes in sample chemical composition over space and time.
- Be more aware of some problems associated with sample stability and contamination.
- Appreciate the merits of simulation experiments at a range of scales.
- Know how to set up a sampling programme for studying element cycling.
- Better appreciate some of the constraints imposed by harsh climatic conditions.
- Be able to calculate element fluxes from raw field data.

17.1 Introduction

So far we have considered how important the cycling of water is at local and global scales (Chapter 2), and how and why the cycling of carbon (Chapter 4) and nitrogen and several other elements (Chapter 5) is essential to every facet of life on earth. We saw that all element cycling is driven by solar energy, either directly or indirectly. We have seen too (Chapter 3) how important the carbon cycle is to global climate and to our understanding of climate change (Chapter 8). Up to this point, though, we have only considered the mechanisms of cycling. We have not examined how appropriate numerical values may be found that can be applied to each step in the cycling process of an element. That is what this chapter is all about.

Lead-in question

What would you need to know to be able to work out how much calcium is being lost each year from the soil of an upland stream catchment?

17.2 Nutrient balances in catchments

Consider a stream in a typical upland catchment in Northern Europe or a wetter region of the USA or Australia. What dictates how much calcium will be carried out of the drainage basin in the stream water? From Chapter 7 it should be clear that the concentration of calcium carried in the river water at a particular point in the stream depends on the sources of water draining into it upstream. It may contain contributions from water draining out of near-surface soils, or from groundwater, or from pollution sources such as drainage outlet pipes from roads or sewage treatment works, etc. To know how much calcium is being transported in the river, we need to know not just the concentration of calcium, but also the amount of water flowing per unit time. For example, if we know the concentration in grams per litre, and the volume of water flowing past us in litres per second, we can calculate the flux of calcium flowing past us in grams per second simply by multiplying the two values together.

If we know the fluxes (g s^{-1}) coming into the river from pollution sources too, and subtract these from the above product, we have a value for how much calcium per second is being lost from soil and the underlying weathered rocks that may provide a groundwater component. Of course, this value only applies at the particular time the sample was taken and the flow of water was measured. We may need many such values to find out how much calcium is being lost per year. We will consider how best to do this a little later. First there is another input flux that we need to consider if we want to quantify how fast the soils in the catchment are losing calcium, namely that from the atmosphere, either in rain and snow (precipitation) or as terrestrial dust or aerosols.

17.2.1 Monitoring discharge in streams

Accurately monitoring discharge (the amount of water flowing in m^3 s^{-1}) in streams can be quite a challenging problem. If the water velocity was uniform across a river channel and with depth, and the stream had a nice uniform cross-section along its length (that you could easily measure), all you would need to do is measure the water flow velocity (m s^{-1}) at a single point and the cross-sectional area (m^2). Multiply the two (metres per second × square metres) and you would have the discharge in cubic metres per second (m^3 s^{-1}). You can estimate flow velocity by timing the movement of small floating sticks between two points a measured distance apart (apples are easier to spot in a bigger river, but you can lose part of your lunch!).

In nature, things are never that simple. River channels are often highly non-linear and cross sections usually very irregular. Therefore generally a flume or weir has to be installed to measure discharge accurately. Figure 17.1 overleaf illustrates a concrete flume at the Glen Dye catchment in NE Scotland, an area studied extensively by the author. Such structures are massive and would be very expensive to install for experiments lasting

Figure 17.1 A concrete flume at Glen Dye in NE Scotland, under very high flow conditions. Flow passes over the outer limbs, so it can still be accurately measured. Under low flow conditions the water is confined to the narrower, central channel.

Figure 17.2 The concrete flume at Glen Dye under low flow conditions, with flow confined to the narrower, central channel. Note the presence of the autosampler on the river bank.

only a year or two. The structure here had been installed already because the river was used to feed a potable water supply. If you look at the sizes of some of the granite boulders carried by the river following intense storm events it is immediately obvious why the structure needs to be so robust. Once installed, the discharge can be found from calibration graphs relating discharge to water height. Figure 17.2 shows the same flume under low discharge conditions in the summer period. In this picture you can see that an automatic sampler has been installed in a small hut beside the flume to collect water samples automatically. The need for this is explained in the next section.

The River Dye is a quite substantial stream, and is subject to very substantial swings in flow; hence the need for such a substantial flume. In much smaller streams such as that in Figure 17.3, more modest flumes may be used. The one shown here was constructed from fibreglass by a company that usually makes small boat hulls. It was designed to allow monitoring throughout the year, and so that water flow should only exceed the monitoring capacity of the flume about once every hundred years. You can see in the picture that it has been strengthened with substantial timbers, to stop the deformation of the structure by ice in mid-winter.

Figure 17.4 shows the stilling well beside the flume, which is there to allow water height to be accurately measured. At the time of installation the height measuring device chosen was clockwork, because of lack of electrical power at the remote field site. Nowadays alternative measurement electronic systems are more often used, but careful consideration needs to be given to snow and icing problems if solar power is used, and to effects of extreme temperatures if battery power is chosen. We will return to the problems of dealing with extreme weather conditions later in this chapter. The author learned the hard way that frozen ink doesn't flow well across chart paper!

It takes considerable effort to install even quite small flumes such as that shown in Figures 17.3 and 17.4. Many researchers therefore favour

Figure 17.3 A fibreglass flume at the Glenbuchat catchment in NE Scotland in the UK.

Figure 17.4 The stilling well and discharge height-measuring device attached to the fibreglass flume shown in Figure 17.3.

devices known as V-notch weirs to record discharge. These are easier to install, less expensive, and commercially available. A typical system, from a Norwegian catchment study, is shown in Figure 17.5.

Figure 17.5 A typical V-notch weir. This weir was photographed by the author at a research catchment in Norway.

17.2.2 Using autosamplers

Monitoring changes in water chemical composition over periods when discharge is changing rapidly is very important if calculated fluxes of elements being transported in rivers are to be accurate. This is because the relative contributions of water following different hydrological pathways in drainage basins may change substantially during rainstorm events and during periods of rapid snow melt. This aspect has been discussed fully by Cresser, Killham and Edwards (1993).

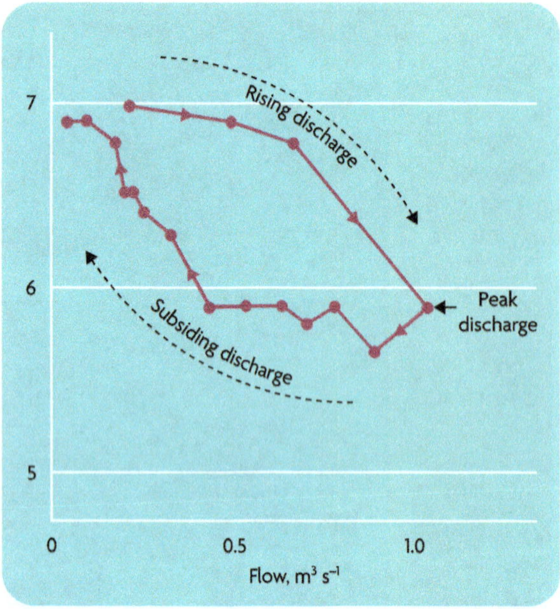

Figure 17.6 Sketch showing the typical relationship between water pH and discharge during, and immediately after, a storm event in a minor upland catchment in northern Europe.

Water draining from surface soils often becomes relatively much more important at high flow, for example. Soils acidify from the top down, and as a result water draining from near surface soils in upland areas with moist climates is more acidic and contains less calcium, silicon and hydrogen-carbonate (alkalinity), as these species are strongly associated with mineral weathering. The surface soils generally have a higher organic matter content, however, so water at high discharge tends to contain much more dissolved organic carbon (DOC). Figure 17.6 shows a plot of pH of river water against discharge during a minor storm for a small upland stream in Scotland. Water pH falls as flow increases, and declines after the rain stops as the flow subsides. Note that the plot does not follow the same path during the subsiding flow phase. It is said to display a hysteresis effect.

Lead-in question

Why does the plot of pH versus discharge in Figure 17.6 follow different paths during the rising and falling discharge?

During rising discharge, unless the soil is already very wet, the soil will be wetting up to depth. Initially therefore, relatively little water will drain laterally down slope through or over surface soil horizons. Thus at first relatively little water will emanate from the surface soil layers, and water from lower in the profile, which is less acidic (has a higher pH), tends to be more dominant. More acidic water from near the soil surface becomes progressively more important over time if heavy rain continues. Once the rain has stopped, however, drainage from saturated surface horizons often remains dominant initially, but not for long. Less acidic water from lower, more mineral-rich soil layers starts to become relatively more important, so river water pH starts to increase. Eventually it will get back to its pre-storm value.

In upland areas where precipitation significantly exceeds evapo-transpiration, the onset of significant amounts of lateral drainage is generally quite rapid. This means that river discharge often rises quite rapidly in response to precipitation events. Figure 17.7, for example, shows how quickly flow in the River Dye in Scotland rises and subsides in response to a double storm event. The scale of the initial response depends upon rainfall intensity and how wet the soil is when the rain starts. Note that the river, however, is close to baseflow conditions for much of the time. It is tempting to assume that changes in water chemistry during events do not matter very much in the context of biogeochemical cycles. However, this is far from the truth. Even although the duration of storms may be relatively short, and the concentration of ions such as Ca^{2+} and HCO_3^- falls at high discharge, the proportional decline in their concentration is generally much lower than the proportional increase in discharge (m^3 of water s^{-1}). Thus the fluxes of these ions ($g\ s^{-1}$) may increase dramatically during storm events. Not taking storm events into account may thus lead to serious systematic underestimates of annual fluxes.

The fact that rivers may be at or close to baseflow for most of the year, sometimes for > 90 per cent of total time, poses problems for those who want to know how water chemistry changes during precipitation events. Unless the scientist is very close to the site, it is difficult to sample

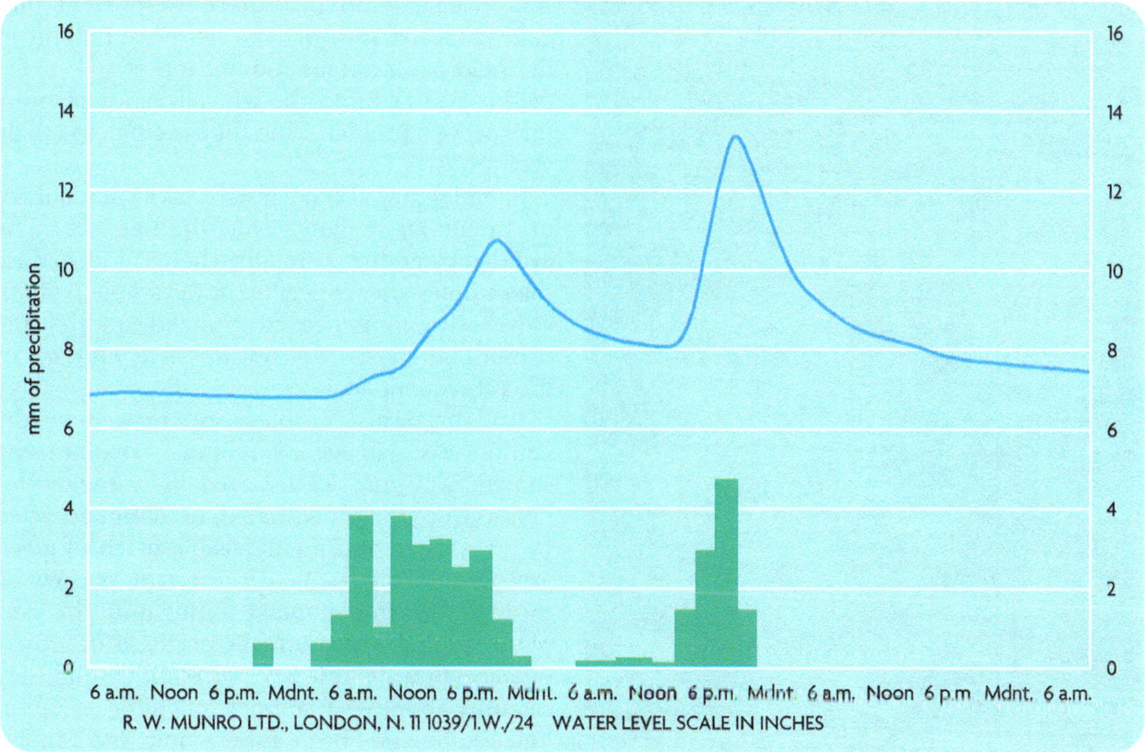

Figure 17.7 Relationships between water discharge and time (solid line) and rainfall and time (vertical bars) before, during and immediately after a double storm event at the Glen Dye upland catchment in north-east Scotland. The vertical scale shows precipitation in mm. Mdnt. indicates midnight.

manually through storm events. Even if he or she lives 'on the doorstep', sampling in the dark every 15 minutes for 12 hours, starting with the onset of a storm at one in the morning may not be either an attractive proposition or a safe thing to do. Autosamplers, such as that shown in Figure 17.8, are an obvious answer.

The sampler shown is battery powered for use at remote sites. It contains 48 sample bottles in three racks each containing 16 bottles. The bottles are fed in turn by a rotating chute from a sample uptake tube. A timer allows the sampling interval to be pre-set, for example for every hour for intensive event sampling or possibly every four hours for reasonably representative sampling of storm discharge events. At the latter sampling frequency the bottles would need to be changed every eight days, since six of the 48 bottles would be used every 24 hours. For more intensive sampling frequency, it is best if the autosampler is activated remotely when heavy rain commences or is forecast. Before 'taking a sample', the autosampler is designed to flush out the sample uptake tube with river water. Thus the uptake pump is switched on until water just reaches the sample discharge funnel, then reversed to back flush the uptake tube before taking the actual sample.

You might think that using autosamplers solves all our sampling problems for events, but sadly this is not the case.

Lead-in question

What problems do you think you might face with the type of autosampler described above in routine use?

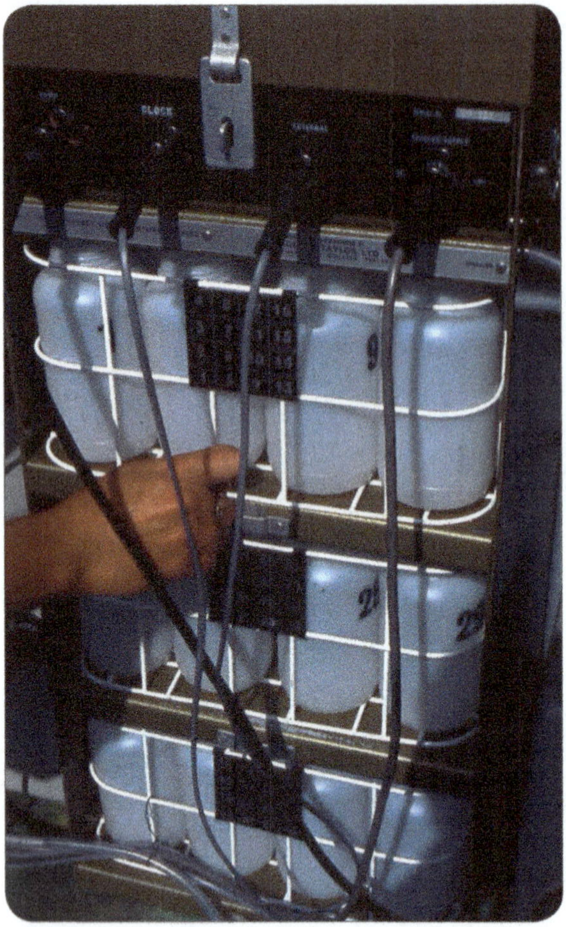

Figure 17.8 A typical autosampler of the type useful for studying changes in water chemistry over time. The housing for the sampler was shown in Figure 17.2.

The samples collected by autosamplers such as that described here are stored in open bottles between collections by the researcher. That is fine if they are perfectly stable, but they may not be in practice. For example, river water is often supersaturated in dissolved carbon dioxide. This is because water draining from soils may have equilibrated with the soil atmosphere containing 1 per cent or more of CO_2. Normal air of the atmosphere contains much less CO_2, so the water re-equilibrates, losing CO_2, and hence losing carbonic acid, H_2CO_3. Thus the water pH will rise until a new equilibrium is reached. When spot sampling we get around this problem by completely filling sample bottles or syringes under water, leaving no headspace, to prevent outgassing. Better still, pH can be measured out in the field. Continuous monitoring of pH in the field is also possible, though calibration stability may prove a problem, and the system needs to be very robust.

Samples may also be unstable as a consequence of microbial activity. This problem may be avoided by adding a suitable chemical to sterilise the samples after collection, for example thymol. Obviously care is needed to make sure that the compound chosen will not interfere with any of the subsequent analyses.

An unexpected problem experienced by the author was damage by animals. The housing shown in Figure 17.2 served as a shelter for rabbits once the snow started, no doubt attracted by an isolated and highly visible patch of green vegetation! This resulted in the rabbits chewing through the sample uptake tubing until the inert plastic tubing was eventually enclosed in protective steel pipe. It is always important to think very carefully about what animals might do to sampling equipment. Sheep, for example, can damage ground level equipment by random grazing, and cows by unintentional trampling or using posts for the occasional vigorous scratch.

Excluding grazing animals such as rabbits may be achieved with wire net fencing, but remember that many small animals will rise to the challenge and attempt to tunnel under the fence. The wire netting should be buried below ground to around 20 cm depth, but also bent outwards for about 25 cm. The rabbits etc. are invariably discouraged when they conclude that the wire extends horizontally all over the landscape below the soil surface.

The final problem that we need to consider is inclement weather, especially snow and freezing conditions.

Lead-in question

Look at Figure 17.9, which shows a small Scottish upland catchment in January. What sampling problems do you think that you would you encounter?

Figure 17.9 A typical small Scottish highland catchment (Peatfold, Glenbuchat) in January.

Source: Courtesy of Dr John Creasey.

17.2.3 Practical problems with inclement weather – need for risk assessments

Monitoring changes in water chemical composition over winter periods can pose a number of problems. If water samples are taken using autosamplers, either the sample uptake probe may act as an ice nucleation centre, or samples themselves may freeze after they have been collected. When water freezes over a few hours, solute species may become concentrated in the initially unfrozen core, resulting in chemical precipitation. The latter may not always be reversible when the samples thaw back in the warmth of the laboratory, so even if a set of replacement bottles is available, problems remain. Ice nucleation may also be problematic if continuous monitoring is used. If preventative heating is used, care is then needed to make sure that it does not modify samples anyway.

The other potentially serious problem is safety and accessibility. No fieldwork programme should ever be embarked upon without a careful and very thorough risk assessment being completed first. Working individually in winter conditions should be avoided if at all possible, and careful consideration should be given to what would be done in the event of an accident, getting trapped by sudden excessive snowfall, etc. As a result of hypothermia, survival time may be very short if inadequate precautions are taken. It should also be remembered that in the northern hemisphere in mid-winter, day length may become very short, limiting what can be safely achieved in a working day.

Lead-in question

How would you set about measuring fluxes of elements deposited in rain and snow?

17.2.4 Measuring inputs in precipitation

Monitoring changes in the chemical composition of precipitation water, in theory at least, seems simpler than measuring river water chemistry. Rainwater composition varies throughout any individual storm as the first rain falling scrubs gaseous and particulate species from the atmosphere. However, this fact is generally ignored by those studying biogeochemical cycles. This is because if a rain gauge has a collecting area of, say, 1 m^2, the flux of an element it collects in g in a year simply has to be multiplied by 10^4 to calculate the deposition flux on a g ha^{-1} yr^{-1} basis. Typically the volume of water collected by a rain gauge would be measured weekly or bi-weekly, and a sub-sample taken for analysis. In other words we assume that the gauge collects a truly representative sample for the sampling area and interval chosen. Thymol may again be added to sterilise samples, but CO_2 out-gassing is not a problem.

Measuring volume of precipitation requires remembering to take a measuring cylinder out to the field, and this should be of appropriate volume and, preferably, unbreakable, as well as chemically inert. Modern laboratory plastic ware fits the bill well, as shown in Figure 17.10, for example. Here a small sub-sample has been taken for analysis in a plastic bottle.

Note that the rain gauge funnel in Figure 17.10 is not a simple, horizontal funnel. It has a fine nylon mesh attached vertically above the funnel. This was done on this gauge in an attempt to collect deposited aerosol as well as simple precipitation, as vegetation catches such aerosols. The problem, however, is knowing just how

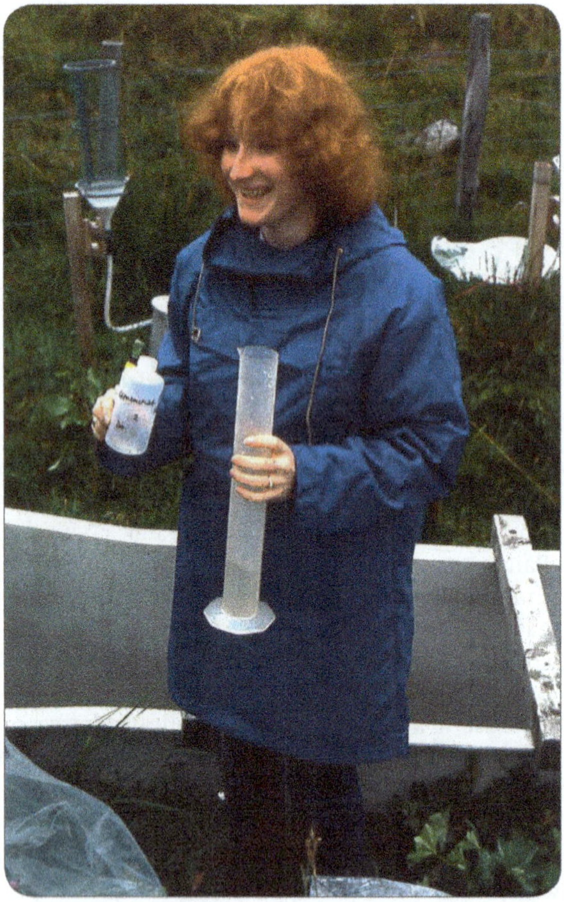

Figure 17.10 Taking a rainwater sub-sample after measuring the volume collected in the rain gauge (shown in the top left-hand corner).

much of the naturally collected aerosol is caught by such a gauge. A crude calibration can be made by comparing extra chloride collected by the modified gauge with the extra chloride found under vegetation, but this is not really a totally reliable method. It may come as a surprise to the reader that in some instances there are no absolutely correct or incorrect methods. This undoubtedly contributes to the uncertainty associated with calculated fluxes.

Another problem with simple funnel-based rain gauges is that they seem to be very attractive to local inquisitive bird populations. Unfortunately, realisation that they have not found a bird-bath sometimes results in them putting the gauge to

Figure 17.11 Coarse filter used to minimise effect of decomposing, wet litter in a gauge collecting water at a forest site.

alternative use – for defecation – with unfortunate consequences for nutrient balance calculations. For this reason often a series of spikes may be attached around the gauge perimeter to discourage perching birds. Small trapped insects are another common problem, and often a fibre filter is used to prevent this from happening. This is essential if precipitation samples are collected in a forest canopy anyway, to minimise the effects of contributions from decomposing litter caught in the gauge. Figure 17.11 shows the type of system used.

Birds and insects are not the only problem with respect to contamination. Figure 17.12, shows a precipitation sample from a bin gauge on the Balmoral Estate in Scotland. To deal with freezing and prevent metal contamination, samples for analysis were collected in clean, polythene bags. Unfortunately on this occasion well-meaning hill walkers had mistaken the rain gauge for a deluxe litter bin – hence the orange peel and chocolate wrappers. The 'vandalism' on this occasion was almost certainly unintentional. Contamination sadly is not always unintentional, and the risk should be considered carefully when planning field experiments.

Plastic bags allow frozen samples to be melted and measured. However, they do not totally deal with the problem of reliable sampling of snow volumes. Snow blows on and over the ground, and it is therefore difficult to quantify its inputs

Figure 17.12 A rainwater sample contaminated by hill walkers, who mistook the gauge for a litter bin.

Figure 17.13 A roofed-sampler at a Scandinavian site, designed to measure aerosol deposition.

precisely. Usually snow gauges are sunk into the ground, below a criss-cross grid mesh rather like those frequently used in packaging to separate glass bottles and jars. This allows snow when appropriate to blow over the top of the collector system, but collects what is intercepted vertically or at an appropriate angle. It goes without saying that the chemical inertness of any materials used to construct sampling systems should be tested prior to their use in the field.

Sometimes, in an attempt to measure aerosol and dry deposition inputs more accurately, filter-type gauges are used under a roof, as in Figure 17.13. At appropriate intervals, the fine lines that trap the aerosols etc., are carefully rinsed with water and the rinse water volume and chemical composition are measured. Calibration is still a problem when it comes to interpretation of results, and also such structures may need to be very strongly built to withstand winds often encountered at remote upland sites.

If rain gauges are to measure the amount of precipitation accurately, it is important to make sure that the presence of the gauge does not significantly modify patterns of air flow in its immediate vicinity. Therefore a standard style of meterological gauge should be used at least to check the volume of water being collected for analysis. Localised eddy currents can significantly modify the volume of water per unit area collected by non-standard gauges.

A particular problem may be experienced if it is deemed necessary to measure changes in throughfall composition and fluxes at various heights under a forest canopy. Sampling towers may be constructed as in Figure 17.14, but obviously sampling becomes much more expensive, and careful risk assessment needs to be made to ensure safety.

Lead-in question

What effect do you think the interception of rainfall by plant foliage might have on the chemical composition of throughflow water?

Figure 17.14 A precipitation sampler designed to measure input fluxes at various heights within a forest canopy.

Figure 17.15 A sampler designed to measure stemflow fluxes to the forest floor.

17.2.5 Measuring effects of interception by vegetation

When precipitation is intercepted by vegetation, its chemical composition may be modified to a substantial extent. One particular effect that has attracted much attention in the context of acid rain impacts on biogeochemical cycling is the effect of water running down the trunks of trees (known as 'stem flow'). Tree bark may retain gaseous pollutants and particulates from the atmosphere. It is therefore often much more acidic than the precipitation, and can contribute to localised soil acidification around tree trunks. Its chemistry is also more closely aligned to that of the throughflow dripping through the vegetation canopy, because the latter also may significantly alter precipitation chemistry. Much of the water flowing along branches towards the trunk originates from interception by foliage. Throughflow is generally enriched in DOC by foliar leaching (Tukey, 1970), and often in the nutrient elements potassium and magnesium, and the trace element manganese. Figure 17.15 shows a sampling device for monitoring stem flow. Care must be taken with such devices to ensure that the materials used do not interfere with the stemflow chemistry, and that they form a water-tight seal to the tree, even when the tree is slowly growing over a long sampling period.

Measuring throughfall chemical composition is a little easier, and conventional gauges with

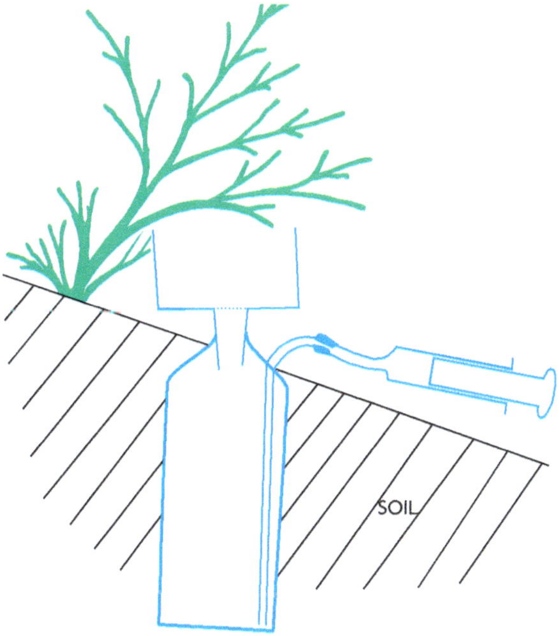

Figure 17.16 A sampler designed to measure input fluxes to the top of a moorland soil under shrubby vegetation.

filters (Figure 17.11) may be used. However, a high degree of replication is needed because of the spatial variation in canopy drip in forests. Ideally the sampling interval should be short, to minimise the effects of decomposition of any forest litter trapped in the gauge filter.

It is also possible to measure throughflow under smaller, shrubby plants such as heather (*Calluna vulgaris*) or bryophytes. Figure 17.16 shows an assembly used for this purpose by the author. The water is collected in a plastic wash-bottle sunk into the soil, and sampled at a reasonable distance using a syringe to prevent excessive localised trampling over a long experiment.

Such devices can give very useful results. For example, it has been shown that most of the nitrate input to the canopy may be retained by foliar interception, either by foliar uptake or by leaf microflora (Edwards *et al.*, 1985). This calls into question the value of experiments done with isolated soil to look at the fate of nitrate inputs to ecosystems.

17.2.6 Measuring changes as water passes through soil

Lead-in question

When you water a potted plant, does the water always drain out of the bottom of the pot, and if not, why not?

Three types of approach may be used to study changes in water solute composition as the water drains through soil. However, we will see later that none is ideal if we wish to know the fluxes of elements passing down the soil profile at various points. The three approaches are to use soil microcosms and collect drainage water, to insert a collector at the depth of interest to collect water draining from the soil under the forces of gravity (a 'zero-tension' lysimeter), or to use tension lysimeters to 'suck' the water out of soil pores.

Figure 17.17 shows a microcosm system being employed in a Swedish forest to assess the effects of throughfall chemistry and forest floor and associated vegetation on drainage water chemistry. The system excludes natural precipitation, but reapplies modified precipitation via a sprinkler device, fabricated from transparent and inert plastic material. The microcosm is shown here reinserted into the soil to maintain an appropriate temperature. Underneath the column of soil is a funnel and collector bottle to collect drainage water (i.e. well below the ground surface).

Figure 17.18 shows the bulk precipitation collector used to create the modified throughfall applied to the microcosm shown in Figure 17.17. It is not always convenient to transport large amounts of artificial rain to remote sites, so use is often made of bulk collectors in the field. On this occasion the system has the advantage of including appropriate amounts of DOC in the water being applied.

Think back now to the lead-in question asked at the start of this section. Unless the potted plant is being over watered (or the compost is so dry it's gone hydrophobic!), no water will drain out of the pot. The water is retained within soil pores.

Figure 17.17 A microcosm system designed to measure output fluxes from the upper layers of a forest soil after applying manipulated precipitation.

This explains why it is difficult to measure fluxes of water passing through soils. Water drains out of microcosms under gravity only when the soils are very wet, and field capacity is exceeded. In the field, however, water flow may occur under drier conditions, so the flow of water and elements is likely to be significantly underestimated. Moreover at some times of the year, water and elements may flow upwards as well as downwards in the 'real' world.

The same limitation applies to zero-tension lysimeters. Figure 17.19 shows a section of guttering, 1 m long, inserted into a forest surface soil at a slight downhill angle to serve as a zero-tension lysimeter. Water draining is collected into a bottle sunk into a soil pit further down slope. Again only water flow when field capacity is exceeded is being collected, not the total downward flow.

Tension lysimeters remove water from soil by the application of suction to the inside of a porous ceramic or Teflon cup that is inserted in soil to the required depth. The size is quite variable, from Rhizon samplers that are about the thickness of the lead in a pencil up to a few cm in diameter. The sampler in Figure 17.20 falls into the latter category. The cup is connected to a plastic pipe, to the top of which a vacuum may be applied with a small hand pump. The other tube from the top of the device passes right to

Figure 17.18 A bulk water collector used to provide water to make manipulated rainfall treatments at a remote forest site.

Figure 17.19 A zero tension lysimeter inserted into the surface horizons of a forest soil.

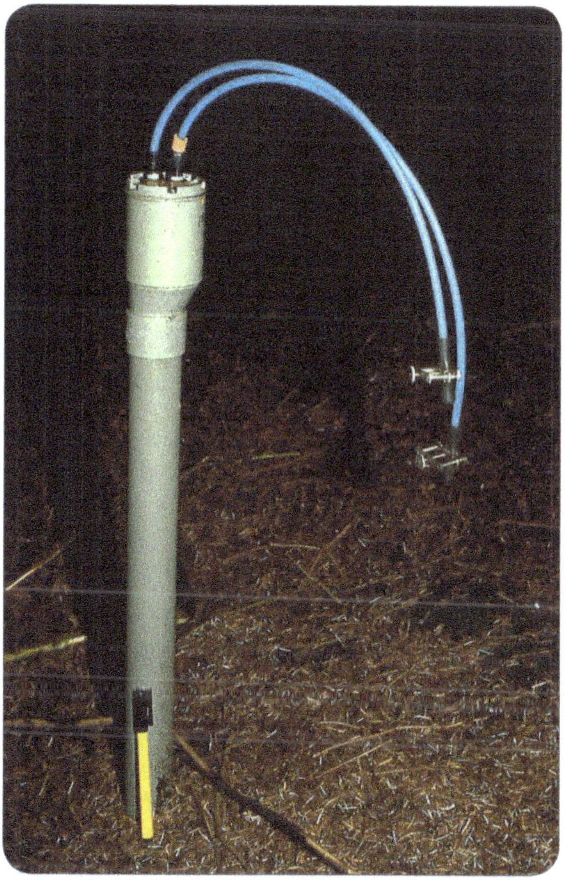

Figure 17.20 A tension lysimeter inserted into a forest soil.

the bottom of the cup, and is used to remove the sample solution with a syringe. While tension lysimeters are excellent for showing how the composition of soil solution changes with depth, it is difficult subsequently to convert the numbers obtained to water or element fluxes, for reasons already discussed.

17.2.7 Measuring litter accumulation rates

Lead-in question

What are the main difficulties that you might face in quantifying the amount of litter per year arriving at a forest floor?

On first thoughts it would seem that all that is required to sample litter fall is installation of collecting bowls of known diameter, and the litter can be collected for drying, weighing and sub-sampling and analysis monthly or biweekly. However, a high level of replication is needed because of spatial variability. Also the bowls need drainage holes to avoid anaerobic decomposition or flooding. On the other hand if rainfall wetting the litter at regular intervals is freely drained, are significant amounts of nutrient elements being leached out from the litter? If they are, it would be possible to measure the leaching losses too, but this is rarely done in practice, the assumption being made that such losses are too small to worry about. Figure 17.21 shows an example of a litter trap. In the author's opinion this is a little small. The larger nylon net mesh bags fixed to hoops of known diameter shown in Figure 17.22

Figure 17.21 A small litter trap based upon a plastic bowl with a drainage hole, pegged to the forest floor.

are probably better, sampling over a larger area. Even they could pose some problems, however. Litter includes not just falling foliage, but also twigs and even quite large branches during periods of high wind or broken off by the weight of accumulating snow and ice. At sites where this is a problem, Skiba from the Centre for Ecology and Hydrology in Edinburgh has used sheets of tough black polythene pegged to the forest floor, with slits to allow drainage.

17.2.8 Measuring litter decomposition rates

Lead-in question

Why is it important to measure litter decomposition rates and how could you measure them in practice?

It is very important to be able to understand how litter composition changes over time as it decomposes, because carbon dioxide is released to the atmosphere and nutrient elements are recycled for plant and microbial uptake. For many years litter bags have been used to measure litter decomposition rates. These are fine mesh nylon bags containing appropriate litter, for example a known mass of air-dry pine or spruce litter needles. Separate samples are oven dried so that results can all be expressed on an oven-dry weight basis. The mesh size used is such that the needles may be readily attacked by bacteria and fungi, and smaller soil fauna, but not so large that litter can be lost. The bags are placed in the litter horizon, and replicate bags are taken back to the lab after appropriate intervals, say 3, 6, 9 and 12

Figure 17.22 Nylon mesh bags attached to horizontal hoops of known diameter used as litter traps. Note the high degree of replication at this site in Sweden.

Figure 17.23 A litter bag set in a forest floor. Usually less of the bag would be visible than in this photograph, which was selected to give a clear impression of the bag's appearance.

months. The mass loss after drying allows decomposition rates to be quantified, and if the litter residues are subjected to chemical analysis, rates of nutrient element release can also be computed. A typical litter bag is shown in Figure 17.23. Careful recording of the bag positions is required, as they can be difficult to relocate after many months.

17.2.9 Use of manipulation experiments

Lead-in question

Why might it often be difficult to make unequivocal interpretations of the results of field experiments that aim to prove the effect of a single variable on ecosystem behaviour?

Suppose that we want to investigate the influence of enhanced nitrogen deposition in rain upon the functioning of a particular type of ecosystem. If we wish to make use of natural pollution gradients in the field we immediately run into a serious complication. Basically we may need to find three or more sites where the only difference between the sites is the amount of N deposition. In other words, the vegetation community, soil type, soil depth, soil parent material, aspect, day length, daily temperature, precipitation amount and type, precipitation chemistry apart from N, position of slope (if all sites are not flat), gradient, etc. must all be identical for all sites. This is invariably almost or completely impossible in the author's experience. So what is the alternative?

We can divide a uniform site into sub-plots, and then, in the above case, control the rainfall chemistry artificially. This may mean excluding precipitation, which can be a challenge on anything but the smallest scale, but it can be done. Figure 17.24, for example, shows one of a series of covered mini-catchments in Norway, for the RAIN (Reversal of Acidification in Norway) experiment. The structures are very substantial, and because of their mountain location, helicopters were used to facilitate their installation. Much thought went into their design because of the high expense, but even then there were problems. The lower walls were open to the atmosphere to avoid excessive cooling and allow some natural dry deposition. Rainwater from the roof was collected and deionised, then reapplied via sprinklers close to the roof after appropriate levels of pollution had been 'put back'. Snow was to be created using snow making devices of the type used on ski-slopes when required. Unfortunately these require a temperature of around five degrees below zero Celsius for successful operation. This ran the risk of exposing large patches of greenery on the hillside when most of it was snow-covered, a magnet for small mammals! This in turn meant researchers having to spend an unreasonably long amount of time scaring off such animals.

The scale of the experiment can also be seen in Figure 17.25. The roof was large enough to enclose mature Scots Pine trees. In spite of the problems, the results were very successful, and the scope for recovery from acid rain if emissions were reduced was clearly demonstrated. Other covered catchments have used transparent plastic sheeting fitted around trees, such that the tops of the trees are outside the covered volume (Figures 17.26 and 17.27). This compromise allows work to proceed with mature trees on a larger scale and with a more typical forest soil, but does leave a number of problems. For example, leaching of organic matter and nutrients from foliage still occurs, but does not reach the ground flora or soil surface in the normal way. This encourages algal growth on the roof surface, and the soluble organic substrate loss may disrupt element cycling in the surface soil horizons

Figure 17.24 One of the RAIN experiment covered mini-catchments on the hillside at Risdalsheia in Norway.

Figure 17.25 A view inside the covered mini-catchment shown in Figure 17.24. The sprinkler pipes are at roof height.

especially. Nevertheless, provided these limitations are borne in mind, such large-scale covered systems allow large numbers of scientists to conduct inter-disciplinary research into the effects of atmospheric pollution. Because of the number of researchers working at the experimental site, walkways pass through between the trees to prevent effects of excessive trampling. A major advantage is that the approach allows use of fully mature trees.

Many experiments on element cycling are done with smaller, younger plants rather than those used in the above major experiments. For example, a

Figure 17.26 A view inside a covered catchment area in which the tree tops are above the roof.

Figure 17.27 A further view inside the covered catchment area shown in Figure 17.26, showing the walkways designed to prevent excessive trampling of the soil.

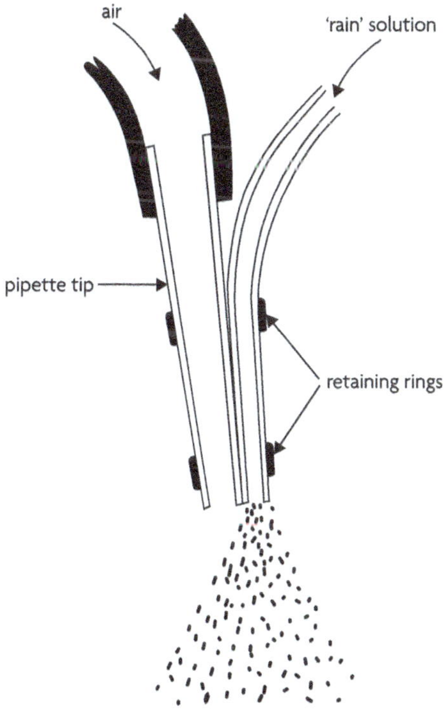

Figure 17.28 A simple and inexpensive pneumatic nebuliser suitable for spraying artificial rain.

set of simple pneumatic nebulisers may be made from plastic pipette tips and lengths of peristaltic pump capillary tubing (Figure 17.28) and used to apply 'rain' of known composition over, for example, pot-grown spruce trees (Figure 17.29).

If the throughfall under the trees is collected (Figure 17.30), it is possible to investigate, for example, how magnesium and potassium leaching

Figure 17.29 An array of nebulisers used to apply 'acid rain' treatments to spruce trees.

from foliage varies with the acidity of the rainfall, or whether sulphuric and nitric acid cause the same amount of leaching (in fact nitric acid causes less leaching).

17.2.10 Assessing effects of changes in the atmosphere

There has been increasing interest over the last decade or so on the influence of changing gas concentrations in the atmosphere on vegetation of various types and ecosystem behaviour, for example the effects of increasing carbon dioxide, sulphur dioxide, ozone or oxides of nitrogen concentrations. To maintain conditions as close as possible to those in the natural environment, often open-top growth chambers are employed, such as those shown in Figures 17.31 and 17.32.

Figure 17.30 Collecting throughfall under a spruce tree during application of artificial acid rain.

Figure 17.31 An open-top growth chamber being used to study the effect of ozone upon young spruce trees.

Figure 17.32 Several open top-chambers are needed, as shown here, to allow adequate replication.

Figure 17.33 Solar domes used at the University of Lancaster for studying mini-ecosystems under controlled atmospheres.

These particular chambers were being used to study the extent of ozone damage on young spruce trees, and a different ozone dose was being applied in different chambers.

Some researchers prefer to have fuller control over experimental conditions than is possible using open-top growth chambers, and use enclosed systems in which the atmosphere can be controlled. For example, if the effects of elevated carbon dioxide, CO_2, concentrations on nutrient cycling are being studied, the gas concentration in the atmosphere of the glasshouse can be monitored, and additional CO_2 added as required to maintain the desired concentration. Figure 17.33 shows the solar domes used at the University of Lancaster to study the effects of increasing CO_2 concentration and of gaseous pollutant concentrations in controlled atmospheres. Care is needed, of course, to make sure that temperature control is appropriate, and the plants being studied are growing in appropriate soil and receiving appropriate precipitation.

The running costs for such solar domes are quite high for long-term experiments. Many researchers therefore work with individual shoots, for example using apparatus such as that shown in Figure 17.34. The concentrations of CO_2 and ozone (O_3) to the chamber inlet are measured and controlled, and the concentrations flowing

Figure 17.34 Chamber system used to study effects of ozone or sulphur dioxide on rates of CO_2 assimilation by individual shoots.

slowly out of the inner chamber are also measured. The assimilation rates may be found by difference if the gas flow rates are known. The outer chamber is flushed with a much faster flow of ambient air to control the temperature of the inner chamber, to avoid generation of unrealistically hot conditions.

Simulation experiments are very attractive in so far as they allow one parameter at a time to be changed in isolation. The down side is the timescale of the experiments. In the natural environment some changes occur over minutes or hours, but others may take several years. This point requires careful consideration when designing environmental experiments.

CASE STUDIES

A case study in Greece – Quantifying biogeochemical cycling

Dr Panagiotis Michopoulos and his colleagues from the Forest Research Institute in Athens have been studying element cycling in Greek forests for a number of years. They have long been interested both in the sustainability of current forest management and utilisation practices and in the impacts of acid deposition throughout the 1990s and beyond (among other reasons to see if they can provide evidence of environmental improvement as sulphur emissions to the atmosphere have been reduced). They are therefore interested to show whether soil organic matter content and composition and soil acidity and pools of base cations stored in soils are changing over time. The photographs below show some of the techniques they are employing.

(a) (b) (c) (d)

Falling litter is sampled in inert plastic barrels down vertical transects (a) to assess any slope effects. A fine mesh plastic net in the barrel bottom (b) allows drainage and keeps the litter reasonably dry prior to analysis. Litter from the forest floor is sampled (c) with a specially made heavy iron box of known dimensions (d). As the box is inserted, the surrounding forest floor litter is scraped away, so that a block of litter of known area and depth can then be bagged, weighed and analysed.

(e)

(f)

(g)

(h)

Rainfall is measured under an open part of the canopy, and collected via a filter (e). Wind speed and direction, air and soil temperature and relative humidity of the atmosphere are also measured (f), to see what effect they have on the chemistry of soil solution as measured with zero tension lysimeters (not shown here). Stem flow from the tree trunks is also collected (g) and analysed at intervals. Note the overflow bottle on the stem flow collector to make sure no excess of sample is lost (h).

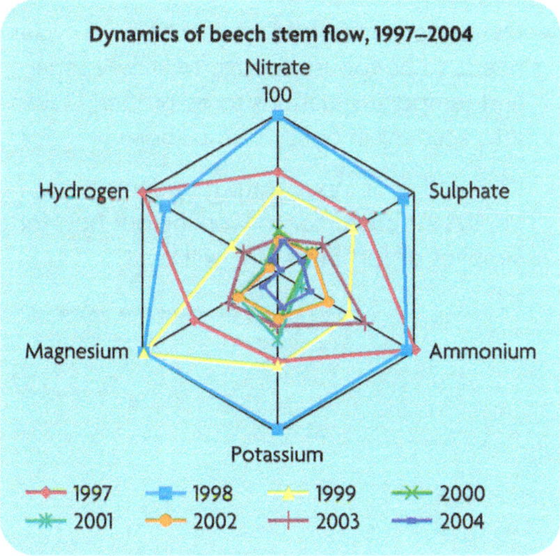

The team from Athens wanted a simple way of showing how the quality of water flowing down the tree trunks was improving steadily since the end of the 1990s. Acid rain can increase the amounts of hydrogen ions (H^+), nitrate, sulphate and ammonium in stem flow, and also the amounts of potassium and magnesium, which are leached from foliage by acid rain and find their way, via branches, to the trunk. Working with Malcolm Cresser at the University of York, they came up with the idea of pollution impact polygons. Observed pollutant concentrations are converted to a scale of 0 to 100, zero being the best possible scenario for each pollutant, and 100 per cent the worst observed scenario. The 100 per cent values are set at the apices of the polygon, in this instance a hexagon, and zero in each case at the centre of the hexagon (this is quite easy to do in Excel). Thus if stem flow quality is improving, the area of the polygon starts to shrink. The improvement in the example above is fairly obvious from 1999 onwards. However, the pollution impact polygon does not, on its own, prove that the improvement is evidence for recovery due to decline in sulphur emissions. That requires alternative methods of statistical analysis of the data.

POLICY IMPLICATIONS

Biogeochemical cycling

- Policy makers need to know how national- and global-scale carbon budgets are changing, especially in the context of assessing risk of climate change as a consequence of rising carbon dioxide and methane concentrations in the atmosphere.
- Knowing the input/output balances for base cations and nutrient elements allows assessment of sustainability of natural ecosystems and of soil use and management practices over mid- to long-term timescales. It is essential for the reliable assessment of the sustainability of organic agriculture compared with traditional agriculture.
- Data from weathering rate studies provides essential inputs to the models that policy makers rely upon when formulating policies to protect the environment over the long term.
- Field studies of biogeochemical cycling are needed to validate predictions from the models that policy makers often have to depend upon. For example, do acidified soils recover at predicted rates from acidification steps? Governments need to be able to demonstrate cost effectiveness of pollution abatement and waste-management measures that have cost a lot to implement!
- Field measurements of changes in weathering rates and rates of element cycling are an essential part of assessing pollutant impacts upon various components of the environment, especially soils and surface fresh waters. For example, we may wish to estimate long-term rates of soil acidification or recovery. They are equally important if policy makers want to know the probable impacts of predicted climate change scenarios.
- Knowledge of element input/output budgets is essential in the assessment of some aspects of waste and effluent disposal practices (e.g. are potentially toxic elements accumulating or is phosphorus accumulating at rates that are a cause for concern?).
- Our understanding of biogeochemical cycles needs to be robust enough to allow spatial and temporal variations to be properly taken into account in policy formulation.
- Knowledge of input/output budgets is an important precursor to working out how long supplies of fertilisers, etc. will last.

CHAPTER REVIEW EXERCISES

Exercise 17.1

The table below shows how discharge ($m^3\ s^{-1}$) and calcium concentration ($mg\ l^{-1}$) change over time throughout a storm event in a small English upland river.

Time	Discharge	Calcium concentration
16.00	0.060	5.05
18.00	0.062	5.04
20.00	0.060	5.05
21.00	0.081	5.11
22.00	0.165	5.08
23.00	0.280	4.60
24.00	0.300	4.48
01.00	0.285	4.40
02.00	0.257	4.38
03.00	0.218	4.24
04.00	0.175	4.22
05.00	0.140	4.33
06.00	0.120	4.41
07.00	0.105	4.48
08.00	0.088	4.61
10.00	0.067	4.78
12.00	0.060	4.95
14.00	0.059	5.00
16.00	0.058	5.00
18.00	0.058	5.02
20.00	0.058	5.05
22.00	0.058	5.04

Enter the data into a spreadsheet (e.g. Excel). Calculate the flux of calcium, in $g\ s^{-1}$, flowing past the sampling point at each sampling time. Remember that there are 1000 litres in 1 m^3 of water. Plot graphs of discharge, calcium concentration and calcium flux as a function of time (time on the horizontal axis as the independent variable in each case). Does the flux increase or decrease as discharge rises? Plot three graphs of calcium versus discharge, one using the entire data set, one using the data for rising discharge only, and one using the data when discharge is falling only. Is calcium concentration significantly correlated to discharge in each of the three graphs? Are the three regression equations different?

If you wished to improve the estimate of the flux of calcium leaving the catchment over the year, how might concentration versus discharge relationships be best used to do this (assuming continuous records of flow are available throughout the year)? Explain the possible reasons for any hysteresis observed in the calcium concentration versus discharge plot using the entire data set.

Exercise 17.2

The funnel of a rain gauge has a radius of 10 cm^2. The table shows the volumes of water collected in litres per month in the gauge at a particular site over a 12-month period. It also shows the concentration of calcium in micrograms per ml (µg Ca ml^{-1}) measured in each monthly sample. Calculate and tabulate:

(i) The monthly precipitation in mm for each month.

(ii) The annual precipitation at the site in mm.

(iii) The mass of calcium in mg deposited in the gauge each month.

(iv) The mass of calcium in wet deposition deposited per hectare each month.

(v) The annual deposition flux of calcium via wet deposition, in kg ha^{-1}.

Useful reminders: 1 ha = 10,000 or 10^4 m^2 or 10^8 cm^2; the area of a circle is πr^2; 1 litre is 1000 ml or 1000 cm^3; there are a million micrograms (10^6 µg) per gram.

Month	Volume of precipitation (litres)	Calcium concentration (µg ml^{-1})
January	2.988	0.25
February	3.200	0.18
March	2.766	0.29
April	3.098	0.30
May	2.880	0.56
June	2.454	0.58
July	2.686	0.42
August	2.990	0.39
September	3.176	0.50
October	2.890	0.40
November	3.418	0.42
December	2.792	0.35

REFERENCES

Cresser, M., Killham K. and Edwards, A. (1993) *Soil Chemistry and its Applications*. Cambridge University Press, Cambridge, 192 pp.

Edwards, A.C., Creasey, J. and Cresser, M.S. Factors influencing nitrogen inputs and outputs in two Scottish upland catchments. *Soil Use and Management*, **1**, 83–87.

Tukey, H.B. (1970) The leaching of substances from plants. *Annual Reviews of Plant Physiology*, **21**, 303–329.

CHAPTER 18

Renewable and non-renewable energy

Craig Adams

Learning outcomes

By the end of this chapter you should:

- Understand the direct and indirect sources of energy utilisable by humankind.
- Appreciate the direct and indirect costs associated with energy sources.
- Be more aware of renewable versus non-renewable energy sources.
- Better understand the relationship between energy use and global warming.
- Understand the issues associated with carbon sequestration.
- Understand the major sectors of energy demand, and how energy need characteristics vary.
- Appreciate the role of energy conservation as a 'source' of energy.
- Understand the key environmental impacts of each energy source.
- Be able to assess renewable and non-renewable energy sources.

18.1 Introduction

There are many sources of energy. The earliest major sources of energy were the sun (e.g. passive solar heating) and biomass (e.g. burning of wood). Industrialisation in the European Union countries (especially England) and other regions led to the massive expansion of the use of various forms of *fossil fuels* (e.g. coal, oil, natural gas and related forms of reduced organic matter) rather than wood burning. These fossil fuels required hundreds of thousands (or millions) of years to produce and are, therefore, *non-renewable* (at least not in periods relevant to human history on Earth).

Nuclear energy may potentially be derived from either fission or fusion processes. Today's nuclear power is *fission* based, in which atoms of an element such as uranium are split, releasing protons and energy, and thereby creating new elements with lower atomic number in the process. Because nuclear fuels needed for fission power generation are a limited resource, current fission-based nuclear power can be considered non-renewable along with the burning of fossil fuels. *Fusion*, on the other hand, can use an isotope of hydrogen (tritium), found in natural water, as its fuel. Tritium is essentially limitless on Earth, thereby potentially providing an inexhaustible fuel supply via fusion power generation. Unfortunately, overwhelming technical difficulties exist in containing the hot sun-like plasma at the heart of the fusion process. At least for the foreseeable future, the axiom related to fusion may have some truth; that is that 'fusion is the energy source of the future, and always will be.'

Renewable energy sources ultimately derive their power from the sun's radiation, from gravity, or from the Earth's rotation. Techniques that harness *solar energy* are more varied than you might at first think. They include photovoltaics, solar thermal heating, biofuel use, windpower, ocean wave power, and hydroelectric power. The Earth's rotation causes the major transoceanic ocean currents (e.g. the gulf stream in the Northern Atlantic Ocean). *Gravitational forces* are at the heart of geothermal energy and of tidal energy. Thus, there are many sources of truly renewable energy that can be tapped to fuel the energy needs of our growing societies and economies. The direct costs of each of these renewable energy sources vary significantly between sources, and can change rapidly as new technologies are developed and/or as their use increases. Each renewable source has great potential, though each comes at some indirect costs to the environment and/or to people living near them.

18.2 A review of non-renewable energy sources

In this chapter, we will review renewable energy sources and why they are critically needed. The merits and issues associated with all energy sources are relative; that is, while all have problems, the nearly 7000 million people on Earth require significant energy for transportation, heat, communication, lighting, and other uses. In that context, we will begin this chapter by reviewing non-renewable energy sources, namely fossil fuels and fission nuclear energy.

18.2.1 Fossil fuels

According to analysis by the US Department of Energy, the contribution of fossil fuels in 2007 corresponded to about 85 per cent of the total energy consumption, with oil (35 per cent) leading both coal (27 per cent) and natural gas (23 per cent). Oil is primarily used for transportation and heat, while coal is primarily used for electricity production. The US Department of Energy predicts steadily increasing worldwide extraction of coal, natural gas and oil in the future, with 41 per cent more fossil fuels being extracted in 2035 (versus 2007) with a 49 per cent increase in overall energy use.

Oil

Oil is formed deep in the Earth where there is sufficient heat and pressure to convert organic matter to liquid over thousands to millions of years. The largest oil reserves in the world are (in order of decreasing magnitude) in Saudi Arabia (19 per cent), Canada (13 per cent), Iran (10 per cent), Iraq (8.5 per cent), Kuwait (7.5 per cent), Venezuela (7.3 per cent), UAE (7.2 per cent), Russia (4.4 per cent) and Libya (3.3 per cent). This compares with the United States (1.4 per cent), China (1.5 per cent), and Norway (0.5 per cent). Earlier in the twentieth century, the idea that oil was a finite (non-renewable) resource was ridiculed by many because discovery of new oil fields outstripped demand. M. King Hubbert, however, predicted in 1956 that oil extraction in the lower 48 United States (i.e. excluding Alaska and Hawaii) would peak in approximately 1970. When his prediction proved accurate, individuals and industry began taking his prediction methods more seriously. Hubbert had several methods for predicting rates of extraction. His third (and most well known) method was based on his observation that maximum extraction of oil tended to lag behind discovery of new reserves by only a few decades. Predictions are that there were initially about two trillion barrels of oil beneath the Earth's surface, and that we have extracted about half of this amount. In fact, after increasing about 12 per cent per year through 2004, world oil production reached a plateau after 2005. In any event, it is the current generation that will begin seeing a rapid decline in the amount of oil produced, necessitating a radical change in the fuel of choice for transportation in most countries.

Another aspect of the diminishing oil reserves is that oil companies now exploit much more difficult (and dangerous) oil fields. Take, for example, the vast amount of deep ocean drilling that oil companies currently use (Figure 18.1). One must only look to the massive BP Deepwater Horizon oil spill in the Gulf of Mexico to understand the implications of drilling for oil in ocean depths a mile or more below the surface of the sea.

We sometimes hear comments suggesting that oil won't 'run out' for at least 100 years. The issue, however, is that the crisis does not originate when the last drop is pumped, but begins when extraction cannot keep up with the ravenous demands for oil in the world. With the world willing to fight and go to war over oil, it is clear that new alternative energy sources are vitally needed.

Natural gas

Natural gas is formed at greater depths and higher temperatures than are optimal for oil, with the extra heat leading to breaking of more chemical bonds leaving smaller organic molecules such as

Figure 18.1 The Deepwater Horizon off-shore oil rig on 21 April 2010, in the Gulf of Mexico.
Source: Photograph 100421-G-XXXXL-003, US Coast Guard Photograph. From http://imagespublicdomain.wordpress.com/2010/05/08/bps-filthy-mess-public-domain-images-of-the-deepwater-horizon-oil-spill-in-the-gulf-of-mexico/

methane. Natural gas is sometimes called the 'clean' energy source, but emits just as much carbon dioxide as other fossil fuels when burned. Natural gas may be a very good short-term transition fuel between the dominant use of oil and gasoline currently, to fuels of the future.

Natural gas can also be converted to a liquid form for use (though the infrastructure to take advantage of this may be problematic). Some estimates are that the peak extraction (production) of natural gas may lag behind that of oil by only several decades, peaking within 20 years in around 2030.

Coal

Coal is the third major fossil fuel in use today. Coal is used primarily for electricity production and accounts for nearly half of the world's electricity production. China is building coal-burning power plants faster than any other nation in the world. Between 2002 and 2006, two of every three coal-burning power plants were being built in China (for a total of almost 400 new plants).

Besides being a major contributor to carbon dioxide emissions into the atmosphere, a problem with coal as an energy source is that it is a very dirty material. Burning of coal causes the emission of sulphur and nitrogen oxides that cause dry and wet acid deposition to fall on the Earth's surface. This acid deposition leads to acidification of surface waters in areas such as the north-eastern United States. Lower sulphur coal from the western United States is transported via railroad in massive amounts to the east coast to help alleviate the amount of sulphur being discharged, or needing to be scrubbed from flue gases via expensive treatment processes. Coal also contains mercury, a potent human toxin especially detrimental to developing children.

Coal extraction itself causes major environmental damage in many regions of the world through open pit strip-mining operations. Similarly, a technique called 'mountain top mining' is also common in places such as the Appalachian Mountains in the United States. In this technique, coal companies remove entire tops of mountains in pristine forests just to get at a narrow band of coal. Much of the debris is then placed in valleys, causing permanent destruction of the mountain regions.

Coal can be burned directly or can be converted to transportation fuels such as hydrogen (or Syngas) or to liquefied synthetic fuels. While synthetic fuels are promoted as a means to bridge between oil/gasoline and other transportation fuels in the future, a major problem is that coal-based synthetic fuels have twice the carbon footprint compared with oil and gasoline due to carbon dioxide being produced during fuel production and again when the fuel is burned.

Other fossil fuels

Hydrocarbons other than in the form of oil, natural gas and coal can be extracted from the Earth including high molecular weight (heavy) tar sands or oil shale. Oil shale is mostly a waxy substance called kerogen which is mined and then converted into oil. Methane hydrates are another possible source of fossil fuel. While there is still a lot unknown about methane hydrates, some believe that methane hydrates may be extractable from the ocean floor, though many technical difficulties would need to be overcome. With the ongoing global warming and increase in the Earth's mean temperature, vast amounts of methane may be also released from the melting permafrost, leading to potentially massive release of methane into the atmosphere. Unfortunately, methane is a powerful greenhouse gas, much stronger than carbon dioxide, and its release can therefore contribute to even greater warming (see Chapter 8).

18.2.2 Nuclear fission

Beside fossil fuels, a second source of non-renewable energy is nuclear fission power generation (Figure 18.2). Nuclear fission accounted for about 5.5 per cent of the total energy produced in 2007. France, Lithuania and Belgium are leaders in the use of nuclear energy, with the majority of their electricity needs met by nuclear power generation.

Thermal fission reactors work by splitting atoms releasing heat used to generate steam and then electricity (via spinning turbine generators

Figure 18.2 Typical nuclear power plant generating continuous power for the electrical grid.
Source: Photograph and permission provided by Duke Energy.

just as in a coal-burning power plant). There are a variety of different types of thermal fission reactors including light water reactors, gas-cooled reactors, the CANDU reactor (Canadian–Deuterium–Uranium), and others. Each type of thermal fission reactor has advantages and disadvantages.

The long-term storage of spent fuel is a major technical and political challenge for countries using thermal fission power generation. An alternative to this open fuel cycle is the closed fuel cycle in which spent fuel is reprocessed to re-enrich it for reuse. Unfortunately, nuclear weapons grade materials that have the potential for use in weapons of mass destruction are also generated in closed fuel systems.

A significant problem with nuclear energy is radioactive waste. High-level radioactive waste is generally in the liquid form, but can be vitrified (or glassified) into a solid form to make transportation, storage and monitoring safer. Other forms of waste include low- and intermediate-level nuclear waste, both of which have their own problems related to storage.

Another issue that is of great concern is the safety associated with nuclear power plants. Notably, large releases of radioactivity from nuclear power plants (e.g. Chernobyl, Russia; Fukushima, Japan) have raised concern regarding the risks associated with this common form of energy. To further exacerbate the issue, in the United States, the average age of its nuclear power plants is now approaching design lifetimes, with few new reactors planned.

The nuclear fuel cycle for uranium involves discovery, mining and extraction, enrichment, fuel fabrication, power generation, and the processing and storage of spent fuel. It is not well recognised that the reserves of uranium-235 used for thermal fission reactors are finite, and will run out at some point in the future. Alternative energy sources to replace these non-renewals (fossil fuels and fission-based nuclear power) are clearly needed, and soon.

18.2.3 Non-renewable energy future

Analyses by the various agencies and governments do not predict with certainty when the production of various sources of fossil fuels and fission-based nuclear materials will peak, but it is clear that it is very soon, within this or the next generation or two. It is clear that there is a critical and immediate need for developing energy policy and technologies to move our societies away from non-renewable energy sources and towards renewable energy. There is, in fact, no choice but to do so.

18.3 Energy costs

It is critical to think of both the *direct* and *indirect* costs of renewable energy sources, as well as those of non-renewables. In fact, it is by neglecting the indirect costs of fossil fuels that these energy sources appear to be so inexpensive. If the indirect costs that ultimately must be paid by society were factored in, fossil fuels would be a much more costly source of energy than they are often currently considered. For example, in the United Kingdom and some other European Union (EU) nations, petrol prices are several times greater than prices in the United States (US), China, South America and other nations. This is due to the factoring in of estimated indirect costs into the sales prices through petrol taxes.

Some of the costs associated with the use of fossil fuels come in terms of damage to the environment, adverse impacts on human health, and contributions to global warming. When fossil fuels are burned, toxic and environmentally-harmful compounds (such as lead, mercury, cadmium, sulphur dioxide, nitrogen oxides, and other compounds) are released into the air and water, and in residual solids. Each of these species can have profound impacts on human health and/or on the environment. For example, massive amounts of heavy metals are released from coal when it is burned and end up in the solid fly ash residual. These heavy metals then may leach into the environment and accumulate in sediments or enter the food chain. Coal-burning power plants specifically are a major cause of the toxic mercury found in fish species (such as tuna) worldwide so are diminishing the value of the ocean's fisheries. Similarly, the sulphur dioxide released during the burning of fossil fuels causes acidification of lakes from the resulting acid rain or acid deposition. Another example is the release of nitrogen oxides from automobiles that cause the formation of ground level ozone through reactions with sunlight, thereby causing major respiratory impacts on humans (see Chapter 8).

Recently, hydraulic fracturing or 'fracking' has become increasingly used to enhance the natural gas (and oil) extraction for gas fields in the United States, Canada, and more recently, in Great Britain. Fracking is implicated in contamination of groundwater by inorganic and organic chemicals and gases.

Lead-in questions

Why are only direct costs of energy often considered by industry and others? How does this impact upon energy policy?

18.4 Energy use and global warming

Burning of fossil fuels is also contributing to global warming. There will be some winners and many losers in the world due to the impacts of global warming on the Earth (especially in areas that can least afford to lose, such as in sub-Saharan Africa). Presently, there is legitimate discussion of the various causes of the increase in the mean temperature of the Earth (or global warming). The scientific consensus is that anthropogenic sources of carbon dioxide, and other greenhouse gases, are causing the majority of the global warming that is currently being observed.

What appears absolutely clear is that natural causes and anthropogenic causes of global warming are both contributing to the Earth's recent temperature rise. What is also clear is that the impacts of global warming will have significant, and mostly unpredictable, consequences for today's society (see Chapter 8). Some individuals and governments argue that the amount of the observed global warming that is caused by anthropogenic sources of carbon dioxide (and other greenhouse gases) must be ascertained with certainty prior to acting to reduce greenhouse emissions. To address this conundrum, one might consider a risk–benefit analysis for two assumptions: (1) global warming is primarily (or completely, as some argue) caused by natural causes; or (2) global warming is significantly increased by anthropogenic releases of carbon dioxide and other greenhouse gases. Detailed analysis of these two opposing assumptions may lead to the same conclusion; that is, the massive use of fossil fuels is detrimental to society, and humankind would benefit by switching rapidly to alternative energy sources.

Specifically, if one considers the extreme costs associated with the focus on fossil fuels as our primary energy sources (when alternatives are available), it is clear that alternative energy sources should, and will, be utilised extensively in the future. For example, these costs for coal usage include massive environmental and human health impacts due to the mining and combustion of coal. Another example is oil, for which political instability, threats to national security, and the pressure to fight wars in oil-producing countries are at least partly due to the heavy reliance on the massive use of oil worldwide. Even if one holds to the thought that global warming is not impacted by carbon dioxide emissions, one might logically come to the conclusion that implementation of alternative energy sources is vital for the protection of civilisation worldwide.

Natural causes of global warming include natural variation in the sun's energy output and changes in geometry with respect to the Earth's position (see Chapter 8). The resulting increase in the Earth's temperature can cause a shift in equilibrium of carbon dioxide between the oceans and the atmosphere, thereby increasing the atmospheric carbon dioxide levels.

Anthropogenic sources of global warming are primarily based on the release of greenhouse gases (GHG) due to activities of humans on Earth (including the massive amounts of carbon dioxide released into the atmosphere resulting from the burning of fossil fuels). For example, driving 100 km could consume about 8.4 litres of petrol in an automobile achieving 12 km l^{-1}. In driving the 100 km, the 6 kg of petrol used would generate about 21 kg of carbon dioxide (a greenhouse gas).

The sun's energy comes to Earth primarily at visible wavelengths due to the sun's temperature. Harmful ultraviolet radiation is mostly absorbed by the atmosphere, as discussed in Chapter 3, thereby protecting us from its adverse health effects on Earth. The atmosphere is mostly transparent to visible light, so that the energy that is absorbed by the Earth heats the Earth's surface. The Earth re-radiates this same energy back towards space, but at much higher infrared wavelengths due to the Earth's mean temperature of about 15 °C. A portion of this infrared radiation is reabsorbed by greenhouse gases in the atmosphere that, in turn, re-radiates the energy equally into space and back to Earth. As the amounts of greenhouse gases increase in the atmosphere, so does the amount of energy trapped by the atmosphere and re-radiated back to Earth. The result is a net increase in the amount of energy trapped on Earth, thereby causing increased temperatures that create global warming (Figure 18.3).

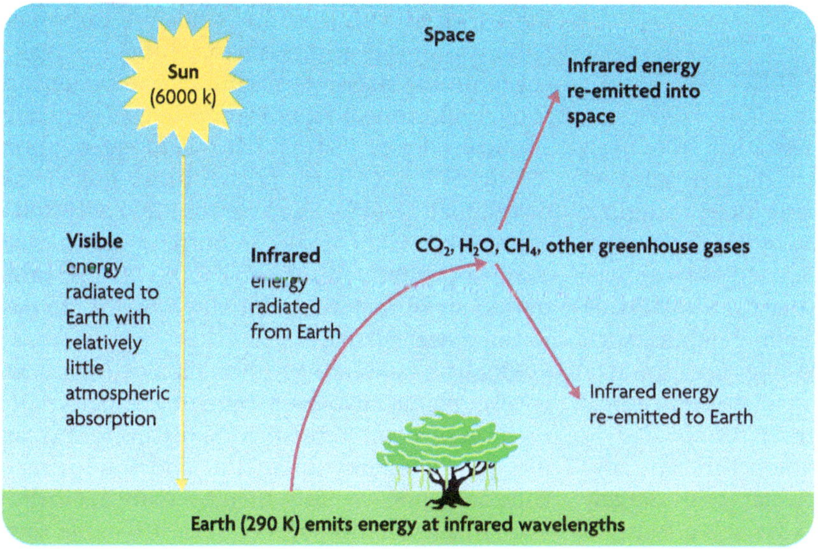

Figure 18.3 Diagram of primary driving forces for global warming including solar radiation (centred in visible spectrum due to the sun's 6000 K approximate temperature), terrestrial radiation (centred in the infrared spectrum due to Earth's 290 K approximate surface temperature), absorption of infrared gaseous components of the atmosphere, and re-radiation of a portion of the absorbed atmospheric heat back to Earth.

The relative contributions of natural and anthropogenic causes of global warming are the subject of intense discussion and debate. These discussions are conducted in the presence of uncertainty and, often, with considerable predisposed bias. An in-depth analysis by scientists and others on the Intergovernmental Panel on Climate Change (IPCC) (established by the World Meteorological Organization (WMO) and by the United Nations Environment Program (UNEP)) studied the best available data, and determined that the anthropogenic causes of increased global warming are about 10 times greater than the natural causes. In any event, the contribution to increased global warming due to massive use of fossil fuels is likely to have tremendous net negative impacts on most of Earth's inhabitants. Both the impacts on global warming, as well as the deleterious effects of the heavy metals and oxides released during the combustion of fossil fuels, dictate that alternative sources of energy must instead be utilised by society in the future.

Lead-in question

Why do many people assume that global warming must be due to either anthropogenic or natural causes, rather than a combination of both?

18.5 Carbon sequestration

Various forms of carbon sequestration have been proposed as sinks for carbon dioxide released during the combustion of fossil fuels, including sequestration in the oceans, geologically, and in biomass. One means of ocean sequestration of carbon dioxide is to pump the carbon dioxide into the ocean at relatively shallow depths, such that it simply dissolves into the ocean water. Unfortunately for this scheme, dissolved carbon dioxide is an acid (specifically, carbonic acid or H_2CO_3), and dissolving it in the ocean can cause acidification of the ocean water with largely unknown (but presumably adverse) effects on marine biota. For example, the shells of sea creatures including shellfish will be more soluble at lower pH levels. Society can ill afford to adversely impact the ecosystem and fish populations in the oceans as these fisheries are a primary (though currently poorly managed) food source for the world's population (see Chapter 15).

Another ocean sequestration method being proposed is to pump (at great cost) the carbon dioxide to great depths in the ocean so that it forms a supercritical liquid that is denser than water so that it will drop to the ocean floor. However, not much imagination is required to anticipate the potential deleterious effects on the ocean's food chain and ecology by replacing the oxygenated water on the ocean floor with a layer of deoxygenated carbon dioxide.

One means of geological sequestration of carbon dioxide is to pump the carbon dioxide into depleted oil formations (and other geological formations), into brine solutions, or (as a compressed gas or supercritical liquid) into voids. Carbon dioxide injection into oil formations is, in fact, a viable tertiary oil recovery technology to enhance the oil extraction yield from an aquifer. Potential issues with geological carbon dioxide sequestration proposals (besides costs) include the slow release and/or the disturbingly rapid release of the carbon dioxide back into the atmosphere, and acidification of aquifers.

Another proposed geological sequestration method is to tie up the carbon dioxide in mineral form such as with calcium as calcium carbonate. However, simple arithmetic may provide a reality check on the massive amount of carbonate rock formed. Burning 1 kg of carbon in coal generates about 3.7 kg of CO_2 which would create about 8.3 kg of $CaCO_3$ rock (or >8 times more mass of residual material than that initial coal). On a volume basis, the coal combustion would yield a much greater volume of $CaCO_3$ rock (or more trainloads of rock than initially brought the coal to the power plant).

Finally, sequestering carbon dioxide in biomass is also being advocated by some experts. One method involves fertilising the oceans to increase the growth of phytoplankton and other organisms on a massive basis, thereby 'tying up' carbon as biomass. The impacts of such a scheme seem dubious at best, however, based on the known impacts of ocean fertilisation, such as the 'dead-zone' created by algal growth in the Gulf of Mexico. This was created by fertilisers flowing into the Gulf from the Mississippi River in the United States. While carbon sequestration is receiving a lot of public interest, research funding and discussion, it is clear that unintended consequences and costs must be carefully weighed against potential benefits.

Lead-in questions

Why is the idea of carbon sequestration popular with many people? What are the risks associated with each proposed means of carbon sequestration (especially for the world's oceans)? Are other approaches more reasonable?

18.6 Energy use sectors

It is useful to differentiate uses of energy into two major categories: *electricity generation*, and *transportation fuels*. These two categories encompass the vast majority of power usage by our societies. Other categories exist, of course, including solar thermal energy (without electricity generation) as used throughout history to heat structures and water, and the direct conversion of river currents to turn mills or wind to pump water and for other processes. For both electricity generation and transportation energy, both renewable and non-renewable energy sources exist.

Transportation fuels include non-renewable fossil fuels (e.g. gasoline, diesel, natural gas and coal), electricity (from either renewable or non-renewable sources), and hydrogen. Hydrogen is not, of course, a source of energy although we sometimes hear incorrectly that hydrogen technology provides a 'limitless source of energy that creates only water as a byproduct'. Hydrogen is, more accurately, simply a means of storing energy in a mobile form, much like a battery. Nonetheless, hydrogen, when generated from electricity (from either renewable or non-renewable sources), or solar-thermally from renewable energy sources, is a transportation fuel with great potential and, presumably, a substantial role for transportation in the future.

Similarly, vehicles running wholly or partially on electrical energy have been, and continue to be, produced. These include exclusively electric automobiles (e.g. the GM EV1 in 1996 and the Toyota RAV4-EV in 1997), as well as hybrid automobiles in which power is generated from a gasoline engine with excess power used to generate electricity stored in batteries, and used in an electric motor to power the vehicle. To store the energy in electrically-powered vehicles, electric automobiles must have batteries which themselves have environmental costs in both their manufacturing as well as when they are beyond their useful life. Some forms of electric vehicles, however, require no batteries such as electric trains or buses used for mass transit. When the electricity used for electric vehicles is derived from renewable sources, electric vehicles represent an ultimately renewable-energy-based means of transportation.

Lead-in questions

What percentage of the population do you think would have their standard daily driving needs wholly met by fully electric cars with a limited range per battery charge (e.g., 100–200 km)? Can you envisage a transportation system in which cars could pull into a battery-replacement station on the highway, and have their batteries swapped out quickly, allowing them to be back on the highway within minutes?

18.7 Energy conservation

A final major, and often neglected, approach to providing energy needs is *energy conservation*. The world average per capita energy consumption (PCEC) in 2006 was about 58 kW h day^{-1}. Many countries or territories in the developed world exceed this consumption by many times. In fact, per capita energy use spans from 1660 kW h day^{-1} in Gibraltar down to just 0.3–0.6 kW h day^{-1} in Cambodia, Afghanistan and Chad. By contrast, usage in the Canada, the United States and the United Kingdom is 343, 269 and 130 kW h day^{-1}, respectively. While increasing criticism of China's rapidly increasing energy consumption is heard, the per capita usage there is just 45 kW h day^{-1}.

For every kilowatt (kW) of energy not consumed by a home or business, the money saved is real money that the owner now has available to spend in some other manner. Many energy conservation methods have payback periods ranging from no time at all (e.g. by turning off lights when not in a room, and/or by not over-illuminating rooms or businesses) to several years (structures or vehicles that last many years). The costs associated with many energy conservation measures are rapidly dropping due to the economy of scale, such as for fluorescent and LED (light emitting diode) light bulbs that are becoming common worldwide as low-energy use replacements for incandescent light bulbs. In the final analysis, of all energy sources (whether fossil, nuclear or alternative), energy conservation is the best and cheapest 'form of energy' that has real monetary value. It immediately reduces carbon dioxide emissions, and should be pursued aggressively and extensively employed.

In earlier sections we briefly reviewed fossil and nuclear energy sources. Now we will focus on the major forms of renewable energy, including those ultimately derived from the sun (i.e. solar thermal, solar photovoltaics (PV), wave energy, hydroelectric, wind, and biomass), from gravitational forces (e.g. tidal power, geothermal), and from the Earth's rotation (ocean currents). Basic principles are discussed for each form of renewable energy, followed by the technologies themselves, and associated environmental impacts. It is clear that there is not one panacea, or universal optimal choice, of alternative energy. It is important to understand how each technology fits into an overall plan of an integrated energy system for transportation and non-mobile energy needs. Each alternative energy source has merit, as well as constraints, related to environmental impacts and energy availability.

Lead-in questions

Why has energy conservation not received more focus in the debate about alternative energy? What factors contribute to some individuals and industries not focusing more on energy conservation opportunities?

18.8 Solar thermal energy

Solar energy ultimately provides most of the energy at the Earth's surface. Solar energy can be used in many ways, including for heating water and air, and for lighting buildings. Solar thermal energy can be used in an active mode with solar collectors or in a passive mode through, for example, well-designed buildings that naturally use the sun's heat and light.

The sun itself is a fusion reactor in which hydrogen is converted to helium releasing massive amounts of energy in the process. Due to the sun's surface temperature of 6000 K, the solar spectrum peaks in visible wavelengths (i.e. 400–700 nm) but with appreciable energy at both the ultraviolet (UV) and infrared (IR) wavelengths. Much of the low wavelength UV energy is absorbed in the Earth's atmosphere allowing the near UV, visible, and IR wavelengths to reach the Earth's surface (see Chapters 3 and 8). The sun's energy that reaches the Earth is approximately 1 kW m^{-2}. This energy can be harnessed to heat air, water, solids, and other materials to provide heating and light, to generate electricity, and for other purposes.

18.8.1 Solar space heating

Passive solar heating of homes and buildings can be efficiently achieved or augmented by properly designed structures that are, for example, oriented with south-facing windows (in the northern hemisphere), have low blockage of sunlight by trees, are thermally massive to absorb and re-emit heat, and which allow circulation of heat. In general, solar energy is absorbed by materials within the building, and is then re-emitted at infrared

(IR) wavelengths. The most effective windows are double- or triple-paned which allow passage of sunlight, but minimise the loss of heat back to the outside. Thermally massive materials (such as concrete) are able to store large amounts of heat that can then be used to heat the structure overnight.

Various passive solar space heating approaches are commonly utilised. First, the simple use of multi-paned windows on south-facing windows can be effective, especially if the blinds are closed to reduce heat loss during periods of no sunlight. Second, the use of a solar-heated atrium or greenhouse attached to the south portion of the structure can capture heat that is allowed to circulate into the structure for heating. Other, more technically sophisticated, designs have also been developed for solar space heating, including active, rooftop-mounted systems.

An inherent problem with solar space heating is that, in both northern and southern latitudes away from the equator, the net sunlight energy (kW m^{-2}) reaching a given location is much lower in the winter than in the summer. The least amount of energy for heating is, therefore, available during the months of the year when it is most needed.

Solar space heating has been used historically by many cultures even when transparent glass was not available. For example, adobe buildings used by Pueblo Indians in the desert of southwest USA and other cultures have used structures with thermally-massive walls to provide both heating during the night, and cooling during the day. The thickness of the walls was designed so that the outside of the walls is heated during sunlight hours, and so that the heat diffuses to the inside by night time. This heat then radiates into the structure's interior all night providing heating. Simultaneously, the outside of the structure's walls are cooled by the cold, clear desert nights so that, by morning, the interior of the structure's walls are relatively cool. The cool interior portion of the walls then provides cooling of the structure's interior throughout the daylight hours. More recently, phase change media (PCM) have been used to utilise materials that change between liquid and solid state to store and release heat in various building materials.

18.8.2 Solar lighting

The use of sunlight for interior lighting can be a highly effective passive method of reducing energy consumption in homes, businesses and other structures. The key to efficient solar lighting is to design a structure that has sufficient light reaching appropriate use areas within the house, while preserving the other functionalities of the building such as noise mitigation, heat transfer, privacy considerations, and related matters.

18.8.3 Solar water heating

Water in homes is most commonly heated by electricity, propane or natural gas and stored in water tanks until use. Recently on-demand, tankless water heating systems have become popular in both the European Union and the United States. These systems can use electricity or gas, and come in whole-house and point-of-use varieties (e.g., see http://www.cpotanklesswaterheaters.com).

Active solar water heating is a highly efficient and proven technology for providing or augmenting water heating needs for homes and businesses. Systems are generally rooftop mounted, and may be more or less complex, with water being circulated through tubes on the roof for heating, and then stored until used. Temperature rises of from 10–50 °C are achievable in typical systems. Rooftop solar collectors for water heating are often used alone, or in conjunction with solar photovoltaic panels on the same structure (e.g. see Figure 18.4).

18.8.4 Solar thermal power generation

Electricity can be generated from sunlight in several ways, including photovoltaics (discussed below) or by active systems that directly utilise solar energy to generate high temperature steam to drive turbines for electricity generation. Such systems generally use linear- or point-parabolic mirrors to focus the sun's energy on tubes containing water or another working fluid used to boil water creating high-enthalpy steam. A potential disadvantage, in terms of return-on-investment of solar thermal power generation systems, is

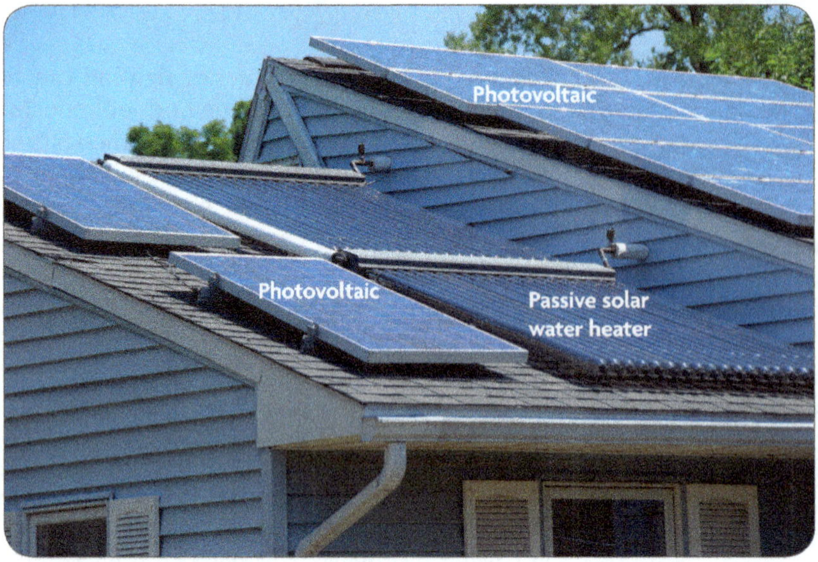

Figure 18.4 Solar house built by the Missouri University of Science and Technology (USA) Solar Decathlon Team in 2004 showing both passive solar water heating collectors (longer tubes) and active photovoltaic solar panels for electricity generation.

that they only produce electricity when the sun is shining (unless sufficient heat storage capacity is integrated into the system). However, this daytime power can be used to provide the additional power production needed during peak demand periods in conjunction with a more continuous power source (e.g. nuclear power or hydroelectricity).

The quantity of solar energy available on Earth varies markedly geographically and temporally. The more distance one moves from the equator, the less solar energy there is available per square metre, on average. In fact, above and below the Arctic and Antarctic Circles, respectively, no direct solar energy at all is available for a portion of the year. Additional geographical variations in solar energy may be due to short- or long-term weather patterns specific to a given region.

Dominant temporal variations in solar energy at a location include daily variations (e.g. the sun only shines during the day), and seasonal variations (e.g. the sun is lower in the sky during the winter than in the summer so that less energy is available per square metre). Furthermore, there are shorter daylight hours during the winter than in the summer. Thus, the net amount of solar energy available for use can be many times less in the winter than in the summer months.

18.8.5 Environmental impacts

A tremendous advantage of solar thermal energy systems is their very small impact on the environment. Rooftop-mounted solar systems have little or no visual impact, and are noiseless. Passive solar heating and lighting can be incorporated into building designs in manners that are highly appealing to most people. Furthermore, pollutants or greenhouse gas emissions are nearly non-existent for these systems. For these reasons, and because of the economical advantages associated with offsetting conventional energy use, the use of solar thermal energy systems in new building design and construction is being rapidly expanded.

Lead-in question

How could solar thermal energy be used more effectively in the building that you are in?

18.9 Solar photovoltaic (PV) energy

In the previous section, the use of solar thermal energy for heating air, water and solids was discussed for various applications, including electricity production. Solar energy can also be used to generate electricity directly by using solar photovoltaic solar cells.

Solar cells use a semi-conductor, usually silicon, to create electricity from solar energy. When sunlight is absorbed by a solar cell, electrons jump from the valence band to the conduction band allowing an electrical current to flow. Solar cells are usually protected by a transparent cover coated with an anti-reflective coating to minimise light reflection and maximise energy absorbed.

18.9.1 PV systems

The most common photovoltaic cells use monocrystalline silicon, although other variations are possible, and many new types are being developed to reduce costs. Photovoltaic cells are generally purchased in panels that are ready to mount on a rooftop (or other location) and connect electrically. A typical roof-mounted system on a house was shown in Figure 18.4.

Photovoltaic systems may be used either on or off the electrical power grid. In off-grid systems, batteries or other means are needed to store electricity when the sun is not shining. Batteries, however, are expensive, have a fixed lifetime, and usually contain toxic metals and acid. An important application for off-grid photovoltaic systems is in remote locations or for systems for which it is not convenient or economical to provide grid power (e.g. road signs or space stations). On-grid systems are connected to the power grid and can take power from the grid or sell power back to the grid when photovoltaic power is less than, or greater than, immediate demand, respectively.

The quantity and variability in the available energy was discussed above for solar thermal technology, and the same concepts apply to solar photovoltaics. Both geographic and temporal variability must be considered in system design.

18.9.2 Environmental impacts

The operation of solar photovoltaic systems is essentially non-polluting and with no environmental impacts. However, there are some concerns about the manufacturing and use of some solar cells (such as the cadmium telluride variety) due to the toxicity of cadmium.

Lead-in questions

For heating water or air in a home, would thermal solar energy or a PV-based system with an electric water heater and furnace be more efficient? Why?

18.10 Oceanic sources of energy – tidal power and power from other ocean currents

Ocean currents are caused by a multitude of factors, including the gravitational effects of the moon, Earth and sun, by solar radiation, and by Earth's rotation on its axis. In this section, the primary causes of ocean currents and tides will be explained, followed by discussion of the two key methods of extracting that energy. The first general method involves tidal barrages (or dams) that extract energy based on differences in the hydraulic head (or elevation, or potential energy difference) between the two sides of the barrage (much like traditional hydroelectric power). The second general method uses turbines and related devices that directly extract energy from the kinetic energy of the currents within the oceans.

Oceanic surface currents are primarily wind driven. Other major forces driving the ocean currents are thermal- and salinity-based density differences, gravity, and the Coriolis Effect (due to the Earth's rotation). These forces combine to form circulating 'gyres' of water that transport heat and energy across vast regions. In the Northern Atlantic Ocean, the North Atlantic Gyre consists of the North Equatorial Current (flowing east to west), the Gulf Stream (flowing north along the North American Coast), the North Atlantic Current (flowing from North America to Europe), and the Canary Current flowing south along Europe and Africa to the equator. Similar currents exist in the other oceans of the world.

Gravitational forces also play a major role in ocean currents worldwide. Simply by visiting most ocean beaches and estuaries in the world, we can observe the tides rising and falling throughout the day. When asked, most people will tell us that the cause of a high tide is due to the water being pulled towards the moon by its gravity. This is true, but is only part of the story. Specifically, it does not explain the reason for there being approximately two high tides (and two low tides) each day, while the moon orbits the Earth only about once a day.

To understand this observation, consider the following. Two identical objects will orbit each other with the centre of rotation being equidistant between the two objects. If one object is more massive, the centre of rotation shifts toward the more massive object. This is the case for the Earth and the moon. While the moon is indeed orbiting the Earth, the Earth is similarly orbiting the moon, with the centre of rotation for this co-orbit being deep within the Earth, the much more massive object. This orbiting of the moon by the Earth causes the Earth to oscillate back and forth slightly (or wobble), thereby causing two simultaneous high tides on opposite sides of the Earth. In general, the lunar tide (when the moon is overhead) is the greater of the two tides due to the combined effect of the Earth's orbitally-derived oscillation, and due to the moon's gravity.

Another factor in the magnitude of the tides is the pull of the gravity of the sun. For example, when the sun is aligned with the moon (or out of phase by 180 degrees), the sun's gravity enhances the magnitude of a high tide. Overall, the causes of the tides and currents are well understood, but are highly complex. Tides can be (and are) accurately predicted, although the resulting currents can be more difficult to predict.

18.10.1 Tidal barrages

Tidal barrage technology

Tidal barrages are a technology with which to harness the potential energy difference between high and low tides in bays (or estuaries) on oceans or seas. Alternatively, offshore barrages are also being considered to reduce (or eliminate) the environmental impacts of damming a bay or estuary, albeit with significantly greater costs.

Figure 18.5 A tidal barrage used to generate power near Saint-Malo, France, on the English Channel.
Source: http://en.wikipedia.org/wiki/File:Barrage_de_la-Rance.jpg
Released into the Public Domain.

In its simplest form, a dam is built across a bay that can trap the tide, preventing it from entering or leaving the bay, so that the water level is allowed to build up between the two sides of the barrage. Once the hydraulic head is sufficiently high, the water is released through conventional turbines, thereby generating electricity. Existing examples of tidal barrages include the SeaGen (County Down, Northern Ireland) and the Rance tidal power plant (Bretagne, France) (Figure 18.5).

The power that can be generated from a tidal barrage is dependent on the volume of water being trapped by the barrage, and by the height of the tide. Thus, the geographic location of the barrage within a given bay is a critical factor in establishing the amount of power that will be generated.

Due to the known and unknown environmental (and economic) impacts that will be caused by tidal barrages, there has been significant (and rational) opposition from some sectors to damming these bays and estuaries. For example, damming of bays may have major negative economic impacts on fish populations, the fishing industry, recreation and tourism.

Building off-shore barrages, on the other hand, would offer nearly unlimited quantities of energy while eliminating the significant environmental (and economic) impacts of damming bays and estuaries.

Temporal power variations

Because of the nature of tides, the use of a single tidal barrage in a bay will generate highly varying amounts of power throughout the day, ranging from zero (twice per day) to its maximum generation capacity (for that day's tide differential). Because electricity cannot be efficiently stored (e.g. with batteries or otherwise) for distribution on a grid, electricity generation must match energy consumption at all times. Therefore, the continuously varying nature of electricity generation from a single tidal barrage can be problematic for a utility with respect to integration within its energy generation system.

One method to reduce the power fluctuations from a tidal barrage is by using a multiple barrage system. Different varieties of multiple barrage systems are being considered. However, each type will be significantly more costly in terms of capital and operational costs and environmental impacts than a single tidal barrage system.

Environmental and related impacts of tidal barrages

Damming a bay or estuary, and controlling the rates at which tidal elevations change, will clearly have significant effects on the ecosystem of the bay or estuary. Estuaries are homes to vast populations of many birds, mammals and fish species, whose food supplies, habitation and reproduction could be significantly impacted. With increasing pressure on the world's food supply, and an already diminishing fishery (fish population) in the oceans, the estuary ecosystem will likely play an increasingly important role in the fishing industry.

An additional major impact is the potential for silt deposition and build-up within a bay or estuary behind the barrage. This additional silt could have a significant negative impact on the ecosystem by covering the natural marine environment on the seafloor. Another impact of a barrage is the inhibition of plant, fish and mammal migration in and out of the estuary. Fish ladders may be installed to partially alleviate the migration of some fish species, although many species of plants and animals might no longer be able to migrate at all. Additionally, with a barrage system, a lock system is required for personal and commercial boats and ships to enter and exit the bay or estuary. This could have a significant impact on the operational costs of the commercial fishing industry, as well as adversely affecting tourism.

Overall, the impact of tidal barrages on the environment and local economies is the subject of current study. It is clear, however, that the impacts will likely be significant, highly site specific, complex, and difficult or impossible to predict fully.

Summary on tidal barrages

The future of tidal barrages is unclear. While significant power generation capability exists in barrages, the ecological and local economic impacts on bays and estuaries may prevent their widespread adoption and use. Offshore tidal barrages, on the other hand, eliminate most or all of the potential for these adverse impacts, albeit at an increased cost. While most of the technology currently exists for offshore barrages, more study and system optimisation of these offshore systems is needed to determine how power generation costs and variability can be optimised.

18.10.2 Ocean current extraction

Ocean current technology

Instead of tapping the potential energy difference of the oceans with tidal barrages, an alternative energy source is tapping the kinetic energy of ocean currents. As discussed above, these currents may be due to the Earth's rotation (fairly constant), to temperature and salinity differences (more variable), and/or to tidal effects (continuously variable, with four maximum-flow- and four zero-flow-periods per day). Submerged marine turbines are similar in concept to wind turbines in which horizontal velocity of the fluid is converted to circular motion to drive a conventional turbine. However, there are many differences between submerged marine turbines versus wind turbines.

First, the density of water is about 1000 kg m^{-3} as compared to just 1.2 kg m^{-3} for air. Thus, water flowing at an equal speed as wind contains more than 800 times more kinetic energy than the wind. Another advantage of submerged marine turbines is that ocean currents in many locations are relatively constant, or at least follow a constant daily pattern (such as tidal currents). Submerged marine turbines are mostly out of site, although floating or stationary platforms may be used instead.

Various types of submerged marine turbines have been proposed, or are currently in use. One design has a hydrofoil blade being driven by the passing fluid rotating on a horizontal axis (much like a traditional wind turbine), with the turbine mounted either submerged or, alternatively, on the surface of the water. Another design has the ocean current spinning a rotor on a vertical axis as for Gorlov's helical turbine or the Polo turbine. Another submerged marine turbine, called the Stingray, uses a blade designed to oscillate up and down driven by the current, with the oscillating motion then converted to circular motion to drive the turbine. Each of these designs (horizontal and vertical rotating shaft, and oscillating) must, in most scenarios, have some means of swivelling so as to remain properly oriented to the ocean current. A highly innovative design called the Rochester Venturi does not spin at all but, rather, has the current pass through a Venturi to create suction on the down-stream side of the orifice used to drive water through a turbine. All of these different designs are being tested and developed at different scales and locations.

Temporal power variations

The temporal power generation from an ocean current electricity generation facility is highly dependent on the nature of the current. Tidal currents are maximised four times daily (twice each in opposite directions). Thus, as with tidal barrages, issues of integration of the ocean current generated power with the over power grid must be addressed appropriately. An issue regarding currents that are driven by the Earth's rotation, temperature differences or salinity differences, is that currents may not vary appreciably in direction but may shift significantly in strength at a specific location. Clearly, the siting of an ocean current generation facility must be chosen carefully based on a firm understanding of these ocean currents. For example, positioning a facility through a constriction such as in channels on the ocean floor, or between land masses, would minimise the shifting of the current.

Environmental and related impacts of ocean current generators

Impacts of submerged marine power generation on the marine ecosystem would likely be minimal. Negligible impacts on the currents themselves are caused by the systems, and the migration of animal and plant species should generally not be impacted. One concern that should be mentioned is the potential for these submerged rotating (or oscillating) blades to attract and/or strike passing animals and fish, thereby maiming or killing them. This concern requires careful study for each design so that technological remedies can be engineered.

Few negative impacts from ocean current generators are likely on the local economy as compared with those of tidal barrages. Clearly, commercial fishermen near generation facilities would need to keep their nets and lines clear of the generators. This would point to the desirability of concentrating marine power generation in specific designated areas rather than having them dispersed across wide regions.

18.10.3 Summary on ocean current extraction

Due to the massive power available in ocean currents, and the relatively small environmental impacts, it is likely that power generation by using ocean currents will, and should, see greatly increased use in the future. Issues with fouling due to drifting vegetation, preventing damage to migrating fish and animals, and shifting location (not direction) of currents all need to be considered and appropriately addressed. Each of these issues appears addressable in the design process. For example, up-current fences or rakes can be implemented to catch or divert drifting seaweed that is large enough to cause a problem with the turbines. Furthermore, the ability to raise a generator out of the water for maintenance can be incorporated into a system design. Field testing of various systems at a moderate scale clearly appears warranted, prior to developing large-scale facilities. Overall, one should expect the role of ocean current generation in the total power generation infrastructure to increase rapidly as the technology matures.

Lead-in question

What advantages and disadvantages are there for wind versus ocean-current power generation?

18.11 Oceanic sources of energy – wave power (from solar-driven forces)

Another source of energy from the oceans and seas is to extract power from their waves. While this technology has been discussed for many years, it is only recently that significant resources have been invested in technology development and prototype testing. Due to its many advantages and few disadvantages, the prospect for wave energy as a renewable energy source is immense.

18.11.1 Basic principles

Ocean waves are formed by the wind in a complex fluid dynamic system. Initial waves are formed due to horizontal stress and frictional forces between the wind and a (smooth) ocean surface. Once a small wave is formed, wind blowing across the surface of the water causes differential pressures and stresses that significantly increase the height and motion of the waves. Because of the relationship between wind velocity and a wave's height and energy, storms generate the most massive and energetic waves. Once formed, a wave can travel for very long distances across the ocean, with little energy loss, causing an ocean swell even in areas with little wind. Because wind is solar driven, wave energy is yet another form of indirect solar energy.

18.11.2 Ocean wave energy technology

A wide array of devices exists for harnessing the energy of waves, and more are constantly being developed and tested. Wave power generators may be sitting on the seafloor in shallow waters, or tethered or floating in deeper waters.

Most devices that are resting on the seafloor are designed as some variety of an oscillating water column (OWC) that can be used to spin generators. An advantage of an OWC is that most of its hardware is above the water surface which facilitates easy maintenance. An additional advantage of OWCs seated on the seafloor is that they are often near shore.

Many floating wave energy generators have been designed, tested and used in practice. Offshore devices have the advantage of having bigger waves (with more energy) available than from the smaller waves that are found in shallow waters. Additionally, vast arrays of floating wave energy facilities can be imagined that would have minimal impacts, with respect to the ecology of the ocean, shipping or fishing.

18.11.3 Temporal variation with ocean wave generators

The power available for energy generation is proportional to the magnitude and frequency of the waves, as well as the efficiency of the device under varied conditions. Clearly, weather and other factors affect wave conditions and, hence, the power output from a wave generator facility. As for tidal-, ocean current-, and wind energy-power generation systems, the varying power from a wave generator power facility must be accounted for within the design and operation of an integrated power network.

18.11.4 Environmental impacts

There are relatively few negative environmental impacts from wave generation systems. Noise and visual impacts are negligible because they tend to be offshore. Impacts on migrating fish and animals would almost certainly be minimal. Similarly, the fishing industry would not be impacted, except in the immediate vicinity of a wave generation facility, especially a floating system. Systems fixed to the seafloor and those that are tethered could pose a problem for nets and lines associated with commercial fishing. Floating and submerged vegetation would generally not be a problem for most systems, especially as compared with some tidal barrage or ocean current systems.

Lead-in question

How much potential do you see for energy generation using wave energy?

18.12 Hydroelectric energy (from solar energy)

Many medium- and large-scale hydroelectric power plants were built in the twentieth century in the United States until public pressure and other sources of energy slowed the construction of new dams. A typical medium-scale plant is the Bowersocks Facility in Lawrence, Kansas, USA which provides nominally 54 kW via a flow of 15 million m^3/day (6300 cfs) and a low-head drop of 5.3 m (17.5 ft) across the Bowersocks Dam on the Kansas River (Figure 18.6). Examples of large-scale, high-flow and higher-head hydroelectric power plants are Bonneville Dam across the Columbia River (Oregon, USA), and Hoover Dam (Nevada, USA) that generate nominally 1.1 and 2.2 GW, respectively.

Figure 18.6 Exterior of the Bowersocks Mills and Power Co. low-head hydroelectric power plant in Lawrence, Kansas, USA. Source: Photograph by kind permission of Susan Adams, copyright Craig Adams.

New hydroelectric facilities are being built internationally. For example, China recently completed construction of the Three Gorges Project which was built ostensibly to control flooding as well as to provide electricity for China's burgeoning economic growth. The Three Gorges Project hydroelectric facility is rated to provide power generation of 18 GW.

The basic principle behind hydroelectric power generation is as follows. Energy from the sun causes evaporation of water from the oceans and lakes, which is then transported overland. Cooling of the atmosphere causes the water to condense and rain to fall. This water then flows down streams and rivers back to the ocean. The potential energy of the water from this solar-driven hydrologic cycle can be tapped by forcing the flowing water to pass through turbines to produce electricity. Hydroelectric power can be harnessed at a large or small scale; both are discussed below.

18.12.1 Hydroelectric technologies

There are three primary means of medium- and large-scale hydroelectric power generation. First is the damming of a river creating a reservoir and providing a potential energy difference between the water levels of the reservoir above and below the dam. A portion of the water then passes through turbines, thereby generating electricity. A second means of tapping hydroelectric power is through diversions, where a portion of a river is diverted through turbines using pipes that parallel the river, and leaving the remaining water flowing along the natural river course.

A third application of large-scale hydroelectric power is through pumped-storage facilities. These facilities pump water from a lower reservoir to a higher reservoir using excess power during times when power generation from other sources (e.g. nuclear or coal-burning power plants) exceeds the immediate demand (e.g. during the night-time hours). When power demand is high (e.g. during daytime hours), the water is transferred from the higher reservoir to the lower reservoir through turbines and used to generate electricity. In pumped storage facilities, reversible (or combination) pump/turbines are used to both pump the water (e.g. at night) and to produce electricity

Figure 18.7 Jocassee Dam and Power Plant associated with the Oconee Nuclear Power Plant in South Carolina, USA, which uses pump-storage to store excess nuclear-generated energy at night for hydroelectric generation and distribution during high-energy demand daytime hours.

Source: Photo and permission provided by Duke Energy.

(e.g. during the day). An example of a pumped storage facility that is used to match continuous power generation to varying demand is the Oconee Nuclear Power Plant (Duke Power/Duke Energy Corporation) in northwestern South Carolina. The 2.5 GW facility operates in concert with the nuclear power plant to store excess power at night, and to generate power during higher demand periods during the day (Figure 18.7). Pumped storage facilities can also be used to store varying power generation, such as that from tidal barrages to a more constant electricity demand.

The energy available from a hydroelectric facility is proportional to the quantity of water flowing through the turbines, and the height differential between the higher and lower water levels across the impoundment (or the hydraulic head). Thus there are two major classes of hydroelectric facilities, low-head and high-head systems. In low-head systems, the head across the dam is relatively small and, thus a very large flow through the turbines is required for significant power generation. High-head systems, usually built in mountainous regions, can achieve large levels of power generation with much lower flow rates.

Due to public opposition to damming rivers and the associated significant environmental impacts, only limited additional hydroelectric power generation facilities are likely to be found in either the USA or the UK, though some countries (e.g. People's Republic of China) are projected to increase hydroelectric power development.

18.12.2 Power variation

Because the power generated is dependent on the flow rate, seasonal variations in power generation are common, with less generation in drier winter or summer periods and more generation during spring runoff periods. Additionally, short-term or long-term droughts also may directly affect or control the amount of power generated from a specific hydroelectric power generation facility.

Short-term power generation can be adjusted to meet demand by varying the amount of water passed through the turbines. For example, hydroelectric power generation is an effective means of matching power generation to demand during daily power generation cycles.

18.12.3 Environmental impacts

There are many environmental impacts from hydroelectric power plants. Damming of rivers floods land that was, or could be, used for other activities such as for towns, agriculture or recreation. For example, fantastic natural environments such as desert canyons in the western US were flooded for the creation of Hoover Dam and Glen

Canyon Dam in 1936 and 1964, respectively. Dams are also subject to potential failure causing catastrophic consequences for downstream regions. They can also impact the migration of fish upstream and downstream, though this can partially be addressed through the use of fish ladders or by other means. Damming of rivers can cause silt and nutrients that are normally transported downstream to be captured behind the dams. Siltation effectively reduces the storage capacity of a reservoir and also deprives downstream regions of sediment and nutrients that were previously naturally replenished by the river.

Finally, while hydroelectric power is normally considered to emit no greenhouse gases, reports to the contrary have indicated otherwise (World Commission on Dams, 2000). Specifically, recent reports have shown that in reservoirs that contain significant anaerobic activity, significant methane production from the anaerobic biodegradation of organic matter can occur. Because methane is a much stronger greenhouse gas than even carbon dioxide (by a factor of 25), hydroelectric power generation (especially in warmer climates) may be considered to create and release greenhouse gases. Reservoirs are also subject to significant algae and cyanobacterial growth. Algae and cyanobacteria can create significant levels of organic and inorganic compounds that have a strong undesirable taste and odour (e.g. MIB or geosmin) and/or significant human toxicity (e.g. microcystines, saxitoxin or cylindrospermopsin). These organic compounds can degrade the quality of water for purposes that include drinking and bathing.

Lead-in question

Where would you expect to see hydroelectric plants being built?

18.13 Wind energy (from solar energy)

Wind energy is a clean source of electrical energy that is rapidly expanding in use worldwide. Wind energy is generally 'environmentally friendly' with a relatively low carbon footprint, and no pollution generated during operation of a wind turbine. Some negative impacts cited by critics include noise pollution and visual impairment of the landscape. Wind energy is abundant, and may be harvested either on shore (on land) or off shore using individual wind turbines or large wind generation farms with many wind turbines.

Winds blow from areas of high atmospheric pressure to low atmospheric pressure that are caused by differential heating of the Earth's surface. The Earth's surface near the equator receives a greater amount of energy per square metre than in higher or lower latitudes due to the greater angle of incidence of the sun's rays on the more northern or southern latitudes. This greater heating of the air mass causes the air near the equator to rise, creating a low pressure zone. The risen air then blows north or south and cools in the process, eventually causing its density to increase, and the air to drop in elevation thereby creating high pressure zones. The wind cycles at the surface from the high pressure zones (nominally oriented at 30 degrees north or south latitude) back towards the equator (or alternatively towards the poles). It should be remembered though that, because of the tilt of the Earth's axis, the maximum solar radiation oscillates between the Tropics of Cancer and Capricorn. Other sources of wind (e.g. the rotation of the Earth) also contribute significantly towards the complex set of forces that drive the meteorology of wind.

18.13.1 Technologies

The wind has been harvested for work for thousands of years. Early windmills were used to grind grain in mills. Later, wind turbines were used to pump water and generate electricity. Modern wind turbines tend to have many fewer blades (and solidity) than early versions. Their designs are based on a fundamental understanding of aerodynamics and are executed in both vertical and horizontal axis configurations.

Vertical axis wind turbines spin about a vertical axis and hence can harvest wind coming from any direction equally without the need for reorientation. Vertical axis wind turbines are sometimes

Figure 18.8 Single horizontal wind turbine located near the Centre for Alternative Technologies (CAT), Machynlleth, Wales, UK.

Figure 18.9 Offshore wind turbines located at the Middelgrunden wind generation facility in Copenhagen, Denmark.

housed in a shroud that increases the velocity of the wind over the blades, thereby increasing the rotational speed of the turbine and increasing power output.

Horizontal-axis wind turbines, on the other hand, must be continually positioned towards the wind to optimise the power generation efficiency. There are typically three blades in a horizontal axis wind turbine. The blades are shaped as air foils which create lift as the wind blows over them. Optimisation of the power generation is complex and involves matching wind turbine generator characteristics to the wind velocity and other parameters. Examples of several horizontal-axis wind turbines are shown in Figures 18.8 to 18.10.

A key issue associated with the rapid expansion of wind energy is the lack of an existing power grid in the areas of highest winds. Development of this power grid system will require significant investment and time, and will need to be initiated soon if wind energy is to reach its potential in the foreseeable future.

18.13.2 Temporal variation in wind power

Power generation from a given wind turbine, as a function of wind speed, is called a power generation curve. Estimation of the net power production from a single wind turbine or an array of wind turbines requires matching the power generation curve to the wind speed frequency curve for the wind generation site. Such wind speed data can be difficult to estimate for a specific site, but can be made by direct measurement over a long period of time, by obtaining data from nearby sites, or by computer modelling.

Wind velocity and, hence, power generation from wind generators, varies frequently. Because energy cannot be easily stored, the net energy consumption must match the power production in large-scale systems. As the electrical energy generated from wind and placed on the electricity distribution grid varies, other power generation sources must be adjusted accordingly. Small-scale systems in remote areas may use batteries to store electricity for low wind periods though this approach can be prohibitively expensive.

18.13.3 Geography

Offshore wind speeds tend to be higher than on-shore speeds due to fewer obstructions to the wind along the Earth's surface. The costs associated with offshore wind generation tend to be higher than those for the equivalent on-shore facilities due to difficulties in construction, the need for more and higher quality materials, and power distribution issues. Offshore facilities have

Figure 18.10 Smoky Hills Wind Farm in Kansas (USA) generates over 250 MW with more than 150 Vestas and GE turbines.

greater power generation potential, however, that can offset the greater initial costs. Furthermore, key environmental impacts associated with the noise and visual aspects of wind generation are generally negated by their location off shore (as discussed below). An example is the offshore 40 MW Middelgrunden wind generation facility located in the harbour at Copenhagen, Denmark, that was built in 2000 (Figure 18.9).

18.13.4 Environmental impacts

Wind energy is gaining in popularity due to its potential abundance and its lack of carbon dioxide and other pollutant emissions (e.g. NO_x, SO_x, etc.). However, there are environmental impacts that have hampered the more rapid increase in the use of wind energy. First, wind turbines generate both mechanical noise from gears and other parts of the turbine mechanism, and aerodynamic noise from the blades passing through the air, especially when located in the countryside. Mechanical noise is usually not an issue at reasonable distances from the wind turbine. New, innovative turbine designs with low-rpm electrical generators have eliminated the gearbox, greatly reducing mechanical noise. Appropriate location of new wind installations at sufficient distances from houses and businesses is required if wind energy is to continue to gain strong support. Aerodynamic noise from the blades moving through the air is generally not an issue and is usually not louder than the wind itself.

A second environmental factor is the visual impact of wind turbines on the landscape or seascape. This is a highly subjective matter, in that many people think that wind turbines are a sign of sustainability and are a positive addition to a landscape. Others, however, object to the sight of wind turbines. In any event, it is prudent to consider visual impacts when selecting a site for wind turbines. Sparsely populated prairies and offshore locations clearly will generally have fewer overall visual and noise impacts.

A third environmental impact is the interference with electromagnetic signals used for communication and navigation. The rotating blades of a wind turbine can reflect electromagnetic signals in a pulsed manner, causing potential problems with radio and related signals. Again, proper siting of the turbine can prevent this issue, and technological fixes may soon be available.

Finally, there has been a concern related to birds being killed by the rotating blades of wind turbines. Overall, the rate of bird fatalities reported in studies on the subject is very low (and of the order of fewer than two birds per year per turbine). Methods of reducing bird fatalities are well understood. These include using the slower rotating blades of modern wind turbines as compared with the much more rapidly rotating older variety. Modern wind turbines also have fewer places for birds to land and roost, and so do not attract birds as much as older turbines. Furthermore, proper location of multiple turbines in relation to each other, and in relation to known avian flyways, is also important in preventing bird fatalities.

Lead-in question

How does having multiple wind farms located in different regions help to provide a more continuous and reliable source of power than a single wind farm?

18.14 Geothermal energy (from gravitation and nuclear forces)

Geothermal (or Earth-based heat) energy may be exploited for direct heating of homes, buildings, structures and processes, or for production of electricity. Geothermal energy can be an important environmentally-friendly energy source, though care is required to prevent negative environmental impacts.

The geothermal gradient is approximately 25 °C per km of depth into the earth. For example, the Earth has a temperature of 260 °C at a depth of 10 km. These high temperatures exist at various depths everywhere in the world, and make geothermal energy potentially exploitable worldwide.

There are several origins of this heat. When the Earth was formed, it was molten, and it continues to possess some of the original heat. Further, gravitational forces cause very high pressures at great depths, causing heating and increased temperatures. More importantly, however, may be the heat released in most locations by the radioactive decay of natural isotopes in the Earth, especially K^{40}, U^{238} and Th^{235}. In regions along the tectonic plates of the Earth (as well as other regions), the molten rock or magma below the Earth's crust reach near (or even to) the Earth's surface, causing effects that range from natural hot springs to active volcanoes (see Chapter 1). Thus, both high-temperature (or high enthalpy) and low-temperature (or low enthalpy) geothermal systems exist and can be exploited as energy sources.

Heat may be stored in porous rock containing significant water or in relatively impermeable rock termed 'hot dry rock'. Water in porous geothermal sources is normally under high pressure, and as it is pumped to the surface it may flash into steam as the water's boiling point at reduced pressure is surpassed. Water from geothermal aquifers often has high amounts of dissolved solids and gases, complicating the process of utilising its heat without polluting the environment or damaging equipment.

The heat from the Earth is not technically renewable in that most heat comes from isotopic decay or from tapping the original heat from the Earth's origin. In fact, however, the Earth's core is cooling at a very slow rate. Furthermore, heat can be extracted from a reservoir more rapidly than it can be replenished, thereby reducing the extractable geothermal energy. Thus, clear understanding of the dynamics and operation of a geothermal system is a prerequisite to utilisation of this energy source.

18.14.1 Direct heat utilisation

Heat from geothermal sources can be used directly, or it may be used to generate electricity. A common method of utilising low enthalpy heat directly is with ground-source heat pumps. These systems are common for homes and businesses, and involve pumping air from the structure through pipes buried beneath the ground. Natural heat from the Earth then warms the cooler air from the building, which is then recirculated back through the building, thereby providing heating. Heat extraction does cause cooling of the ground during the winter season, but this heat is restored naturally (or otherwise) during the summer season. Ground-source heat pumps have proven to be effective in many regions, and are commonly used. In some regions, these systems may also be used for cooling in the hot summer months by dumping heat from the building into the cooler ground, thereby cooling the recirculating air (and storing the heat for use in the following winter months).

Steam and hot water from geothermal sources can also be used directly to heat buildings and other structures. Countries such as Iceland use direct geothermal energy on a large-scale basis to keep roads and sidewalks ice free year round, to heat buildings, and for recreation. Direct use of geothermal energy offers the advantage of being able to utilise both high- and low-enthalpy heat sources.

18.14.2 Electricity generation

Sustainable generation of electricity from geothermal sources is complex and highly dependent on many factors including the nature (e.g. porosity, fraction pattern, material, depth) of the geologic formation, the water content at the source, the temperature and pressures, salinity, spatial requirements, and other factors. The development of geothermal energy can be relatively costly due to requirements for exploration and characterisation of geological formations, drilling costs, and environmental protection costs. Nonetheless, geothermal energy provides an outstanding, renewable source of energy that is able to provide significant power to the grid. There are many types of geothermal-based power generation systems. The simplest system utilises dry steam from geothermal sources to spin turbine generators, but these systems are fairly inefficient, and are not as common as flash systems. In dry steam systems, the resulting condensed water is usually re-injected into the geothermal source to prevent rapid depletion of the steam supply.

A second geothermal power generation system is the single flash steam system. In these systems, pressurised hot water brought up from the extraction well is vaporised to steam in a boiler at the surface. The steam is then used to generate electricity, while the remaining brine solution is pumped back into the geothermal formation.

A third option is to use a binary-cycle system in which the water from the well operates in a closed loop. Specifically, hot water is brought up from the well, and then passes through a heat exchanger in which the heat is transferred to a working fluid (normally a volatile organic compound with optimised vaporisation characteristics). The high-pressure gaseous working fluid is used to spin the turbines, after which it re-condenses to a liquid and is reused in a closed loop. The water from the well similarly condenses after transferring its heat to the working fluid, and is then reinjected back into an injection well. Binary-cycle systems have a great advantage over either the dry steam or the single flash systems because the water from the geothermal source is used in a closed loop. Specifically, there are no emissions to the atmosphere and no high salinity water to dispose of as in the other systems.

18.14.3 Quantity and variability

The quantity of geothermal energy that is potentially available is enormous. Heat for direct use via ground-source heat pumps is available almost anywhere worldwide in quantities sufficient for distributed use in homes and businesses. The variability in this low-enthalpy heat is mostly seasonal; that is, it is primarily based on the heat extraction rate relative to the reheating rate from natural sources (or from use of the ground source system for cooling structures in the summer months).

Geothermal heat for power generation is also available worldwide because high temperatures are everywhere present at sufficient depth within the Earth. However, economical utilisation of this geothermal heat requires it to be sufficiently accessible with respect to depth and geological formation characteristics. The variability of these geothermal heat sources is also related to the heat extraction rate relative to the reheating rate. Modern geothermal operations have learned that they must properly manage the net heat mining or else geothermal fields are unsustainable.

18.14.4 Environmental impacts

Exploitation of geothermal energy has significant potential for negative environmental impacts, if not properly managed. The first concern is that gases from geothermal sources can contain hydrogen sulphide, carbon dioxide, methane, hydrogen and other polluting gases. Hydrogen sulphide has a strong 'rotten egg' smell that is often associated with older geothermal power production facilities. Carbon dioxide, methane and hydrogen are all strong greenhouse gases and, therefore, must be captured to prevent release into the atmosphere to protect the environment.

A second concern is that condensed water in single flash systems contains highly concentrated dissolved solids because of both the initially high mineral content, and the concentrating effect of distilling off water in the flash process. This brine solution can contain not only high concentrations of calcium, magnesium, chloride and sulfate, but also toxic metals such as cadmium or lead. In modern geothermal power production systems, the brine stream must either be treated to remove pollutants or be reinjected into the geothermal formation.

A third issue with geothermal power production is the potential for excessive heat extraction to cool a formation to such a degree that the local environment is negatively impacted when hot springs are cooled and geysers are adversely affected. This can result in a significantly negative impact on regional tourism that is so important to many areas of the world.

Each of these potential impacts can be managed to a greater or lesser degree with sufficient planning and economic investment. Nevertheless, these potential environmental impacts are real, and form the basis for much of the public resistance to geothermal power production.

Lead-in question

How and where can geothermal energy be developed in a manner that would have minimum impact on the environment?

18.15 Biomass energy (from solar energy)

Plants grow and create biomass using carbon dioxide as the carbon source and solar energy as the energy source, through a process called photosynthesis (see Chapter 4). In essence, the energy of the sun is stored in the chemically-reduced molecules within the biomass. Fossil fuels come from prehistoric plants that were buried beneath the Earth and chemically reduced to form coal, oil and natural gas. Biomass may be from natural sources (e.g. forests), cultivated crops (e.g. soybean, switchgrass or algae), or from wastes (e.g. swine manure). The energy content of biomass feedstocks varies greatly. Proper selection of feedstocks is the focus of much study in the EU, US and worldwide. Unfortunately, selection of feedstocks for biofuel production has often been based on political, rather than sound scientific, considerations.

Biomass can be thermocatalytically converted to heat (or electricity) through combustion (or gasification or liquification), such as through burning of wood or wood products (e.g. charcoal). Biomass can also be chemically converted to biodiesel through a trans-esterification process. Finally, biomass can be biochemically converted to fuel gas (i.e. methane) through anaerobic digestion, or to ethanol through fermentation.

Most biofuels offer a tremendous advantage over fossil fuels in that there is usually a smaller *net* carbon dioxide footprint. Specifically, the carbon dioxide released during the burning of biofuels was originally sequestered into the biomass during its production. One must also consider, of course, the carbon footprint involved in growing, harvesting, transporting and manufacturing the biofuels.

Biomass can also be used as an energy source in many different ways. In this chapter, we focus on the use of biomass for biodiesel and for ethanol production due to the current emphasis on these processes as renewable energy sources.

18.15.1 Biodiesel

Biodiesel is used essentially like petroleum diesel products for transportation and the operation of other engines. Biodiesel is produced through trans-esterification of fats and vegetable oils to produce methyl esters (i.e. the biodiesel) and glycerin (which has commercial value for production of cosmetics and other products). The process requires approximately 1 kg of alcohol for every 10 kg of oil or fat consumed (www.biodiesel.org). Feedstocks for biodiesel include oil from crops, including coconut, jatropha, rapeseed/canola, peanut, sunflower, safflower, soybean and corn. Additionally, used cooking oil and animal fat may be used as a feedstock. Another feedstock showing great promise is algae which can be grown rapidly and can have high oil content. Currently in the United States, the mix of biodiesel feedstocks includes about 40 per cent soybean, 20 per cent used cooking oil and animal fat, and 10 per cent canola (rapeseed) and camila oil, with the remainder being other sources.

All feedstocks are not equivalent, but vary greatly in their energy content, growth rates, and impacts on the environment. For example, of the 'fuel crops' most commonly used, corn and soybean, produce the least amount of oil per year per hectare of land farmed. Rapeseed (canola), peanut, and sunflower oils can be produced at intermediate rates, while coconut and palm oils can be produced at higher rates. Several crops have gained special attention lately for potential use as feedstock for biodiesel products.

Corn-based ethanol production has been widely promoted, although it has many issues affecting its viability (described below). Jatropha is also problematic as a biofuels feedstock. Jatropha is a woody plant that can grow in many environments, but it is highly toxic to the extent that workers must take special care in its handling. This is also an invasive species, which raises concerns about its importation and use in some countries. Switchgrass, on the other hand, is not toxic and can grow in many environments. It is relatively high in oil content and is reported to have yields greater than each of the crops mentioned above. Much research is ongoing in optimising switchgrass, which shows great promise as a core feedstock for biodiesel. A further advantage of switchgrass is that it is not a food source and, therefore, fuel production does not compete in the marketplace with food production for people (although of course it requires land).

Algae may show the most promise of all feedstocks for biodiesel production. They can be grown easily in many climates. Algae's oil yields are projected to exceed those of even switchgrass because of algae's rapid growth rate and relatively high oil content (i.e. 15–45 per cent). For these reasons, small biodiesel producers as well as major industries are studying algae as possible feedstock for biodiesel production. For example, Chevron and the US Department of Energy National Renewable Energy Laboratory recently began collaborative development of algae-based biodiesel production (http://www.nrel.gov/news/press/2007/535.html). Algae farms could be placed next to large carbon dioxide emitters, such as coal-burning power plants, to sequester carbon from burning coal into the algae biomass. More research is needed to fully optimise algae-based biodiesel. Algae-based biodiesel does not cause market competition between energy and food, and algae are easy and inexpensive to grow, and give high yields.

18.15.2 Ethanol

Ethanol can be produced from fermentation of crops such as corn or sugarcane. Alternatively, cellulosic ethanol can be produced from grass, wood waste and other agricultural wastes. In essence, cellulosic materials are broken down enzymatically to simple sugars that are then fermented and distilled into ethanol. The advantage of ethanol over biodiesel is that it can be blended into gasoline mixtures for use in automobiles (including as a fuel oxygenating agent). A disadvantage of ethanol-based biofuel is the high energy costs required for processing, especially for distillation.

18.15.3 Environmental impacts

In the USA, corn-based ethanol production has been promoted for years as the environmentally-friendly replacement for gasoline, and has been made possible by tax subsidies from the US Congress to promote its use (which were being phased out in 2011). However, corn-based ethanol has many problems that should not be overlooked. First, it may take approximately the same amount of fossil fuel energy to produce corn-based ethanol as the product contains due

CASE STUDIES

Biodiesel Case Study – Feedstock shifts by largest US-based biodiesel producer

Renewable Energy Group (REG) is the largest biodiesel producer and marketer in North America (and is located in Ames, Iowa, USA). REG currently owns/operates five biodiesel production facilities (in Iowa, Texas and Illinois, USA) with a total production capacity of about 680 million litres/year. Economic forecasting of the increasing costs of soybean oil led REG to consider alternative feedstocks for the biodiesel production. To support this approach, REG built multi-feedstock plants, and broke ground on the first one in 2004. This case study is illustrative of how shifting economics can profoundly affect the production and use of biodiesel (and other alternative energy sources). It also demonstrates that technological feasibility and considerations are just one component of bringing an alternative energy source to market.

As part of their business model, the purchasing plan of REG was originally focused on the use of soybean oil as its primary feedstock for biodiesel production. REG marketed its soybean-based biodiesel under the tradename SoyPOWER. As soy-based biodiesel production rose, the demand for soybean oil also rose dramatically. This increased demand caused the price of soybean oil to double or triple in market prices in the last decade (between 2000 and 2010).

REG had to develop new production technology to allow the use of lower cost alternative feedstocks such as animal fat and used cooking oil. The source of animal fat is generally from slaughterhouses where cattle, swine and poultry are processed for market. Historically, market costs for animal fats had been significantly lower than for soybean oil but could be expected to rise with increasing demand (and have risen in recent years). REG also uses used cooking oil as a biodiesel feedstock. Restaurateurs used to have to pay to have their used cooking oil taken away, but now are paid as the demand for used cooking oil has increased. As all of these feedstocks continue to increase in price, the biofuel industry will continue to look for new, lower cost alternatives.

The specific technological developments and production facility modifications used by REG required to use animal fats and used cooking oils instead of soybean oil included a feedstock purification step that takes out impurities in the raw materials before producing the diesel fuel. REG implemented these changes in its production facilities allowing the successful changeover in feedstocks in response to market fluctuations in raw materials prices. Currently, REG is positioned technologically and structurally within its various production facilities to utilise new feedstocks including algae, jatropha oil, and inedible corn oil depending on availability, pricing and other factors. In fact, REG has studied 36 different alternatives for their use as biodiesel feedstocks. They found that 34 of these feedstocks were viable, and all of their production facilities are designed to use any of these feedstocks with only minor processing changes. Thus, when the marketplace has large volumes of a specific feedstock (and hence, relatively lower costs), REG is able to utilise those raw materials directly in their current production facilities.

While algae-based biodiesel continues to provide great promise as a high-volume biodiesel feedstock, the costs are currently too high for the biodiesel industries use. Key technology development needs include more economical algae separation/dewatering and lipid-extraction technologies. With these technological advances, it is anticipated that algae may become a major feedstock alternative for biodiesel production.

REG's significant and major response to economic factors for selection of their feedstock provides a unique glimpse at the challenges of operating a successful biodiesel production and marketing company. As with all alternative energy technologies, the economics have to work for the technology to be sustainable, that is, used long term. Rapidly changing economics can have profound effects on the economic viability of green energy businesses, in general, which stresses the critical importance of outstanding business management as well as technological leadership.

to the energy intensive farming, transportation and production processes involved. Second, corn farming for fuel increases the amount of land under cultivation, and thereby may increase soil erosion and loss of top soil. Third, corn farming is pesticide intensive (e.g. with atrazine) and thereby contributes to increased water pollution, which is associated with increased water treatment costs and human health impacts. Fourth, corn farming is also fertiliser intensive, leading to increased use of nitrogen and phosphorus fertilisers (with high carbon footprints). Runoff of N and P into surface waters causes their degradation and enhances eutrophication and algae growth. Fifth, use of food crops (corn as well as soybeans and other crops) as energy fuels increases food prices, including the cost of meats and poultry (which are fed corn, soybeans etc.). One could argue that taxpayers may be subsidising food price increases and environmental degradation through tax incentives provided to the corn-based ethanol industry.

Lead-in question

Why has algae- or switchgrass-based biodiesel not had the industry or governmental support enjoyed by corn-based biofuels production?

18.16 Integrated power systems

Integration of more or less temporally varying power supplies to continuously match supply with demand is always a constraint that must be met, even when energy storage (e.g. hydraulic) is employed to somewhat dampen the supply/demand differential. A scenario for a hypothetical power utility would be to utilise the following mix of energy sources:

1 Constant coal or nuclear power generation for the base demand.

2 Variable wind or tidal power generation.

3 Diurnally-variable photovoltaic solar power generation.

4 Natural gas power generation to continuously match power generation with consumption.

A hypothetical mix for a one-day period is presented in Figure 18.11. Natural gas power plants can very rapidly adjust energy output, making them a good component of an integrated power system. Natural gas, of course, is a fossil fuel and emits carbon dioxide, a greenhouse gas. Thus, as society moves farther from fossil fuels to truly renewable energy, more innovative means will need to be devised to match power generation with instantaneous demand. For example, having many multiple wind- or oceanic-based power generators within a system also has the benefit of damping power generation. Finally supplementary energy may come from waste exploitation, for example as discussed in Chapter 12 for disposal of food waste.

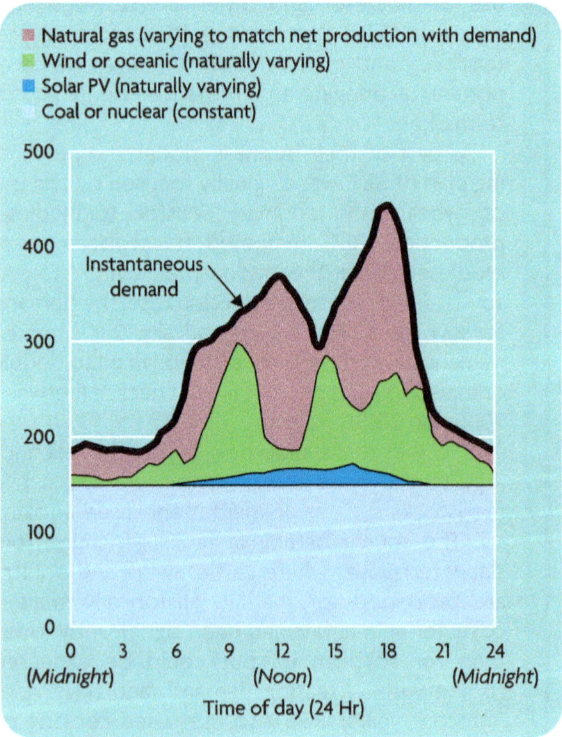

Figure 18.11 Hypothetical mix of coal or nuclear power (base load), augmented by naturally varying wind or ocean energy, diurnally varying solar photovoltaic (PV), augmented by systematically-varied natural gas power production to match net production with instantaneous consumption.

18.17 Summary

Society in both developed and developing countries will eventually move almost completely to renewable energy sources. The only real questions are at what rate will the transition occur, and whether or not the transition will be done in a manner that is smooth and relatively painless, or will it be abrupt with major impacts on society. There are many truly renewable energy sources, each with distinct advantages and disadvantages, as has been discussed in this chapter. It is critical that engineers, scientists, policy makers and economists make sound decisions regarding the appropriate mix of energy sources to utilise in a given region. This requires full consideration of direct and indirect costs in terms of economics, and social and environmental impacts. Most importantly, decisions must be made now and cannot be put off. Energy infrastructure is costly, and takes many years between system design and its construction. Society must move forward in an orderly and rapid manner in designing and building the energy infrastructure of the future.

POLICY IMPLICATIONS

Renewable and non-renewable energy

- Policy makers need to understand and incorporate both direct and indirect costs of energy sources into policy decisions so that sustainable means of powering our society and transportation infrastructure may be realised.
- The real economic value and environmental benefits of energy conservation as an energy source should be recognised for western societies consuming energy at many times the Earth's carrying capacity.
- The economic impact on alternative energy development of frequent and radical fluctuations in the price of fossil fuels should be managed through appropriate policies.
- Energy policy should embrace the reality of integrated systems that utilise many different alternative and base-load energy sources, each with different merits and issues.
- Issues of base-load electricity generation and transportation fuels for future generations must be addressed urgently in the light of future declines in extraction of non-renewal energy sources (due to either scarcity or their high impact on the environment).

CHAPTER REVIEW EXERCISES

Exercise 18.1

List five to ten alternative and non-renewable energy sources. Discuss the advantages and problems with each energy source and how it might play a role in an overall energy strategy.

Exercise 18.2

Calculate the mass (kg) of carbon dioxide (CO_2) that would need to be replaced into a depleted oil aquifer to compensate for one kg of crude oil pumped (assume an empirical formula of

$C_{85}H_{12}O_1S_1N_1$). If equal densities are assumed, what volume of liquefied carbon dioxide must be injected for each cubic metre of oil extracted?

Assuming a vehicle achieves 50 km/US gallon of gasoline (C_6H_6), estimate the volume (m^3) of $CaCO_3$ that would be generated for each m^3 of gasoline burned. What volume of $CaCO_3$ would be generated for each mile a car travels down a highway?

(1 U.S. gallon = 3.79 litres; density of gasoline = 0.877 g cm^{-3}; density of $CaCO_3$ = 2.71 g cm^{-3}.)

Exercise 18.3

Imagine that you are now a staff aid for a leading legislator/politician. You have been assigned to draft the key points for new energy legislation. The legislator has specifically required that you draft a bill that makes good rational sense for the country, and that sets the country on a firm path towards sustainable energy policy. The bill is to be based on the best data available including fossil fuel projections, environmental and economic impacts, safety, alternative technologies, etc. The legislator wants the policy to make sense for the short- (5 year), medium- (20-year), and long-term (100-year) time horizons. Write a short report that includes:

1 Statement of need (1 paragraph).
2 Facts, factors and assumptions considered. (Bullets with introductory paragraph.)
3 Key points of the bill. This is the most important section. (Bullets with introductory paragraph.)
4 Measures or incentives for successful implementation of the bill's goals. How will goals of the bill be encouraged or enforced? (Bullets with introductory paragraph.)
5 Anticipated criticisms of the bill with your (or the legislator's) rebuttal response. Do not miss the obvious and most important anticipated criticisms of each aspect of your proposed legislation.

REFERENCES

Alley *et al.* (2007) Report of Working Group I of the Intergovernmental Panel on Climate Change.

Boyle, G., Everett, B. and Ramage, J. (2004) *Energy Systems and Sustainability: Power for a Sustainable Future*. The Open University, Oxford (UK).

Boyle, G. (2004) *Renewable Energy: Power for a Sustainable Future*, 2nd edn. The Open University, Oxford (UK).

Fanchi, J. (2004) *Energy: Technology and Directions for the Future*. Elsevier Academic Press, London/New York.

Heinberg, R. (2004) *Understanding Renewable Energy Systems*. Earthscan, London/Sterling (VA, USA).

Komer, P. (2004) *Renewable Energy Policy*. iUniverse, Lincoln (NE, USA).

Morgan, S. (2003) *Alternative Energy Sources*. Heinemann Library, Chicago.

Pahl, G. (2005) *Biodiesel: Growing a New Energy Economy*. Chelsea Green Publishing, White River Junction (VT, USA).

Smil, V. (2005) *Energy at the Crossroads: Global Perspectives and Uncertainties*. MIT Press, Cambridge (MA, USA)/ London (UK).

Smith, K. (2005) *Powering the Future: An Energy Sourcebook for Sustainable Living*. iUniverse, Lincoln (NE, USA).

Sorensen, B. (2004) *Renewable Energy: Its Physics, Engineering, Environmental Impacts, Economics and Planning*. Elsevier Academic Press, London/New York.

Tester, J., Drake, E., Driscoll, M., Golay, M. and Peters, W. (2005) *Sustainable Energy: Choosing Among Options*, MIT Press. Cambridge (MA, USA)/London (UK).

US Department of Energy, US Energy Information Administration/International Energy Outlook 2010.

Walisiewicz, M. (2002) *Alternative Energy: A Beginner's Guide to the Future of Energy Technology*. Series Ed. J. Gribbin, Essential Science, London/New York.

World Commission on Dams (2000) 'Dams and Development: A New Framework for Decision-making' (www.dams.org).

CHAPTER 19

Soil pollution and abuse

Malcolm Cresser, Sophie Green and Clare Wilson

Learning outcomes

By the end of this chapter you should:

- Be aware of how soil has served as a repository for waste from anthropogenic activities even from prehistoric times.
- Be aware that the capacity of soils to deal with such waste is finite, and is sometimes exceeded.
- Realise the extent to which scientific, technological and industrial development has added to soil pollution burdens.
- Realise that supposedly beneficial management practices such as irrigation and fertiliser and manure applications may contribute to soil pollution burdens.
- Understand how policy makers attempt to deal with this finite capacity by setting critical loads or target loads for specified pollutants.
- Be able to discuss examples of the consequences of exceeding the critical loads for soils and associated fresh waters.
- Understand the problems that may arise in soils and associated surface waters as a consequence of road salting and gritting of roads to reduce accident risk in freezing wintry conditions.
- Be better able to discuss the significance of global food markets to soil pollution, with particular reference to sewage sludge disposal.

19.1 Soil pollution in pre-historic and historic times

19.1.1 Prehistory

Most people's concepts of early man conjure up images of primitive beings with little or no sense of personal hygiene, happily hunting and gathering and relieving themselves locally as and when the need arose. As we saw in Chapter 12, eating locally, paired with very local defecation, should at least be regarded as more environmentally friendly and sustainable than many of our present day practices which involve flying food and associated plant and animal nutrient elements (and potentially polluting elements) round the world and spreading them far from their sources of origin.

Shelter in some form, food and drinking water that looked and tasted not too bad would have been basic requirements of human life, just as they are today. It can't have been too long before excrement was deposited outside places of shelter and from then to the realisation that it had the beneficial effects of manure on crops. As we saw earlier in this book, manure use was widespread by the time of the ancient Egyptians and Greeks.

However, one author has worked on a stone-age archaeological site on Arran off the west coast of Scotland where there was clear evidence of shell sand being moved over considerable distances, presumably to be used as a fertiliser through its beneficial liming effects in local acidic soils. Recognition of the fertiliser effect was no doubt also a consequence of chance observation, when residues of sea-food shell fish meals were dumped over extended periods. The inference has to be that these early humans had a fairly high level of intelligence. They can be forgiven, though, for not recognising that every time they added shell sand they added a selection of desirable and possibly some undesirable elements too; forgiven because undesirable effects might have taken decades or even centuries to manifest themselves. So unwittingly they were starting to use the soil as a global dustbin for elements possibly serving no useful purpose.

19.1.2 Soils as waste repositories in more recent history

As long as human populations were small, and just one of many locally functioning natural ecosystem component parts (see Chapter 6), their activities remained a relatively insignificant problem to ecosystem sustainability. However, the development of urban populations soon started to increasingly separate food production and waste generation. At first of course the human waste would be removed manually in carts and at night (not a job in high demand!) and dumped in ditches or on soils relatively locally, replenishing local soil fertility to some extent, albeit in a patchy mosaic. Much undoubtedly found its way to local water resources too, as it still does sometimes in poorer parts of developing countries, for example from squatter communities using the only water supply available to them for sanitation. But as transport improved, the separation of food production site and waste disposal became more of a problem.

The next problem that arose was the development of industries such as cloth dyeing and leather tanning within towns and cities. To some extent waste was simply waste to most people, so it is easy to see how potentially phytotoxic elements that were not even known to exist at the time could become mixed with the more recyclable nutrients from human urine and excrement, especially when stale urine itself was regarded as a useful industrial starting material.

At this point it is worth considering a few examples of materials that would have been used in building or in buildings at that time, to see how the relative concentrations of some important elements in such materials varied. We'll include bone as one of our materials, not so much because the authors object to the burial of human dead (though it is another example of anthropogenic element redistribution in a very patchy way!), but because there was a need to dispose of animal bones over a range of sizes.

It will come as no surprise to most readers that calcium and phosphorus concentrations are high in bones because of their chemical nature. Phosphate in soils is often used by archaeologists

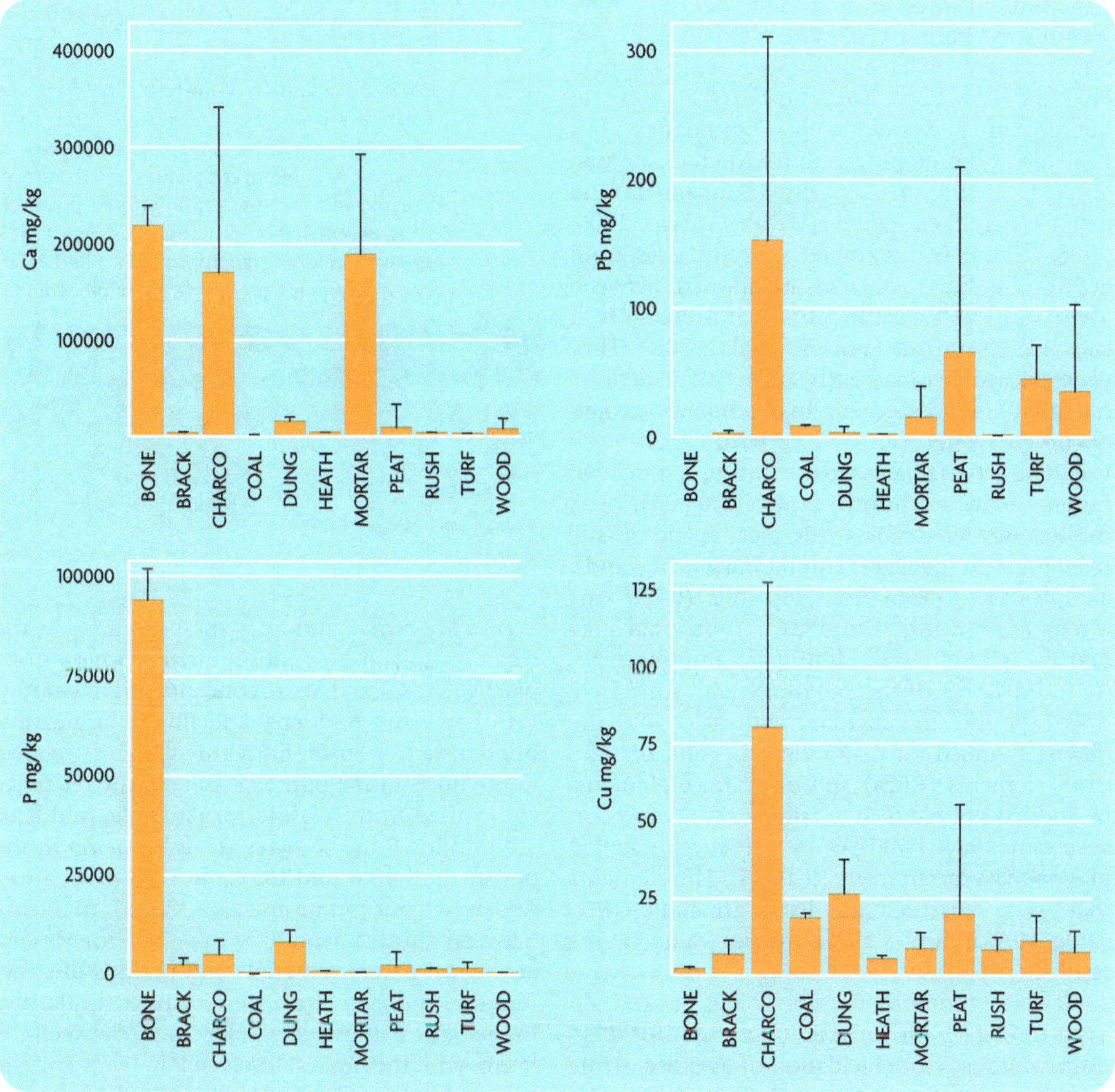

Figure 19.1 Comparisons of concentrations in a range of materials used in the construction of early buildings or that might have been used within buildings for diverse purposes a few hundred years ago, that could eventually find their way to local soils. From left to right, the replicate test materials were animal bone, bracken, charcoal, coal, dung, heath vegetation, mortar, peat, rushes, turf and wood. Mean vales are plotted and errors are shown as least significant differences.

Source: Based on some of the data published by the authors in Wilson *et al.* (2008).

to identify impacts of past human habitation and animal consumption. In charcoal, both the elements (and others measured by Wilson *et al.* (2008) such as Ba, Sr and Zn) are concentrated from wood during the formation of the charcoal. It should come as no surprise that calcium (Figure 19.1) and strontium were also present at high concentrations in old mortar samples collected from walls. The main point here though is that old buildings fall down under the ravages of time, so these and other elements find their way into local soils.

Several studies over recent decades have investigated patterns of multi-element soil contamination on archaeological sites. Multi-element analyses have been used around the world to identify sites and establish their boundaries. They have also been employed to try to elucidate patterns of former human activity within and around archaeological structures (Wilson *et al.*, 2005, 2008, 2009). For example, six farm sites from across UK that had been abandoned between 1890 and 1940 were identified and the soils at each were sampled (Wilson *et al.*, 2005). They were chosen because each farm had a definite history of use based on living memories and documentary records, and each provided equivalent functional areas representing a range of human activities; hearth, house (kitchen), byre, midden, garden, arable fields, pasture and off-site reference soils (generally communal grazing land). Samples were taken from top-soils (0–20 cm) across a 1 m grid within each functional area and test pits allowed additional sampling from the floor layers of the buildings. The soils were dried, ground and digested with acid, and the digests analysed for major elements and the rare earth elements (REEs) and other trace elements by inductively coupled plasma atomic emission spectrometry (ICP-AES) or by inductively coupled plasma-mass spectrometry (ICP-MS). The elements that were measured are listed in Table 19.1. Table 19.2 indicates the potential value of the results.

Concentrations of Ba, Ca, P, Pb, Sr, and Zn seem to display the most useful pattern for diagnostic purposes for elucidation of area use, while most additional elements, including Co, Cr, Cu, Eu, Fe, La, Mn, Y and Rb, provide this level of information at one or more farm sites, and particularly at Olligarth and Grumby. Both of these farms have an igneous geology and occupy well-drained sites. However, Ti, Cs, Na, V, Zr and some REEs failed to show significant enhancement between functional areas at any of the six sites.

The main point to take away from these results in the present context, however, in spite of the great interest of their diagnostic potential to archaeologists, is that in the fullness of time abandoned buildings leave complex contamination element fingerprints in the soils that eventually evolve and bury their remains. In the resultant soils the pollution distributions will be patchy. The Ca and Sr are concentrated in hearths, P in byres and middens, and Pb in the hearths, middens and houses. Enrichment factors relative to the unmanured outfields and off-site reference soils varied from 3.7-fold for Ba in the middens, 4.7-fold for P in the byres, 12-fold for Zn in the houses, and 48.6-fold for Ca in the hearth areas. Absolute concentrations also varied markedly between sites, as would be expected for varying geological parent materials and periods of structure use; however, there were distinct similarities in the basic patterns of enhancement between the farms, with the highest levels of P in byres and the highest levels of Ca, Sr and Pb in the hearths and houses.

Table 19.1 Elements determined by Wilson *et al.* (2005) using ICP-AES and ICP-MS techniques. Note that some elements were determined by both methods, to provide an indication of the reliability of the two analytical procedures. It is unusual, though not impossible, to get the same incorrect answer by two or more different techniques, and the procedure showed that calcium interfered positively in the determination of neodymium in ICP-AES. However, analysing appropriate certified reference materials is a preferable check procedure.

ICP-AES	ICP-MS
Al, Fe, Mg, Ca, Na, K, Ti, P, Mn, Ba, Co, Cr, Cu, Li, Ni, Sc, Sr, V, Y, Zn, Zr, La, Ce, Nd, Eu, Dy, Yb, Pb	Zn, As, Rb, Sr, Mo, Cd, Sn, Cs, La, Ce, Pr, Nd, Sm, Eu, Gd, Tb, Dy, Ho, Er, Tm, Yb, Lu, Hf, Ta, Pb

A second important point to remember is that the data were for isolated farms that had not been subjected to heavy loads of metals from disposal of more contemporary sewage sludge. The problems of contamination from sewage sludge were considered in Chapter 12, in section 12.3.3; you may remember that there are still problems today from the mixtures of waste streams that enter the sewage system in many countries, and therefore there have to be tight constraints on the extent of disposal of sewage sludge onto agricultural soils especially.

Table 19.2 A simplified schematic representation of the potential to use concentrations of specific elements to help differentiate between functional areas based on results of Wilson *et al.* (2005). Underlying geology and the dates farms were abandoned are also shown.

Farm	Auchindrain	Balnreich	Cwm	Far	Grumby	Olligarth
Geology	Schists	Schists	Shale	Oolite	Gneiss	Rhyolite
Abandoned	1890s	1900s	1917	1938	1940	1940
Element						
Al					•	•
As	••				•	•
Ba	•	••	••	•	••	••
Ca	•	••	••	••	••	••
Ce					••	•
Co		•		•	••	••
Cr	•				••	••
Cs					•	
Cu	•	••	••		•	••
Dy				•	•	••
Eu				•	••	••
Fe					••	••
K		••				•
La		••			••	•
Li				•	•	••
Mg			•	••	•	•
Mn		•		••	••	••
Mo	•					•
Na				•		
Nd		••				•
Ni	•	•	•	•	•	•
P	••	••	•	•	•	••
Pb	•	••	••	•	••	••
Rb		•		••	•	•
Sc			•	•	••	•
Sr	•	••	••	••	••	••
Sm		•		•	•	•
Sn		••	•		•	•
Ti					•	
V					•	•
Y		•		•	••	••
Zn	••	••	••	••	••	••
Zr					•	•

• Indicates significant overall differences in elemental concentrations between soils associated with buildings (houses, byres, middens and gardens) and surrounding fields. •• Indicates significant differences in elemental concentration between functional areas. Far and Cwm stand for Far House and Cwm Eunant.

19.1.3 Soils as waste repositories for lead

In the previous section we saw that one of the elements that provided useful indication of human activities is lead. Table 19.3 shows the distribution of lead concentrations between different parts of a croft in Shetland, and Table 19.4 shows lead concentrations in potential source materials, based upon data in Wilson *et al.* (2006). However, although these values suggest that the midden and house remains especially will have input lead to the resulting soil, they tell us nothing about the origins of the lead within those areas.

The element lead exists as a number of stable isotopes (atoms of a given element with the same atomic number but differing atomic mass because they have different numbers of neutrons in their nuclei). Lead originating from different sources has very small differences in the ratios of lead isotopes.

Table 19.3 Lead concentrations in top soils or floor layers from areas of different past known uses at an abandoned croft in Shetland, a group of islands to the North of Scotland.

Context	Mean Pb mg kg^{-1}	Minimum Pb mg kg^{-1}	Maximum Pb mg kg^{-1}	Standard deviation
Hearth	179	111	246	55.2
House floor	203	141	396	109
House overburden	153	89.4	300	53.4
Byre floor	100	97.5	103	2.40
Barn floor	67.2	35.9	82.4	19.4
Midden	232	81.2	483	182
Kailyard	71.3	65.9	86.9	5.02
Arable fields	58.5	51.4	65.7	4.03
Grazing	25.4	18.0	32.6	4.65

Table 19.4 Mean lead concentrations in reference materials.

Reference material	Mean Pb mg kg^{-1}	% loss on ignition	Pb mg kg^{-1} mineral matter
Coal	9.26	95.5	206
Turf	61.6	58.5	148
Seaweed	1.09	67.4	3.34
Plaster	10.4	2.62	10.7
Dung	1.78	82.5	10.2
Wind blown sand	3.91	.513	3.93
Till	4.24	9.97	4.71

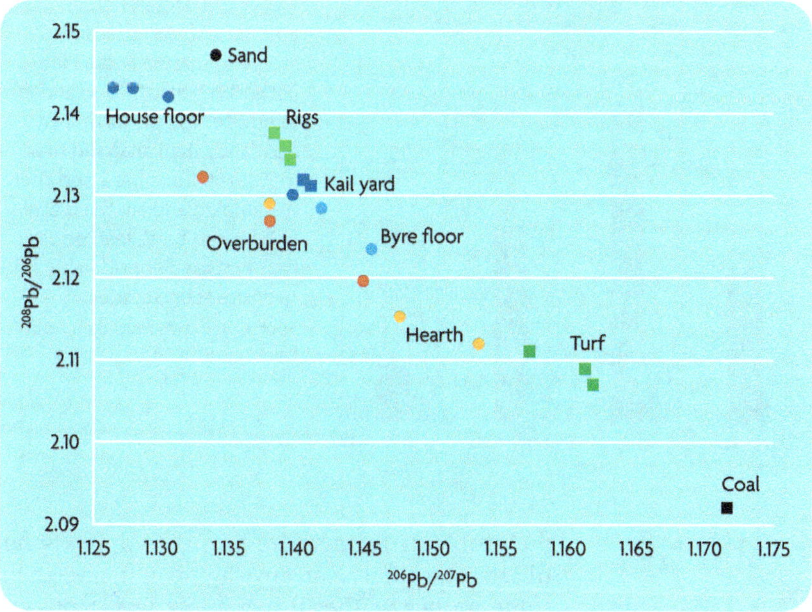

Figure 19.2 Relationships between pairs of lead isotope ratios for diverse source materials collected from an abandoned croft in the Shetland Isles showing typical clustering. The results suggest that position on such a plot for an unknown sample might be useful for diagnosing prior use of an abandoned area of a farm.
Source: From results in Wilson *et al.* (2006).

Figure 19.2 shows that when pairs of isotope ratios are plotted against each other, distinctive groups are found for lead isotope ratios from different source areas.

Note that the differences in isotopic ratio between sample types are very small in Figure 19.2, and the precision of ICP-MS is insufficient for this sort of study; instead thermal ionization mass spectrometry (TIMS) must be used; it has better precision because it measures the isotopes of interest simultaneously rather than sequentially.

Lead contamination of soils rose to prominence as an issue in the 1980s and 1990s, not because of problems from lead in soils from archaeological remains but because of the realisation that its use as an anti-knock agent in fuels was resulting in serious atmospheric contamination with volatile organo-metallic alkyl lead compounds. This had very worrying neuro-toxicological consequences for humans, especially those that lived in inner city areas subject to heavy vehicle congestion. It was recognised also that deposition from the atmosphere was occurring even in very remote rural areas, leading to significant amounts of this potentially toxic heavy metal accumulating in soils (Billett *et al.*, 1991). Figure 19.3, for example, illustrates the amounts of lead that had accumulated in surface organic layers of remote forest soils in north-eastern Scotland over a period of 40 years. Zinc and copper had also accumulated in the organic soils, though to a lesser extent.

An important point to note when looking at results such as those shown in Figure 19.3 is that the authors expressed their results as the total pools of heavy metals stored in the organic soil layers at the sites on a kg of lead per hectare basis, and not just as lead concentrations. This was because organic matter had been accumulating at the site over the test period too and this thickens the surface soil horizons and may have a significant dilution effect. Thus it is possible when such changes are occurring for pools of stored pollutant metals to increase while their concentrations actually decrease.

The lead used as an anti-knock agent in petrol had very characteristic isotopic signatures and this made it much easier to demonstrate the origins of the lead pollution input. Such unequivocal evidence was an important aspect of the debate that finally resulted in the banning of lead in petrol in the United Kingdom and other countries. It is perhaps worth noting that surface soil lead concentrations at the remote site were up to 100 mg kg^{-1}. This is close to the critical concentration value at which microbial activity in soils is often said to be adversely affected.

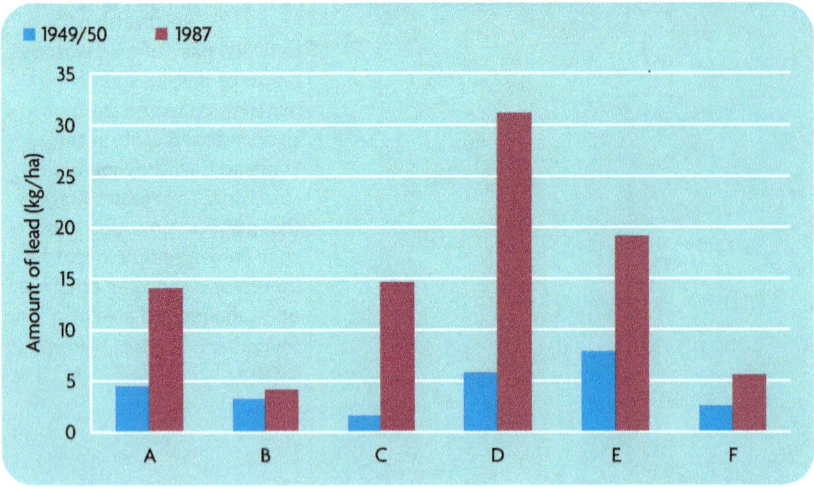

Figure 19.3 Differences in amounts of lead of lead stored in mature Scots Pine (*Picea sitchensis*) woodland in a remote area of north-east Scotland when analysed first in 1949/50 (blue bars) and then re-sampled and analysed again in 1987 (red bars); lead pools had increased considerably over the four decades.

Source: Redrawn from data from Billett *et al.* (1991).

19.2 How does soil deal with nitrogen deposition?

In Chapter 5 we looked at the origins of nitrogen found in soils at the current time, and saw that over centuries to millennia reserves of up to around 10–15 tonnes of organic N may build up, all originating initially from the atmosphere. Later (Chapter 16) we considered how anthropogenic activities have massively increased the amount of deposition that soils receive from the atmosphere, especially via the deposition of nitrate and gaseous NO_x species and reduced N species such as ammonium (NH_4^+) and ammonia (NH_3). You might be forgiven for thinking: 'What's the problem? Farmers are having to apply N fertilisers year after year anyway, so it'll save them some money as they'll just apply less.' However, farmers don't apply fertiliser annually in remote upland moorland and forest sites, so what happens there?

19.2.1 Evidence for N accumulation in soils

Let's return to the remote Scottish upland forest site where we saw lead, zinc and copper were accumulating and see how much of the 20–30 kg of pollutant N being deposited there from the atmosphere is also accumulating in the organic soil layers. The results are shown in Figure 19.4. When Billett *et al.* (1990) first re-sampled and

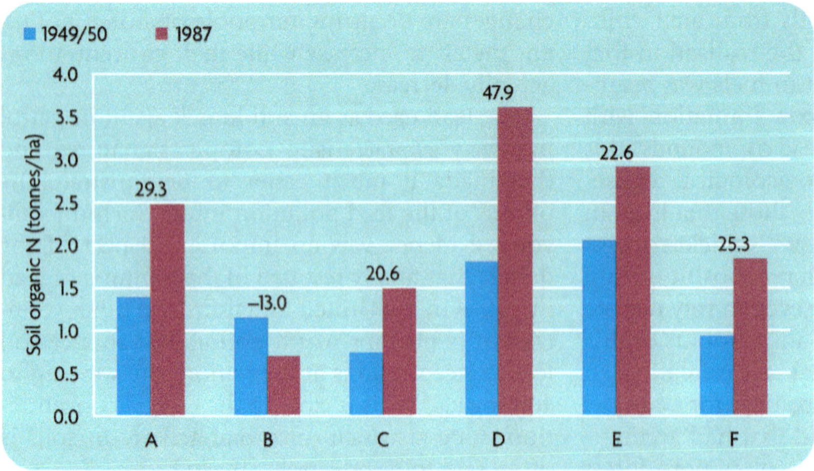

Figure 19.4 Bar chart showing the pools of nitrogen stored in the organic surface soils at the six Scots Pine forest sites in a remote area of north-east Scotland, as mentioned in Figure 19.3. The numbers shown are rates of N accumulation (loss for site B) in kg ha^{-1} yr^{-1} in the surface organic horizons.

Source: Data from Billett *et al.* (1990).

re-analysed soils from this site they had anticipated that they would find a sharp decline in the soil C:N ratio, which they thought would provide evidence of how the N pollution was changing the soil. To their initial surprise they found C:N ratio had increased, not decreased, at 4 of their 6 sites. It did not take them long to realise that, because of acidification at the sites, carbon was accumulating much faster than nitrogen, thereby diluting the N concentration and explaining their results. That soon led to their realising that to detect N accumulation they needed to plot their results as amount of soil N per hectare rather than N concentration.

Lead-in question

How would you plan an experiment to see if atmospheric N deposition was changing the N stored in upland heather-moorland soils?

As is often the case in environmental science research, finding the answer to one question often leads to another question. In this case the new question was: 'How long can such soils continue to accumulate organic N before something else new starts to happen to the N cycle, and what will that be?' We will return to that question later. Meanwhile, here, for now, we will look at the answer to another question. If N is accumulating at coniferous forest sites, is the same thing happening at moorland sites in the uplands? If it is, and such sites eventually become N-saturated and start leaking mineral N species such as nitrate into local headwater streams, could this have serious implications for river water quality downstream in the longer term?

At this point it is worth mentioning one of the author's most common statements to students planning research projects, whether small or large, namely: the wisdom of forethought is much more valuable than the wisdom of hindsight. The topography of upland landscapes is often complex and the geology may be highly variable, as may the soils that evolve as a consequence. Even over a relatively small regional area climate too may vary considerably because of altitude and aspect. These factors may, and do, interact to influence the vegetation that becomes established. Both soluble organic-N species and soluble inorganic N species may move down slope in drainage water at various depths, acting as inputs of N down slope.

So when Crowe *et al.* (2004) decided to examine the effects of atmospheric N pollution and precipitation of soil profile nitrogen storage (SPNS) in Scottish uplands, they confined their sampling strategy to a single parent geology (sandstone/quartzite), a limited altitude and aspect range, and readily recognisable distinctive vegetation cover (dominant *Calluna vulgaris*) and a distinctive soil profile (a well-developed podzol, as shown in Figure 19.5). Even these constraints did not eliminate all variables; for example position of profiles on a slope was not pre-determined, so amount of up-slope area was still variable. At least it represented a reasonable level of awareness of all the factors likely to be influencing the dynamics of N species transformations in soil.

The predicted values of soil profile nitrogen storage (SPNS) in Figure 19.5 were obtained from the equation:

$$\log_{10}(\text{SPNS}) = 2.271 - 0.636 \log_{10}(\text{Precipitation in mm}) + 0.464 \log_{1}(\text{N Deposition in kg N ha}^{-1}\,\text{yr}^{-1})$$

Always remember, by the way, to include units in (or under) tables when writing equations. If you don't it is impossible for anyone else to apply your equation. Always remember, too, to look at any equation that you derive to check that it makes intuitive sense. This one does seem to, as SPNS is higher when N deposition is higher and lower in wetter areas. The latter could reflect greater probability of either higher N leaching losses or higher losses of N via denitrification in more persistently wetter soils. More experiments would be needed though to confirm which was the more important.

We have now seen that soils initially start to store more nitrogen when subjected to higher levels of N pollution from the atmosphere, though it is not clear how sustainable this storage would be in the longer term. We have also seen

Figure 19.5 A typical *Calluna vulgaris*-dominated moorland podzol profile (a) as sampled by Crowe *et al.* (2004) and (b) the relationship that they found between SPNS predicted from a combination of annual precipitation amount and atmospheric N deposition and directly measured SPNS.

Source: Fig. 19.5(a) Malcolm Cresser. Fig. 19.5(b) courtesy of Springer Science + Business Media. Note that the one point substantially away from the 1:1 line corresponds to a site with exceptionally high N deposition, so it is possible that this represents a breakdown in normal functioning rather than just a random error.

that it is not advisable to consider the N cycle in isolation as element biogeochemical cycles are interactive. This was particularly true for the carbon cycle, when additional carbon storage far outstripped the additional nitrogen storage in organic soil layers in forest. In part this is likely to be due to an initial N fertiliser effect from moderate levels of N pollution. Remember that we saw in Chapter 7 that vegetation is a factor of soil formation too, and all factors of soil formation are interactive over diverse timescales.

Timescales are important when investigating the effects of pollutant deposition to soils. For example, if we collect samples of peat soils in the UK from along a known N deposition gradient, and measure their C:N ratio, there is a distinctive (and to some extent predictable) relationship between peat C:N ratio and N deposition flux, as shown in Figure 19.6. If we collected peat from a relatively pristine site, and started applying artificial N deposition treatments to simulate the effects of atmospheric pollution, it might take many years or decades to produce the relationship shown. That makes using spatial separation as a surrogate for time seem like a very attractive proposition. However, a note of caution is necessary. For interpretation to be totally unequivocal we need to be sure that N deposition is the only variable at our selected sites. Unfortunately pollution gradients for N often have parallel gradients for S deposition, day length and climate too, so interpretation must be done with great care.

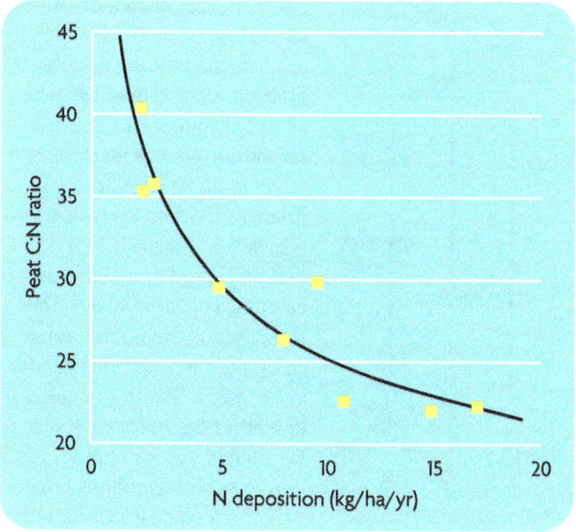

Figure 19.6 A plot indicating that C:N ratio falls with increasing N deposition for peat soils from along a pollution gradient.

Source: Based upon the results of Yesmin *et al.* (1995).

19.2.2 Does soil N accumulation matter? – N leaching and N critical loads

Ecosystems evolved naturally in relatively pristine environments to conserve the small annual inputs of atmospheric N from biological fixation of N_2 gas by algae or symbiosis with plants such as clover, or from naturally occurring ammonia or nitrate in the atmosphere, and by tight biogeochemical N cycling (see Chapter 5). Leaching losses to nearby streams would have been minimal, though small losses would have eventually started to occur in winter months when N uptake by biota was minimal. The N losses would have been as organic and inorganic N species. This can be confirmed by monitoring seasonal variations in N species concentrations in streams in the least polluted parts of the planet.

Lead-in question

What happens when soils become N saturated?

Farmers in agricultural areas often apply nitrogen fertiliser at rates well in excess of atmospheric N deposition rates, even in N polluted areas. Where they do it is common to see significant amounts of nitrate leaching into surface waters over autumn and winter months, because crop uptake is lower when crops have been harvested and often there are recent inputs of readily mineralisable plant litter to the soil too. Such leaching also often follows felling of forest trees in catchments. It is a cause for concern because nitrate concentrations above 50 μg ml^{-1} are generally deemed unacceptable for potential potable water supplies.

In minimally managed upland areas too, however, strong seasonal trends are seen where N deposition is high. The soils can no longer continue to accumulate the nitrate-N and ammonium-N being deposited from the atmosphere and nitrate leaching becomes substantial, even in the summer months. The soils are then described as N saturated. The concept was advanced by Ågren and Bosatta (1988) and by Aber *et al.* (1989), who were concerned about increases in nitrate leaching from forest ecosystems. It is now widely referred to as the nitrogen saturation hypothesis. Figure 19.7 illustrates the seasonal variations in the extent of nitrate leaching in N impacted areas of the UK. Even in the relatively less N-polluted north-east of Scotland nitrate leaching is now observed in upland catchments even in the summer months. The River Etherow catchment in the South Pennines, between Barnsley and Manchester in England, has been heavily polluted with N and S for decades. It feeds a series of reservoirs used to supply potable water, but note that the waters in the catchment are still well below the critical 50 μg nitrate ml^{-1} threshold value (100 micromoles of nitrate equates to 6,200 micrograms or 6.2 mg of nitrate, so even 100 μmoles nitrate l^{-1} only equates to 6.2 μg nitrate ml^{-1}).

In upland drainage basins such as that of the River Etherow, it is common practice in Britain periodically to burn the *Calluna vulgaris* (heather) vegetation every few years to encourage the regrowth of the plants (Figure 19.8). This is to provide a better environment for grouse especially. Much of the income in such rural communities is often generated by sporting activities such as shooting and angling. The practice also returns much nitrogen back to the atmosphere as nitrogen gas (N_2), although some oxides of nitrogen may also be generated. However, unfortunately destruction

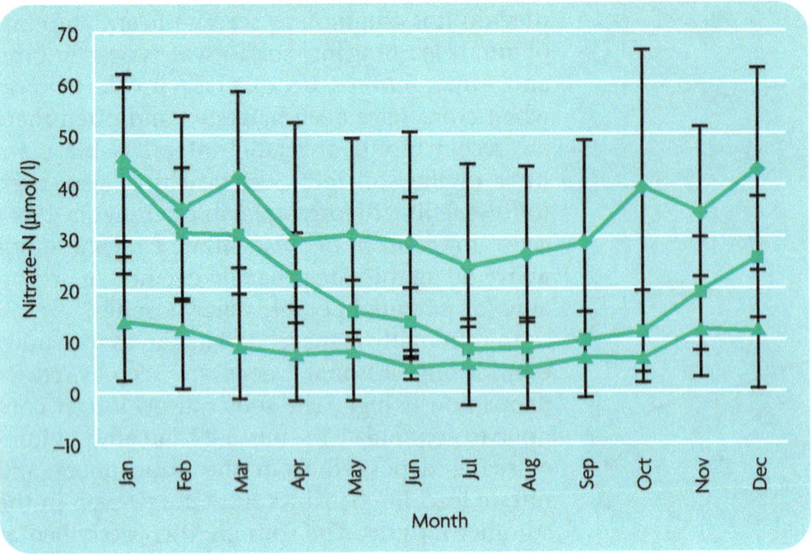

Figure 19.7 Examples of the strong seasonal trend in nitrate leaching, and strong pollutant-N effect in upland river systems as a consequence of soil N saturation. Nitrate leaching is lower in north-east Scotland (triangles), where N deposition is lower, and highest in the most polluted River Etherow catchment (diamonds), with the data for the Nether Beck catchment in Cumbria (squares) in between. Data shown are for 16 Nether Beck sub-catchments, 13 Etherow sub-catchments and 16 Dee sub-catchments. Error bars show standard deviations of the means.

Source: Redrawn from Cresser *et al.* (2004).

Figure 19.8 Typical appearance of heather remains after burning at the River Etherow catchment in the UK.

Source: Photographed by Richard Smart Ph.D.

of the vegetation may also facilitate the transport of nitrate to adjacent streams (Cresser *et al.*, 2004).

During the past two decades scientists have given much thought to quantifying critical loads of N deposition for soil ecosystems. The approach that has found most favour is known as the steady state mass balance approach. Basically in this approach scientists make informed guesses of values for how much N can be safely stored in a maturing forest (if appropriate), how much can be stored sustainably each year in soil as soil evolves, how much N can be leached safely in organic or inorganic forms and how much can be disposed of by denitrification. Add those figures up and you supposedly get the amount of N that can be deposited each year on that particular soil 'without causing any damage to the functioning of the soil system'. The approach is attractive to policy makers as it allows maps to be drawn and abatement policies to be formulated. The present authors though are highly sceptical since the concept apparently assumes, in its present form, that the Earth is effectively flat and no water containing solute N species ever moves down slope. Moreover it is not easy to justify allowing continued accumulation of N in soil over the longer term.

Another problem in assessing the effectiveness of critical loads as a tool for design of pollution abatement policies is in deciding what constitutes 'damage'. Ideally there should be some plant bio-receptor that can act as an indicator species. If however the effect of the receptor is just reduced growth rate this may be very difficult to detect under field conditions; no controls will be available on site that can be used for reference purposes.

In this chapter we have concentrated upon the soil/water link as the indicator of soil N saturation and possibly the greatest cause for concern. However, it should also be pointed out that shifts in soil C:N ratio are bound ultimately to impact upon biodiversity of plant communities and associated biota that, in turn, depend upon them. Nor should we ever forget the inextricable link between the carbon and nitrogen cycles at a time when soil carbon storage and carbon feedback to the atmosphere are a cause for such concern.

19.3 How does soil deal with sulphur deposition?

As well as functioning as a global dustbin for much of the N pollution in the atmosphere, soil has also been tacitly expected for centuries to safely handle high loads of sulphur deposition, initially from combustion of wood, but more recently from combustion of fossil fuels. In Chapter 16 we saw the problems associated with human activities releasing vast quantities of sulphur dioxide and some sulphur trioxide into the atmosphere, and the acidifying nature of this pollution. Figure 19.9 illustrates an early critical loads map for acid deposition to soils, as generated at CEH by Keith Bull and Jane Hall for the Department of the Environment (now Defra) in the UK.

At the time that map was generated, acidity associated with sulphur deposition was the main cause for concern in the context of potential soil acidification. Therefore the map was based upon best-estimates of soil mineral weathering rates, and it was assumed that weathering rate provides a robust indication of the alkalinity available in soils to neutralise acid inputs from the atmosphere. To the authors this is highly questionable. It presupposes that the alkalinity generated would otherwise not be having any impact upon soil physico-chemical properties. This cannot be true, since it was the replenishment of base cations on cation exchange sites via biogeochemical weathering that gave the soil its pre-pollution pH value. It seems to be a case of double accounting therefore assuming that the associated HCO_3^- is available to neutralise acidity inputs.

Sulphur deposition to soils in agricultural areas was not regarded as a problem, because it helped meet crop sulphur requirements. Sulphur dioxide was only a problem if the gaseous concentrations were high enough to damage crops directly via stomatal uptake of the gas. The associated acidity when sulphuric acid was deposited on agricultural soils was compensated for, often unwittingly but quite simply, by farmers' tweaking the lime additions needed to maintain soil pH at the desired value. Note that critical loads of acidity in Figure 19.9 are much lower in the highland regions of Scotland, Wales and the English Pennines, and higher in more agricultural lowland areas (Hornung *et al.*, 1995). Ironically, once sulphur emissions were dramatically reduced in the UK in the 1990s, sulphur deficiency in crops started to become more of a problem. In part, however, this was because ammonium nitrate and urea became more widely used as nitrogen fertilisers, rather than ammonium sulphate. The latter often had been a bi-product of gas works that used to produce gas as a fuel from coal.

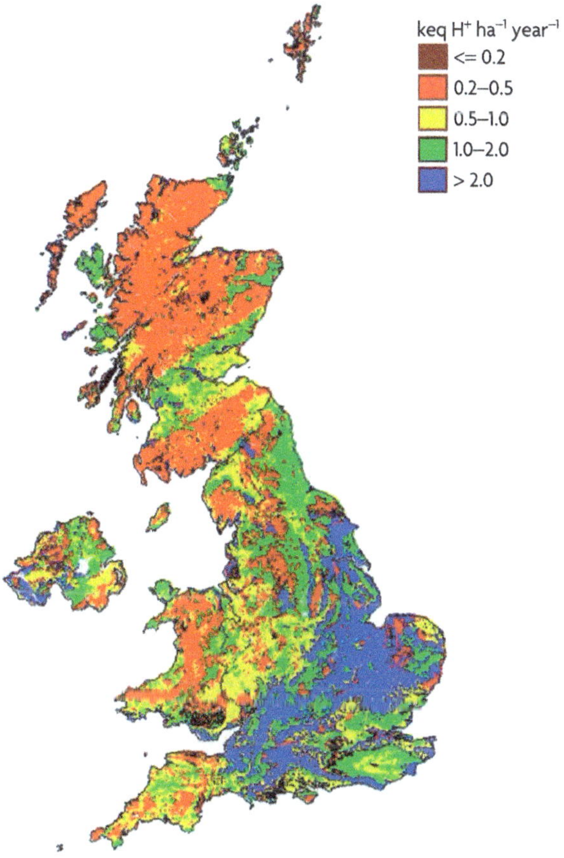

Figure 19.9 An early critical loads map for acidity.
Source: Centre for Ecology & Hydrology, www.ceh.ac.uk.
Note that the units used were keq ha^{-1} yr^{-1}. H^+ is monovalent, so a kiloequivalent is the same as a kilomole. (May not reflect most up to date data/information.)

In minimally managed soils in uplands, such as the podzol in Figure 19.5(a), the iron and aluminium hydrous oxides in the soil mineral horizons have a high capacity for absorbing sulphate anions. The divalent SO_4^{2-} anions displace hydroxide ions from the surface of the hydrous oxides of iron and aluminium by anion exchange reactions. This OH^- neutralises the H^+ associated with sulphate anions in acid rain, so that, at least in the short term, sulphate adsorption helps buffer soil pH. However, the sulphate absorption capacity is always finite, so if sulphate deposition remains high eventually sulphate starts to leak out of the soil in drainage water. In acidic soils the accompanying mobile anions may be H^+ and to a lesser extent, Al^{3+}, which may cause acidification problems in streams and rivers down slope.

Critical loads maps for acidity were steady state maps, which made it difficult to validate their use by direct field measurements. If sulphuric acid mobilised base cations such as Ca^{2+} and Mg^{2+} from upper slopes, the leached base cations were inputs to soils on lower slopes, and helped buffer pH by cation exchange down slope. Therefore, if critical load of acidity was exceeded, the first place to start looking for damage would be close to the tops of hills and mountains. The question then though was what changes should be sought? Finding suitable time zero reference conditions has always been a problem with this type of research.

19.4 Can irrigation water be a potential problem?

As well as supplying much needed water, irrigation also adds to soils whatever solute species are dissolved in the water. For example, in old Aberdeen in Scotland, many of the older houses had lead water pipes until the 1980s; consequently the tap water from the houses could contain 0.05 mg of lead per litre. Suppose a keen gardener adds 10 cm of water to his lawn via a sprinkler over night. Each square metre has received 100 litres of water, and with it 5 mg of lead, which equates to 50 g of lead per hectare. If he does this 20 times a year he's adding lead at a rate of 1 kg ha^{-1} yr^{-1}. Lead is strongly absorbed in upper soil horizons, and subsequently very immobile.

The authors became aware of this issue many years ago when a student project was designed to show the dramatic accumulation of lead close to a house from lead paint scrapings. The conclusion was that the house owner had been much more careful in tidying away his paint scrapings than he had been in watering his garden as lead was fairly evenly distributed across the garden. But the problem can exist on a much wider scale in agriculture where industrially polluted river and canal water is used for irrigation, as has been the case in parts of Turkey (Aydinalp *et al.*, 2005). Figure 19.10 illustrates how irrigation in the

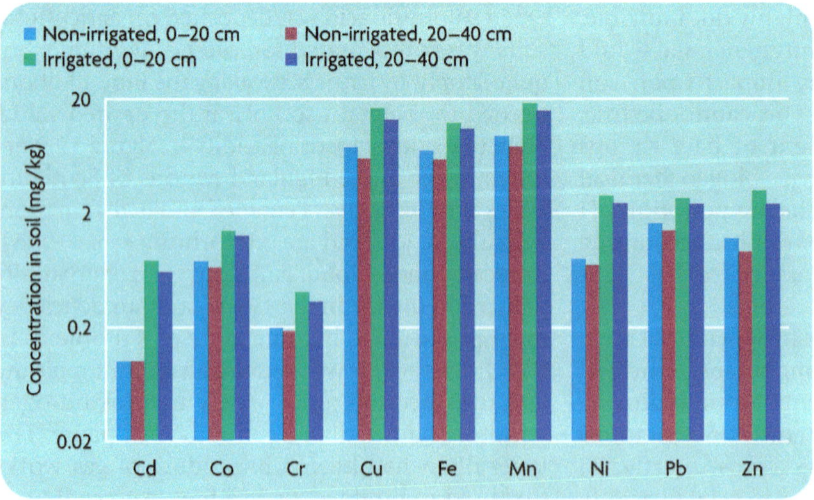

Figure 19.10 The DTPA-extractable heavy metal concentrations in non-irrigated and irrigated soils of the Bursa Plain, Turkey, at 0–20 cm and 20–40 cm depths. Note the logarithmic concentration scale. The extractable concentrations of these elements for irrigated soils were all significantly higher than those for non-irrigated soils. The concentrations of heavy metals found are highest in the topsoil.

Source: Data from Aydinalp *et al.* (2005).

Bursa Plain in Turkey has led to dramatic increases in extractable concentrations of heavy metals in the soils. In this area the soils are vertisols that have many cracks that pass to considerable depth; therefore there is almost as much contamination with heavy metals at 20–40 cm as there is at 0–20 cm.

19.5 Pollution from road salting

In winter it is obvious when the gritters are active on our roads from flashing amber lights, copious warnings, and white salt coated vehicles, but it's not just roads and vehicles that get coated with road salt. Roadside vegetation and soils receive high loadings of the salting agent used, which can have a host of environmental implications. Moreover, roads often run parallel to rivers for long distances in upland regions (Figure 19.11), for sound engineering reasons, so road salting impacts on down-slope soils and associated streams or rivers may be considerable.

Salt and grit are used on roads to maintain traffic flow and prevent accidents under freezing conditions. There is a range of different chemical agents used to keep roads clear (e.g. sodium chloride, calcium chloride, magnesium chloride, calcium magnesium acetate, sodium formate, urea). Local authorities in the UK tend to apply sodium chloride, otherwise known as rock salt. It can be applied as a liquid or solid, and may be applied alone or mixed with grit and sands, with possibly an anti-caking agent.

It has been estimated that 75–90 per cent of the road salt applied enters the roadside environment via several pathways: runoff, splash, spray, aerosol deposition, mechanical malfunction during application, post-application snow-ploughing, or wash off of throughfall from vegetation. Most deposition occurs within the first 10 m from the road, although elevated concentrations have been observed in excess of 100 m away. Once in the soil, salts can undergo physico-chemical interactions with exchangeable base cations, and the products remaining in the solution phase drain to nearby surface waters or groundwaters. Some may be transported directly to surface waters through drainage systems (Figure 19.12). Quite often, though, drains are used to channel road runoff water rapidly onto down-slope soils (Figures 19.13a and b). Road salts may be diluted en-route by fresh rainfall or melting ice/snow. However, some of the highest concentrations of salt observed in surface waters are associated with spring snow/ice melts as salt has accumulated in roadside snow/ice.

Rock-salt affected soils are exposed to elevated concentrations of sodium cations and chloride anions. Chloride is considered a 'conservative' species, and is thought to pass through the soil phase in solution with minimal chemical interaction. However, sodium ions tend to have significant interactions on the cation exchange sites of the

Figure 19.11 A main road in northern England, running parallel to a down-slope river channel for several km.

Figure 19.12 Road surface drainage being piped directly into a river. Road runoff from the drainage pipe can be clearly seen. The river is on the other side of the dry stone wall, passing under the road bridge.

soil organic matter and mineral phase. Introduction of road salts to roadside soils is an excellent example of how cation exchange sites operate in soils and how important they are in driving soil–soil solution dynamics.

Lead-in question

Why do sodium ions displace calcium and magnesium ions from cation exchange sites?

Sodium, potassium, magnesium, calcium are all base cations and are electrostatically attracted to negatively charged interactive cation exchange sites present in soils, as discussed in Chapter 7. The cation exchange capacity (CEC) is in constant instantaneous exchange with any cations in the soil solution phase. For cations to be mobile in solution, charge balance must be simultaneously achieved. Thus each positively charged cation has a counter anion (or anions) with corresponding negative charge (e.g. sodium Na^+ and chloride Cl^-). Cation exchange facilitates movement between the soil solid and soil solution phases.

The extent to which cation exchange displacement reactions occur is highly dependent on the species initially occupying the exchange sites and what is available in solution. Let's take the case of an acidic soil under grassland in an upland area. If it's acidic, hydrogen ions tend to dominate the cation exchange sites.

Figure 19.14 compares the proportional occupation of the cation exchange sites by sodium at 2, 4, 8, 16, 32 and 64 m from a busy road, the A6, in north west England for soil down slope of a road directly affected by road drains, affected by spray from the road or not affected at all (i.e. control transects). What is clear from this figure is the significantly higher proportion of the CEC occupied by sodium close to the road for drainage- and spray-affected transects when compared to the control

(a) (b)

Figure 19.13 A road drain on the A6 road in Cumbria, England (a), channelling road runoff via a tile drain directly onto soils down slope (b).

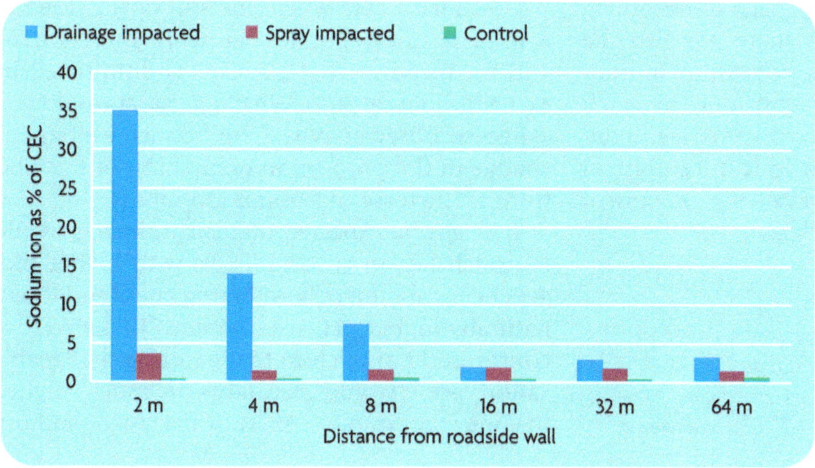

Figure 19.14 Proportion (per cent) of CEC occupied by sodium at 2, 4, 8, 16, 32 and 64 metres from a wall beside the road for direct drainage-affected), spray-only-affected and control transects for a naturally acidic soil under grassland.

Source: Based on authors' original data.

graph. There are big sodium ion effects, with decreasing impacts according to the degree of exposure (drain > spray > control). The road area effectively channels road surface runoff towards the drains and then onto the soils immediately down slope of the drains. In this case the water is piped directly onto the soil surface, concentrating the effects of road salt to specific areas along the roadside and there is high sodium dominance for the drainage-impacted transects over at least 2–8 m. The spray-impacted transects are affected to a much smaller extent, and the control transects demonstrate background levels well below the percentages for the other two scenarios. Most of the sodium for the control transects comes from inputs in rainfall at this site. It was also found that the percentage of the CEC occupied by ammonium ions was lower as a consequence of displacement by sodium cations (Green and Cresser, 2008a).

Clearly if no hydrogen ions have been displaced by the salt for the control transects there will be a higher H^+ percentage occupancy of CEC for the controls, but a lower H^+ percentage of CEC for both drainage pipe transects and spray-contaminated transects between 2 and 16 m (Figure 19.15). The link between Na percentage and H^+ percentage of CEC is also obvious between 16 and 64 m for the three pollution scenarios. The reason is that exchangeable hydrogen ions are competitively displaced by incoming sodium ions. However, on moving down slope for the drain- and spray-affected transects, the Na percentage of CEC declines and the H^+ percentage of CEC increases as competitive displacement declines.

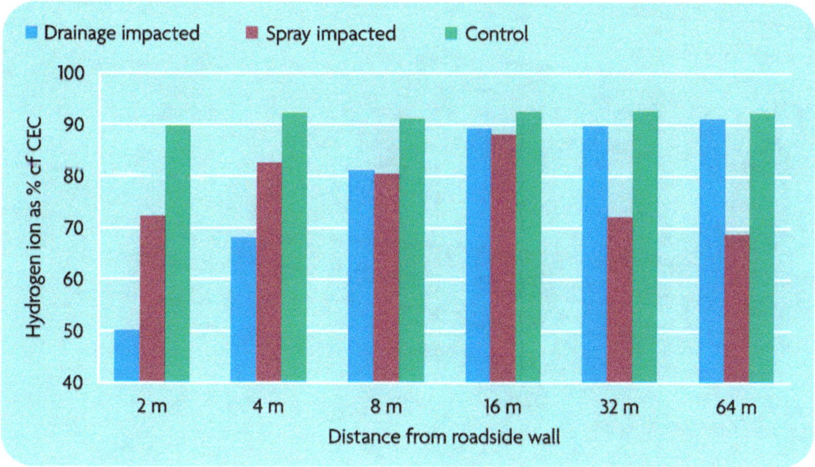

Figure 19.15 Proportion (per cent) of CEC occupied by hydrogen ions at 2, 4, 8, 16, 32 and 64 metres away from the wall for drainage-affected, spray-affected and control transects for the acidic soil.

Source: Based on the authors' original data.

This is a very simplistic approach to observing such a complex system – as there are also the roles of calcium, magnesium, potassium, ammonium and aluminium to consider in such exchanges, and road salt used in this area contains about 2 per cent gypsum (calcium sulphate) as well as magnesium. However, the process of exchange in this example is clear.

Lead-in question

The discussion above is for an acidic upland soil. Would a calcareous soil behave differently?

Consider a case where the soil down slope of a road is calcareous in nature (Figures 19.16a and b), meaning that the CEC is dominated by exchangeable calcium. What *differences* would you expect to observe? Will there be such a dramatic change in the proportion occupation of the CEC from an increase in sodium ion input?

It might be expected that such impacts would be much less in roadside soils with higher biogeochemical mineral weathering rates (and thus a naturally higher pH, see Chapter 7). The relative contribution of calcium to the total exchangeable cation pool would naturally be much higher, making the impact of seasonal high sodium

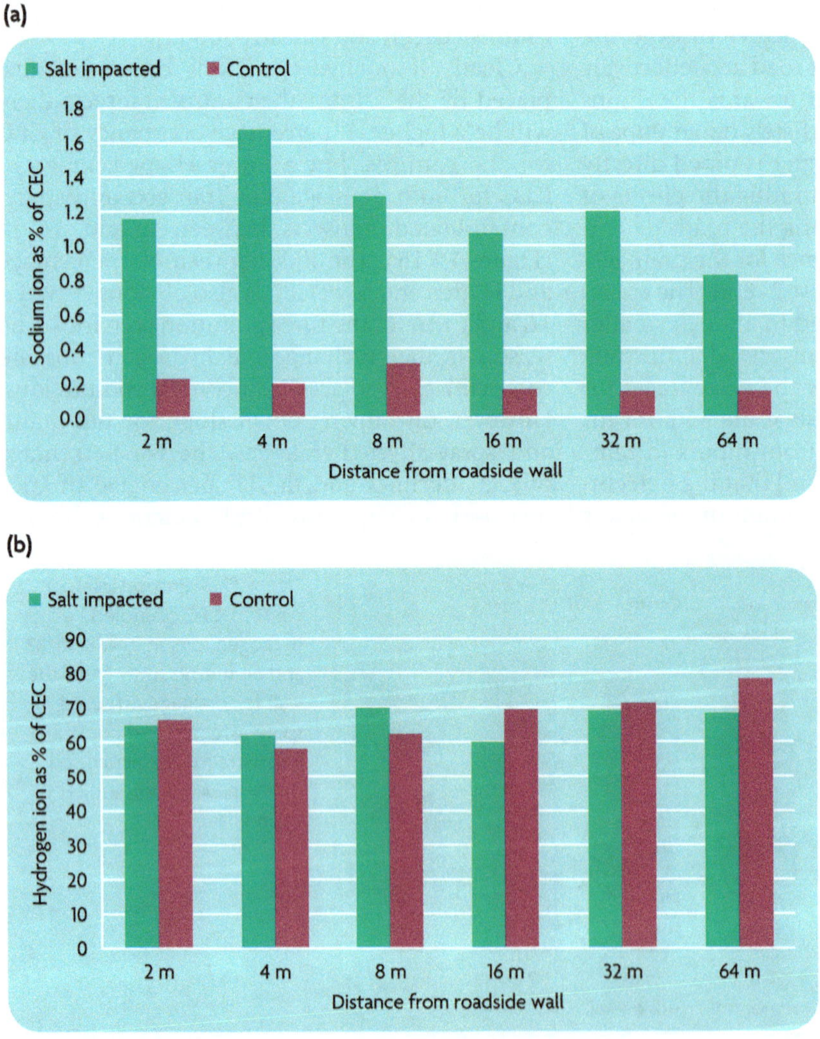

Figure 19.16 Proportion (per cent) of sodium ions (a) and hydrogen ions (b) on CEC at 2, 4, 8, 16, 32 and 64 metres below a roadside wall for spray-affected and control transects on a more base-rich soil.
Source: Based upon results of Green and Cresser (2008b).

inputs less dominant. It's energetically unfavourable for incoming, monovalent sodium ions (Na^+) to displace divalent calcium ions (Ca^{2+}) compared to hydrogen ions (H^+) from the CEC (i.e. smaller hydrated ionic radii and higher charge = greater electrostatic attractive tendency towards negatively charged cation exchange sites). So theory suggests that the effects observed above for acidic upland soils are likely to be dampened on calcareous soils.

Figure 19.16 shows what happens in practice. There is clear sodium elevation for the spray-affected transects as compared to the control over 2–64 m. However, the contribution of Na^+ to exchangeable base cations is lower than at the upland acidic soil site in Figure 19.14. The higher Na percentage occupation of CEC does not directly correspond to a lower H^+ percentage of CEC for the control transects, and thus a lower H^+ percentage of CEC for spray-contaminated transects at the upland calcareous site, in marked contrast to observations for the upland acidic grassland. Note that although this more base-rich soil is evolved from calcareous parent material, it is acidic at this stage in its evolution. However, the H^+ percentage of CEC is substantially lower in Figure 19.16 than for the acid soil in Figure 19.15. It is more likely that the exchangeable hydrogen ions are competitively displaced by divalent calcium and magnesium ions released from biogeochemical weathering for most of the year than by sodium from the road salt. The results in Figures 19.14 to 19.16 were all obtained in April, at the end of the period of heavy road salting in both areas.

The above section highlights the important role that soil parent material plays in soil development, actively defining the soil's features and thereafter the mechanism by which soil responds to incoming pollutants. This is as true for road salt impacts as it was for S and N deposition impacts discussed earlier.

19.5.1 The importance of soil pH to road salt impacts

Soil pH, which is a key regulator in the majority of soil processes (see Chapter 7), is itself driven by the proportional occupation of the CEC by base cations compared to hydrogen and aluminium cations. Measurement of soil pH along down-slope transects on the upland acidic soil demonstrated that the drain-influenced soil at 2 m from the highway had a pH approximately two-and-a-half units higher during April 2005 than spray-affected transects, and three units higher than the control soil (Figure 19.17). This corresponds to the high sodium dominance and low hydrogen content of the CEC (Figures 19.14 and 19.15); the decline in pH to a value of ca. 4.2 below 8 m is clearly associated with the decreasing sodium ion content.

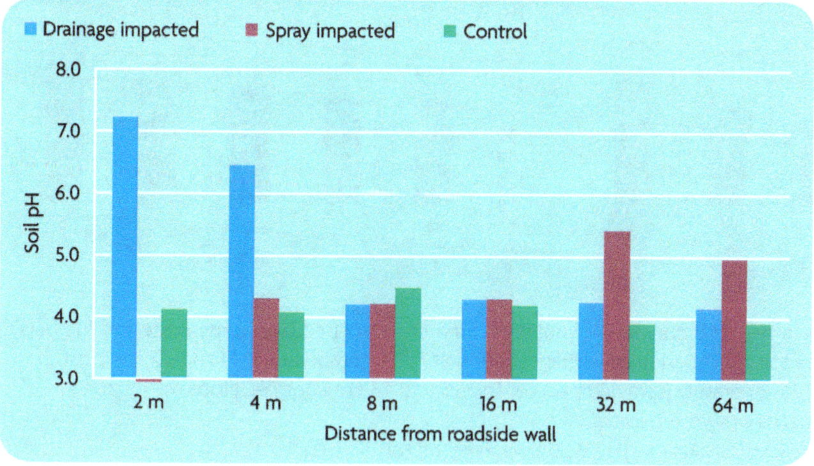

Figure 19.17 Soil pH (measured in water) at 2, 4, 8, 16, 32 and 64 metres from the roadside wall for drain-affected soils, spray-affected soils and control soils.
Source: Redrawn from authors' original data.

Lead-in question

Is such a shift in pH important? What implications does it have?

Changes in pH can have dramatic effects on soil processes. For example, mineralisation and nitrification rates are modified via pH effects on soil microbes. Nitrification (conversion of ammonium-N to nitrate-N) and nitrate immobilisation were studied for the three salt-impact scenarios (drainage-impacted soil, spray-impacted soil and control soil) discussed earlier (Green *et al.*, 2008a). To do this, sub-samples of fresh, homogenised soil were spiked with ammonium-N solution, nitrate-N solution or just the same volume of deionised water (as controls). The spiked soils were incubated at room temperature or at 4 °C, and after 0, 1, 2, 5 and 9 days duplicate samples were analysed for extractable ammonium-N and nitrate-N.

Between days 0 and 1, spike ammonium-N was lost in all soils to a significant extent, but interestingly not all was converted to nitrate-N (Figure 19.18). It is clear from Figure 19.18 that

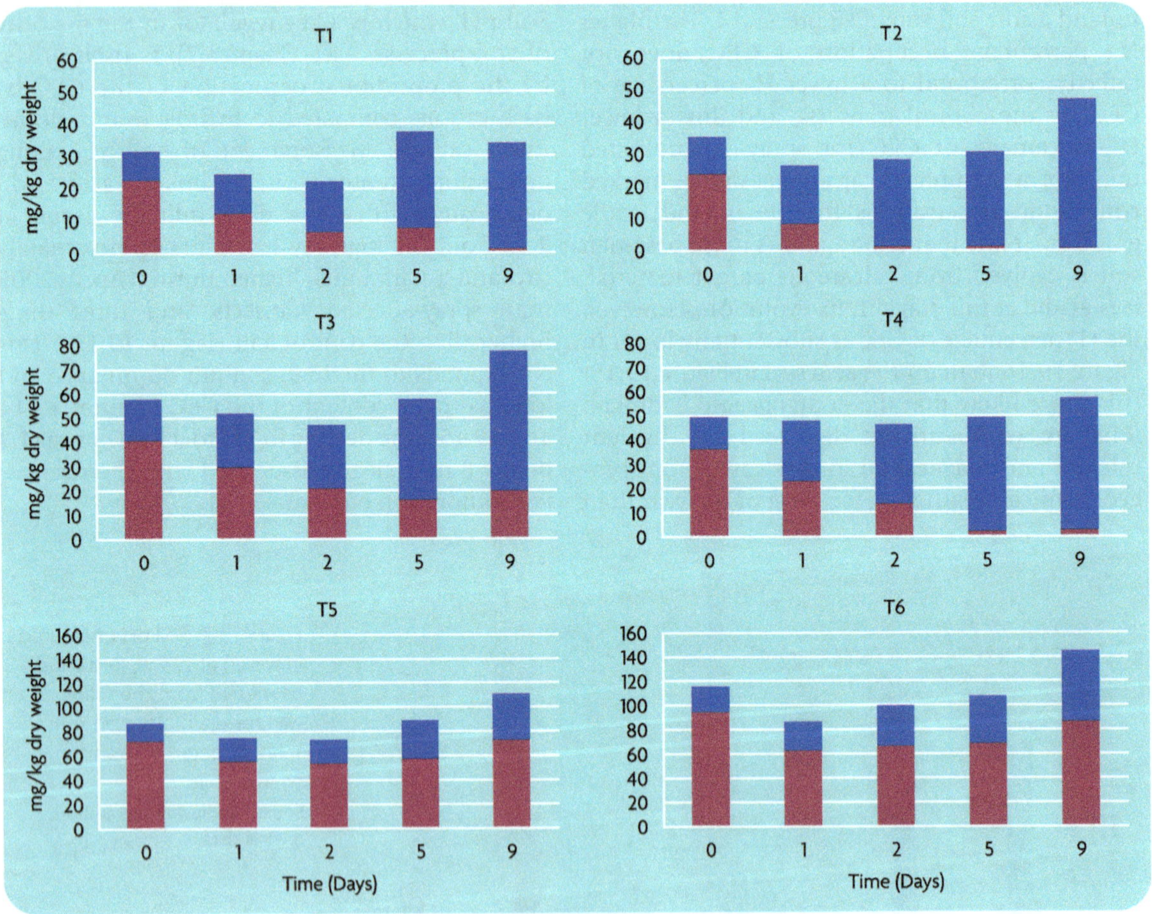

Figure 19.18 Changes over nine days in the total mineral N (total bars), ammonium-N (red bars) and nitrate-N (blue bars) for sites T1–T6 for the ammonium-N spiking experiment with incubation at room temperature. T1 and T2 are most (direct-drainage) impacted, T3 and T4 moderately (spray) impacted, and T5 and T6 are controls. Note the changes in vertical scales used. All results are means of two replicates.

Source: Reproduced from Figure 3 in Green, Machin and Cresser (2008) *Environmental Pollution*, 152(1), 20–31.

if nitrification is occurring at all in the control soils (T5 and T6), it is slower than in salt-affected soils, and/or any nitrate being produced is being immobilised by soil microbial biomass or lost by denitrification. As ammonium-N concentration is changing only slightly over time for the T5 and T6 controls (Figure 19.18) probably low nitrification rate is the cause. It is interesting, however, to note that when the soils were spiked only with deionised water, some nitrification of 'native' soil ammonium-N was apparent for the control soils. If nitrification is being inhibited by low soil pH in the control soils, then the additional chloride added with the ammonium spike would, as a mobile anion, further lower the soil solution pH, reducing nitrification rate even further upon ammonium-N spike addition (see Chapter 7 for discussion of mobile anion effects on soil pH).

The nitrate concentrations changed very little over time following spiking of the soils with nitrate (Green *et al.*, 2008a), which would not be the case if denitrification rate was substantial as suggested above. Moreover, the soils were not particularly wet even after spiking, and the amount of water added with the control deionised water, ammonium-N and nitrate-N spikings was constant between treatments, so if it occurred the denitrification rates would be similar for all treatments. Thus the attribution of low nitrification rate to low pH is by far the most likely hypothesis.

There was no evidence of nitrate immobilisation by soil microbial biomass after spiking with nitrate-N. However, in the road salt-impacted soils (T1–T4), there was rapid decline in ammonium-N over the first 1–2 days. This was almost certainly by nitrification, because total mineral N remained virtually constant over the first five days. This suggests that the higher pH in the salt-impacted soils is again favouring nitrification. Over a longer timescale nitrification was readily apparent in all six soils. The fact that ammonium was starting to accumulate by day 9, especially in the most acidic, control soils (T5 and T6), supports the hypothesis that greater acidity in the control soils inhibits nitrification, or rather that the increase in soil pH caused by the salting impact is greatly favouring nitrification.

Although the results are not reproduced here, nitrification still was significant even at 4 °C in all soils, but the impacts of different long-term salt applications on the soils were much less pronounced at the lower temperature (Green *et al.*, 2008a). At the lower temperature nitrification was more important in the most acidic, control soils (T5 and T6), so it may be concluded that the influence of soil pH upon nitrification rate is highly temperature dependent. It cannot be stated categorically that no residual salinity effects are occurring, but if they have adverse effects upon relevant microbial activity they must be small compared to the acidity neutralisation effect. The results from room temperature incubation in this experiment point to long-term soil pH effects being very important over summer months.

19.5.2 The effects of soil pH increase from road salt on a local surface water

We have seen above that the degree of long-term salt exposure of the soil can control the rates of key microbial N transformation processes, primarily by increasing soil pH. This suggests a risk of increasing amounts of nitrate-N from nitrification of ammonium from atmospheric deposition or produced *in situ* in soils that would naturally be acidic. Thus there is potential for enhanced nitrate loading of waterways in UK uplands due to leaching, which has particular relevance to the Water Framework Directive. Green and Cresser (2008a) investigated how the nitrate concentration in river water varied throughout the year as progressively more and more of a river was affected by road runoff from a road up slope of the river. There was a strong increase in nitrate-N concentrations from where the road and river (Crookdale Brook in Cumbria, UK) converged at S1, on moving downstream towards S6 (Figure 19.19), especially during periods when winter maintenance was prominent (22/11/2005–27/01/2006). This suggests a relationship between the quantity of nitrate-N in Crookdale Brook and road salt application. Spatial trends for chloride were very similar to those for nitrate, and a residual increase in chloride along the river stretch was still seen even in July, well after the end of salt applications. Nitrate concentrations are low in

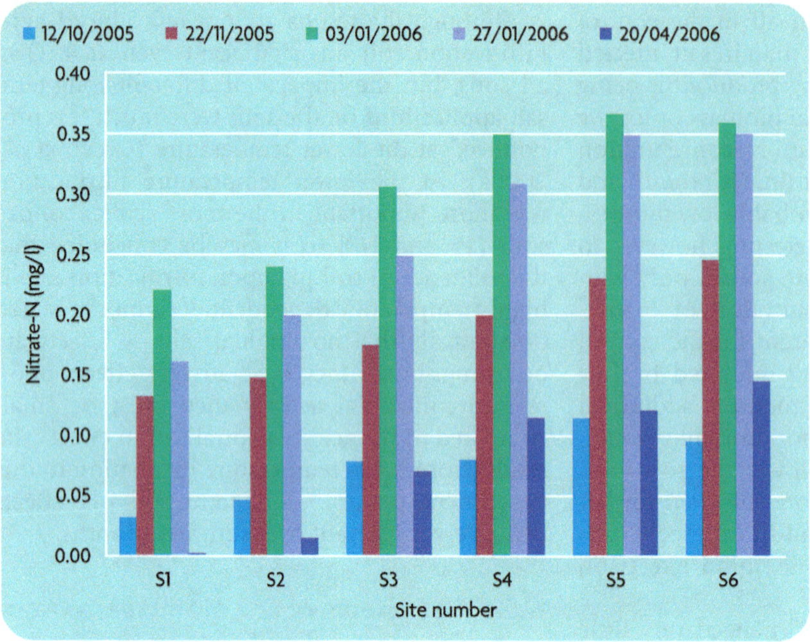

Figure 19.19 The change in nitrate-N concentration (mg l^{-1}) with time and distance along Crookdale Brook for the period 12/10/2005–22/04/2006. The distance between each sampling point from S1 to S6 is approximately 200 metres.

Source: Redrawn in simplified form from authors' original data; a more extensive data set is shown in Figure 8 of Green and Cresser (2008a).

April and July, as expected due to greater plant uptake. Above S1, the catchment area itself spans over 7 km^2; so the extent to which nitrate-N concentration increases in Figure 19.19 is very significant, bearing in mind the relative area of unaffected catchment upstream of S2.

Lead-in question

Why are all these exchange processes important and what are the ecological implications?

At acidic upland sites the introduction of elevated concentrations of road salt has a host of ecological implications following on from the cation exchange reactions and the significant temporal/spatial pH shifts. Changes in pH lead to changes in microbial activity which are highlighted by the enhanced nitrification observed at the upland acidic grassland site; this in turn has reverberations on the N cycle. Displacement of ammonium ions from the CEC may lead to N deficiencies in roadside soils. In addition, toxicity effects of high salinity on microbes may be masked by the pH changes observed. Possible ecological effects have been summarised elsewhere by Green *et al.* (2008a). They include: effects on vegetation of Na^+ accumulating to toxic concentration within plant tissues; increasing osmotic pressure differences causing desiccation; effects of changes in nutrient element balances (e.g. reduced levels of available ammonium and K^+ within the soil); effects of changes in soil structure. In addition, ammonium and nitrate leaching may also suppress vegetation growth. The authors have observed chlorosis very characteristic of nitrogen deficiency in conifers in woodland down slope of salt piles left beside roads in winter for use near bridges or dangerous bends.

Elevated chloride concentrations are known to interfere with photosynthesis in algae in surface waters (Williams *et al.*, 1999), with shifts in population occurring at 12–235 mg l^{-1} due to the varying degree of sensitivity between algae species. In higher organisms they can result in potentially fatal metabolic acidosis and osmotic stress, as well as behavioural changes (Williams *et al.*, 1999). It is possible that changes in population, community structure and/or biodiversity may occur due to acute and chronic toxicity of road salts, in combination with other physical and chemical impacts generated directly or indirectly from the application of road salt.

Raised levels of sodium ions in water can cause high blood pressure and hypertension, so individuals who already suffer from this condition and on salt-restricted diets should not ingest greater than 20 mg l^{-1}. Several states in the north east US and Canada have measured sodium from road salt in well waters at concentrations that are 2–140 times the recommended limit for individuals on salt-restricted diets (Amrhein *et al.*, 1992). Hence, increases in sodium and chloride ions can also cause problems with water balance in the human body. This is a particular concern when considering upland soils and associated fresh water bodies as these tend to be used as potable sources in the UK.

19.5.3 Does road salt flush organic matter into rivers?

It is well known that high salt concentrations in soils lead to dispersal of organic matter. It is also well known that in acid soils high mobile anion concentrations lower the soil solution pH (see Chapter 7). Therefore it is important to consider whether or not road salting over many years is increasing the load of organic matter draining into rivers. Figure 19.20 shows what happens to pH and the concentration of organic carbon (DOC) in the equilibrating water when soils affected by road runoff drainage water or road spray and control soils are treated with increasing concentrations of sodium chloride up to 10,000 mg l^{-1}, as described by Green *et al.* (2008b).

The pH reduction from the increasing salt concentration is very obvious for all three road salting scenarios (Figure 19.20). However, the dispersal effect on DOC of 10,000 mg l^{-1} salt is marked for the control soils, much smaller for spray impacted soils and not significant at all for drainage water directly impacted soils. Green *et al.* (2008b) explained this in terms of a 'when it's gone it's gone' hypothesis. In other words the salt effect has gone on for several decades so any readily mobile organic matter has already been leached from the soil. This concept has important implications for the water industry as removal of DOC is expensive, but expected by consumers.

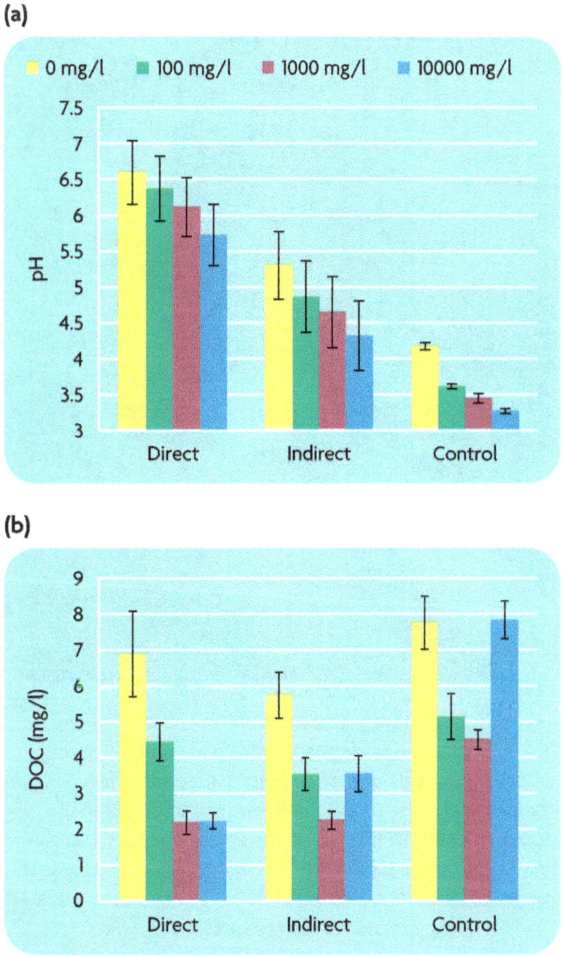

Figure 19.20 The effect of increasing salt concentrations over the range 0–10,000 mg l^{-1} on DOC concentrations in mg l^{-1} in filtrates (b) and on filtrate pH (a) for drain-affected soils, spray-affected soils and control soils. Error bars reflect 95 per cent confidence intervals.

Source: Redrawn from part of Figure 1 in Green, Machin and Cresser (2008) *Chemistry and Ecology*, 24(3), 221–231.

The highest concentration tested seems very high, but is still smaller than the highest salt concentration measured under field conditions.

19.5.4 The first flush effect

Roads do not only acquire pollution from road salting and gritting, of course. They are also subject to deposition of everything ranging from atmospheric terrestrial dust through wear components

Figure 19.21 System set up to monitor the quality and quantity of material deposited on road surfaces during extended periods of dry weather. The road surface is sprayed with a jet of deionised water and dispersed/dissolved material is collected in a container via a suction head. The box has a known area and a seal at the base; it can be used either at the kerbside or further out in the road.
Source: Photograph courtesy of Ana Deletic and David Orr.

from car engines, exhausts, catalytic converters and tyres to mud moved about the country on the underside of vehicles. After several days of dry weather the amount of material flushed away at the start of a rainstorm event may be quite considerable; this first flush too may pass to either rivers via storm drains or to roadside soils, so it is important to know what it consists of (Figure 19.21).

19.6 Using soil to protect surface waters

So far we have been considering at some length how soils may become contaminated by runoff from road surfaces. Sometimes, however, the capacity of soils to retain pollutants is used deliberately to protect surface waters. Deletic and colleagues from Monash University in Australia, for example, have made extensive studies of the use of large tanks of vegetated soil to remove contaminants from the runoff from large areas of tarmac surfaces such as might be found in multi-storey car parks (Figures 19.22a and b and 19.23).

You may well be thinking that the end product of the system considered above is yet more contaminated soil, as indeed it is. However,

(a)

(b)

Figure 19.22 Picture (b) shows a large tank of vegetated soil being used to remove pollutants from runoff from the car park (a) at Monash University in Australia. Water may be sampled at the inlet and outlet of the system so that the efficiency of removal may be monitored for diverse pollutants under a wide range of climatic conditions.

Figure 19.23 Large macrocosms used by the environmental engineers at Monash University to optimise the soil–plant combination for use in their pollution removal systems and to investigate how soil solution chemical composition changes with depth of water infiltration under diverse climatic conditions.

removal of nitrogen, phosphorus and several other potential pollutants by vegetation is a much easier problem to deal with than serious water eutrophication. Moreover, the contained soil may also be treated more readily to deal with the stored pollutants that it retains.

19.7 Soil pollution from catastrophic events

So far we have been considering how soils deal with the anthropogenically generated pollution loads that they are subject to over a range of timescales from daily to seasonal, and year after year. However, soils also have to deal with occasional catastrophic events such as flooding with sea water from a rare tidal surge or from a tsunami. When sea water moves far in land much may drain back via river networks and/or storm drains to the ocean; however, flood waters may also move enormous quantities of both natural material and contaminants from destroyed buildings and their contents, depositing both at large distances from their original source. Deposits may include substances such as oil and petrochemicals from destroyed vehicles or sewage from treatment plants. After a few weeks, rain helps reduce the salinity of soil solutions, but contamination (salinisation) of groundwaters may remain a problem.

Recovery rate will depend upon the original soil mineralogy and land use and the extent and nature of pollution deposited. If eroded subsoils are deposited in deep layers over top soils the lack of organic matter and available nitrogen at the new surface may seriously restrict re-growth of vegetation for many years. Recovery may be accelerated by removal of the surface deposits, but sometimes this can be prohibitively expensive for farmers.

A particular problem comes if the deposited materials include radioactive materials, as happened after the horrifying tsunami incident in Japan in March 2011 when the Fukushima nuclear power plant about 400 km north-east of Tokyo was critically damaged. The wall of water swept ashore at several 100 km per hour, throwing vehicles, ships and buildings inland after an 8.9-magnitude tremor. Many hundreds died and many had to be evacuated from the adjacent area. Under these circumstances radioactive elements accumulate in soils, but accumulation may not necessarily be at the surface, making risk assessment in subsequent months particularly difficult.

After other nuclear accidents, such as the explosion and fire at the Chernobyl nuclear plant in Ukraine in April 1986, large quantities of radioactive contamination may be released high into the atmosphere and transported over large distances, crossing national boundaries. Significant quantities of radioactive elements landed in British uplands, for example, and were retained by cation exchange in surface soils. Where the land was used for rough grazing, sheep, which tug at vegetation and therefore consume some surface soil as well as vegetation, accumulated contamination and many could not be used for meat. Over extended periods the deposited elements such as ^{90}Sr and ^{137}Cs become more dilute at the surface, and are redistributed to greater depth in the soil. Radioactive iodine (^{131}I) is potentially more of a problem as it is more mobile and can be concentrated up in the thyroid gland. However, the soil retains the bulk of the pollution, thereby at least protecting local surface waters and groundwaters to a large extent. This is important for element isotopes with relatively long half lives.

19.8 Soil pollution from agricultural activities

We mentioned earlier that use of sewage sludge as a fertiliser may introduce a range of contaminants to soils as may inappropriate irrigation water use. These may be potentially toxic heavy metals but organic pollutants may also be found from various sources. Soils also have to deal with a whole range of organic agrochemicals such as herbicides, insecticides, and molluscicides. Considerable effort goes into trying to ensure that these chemicals degrade, if not fully, at lease to harmless metabolites. The degradation rate should be sufficiently fast to allow continued use without excessive build up in soil over the long term. The retention of organic agrochemicals in soils also receives much attention, particularly because of the need to restrict movement to surface waters and groundwaters. It should always be remembered too that fertilisers are not high purity chemicals, so careful attention should be paid to what impurities are being unintentionally added to ensure that risk of accumulation of potentially toxic elements is avoided.

POLICY IMPLICATIONS

Soil pollution

- Policy makers need to be aware that road salting disrupts organic matter carbon and nitrogen cycling down slope of upland roads, especially where soils are naturally acidic.
- Policy makers need to be aware that inputs of potentially toxic metals to soils come from atmospheric pollution as well as sewage sludge and sometimes from fertiliser use and from irrigation.
- Policy makers need to consider the consequences of exceeding the retention capacities of soils for pollutants for the quality of surface waters and groundwaters.
- In areas susceptible to catastrophic events such as flooding, policy makers need to have management strategies in place to deal with the subsequent soil pollution problems.
- Special consideration needs to be given to procedures for dealing with nuclear accidents.
- Policy makers need to pay more attention to the finite capacity of soils to deal with pollution loads in the longer term. In particular, they need to reassess many of the limitations of the critical loads approach as a pollution management tool in the longer term.

CHAPTER REVIEW EXERCISES

Exercise 19.1

An enthusiastic gardener waters his garden with a sprinkler every 3 days over a 90-day period each summer, applying 10 mm of water each time. He is in a soft water area and his tap water has an average concentration of 0.05 mg of copper per litre. Assuming that the water is evenly distributed, calculate how much copper he adds per year to his soil in mg per m^2 per year.

If the copper is retained in the top 10 cm layer of soil, and the soil in this layer has a dry bulk density of 1.5 g cm^{-3}, calculate what the annual increase in the soil copper concentration would be in mg per kg of dry soil.

Exercise 19.2

Water taken from the Nilufer River in Turkey has been reported to have the following concentrations of potentially toxic elements in units of mg l^{-1}.

Cadmium	Cobalt	Chromium	Copper
5.2	3.8	11.8	18.1
Manganese	**Nickel**	**Lead**	**Zinc**
8.2	8.5	7.8	21.9

If a farmer abstracts sufficient water from the river to irrigate his crop with the equivalent of 500 mm of rain each year, calculate how much of each element he is adding per year in units of kg ha^{-1}.

Approximately how much of the amount of each element do you think would remain in the soil and why?

Exercise 19.3

Why do sulphate saturation and nitrogen saturation of upland moorland soils contribute to the acidification of the water in rivers that their drainage water flows to? Would you expect the increase to be rapid or slow? Give reasons for your conclusion.

REFERENCES

Aber, J.D., Nadelhoffer, K.J., Steudler, P. and Melillo, J.M. (1989) Nitrogen saturation in northern forest ecosystems. *BioScience*, **39**, 378–386.

Amrhein, C., Strong, J.E. and Mosher, P.A. (1992) Effects of de-icing salts on metals and organic matter mobilization in roadside soils. *Environmental Science and Technology*, **26**, 703–709.

Ågren, G.I. and Bosatta, E. (1988) Nitrogen saturation of terrestrial ecosystems. *Environmental Pollution*, **54**, 185–197.

Aydinalp, C., Fitzpatrick, E.A. and Cresser, M.S. (2005) Heavy metal pollution in some soil and water resources of Bursa Province, Turkey. *Communications in Soil Science and Plant Analysis*, **36**, 1691–1716.

Billett, M.F., Fitzpatrick, E.A. and Cresser, M.S. (1990) Changes in carbon and nitrogen status of forest soils organic horizons between 1949/50 and 1987. *Environmental Pollution*, **66**, 67–79.

Billett, M.F., Fitzpatrick, E.A. and Cresser, M.S. (1991) Long-term changes in the Cu, Pb and Zn contents of forest soil organic horizons from northeast Scotland. *Water, Air and Soil Pollution*, **59**, 179–191.

Cresser, M.S., Smart, R.P., Clark, M., Crowe, A., Holden, D., Chapman, P.J. and Edwards, A.C. (2004) Controls on leaching of N species in upland moorland catchments. *Water, Air and Soil Pollution, Focus*, **4**, 85–95.

Crowe, A.M., Sakata, A., McClean, C. and Cresser, M.S. (2004) What factors control soil profile nitrogen storage? *Water, Air & Soil Pollution, Focus*, **4(6)**, 75–84.

Green, S.M. and Cresser, M.S. (2008a) Nitrogen cycle disruption through the application of de-icing salts on upland highways. *Water, Air and Soil Pollution*, **188**, 139–153.

Green, S.M. and Cresser, M.S. (2008b) Are calcareous soils in uplands less susceptible to damage from road salting than acidic soils? *Chemistry and Ecology*, **24**, 1–13.

Green, S.M., Machin, R. and Cresser, M.S. (2008a) Effect of long-term changes in soil chemistry induced by road salt applications on N transformations in roadside soils. *Environmental Pollution*, **152**, 20–31.

Green, S.M., Machin, R. and Cresser, M.S. (2008b) Long-term salting effects on dispersion of organic matter from roadside soils into drainage water. *Chemistry and Ecology*, **24**, 221–231.

Hornung, M., Bull, K.R., Cresser, M., Ullyet, J., Hall, J.R., Langan, S., Loveland, P.J. and Wilson, M.J. (1995) The sensitivity of surface waters of Great Britain to acidification predicted from catchment characteristics. *Environmental Pollution*, **87**, 207–214.

Williams, D.D., Williams, N.E. and Cao, Y. (1999) Road salt contamination of groundwater in a major metropolitan area and development of a biological index to monitor its impact. *Water Research*, **34**, 127–138.

Wilson, C.A., Davidson, D.A. and Cresser, M.S. (2005) An evaluation of multi-element analysis of historic soil contamination to differentiate space use and former function in and around abandoned farms. *Holocene*, **15**, 1094–1099.

Wilson, C.A., Bacon, J.R., Cresser, M.S. and Davidson, D.A. (2006) Lead isotope ratios as a means of sourcing anthropogenic lead in archaeological soils: A pilot study at an abandoned Shetland croft. *Archaeometry*, **48**, 501–509.

Wilson, C.A., Davidson, D.A. and Cresser, M.S. (2008) Multi-element soil analysis: an assessment of its potential as an aid to archaeological interpretation. *Journal of Archaeological Science*, **35**, 412–424.

Wilson, C.A., Davidson, D.A. and Cresser, M.S. (2009) An evaluation of the site specificity of soil elemental signatures for identifying and interpreting former functional areas. *Journal of Archaeological Science*, **36**, 2327–2334.

Yesmin, L., Gammack, S.M., Sanger, L. and Cresser, M.S. (1995) Impact of atmospheric N deposition on inorganic- and organic-N outputs in water draining from peat. *Science of the Total Environment*, **166**, 201–209.

CHAPTER 20

Risk assessment and remediation of environmental contamination

Ken Killham and Graeme Paton

Learning outcomes

By the end of this chapter you should:

- Understand the value of the concept of risk assessment in evaluating potential environmental pollution problems.
- Be aware of the types of policy and legislation driving demand for risk assessment and for remediation of environmental pollution.
- Recognise the importance of the soil–pathway–receptor model as a risk assessment tool.
- Be more aware of the nature and scale of environmental contamination problems.
- Recognise the scope of bioremediation compared to conventional approaches in remediation of contamination.
- Be aware of the types of bioremediation and of the factors controlling bioremediation success.
- Understand the role of risk assessment in remediation of contaminated sites through case studies.

20.1 Risk assessment – introduction and definition

Risk assessment in the context of uses of contaminated land was adopted from the nuclear industry, where there was a need to develop a quantitative approach to justify statements that using nuclear energy represented an acceptably low risk to the population. The lessons that were learned then have been refined and improved with time.

Risk assessment can be defined as a systematic process for identifying and analysing the risks inherent at a particular site.

20.2 Generic and site-specific risk assessment

Risk assessments can be carried out in two ways, by a generic approach or by a site-specific approach.

Figure 20.1 Intrusive site investigation (using a combination of trial pits and boreholes) coupled to a desk-top study is often the best way to identify hazards.

20.2.1 The generic approach

In this approach, a set of guidelines and standards is applicable to all sites. Here values are developed using defensible scientific approaches that tend to become 'fixed in stone' once set up. This has the advantages of convenience, a modest demand for data, and consistency in use.

20.2.2 The site-specific approach

Criteria in this approach can be developed on a site-by-site basis. It is used when either the relevant generic criteria are missing, or generic criteria would not be sufficiently protective or would be overprotective given the site conditions. This approach may also be relevant when local background levels are high compared to generic criteria. The site-specific approach has a tremendous advantage in that there is greater flexibility in adjusting how the consequences of the values found are used as site knowledge becomes enhanced.

A risk assessment for contaminated land involves the following stages:

- Hazard identification – 'what is the possible problem?' (Figure 20.1).
- Hazard assessment – 'how big a problem might it be?'
- Risk estimation – 'what will be the effect?'
- Risk evaluation – 'does it matter?'

The overall process is summarised schematically in Figure 20.2. For further reading on good practice in contaminated land risk assessment, the text of Rudland and Jackson (2004) is recommended.

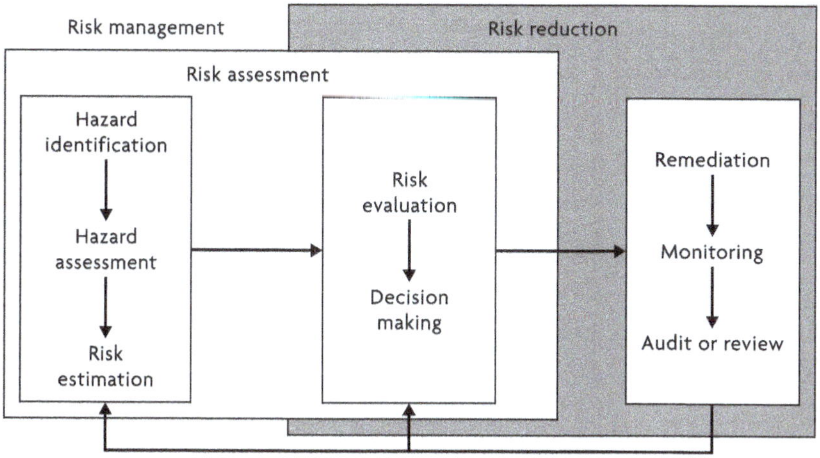

Figure 20.2 Hazard identification, hazard assessment, risk estimation and risk evaluation are the key steps towards decision making prior to remediation of environmental contamination. These steps facilitate design of a monitoring and audit process to demonstrate that risk reduction has succeeded.

20.3 Policy and legislation – the drivers of risk assessment and remediation of environmental contamination

Globally, legislation has developed via several routes. In general, soil protection policy has lagged behind environmental policy for air and water protection. In the UK and several other European countries, legislation now specifically addresses the issues of contaminated land, with regard to assessment, responsibilities and remediation. Under this legislation, there is a need to define a range of clean-up targets or, increasingly commonly, use risk-derived criteria in deciding upon management strategy. The UK legislation for dealing with contaminated land is Part IIA of the 1990 Environmental Protection Act (EPA, 1990, which came into force in England and Wales in April 2000). It represented a fundamental change in the way we think about contaminated land, contamination being defined in terms of causing 'significant harm' or 'pollution of controlled waters', rather than based on some arbitrary contaminant concentration (Nathanail and Bardos, 2004).

20.4 The source–pathway–receptor model of environmental risk

Quantifiable risk assessment needs to be placed in the context of environmental protection. It is not the presence of a chemical of concern *per se* that makes it harmful in the environment but the presence of a receptor and a connection (a pathway) between the contaminant and the receptor.

Risk assessment methodology addresses 'significant pollutants' and 'significant pollutant linkages' within the context of a source–pathway–receptor model of the site (Nathanail and Bardos, 2004). This is represented schematically in Figure 20.3.

'Risk' is defined as the combination of:

- the probability, or frequency, of occurrence of a defined hazard (e.g. exposure to a property of a substance with the potential to cause harm); and
- the extent (including the seriousness) of the consequences.

In order for significant harm to occur, a 'source –pathway–receptor' linkage must be established.

SOURCE	PATHWAY	RECEPTOR
Gaseous emissions via chimneys, vehicle exhausts etc. and particulate emissions	Atmospheric transport, plant uptake; inhalation	Humans/animals, plants
Waste water, effluent	Drainage to surface waters or groundwater	Aquatic organisms, humans, animals soil microbes
Solid waste disposal landfill sites	Leaching, volatilisation, movement by wind	Diverse
Metal corrosion Leakage from storage	Movement in drainage water	Plants, soil microbes, water consumers
Pesticide misuse	Movement with runoff Atmospheric transport	Plants, animals, humans, aquatic organisms

Figure 20.3 The source–pathway–receptor model is fundamental to defining contamination. Note that food chains need to be considered too under 'receptors'.

This requires each component of the source–pathway–receptor model to be linked as follows.

20.4.1 Source

The first component of the model is a source of contaminant in or under the ground, which is bioavailable and has the potential to cause harm to receptors or to cause pollution of controlled waters.

The chemicals of concern (CoCs) (identified through analysis, but likely to be associated with recognised contaminants from former/current land uses) lead to the contamination of soil, water and the atmosphere. The partitioning of the CoCs will define where in the environment the chemical will be found and its relative ability to transfer between environmental matrices.

20.4.2 Pathway

The second component is a pathway with one or more routes or means by which a receptor either:

- is being exposed to, or affected by, a contaminant, or
- could be so exposed to, or affected by, a contaminant.

The migration of CoCs is governed by a number of physical, chemical and biological mechanisms. It may proceed along several pathways:

- CoC release from, and transport through, soil (leaching) – the movement of dissolved substances with water percolating through soil.
- Transport with water in saturated and/or unsaturated zones. The saturated zone encompasses the area below ground in which all interconnected openings within the geologic medium are saturated with water (groundwater was discussed in Chapter 2). The unsaturated zone is that portion of the subsurface in which the inter-granular openings of the geologic medium contain a mixture of both water and air.
- Flow as free phase (e.g. non-aqueous phase liquid (or 'NAPL') migration).
- Sorption, which refers to the action of either absorption or adsorption. Absorption is the incorporation of a substance in one state into another of a different state (e.g. liquids being absorbed by a solid or gases being absorbed by water). Adsorption is the physical adherence or bonding of ions and molecules onto the surface of a solid matrix.

- Biodegradation – biologically mediated breakdown.
- Advection – the bulk movement of solutes with the mean groundwater velocity.
- Mechanical dispersion – the main mechanism that causes solutes to spread and dilute in an aquifer.

20.4.3 Receptor

The third component is a target such as humans, water resources (surface and/or groundwater) and flora/fauna (e.g. on UK sites of special scientific interest – 'SSSIs', livestock, wild animals, birds, landscape, etc.). Targets may also be buildings that are, or may be expected to be, or have been, exposed to or affected by the source on contamination at a site. Multiple receptors can be involved (e.g. aerial contamination that falls back to ground, surface water run-off entering a river system, or an underground plume making its way to the saturated groundwater zone). Examples of links between pathways and receptors are illustrated in Figure 20.4.

A 'significant pollutant linkage' is the relationship between the contaminant, the pathway and the receptor. Unless all three components of a pollutant linkage are identified in respect to a site, then that land should not be identified as contaminated. It is important also to remember that there can be multiple pollutant linkages on any given site.

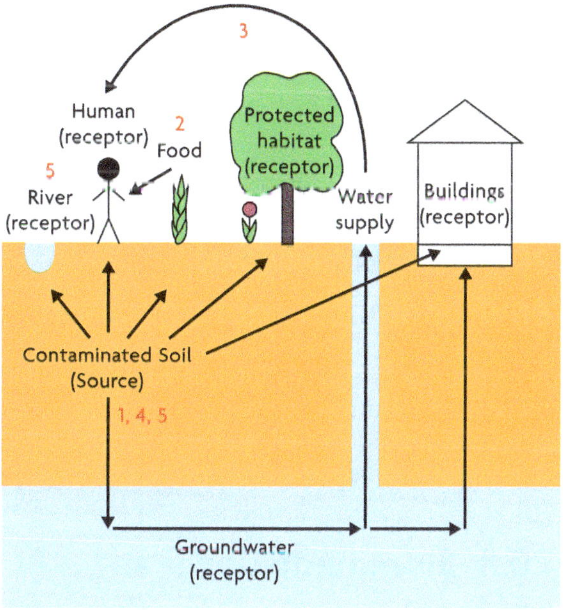

Possible pathways

Ingestion of contaminated soil/dust – 1
Ingestion of contaminated food – 2
Ingestion of contaminated water – 3

Inhalation of contaminated soil particles/dust/vapours – 4

Direct contact with contaminated soil/dust/water – 5

Figure 20.4 Pathways to receptors. Pathways are based on the properties of the CoCs and the nature of the site. A pathway can only be recognised if it is capable of exposing a receptor to an identified contaminant. Likewise, the contaminant should be capable of harming that specific receptor.

20.5 Risk derived remediation targets

It is generally not economically feasible to remove all CoCs from contaminated soils/sites. The objective therefore is to remove sufficient such that the risk is deemed to be acceptable. Coupled to this, is the belief that risk assessment is 'fit for purpose'. This means that residential property with gardens will require more conservative remediation targets than commercial or industrial sites. The risk-derived criteria are receptor defined (in the case of residential end use, the receptors are humans with potential direct contact with the soil via gardens), so a knowledge of the intended end use is essential in agreeing protective remediation targets with an environmental regulator (a local council or national environment agency).

20.6 Environmental contamination – the nature and scale of the challenge

Although contamination by chemical and biological agents affects every part of the environment,

Figure 20.5 Industrial activities such as manufacturing gas and petroleum refining are associated with both point source and diffuse environmental contamination.

this chapter will focus on land contamination, particularly as the science of its remediation is the best developed. While biological contamination is usually localised, chemical contamination is virtually ubiquitous. Even though estimates of the extent of chemically contaminated (by organic contaminants and/or metals) land will vary with definition/standards, as well as with methods of chemical analysis, and there is no reliable global figure, the extent of the problem is very considerable indeed. Estimates of the extent of chemically contaminated land in England and Wales range from 50,000 to 200,000 hectares (Royal Commission, 1996; Environment Agency, 2000), with up to a further 9000 hectares in Scotland (Scottish Executive, 1998). Not surprisingly, the main areas of heaviest chemical contamination are associated with long histories of industrial development such as that in Figure 20.5. However, the problem is worldwide, with chemical contamination from environmentally persistent pesticides and from diffuse atmospheric pollutants found right across the globe.

20.7 Remediation of environmental contamination

Remediation can be defined as action taken to prevent or minimise, or remedy or mitigate the effects of any identified and unacceptable risks due to environmental contamination.

20.8 Bioremediation – an alternative to traditional remediation approaches

Bioremediation refers to the application of biodegradative processes to remove or detoxify contaminants found in water, soil or sediments. In the case of organic contaminants, this process should (ideally) end in the production of carbon dioxide or methane and water, though CO_2 is preferred as a less potent greenhouse gas unless methane is trapped and burned. The effect of contaminants on microorganisms can be stimulatory (necessary for biodegradation), inhibitory or neutral.

20.8.1 Bioremediation requires bioavailable and bioaccessible contaminants

The terms 'bioavailable' and 'bioaccessible' are used in published literature in a number of ways. In this chapter, the definitions used are those proposed by Semple *et al.* (2004).

A bioavailable compound is defined as a 'compound which is freely available to cross an organism's membrane from the medium the organism inhabits at a given point in time'.

A bioaccessible compound is defined as 'a compound which is available to cross an organism's membrane from the environment it inhabits, if the organism has access to it; however it may either be physically removed from the organism, or only bioavailable after a period of time'.

These terms are both usually applied to contaminants in soil. Figure 20.6 indicates the types of process that need to be considered in this context.

20.8.2 Other conditions required for bioremediation

For biodegradation to take place the following must apply:

- An organism (plant or microorganism) with the necessary enzymes or properties must exist.
- That organism must be present with the pollutant.
- The pollutant must be bioaccessible.
- If the enzyme is extracellular, the bonds it acts upon must be 'exposed'.
- If the enzyme is intracellular the pollutant must be capable of penetrating the surface (membrane/wall) of the degrader organisms.
- Environmental conditions must favour the proliferation of the degrading organism.

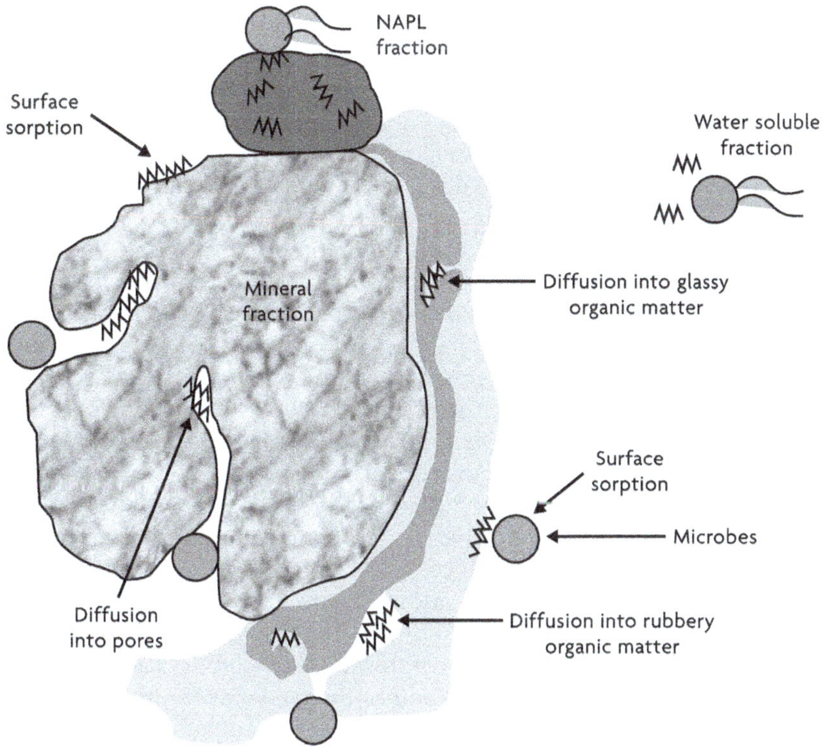

Figure 20.6 Interactions between soil matrix and hydrocarbons determine contaminant bioavailability and bioaccessibility and hence likely bioremediation.

20.8.3 Recalcitrant contaminants

A contaminant is referred to as recalcitrant if there are no microorganisms present in the environment that have the ability to degrade it. This is quite rare as, given time, degrader organisms will generally evolve in the presence of a new organic contaminant.

20.8.4 Why use bioremediation?

There is a strong drive to find increasingly sustainable ways of remediating environmental contamination. Bioremediation exploits the activity of either indigenous or added organisms (microorganisms and plants). It provides an attractive alternative (and complementary approach) to the conventional 'dig and dump' remediation of contaminated land. The attraction is sometimes based on reduced inputs and lower costs. It is also based on regulatory drivers, particularly tied to reduced dependence on landfill as shown in Figure 20.7.

Figure 20.7 Excavation combined with landfill ('dig and dump') is still the most common method of remediating contaminated land, but more sustainable methods are increasingly being developed and applied.

20.8.5 Types of bioremediation

Two broad management approaches to bioremediation can be adopted, depending on whether contaminated material is excavated (*ex situ* bioremediation) or not (*in situ* bioremediation) for biological treatment. Both *in situ* and *ex situ* bioremediation of contaminated land require integrated management of the plant, soil and water system. The key objective is to remove any constraints to the activity of the microorganisms which carry out bioremediative processes, and then optimise the physiochemical conditions that determine the rate of these processes.

Ex situ bioremediation

There are three main technologies used in *ex situ* bioremediation. These are:

- Soil slurry bioreactors and biofilters.
- Composting – composts, biopiles and windrows.
- Land farming.

Soil slurry bioreactors and biofilters

Soil slurry bioreactors (Figure 20.8) were developed from the wastewater and chemical processing industries. An aqueous slurry is created by combining soil, sediment or sludge with water and other additives (Kuyukina *et al.*, 2003). The reactors are mainly aerobic, with oxygen being passed throughout the liquid at a rate sufficient to maintain aerobic conditions – a costly procedure. The bioreactor is stirred mechanically or by air sparging to keep the solids and microorganisms suspended and in contact with the contaminants. Agitation can vary from intermittent mixing to intense mixing. Bacterial growth occurs in suspension, resulting in the breakdown of the contaminants. When the process is completed the water is removed and the treated soil disposed of.

Biofiltration (Figure 20.9) is another type of *ex situ* bioremediation, providing a large surface area for colonisation by biodegrading microorganisms and optimising reaction time between the contaminants and the surfaces of the biofilter.

Figure 20.8 Bioreactors provide a form of *ex situ* bioremediation and are sometimes used to speed up biodegradation of organic contaminants in soil as many of the physical barriers (e.g. soil structure) to biodegradation are removed and conditions can be optimised.

Figure 20.9 A biofiltration system.

Composting – composts, biopiles and windrows

In general, composting differs from other bioremediation processes involving enhanced bioremediation in that soils are mixed with organic amendments. Amendments/bulking agents improve drainage, ventilation and thermal properties and stimulate intense biological activity (Antizar-Ladislao *et al.*, 2004, 2009). They may include:

- Commercial or domestic green waste.
- Animal litter and manures.
- Wood chip or straw.
- Food process wastes.
- Used mushroom composts etc.

Composting can involve biopiling and windrowing. Biopiling is a solid-phase bioremediation technique in which the soil is excavated and mixed with other material (amendments) (Figure 20.10). The mixed material is placed on a lined treatment area which allows for the collection (and often recirculation) of leachate. Aeration and irrigation systems are incorporated into the design. It may be necessary to collect gas emissions.

Windrowing is a solid-phase bioremediation technique in which the soil is excavated and mixed with other material (amendments). The mixed material is placed on a lined treatment area which allows for the collection of leachate (Figure 20.11). Aeration is achieved by regular turning of the composting material, usually using specialised machinery. Irrigation is achieved by adding water during turning. It may be necessary to collect gas emissions.

Landfarming

In landfarming, contaminated material (soil, sludge and/or sediment) is applied to the soil surface in a controlled manner and the indigenous microorganisms are allowed to degrade the contaminants aerobically (Kuyukina *et al.*, 2003). It is a shallow treatment procedure involving a large exposed surface area, usually divided into treatment cells. Treatment cells are lined and incorporate a leachate collection system

Figure 20.10 Biopiling requires even aeration and this constrains design. The component parts of a biopile, which is a static construction, are the base, aeration system, contaminated material, irrigation system and cover.

Figure 20.11 Windrowing of contaminated soil where the base for windrows is similar to biopiles. Windrows usually involve a lined treatment area into which the contaminated soil is excavated and combined with amendments (if needed) and bulking agents. However, the soil/compost is laid out in elongated rows and aeration is achieved by periodic turning.

(Figure 20.12). Aeration is by tilling and moisture is added by irrigation or spraying. Although biodegradation occurs contaminants are also lost by volatilisation.

Landfarming is similar to composting methods such as biopiling, although a much larger area is covered (e.g. 40,000 m^2). The amount required depends on the volume of contaminated material for treatment and the depth of the landfarm soil (commonly 30–45 cm). The contaminated material is applied in layers usually no more than 20 cm deep.

As with all forms of bioremediation, *ex situ* bioremediation proceeds most rapidly when conditions for microbial/plant activity are optimal or at least near optimal. Oxygen can be introduced by turning of windrows and landfarms and by injection of slurry systems. Water can be added to the two former types of *ex situ* treatment by spraying/irrigation, and nutrients (mainly nitrogen) can be added with water. Slow-release fertilisers are the preferred way of adding nutrients and are generally reapplied when the available nitrogen levels fall below 50 mg kg^{-1} of soil/matrix. Monitoring of available nitrogen levels (along with respiration rates in most cases) is usually carried out every two weeks during the first few treatment weeks and then at longer intervals. Liming is sometimes necessary as the pH may be suboptimal and may change during bioremediation.

In situ bioremediation

In order for *in situ* bioremediaiton to take place at a reasonable rate and so remove contaminants, the environmental conditions have to be suitable

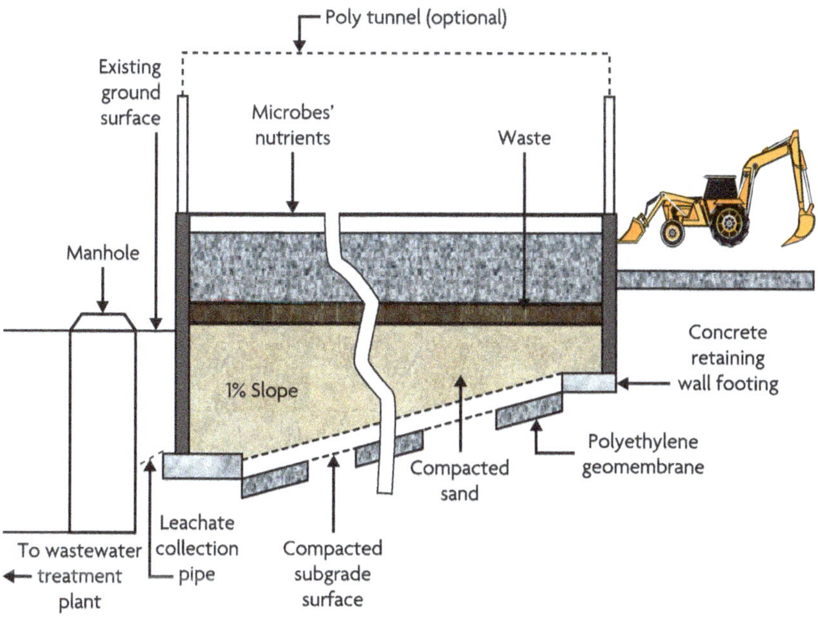

Figure 20.12 Schematic representation of landfarming, showing how contaminated material is applied to the soil surface in a controlled manner and the indigenous microorganisms are allowed to degrade the contaminants aerobically. It is a shallow treatment procedure involving a large exposed surface area.

for microbial activity. The most important conditions affecting activity are:

- water content or moisture (strictly 'water potential');
- pH;
- temperature;
- nutrient status;
- electron acceptor (usually oxygen) status.

Soil porosity is the single most important physical property of the soil that influences the above conditions and the technology employed, and is fundamental to the success or otherwise of *in situ* remediation and bioremediation. The rate of flow of water through soil is governed by its hydraulic conductivity or coefficient of permeability which is defined by Darcy's law: The lower the porosity, the lower the hydraulic conductivity.

There are several technologies used in *in situ* bioremediation. These include:

- pump and treat;
- anaerobic degradation;
- natural attenuation;
- bioventing and sparging;
- permeable reactive barriers;
- phytoremediation (including rhizoremediation).

Pump and treat

Pump and treat technology (Figure 20.13) is a common form of groundwater remediation. Contaminated groundwater is pumped to the surface using a series of extraction wells (Figure 20.14), where it is subsequently treated to remove the contaminants, and then either re-injected into a groundwater aquifer or discharged into a sewer (USEPA, 2005).

Anaerobic degradation

The rates of anaerobic degradation are generally less than those for aerobic degradation, in the case of hydrocarbons, typically an order of magnitude less. Some contaminants may be degraded under anaerobic conditions, as indicated below:

- Aromatic hydrocarbons (e.g. BTEX compounds) and chlorinated solvents may be degraded under denitrifying conditions.

Figure 20.13 Installing pump and treat equipment involves mobilisation costs, but is a well-used technology for contaminants such as volatile chlorinated organics.

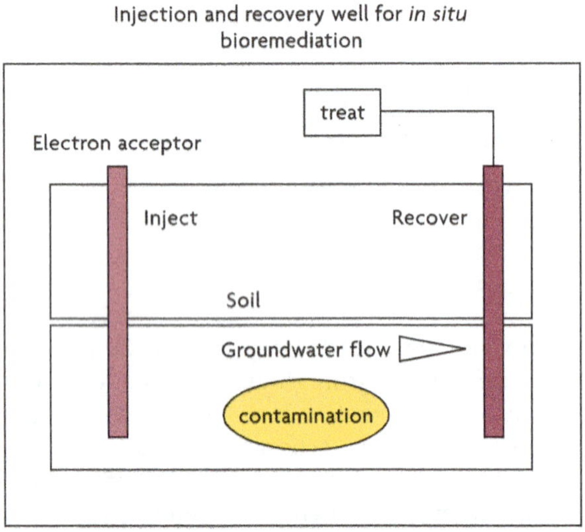

Figure 20.14 Pump and treat sometimes requires provision of an electron acceptor such as oxygen, nitrate or sulphate, the contaminant(s) being the electron donor.

- Nitro-substituted compounds (e.g. TNT) may be degraded.
- TCE and chlorinated phenols can be degraded by methanogens.
- Aromatics and aliphatics can be degraded under sulphate reducing conditions.
- Aromatics can be degraded using Fe(III).

Anaerobic degradation may be limited by the lack of a suitable electron acceptor, but this may be overcome by the addition of metal species, primarily Fe and Mn, sulphur species and carbon species (e.g. humic acids).

Natural attenuation

Natural attenuation (Figure 20.15) relies completely on the naturally occurring biodegradation to remove an environmental pollutant with no intervention at all (Krupka and Martin, 2001). This method, which may be aerobic and/or anaerobic, is often very slow and may often not be fast enough to remove toxic substances before damage to the site and surrounding ecology can occur. This approach has been used, particularly in Europe, for the treatment of marine crude oil spills along coastlines.

The rate of movement of oxygen from the air spaces into the pore water and then to the microorganisms is the most important process governing bioremediation technologies. This movement is controlled by diffusion and convection, which in turn are affected by the water content of the matrix. If the soil pores become saturated with water then the movement of oxygen is slowed down. The microorganisms quickly use up any available oxygen at a rate that exceeds the rate at

Figure 20.15 Natural attenuation of chlorinated solvent contamination in a wetland.

which convection and diffusion can replace it, resulting in anaerobiosis.

It is not only water saturation that can cause this effect. When soils and sediments are contaminated with organic compounds, aerobic and anaerobic areas are present. This is because the contaminants stimulate the aerobic microbial population present which results in the concentration of oxygen being rapidly depleted and anaerobiosis results.

Bioventing and biosparging

Bioventing (Figure 20.16) relies on effective aeration, and to achieve this, air/oxygen is introduced into the soil either by vacuum extraction or by forcing in the air under positive pressure. It is the geology of the site that dictates whether or not this procedure can be used. The permeability of the soil determines the relationship between the applied pressure, or vacuum, and the gas flow rate. The aim is to maximise the biodegradation of the contaminants while minimising losses by volatilisation. A uniform airflow is the best way to do this, but in practice this may not be possible.

Biosparging (Figure 20.17) is similar to bioventing, but air is introduced into the saturated zone, below the lowest point of contamination. The technology is not so well developed as that for bioventing. The criteria for design are similar to those for bioventing however.

Permeable reactive barriers

A permeable reactive barrier (PRB, Figure 20.18) is a created treatment zone incorporating reactive material(s) that is created below the surface to remediate contaminated fluids flowing through it. It has a negligible overall effect on overall fluid flow rates through the subsurface strata. Typically this is achieved by creating a permeable reactive zone, or a permeable reactive 'cell' sandwiched between low permeability barriers so that the contaminant is chanelled via the reactive media.

Permeable reactive barriers can also be called permeable treatment walls (PTW) or active barriers. They are most commonly used *in situ* for the treatment of groundwater. The contaminants pass through the barrier under natural hydraulic gradients and are removed by chemical/biological

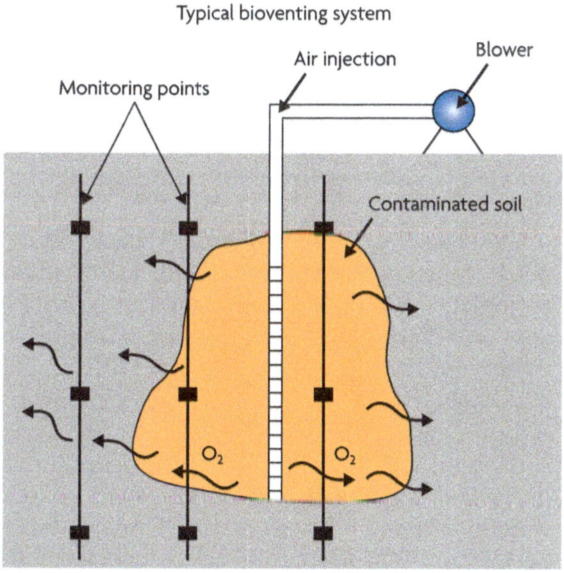

Figure 20.16 Bioventing is used for many types of hydrocarbon contamination, but is not suitable for soils with low permeability or for contaminants with high vapour pressures as these will volatilise and the procedure may further disperse volatile compounds.

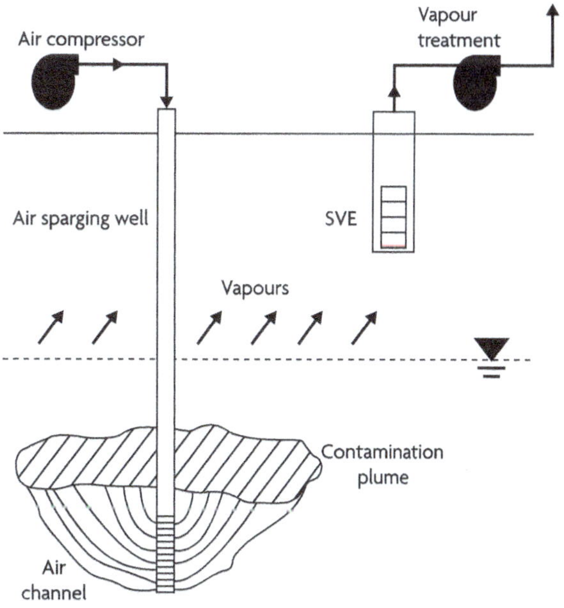

Figure 20.17 Diagram of a biosparging system. Biosparging is best used when contaminants are located in a confined aquifer. Vidali (2001) has reviewed the main features of biosparging and bioventing systems.

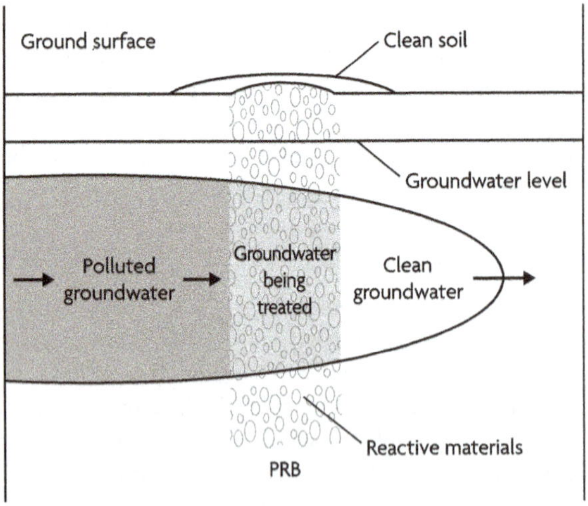

Figure 20.18 Schematic representation of a permeable reactive barrier. These tend to be long-term installations and usually involve considerable installation costs.

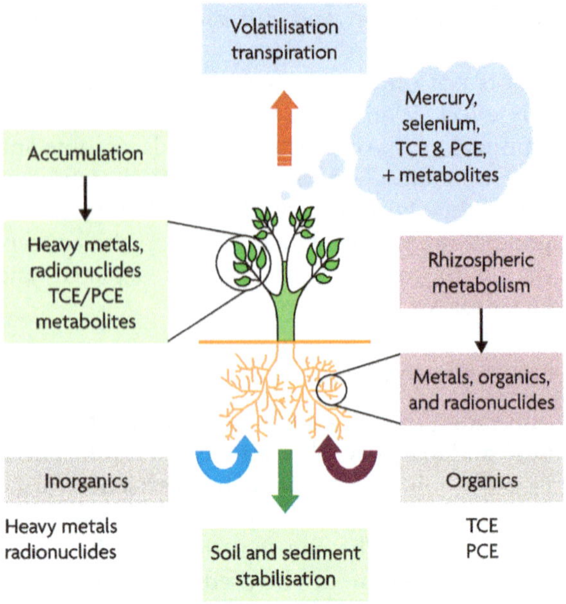

Figure 20.19 Schematic representation of the diverse pathways by which phytoremediation can operate to remove toxic chemicals. TCE and PCE here indicate trichloroethylene and perchloroethylene.

remediation or by sorption and precipitation. The decontaminated groundwater emerges downstream of the barrier. The advantages and disadvantages of permeable reactive barriers have been presented by Striegel *et al.* (2001).

Phytoremediation (including rhizoremediation)

Synergistic enzymatic and chemical activities of plant and microbial metabolism in the rooting zone of contaminated soil can transform and degrade contaminants and hence bioremediate environmental contamination, as well as 'green' a formerly contaminated, and possibly derelict, site.

In the case of metals, the combination of activities of plant roots and associated microorganisms is termed 'rhizoremediation' and usually involves the metal immobilisation in the soil or plant, while in the case of organic pollutants, it usually involves their degradation (Figure 20.19). Rhizomediation of metal contaminants is generally most effective when the metals are taken up by the plant, potentially for removal when the plants are harvested (Figure 20.20), but is constrained by the fact that many of the strongest metal accumulators (particularly the hyperaccumulators) tend to be small plants. Future strategies may involve introducing their key genes involved in transporting and tolerating considerable metal loads (such as the genes encoding the cysteine-rich, metallothionein metal-binding proteins) into other plants (Milner and Kochian, 2008).

In the case of organic pollutants, the aim of rhizoremediation, as introduced above, is usually their degradation, either to non-toxic forms or, through complete mineralisation to carbon dioxide and water.

The terms phytoremediation (Harvey *et al.*, 2002; Morikawa and Erkin, 2003) and rhizoremediation (Kuiper *et al.*, 2004) are commonly applied when plant metabolism contributes to the bioremediation process in some way. In certain cases, such as the elevated rates of hydrocarbon degradation in grassed compared to bare soils with otherwise similar physicochemical properties, it is the degradation of plant residues which is the main key to driving biodegradation or immobilisation of the target contaminants (Gunther *et al.*, 1996). In some other systems, however, the plant and/or its symbionts are more actively/specifically involved. This is exemplified by deep rooted poplars which can access mobile organic contaminants and degrade/detoxify them

Figure 20.20 Rhizoremediation is particularly useful for dealing with large volumes of contaminated soil at low concentrations of contaminant. When the roots have become saturated they can be removed and the area replanted. Radioactively contaminated pools at Chernobyl were successfully treated using this technique via sunflowers (*Helianthus annuus*) (Vanek *et al*., 2010).

Source: Image from Pearson Online Database/Imagestate/John Foxx Collection.

by metabolism by either non-modified plants or plants with introduced pollutant transformation genes, or by the action of microbial endophytes in the tissues of the trees (Doty, 2008).

The mycorrhizosphere – a bioreactor for contaminant bioremediation

Mycorrhizal symbioses involving plant root associations with fungal symbionts (mycobionts) confer a number of properties to the host plant which are key to bioremediation of soil contamination and offer interesting options for management of contaminated sites. Ectomycorrhizal association, common to most trees, offers considerable metal immobilisation potential for example, and enables phytoremediation with trees that can colonise sites to greater depths than hyperaccumulator or other metal-tolerant plants. There are at least two mechanisms for this; one involves complexation of the metals with oxalates released from the mycorrhizal roots of conifers into the 'mycorrhizosphere' and another involves metal uptake via the mycorrhizal fungal hyphae and subsequent immobilisation in the tissue of the tree (Meharg, 2003).

Furthermore, because metal-contaminated sites select the mycorrhizal fungal community that can best cope with the presence of potentially toxic metal ions, the associated fungi are able to confer diverse physiological attributes to the host to enable colonisation of a wide range of contaminated sites (Meharg, 2003).

The zone of soil around the mycorrhizal fungal hyphae, or 'mycorrhizosphere', appears to be a key microhabitat of the plant–soil system where not only metals can be phytoremediated but organic contaminants also, through a variety of possible mechanisms (Joner and Leyval, 2003); it thus offers additional potential for management of the plant/soil system for bioremediation.

There are numerous advantages to rhizoremediation, many of which are associated with the ability of the root system to penetrate to otherwise inaccessible contaminants and make them accessible to plant or microbial metabolism. It is also a process ideally suited to the remediation of contaminants such as radioactive elements (Macek *et al.*, 2009) and explosive compounds (Gerhardt *et al.*, 2009), materials that pose obvious risks in traditional 'dig and dump' approaches. One of the main drawbacks to phytoremediation, however, is that it is usually a relatively slow and seasonal process and so it is often necessary to allow rhizoremediation to proceed for a number of years before clean-up soil/groundwater targets are achieved.

20.8.6 Managing/optimising bioremediation

Monitored natural attenuation assumes that a contaminated system will remediate without intervention and this can be demonstrated through monitoring. In managing all other types of bioremediation, issues of nutrient status/supply, water potential, oxygenation and the presence of an adequate inoculum are of paramount importance to ensure optimal rates of remediation where the aim is degradation of organic pollutants, and particularly for most petroleum hydrocarbons (Piotrowski *et al.*, 2006). Other issues such as toxicity of contaminants and co-contaminants, varying bioavailabilities of target contaminants, and pH-dependent toxicity (particularly for heavy metals) are sometimes poorly understood and often require careful consideration to ensure success.

20.8.7 The 'bioaugmentation versus biostimulation' debate

Unassisted, bioremediation (*ex situ* or *in situ*) will often proceed at rates that are too slow for redevelopment of contaminated sites. It is therefore often necessary to carry out some kind of managed intervention to obtain satisfactory rates of bioremediation. Two key types of bioremediation intervention which offer considerable promise for effective management of contaminated land are biostimulation and bioaugmentation. There is considerable debate about their relative merits however.

Biostimulation

Biostimualtion aims to enhance (stimulate) activities of the indigenous microorganisms to degrade contaminants at a site. It has been discussed briefly earlier under *ex situ* bioremediation and managing bioremediation. It is commonly used in the bioremediation of oil-contaminated sites as a simple and relatively cheap approach to bioremediation that does not change genetic diversity. Stimulation of mineralisation needs the correct balance of carbon to nitrogen to phosphorus (C:N:P), at an appropriate pH. This is achieved by the addition of:

- Nutrients, such as fertilisers and trace elements.
- Liming materials (usually $Ca(OH)_2$ as agricultural lime, $CaCO_3$).
- Optimal physical conditions can be achieved with the addition of:
  - water;
  - oxygen. The levels may be increased simply by adding a bulking material (e.g. bark or woodchips as in Figure 20.21) to increase the volume of air present, by injecting oxygen, or by the addition of hydrogen peroxide.

When the pollutants are hydrophobic, the addition of a surfactant will maximise the contact

Figure 20.21 Woodchips being stockpiled for supplementing *ex-situ* bioremediation (windrow) of hydrocarbon-contaminated soil.

between the pollutant and water (containing dissolved nutrients).

There is more scientific evidence for the efficacy of biostimulation than for bioaugmentation. Biostimulation is based on the concept that suitable organisms/genes are present for bioremediation but that activity is restricted by a constraint that can be alleviated. It therefore involves the addition of electron acceptors, nutrients, or electron donors to increase the numbers or stimulate the activity of indigenous biodegradative microorganisms (Widada *et al.*, 2002).

Bioaugmentation

Bioaugmentation is the addition (inoculation) of 'foreign' organisms into contaminated soil or water to improve the biodegradation capacity of the system. It is most suited for the degradation of single contaminants (e.g. resistant pesticides such as pentachlorophenol and the hormone-acetic acid herbicides such as 2,4-D) (Zuzana, 2009). The procedure can be used for soil composting, slurry bioreactors and surface soils and may involve the addition of white rot fungi or bacteria.

There are three main ways in which bioaugmentation can be approached:

- Increase the genetic diversity of the soil or water by adding microorganisms not usually found at the site.
- Take samples from the site and use them as initial inocula for serial enrichments with the contaminants found at the site as the sole carbon source. Those microorganisms that are capable of degrading the contaminant will increase in number and can then be returned to the site and increase the rate of biodegradation.
- An unknown consortium of microorganisms can be introduced by adding materials such as compost and sewage sludge.

Bioaugmentation is based on the concept that bioremediation cannot proceed because of a lack of suitable organisms and therefore involves the addition of (indigenous or non-indigenous) laboratory-grown microorganisms capable of biodegrading the target contaminant (Widada *et al.*, 2002; Vogel, 1996) or serving as donors of catabolic genes (Top *et al.*, 2002). For most contaminated soils which retain reasonable biological activity, organisms for bioremediation are present, but their activity is constrained. So, biostimulation will be commonly applied as a management tool in bioremediation. However, in contaminated sites where suitable organisms for bioremediation are at low population densities or are absent, real benefits can be derived from bioaugementation.

Combined bioaugmentation and biostimulation

The advantages of bioaugmentation and biostimulation have been combined to remediate contamination from the pesticide atrazine successfully (Silva *et al.*, 2004). Bioaugmentation requires the degrading inoculum to reach the contamination in the soil/fill/stratum as well as carry out biodegradation of the target organoxenobiotic. In the field, this aspect of 'delivering' the degrader microorganisms to their site of action may well be the most challenging part of a bioaugmentation programme, as many organoxenobiotics become bound in micropores of the soil and these may not be pathways for the percolating water which can carry the degraders. This constraint to bioaugmentation will often be an important factor in applying a combined approach of bioaugmentation and biostimulation as indigenous degraders may already have colonised these more inaccessible parts of the soil; therefore the activity of such degraders may be enhanced through biostimulation.

CASE STUDIES

A case study – What happens in practice?

It seems appropriate at this point to consider a typical case study, to see what typically happens in practice.

Site description

The case study involves a large hydrocarbon-contaminated site in the UK. It was originally a railway siding and train fuelling depot (see figure) and so most of the contamination was diesel fuel from trains. The contamination was quite extensive, and had migrated to considerable depth in the alluvial, sandy soils. The amount of soil excavated because of exceedance of regulatory limits was 27,000 tonnes.

The site prior to remediation was a railway marshalling yard with diesel fuelling facilities such as the one shown. The facility was associated with extensive, long-term diesel contamination.

Risk assessment

For this site, the end use was intended to be residential with gardens. This meant that the receptors that required protection from the CoCs were humans (in fact toddlers are deemed the most sensitive group because of their play in gardens) and the water environment. Quantitative risk assessment established that the remediation targets agreed with the local council posed the following risk:

- A greater than 100,000:1 chance of a toddler developing cancer in a 70-year life span from exposure to CoCs through soil ingestion between age 0 and 5 years at the site.
- The water environment would not be impaired beyond environmental quality standards from CoCs associated with the site. The sampling point for this was agreed at 44 m from the site boundary.

Remediation strategy – *ex situ* bioremediation by windrows

Because of the considerable amount of contaminated soil and the presence of diesel range hydrocarbons as the primary contaminant, it was decided to avoid unnecessary landfill and follow the more sustainable option of *ex situ* bioremediation using windrows (see Figure 20.11 and the following figure). In the past, the primary strategy for such sites was to excavate and landfill the contaminated soil. Increasingly, however, bioremediation offers a cost-effective and sustainable alternative to the more traditional 'dig and dump' approach (Alexander, 1999; Atlas and Philp, 2005).

At the case study site, excavated, contaminated material was very considerable and bioremediated through inoculation and intensive management of the hydrocarbon-degrading soil microbial community in windrows such as the one visible behind the excavation.

Windrowing of hydrocarbon-contaminated soil is, along with biopiling, an *ex situ* bioremediation technique that relies on the action of micro-organisms to break down organic compounds (Semple *et al.*, 2001; Khan *et al.*, 2004; Li *et al.*, 2004). The methods involve the excavation and piling of contaminated soils into piles, usually to a height of 2–4 m, in order to enhance aerobic microbial activity through aeration (see following figure), the addition of nutrients and the control of moisture and pH (Jørgensen *et al.*, 2000; Khan *et al.*, 2004). If aeration is forced, then biopile is the correct term. If aeration is simply by turning with passive diffusion of oxygen, then windrow is the appropriate term.

The lined windrows required intensive management such as covering during heavy rainfall to not only prevent slumping, but also to ensure a soil matric potential and aeration/oxygen diffusion commensurate with optimal hydrocarbon degradation activity.

Windrows and biopiles have been effectively used to remediate a wide range of contaminants such as petroleum hydrocarbons, pesticides, PAHs and sewage sludge (Semple *et al.*, 2001; Thassitou and Arvanitoyannis, 2001; Khan *et al.*, 2004).

Non-engineered windrows/biopiles rely mainly on wind-induced pressure gradients (i.e. natural airflows) that are non-uniform and particularly weak in the centre part of the pile, which may lead to a local O_2 deficiency (Eweis *et al.*, 1998; Li *et al.*, 2004). Engineered windrows and biopiles are often covered and lined with waterproof plastic to control water infiltration, run-off and volatilisation as well as to enhance solar heating (Fahnestock *et al.*, 1998; Khan *et al.*, 2004). An impermeable membrane or clay layer may be used to reduce the risk of pollutant leaching into uncontaminated soil.

The distribution of soil characteristics such as texture, permeability, water content and bulk density/porosity which are critical to the activity of the hydrocarbon-degrading microbial population are often non-uniform, and therefore turning the contaminated soil may be required to promote optimal biodegradation conditions (Khan *et al.*, 2004).

Windrows and biopiles containing organic matter (e.g. wood waste, sewage sludge and food waste) are usually referred to under 'composting' (Vidali, 2001). In this type of biopile/windrow, the degradation of organic matter results in an increase in the biopile temperature leading to changes in the microbial community structure during the course of bioremediation (Semple *et al.*, 2001; Thassitou and Arvanitoyannis, 2001). The degradation process in the composting method is initiated by mesophilic bacteria which are active at temperatures between 30 and 45 °C (Thassitou and Arvanitoyannis, 2001). However, the increase in the biodegradation rate results in heat production leading to temperature increases of up to 65 °C (Semple *et al.*, 2001; Thassitou and Arvanitoyannis, 2001). The increase in the pile temperature results in an increase in thermophilic bacteria and a decline in the mesophilic microbial population (Semple *et al.*, 2001; Thassitou and Arvanitoyannis, 2001).

The advantages of windrow/biopile strategies include their cost efficiency (Semple *et al.*, 2001); they are easy to design and they can be also managed on site (Khan *et al.*, 2004). In addition, the area and time required for biopile treatments are less than those required for land farming. Furthermore, vapour emissions can be controlled using a closed system, and they can be designed to fit a range of products and site conditions (Khan *et al.*, 2004). The limitations include space requirements and the evaporation of volatile compounds which often require treatment prior to discharge to the atmosphere (Mueller *et al.*, 1996; Khan *et al.*, 2004).

Optimising bioremediation

On the study site, a simple relationship was used which, coupled to optimal management of water, pH, nutrients and aeration, enabled optimisation of the hydrocarbon degradation by the indigenous soil microbial community (i.e. biostimulation). The *Bioremediation factor* that has to be optimised is a function of [TPH], 'availability', 'degraders' and 'constraints'. Here TPH is the total petroleum hydrocarbon concentration in the soil, 'availability' refers to the fraction of these hydrocarbons that is bioavailable for bioremediation, 'degraders' refers to the population density of hydrocarbon degraders in the soil measured by most probable number techniques, and 'constraints' refers to the limitation imposed on the microbial community by the chemical toxicity of the contaminated soil environment. Availability and constraints were both quantified by using specialised biosensors to assess hydrocarbon bioavailability and toxicity (Killham and Paton, 2003).

Using this relationship to predict the likely success of bioremediation, the windrows of contaminated soil were carefully managed and the hydrocarbon contamination degraded down to compliance concentrations over 6–8 weeks. The windrows were monitored during this period to ensure high rates of hydrocarbon degrader activity in the sandy soils under remediation. This was simply achieved by measuring oxygen uptake and carbon dioxide evolution by the soil, as well as monitoring reduction in hydrocarbon concentration (see following figure). When the O_2 consumption/CO_2 evolution rates decreased markedly, the windrows were turned mechanically to introduce air and oxygenate the microbial degraders throughout the windrow.

Meeting the risk-based remediation targets

On completion of the adopted, *ex situ* bioremediation strategy, the local council regulator required a detailed verification report to confirm that, across the site, the quantitative risk derived criteria had been met. For this site, an approval letter was received within two weeks of submission of the verification report. The letter enabled the developer of the site to gain approval from the National House-Building Council (NHBC) to build the planned housing scheme.

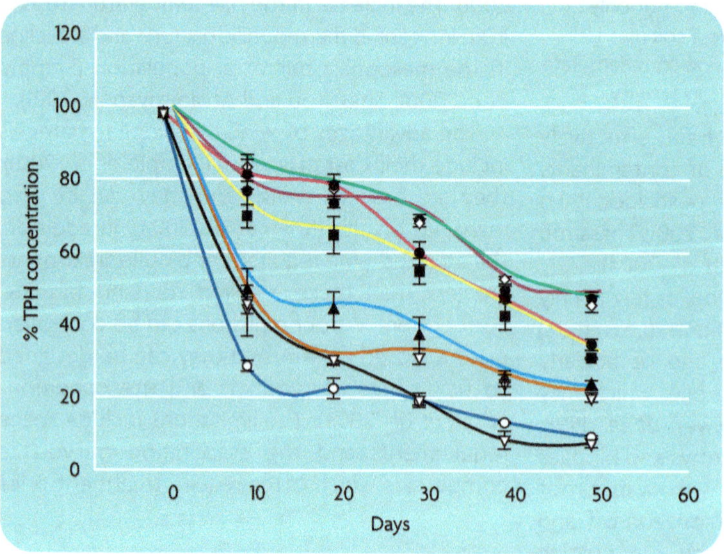

Monitoring of hydrocarbon degradation rates (as a percentage of original hydrocarbon concentration) across the different windrows of the site demonstrates rapid bioremediation. Each symbol represents a different windrow on the site and the data for each point are the means of five observations. TPH refers to Total Petroleum Hydrocarbons.

POLICY IMPLICATIONS

Risk assessment and remediation

- Policy makers need to be aware of how risks associated with any contaminated soil or water resources under their jurisdiction may be, and should be, assessed.
- It is important for policy makers to know how to assess change in risk associated with proposed change(s) in land use.
- Policy makers need to know how to assess the success of remediation prior to allowing change in land use to proceed, and therefore what monitoring data to demand.
- It is important for policy makers to understand the relative merits of diverse remediation approaches.
- Planners need to be aware of risks that may occur during the remediation process.
- It is important to consider the potential consequences of extreme climatic and other potential extreme events during the remediation process when making risk assessments.

CHAPTER REVIEW EXERCISES

Exercise 20.1

Why are the rates of hydrocarbon degradation associated with windrowing and biopiling generally more rapid than those with landfarming?

Exercise 20.2

In order for significant harm to occur, a 'source–pathway–receptor' linkage must be established. Explain how this linkage is used to define contaminated land and subsequently during remediation.

Exercise 20.3

Why is phytoremediation advantageous for treatment of sites contaminated with CoCs such as radionuclides and explosives?

Exercise 20.4

Explain how risk assessment would be used to drive the remediation of environmental contamination using a case study with which you are familiar in your local area.

REFERENCES

Alexander, M. (1999) *Biodegradation and Bioremediation*. Academic Press, New York.

Antizar-Ladislao, B., Lopez-Real, J. and Beck, A.J. (2004) Bioremediation of polycyclic aromatic hydrocarbons (PAHs) contaminated soil using composting approaches. *Critical Reviews in Environmental Science and Technology*, **34**, 249–289.

Antizar-Ladislao, B. and Russell, N.J. (2009) In-vessel composting as a sustainable bioremediation technology of contaminated soils and waste. In: *Composting: Processing, Materials and Approaches* (Chapter 8) (eds Perreira, J.C. and Bolin J.L.), Nova Science Publishers Inc., New York.

Atlas, R.M. and Philp, J.C. (eds) (2005) *Bioremediation: Applied Microbial Solutions for Real-World Environmental Cleanup*. American Society for Microbiology Press, Washington D.C.

Doty, S.L. (2008) Tansey Review: Enhancing phytoremediation through the use of transgenics and endophytes. *New Phytologist*, **179**, 318–333.

Environment Agency (2000) *The State of the Environment of England and Wales: The Land*. HMSO, London.

Environmental Protection Act (Part IIA), UK (1990) HMSO (1990).

Eweis, J.F., Ergas, S.J., Chang, D.P. and Schroeder, E.D. (1998) *Bioremediation Principles*. McGraw Hill series in Water Resources and Environmental Engineering, International Edition, Malaysia.

Fahnestock, F.M., Wickramanayake, G.B., Kratzke, R.J. and Major, W.R. (1998) *Biopile Design, Operation, and Maintenance Handbook for Treating Hydrocarbon-Contaminated Soils*. Battelle Press, USA.

Gerhardt, K.E., Huang, X., Glick, B.R. and Greenberg, B.M. (2009) Phytoremediation and rhizoremediation of organic soil contaminants: Potential and challenges. *Plant Science*, **176**, 20–30.

Gunther, T., Dornberger, U. and Fritsche, W. (1996) Effects of ryegrass on biodegradation of hydrocarbons in soil. *Chemosphere*, **33**, 203–215.

Harvey, P.J., Campanella, B.F., Castro, P.M.L., Harms, H., Lichtfouse, E., Schaffner, A.R., Smrcek, S. and Werck-Reichharts, D. (2002) Phytoremediation of polyaromatic hydrocarbons, anilines and phenols. *Environmental Science and Pollution Research*, **9**, 29–47.

Joner, E.J. and Leyval, C. (2003) Phytoremediation of organic pollutants using mycorrhizal plants: a new aspect of rhizosphere interactions. *Agronomie*, **23**, 495–502.

Jørgensen, K.S., Puustinen, J. and Suortti, A.M. (2000) Bioremediation of petroleum hydrocarbon-contaminated soil by composting in biopiles. *Environmental Pollution*, **107**, 245–254.

Khan, F.I., Husain, T. and Hejazi, R. (2004) An overview and analysis of site remediation technologies. *Journal of Environmental Management*, **71**, 95–122.

Killham, K. and Paton, G.I. (2003) Intelligent site assessment: a role for ecotoxicology. In: *Bioremediation: A Critical Review* (eds Singleton, I., Milner, M.G. and Head, I.M.), Horizon Press, London.

Krupka, K.M. and Martin, W.J. (2001) *Subsurface Contaminant Focus Area: Monitored Natural Attenuation (MNA) – Programmatic, Technical, and Regulatory Issues*. USDE Publication PNNL-13569.

Kuyukina, M.S., Ivshina, I.B., Ritchova, M.I., Cunningham, C.J., Philp, J.C. and Christofi, N. (2003) Bioremediation of crude oil contaminated soil using slurry-phase biological treatment and landfarming techniques. *Soil and Sediment Contamination*, **12**, 85–99.

Kuiper, I., Lagendijk, E.L., Bloemberg, G.V. and Lugtenberg, B.J.J. (2004) Rhizoremediation: a beneficial plant-microbe interaction. *Molecular Plant–Microbe Interactions*, **17**, 6–15.

Li, L., Cunningham, C.J., Pas, V., Philp, J.C., Barry, D.A. and Anderson, P. (2004) Field trial of a new aeration system for enhancing biodegradation in a biopile. *Waste Management*, **24**, 127–137.

Macek, T., Uhlik, O., Jecna, K., Novakova, M., Lovecka, P., Rezek, J., Dudkova, V., Stursa, P., Vrchotova, B. and Meharg, A.A. (2003) The mechanistic basis of interactions between mycorrhizal associations and toxic metal cations. *Mycological Research*, **107**, 1253–1265.

Macek, T., Uhlik, O., Jecna, K., Novakova, M., Lovecka, P., Rezek, J., Dudkova, V., Stursa, P., Vrchotova, B. and Pavlikova, D. (2009) Advances in phytoremediation and rhizoremediation. *Soil Biology*, **17**, 257–277.

Meharg, A.A. (2003) The mechanistic basis of interactions between mycorrhizal associations and toxic metal cations. *Mycological Research*, **107**, 1253–1265.

Milner, M.J. and Kochian, L.V. (2008) Investigating heavy-metal hyperaccumulation using *Thlaspi caerulescens*. *Annals of Botany*, **102**, 3–13.

Morikawa, H. and Erkin, O.C. (2003) Basic processes in phytoremediation and some applications to air pollution control. *Chemosphere*, **52**, 1553–1558.

Mueller, J.G., Cerniglia, C.E. and Pritchard, P.H. (1996) Bioremediation of environments contaminated by polycyclic aromatic hydrocarbons. In: *Bioremediation: Principles and Applications* (eds Crawford, R.L. and Crawford, D.L.), Cambridge University Press, Cambridge, pp. 125–194.

Nathanail, C.P. and Bardos, P. (2004) *Reclamation of Contaminated Land*. Wiley, Chichester, pp. 238.

Pavlikova, D., Demnerova, K. and Mackova, M. (2009) Advances in phytoremediation and rhizoremediation. In: *Advances in Applied Bioremediation* (eds Singh, A. Kuhad, R.C. and Ward, O.P.), Springer, Berlin, pp. 257–277.

Piotrowski, M.R., Doyle, J.R. and Carraway, J.W. (2006) Integrated bioremediation of soil and groundwater at a superfund site. *Remediation Journal*, **2**, 293–309.

Royal Commission on Environmental Pollution (1996) *Sustainable Use of Soils*. HMSO, London.

Rudland, D.J. and Jackson, S.D. (2004) *Selection of Remedial Treatments for Contaminated Land. A Guide to Good Practice*. CIRIA Publication C622, pp. 90.

Scottish Executive (1998) *Scottish Vacant and Derelict Land Survey 1998*. ENV/1999/1, ScottishExecutive, Edinburgh, www.scotland.gov.uk/library2/doc05/lssb-00,htm.

Semple, K.T., Reid, B.J. and Fermor, T.R. (2001) Impact of composting strategies on the treatment of soils contaminated with organic pollutants: review. *Environmental Pollution*, **112**, 269–283.

Semple, K.T., Doick, K.J., Jones, K.C., Burauel, P., Craven, A. and Harms, H. (2004) Defining bioavailability and bioaccessibility of contaminated soil and sediment is complicated – different interpretations create more than a semantic stumbling block. *Environmental Science and Technology*, **38**, 228A–231A.

Silva, E., Fialho, A.M., Sá-Correia, I., Burns, R.G. and Shaw, E.J. (2004) Combined bioaugmentation and biostimulation to clean up soil contaminated with high concentrations of atrazine. *Environmental Science and Technology*, **38**, 632–637.

Striegel, J., Sanders, D.A. and Veenstra, J.N. (2001) Treatment of contaminated groundwater using permeable reactive barriers. *Environmental Geosciences*, **8**, 258–265.

Thassitou, P.K. and Arvanitoyannis, I.S. (2001) Bioremediation: A novel approach to food waste management. *Trends in Food Science and Technology*, **12**, 185–196.

Top, E.M., Springael, D. and Boon, N. (2002) Catabolic mobile genetic elements and their potential use in bioaugmentation of polluted soils and waters. *FEMS Microbiology Ecology*, **42**, 199–208.

USEPA (2005) *Cost-Effective Design of Pump and Treat Systems*. OSWER 9283.1-20FS, EPA Publication 542-R-05-008.

Vanek, T., Podlipna, R. and Soudek, D.P. (2010) General Factors Influencing Application of Phytotechnology Techniques. *NATO Science for Peace and Security Series C: Environmental Security*, pp. 1–13.

Vidali, M. (2001) Bioremediation. An overview. *Pure and Applied Chemistry*, **73**, 1163–1172.

Vogel, T.M. (1996) Bioaugmentation as a soil bioremediation approach. *Current Opinion in Biotechnology*, **7(3)**, 311–316.

Widada, J., Nojiri, H. and Omori, T. (2002) Recent developments in molecular techniques for identification and monitoring of Xenobiotic-degrading bacteria and their catabolic genes in bioremediation. *Applied Microbiology and Biotechnology*, **60**, 45–59.

Zuzana, S., Katarína, D. and Lívia, T. (2009) Biodegradation and ecotoxicity of soil contaminated by pentachlorophenol applying bioaugmentation and addition of sorbents. *World Journal of Microbiology and Biotechnology*, **25**, 243–252.

CHAPTER 21

Pollution swapping

Keith Goulding

Learning outcomes

By the end of this chapter you should:

- Be more aware of the fact that solutions to one environmental pollution problem at a site may create different pollution problems at the same site.
- Be more aware of the need to consider the transfer of pollution problems between soil, aqueous and atmospheric phases when trying to find solutions to specific pollution problems.
- Understand the very limited value of finding solutions for one site that simply transfer the same or a different problem to an alternative site.
- Understand better the importance of valuation of ecosystem services when evaluating, and attempting to solve, a specific pollution problem.
- Be better able to explain the importance of interdisciplinary approaches to solving pollution problems.

21.1 Introduction

The environmental focus that agricultural policy has had over the last 20 years or so has been driven mostly by public concerns over matters such as the impacts of agriculture on water quality and biodiversity. It has resulted in much legislation in the UK and Europe, such as the EU Nitrate Limit, Integrated Pollution Prevention and Control to reduce ammonia emissions to air, and various environment schemes such as the Entry Level (ELS) and Higher Level (HLS) Schemes designed to reverse, or at least halt, the decline in biodiversity, especially for farmland birds. But the search for solutions to single, seemingly simple policy problems risks overlooking 'knock-on' effects and the possibility of 'pollution swapping', solving one problem but causing another.

We saw an example of concentrating too much on a single determinant in Chapter 16. There it was pointed out that when high loads of atmospheric pollutant nitrogen are deposited in remote forest areas the soil C:N ratio does not necessarily fall as might be expected. This is the case when the soil carbon content is increasing much faster than its nitrogen content. What we did not address there was the series of questions that then arise: Is the accumulation of acidic litter a problem? Will it result in more acidic drainage water passing to local streams? Will it increase the dissolved organic matter concentration in local streams? Will it adversely affect uptake of plant nutrients, especially those mediated by mycorrhizae? Will it change biodiversity of ground flora? These problems, however potentially serious they may be, are simply examples of pollution effects. They are not examples of pollution swapping.

Pollution swapping refers to the creation of a new pollution problem as a direct consequence of attempting to *solve* a pollution problem that already exists. The present chapter looks at some specific examples. We will focus on nitrogen and then greenhouse gases (GHGs), and then move on to 'problem swapping' and the wider aspects of developing sustainable agricultural systems, in which biodiversity and the economic and social aspects of sustainability are considered. We will see that too strong a focus on only a part of a system is unlikely to develop a truly sustainable solution and that finding optimal, sustainable agricultural systems is not easy.

21.2 Pollution swapping – losses of nitrogen to air and water

Figure 21.1 shows a simplified nitrogen cycle for farmed land. The cycle is 'leaky', even in natural

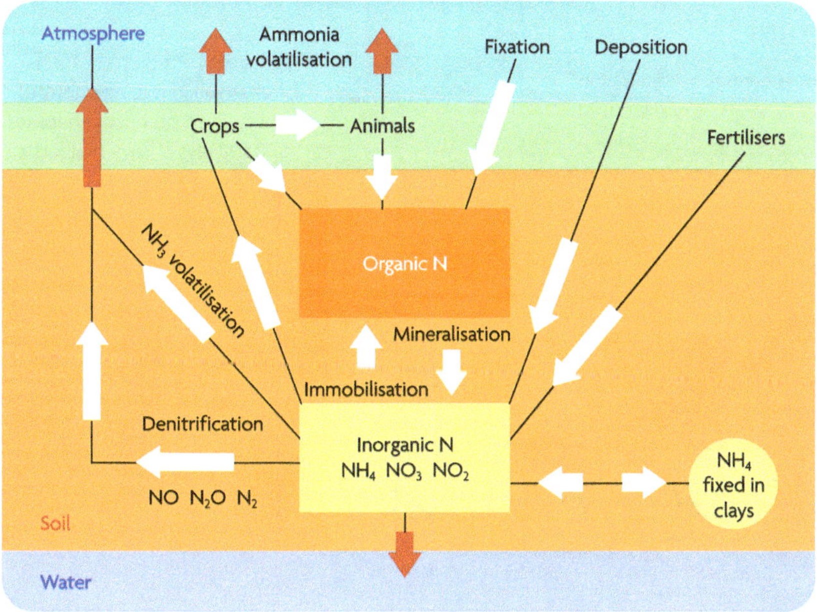

Figure 21.1 A simplified nitrogen cycle for farmed land.

ecosystems, as we saw in Chapter 5: small amounts of nitrate are leached with percolating rainwater from land into surface and groundwaters to sustain, at small concentrations, aquatic life, and small amounts of nitrate are denitrified (essentially reduced to nitrous oxide, N_2O, and dinitrogen, N_2) by ubiquitous microorganisms in soils and fresh water and ocean sediments. Indeed, denitrification of nitrate to N_2 closes the cycle, returning fixed nitrogen to its inert state in the atmosphere (Boyer *et al.*, 2006 and see Chapter 5). It is assumed that denitrification to N_2 and modest mineral and organic N leaching balanced fixation by algae, lichens, legumes and lightning before man perturbed the cycle by planting more legumes and developing the Hãber–Bosch process (Galloway *et al.*, 2004). In natural ecosystems, leaching of N in dissolved organic matter may be significant (see section 5.3.3).

Adding nitrogen to enhance agricultural production, whether via legumes or fertiliser or manures, and enhanced atmospheric deposition from polluted air to natural ecosystems, perturbs all parts of the cycle, thus increasing emissions back to air and water (see, for example, the Nitrogen Cascade on the International Nitrogen Initiative website: http://www.initrogen.org/). Figure 21.1 shows the complexity of the relationships between pools of nitrogen and transfer processes and why simply trying to block one route of loss is only likely to result in an increase in the transfer of nitrogen in another direction. For example, injecting animal slurry into soil reduces ammonia emissions but risks increases in nitrate leaching and the loss of N_2O through denitrification (Thompson *et al.*, 1987).

The first, and possibly the simplest, example of pollution swapping concerns nitrate in water. Public anxiety over increasing concentrations of nitrate in water (not necessarily supported by science, see L'hirondel and L'hirondel, 2001) has led to the EU Nitrate Limit and policy responses such as Nitrate Vulnerable Zones. Much research effort has been spent on trying to find management systems that reduce nitrate leaching: Defra and MAFF, for example, funded a multi-million pound nitrate research programme throughout the 1990s.

Figure 21.2 shows the results of field experiments and computer modelling that looked at the impacts of a number of farm management systems on losses of nitrogen to air and water from a dairy farm (Jarvis, 2000). Total denitrification was mostly as N_2, but a significant amount of N_2O would have been emitted as well. It shows that total losses of nitrogen from an intensive dairy system can be reduced by almost two-thirds. However, no loss can be completely prevented and the most effective system for reducing total losses – a clover-based system – resulted in almost no reductions in ammonia volatilisation. Policy makers need to

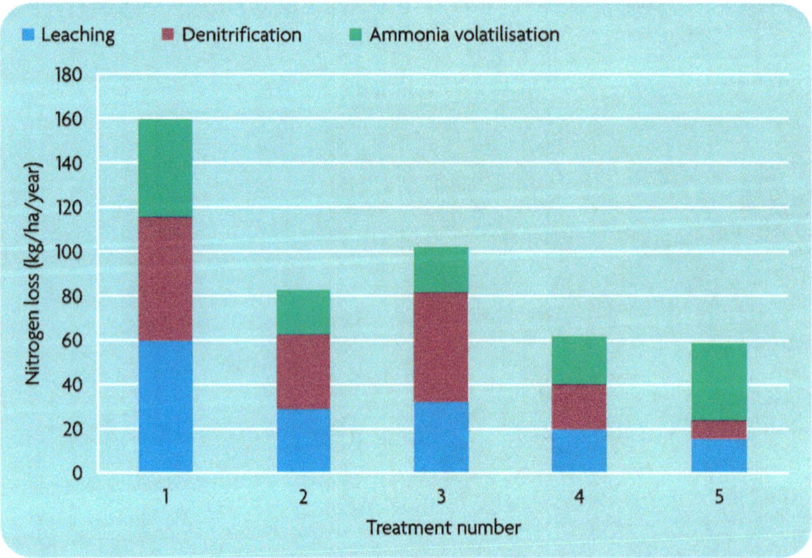

Figure 21.2 Predicted effects of changes in management of a dairy farm on nitrogen losses via pathways shown (IGER data, adapted from Jarvis, 2000). Treatments were: (1) A case study. (2) Tactical fertiliser use with slurry injection. (3) Maize silage. (4) Tactical fertiliser use with slurry injection and maize. (5) No fertiliser, only clover.

Source: Based on data from Jarvis (2000).

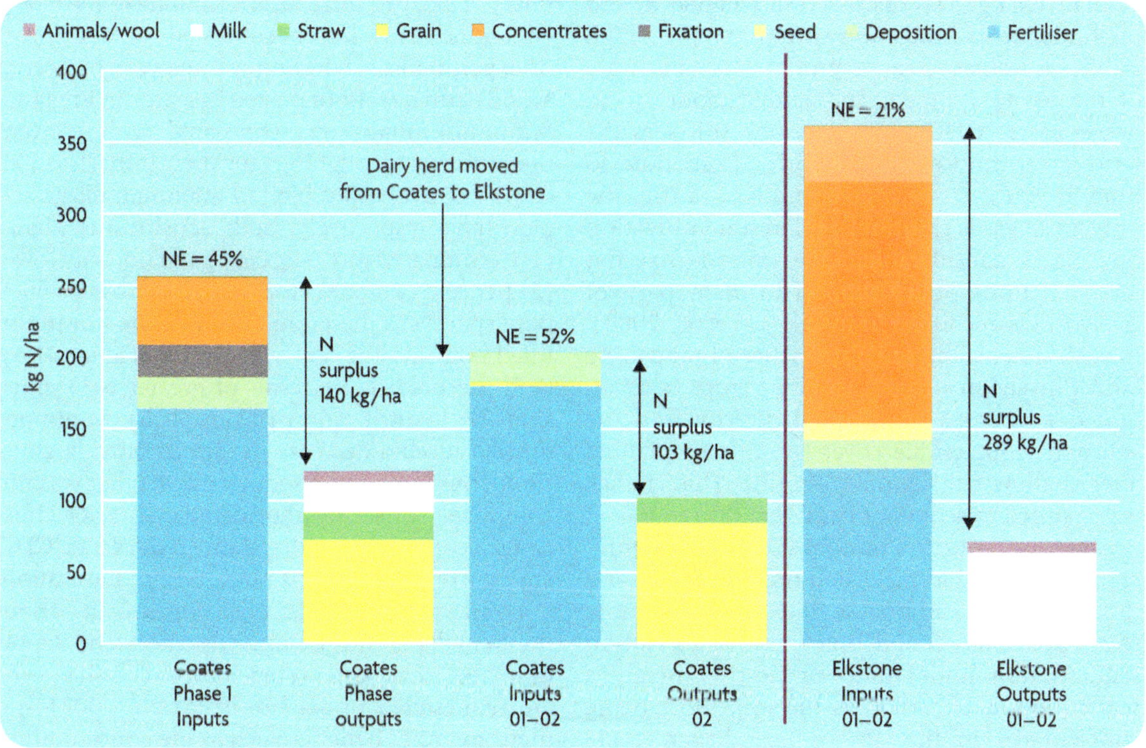

Figure 21.3 Nitrogen budgets for Coates Farm before the dairy herd was moved to the neighbouring Royal Agricultural College farm at Elkstone, and the effect on the Elkstone budget.
Source: Based on Leach *et al.* (2004).

realise that all farm systems will have some impact on the environment and cause some pollution.

Pollution swapping does not only mean swapping between pollutants but includes swapping the location of the source or impact of the pollution. From 1995 to 2003 the Ministry of Agriculture, Fisheries and Food (MAFF), as it was then, funded a farm system study at the Royal Agricultural College's Coates Farm near Cirencester in Gloucestershire, UK. The research made a full and balanced nitrogen budget for the farm and identified the main loss processes (Leach *et al.*, 2004). Part way through the study the College moved the dairy herd from Coates to a neighbouring farm at Elkstone. Figure 21.3 shows the change in the nitrogen budgets at Coates before and after the herd moved and the final budget at Elkstone (unfortunately we do not have a pre-move budget; as is often the case with economically-driven changes in management, this happened quickly and with little forewarning for the research team). The nitrogen surplus (the excess of all inputs less all outputs in saleable produce) at Coates was reduced by almost 140 kg N ha^{-1} to around 100 kg N ha^{-1} but the surplus at Elkstone was almost 290 kg N ha^{-1}. Pollution was reduced at Coates Farm but only by moving it elsewhere. This is a local scale example of the pollution swapping caused by importing food and thereby exporting agricultural production and pollution, a concept developed further in Chapter 22.

21.3 Pollution swapping – climate change

Lead-in question

Can you think of ways in which steps taken to ameliorate, or slow, climate change might bring about pollution swapping problems?

Climate change was discussed in Chapter 8, and methods for its mitigation currently have a high public, as well as research, profile. Most are aware of the Kyoto Protocol and know about greenhouse gases (GHGs, although one still sees the term 'green house gases' in print from time to time; if you ever see a green house gas run; the only green gas is chlorine!). Interest in mitigation options is considerable. One option that has received strong policy support in many parts of the world is 'no-till' or 'min-till' (Lal *et al.*, 2004). In such systems the age-old practice of ploughing land ('inversion tillage') to plant crops is abandoned and crops are planted directly into the soil and crop residue cover (no-till) or planted after shallow cultivation (min-till). This change was claimed to sequester (lock-up) carbon in soil because the reduction in tillage reduced the mineralisation (oxidation) of soil organic matter and thus produced carbon dioxide. However, more recent work has found that there is no carbon sequestration under min- or no-till, merely a redistribution of carbon in the soil, more being concentrated in the surface (e.g. Baker *et al.*, 2007); even past protagonists have changed their view (Christopher *et al.*, 2009).

In addition, Smith *et al.* (2001) suggested that, whatever its impact on carbon dioxide fluxes, reduced tillage is likely to increase N_2O and methane (CH_4) emissions. This has been tested in a number of experiments, including a three-year project at Wood Farm near Rothamsted Research in England. Minimum tillage was compared with conventional tillage (ploughing) for their impacts on all GHG fluxes. In soil subject to minimum tillage and ploughing, respectively, bulk density was higher (1.13 compared to 1.04 g cm^{-3}) and porosity lower (51 per cent compared to 55 per cent). Water-filled pore space was therefore higher in the minimum tilled soil, i.e. the soil was wetter (Figure 21.4), with the predictable result of greater N_2O emissions by denitrification (Figure 21.5). Minimum tillage caused a decrease in carbon dioxide emissions (but not carbon sequestration) that was not completely offset by the increase in N_2O emissions, so minimum tillage did reduce net GHG emissions, but there was no carbon sequestration.

Some reviews have been optimistic about the benefits of no- or min-till; the Stern Report estimates that, on average, 0.14 t C ha^{-1} yr^{-1} are sequestered under no-till (Stern, 2007) and there are other benefits, such as the concentration of organic matter near the soil surface, which is good for soil structure, seedling emergence, and water infiltration and retention (Powlson and Jenkinson, 1981; Baker *et al.*, 2007). However,

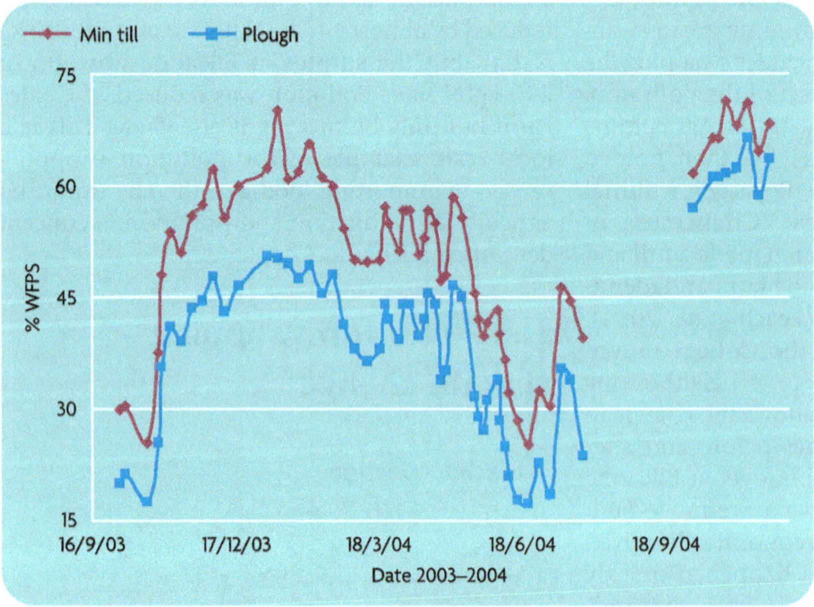

Figure 21.4 The percentage of water filled pore space (per cent WFPS), i.e. the wetness of soil, in adjacent fields at Wood Farm subjected to ploughing or minimum tillage.

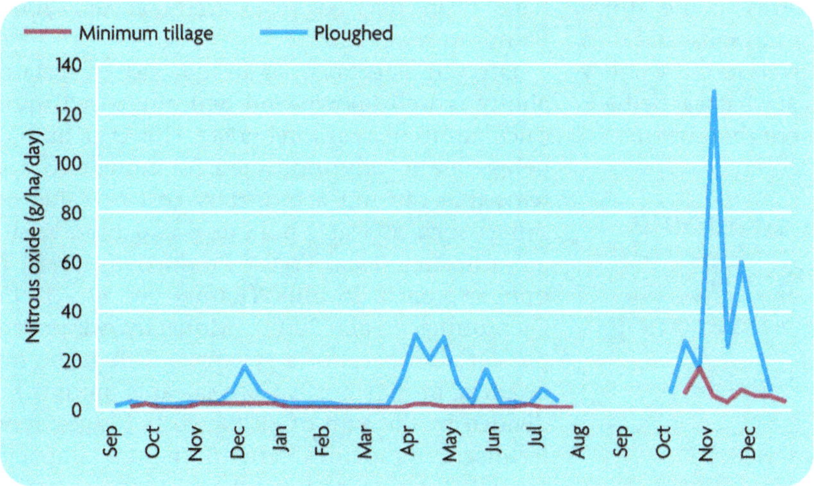

Figure 21.5 Emissions of nitrogen as nitrous oxide (N_2O-N), mostly from denitrification, from soil in adjacent fields at Wood Farm that were either ploughed or minimum tilled.

Johnson *et al.* (2007) calculated that an extra 3 kg N_2O emitted ha^{-1} yr^{-1} could offset carbon sequestration of 0.3 t C ha^{-1} yr^{-1} (an extra 4 kg N_2O ha^{-1} yr^{-1} was measured at Wood Farm) and Almaraz *et al.* (2009) found that increased N_2O emissions completely offset decreased CO_2 emissions in a comparison of conventional and minimum tillage experiment. It is clear then that in any work to develop policies to mitigate climate change the full Global Warming Potential (GWP, i.e. the effect of the change on all GHGs) of any change must be assessed.

21.4 Pollution swapping between air, soils and water

As noted already, government policies tend to be narrowly focused. For some time, the European Union's Nitrate Directive has been a strong driver for much of the UK Government's agricultural policies, such as the Nitrate Vulnerable Zones (NVZs) that cover much of the UK. Such strong drivers and well-developed policies can have a major impact on the pollutant in question (e.g. Nimmo Smith *et al.*, 2007). Unexpected and unwanted knock-on effects can be severe, however. The Silsoe Whole Farm Model simulates the interactions that occur on farms between such unchangeable things as soil type and climate and manageable aspects such as labour, machinery, crops grown and inputs. It models outputs such as profit and also environmental impacts except for biodiversity (Williams *et al.*, 2006a, b).

The model was used to assess the impact on a particular farm of policies that sought to decrease nitrate leaching (Figure 21.6). In summary, the results suggest that the farmer would change the farm system such that nitrate leaching to fresh waters was greatly decreased at minimal cost, but to a new system that greatly increased nitrogen losses by denitrification, and thus probably

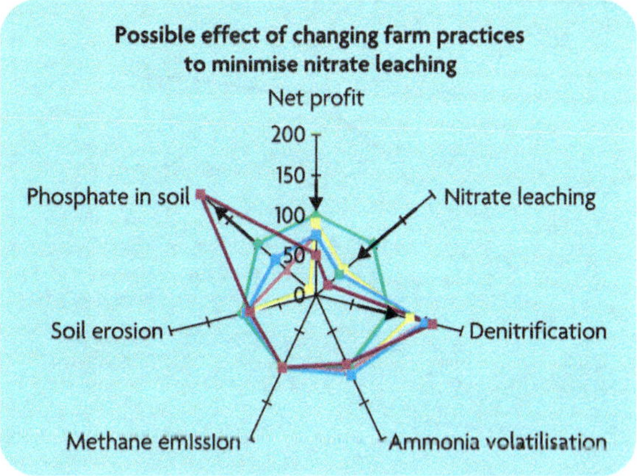

Figure 21.6 Simulations from the Silsoe Whole Farm Model of the impact of policies that caused a farmer to change management practices to reduce nitrate leaching but at a financial cost and with the result of increased phosphorus load in soil and increased nitrous oxide emissions to air from denitrification.

releasing nitrous oxide emissions to the atmosphere, and the phosphorus load in soils, and thus probably phosphate losses to waters too. The policy objective was therefore met but at what was perhaps a greater cost to the environment than before.

21.5 From pollution swapping to problem swapping and identifying sustainable farming systems – Total Factor Productivity

The principal indicator of sustainability on most farms is the profit they make; only a few of the farms that are sometimes called 'hobby farms' can afford to run at a loss. However, the broader aspects of sustainability require an assessment of more than profit. Farmers are expected to produce food but to avoid polluting air and water and reducing biodiversity. Ways in which environmental impact and biodiversity can be included in assessments of sustainability have been the subject of debate for many years now, and various methods of assessing the sustainability of systems have been devised (e.g. Pretty *et al.*, 2000; Balmford *et al.*, 2002).

Lifecycle assessment as an indicator of sustainability is well known and typically based upon calculations of a carbon budget. However, human beings are greatly motivated by money, so systems that can put a monetary cost or value on things such as water pollution and biodiversity are attractive. Total Factor Productivity (TFP) is one approach to this (Barnett *et al.*, 1995; Glendining *et al.*, 2009). Monetary values are put on all aspects of a system and then the net financial profit (mostly from saleable produce) is divided by the sum of all the costs (wages, rent, fertilisers, pesticides, environmental impacts, etc.); if the value is greater than 1 then the system can be said to be sustainable.

Figure 21.7 shows TFP applied to winter wheat production in the UK in 2006. The left hand plot shows yield increasing with the costs of production (effectively with inputs, i.e. increased intensity) but that TFP reaches a maximum of approximately 1.1 (i.e. it is sustainable) at a cost per hectare of around £840. This is because, as the right-hand side of the figure shows, more intensive systems have greatly increased GHG emissions and thus environmental costs, but more extensive systems

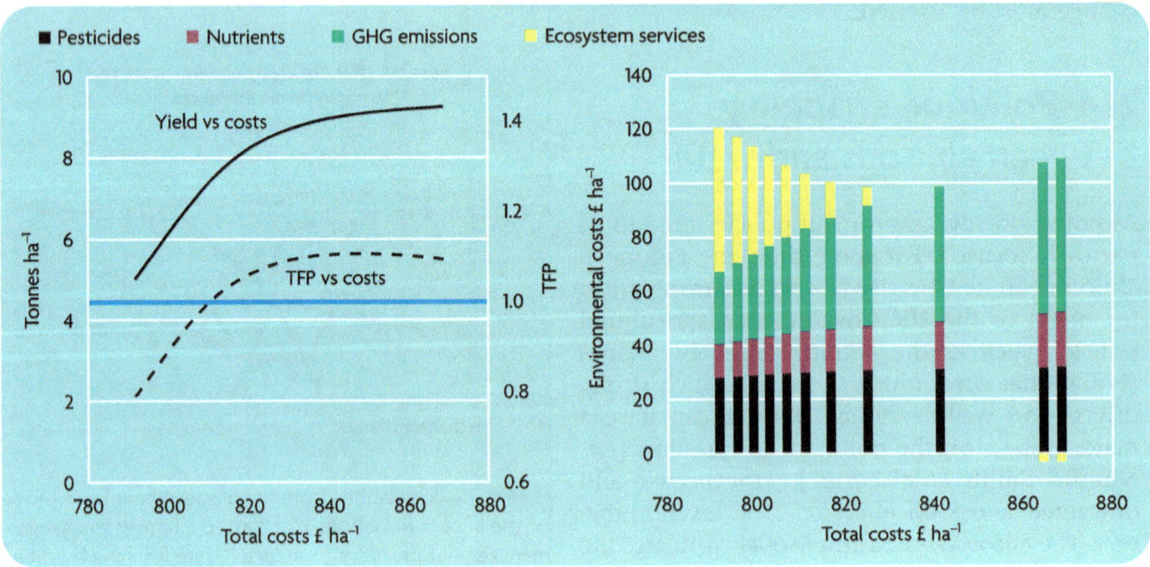

Figure 21.7 Total Factor Productivity analysis of winter wheat production in 2006.
Source: Glendining *et al.* (2009).

require more land and therefore cost more because of lost Ecosystem Services (biodiversity, water storage, landscape, etc.) (Costanza *et al.*, 1997). This shows the dilemma that arises in complex systems in which air and water quality have to be valued alongside biodiversity, landscape and human health and welfare.

21.6 Problem swapping on a real farm

The conscientious reader will have realised that we have moved from discussing pollutants to discussing problems, or rather from the environment to the three pillars of sustainability: economic, environmental and sociopolitical. But some say that there are four or even five pillars of sustainability, adding culture and ethics/belief.

The Allerton Trust's Loddington Farm is managed according to the principles of integrated farming as set out by LEAF (Linking Environment and Farming, http://www.leafuk.org/leafuk/). The farm is managed as a complex mosaic of crops, woodland and field margins (Figure 21.8) (Boatman and Brockless, 1998). It is profitable and has been very successful at increasing biodiversity, but at a 'cost'. Songbird numbers more than doubled between 1992 and 2001 for example (Figure 21.9) but this was at least partly due to

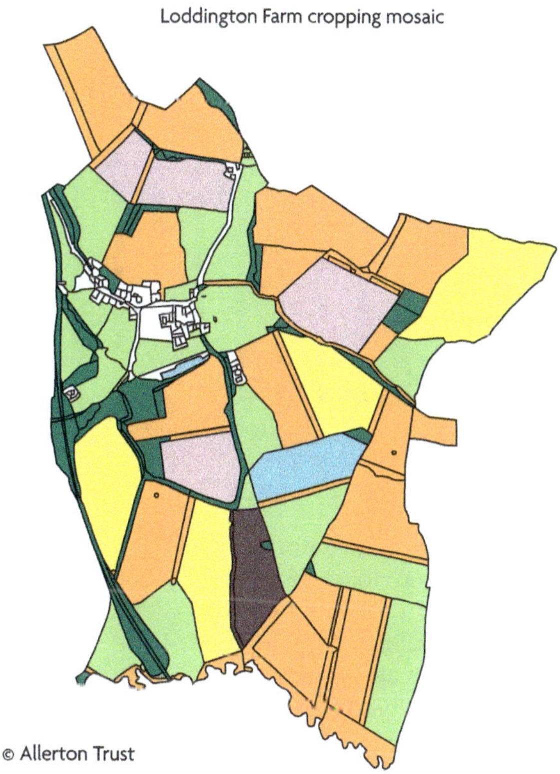

Figure 21.8 The mosaic of crops grown at the Allerton Trust's Loddington Farm in Leicestershire, UK in 2005.
Source: The Allerton Trust.

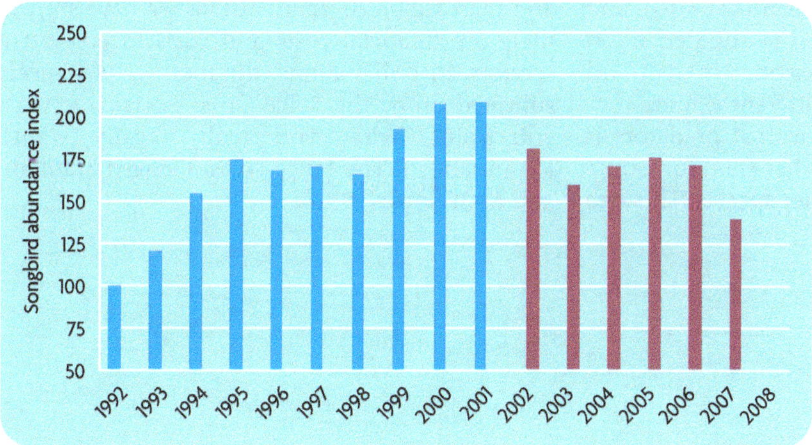

Figure 21.9 Songbird counts at Loddington Farm showing the increase relative to number in 1992 when predator control was initially started in 1993 (blue bars) and the decline after 2001 when control was stopped (red bars).
Source: Based with permission of Alastair Leake on data from the Game & Wildlife Conservation Trust website at: http://www.gwct.org.uk/research__surveys/the_allerton_project/gamebird__songbird_counts/default.asp

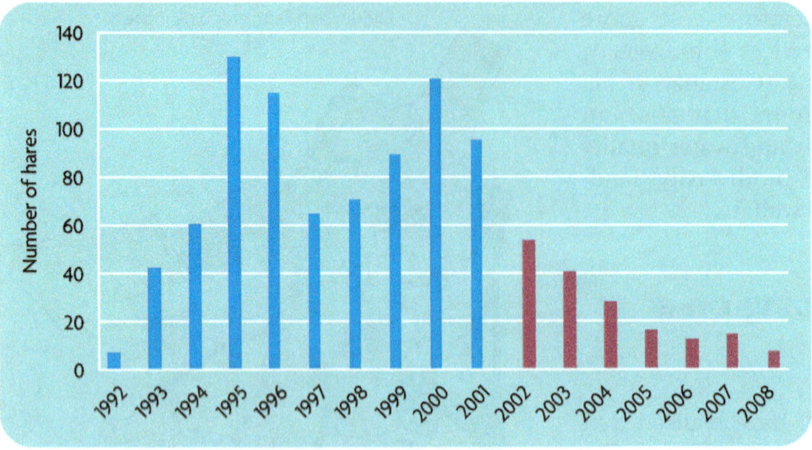

Figure 21.10 Change in the numbers of brown hares at Loddington Farm between 1992 and 2008. Note the increase when predator control was introduced in 1993 (blue bars) and the decline after it was stopped in 2001 (red bars).
Source: Based with permission of Alistair Leake on data from the Game & Wildlife Conservation Trust.

the control of predators – corvids (crows and magpies) and foxes (Sage *et al.*, 2005). When predator control stopped after 2001 songbird numbers declined. The effect of predator control is even more apparent in Figure 21.10 where the change over the years in the numbers of brown hares is shown.

The results in Figure 21.9 need quite careful interpretation, however, as the Game and Wildlife Conservation Trust website explains. Some bird species, such as the song thrush and the spotted flycatcher increased markedly over the early years of the project but declined rapidly once predator control was stopped. The populations of linnets and bullfinches showed a similar but less dramatic trend; however, the small number of reed bunting grew in the later years. Grain in hoppers that were provided initially was not maintained in the two years after predator control ceased, and this probably also contributed to the decline in songbird numbers over this period. Different bird species use feed hoppers to different extents.

For many people, the control of predators is unacceptable, whatever the benefits for songbirds, brown hares or any other species. This was shown very clearly in the lively debate in the correspondence pages of *The Times* in April 2009 (article: 'Charities in dispute over culling Magpies', 18 April 2009; letters: 'Songbirds and culling Magpies', 20 and 22 April 2009).

21.7 Conclusions

In this chapter we have tried to show the complexity of interacting processes in the landscape and the difficulty of solving one problem without causing another; pollution swapping is only part of the problem. There are not likely to be any ideal solutions that are economically viable and satisfy all the very different wishes of the public; more extensive (as opposed to more intensive) farm systems will reduce environmental impact and increase biodiversity but, as we have shown, there are financial costs and costs to ecosystem services. In policy terms, the message was nicely summed up by the Spice Girls: 'So tell me what you want, what you really, really want!' ('Wannabe'; © The Spice Girls, Richard Stannard and Matt Rowe, 1996).

POLICY IMPLICATIONS

Pollution swapping

- Policy makers need to be aware of what pollution swapping means.
- Policy makers need to assess carefully when formulating pollution control policies whether any new pollution problem may be caused or exacerbated when the policy is implemented.
- When formulating and implementing pollution abatement policies, policy makers need to be aware of, and carefully consider, risks of moving the pollution problem to another site.
- Policy makers need to possess and integrated overview of how ecosystem components interact and to be aware of the risks of abatement policies that transfer pollution problems from soils to fresh waters and/or the atmosphere.
- Policy makers need to be aware of the risks in compartmentalising pollution problems. For example, they also should consider potential climate change impacts via greenhouse gas emissions when implementing policies to control nitrate leaching.

CHAPTER REVIEW EXERCISES

Exercise 21.1

Critically discuss the pollution swapping implications of the change to disposal of treated sewage sludge on agricultural and forest land rather than dumping it at sea as tended to happen more in the past.

Exercise 21.2

Discuss the potential pollution swapping problems associated with liming upland organic matter-rich soils to combat the effects of acid rain.

Exercise 21.3

To what extent do you think that the often slow timescales of environmental recovery following implementation of policies to manage pollution impacts might increase the risks of pollution swapping occurring?

REFERENCES

Almaraz, J.J., Mabood, F., Zhou, X., Madramootoo, C., Rochette, P., Ma, B-L. and Smith, D.L. (2009) Carbon dioxide and nitrous oxide fluxes in corn grown under two tillage systems in south-western Quebec. *Soil Science Society of America Journal*, **73**, 113–119.

Baker, J.M., Ochsner, T.E., Venterea, R.T. and Griffis, T.J. (2007) Tillage and soil carbon sequestration – what do we really know? *Agriculture, Ecosystems and Environment*, **118**, 1–5.

Balmford, A., Bruner, A., Cooper, P., Costanza, R., Farber, S., Green, R.E., Jenkins, M., Jefferiss, P., Jessamy, V., Madden, J., Munro, K., Myers, N., Naeem, S., Paavola, J., Rayment, M., Rosendo, S., Roughgarden, J., Trumper, K. and Turner, R.K. (2002) Economic reasons for conserving wild nature. *Science*, **297**, 950–953.

Barnett, V., Johnston, A.E., Landau, S., Payne, R.W., Welham, S.J. and Rayner, A.I. (1995) Sustainability – the Rothamsted experience. In: V. Barnett, R. Payne and R. Steiner (eds) *Agricultural Sustainability: Economic, Environmental and Statistical Considerations*. John Wiley, Chichester, pp. 171–206.

Boatman, N.D. and Brockless, M.H. (1998) The Allerton Project: farmland management for partridges (*Perdix perdix, Alectoris rufa*) and pheasants (*Phasianus colchicus*). In: *Perdix VII: Proceedings of the VIIth International Symposium on Partridges, Quails and Pheasants* (eds M. Birkan, L.M. Smith, N.J. Aebischer, F.J. Purroy and P.A. Robertson). Gibier Faune Sauvage, **15**. Office National de la Chasse, Paris, 563–574.

Boyer, E.W., Alexander, R.B., Parton, W.J., Li, C., Butterbach-Bahl, K., Donner, S.D., Skaggs, W. and Del Grosso, S.J. (2006) Modeling denitrification in terrestrial and aquatic ecosystems at regional scales. *Ecological Applications*, **16**, 2123–2142.

Christopher, S.F., Lal, R. and Mishra, U. (2009) Regional study of no-till effects on carbon sequestration in the Midwestern United States. *Soil Scence Society of America Journal*, **73**, 207–216.

Costanza, R., D'Arge, R., de Groot, R., Farber, S., Grasso, M., Hannon, B., Limbuirg, K., Naeem, A., O'Neill, R.V., Paruelo, J., Raskin, R.G., Sutton, P. and van den Belt, M. (1997) The value of the world's ecosystem services and natural capital. *Nature*, **387**, 253–260.

Galloway, J.N., Dentener, F.J., Capone, D.G., Boyer, E.W., Howarth, R.W., Seitzinger, S.P., Asner, G.P., Cleveland, C.C., Green, P.A., Holland, E.A., Karl, D.M., Michaels, A.F., Porter, J.H., Townsend, A.R. and Smarty, C.J.V.R. (2004) Nitrogen cycles: past, present, and future. *Biogeochemistry*, **70**, 153–226.

Glendining, M.J., Dailey, A.G., Williams, A.G., van Evert, F.K., Goulding, K.W.T and Whitmore, A.P. (2009) Is it possible to increase the sustainability of arable and ruminant agriculture by reducing inputs? *Agricultural Systems*, **99**, 117–125.

Jarvis, S.C. (2000) Progress in studies of nitrate leaching from grassland soils. *Soil Use and Management*, **16**, 152–156.

Johnson, J.M-F., Franzluebbers, A.J., Weyers, S.L. and Reicosky, D.C. (2007) Agricultural opportunities to mitigate greenhouse gas emissions. *Environmental Pollution*, **150**, 107–124.

Lal, R., Griffin, M., Apt, J., Lave, L. and Morgan, M.G. (2004) Managing soil carbon. *Science*, **304**, 393.

Leach, K.A., Allingham, K.D., Conway, J.S., Goulding, K.W.T. and Hatch, D.J. (2004) Nitrogen management for profitable farming with minimal environmental impact: The challenge for mixed farms in the Cotswold Hills, England. *International Journal of Agricultural Sustainability*, **2**, 21–32.

L'hirondel, J. and L'hirondel, J.-L. (2001) *Nitrate and man. Toxic, harmless or beneficial?* CABI, Wallingford.

Nimmo Smith, R.J., Glegg, G.A., Parkinson, R. and Richards, J.P. (2007) Evaluating the implementation of the Nitrates Directive in Denmark and England using an actor-orientated approach. *European Environment*, **17**, 124–144.

Powlson, D.S. (2000) Tackling nitrate from agriculture. *Soil Use and Management*, **16**, 141–141.

Powlson, D.S. and Jenkinson, D.S. (1981) A comparison of the organic matter, biomass, adenosine triphosphate and mineralizable nitrogen contents of ploughed and direct-drilled soils. *Journal of Agricultural Science*, **97**, 713–721.

Pretty, J.N., Brett, C., Gee, D., Hine, R.E., Mason, C.F., Morison, J.I.L., Raven, H., Rayment, D. and van der Bijl, G. (2000) An assessment of the total external costs of UK agriculture. *Agricultural Systems*, **65**, 113–136.

Sage, R.B., Parish, D.M.B., Thompson, P.G.L. and Woodburn, M.I.A. (2005) Songbirds in game crops. In: *Proceedings of the XXVth International Congress of the International Union of Game Biologists – IUGB and the*

IXth International Symposium Perdix, Vol. 2 (ed. E. Hadjisterkotis). Ministry of the Interior, Nicosia, 173–181.

Smith, P., Goulding, K.W.T., Smith, K.A., Powlson, D.S., Smith, J.U., Falloon, P. and Coleman, K. (2001) Enhancing the carbon sink in European agricultural soils: including trace gas fluxes in estimates of carbon mitigation potential. *Nutrient Cycling in Agroecosystems*, **60**, 237–252.

Stern, N. (2007) *Stern Review on the Economics of Climate Change*. HM Treasury, London: http://www.hm-treasury.gov.uk/sternreview_index.htm.

Thompson, R.B., Ryden, J.C. and Lockyer, D.R. (1987) Fate of nitrogen in cattle slurry following surface application or injection to grassland. *Journal of Soil Science*, **38**, 689–700.

Williams, A.G., Audsley, E. and Sandars, D.L. (2006a) *Determining the environmental burdens and resource use in the production of agricultural and horticultural commodities*. Main Report. Defra Research Project IS0205. Bedford: Cranfield University and Defra. Available on www.silsoe.cranfield.ac.uk and www.defra.gov.uk.

Williams, A.G., Audsley, E. and Sandars, D.L. (2006b) Energy and environmental burdens of organic and non-organic agriculture and horticulture. In: *What will organic farming deliver? COR 2006* (eds Atkinson, C. *et al.*) Association of Applied Biologists, Wellesbourne, Warwick CV35 9EF, UK, pp. 19–24.

CHAPTER 22

The trouble with man is . . . or 'what have you damaged today?'

Elena Dawkins and Anne Owen

Learning outcomes

By the end of this chapter you should:

- Understand the difference between territorial and consumption based environmental impact assessment and emissions accounting.
- Appreciate how impacts from consumption may be calculated.
- Be aware of the indicators used for consumption-based environmental impact assessment.
- Consider how and why environmental impacts from consumption may vary between countries and within countries.
- Appreciate what could be meant by sustainable lifestyles.
- Understand the policy implications of using a consumption-based approach.
- Be able to review policies to reduce the footprint at a national, local and individual level.

22.1 Introduction

Up until now in this book we have considered the pressures that various natural processes and human activities put on the environment and looked at how we rely on the natural environment for the provision of resources and for the absorption of wastes. In the UK and many other countries numerous environmental pressures are well documented and understood and there are legislative systems in place that attempt to deal with any perceived negative environmental impacts. The Clean Air Act of 1956, was brought in by the UK government in response to London's Great Smog of 1952. The Act, often cited as the first environmental law, legislated for zones where smokeless fuels had to be burnt (Guissani, 1994).

In many countries of the world environmental regulations are not very well developed, leading to detrimental effects on the local environment. Just as regulation is lacking in some countries of the world, international legislation for some potentially very damaging global pollutants such as greenhouse gases, is also absent. A lack of legislation to deal with harmful pollutants, when combined with the ever increasing demands for natural resources, is having detrimental effects on the environment at both local and global scales. These present day problems may be exacerbated in the future as the global population and affluence grow (Peters, 2008).

Many of the Earth's resources, and its capacity to absorb waste, are finite, meaning that once environmental tipping points are reached it is impossible to recover resources or reverse any negative effects. This is a global problem; we have one planet with limited environmental capacity and growing pressure on its resources. While we can focus on our local environment and ensure that our domestic resources are used sustainably, we must recognise that with global trade we use goods and resources from all over the world. It is important that we take a holistic view of our activities and any local or global environmental implications they may have (Rockstrom *et al.*, 2009).

If we look at everything required to provide a good or service, whether it is lighting and heating of an office or an entire manufacturing chain with the extraction of raw materials, processing, transportation and then selling of a product, it is clear that most goods we buy will have some environmental impact.

22.1.1 What do we mean by impact?

In the early 1970s, Ehrlich and Holden (1971) devised a simple equation to describe environmental impact. Impacts (I) were expressed as a product of population (P), affluence (A) and technology (T).

$$I = PAT$$

Impact can be taken to mean resource depletion or waste accumulation; population is a count of human population, affluence is the level of consumption of the defined population and technology refers to the process of obtaining resources and turning them into goods and waste.

The IPAT equation was initially used to highlight the contribution of a growing population on the environment. Later commentators have used the equation to demonstrate that environmental impact is a function of more than pollution; multiple drivers act together and produce a compounding effect. We will consider the role and influence of each driver with reference to greenhouse gas emissions later in this chapter.

Lead-in question

When goods are produced, who should be responsible for the environmental impact of their production?

22.1.2 Taking responsibility

This brings us to an important question: who should be responsible for resource use and the environmental impacts of producing goods and services? Traditionally, the responsibility is attributed to people who extract the resources or the people who process them and add value for sale. The responsibility tends to fall at the point of production where the impact occurs. The people who consume the good or service have little responsibility for, or knowledge of, the environmental impacts that have occurred in their products' manufacturing processes. If goods and services are produced as a result of consumer demand then should the consumers take some of the responsibility for any environmental impacts?

By extension, if goods are produced in other countries then exported for consumption within the UK, should the UK take some of the responsibility for environmental impacts of production in other countries? Or is this the sole responsibility of the country where the good (or service) is produced? The two cartograms below (Figures 22.1 and 22.2) help to illustrate these questions. Cartograms are maps where the land area is distorted and morphed to illustrate some other mapped variable. The cartograms below have their land areas morphed according to the size of their import and exports of toys. It is clear to see that Asia is the greatest exporter of toys and Europe and the USA are the greatest importers.

Such a large export industry in Asia must have some environmental impacts, whether this is on local air or water pollution, emissions of global pollutants or degradation of land. Some of these impacts may be well regulated and controlled in their country of origin, but others may not be. Would it be right to attribute all responsibility of the environmental impacts to where the goods are produced? It is ultimately consumer demand that drives the production of goods, so should the countries that consume the goods take on some responsibility for the environmental impacts of the production of those goods?

Figure 22.1 Export of toys. Territory size shows the proportion of worldwide net exports of toys (in US$) that come from the country shown. Net exports are exports minus imports. When imports are larger than exports the territory is not shown.

Source: © Copyright SASI Group (University of Sheffield) and Mark Newman (University of Michigan).

Figure 22.2 Imports of toys. Territory size shows the proportion of worldwide net imports of toys (in US$) that come to the country shown. Net imports are imports minus exports. When exports are larger than imports the territory is not shown.

Source: © Copyright SASI Group (University of Sheffield) and Mark Newman (University of Michigan).

Lead-in question

How can we measure what the impacts of consumption are?

22.2 Understanding and measuring the impacts of consumption

If we recognise that it is our consumption or demand that drives production and any associated environmental degradation, we must next think how we could measure this and then how we can address it. First, let's consider measurement. Somehow we need to link the environmental consequences of production to the consumption of the final good. In the UK we have a relatively good understanding of both; we measure the pollutants emitted during production processes and collect information on what people are buying. However, this information is collected separately and for different purposes. The link between them is the supply chain. By looking at supply chains we know what is required to produce each good.

If, at every stage of the supply chain, we allocate any impacts to the purchaser, rather than the producer of a good, all of the impacts can be accumulated and allocated to the point of final demand. Take the greenhouses gas (GHGs) emissions that could be associated with purchasing a carton of yoghurt as an example. GHGs emitted at the first stage of production (from the dairy farming) are allocated to those who buy the products (the milk), rather than the producers. These buyers could be another industrial sector; a yoghurt manufacturing company. At this point we need to consider environmental consequences of packaging too. If these goods are then bought by others (supermarkets) the impact can be allocated to those buyers and so on (intermediate demand). Further emissions from transportation, processing and refrigeration are added to the total impact. Eventually a good or service is bought by an end consumer (final demand) and all of the cumulative supply chain impacts of that good or service are allocated to them.

There are two main methods for allocating impacts to the point of final demand: Process Life Cycle Analysis (PLCA) and Environmentally Extended Input-Output Analysis (IO). LCA can be described as a 'bottom up' approach that starts from a single product and attempts to measure the impacts generated at each stage of the product's supply chain. To measure the impacts associated with a consumer's total consumption of all goods and services using this PLCA method would be an impossible challenge. There are far too many different products to examine and maintaining a consistent approach across all of them would be extremely difficult. Calculating the impacts embedded in the consumption of goods and services of a whole country or region requires a 'top down' approach provided by Environmentally Extended Input–Output (EEIO) methods (Miller and Blair, 2009). This is a macro-economic modelling technique that combines an economic modelling framework with data from environmental accounts (see Figure 22.3). The economic component describes the links between different sectors of the economy and hence the supply chains of products produced. Added to this are data about the total environmental impacts associated with the sectors of the economy. The model in this initial state describes the environmental impact of sectors, and how they link to other sectors, in terms of what they buy and sell to each other. In order to view this data from the perspective of the products produced at the end, rather than the sector inputs and environmental impacts at the beginning, the model is turned on its head. This process was first developed by Leontief (1970) in the 1930s and has been used ever since to answer the question of how a change in demand for the product might change the resulting impacts from all of the sectors involved in its manufacturing, along the whole supply chain of that product.

EEIO techniques can provide an estimate of 'T' for the IPAT equation; a measure of environmental impact per unit spend or an impact conversion factor where resource intensive products will have a higher factor than others. It can also provide an estimate of the 'A' component – affluence – in

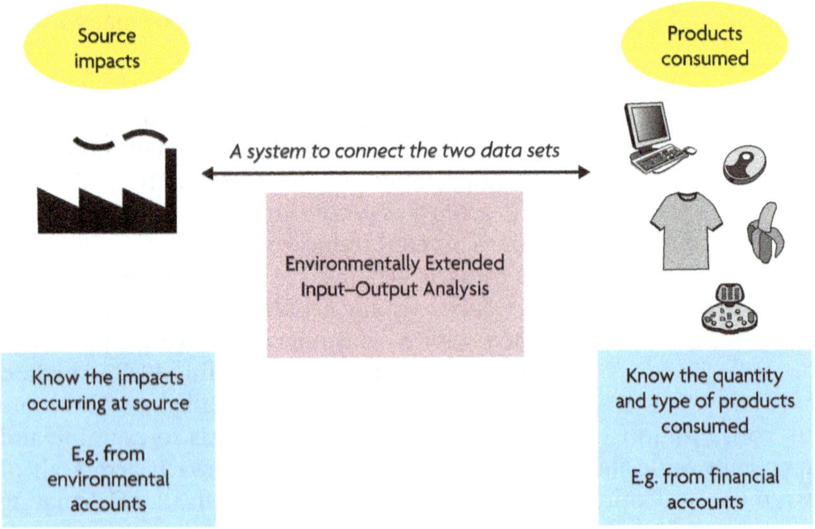

Figure 22.3 From production to consumption – how emissions at source relate to impacts from products.

terms of total consumption (or demand) across different sectors of the economy, from householders for example, or government.

22.2.1 Issues with the consumption approach

Supply chains and international trade are growing in complexity and goods and services are being consumed further and further away from where they are produced. Consequently any environmental impacts of production are becoming increasingly distanced from where goods are consumed.

Further to this, local environmental impacts can vary considerably. One manufacturing process may have serious implications for water use in a local area; in another country the same process may have no water implications, but may have adverse effects on the soil quality. For many goods, the importers or consumers will have to rely on national governments to prevent local environmental damage in the producing countries. Consumers can make purchasing choices based on environmental standards and there are a number of labelling schemes to help with this, such as the Forest Stewardship Council (FSC) certification for wood products. These standards exist for some products, but not all, and consequently it remains difficult for consumers to make informed choices about the full environmental impacts of an individual good or service.

Some specific environmental impacts, such as greenhouse gas or carbon emissions, are easier to track than local environmental impacts, because they can be measured consistently and have the same global impact wherever they originate. They still require consistent data and a way to attribute this to the consumer, but there are a growing number of studies and methodologies that have started to look at attributing these types of impacts to the consumer. The methodologies vary, but in principal they have the same aim of attributing supply chain impacts to the point of final demand (consumption) rather than the point of production.

This chapter explores a number of indicators that we can use to measure the environmental impact of our consumption; how we can quantify it, how it varies and how we can address the issues that arise from increasing consumption now and in the future.

22.3 Environmental indicators of consumption

In general, an indicator that allocates all of the environmental impacts of producing a good or service to a consumer is called a footprint. The footprint methodology can be used to assign resource use, measured in hectares of land or litres of water, or emissions of pollutants, such as carbon dioxide and other harmful greenhouse gases to the consumption of goods and services.

The carbon footprint has become a popular indicator when discussing issues of climate change and has become widely used. We will look at the carbon footprint in more detail later. First, we will look at the original footprint indicator, the 'ecological footprint', which attempts to estimate resource use as a whole and attribute it to the goods and services that we consume.

22.4 The ecological footprint

The concept of the *ecological footprint* was first discussed by Wackernagel and Rees in the mid-1990s. Their book *Our Ecological Footprint, Reducing Human Impact on the Earth* begins by reminding people of their dependence on nature and the resources of the world. This is the essence of the ecological footprint concept – looking at what the world can provide and comparing this to what humans demand (Wackernagel and Rees, 1996).

The most powerful message that arises from the concept of ecological footprinting is that we are currently using up more of the world's resources than we have available. We are placing demands on nature to provide us with food, materials, energy and waste absorption at a faster rate than they can be provided or renewed.

You may wonder whether this is actually true. Are we really using up our natural resources faster than they can be renewed? Surely the planet can cope or is coping with our demands, can't it? But you will also be aware of many of the environmental problems that we face today; the depletion on fisheries (Chapter 15), the extinction of species (Chapters 6 and 25), loss of biodiversity (Chapter 25), increases in greenhouse gases (Chapter 8), growth of waste, land/soil degradation (Chapter 19), desertification…When you begin to consider all of these things together the possibility that we are using up resources faster than they can be renewed does not seem so far fetched.

It is important that we consider our use of natural resources, particularly if they are undervalued economically. Some services that the environment provides are undervalued or not given any monetary value at all. Part of the reason for this is that it is very difficult to value the environment. How do we place a value on the existence of trees absorbing carbon dioxide, or preventing soil degradation or flooding; or on biodiversity? Traditionally it is only the value of the wood, or the land that is counted. This has led to degradation or exploitation of many environmental resources that are not valued economically. By attempting to measure all of the natural resources required to meet our demands we can start to understand what is available and compare this to how much we use. The ecological footprint is a useful tool for doing this. It does not place a monetary value on environmental resources, but it does try to quantify demand in terms of bio-productivity required to meet that demand.

The ecological footprint for a particular population is defined as the 'total area of productive land and water ecosystems required to produce the resources that the population consumes and assimilate the wastes that production produces, wherever on Earth that land and water may be located' (Rees, 2000).

The ecological footprint is usually described in terms of global hectares (gha). This is a hectare of land with world average bio-productivity – a world-average ability to produce resources and absorb waste. An important point to note, and one often subject to confusion, is that a value in global hectares is not a measure of land area as such; it should be viewed instead as a measure of potential bio-productivity.

Since its original definition and calculation in the 1990s, the ecological footprint has come under much scrutiny and gone through a number of methodological transformations (Wiedmann and Barrett, 2010), both of which we will discuss in more detail later in the chapter. For now, we will take a look at the global ecological footprint, how it has changed over time and how ecological footprints vary across the world.

22.4.1 The global ecological footprint

In 2006, the global ecological footprint was calculated to be 17.1 billion global hectares (gha),

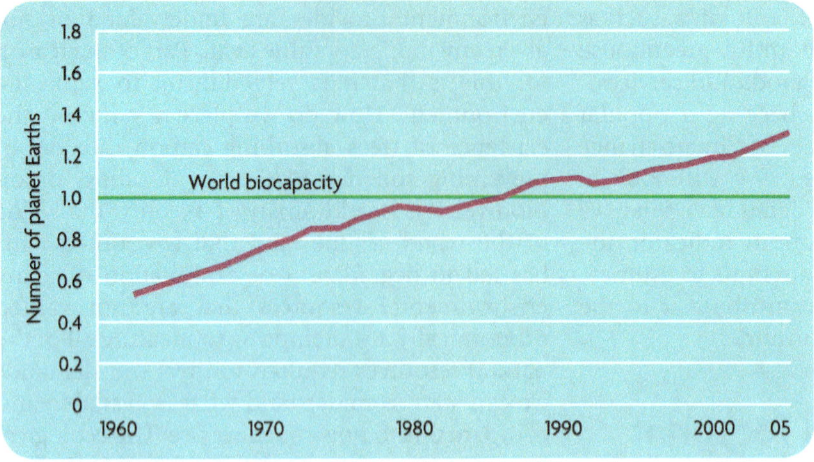

Figure 22.4 Change in humanity's ecological footprint from 1960–2005.
Source: © 2008 WWF.

or 2.6 gha per person. The total available supply of bio-productive land area was calculated to be 11.2 billion gha or 1.8 gha per person. We are using up the planet's resources faster than they can be replenished. In fact our footprint now overshoots the Earth's biocapacity by more than 40 per cent. What the global population uses in one year takes the planet 17 months to replenish (The Global Footprint Network, 2010).

This hasn't always been the case. Historically, pre-industrialisation and with a smaller global population, we used fewer resources and the share of bio-productive land would have been higher. Over the past 40 years alone bio-productivity is estimated to have dropped from 3.4 gha per person in 1961 to the 1.8 gha per person of 2006. As consumption increases as a result of development and population growth, and more resources are degraded or lost, this situation may worsen. Humanity's ecological footprint grew larger than global bio-capacity in the 1980s. Since then demand has continued to exceed supply (The Global Footprint Network, 2010).

Clearly this cannot go on forever – if we are using resources faster than they can be replenished eventually they may run out. Overuse is possible now because the Earth has ecological assets (forest, fisheries and fossil fuels). The message from the ecological footprint is that if consumption continues at this rate, or increases even further, we run the risk of depleting the planet's biological resources and interfering with its long-term ability to renew them (WWF, 2008).

This message is a very powerful one and as a result the ecological footprint has become a widely used tool to highlight the impacts of human activities on the planet. The concept is often praised as a very effective communication tool for describing total human resource use in a way that people can easily understand. The term 'One Planet Living'™,[1] for example, is a popularised term used to communicate the problem of resource overuse and living beyond our environmental means. The campaign for 'One Planet Living' emphasises the problem of our current levels of resource use. If everyone used resources at the same rate as a European we would need two-and-a-half planets to meet demand and if everyone lived like a North American we would need five planets (The Global Footprint Network, 2010). These statements are based on a comparison of the ecological footprints across the world. Figure 22.4 summarises how the overall situation is changing. The next section looks at this in more detail.

---

[1] © 2008 WWF International and BioRegional One Planet Living™ is a jointly owned trademark of BioRegional and WWF International. BioRegional Development Group ('BioRegional') is a registered charity no. 1041486 and company limited by guarantee registered in England and Wales no. 2973226 whose registered office is at the BedZED Centre, 24 Helios Road, Wallington, Surrey SM6 7BZ UK. WWF – World Wide Fund for Nature (formerly World Wildlife Fund) ('WWF International') is a foundation established under section 80 *et seq.* of the Swiss Civil Code whose registered office is at Avenue du Mont-Blanc, 1196 Gland, Switzerland.

22.4.2 Comparing the ecological footprint of different countries

The global average ecological footprint is 2.6 gha per person and the world capacity is 1.8 gha, but what is the range? National Footprint Accounts are produced for more than 120 nations and provided through the Global Footprint Network. The Global Footprint Network serves as the steward of the National Footprint Accounts, as the calculation system that measures the ecological resource use and resource capacity of nations over time. Based on approximately 4000 data points per country per year, the Accounts calculate the Footprints of over 120 countries from 1961 to the present. The highest reported ecological footprint in the 2006 accounts was 10.3 gha per person in the United Arab Emirates; the lowest was 0.5 gha per person in Haiti. The UK has the eighth highest per capita ecological footprint in the world at 6.1 gha per person. The average ecological footprint for Europe is 4.5 gha per person. In North America it is 8.7 gha per person and in Africa it is 1.4 gha per person (The Global Footprint Network, 2010).

When ecological footprints are compared at a national scale (Figure 22.5) there are a number of patterns that emerge. First, developed countries tend to have higher ecological footprints than less developed countries as their consumption levels are higher. Secondly, it is only the countries with a higher average income per person that have had large increases in their ecological footprints since 1961. Middle and low income countries have recorded very little, or no, increase in their ecological footprint over this time. Thirdly, when you compare a country's individual bio-capacity to their ecological footprint there are a number of countries which are using more bio-capacity than they control within their borders. Equally, there are some countries where capacity is greater than demand. There will always be areas of the world that are more productive than others and this is not necessarily where the largest populations are. If the majority of countries use more bio-capacity than within their borders, with some using considerably more, then we end up with the situation we have now; 2.6 gha per person being used but only 1.8 gha per person available.

These ecological footprint patterns are often linked to issues of development; the more developed a country is regarded in terms of economy or incomes, the higher their ecological footprint. These results tend to raise important and difficult questions about development and its consequences. There are countries such as Poland, Hungary and Japan that have lower ecological footprints of 3.3, 3.5 and 4.4 gha per person respectively, but would also be classed as developed or emerging economies, with relatively high standards of living. But let's not forget, all of these countries still far exceed the global capacity footprint of 1.8 gha per person (The Global Footprint Network, 2010).

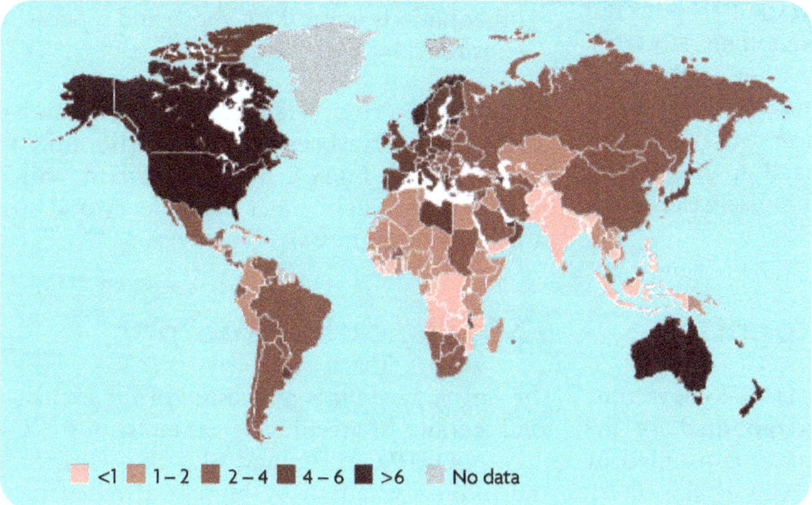

Figure 22.5 Ecological footprint by country, measured in global hectares per person.
Source: © 2008 WWF.

Lead-in question

Does the ecological footprint as a policy formulation tool tell us everything that we need to know?

22.4.3 Criticisms of the ecological footprint

The EF is limited in scope and should not be over-interpreted – there are some things it can not tell us. It does not account for our consumption of non-renewable (abiotic) resources or the damage that we cause to ecosystems through over extraction and pollution, so it does not show the full impact of our consumption. We would need to consider other complementary indicators to get the full picture of our ability to sustain natural resources to meet our needs indefinitely (Wiedmann and Barrett, 2010).

The EF is an aggregated indicator; it includes resource use and land use in one indicator. This can make it difficult to use when making detailed decisions about which resources are most over-exploited, which ecological limits are most threatened or which land should be protected. This can mean it has limited use in policy development.

The most significant message communicated by the ecological footprint is the concept of overshoot; we are using more resources than the earth can produce and producing more pollution than the earth can absorb. This is an extremely powerful message that can motivate stakeholders. However, the main cause of this overshoot is the land required to absorb greenhouse gas emissions from fossil fuel combustion. The concept of overshoot is primarily telling us that human activity is responsible for excessive greenhouse gases in the atmosphere.

22.5 The water footprint

The *water footprint* of a country is the total volume of fresh water consumed and polluted for the production of goods and services consumed by citizens in the country (Hoekstra *et al.*, 2009). Consumption is defined as water permanently removed from a water body in a catchment, which happens when water evaporates, returns to another catchment area or the sea or is incorporated into a product. The pollution element of the footprint is the water required to dilute pollution so it can be returned to the environment.

The water footprint can be broken down into different components as described below:

- The blue water footprint refers to consumption of blue water resources (surfacewater and groundwater) along the supply chain of a product.
- The green water footprint refers to consumption of green water resources (rainwater stored in the soil as soil moisture/groundwater).
- The grey water footprint refers to pollution and is defined as the volume of fresh water that is required to assimilate the load of pollutants based on existing ambient water quality standards.

The water footprint of a country includes the internal footprint and the external footprint:

- The internal footprint is the use of domestic water to produce goods and services consumed within the country (excluding water used to produce goods that are exported).
- The external footprint is the use of water in other nations to produce goods and services consumed by the country.

This gives us a complete picture of how national consumption translates to water use not only in the country of interest but also abroad, which allows us to analyse water dependency and sustainability of imports.

22.6 The carbon footprint

The *carbon footprint* is a measure of the exclusive total amount of greenhouse gas emissions (CO_2, CH_4, N_2O, HFC, PFC, and SF_6) that is directly and indirectly caused by human activities or is accumulated over the life stages of a product (Wiedmann and Minx, 2008). This includes activities of individuals, populations, governments, companies, organisations, processes, industry sectors etc. Products include goods and services. In any case, all direct (onsite, internal) and indirect emissions (offsite, external, embodied, upstream, downstream) need to be taken into account. The carbon footprint is expressed as a measure of CO_2e (carbon dioxide equivalence). Sometimes we consider emissions of carbon dioxide (CO_2) and omit the contributions of other greenhouse gases in a measure of carbon footprint.

22.6.1 A comparison of carbon measurement at the national level

At the national level total emissions are most often calculated from a territorial perspective. This is when any emissions produced within a country's territory are collated. This would include emissions from factories, power stations, transport, agriculture and householders' fuel use etc. within that country. The carbon footprint, as we have mentioned previously, requires an estimation of emissions embedded within products consumed, accumulated through the supply chains of those products. At the national level, this requires an estimate of the emissions embedded within products consumed regardless of where they are made – at home, or abroad. A method such as the Input–Output macro-economic framework described in section 22.2 can be used to estimate emissions embedded within products made domestically and imported, by incorporating trade data into the national accounts. This allows the calculation of a national carbon footprint, or the carbon emissions associated with the consumption of a population. Figure 22.6 explains the data that are used to construct the two types of

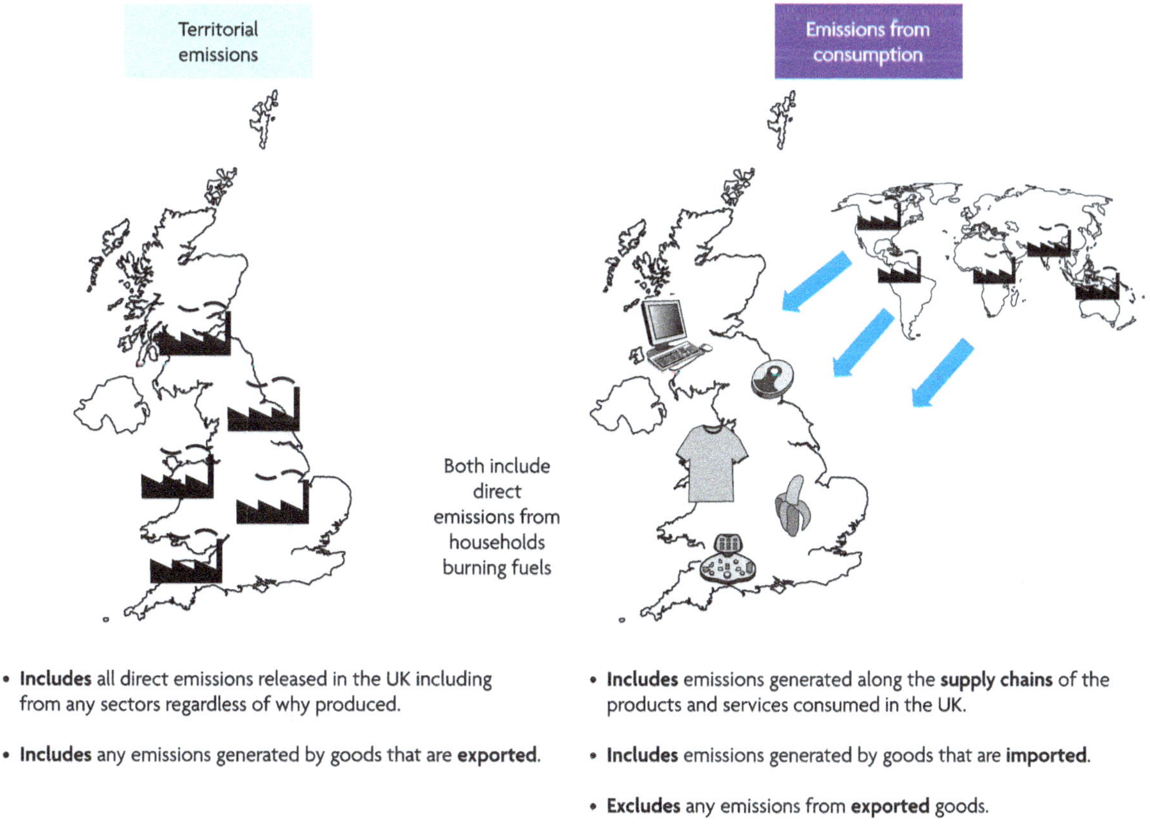

Territorial emissions
Emissions from consumption
Both include direct emissions from households burning fuels
• **Includes** all direct emissions released in the UK including from any sectors regardless of why produced.
• **Includes** any emissions generated by goods that are **exported**.
• **Includes** emissions generated along the **supply chains** of the products and services consumed in the UK.
• **Includes** emissions generated by goods that are **imported**.
• **Excludes** any emissions from **exported** goods.

Figure 22.6 Territorial emissions data and consumption emissions data.

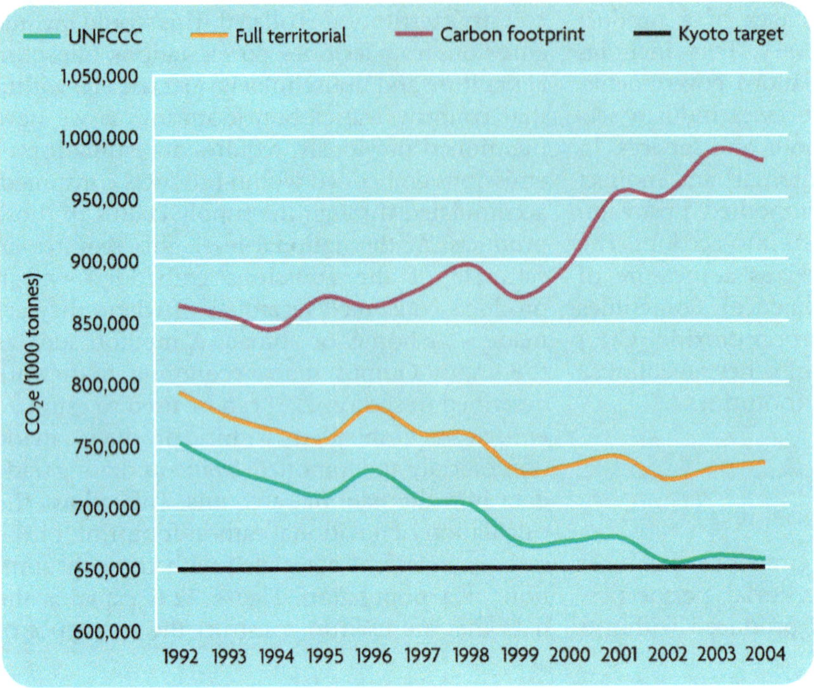

Figure 22.7 UK carbon emissions between 1992 and 2004 under differing measuring methods (Weideman *et al.*, 2008).

emissions accounts – territorial and consumption based. On a global scale, the total emissions from territories or by consumption would be the same; it is the allocation between countries that differs in the two methods.

22.6.2 Comparing the carbon footprint over time

Figure 22.7 shows the difference between the UK carbon emissions related to production and those related to consumption. The emissions that the UK reports to the UNFCCC and those that relate to the Kyoto target[2] include emissions from UK factories, businesses, heating and powering homes and domestic travel. If we include emissions from aviation and shipping – the full territorial impact – the UK is not as close to achieving the Kyoto target. Using the consumption approach, we see that the carbon footprint actually increases by around 13 per cent within the time period. It can be argued that much of the UK's perceived reduction in territorial emissions may come from shifting the production of our goods to factories abroad. But how can we fully understand why the UK's carbon footprint has increased? Is it a function of population growth, the emissions intensity of the products or the volume of the products bought? The IPAT equation described earlier can aid our understanding of changing footprints.

22.6.3 Carbon footprint and the IPAT equation

As discussed previously, the IPAT equation is a simple and useful method for understanding the drivers behind changes in environmental impact and the same approach can be used for analysing changes to the carbon footprint. Figure 22.8

[2] The Kyoto Protocol is an international agreement linked to the United Nations Framework Convention on Climate Change. The major feature of the Kyoto Protocol is that it sets binding targets for 37 industrialised countries and the European Community for reducing greenhouse gas (GHG) emissions. These amount to an average of 5 per cent against 1990 levels over the five-year period, 2008–2012. (Extract taken from http://unfccc.int/kyoto_protocol/items/2830.php, July 2010.)

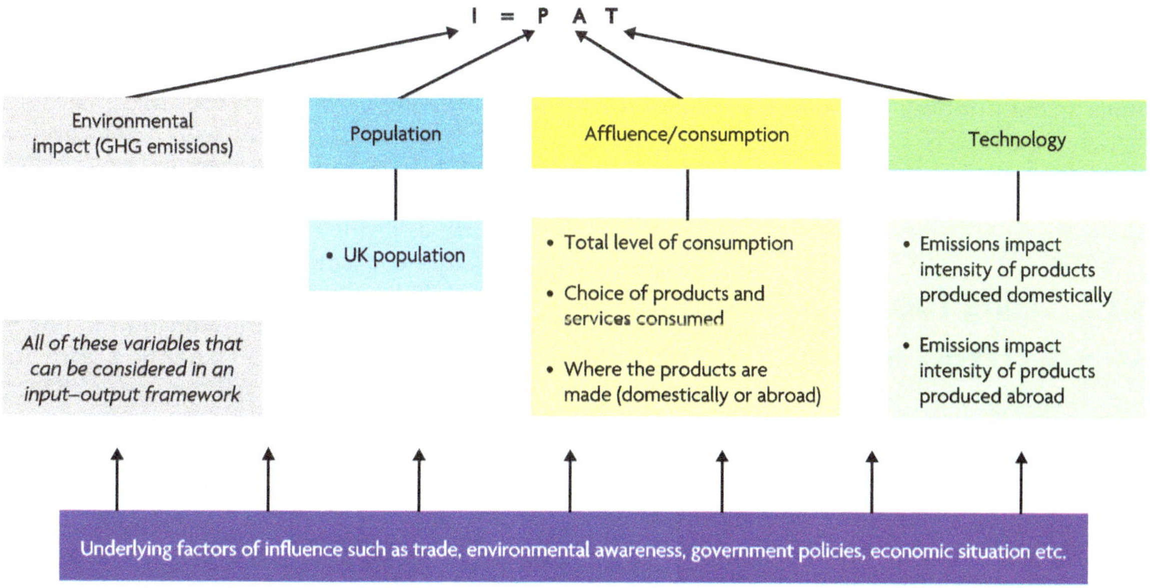

Figure 22.8 Carbon footprint and the IPAT equation.

shows how population, affluence and technology can all be thought of in terms of a carbon footprint and be analysed within an input–output framework.

Each of the drivers of impact in the IPAT equation will have a varying magnitude of influence. The change of one driver may also outweigh an opposing change in another. Technology may reduce the overall impact as efficiency improves, but a growing level of consumption may counteract those improvements, leading to increased impacts overall. One way to assess the individual influence of each of the drivers is to assess their impact on a data trend individually. This has been completed for the carbon footprint using a method called structural decomposition analysis. The results, displayed in Figure 22.9, demonstrate that the decline in emissions intensity in the UK between 1992 and 2004 has been completely outweighed by an increase in consumption levels, leading to an overall increase in the total carbon footprint of that period. Population and household size changes would have alone resulted in a small increase in the carbon footprint, but changes in the choice of products consumed (consumption basket) had a small negative impact. Changes to the structure of the economy (input structure) had little impact on the overall

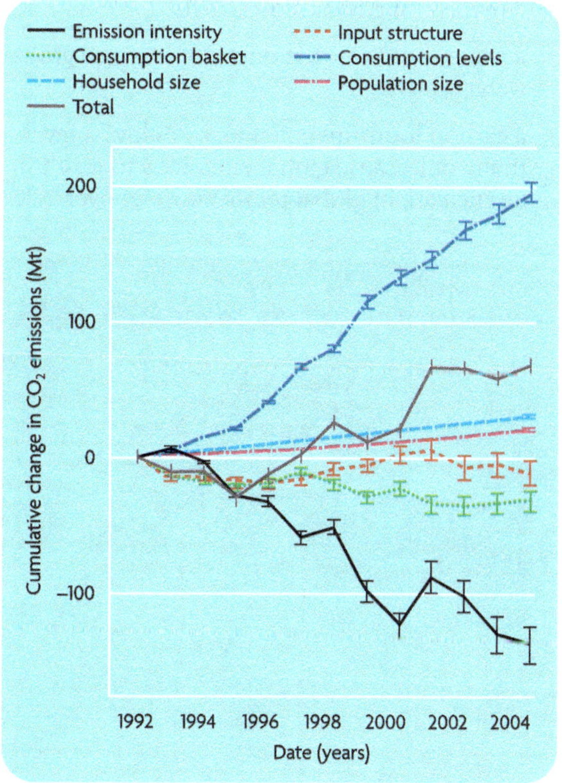

Figure 22.9 Drivers of a rise in consumption emissions.
Source: Minx *et al.*, 2009.

total. Looking at each driver individually is a valuable technique for understanding the role that each one plays in changing an indicator over time. It does not tell us about the relationship between the drivers, but is a useful starting point in an analysis of historical trends, which can then be used to inform future scenarios.

22.7 Using footprint indicators to set targets and budgets

Footprint indicators are used to assess the environmental impact of a country, region or individual. They can also be used to explain sustainable levels of consumption, help set reduction targets and provide impact budgets over a period of time. It is suggested that sustainable levels of consumption could be:

- an ecological footprint equal to or below its fair share of global bio-capacity;
- a water footprint that does not increase global water scarcity;
- a carbon footprint equal to, or below, a level that would cause global emissions to result in an increase in global temperature of over 2 °C.

22.7.1 Allocating carbon reduction responsibility

Footprint analysis can help countries plan their carbon budgets for staying within the level of between 445 to 535 parts per million for CO_2 in the atmosphere in 2050 as recommended in the fourth assessment report of the IPCC (IPCC, 2007). This concentration level gives a carbon budget of 750 Gt between 2010 and 2050. Figure 22.10 shows three emission reduction pathways each keeping within the 750 Gt limit. Clearly the earlier global emissions reach their peak, the less drastic the reduction rate needs to be and the more 'budget' there is left in 2050. But which countries need to make the most reductions to the amount of carbon they emit? How do we decide how much of the responsibility each country has in making carbon reductions?

There are a number of ways of allocating emission reduction responsibilities to countries. Traditionally, responsibility lies with producer or territorial emissions; those emissions from factories, businesses, homes and traffic within the country's borders. However, there are calls from leading researchers to start to consider other ways of assigning responsibility. One method may involve considering past emissions. You could argue that countries who have traditionally

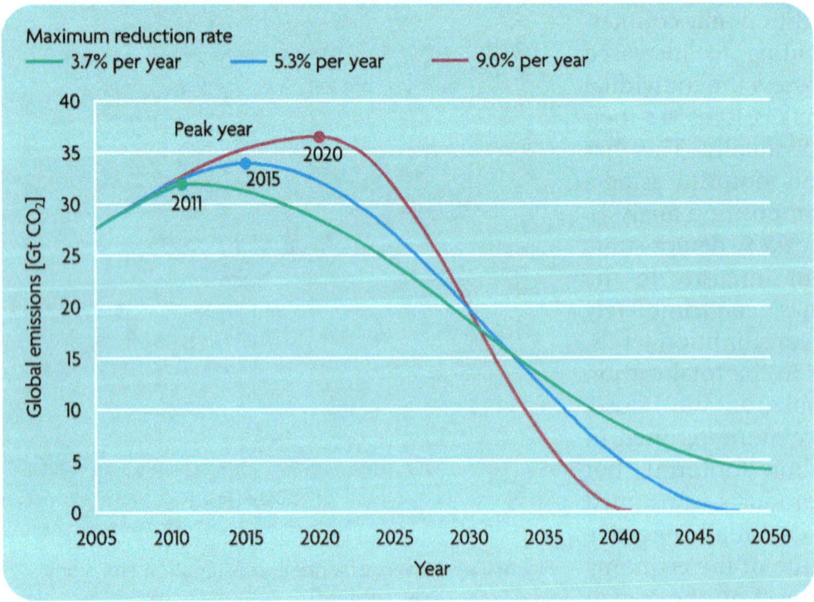

Figure 22.10 Exemplary emission pathways in order to remain within a budget of 750 Gt between 2010 and 2050. At this level, there is a 67 per cent probability of staying below a warming of 2 °C (Messner *et al.* (2010) The budget approach: A framework for a global transformation toward a low-carbon economy. *Journal of Renewable and Sustainable Energy*, 2(3), American Institute of Physics).

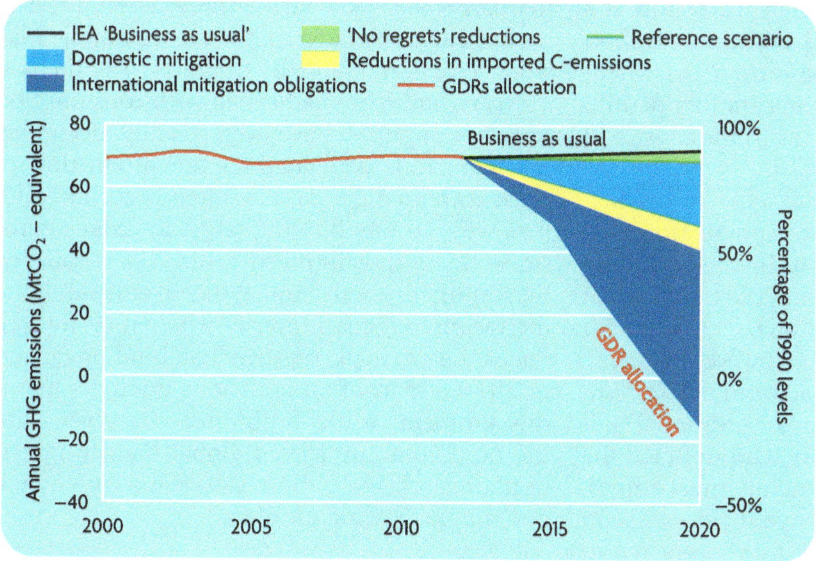

Figure 22.11 GDR allocation for Sweden (Baer *et al*., 2007).

contributed most to greenhouse gas emissions should shoulder the most blame for the planet approaching dangerous levels of climate change. Another consideration may be to consider countries' development pathways between now and 2050. How might countries grow in terms of their wealth, population and consumption habits? How can developing countries continue to grow in a climate constrained world?

The Greenhouse Development Rights (GDR) Framework (Baer *et al*., 2007) attempts to allocate emission reduction responsibility in a fair manner taking account of countries' past emission histories and the need to allow all people to reach a level of human development. The framework states that:

- The south will not prioritise rapid emissions reductions above its goal of human development for its people.
- The right to development must be recognised and protected by any climate regime that has even a hope of being viable.

The GDR framework defines and quantifies the burdens that should be borne by relatively well-off populations. It assesses capacity and responsibility of individuals – taking account of the unequal distribution of income within countries. It then defines national obligations under a climate regime accordingly. Consider a relatively well-off country such as Sweden. What does the GDR Framework describe for Sweden's emissions reduction pathway? Figure 22.11 shows that under the GDR framework, if you take into account Sweden's affluent population and its historical emissions record, Sweden should be reducing emissions by over 100 per cent. But how can a country reduce emissions by over 100 per cent and does this even make sense? This target only becomes meaningful if Sweden invests in reductions internationally. This would mean investing in schemes in other countries to help them move towards a low-carbon economy.

22.7.2 Emissions trading

The Greenhouse Development Framework described above is one method for allocating emissions budgets, but this or anything similar is yet to be widely accepted and agreed across the globe despite numerous meetings and summits. Another method for reducing emissions is emissions trading. This is when a limited number of carbon emission permits are shared out across business and industry and then bought and sold on an open market. The permits to pollute can be sold by a company that doesn't require them to a company that does and, over time, those permits are reduced, leading to a steady reduction in the total carbon emissions released by the sectors as a whole. In theory this should encourage carbon mitigation technologies to be introduced whenever the cost is lower than buying another permit, generating emissions reductions in the most cost effective manner.

Some would argue that this type of market-based solution is a more acceptable form of carbon budgeting because it involves trading carbon as a good on an open market – a system that business and industry would be familiar with and could more easily adopt. However, this type of system does not take into account any responsibility for historical emissions. Additionally, it may adversely impact on those who polluted the least historically and are potentially most vulnerable to affects of climate change if prices rise in accordance with carbon costs. As a recent report by Ackerman (2009/2010) concluded: 'while carbon prices will change energy costs, energy consumption and carbon emissions, relying on this mechanism alone would be both ineffective and inequitable. Other policies are needed to offset the equity impacts of higher fuel costs and to launch the development of new, low-carbon energy technologies of the future.' This type of emissions trading system can also suffer from technical problems of large price fluctuations, influential permit allocation decisions and carbon leakage.

Carbon leakage is a term used to describe the transfer of emissions from one location to another. It is called leakage because it usually refers to the movement of carbon emissions from a constrained or limited system, to an area where there are no limits (Peters, 2008). This could occur when an emissions trading scheme is set up in one location, causing industry and business to move to a place where there is no price on carbon. The same argument could be made for any type of legislation that increases the costs of trading in one country compared to another. Historically, however, this type of industrial shift has been more attributed to labour costs above anything else. Carbon leakage can also be used to describe the impact of off-shoring polluting industry from a country with binding targets for emissions reduction to one without. Whether intentional or otherwise, this problem is evident in many countries. Take the UK, for example; here we have demonstrated that while territorial emissions have been gradually declining over the past 15 years, emissions associated with consumption of goods imported have been steadily increasing. The emissions generated in the production of those imported goods may often occur in countries without any binding target to reduce emissions. So as consumption in the UK is satisfied by imported goods, unregulated emissions are increasing in those countries where they are produced. Again, if all countries had binding targets, or all were part of an emissions trading scheme, this problem would be limited. However, with the no commonly agreed global budget, targets or trading scheme carbon leakage is a current and potentially growing problem.

Lead-in question

What can be done at a local level?

22.8 Investigating footprint reductions in more detail

It is possible to assess and budget at a national level, but also at the individual or community level. Understanding carbon emissions at this smaller scale is necessary for tackling both territorial and consumption based emissions; direct emissions from household activities such as heating or travel are a key contributor to territorial emissions alongside businesses and industry. In addition to these direct emissions, a consumption based approach helps to inform how lifestyles, choices and consumption patterns influence carbon emissions nationally and globally through supply chains. The next part of this chapter looks at individual and household footprints in the UK from a consumption perspective; what contributes to a high footprint and how it might be reduced.

Before we can start to consider how footprints can be reduced, we need to consider what factors contribute towards having a large impact. Is it the type of house we live in or our diet or the way we get to work which has the greatest impact on our footprint? What products have the highest emissions intensity or, put another way, if we had a pound to spend on a product, what product has the highest impact per pound spent? Once we understand the drivers of a high impact lifestyle, we can then start to consider the policies that can be put in place to reduce footprints.

22.8.1 Why is the UK carbon footprint high?

By breaking down the carbon footprint to its component themes, we can begin to discover which elements of our lifestyles have the highest impact and identify methods for reducing the UK's footprint. The average UK citizen has a carbon footprint of 16.24 tonnes CO_2e (SEI, 2010) of which almost a half is related to emissions associated with heating and powering the home and using private and public transport (Figure 22.12). The remaining half comprises the emissions related to consumption of food, goods and services.

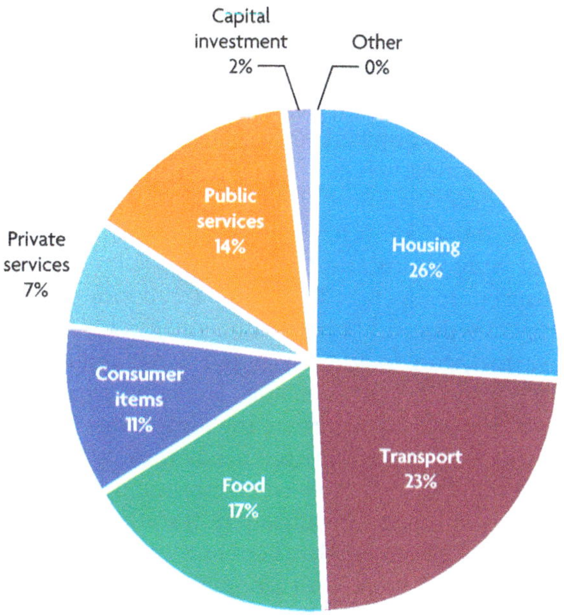

Figure 22.12 UK carbon footprint (including all GHGs) broken down by theme (SEI, 2010).

The Family Expenditure Survey reveals how the average family in the UK spends their money in a typical week – i.e. what is being consumed. When we group this consumption into the categories of home and energy, transport, food, consumables and services we see that spend is fairly equal across the five categories. This is shown by the height of the red bars in Figure 22.13. Of the £350 spent per household per week, 30 per cent is spent on consumable items such as clothes, furnishings and electrical gadgets, compared to less

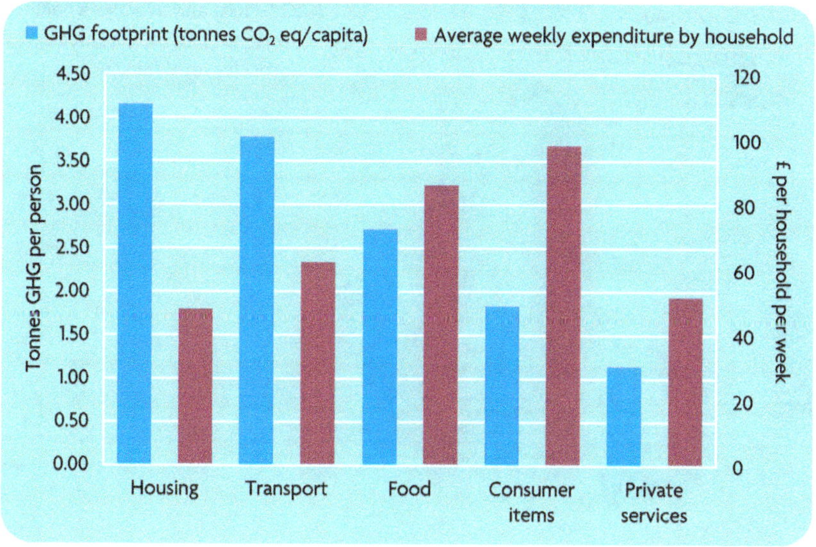

Figure 22.13 UK carbon footprint (including all GHGs) compared with average weekly household expenditure (SEI, 2010).

than a fifth of the weekly budget on running a car and public transport.

The carbon footprint associated with this spend is not as uniform. The graph reveals that the highest footprint is associated with fuel burning activities such as heating the home and transportation. We know that home and energy and transport account for half of a person's footprint but only contribute to one-third of the weekly household spend. Therefore, by concentrating on activities with a high impact per pound spent we can start to identify where large reductions in footprint can be made. It seems clear that housing and transport are the areas where action needs to be taken. Energy needs to be produced and transport run more efficiently and the amount of energy used and distances travelled need to reduce. Perhaps residents could be encouraged to spend their spare cash on shared service based activities such as the cinema or sports rather than buying products as the impact per pound spent on services is considerably lower than that spent on consumables.

We will discuss ways in reducing the footprint in more detail later on in this chapter but first we need to discover who has a large footprint and where they are found.

22.8.2 Who has large footprints and where do they live?

The carbon footprint measure takes into account the CO_2e associated with heating and powering the home, all the goods and services purchased by the household (including food and consumables) and the impact of using and running private cars and public transport. Obviously, different households consume in different ways. Some people will live in large, inefficient houses and leave their lights and appliances on. Some people have to travel long distances by car to get to their place of work. Diets vary vastly between different people and, finally, some families will be able to afford many more goods and services than others. But which of these factors specifically contribute to a high footprint lifestyle? Or rather, what type of expenditure will result in a high footprint? Very simply, footprints are large where spending is greatest or if the items bought have a high impact per pound spent.

If we know the spending habits of the families living in different parts of the UK we can start to break down the footprint to a localised picture. Figure 22.14 shows footprint by English and Welsh Local Authorities; the London commuter

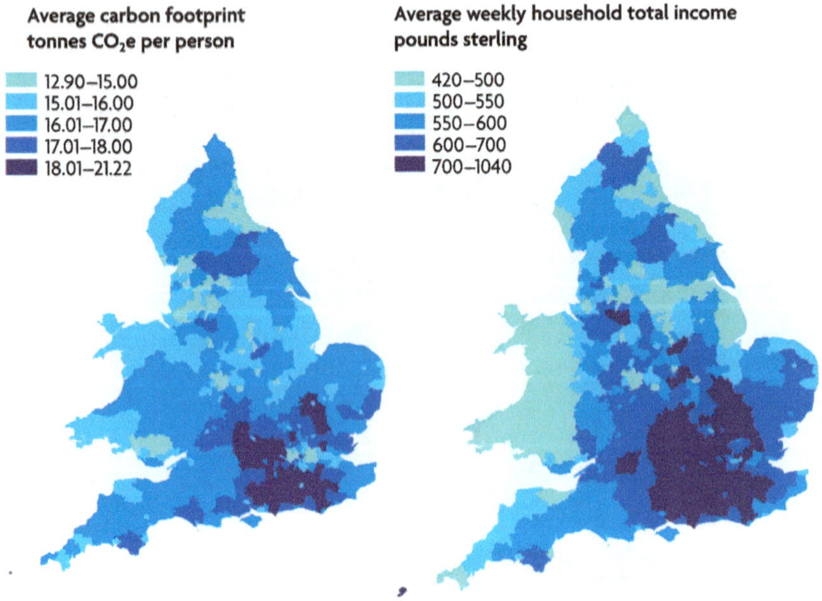

Figure 22.14 Maps comparing UK carbon footprint (including all GHGs) and average weekly household income at local authority level (SEI, 2010).

belt contains the area with the largest carbon footprint. The lowest footprints are found in the southern Welsh valleys and the north-east of England. Comparing the footprint map with the average weekly household income map for England and Wales confirms a clear link between the two measures. Generally speaking, a high income household is likely to have higher rates of consumption and a higher footprint.

This simplicity in the methodology is one of the major flaws in footprint calculation. By using expenditure to estimate size of footprint, a family who spend £200 on a new fridge will be given a higher impact than the family who spend £100. The more expensive fridge might be an energy efficient A-rated appliance but the methodology has no way to make this distinction between different qualities of the same product type.

22.8.3 Why do different people have different footprints?

If we look in more detail at what people spend their money on, what their footprint is and where they live, interesting patterns can be observed. Copmanthorpe is a small commuter village outside of York. The pattern of residents with a high car fuel footprint is almost a negative image of those with a high impact associated with public transport (Figure 22.15). This means that people with a high car footprint tend to have a low public transport footprint and vice versa. When you compare the impact per kilometre travelled of cars compared to public transport, public transport is nearly always the better option. You would expect the footprints associated with public transport to be considerably lower but this is not the case in Copmanthorpe; the public transport footprints are about the same size. Some local knowledge is needed to interpret these maps. In Copmanthorpe many of the public transport users are commuting by train to Leeds and London, whereas the majority of the car commuters will be travelling the few miles up the road to York City Centre or suburbs.

By identifying those high impact footprint activities, such as fuel and heating, along with high volume of use, such as the many miles travelled by train in the example above, and then considering local factors such as access to services, a picture of the consumption behaviour of residents can be painted. If we know how and why residents consume like they do, policy makers can have a better understanding of how to target homes to bring about reductions. Different groups in society are likely to respond better to different actions.

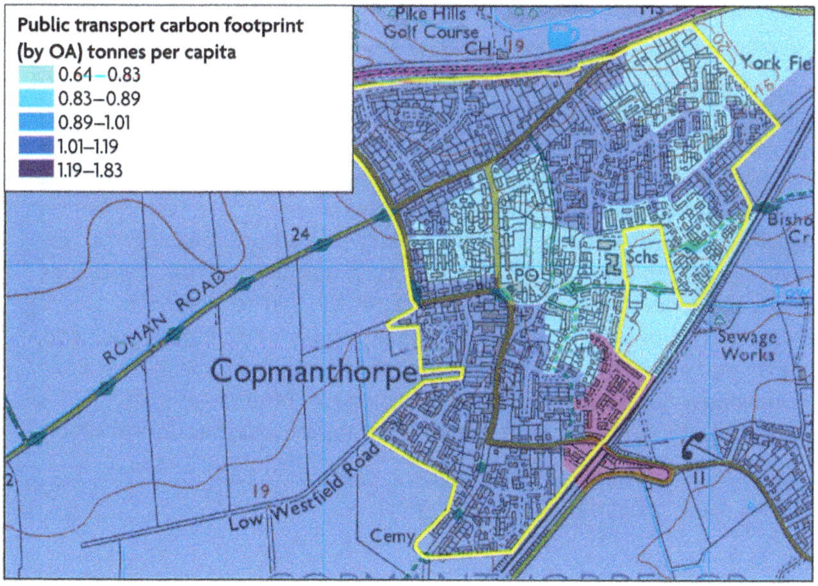

Figure 22.15 Maps comparing impact of using personal and public transport in a commuter village outside of York (SEI, 2010).

22.8.4 How can we reduce our footprint?

As discussed, determining who is responsible for ensuring that the UK's footprint halts the current trend for increasing and starts to decrease towards sustainable levels is not a simple decision. Some responsibility lies with those who are actually generating the emissions, the companies owning the factories and power stations that provide us with our goods and energy. If the products could be produced more efficiently using cleaner technologies, environmental impact is reduced. Businesses must be responsible for all emissions associated with their products, which means that emissions all the way along the supply chain need to be accounted for and considered.

The governments of countries need to ensure that legislation is in place to ensure that factories and power stations reduce their emissions. All levels of government have a role to play to make sure that infrastructure is in place that allows communities to behave in a more sustainable manner, whether this is a grant to improve the energy efficiency of homes, provision of local services to reduce the need to travel, or space for local allotments.

Finally, the consumers have a role to play by making informed choices over the way they live their lives. Individuals could choose to buy products that are locally produced, from sustainable sources and made by companies who are attempting to become environmentally responsible. They can also choose not to consume so much, spending their money on goods that last and need replacing less frequently. They can take action to make homes and cars more efficient and reduce food wastage (see Chapter 12). In the next section we provide case studies of how footprints can be and have been reduced at the national and local level.

22.8.5 Action at a national level

CASE STUDIES

Code for Sustainable Homes

The Code for Sustainable Homes is a guide for industry on the construction of more energy efficient houses. Any homes built to the new standards will be more energy efficient than a house built to current (2006) UK building regulations. The code is progressive, with each level demanding, among other things, higher levels of energy efficiency. The figure on page 525 displays the improvements in energy efficiency (in terms of carbon emissions) for an individual house at each level of the code (Owen *et al.*, 2008). The 2006 Building Regulations represent a significant shift in performance (nearly 50 per cent lower emissions) over the 'average house' in the UK housing stock. By Code level 4, the house is 44 per cent improved from the 2006 building regulations in terms of energy use. Level 6 homes are described as 'zero carbon', meaning that any energy used must either be from renewable sources or 'offset' by the generation of onsite renewable energy paid back to the national grid. However, the construction of the home along with the provision of renewable energy does have some carbon output further up the supply

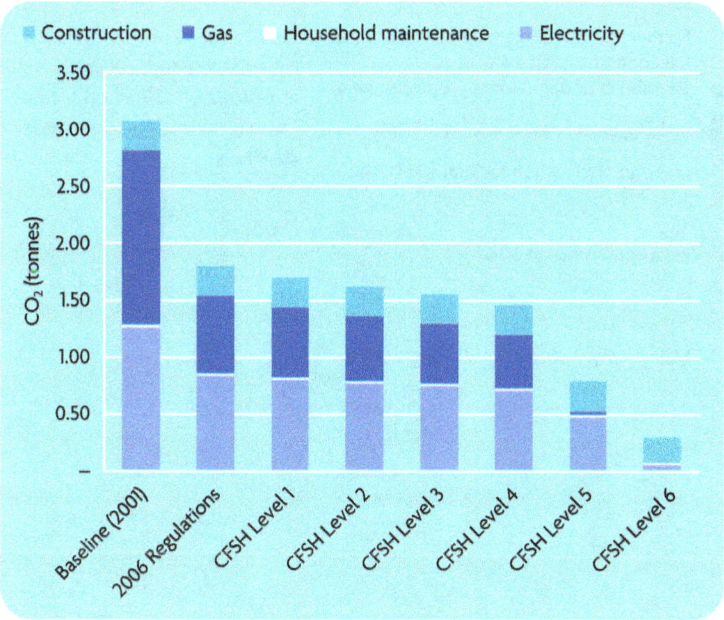

Impact of Code for Sustainable Homes policy (Owen *et al.*, 2008).

chain which is not incorporated into the assessment. The UK government has implemented a timetable which states that all homes should be built to Code Level 3 by 2010, Level 4 by 2013 and Level 6 by 2016.

Is this enough?

At first glance, this government policy sounds as though it could significantly reduce the carbon footprints of residents by tackling the footprint associated with heating and powering the home. However, this policy only affects new housing stock and with replacement rates being less than 0.1 per cent of the housing stock in some parts of the country, it could take 2000 years for every home to be built to the highest standard of efficiency!

What more can be done? – Leeds City Region 80 per cent reduction case study

The Leeds City Region (LCR) area of the North of England considered the scale of change required within the housing sector in order to achieve the UK government's 80 per cent carbon dioxide emissions reduction target by 2050. With a growing population and an additional 263,000 housing units to be built within LCR by 2026, the housing sector would need to reduce its expected total carbon dioxide emissions by 38 million tonnes between 2010 and 2026 to be on track for 80 per cent savings in 2050 (Dawkins and Barrett, 2008). The study demonstrated that the majority of the emissions savings can come from improving the energy efficiency of existing homes by insulating lofts, cavity walls and boilers and fitting double glazing. The figure overleaf shows how the 38 million tonne reduction would be achieved.

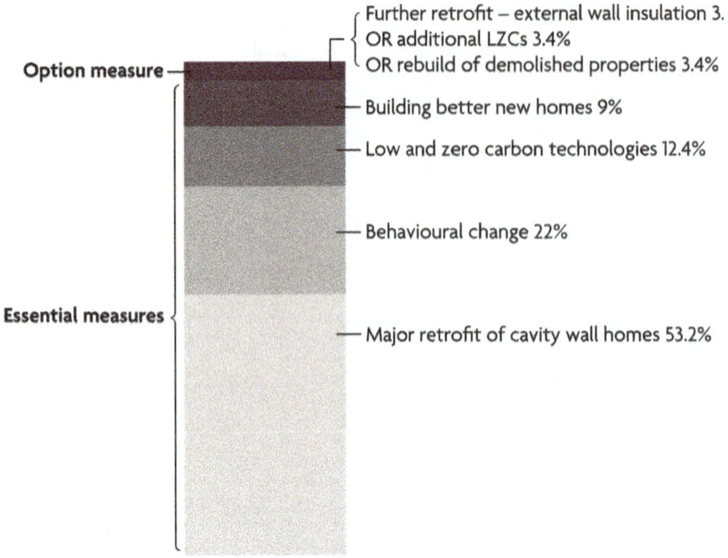

Contribution of measures towards a reduction in emissions of 38 million tonnes CO_2e (Dawkins and Barrett, 2008).

22.8.6 Transport at a local level

At first glance there are more measures available to encourage low footprint travel than there are to encourage low footprint housing. Tackling energy use in the home can only be achieved by making homes more efficient, encouraging households to consume less and supplying cleaner energy. The footprint of personal travel can be addressed by looking at fuel efficiency, modal choice, spatial planning, integrated personal transport, infrastructure efficiency, journey purpose, financial incentives . . . and the list goes on. But in many ways personal travel is much more difficult to address. There are many simple changes which can be made in the home without fundamentally changing the way people live: insulating a loft for example. A shift in transport behaviour often requires a more radical change; persuading people to change an everyday habit, such as commuting by car, is often difficult, especially if it is perceived to be more convenient, cheaper and quicker. Transport demand trends can often be less reactive to changes in price compared to other consumption areas. This makes it difficult to influence with minor price incentives.

CASE STUDIES

Sustainable Travel Towns

In 2004 the UK government launched a Sustainable Travel Towns project where three demonstration towns (Peterborough, Worcester and Darlington) were selected to receive support for the sustained implementation of a package of Smarter Choices (an initiative to encourage people to use less carbon intensive forms of transport) measures and infrastructural improvements over a period of five years (Sloman *et al.*, 2010). One of the key methods for changing travel behaviour was an individualised marketing initiative. The impact of the Sustainable Travel Town initiative was measured and there was found to be a significant change in travel behaviour across all demonstration towns, with increases in walking, cycling and the use of public transport and decreases in car use. The following figure shows the modal shift away from cars to greener methods of transport in Darlington.

Mode	Change in number of trips following implementation of individualised travel marketing scheme
Walking	+ 14 %
Cycling	+ 19 %
Car driver	− 6 %
Car passenger	− 16 %
Public transport	+ 9 %

Modal shifts in Darlington Sustainable Travel Town.

But what might the effect be on the transport footprint if sustainable travel towns are implemented across the country? Would the Sustainable Travel Town initiative work in every town in the UK?

Is this enough?

Sustainable Travel Towns, taken as a policy in isolation, will reduce the impact of personal travel, but the concept may not work in rural areas where there is a greater need for travel. In addition, the policy does little to counteract the trend for increased trip distance; it focuses instead on the number of trips made. Year on year, people are travelling 0.7 per cent further by car and 4 per cent further by train (Sloman *et al.*, 2010).

The examples above show how reductions can be made in the housing and transport sectors. If we are to reduce footprint to sustainable levels, consumption of food, goods and services must also be tackled. The next case study shows how a community initiative attempted to tackle residents' entire footprint.

CASE STUDIES

Green Neighbourhoods: What more can be done?

National, regional and local government can influence the housing and transport footprint, but have limited influence over the food we eat and the goods and services we buy. Reducing the impact of food, goods and services might be better tackled at a community and individual level. The aim of the York Green Neighbourhoods Initiative was to identify the streets and communities in the City of York which have the greatest potential to reduce their environmental impact (Haq and Owen, 2009). The Green Neighbourhood framework used spatial data to find the streets in York which contained residents who, despite being sympathetic towards green issues, also had some of the largest footprints in the city. This

➤

selection was then refined further to concentrate on those areas where households had the potential to make changes (see the following figure). For example, the homes might be in need of energy efficient measures so could potentially use less energy. Or an area might have lots of local facilities and good bus and cycle lanes, reducing residents' needs for travel by car.

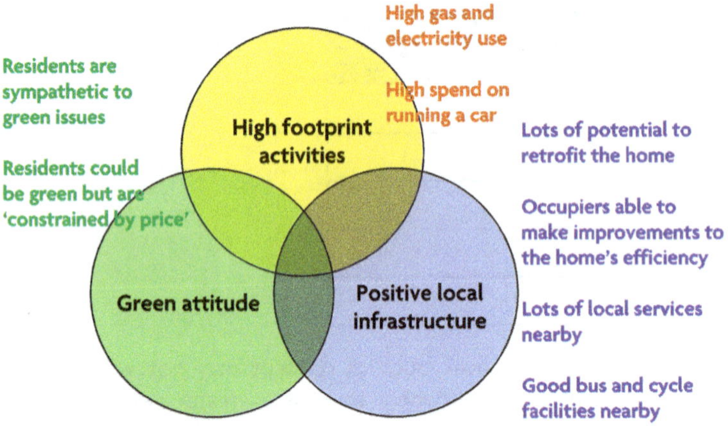

Selection procedure for York's Green Neighbourhood challenge (Haq and Owen, 2009).

Once the streets were selected, residents were surveyed to find out the size of their carbon footprint. Community meetings were then held where residents discussed actions they could take to reduce their impact. Actions ranged from setting up community allotment schemes, sharing tools, cars and baby clothes to applying for grants to make homes more energy efficient.

How much is possible?

The figure shows the carbon footprint measured in tonnes of CO_2e of a typical respondent in the Green Neighbourhood Challenge. The first bar shows their impact before the challenge and the second bar shows what is possible if they signed up to every single carbon reduction pledge (such as installing loft insulation or using renewable energy). We see that the resident is only able to reduce their footprint by around 40 per cent in total. In some areas, such as housing, reductions of around 80 per cent are possible, but for other categories, such as activities, there are few pledges to be made. For this part of the footprint, business and industry need to make their goods and services more efficient, which will in turn lessen the consumer impact. The category of 'other' is the impact of government spending on health, education, defence etc. shared by every person in the country. It is the government's responsibility to provide these services more efficiently, by being more resource efficient or by purchasing less carbon intensive goods.

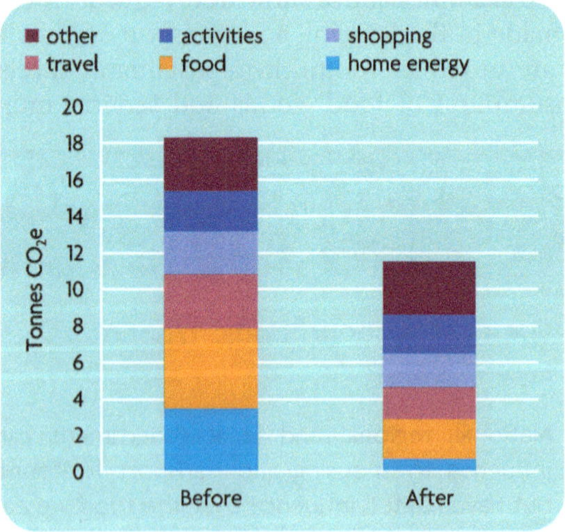

A typical Green Neighbourhood Challenge member's footprint before the challenge and the potential savings that could be made (Haq and Owen, 2009).

The next figure reveals how the community reduction pledges can reduce the impact of an individual's footprint. Supporting the individual pledges with the decarbonisation of the national electricity grid can reduce the impact further. The remaining component of the individual footprint is the part allocated to each person for government spend on services; this could also be reduced to take the total footprint closer to the 80 per cent reduction target.

Carbon footprint reductions possible through behaviour change and decarbonisation of electricity supply.

Footprint theme	Footprint before (tonnes CO_2e)	Emission reduction pledges	Footprint after pledge (tonnes CO_2e)	Footprint after resource efficiency and grid decarbonisation (tonnes CO_2e)
Home energy	3.52	Use less energy; all energy efficient lights and appliances; green tariff electricity; well-insulated home; condensing boiler; ground source heat pump; solar panels	0.75	0.24
Food	4.36	Reduce food waste; eat a low meat diet; grow some of my own fruit and vegetables	2.09	0.89
Travel	2.92	Cycle rather than drive journeys under 3 miles; replace car with A rated emissions one; travel by train for journeys over 100 miles; never fly domestically; holiday in the UK rather than abroad; join car sharing club	1.83	0.85
Shopping	2.37	Only buy good quality items that last; join the local library Download music; Use eBay and Freecycle; Share tools and DIY equipment	1.77	0.76
Activities	2.27	Jogging rather than using a gym	2.19	0.89
Other – mainly government spending	2.85	It may be possible to reduce emissions further by reducing the impact of government spending on services.	2.85	0.12
Total	18.29		11.48	3.75

22.9 Conclusion

To successfully reduce GHG emissions across the globe action is required at all levels of society. Businesses must strive to become more resource efficient, governments must encourage and support decarbonisation of the economy as a whole and individuals need to be aware of the choices they can make to move towards low carbon lifestyles. The footprint approach provides a framework for considering how the role of consumption and demand for products is driving environmental impact globally. The impact that government policy has on territorial emissions from and resource use by businesses and industry is well documented and understood. Government policy can also influence consumer choice and lifestyles. The footprint approach can, for the first time, quantify this influence and reveal the role consumers must play in reducing environmental impact.

There are numerous indicators that can be used to measure humanity's environmental impact. But is considering our resource use in terms of land area and water volume and polluting impact in terms of emissions of harmful greenhouse gases really telling the full story? The footprint approach does not consider Earth's resources required for biodiversity, for example. In addition, progress towards sustainability may also be considered in terms of material living standards, health, education, personal activity (including work), political voice and governance, social connections and relationships, environmental conditions (present and future) and insecurities (economic and physical) (Stiglitz, 2009). Reducing GHG emissions will be vital to ensure that the potential impacts of global warming are either prevented or minimised as far as possible, but the wider environmental impacts that our consumption has on the planet will need to be considered as well. Perhaps a new look at a system which has the ever increasing consumption of finite resources on a finite planet as a fundamental component for success might be required?

POLICY IMPLICATIONS

Accounting for impacts from a consumption perspective

- Policy makers need to understand how impacts are accounted for from both consumption and territorial perspectives.
- Understanding the difference between direct source emissions and supply chain emissions from products is paramount for decisions to be made as to where responsibility for reduction lies.
- Accounting for emissions from a consumption, rather than a production, approach has implications when thinking about how to allocate carbon reduction budgets to countries.
- Policy makers need to be aware of where the largest impacts lie so that they can prioritise where to take action in order to reduce impacts and move towards a sustainable future. For example, should policy makers concentrate on modal shift to sustainable forms of transport, concentrate on food waste, or concentrate on decarbonisation of the electricity sector as a strategy?
- In order to keep within global concentrations of CO_2e by 2050 that avoid dangerous levels of climate change, policy makers need to understand that businesses, government and individuals all have a role to play.
- Impact is a combination of population, affluence and technology. An increase in technological efficiency, meaning that the products we buy are made more using less energy, will not bring about sufficient reductions in impact if populations continue to increase their rate of consumption. Policy makers need to tackle both resource efficiency and question our reliance on economic growth.

CHAPTER REVIEW EXERCISES

Exercise 22.1

How would you explain the concept of 'One Planet Living' to someone unfamiliar with the ecological footprint? Briefly explain how a 'global fair share' is calculated and what we mean when we describe an American citizen as living a five planet lifestyle.

Exercise 22.2

Consider the following table. Decide how the emissions associated with each spend should be allocated.

Spend on	UK production or territorial emissions	UK consumption emissions	Both UK production and consumption	Neither UK production nor consumption
A television manufactured in the UK and bought by an American citizen				
A fridge manufactured in Japan and bought by a UK citizen				
A UK citizen's car journey from London to Manchester				
A UK student taking a flight from Sydney to Los Angeles				
A German holiday maker staying in a hotel in Brighton				
A UK family heating their home with gas				
A car with parts flown in from China, assembled in Newcastle and bought by an Australian				

Exercise 22.3

Plot a scatter graph of ecological footprint, measured in global hectares against GDP measured in US dollars per person. Explain the relationship between these two variables and suggest reasons for this (data from The Global Footprint Network, 2010).

Country	Ecological footprint (gha/person)	GDP (US $/person)
Argentina	3.00	15,119
Mexico	3.25	11,370
Canada	5.76	36,584
United States	9.02	44,005
Egypt	1.40	5,587
Tanzania	1.03	886
Sierra Leone	0.77	1,817
China	1.85	7,303
India	0.77	3,712
Thailand	1.72	9,424
United Arab Emirates	10.29	53,496
Japan	4.11	31,236
Italy	4.94	29,048
France	4.60	30,119
Germany	4.03	31,291
United Kingdom	6.12	32,103

Exercise 22.4

Consider the supply chain of emissions associated with a consumer buying and using a book. Put this supply chain into the correct order (ending with 'final consumer').

Paper mill 0.4 kg	Timber growing 0.2 kg	Printing 0.1 kg	Final consumer 0 kg	Book distribution 0.5 kg

What is the full impact of buying and using a book?

Do the same as this for a supply chain of emissions for a consumer buying and using a car.

Why are there emissions associated with the 'final consumer' stage of this supply chain?

Lighting the showroom 10 kg	Iron ore extraction 2000 kg	Steel production 1000 kg	Car assembly 1500 kg	Final consumer 2000 kg

REFERENCES

Ackerman, F. (2009/2010) *Trade and Environment Review: Promoting poles of clean growth to foster transition to a more sustainable economy, Carbon Markets are Not Enough*. Stockholm Environment Institute and Tufts University, United Nations.

Baer, P., Athanasiou, T. and Kartha, S. (2007) *The right to development in a climate constrained world*. Berlin, Germany, Heinrich Boll Foundation, ChristianAid, Stockholm Environment Institute.

Dawkins, E. and Barrett, J. (2008) *Making homes more energy efficient. How to reduce domestic carbon emissions while delivering a growth-based housing strategy*. Report to the Environment Agency by Stockholm Environment Institute, York, UK.

Ehrlich, P.R. and Holden, J. (1971) Impact of Population Growth. *Science*, **171**, 1212–1217.

The Global Footprint Network (2010) *The Ecological Wealth of Nations*. Global Footprint Network, Oakland, California, USA.

Guissani, V. (1994) *The UK Clean Air Act 1956: An Empirical Investigation*. Centre for Social and Economic Research on the Global Environment, University College London and University of East Anglia.

Haq, G. and Owen, A. (2009) *Green Streets: The Neighbourhood Carbon Footprint of York*. Stockholm Environment Institute, York, UK.

Hoekstra, A.Y. Chapagain, A.K., Aldaya, M.M. and Mekonnen, M.M. (2009) *Water Footprint Manual*. State of the Art Water Footprint Network, Enschede, The Netherlands.

IPCC (2007) *Climate Change 2007: The Physical Science Basis*. Contribution of Working Group I to the Fourth Assessment Report of the Intergovernmental Panel on Climate Change (eds Solomon, S., D. Qin, M. Manning, Z. Chen, M. Marquis, K.B. Avery, M. Tignor and H.L. Miller). Cambridge University Press, Cambridge, UK and New York, USA.

Kriegler, E., Hall, J.W., Held, H, Dawson, R. and Schellnhuber, H.-J. (2009) Imprecise probability assessment of tipping points in the climate system. *Proceedings of the National Academy of Sciences USA*. 10.1073/pnas.0809117106.

Leontief, W. (1970) taken from Kurz, H.D., Dietzenbacher, E. and Lager, C. (1998), *Input-output analysis Volume II*. Edward Elgar Publishing Ltd, Cheltenham, UK.

Miller, R.E. and Blair, P.D. (2009) *Input–Output Analysis: Foundations and Extensions* 2nd edn. Cambridge University Press, Cambridge UK.

Minx, J.C., Baiocchi, G., Wiedmann, T. and Barrett, J. (2009) *Understanding Changes in CO_2 Emissions from Consumption 1992–2004: A Structural Decomposition Analysis*. Report to the UK Department for Environment, Food and Rural Affairs by Stockholm Environment Institute at the University of York and the University of Durham, Defra, London, UK, in press.

Owen, A., Paul, A. and Barrett, J. (2008) *Ashford's Footprint: Now and in the Future*. Stockholm Environment Institute, York, UK.

Peters, G. (2008) Reassessing carbon leakage (cited in: Rockstrom, Steffen, *et al.* 2009) A paper for the Eleventh Annual Conference on Global Economic Analysis, 'Future of Global Economy', Helsinki, Finland, 12–14 June 2008.

Rees, W.E. (2000) Discussion: Eco-footprint analysis: Merits and brickbats. *Ecological Economics*, **32**, 347–349.

Rockstrom, J., Steffen, W., Noone, K., Persson, A., Chapin, F.S., Lambin, E.F., Lenton, T.M., Scheffer, M., Folke, C., Schellnhuber, H.J., Nykvist, B., de Wit, C.A., Hughes, T., van der Leeuw, S., Rodhe, H., Sorlin, S., Snyder, P.K., Costanza, R., Svedin, U., Falkenmark, M., Karlberg, L., Corell, R.W., Fabry, V.J., Hansen, J., Walker, B., Liverman, D., Richardson, K., Crutzen, P., Foley, J.A. (2009) A safe operating space for humanity. *Nature*, **461**, 472–475.

Sloman, L., Cairns, S., Newson, C., Anable, J., Pridmore, A. and Goodwin, P. (2010) *The Effects of Smarter Choice Programmes in the Sustainable Travel Towns: Research Report*. A report to the Department for Transport.

SEI (2010) Access to methods for calculating footprints is available via the SEI site at: http://www.resource-accounting.org.uk/ accessed in April 2012.

Stiglitz, J. (2009) *Report by the Commission on the Measurement of Economic Performance and Social Progress*, accessed in April 2012 via http://www.footprintwork.org/en/index.php/GFN/blog/stiglitz_report.

Wackernagel, M. and Rees, W. (1996) *Our Ecological Footprint: Reducing Human Impact on the Earth*. New Society Publishers, Gabriola Island, BC.

Wiedmann, T. and Barrett, J. (2010) A Review of the Ecological Footprint Indicator – Perceptions and Methods. *Sustainability*, **2**, 1645–1693.

Wiedmann, T. and Minx, J. (2008) A Definition of 'Carbon Footprint'. In: C.C. Pertsova, *Ecological Economics Research Trends*, Chapter 1: 1–11, Nova Science Publishers, Inc., Hauppauge NY, USA.

Wiedmann, T., Wood, R., Lenzen, M., Minx, J., Guan, D. and Barrett, J. (2008) *Development of an Embedded Carbon Emissions Indicator – Producing a Time Series of Input–Output Tables and Embedded Carbon Dioxide Emissions for the UK by Using a MRIO Data Optimisation System*. Report to the UK Department for Environment, Food and Rural Affairs by Stockholm Environment Institute at the University of York and Centre for Integrated Sustainability Analysis at the University of Sydney, June 2008. Defra, London, UK.

WWF (2008) *The Living Planet Report 2008*. Gland, Switzerland.

CHAPTER 23

The nature and merits of green chemistry

Andrew Hunt and James Clark

Learning outcomes

By the end of the chapter you should:

- Know that the use of traditional chemical technologies can give rise to polluting and unsustainable practices or methods.
- Become aware that 'chemicals' is currently seen by many as a dirty word.
- Understand what the term 'green chemistry' means and how the ethos of this movement can be applied to the manufacture of chemicals and materials essential for modern society.
- Be aware of the issues surrounding elemental sustainability and the depletion of resources for many elements that are commonly used in consumer goods.
- Be more aware that societies' reliance on petrochemicals or fossil fuels is highly unsustainable and switching to renewable materials should be of great environmental benefit.
- Understand the concept of a biorefinery and its potential to sustainably make the chemicals, materials and consumer products of the future.

23.1 Introduction – 'chemistry is a dirty word'

Chemicals are an essential part of modern society (Figure 23.1). It is difficult to think of any articles that we use that do not contain chemicals and which did not involve chemistry in their manufacture. If you look around the location you are in when you read this (room, bus, train etc.), try to identify an item for which this isn't true.

This seems to be good news for chemistry and the chemical manufacturing industry, which supplies chemicals to an enormous range of chemical-user companies including those that manufacture pharmaceuticals, home and personal care products, electronic components, furniture and construction materials, plastics, food and clothing, as well as aerospace and other transport industries. Yet chemical manufacturing is an industry that suffers from one of the lowest reputations of all manufacturing, and is subject to an exponentially growing amount of legislation (Figure 23.2). In fact it is struggling with economic, environmental, resource and social pressures at all stages across the lifecycle of chemical products (Figure 23.3).

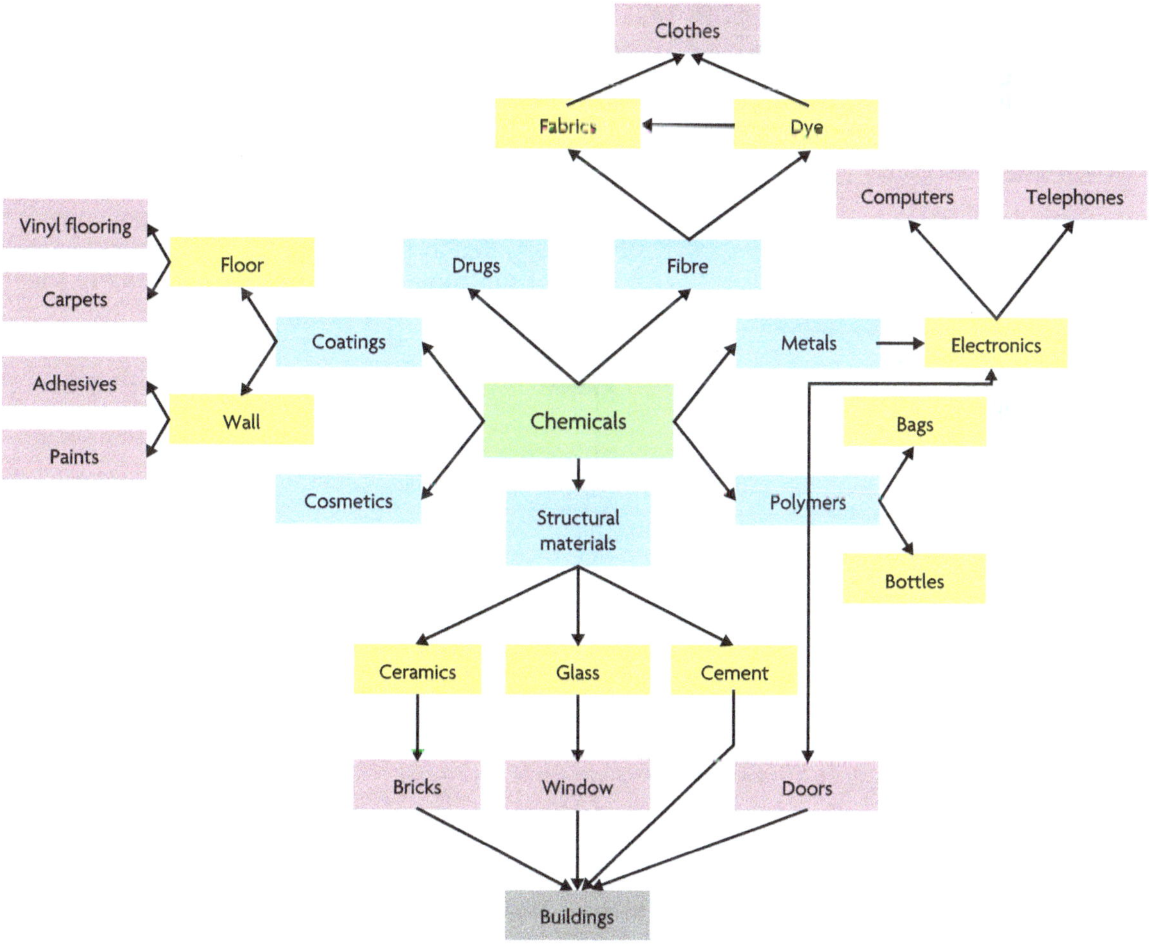

Figure 23.1 The uses of chemicals as an essential part of modern society.

Figure 23.2 Chemical manufacture and energy production are viewed as dirty industries by society.
Source: marmit at rgbstock.com

Lead-in question

Chemicals will remain essential to modern society, so what can be done to improve the public perception and sustainability of the chemical industry?

Green chemistry is the design of chemical products and processes that reduce or eliminate the use and generation of hazardous substances. Key to this movement's ethos is the discovery and application of new chemical/technologies leading to prevention/reduction of environmental, health and safety impacts at source. Both environmental and human tragedies such as the oil spill in the Gulf of Mexico and the Bhopal disaster in India can be viewed as clear drivers for the implementation of green, clean and safe chemical technologies including the use of alternatives to oil.

Another key aspect of the green movement is sustainable development. This is where we as a society meet our needs without impairing the needs of future generations. Sustainable development is achieved in a business context at the convergence of the triple bottom line (Figure 23.4). This is

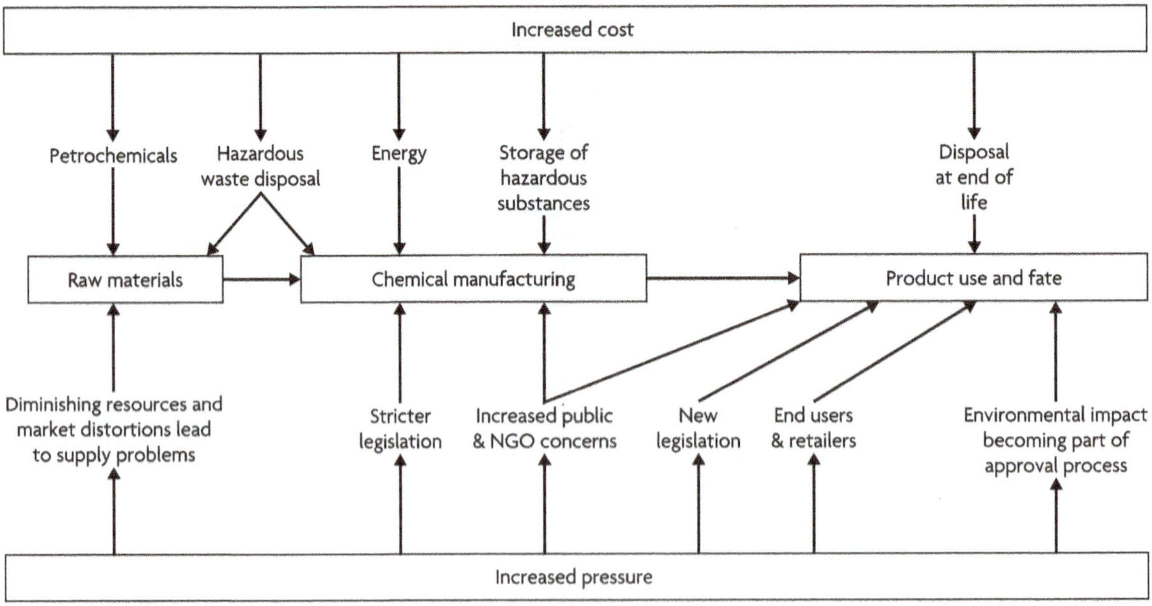

Figure 23.3 Summary of the pressures on the chemical industry across the lifecycle of a chemical.

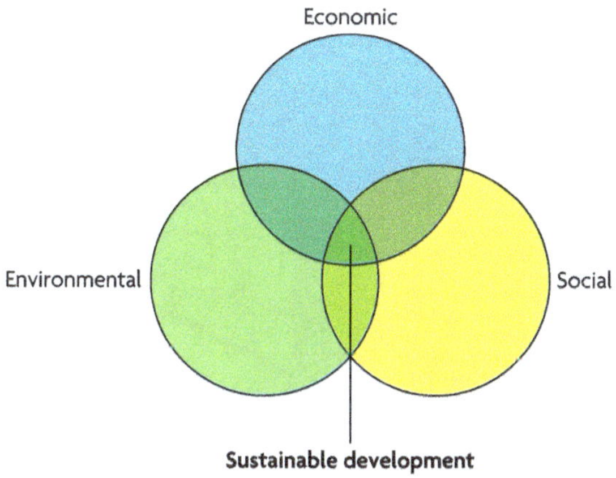

Figure 23.4 Sustainable development as commonly portrayed in terms of the triple bottom line.

where the best balance is achieved between economic constraints, social acceptability and environmental protection.

The development of green and sustainable strategies for the chemical industry will ensure long and healthy lives for humanity and planet Earth. Frequently, some sectors of industry can be reluctant to change from well known processes to new green technologies. For these sectors, for green chemistry processes to be implemented they cannot just be clean, efficient and a benefit to society; they must also be cost effective and at least economically comparable to, or better than, the existing processes. Wherever and whenever possible it is important that a holistic approach is applied to the development of green chemistry processes.

Lead-in question

Are element resources likely to run out at a rate that soon might start to pose problems to industrial companies?

23.2 Elemental sustainability

At the beginning of the lifecycle of any chemical product we must have a natural resource or natural resources. It can be surprising how much resource is consumed in making chemical products – the elements that go into the chemical, the other compounds used to help extract these elements from their origins (minerals, oil, gases etc.) and therefore the chemistry needed to make the product. There's also lots of energy involved; for example, about 10 per cent of all oil that is imported into the European Union goes for energy to make chemicals (another 10 per cent is used for the carbon that goes into organic chemicals such as paints, solvents and plastics). Large volumes of water are also used in making chemicals (leading to large volumes of dirty water at the end of the process).

A fundamental resource problem is that we do not have an infinite supply of elements; the world is a closed system and, short of going out to mine the other planets or asteroid belt (an idea often used in science fiction!), we have to make do with what we've got. But surely the Earth is such a big place that this isn't a problem? The answer to this rhetorical question is that there is a problem – we are beginning to run out of many elements that are vital to our industries and the articles they make and that we want to use in a modern society.

Many new low carbon technologies that are viewed as potential sources of clean energy, including wind turbines, photovoltaics and fuel cells, require rare metals for their production, as do electric cars and catalytic converters. However, traditional supplies of these elements are running out. It may be surprising to note that many of the elements we would also consider to be common such as phosphorus, aluminium and copper are being consumed at a remarkable rate and also running out, at least in terms of traditional sources (Figure 23.5). This is also reflected in rising prices of elements and associated social consequences. At the time of writing, for example, theft of copper and other metals has become a growing problem in the UK.

We are all becoming familiar with the concept of living sustainably and trying to be 'carbon neutral' (Chapter 22). However, as a species we are still highly unsustainable when it comes to our uses of the vast majority of elements. Increasingly in the future we will also need to

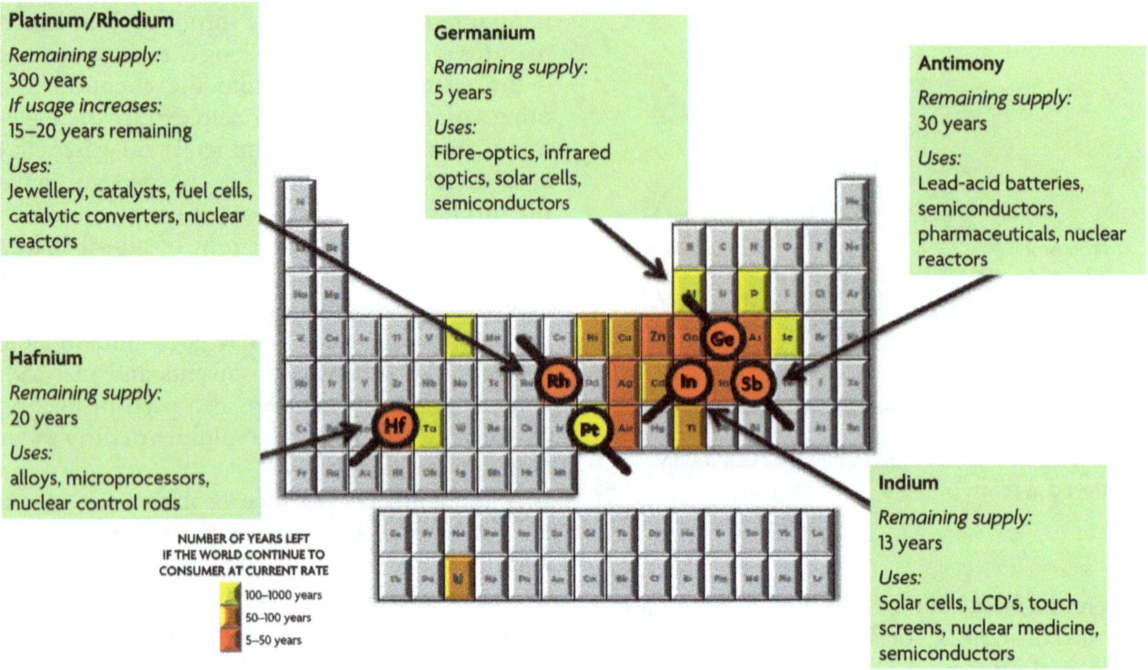

Figure 23.5 Estimated numbers of years remaining for some rare and precious metals if consumption and unsustainable disposal continue at present rates.
Source: Courtesy of Helen Parker.

be 'neutral' for a much larger range of elements. Elemental sustainability is a concept that is likely to become increasingly important over the coming decades. In order to maintain the consumer lifestyle that those of us in the developed world have come to expect, we must increase the number and amounts of elements that we reuse or recycle.

The recovery and recycling of elements from waste streams must become cost effective and environmentally beneficial. Many benefits already exist in terms of metal recovery; for example, the recycling of aluminium typically saves around 95 per cent of the energy compared with that needed for production from ores, and as such only generates 5 per cent of the CO_2 compared to the mining and electrolysis of alumina from bauxite ore. These energy savings are also significant for steel recovery, where 74 per cent of the energy is saved; equally important are a 90 per cent saving in virgin materials, an 86 per cent drop in air pollution, a 40 per cent reduction in water use and a 76 per cent reduction in water pollution. However, the largest saving in percentage terms is a 97 per cent reduction in mining waste generated.

Lead-in question

What 'waste' products could become resources for the chemical industry?

Waste electrical and electronic equipment (WEEE) is a rapidly growing waste stream that contains vast amounts of rare and valuable elements (Figure 23.6). However, the recovery of elements from such wastes is currently underexploited commercially and generally we only see such waste as a problem and not as a resource. This must change!

Figure 23.6 A typical mountain of waste electrical and electronic equipment.
Source: iStockphoto

23.3 LCD waste and the potential for elemental recovery

LCD (liquid crystal display) screens are one example of WEEE that has great potential for elemental recovery. These devices contain the scarce element indium, a silvery-white rare metal with an estimated abundance of 0.24 ppm in the Earth's crust. It has been predicted that, at its current rate of use, reserves of this element will be exhausted within 13 years. Greater than 65 per cent of the total globally extracted indium is used in the manufacture of indium tin oxide (ITO) for LCDs. An estimated 2.5 billion LCDs are approaching their end of life, with LCD waste being the fastest growing waste stream in the EU (Hunt *et al.*, 2009). The growing production of electrical and electronic equipment (EEE) means that increasing amounts of these devices (including the valuable elements contained within them) are finding their way into waste streams. The WEEE Directive of the European Parliament and the Council on Waste Electrical and Electronic Equipment requires the disassembly of all LCDs with an area greater than 100 cm^2 and those containing mercury backlights. Once the backlight has been removed an LCD is rendered 'safe' and may be sent for incineration or landfill. Both options are wasteful and potentially hazardous to the environment. In the UK alone it was predicted that over 10,000 metric tonnes of LCD would be available for recycling in 2010, containing 9 tonnes of liquid crystals, 900 kg of indium and 8000 tonnes of optical quality glass.

The development of an innovative holistic strategy for the recovery and reuse of valuable materials from LCD panels is important to successfully reducing WEEE and preserving indium reserves for future generations, an important facet of green chemistry and sustainable development.

The current processes used for the recovery of elements from WEEE are pyrometallurgy and hydrometallurgy. Pyrometallurgy is the most utilised industrial method for metal recovery and involves the thermal treatment of crushed WEEE by incineration, smelting or high temperature gas phase reactions. These processes generate large quantities of unwanted slag, containing impurities such as zinc, iron, aluminium and lead and ceramic or glass components. Only partial separation of metals can be achieved, and further refining is necessary to remove precious metals such as platinum and palladium. In comparison, hydrometallurgical processes are more precise and easily controlled. They typically use strong acids or bases to leach metals out of WEEE. The resulting metal solution undergoes multiple separation, purification and recovery steps in order to isolate and concentrate the elements. Unfortunately, this method could have damaging toxicological impacts due to the large amounts of reagents used and the content of the wastewaters formed.

An ideal recovery solution would combine green technologies such as bioleaching and biosorption of valuable elements from WEEE. This would allow for the selective recovery of elements without the need for toxic reagents or energy intensive processes. Bioleaching is the extraction of specific metals from their ores or WEEE through the use of bacteria; biosorption is the use of renewable or bio-derived materials for the adsorption of specific metals from solution. However, these technologies are still in their infancy and further research into these areas is vital to fully realise the potential of these methods for metal recovery from WEEE both individually and in combination.

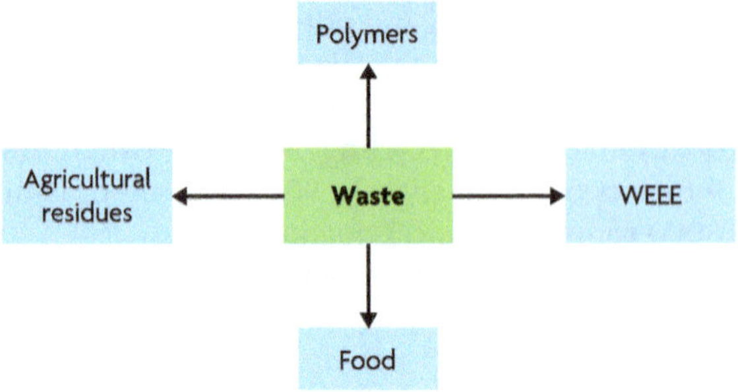

Figure 23.7 Resources of the future.

The example of indium recovery from LCDs highlights the potential and necessity for new approaches to our waste. We must attempt to recover all elements and reuse them in close-looped systems either by recovery from landfill sites, incineration ashes, wastewaters or new sources of elements including the disassembly of WEEE at their end of life (Figure 23.7). These recovery methods should use novel and benign methods that can reduce the environmental burden of mining, selectively recover all elements, limit the demand for new supplies and increase the lifetime of our reserves.

23.4 Starbon® technologies and their applications

One type of material that has demonstrated some promise as a biosorbant and could be used to recover elements from hydrometallurgical processes is known as Starbon® (Budarin *et al.*, 2006). These are a novel family of mesoporous materials (with pore diameters of 2–50 nm) derived from polysaccharides. Polysaccharides, including starch, are widely available, relatively inexpensive, non-toxic, biodegradable, possess polyfunctionality, have great potential for chemical modification and are found in nearly every geographical location on the planet. The extensive use of polysaccharides as an absorbent of metals and valuable elements has been restricted by low surface area ($< 1\ m^2 g^{-1}$) and low degree of porosity.

The development of polysaccharide-derived mesoporous materials with large pore volume and surface areas has opened new doors to their use as absorbents. Through controlled pyrolysis it is possible to form tuneable, nano-structured, graphitisable and mesoporous carbons (Starbons®) using no templating agents. The Starbon materials typically have surface functional groups ranging from predominantly hydroxylic at low temperatures, through C=C and C=O functions (which may be conjugated) at medium temperatures, to aromatic and graphitic at high temperatures (Figure 23.8).

The surface chemistry, functionality and surface polarity of these materials can be controlled by varying the temperature of preparation and selection of the polysaccharide precursor. Incorporation of polysaccharides that are abundant in regions of developing countries, including starch, alginic acid, okra, chitin and chitosan, would be of great benefit for potential water treatment applications in such regions.

These materials have surface functionalities ranging from hydrophilic to hydrophobic. The mesoporous Starbon® family includes a continuum of materials from polysaccharides to activated carbon in nature. This novel Starbon® technology utilises the natural ability for polysaccharides to retain their organised structure during pyrolysis. Starbons® have been demonstrated to be effective at removing metals from potable drinking water (Table 23.1). Adsorption properties of Starbons® towards metals are enhanced with increasing preparation temperature.

These materials are ideal candidates for the biosorption of metals extracted by hydrometallurgical processes. These Starbons® have already demonstrated great promise as catalytic supports, catalysts, nanoparticle delivery systems and chromatographic stationary phases. Starbons® have great potential for their use as catalysts for the preparation of 'platform molecules' from biomass as part of a 'biorefinery'. Platform molecules are a group of organic molecules that can be used as the building blocks for fine chemicals, pharmaceuticals, personal care products and solvents.

Lead-in question

What will happen when we run out of natural petrochemical resources?

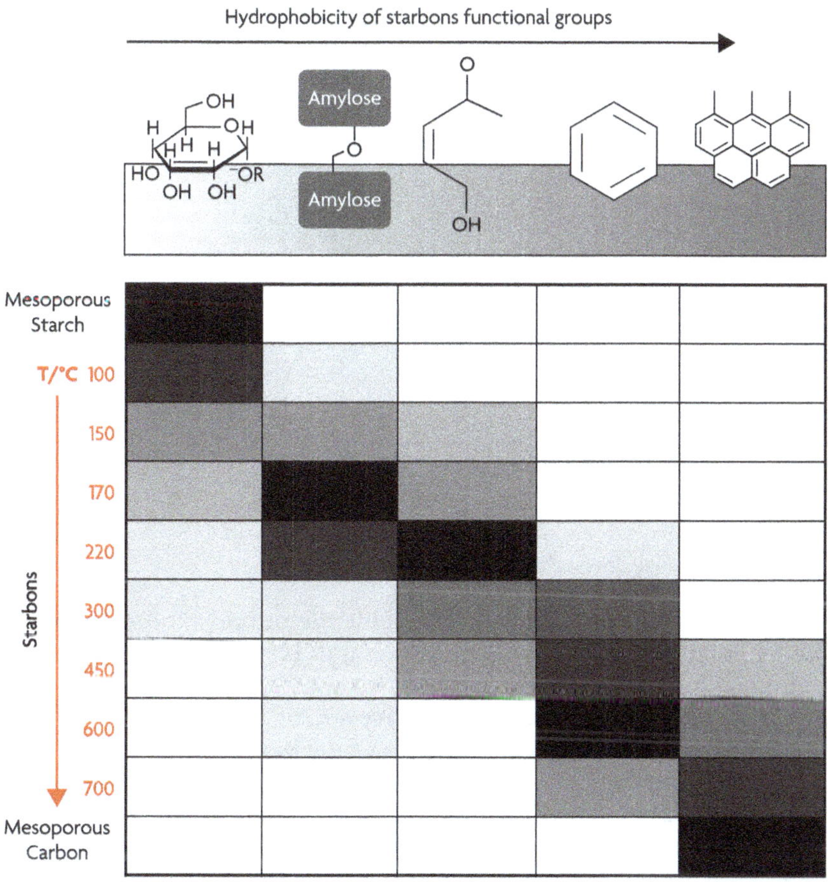

Figure 23.8 Starbon® consists of a continuum of materials from polysaccharides to activated carbon.

Source: Courtesy of Dr Vitaliy Budarin.

Table 23.1 Adsorption (percentage) of metals from potable drinking water at specified concentrations.

	Metal concentration ($g\,l^{-1}$)	Adsorption (%) by: S350	S700	S1000
Mg	4.67	50.8	81.6	83.2
Ca	14.9	74.4	88.7	95.5
Ba	0.14	99.0	99.1	99.4
Fe	0.12	77.5	88.8	90.0
Ag	0.11	99.1	99.1	99.1
Zn	0.039	53.8	84.6	97.4
La	0.024	66.7	83.3	91.7
Cu	0.008	75.0	87.5	75.0
	Average:	74.5	89.1	91.4

Source: Starbon® water purification data by Dr Vitaliy Budarin.

23.5 The biorefinery and its potential for replacing the petrochemical industry

The twentieth century saw a boom in the chemical industry with the emergence of an organic chemical manufacturing industry based on a cheap carbon feedstock, oil (Figure 23.9). This revolutionised the industry and shifted the main energy source away from bioresources, thereby creating the basis of the petroleum refinery we know today.

Environmental and political concerns over the impact of continued fossil fuels use, their depletion and security of supply, combined with a growing population, have created a need for renewable sources of carbon. Over the last two decades there has been a global policy shift back

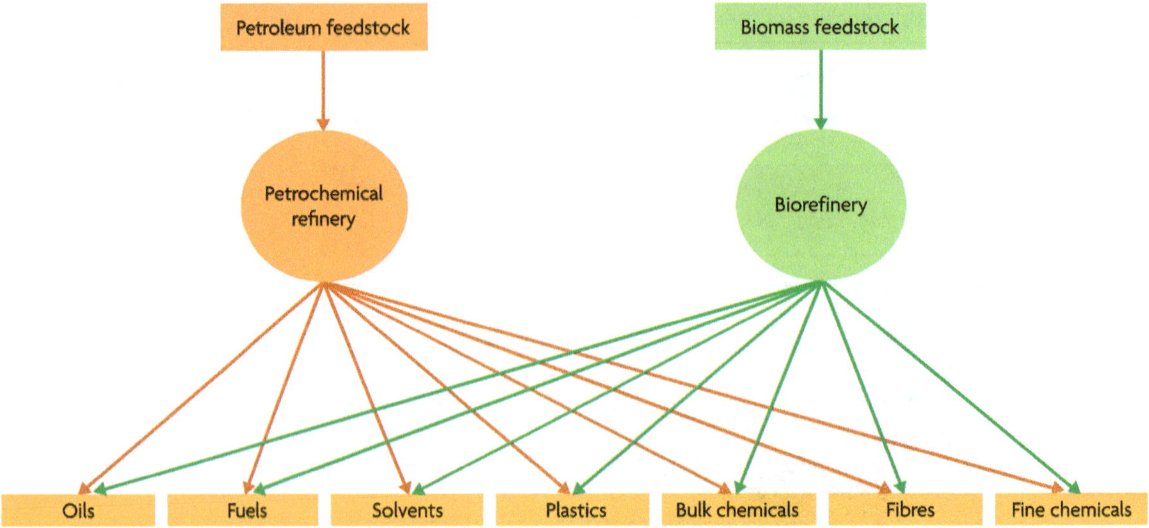

Figure 23.9 A typical petrochemical refinery and an analogous biorefinery.

towards the use of biomass as a local, renewable and low carbon feedstock. The 'biorefinery' concept is a key tool in utilising biomass in a clean, efficient and holistic manner, while maximising value and minimising environmental impact. However, the use of biomass as a source of energy, chemicals and materials is not new and has been taking place for millennia. The biorefinery concept is analogous to today's petroleum refineries. Biorefineries are ideally integrated facilities for conversion of biomass into multiple value-added products, including energy, chemical and materials (Figures 23.9 and 23.10). It is important that biorefineries utilise a range of low value, locally sourced feedstocks that do not compete with the food sector, including low value plants such as trees, grasses and heathers, energy crops and food crop by-products (e.g. wheat straw), marine resource wastes, seaweeds and food wastes.

The main transformations available to the biorefinery can be classified as extraction, biochemical and thermochemical processes. The application of green chemical technologies (including supercritical fluid extraction, microwave processing, bioconversion, and catalytic and clean synthesis methods) are all utilised with the aim of developing new, genuinely sustainable, low environmental impact routes to important chemical products, materials and bioenergy (Figure 23.10) (Clark *et al.*, 2006). These methodologies are usually studied independently of one another; however, the integration and blending of technologies and feedstocks is a way to increase the diversity of products and the social, economic and environmental benefits of the biorefinery.

An integrated zero waste biorefinery that sequentially exploits an extraction, followed by biochemical and thermal processing, with internal recycling of energy and waste gases is viewed as a model system (Figure 23.10). Extraction of secondary metabolites prior to their destruction in subsequent processes can significantly increase the overall financial returns.

23.5.1 Supercritical fluid extraction

Extraction of secondary metabolites can be achieved through the use of traditional organic solvents, including hexane and dichloromethane. However, such solvents are environmentally damaging, often toxic, may be highly flammable, leave solvent residues in the product and typically the extracts require further purification prior to use. More environmentally benign solvents such as liquid and supercritical carbon dioxide are effective at extracting metabolites from biomass.

A supercritical fluid is defined as a substance that is above its critical temperature and critical pressure. The critical point represents the highest temperature and pressure at which the substance

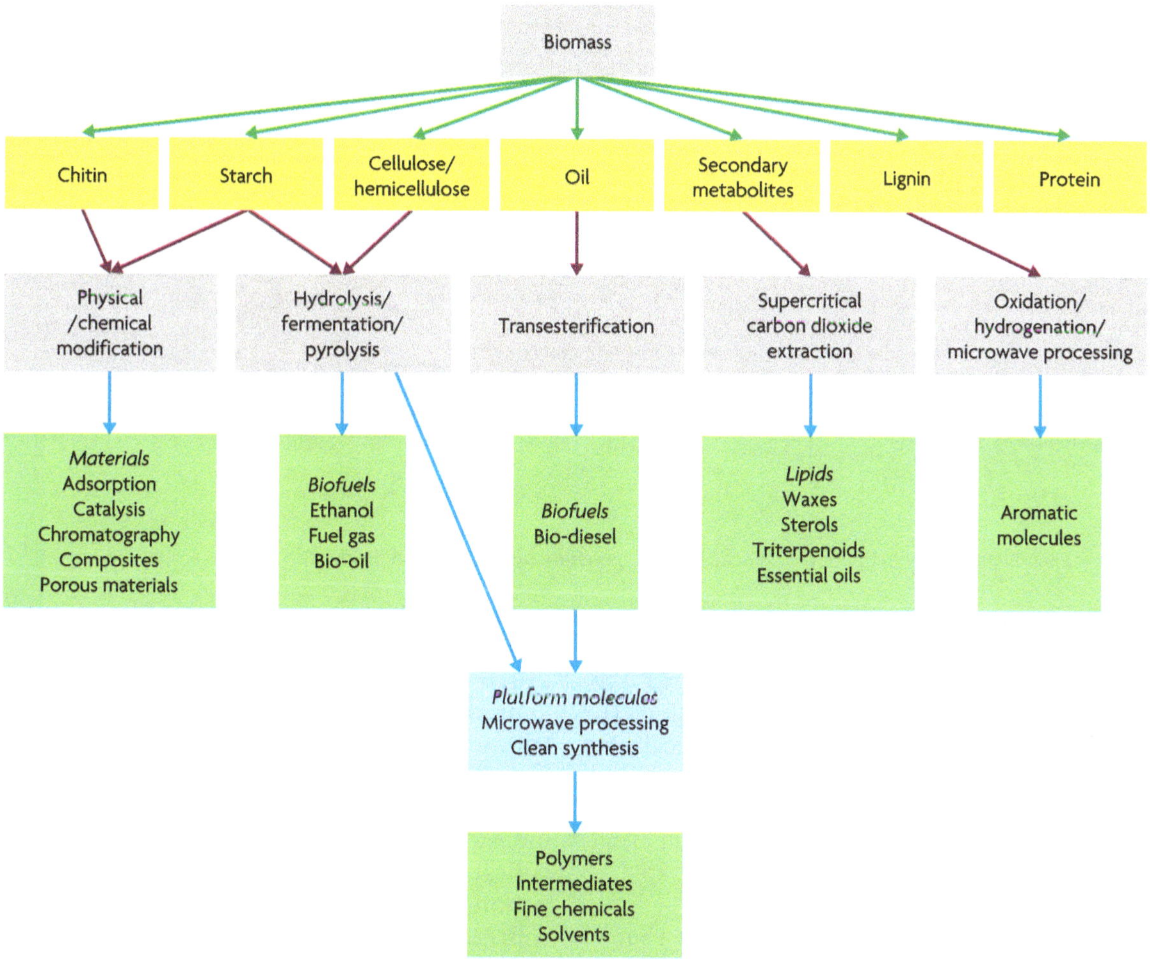

Figure 23.10 Summary of potential production of useful chemical resources from a biorefinery.

can exist as both a liquid and gas in equilibrium. A phase diagram can be used to explain how the substance changes with variations in temperature and pressure (Figure 23.11). Carbon dioxide can exist as a gas, liquid, solid or supercritical fluid. The critical point for carbon dioxide occurs at 73.8 bar and 31.1 °C. A variety of units are used to represent pressure in the literature and conversion factors of these units are as follows: 1 atmosphere = 1.01 bar = 14.7 psi (pounds per square inch) = 0.101 MPa (megapascals).

The physical properties of a supercritical fluid are generally between those of gases and liquids, although these properties can vary significantly as temperature and pressure are changed within

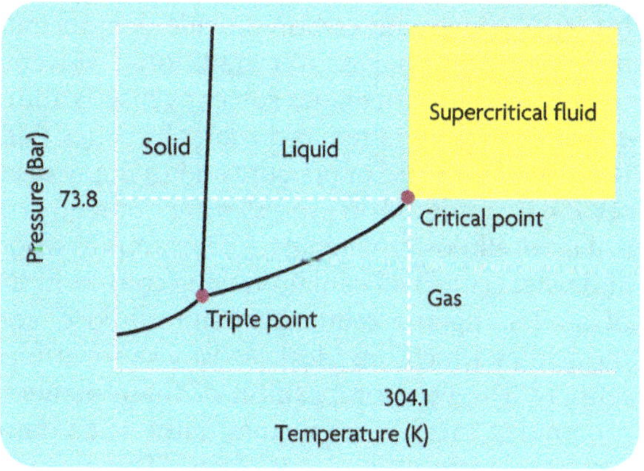

Figure 23.11 Phase diagram for carbon dioxide.

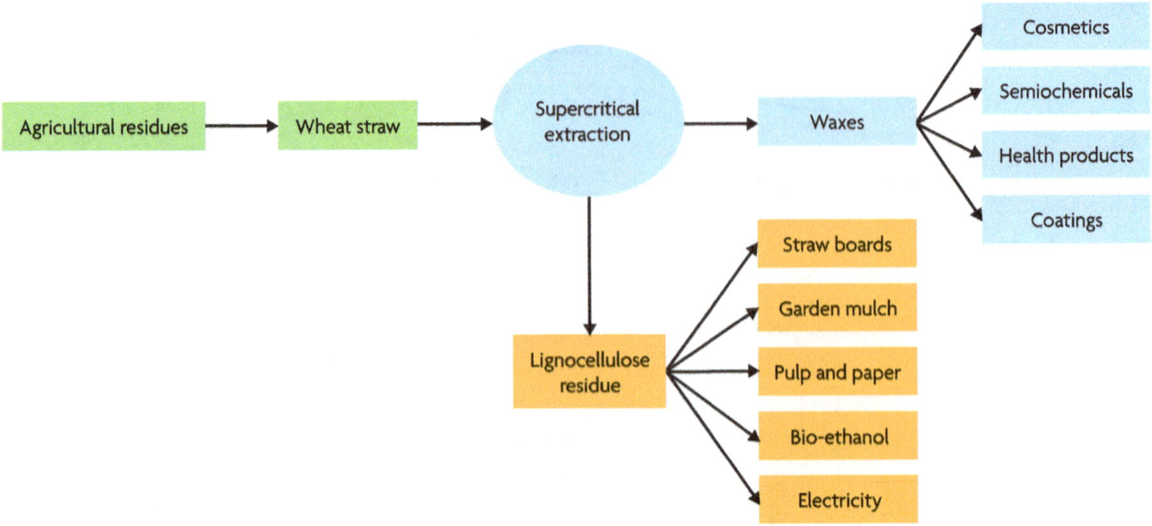

Figure 23.12 Supercritical extraction of wheat straw to generate high value lipid products.

the phase. Varying the temperature and pressure (and therefore density) of the supercritical fluid means the solvent properties can be tuned, providing an enhanced selectivity of extraction. The viscosity of a supercritical fluid is typically an order of magnitude lower than that of a liquid, whereas its diffusivity is at least one order of magnitude higher. The enhanced diffusivity leads to improved extraction times as compared to those for traditional solvent extraction.

Product isolation post extraction from a supercritical carbon dioxide solution is achieved to total dryness simply by pressure release and evaporation. Supercritical carbon dioxide is employed on an industrial scale for extraction of hops, treatment of wastewater, decaffeination of coffee and dry cleaning. Extractions with carbon dioxide demonstrate numerous advantages over traditional solvents such as dichloromethane; solvent reuse is more efficient, toxicity is reduced, no solvent residues remain after release of pressure and in many cases secondary purification processes are not required.

Lignocellulosic materials (for example, wood and wheat straw) all contain varying compositions of waxes, lignin, cellulose, hemicellulose and inorganics which are ideal feedstocks for a biorefinery. The wax composition of these plants is a complex mixture of alkanes, fatty acids, fatty alcohols, sterols and wax esters. These waxes are increasingly finding applications in health and personal care products, as well as possible use as semiochemicals to prevent crop damage by insects (Figure 23.12).

As previously mentioned, extraction with supercritical carbon dioxide leaves no solvent residue and, as such, makes the recovered products ideal for applications in food, pharmaceuticals and cosmetics industries. Current opinion is that using supercritical carbon dioxide as a solvent for extraction is prohibitively costly and, when compared directly with use of organic solvents, this view would appear to be justified. However, the ability to carry out highly selective extractions with little subsequent purification of recovered products is of sufficient benefit to outweigh the expense. As legislation and consumer demand move towards cleaner, more natural processes the use of supercritical carbon dioxide becomes more attractive as solvent residues cease to be an issue (Hunt *et al.*, 2009). Also, in many jurisdictions, the extracts or products can be classified both as natural and organic.

23.5.2 Biochemical conversion

Biochemical routes offer advantages of low processing temperatures and high selectivity. However, they generally require pre-treatment of biomass, long processing times, large amounts of space for batch systems, difficult lignin treatments and downstream processing. However,

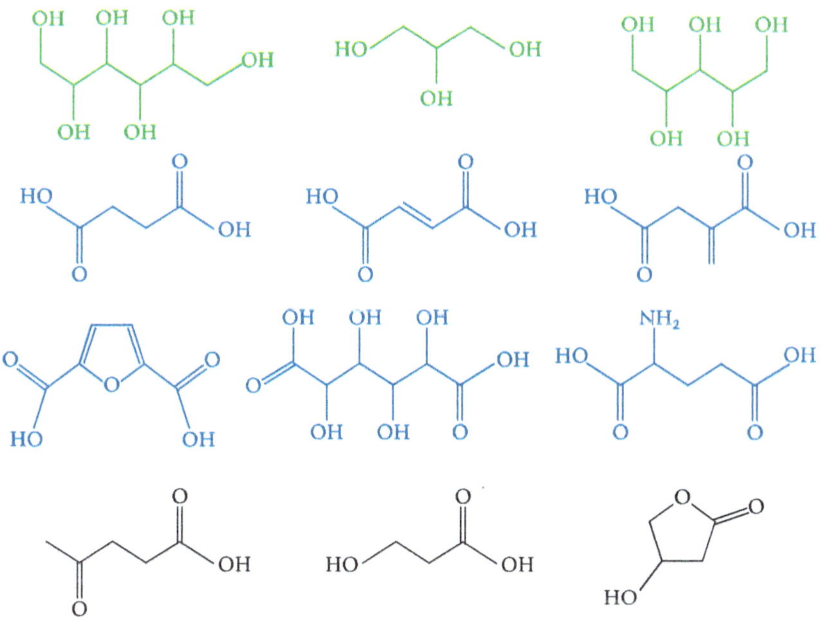

Figure 23.13 Biorefinery platform molecules produced through fermentation.

fermentation broths are a future source of platform molecules that are the renewable building blocks for the production of polymers, solvents and fine chemicals (Figure 23.13).

New solvents derived from biomass through fermentation and biochemical conversion (Figure 23.14) are now industrially available and coming onto the market. However, a greater user industry

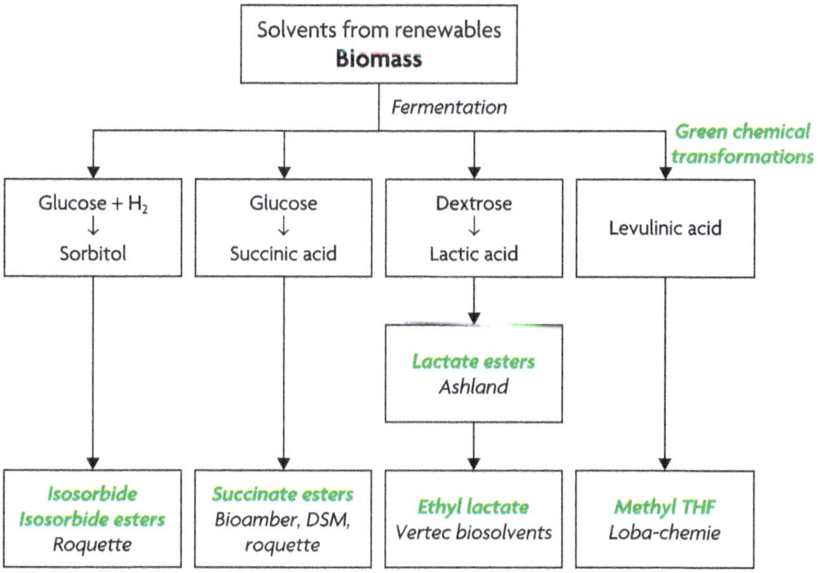

Figure 23.14 Solvents generated from biomass.

pull is required for the large scale uptake of these products.

23.5.3 Microwave processing

Biomass residues post metabolite extraction and biochemical conversion to platform molecules are still a sustainable source of carbon. There is considerable interest in the use of agricultural residues for the production of renewable energy to reduce carbon dioxide emissions. Thermochemical conversions of biomass to valuable chemical products, including gasification, pyrolysis and direct combustion to produce oils, gas, char or ash, are both fast and typically continuous. However, these methods are non-selective and require high operating temperatures (>500 °C), which reduces energy efficiency. Moreover the transportation of biomass to processing facilities can be costly due to the high water content and low density. The efficient use of biomass by thermochemical conversion is viewed as a possible route to revolutionise the long-term manufacture of organic chemicals, materials and energy.

Energy efficient, low-temperature microwave activation of biomass can be used to enhance the material's calorific value, as well as reduce transportation cost and associated pollution. The process converts biomass into an energy-concentrated solid fuel with a minimum carbon dioxide burden depending on biomass sources (Budarin *et al.*, 2010). The process also yields oils with potential applications in transport fuels (Budarin *et al.*, 2009). Initial studies on the microwave processing of marine residues post extraction of valuable components demonstrate the potential for enormous economic and environmental benefits with the utilisation of biomass for a variety of higher value energy products with flexible and controllable technologies that can be installed close to source.

23.5.4 Inorganic residues after combustion

In the true spirit of an integrated holistic biorefinery there is still potential value to be gained from the ash after combustion of the residues. Never accept that what you have is really waste! This so called 'waste' can be turned into a valuable product, bio-silicate solution, without the need for additional chemicals and with a less energy intensive process than that in current use to produce silicate solutions. This bio-silicate solution is a good adhesive and can be utilised to form new, entirely bio-derived, fire resistant, moisture resistant composite boards. The inherent alkali in pot ash from the combustion of seaweed could be used to extract silica from other biomass sources such as wheat straw. In addition, this process should enable more efficient recycling of all elements in ash. Two wastes can be better than one!

As demonstrated in this chapter, there are clear advantages to an integrated biorefinery, including the range and diversity of products, flexibility and efficiency. Figure 23.15 illustrates the potential of an integrated seaweed biorefinery. Using some of the innovative green technologies previously described can lead to an intricate web of products for different markets.

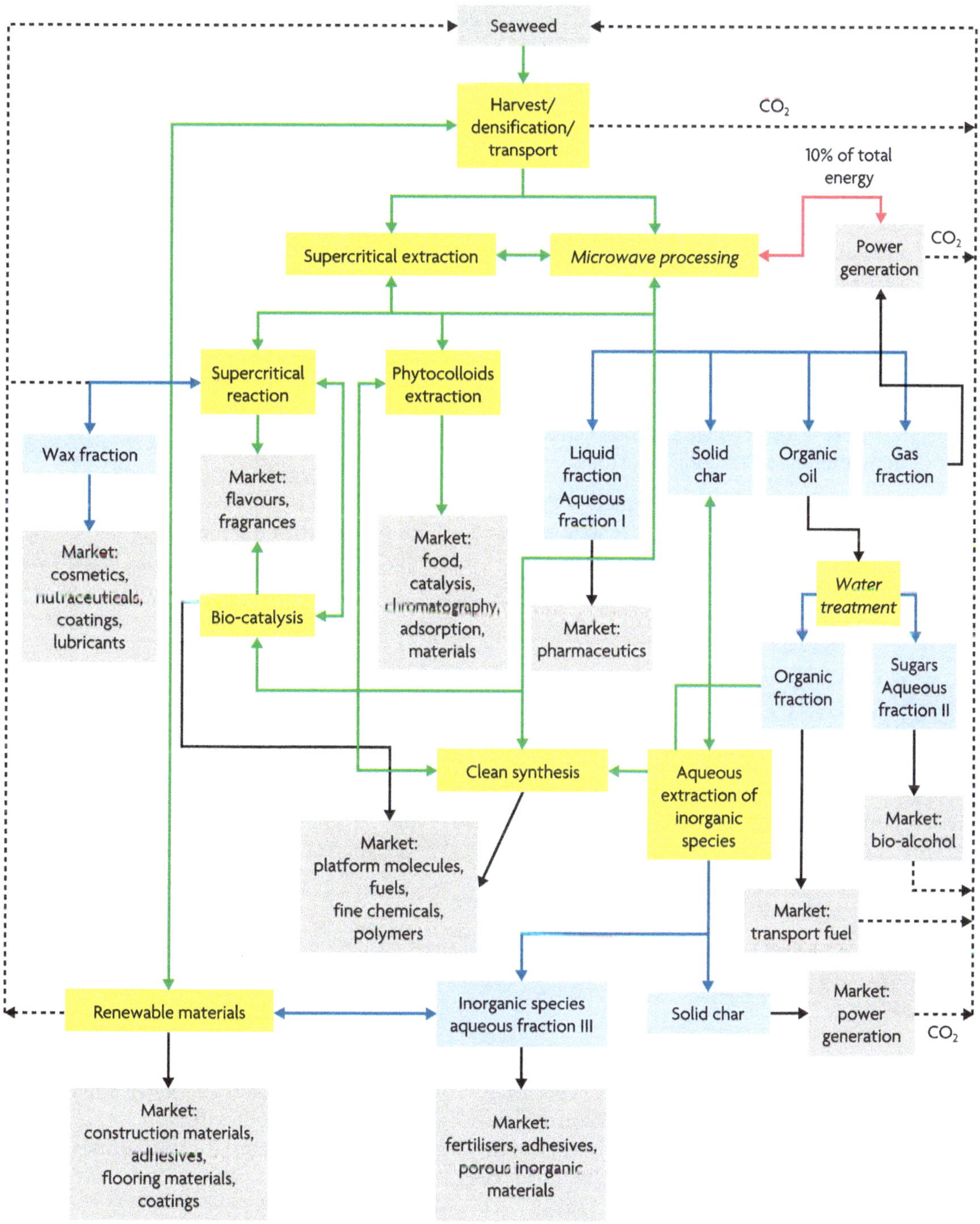

Figure 23.15 The potential functioning of, and products from, a seaweed biorefinery.

POLICY IMPLICATIONS

Green chemistry

- Green chemistry is still a relatively new branch of science. Policy makers need to keep abreast of developments in this area of science to make sure that potential opportunities are not missed.
- Policy makers need to be aware of the finite nature of fossil fuel and element natural resources and the ultimate need to find sustainable replacements in the chemical and other manufacturing industries.
- Sustainable use of waste products within the chemical industry is a highly attractive alternative to most other forms of waste disposal.
- Policy makers should give serious consideration to the potential uses of agricultural residues in chemical and energy production.
- Exploitation of products from green chemistry offers considerable scope for reduction of potentially damaging chemical discharges to the environment compared with some current disposal processes.
- It is important that policy makers are aware of the incredibly broad range of products regarded as essential to everyday life that involve one (or several) chemical processing steps that may be consuming finite natural resources.

CHAPTER ACKNOWLEDGEMENTS

The authors would like to thank Dr Vitaliy Budarin for the use of his Starbon water purification data (Table 23.1) and Starbon diagram (Figure 23.8). We also thank Helen Parker for the use of Figure 23.5.

CHAPTER REVIEW EXERCISES

Exercise 23.1

Look at the list of ingredients (chemicals) on the label of a shower gel or shampoo bottle. How many can you work out the function of in the product? [Hint: Start by looking for any surfactants.]

Exercise 23.2

List ten industrially important metallic elements and outline briefly why each is important. Try to locate the main sources of the elements that you chose and produce a sketch illustrating their global geographical distribution. [Hint: Concentrate on where there are established mines.]

Exercise 23.3

Compare critically the advantages and disadvantages of extractions with supercritical carbon dioxide compared to those with dichloromethane. [Hint: Think about environmental, sustainability and cost issues.]

REFERENCES

Budarin, V., Clark, J.H., Hardy, J.J.E., Luque, R., Milkowski, K., Tavener, S.J. and Wilson, A.J. (2006) Starbons: New starch-derived mesoporous carbonaceous materials with tunable properties. *Angewandte Chemie International Edition*, **45**, 3782–3786.

Budarin, V., Clark, J.H., Lanigan, B.A., Shuttleworth, P., Breeden, S.W., Wilson, A.J., Macquarrie, D.J., Milkowski, K., Jones, J., Bridgeman, T. and Ross, A. (2009) The preparation of high-grade bio-oils through the controlled, low temperature microwave activation of wheat straw. *Bioresource Technology*, **100**, 6064–6068.

Budarin, V., Clark, J.H., Lanigan, B.A., Shuttleworth, P. and Macquarrie, D.J. (2010) Microwave assisted decomposition of cellulose: A new thermo-chemical route for biomass exploitation. *Bioresource Technology*, **101**, 3776–3779.

Clark, J.H., Budarin, V., Deswarte, F.E.I., Hardy, J.J.E., Kerton, F.M., Hunt, A.J., Luque, R., Macquarrie, D.J., Milkowski, K., Rodriguez, A., Samuel, O. Tavener, S.J., White, R.J. and Wilson, A.J. (2006) Green chemistry and the biorefinery: a partnership for a sustainable future. *Green Chemistry*, **8**, 853–860.

Hunt, A.J., Budarin, V.L., Breeden, S.W., Matharu, A.S. and Clark, J.H. (2009) Expanding the potential for waste polyvinyl-alcohol. *Green Chemistry*, **11**, 1332–1336.

Hunt, A.J., Sin, E.H.K., Marriott, R. and Clark, J.H. (2010) Generation, capture, and utilization of industrial carbon dioxide. *ChemSusChem*, **3**, 306–322.

CHAPTER 24

Doing environmental science at the right scale

Dave Raffaelli

Learning outcomes

By the end of this chapter you should:

- Be more aware of the nature of patchiness and extent of spatial variation in many of the components of ecosystems that scientists wish to study, from areas of a few cm^2 up to landscapes or whole regions.
- Realise that temporal variations occur in natural systems over time intervals that may vary from seconds through days, weeks, months, seasons and decades to centuries or millennia.
- Understand how and why these spatial and temporal variations need to be considered when planning experiments to investigate changes over time or in response to some new external pressure.
- Be able to design experiments that unequivocally answer explicit questions, or test a clearly stated hypothesis, and have incorporated appropriate control treatments to give robust baseline data.
- Be aware of the problems of trying to use matched pairs of areas to study an external pressure effect as a surrogate for long-term studies.
- Understand and know how to avoid risk of autocorrelation.
- Be aware that management based upon experimental findings at the wrong scale can lead to surprises in the real world.

24.1 Why scale is important

Environmental patterns and processes operate over a broad range of space and timescales, from less than a few square mm, to hectares, to thousands of km^2, and from seconds, to seasons, to centuries (Figure 24.1). Patchiness occurs at all of these scales, creating environmental heterogeneity, so that thought and care are needed when choosing the time and spatial scales for investigations in ecology and environmental science. The appropriateness of choices will affect how well changes can be detected or important processes understood, and, most importantly, whether you will be able to persuade others of the importance of your findings. For instance, you are unlikely to be able to convince policy makers of the global or even regional significance of measurements of greenhouse gas emissions from wetlands if your estimates are based only on one or two small diameter cores. Answers to landscape-scale questions are best provided by measurements made at those large scales. Samples analysed must be adequately representative with respect to both space and time for the specific hypothesis that is to be tested.

Ensuring that all measurements are made at appropriate similar scales when testing for inter-relationships between two or more determinants can be very difficult. For example, soil pH is known to be an important driver of plant community structure and composition, but, whereas pH is typically recorded using a few g of soil, plant communities are recorded at 1–100 m^2 scales. Ideally, pH should be measured at the same spatial scale as the plants, but this would involve a very large beaker! Instead, we might take many replicate soil cores throughout the plant community, measure pH for each core and take an average value from the replicates (note though that soil pH is a logarithmic function, as discussed in Chapter 7, so the mean of the H^+ concentrations should be calculated and converted back to a pH value). We then would assume that the calculated value is representative of the larger area. In practice, this is often the best we can do, but, as we shall see later, we may get a different answer to our question, depending on the spatial and temporal scale of our study.

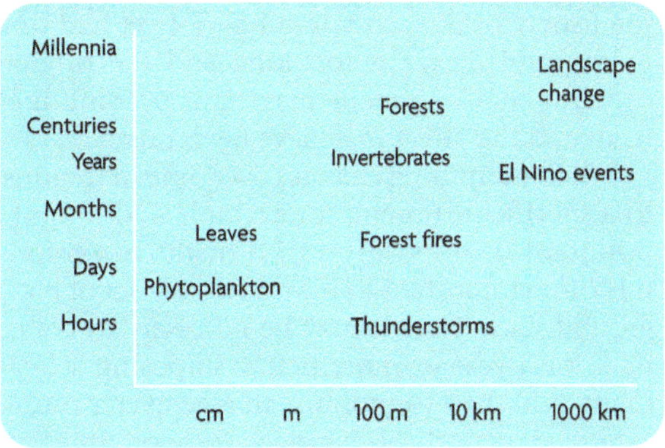

Figure 24.1 Ecological processes operate at distinct spatial (*x*-axis) and temporal (*y*-axis) scales. Studying these processes requires working at the correct scale.

24.2 Components of scale

Choosing the scale correctly for the explicit question being asked is therefore a fundamental starting point in any environmental investigation. To appreciate this fully, we need to consider the different components of scale: *grain*, *lag* and *extent* (Figure 24.2). For spatial scale, grain is the size of the unit from which the recordings are made, for

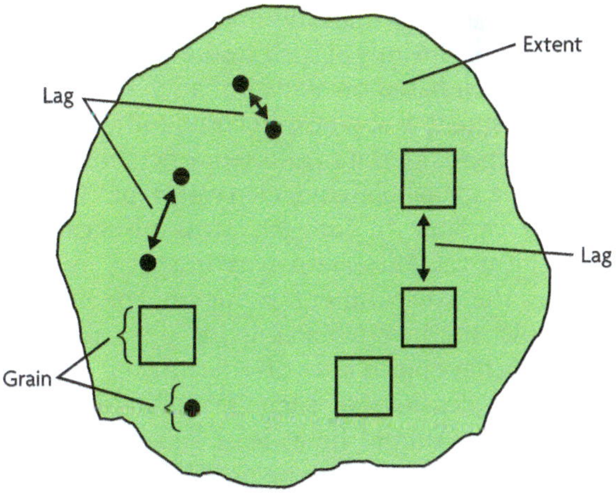

Figure 24.2 The three main components of ecological scale: extent (the larger area in which the study is carried out), grain (the size of the sampling unit used to record data, here a quadrat or a small core) and lag (the distance between sampling units).

example the area of a quadrat, the volume of soil in a core of specified cross-sectional area or the size represented by a pixel on a satellite image. Defining the appropriate grain is the first thing one should do when starting a piece of work and usually involves a pilot study exploring the effect of different sizes of grain on the reliability of measurement of the variable under study. This often involves critically comparing the efficiency and cost of resources if taking many small samples rather than fewer larger samples (Sutherland, 2007).

Lag is the distance that separates replicate sample units. Apart from the obvious need to ensure that lag is larger than the grain and the need to avoid re-sampling areas that you have already disturbed, few researchers explicitly consider the effects of lag in their sampling or monitoring programmes. Getting lag right can be as important as deciding on the correct grain, because of the well-documented phenomenon of spatial autocorrelation. This occurs when measurements made at one location are not totally independent of measurements made at an adjacent, nearby location. Since the two measurements are not independent, the researcher is effectively measuring the same thing twice, and these measurements cannot therefore be treated as separate replicates in any statistical analysis. Lack of awareness of spatial autocorrelation means that many environmental investigations have far fewer independent samples than the investigators thought, and the conclusions of such studies may therefore be invalid if based on statistical analysis.

To ensure that spatial autocorrelation does not compromise the data collected, one should first assess the spatial pattern (often called 'patchiness') of the variable of interest. Most physical and biological variables are patchy in space, in other words the environment is heterogeneous. For example, the invertebrates living on a tidal flat typically have patch sizes of $<10\ cm^2$, and estimating local densities of such invertebrates means that the lag should be at least greater than this patch size. It should also be noted that patchiness occurs at a hierarchy of scales, with smaller patches nested within larger patches and so on, as shown for tidal flat invertebrates (Figure 24.3). Depending on the question posed, different patch sizes need to be considered within this hierarchy.

Extent is the total area over which the samples are collected or measurements made. Spatial extent is often what most people refer to as 'spatial scale', even though it is only one of three separate components, and extent is what is usually represented in diagrams such as Figure 24.1. Working over large extents is not necessarily the same thing as working at large, or even appropriate, scales if considerations of grain and lag are not taken into account. For instance, can we be sure that soil carbon flux measurements based on twenty 10-cm diameter replicate cores scattered across a 20 ha site truly represent landscape-scale processes?

Spatial scale is clearly an important factor in environmental science. Spatial scale needs to be defined and the appropriate scale selected for the question being addressed. Similar arguments apply to temporal scale, since physical and biological patterns and processes are also patchy over time. Interpretations of investigations are likely to be just as problematic if the wrong timescale is chosen. For instance, impressions of the impact of a contaminant on the environment might be different if the investigation was short term, because the effects on the biology of that environment may only be apparent in the longer term, for example, through effects on reproduction or longevity of species. Similarly, experiments aimed at estimating the effects of harvesting commercial fish or re-introducing species into ecosystems may provide different outcomes depending on how long the experiments are run for. If the experiment is too short, no effects might be found because the fishery or the ecosystem has not yet had time to respond. Leave it too long and the original experimental treatment may have become ineffective so the effect may have been missed.

When examining changes over time in response to a specific treatment, for example addition of a pollutant or imposition of drought conditions, it is often important to have a reliable set of background data. Thus it may be necessary to collect data for a year or more before imposing a treatment, and then to monitor change over a further extended period. Sometimes, to save time and money, scientists try to circumvent this problem by using matched pairs of areas such as adjacent fields or catchments.

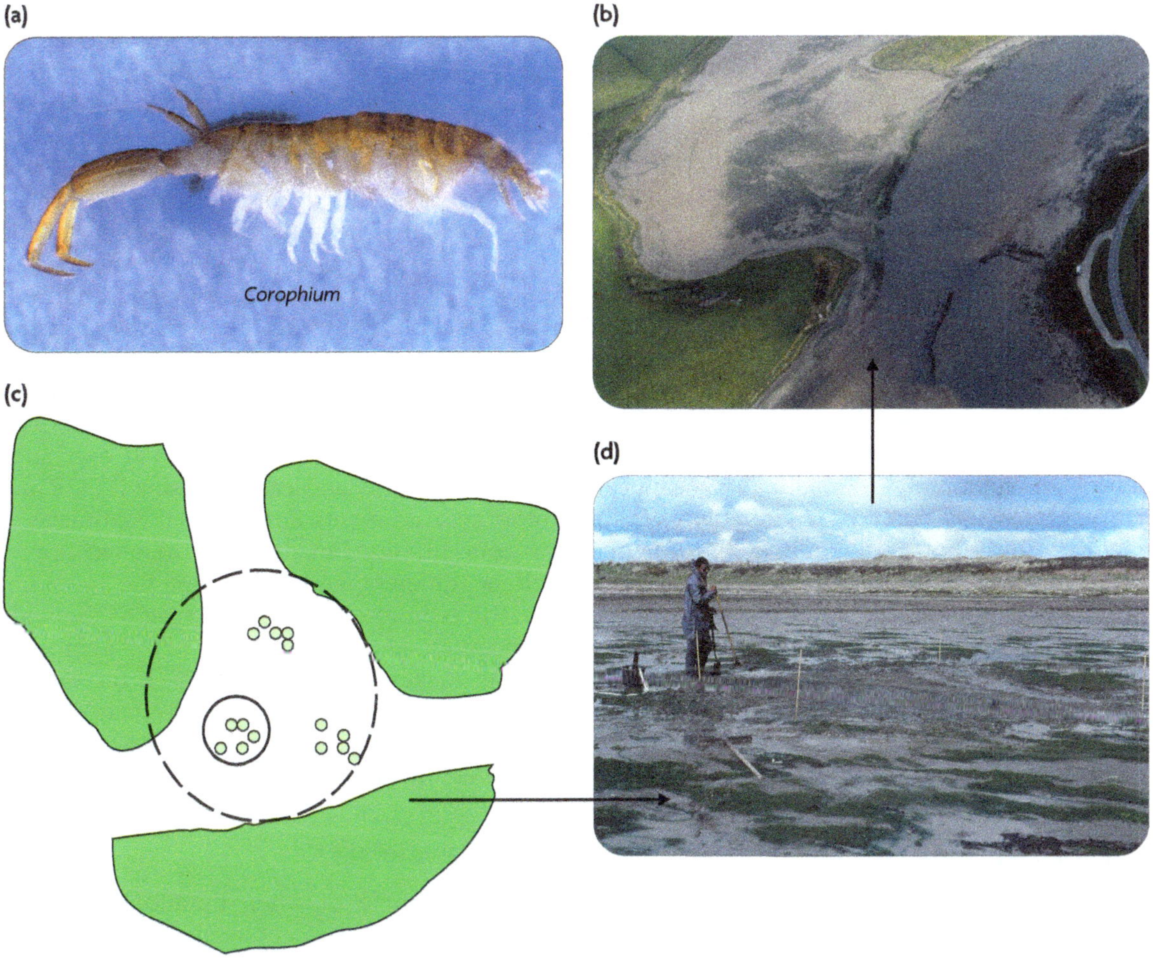

Figure 24.3 The spatial patchiness of organisms such as the amphipod shrimp *Corophium* (top left) and the green seaweeds *Ulva*, on the Ythan estuary, Scotland, changes across a nested hierarchy of scales, in relation to mats of green weed which the shrimp cannot tolerate. Top right: at the highest level in the hierarchy, weed mats and the shrimp are distributed in a mosaic of patches at the 100s–1000s of metres scale, most shrimps occupying the areas between the large weed patches. Bottom right: within these larger patches of the order of 10s of cm in size, are smaller clumps of weed and *Corophium* also lives in the gaps between these. Bottom left: even at the finest scale of only a few cm, *Corophium* distributions are still patchy, this time at the scale of mm to cm, due to interactions between the individual shrimps that occupy permanent burrows.

Lead-in question

What are the risks in using matched pairs of experimental sites as a surrogate for obtaining an extended set of background data and then monitoring change(s) in response to imposition of an experimental treatment at a single site?

To test hypotheses quickly, matched pairs of sites have sometimes been used; for example, adjacent forested and heather moorland catchments have been used in attempts to establish the effects of mature coniferous forestation upon river water quality. However, it is crucial that the two adjacent catchments would have been identical in every significant respect prior to the trees having been planted. That means that the underlying geology and the soils evolved from it must be identical, and so must surface and sub-surface topographic characteristics. Aspect must be identical. Prior land use and management history must have been identical for many decades or even centuries. The catchments must have experienced identical climates. While it is easy to suggest this perfect pairing, the perfect match is much harder to prove in practice, especially as several of the key features may be well below the ground surface. Moreover, just using a single pair of matched catchments would undoubtedly leave the results open to doubt because of inadequate replication.

At small scales, of course, many of these problems can apparently be made to disappear. Homogenised soil can be used effectively in replicate flower pots to unequivocally study one or more treatment effects on seedling plants. The only question that then remains is: 'How well does the confined, massively disturbed soil in pots represent field conditions?' This takes us back to the explicit question being asked.

24.3 Different answers at different scales?

24.3.1 Spatial scale

One of the clearest illustrations of the effects of scale comes from the use of different size quadrats for counting organisms, such as terrestrial flowering plants in meadows or invertebrates on rocky shores. The larger the quadrat, the easier it is to miss the odd individual present, so that density estimates, as well as counts of the number of different species, tend to be slightly lower when made using larger quadrats; it is easier to ensure all the area is searched when using a 10 cm × 10 cm quadrat count compared to a 1 m × 1 m quadrat. Try it and see!

A different scale-effect can occur when measuring physico-chemical variables at different scales, and where different techniques have to be employed at those scales. As an example, the temperature of a bucket of sea water can be measured using a simple thermometer, but the temperature of the ocean is best measured from space using quite different technology. Sea temperature can be assessed by satellite by sensing the ocean radiation at two or more wavelengths in the infrared part of the electromagnetic spectrum. Scientists have to go to great lengths to inter-calibrate these different techniques so that they provide the same information, but sometimes this isn't possible or even desirable, especially in rapidly emerging science areas. Consider the emissions of carbon dioxide (CO_2), sulphur dioxide (SO_2) and hydrogen sulphide (H_2S) from volcanic areas that pose a health risk to people living nearby. Gaseous emission fluxes can be accurately measured using small cores only a few cm in diameter and the results can be scaled up to the landscape. Such estimates can also be made by flying small aircraft over large areas of the same landscape, by using LICOR and Infrascan instruments. Sometimes the estimates of gas concentrations made using these techniques will be different and this poses the question: 'Are the estimates different because of the scale of measurement or because of the different tools used to make the measurements?' If the former, then by forcing inter-calibration we may obscure interesting and important processes operating at the different scales.

24.3.2 Temporal scale

A good example of when duration (temporal extent) of an experiment really does matter is provided by Robert Paine's classic study of the effects of removal of predatory starfish from rocky shores in the Pacific (Figure 24.4). The most popularised result from these experiments in ecological textbooks is the dramatic decline in the diversity of the shore community when the starfish were removed, leading to the now classic concept of keystone predators in ecological systems. A keystone predator normally keeps the superior competing species (in this case, a mussel) in check, thereby allowing many other species to co-exist. If the predator is removed, the other species are rapidly outcompeted for space. If starfish are allowed back onto the shore, the system reverts to its original state. However, if the experiment is continued (starfish are excluded) for many years, then the mussels grow too large to be eaten when starfish are re-introduced, and the system becomes locked into a new state of no starfish and continued

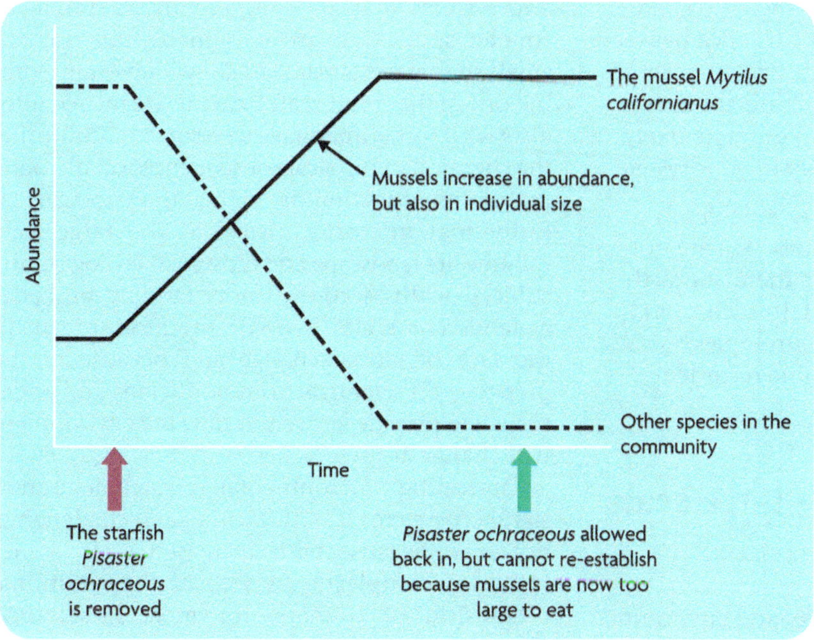

Figure 24.4 Schematic representation of alternate states observed in a starfish-mussel interaction study by Robert Paine (2004).

dominance by the mussel. Our understanding of the role played by starfish in this system would be very different if we adopted only a short-term perspective. We would never have imagined that system could 'flip' into a different state if the starfish was removed for long enough.

These observations can be used to make a more general point. The outcomes of 'pulse' experiments, where the treatment (e.g. nutrient addition, species removal) is applied once at the start, and 'press' experiments, where the treatment is applied continuously throughout the course of the experiment, are likely to yield quite different outcomes (Raffaelli and Moller, 2000).

Similar considerations of patchiness in time apply to the measurement and monitoring of physical variables. Getting the temporal scale right is therefore as important as getting the spatial scale right.

24.4 How do I select the appropriate scale?

The short answer to this question is that it all depends on the exact nature of the question you are asking. There is no substitute for thinking through long and carefully what you wish to achieve before embarking on an investigation. If you get it wrong, much time and effort may be wasted. Occasionally, this reflection might mean not attempting the study at all, as described below with reference to the management of disease and rabbits in New Zealand. Usually, though, there is a way around scaling issues and often the most important thing is that you are aware of such issues. Here are some suggestions:

- Make sure you clearly understand the scales (spatial or temporal) over which the answer to your question is needed. If it is at the small scale, then work at that scale, and if it is the large scale, then try to work at that scale.
- You only have finite time, energy and resources, so there may well need to be a trade-off, for instance, between sample size (grain) and number of samples. From a statistical point of view, it is always best to take many small samples rather than a few large samples, because this increases the power of your analysis.

- Be aware of the patch size of what you are measuring so that your lag (distance between samples) exceeds that patch size, in order to avoid the problem of auto-correlation. It is always worth taking time to estimate likely patch size and heterogeneity in your system.
- If you are relating biological variables to physical or chemical variables, you are not likely to be able to measure them at exactly the same scales (see above). In such cases, make sure you at least take multiple measurements for the variable recorded at the smaller scale.

24.5 Working at the large scale: landscapes

Much of what we do in ecology and environmental science is at the relatively small scale. Most of us tend to investigate patterns and processes at scales of time and space which fit in well with ourselves. This usually involves 3–5-year time spans and systems that are conveniently at similar or smaller scales than us as humans, but not too small so that we can at least see what we are working on! At the small-scale extreme are microbiologists who sometimes have to resort to monitoring the chemical or physical consequences of microbial population functioning, rather than looking at individual microorganisms; at the large-scale extreme are landscape and ecosystem ecologists, the subject of this section. Understanding processes at landscape scales is becoming more urgent in the face of global change and because of the pressures on scientists from politicians and wider society to provide answers to how the environment will change at these scales.

Up-scaling environmental research to understand ecosystems is both exciting and challenging, not least because with increased scale comes increased complexity, as well as the additional dimension of humans for most landscapes. Complexity doesn't simply mean more complicated. The distinction is best seen by thinking of two kinds of world: cog-world and bug-world (Figure 24.5). In cog-world, the system is often

Cog-world

Inter-connected cogs.
Cogs of different sizes drive one another.
Cogs (components) never change.
Response to environment is *linear* and *predictable*.

Bug-world

Bugs are connected and they *interact*.
Connections can change, and groups of bugs can be re-arranged.
System is *self-organising* and *adapts* to change.
Response to environment may be *non-linear* and *non-predictable*.

Figure 24.5 The contrast between cog-world and bug-world systems.
Source: Cog photograph Adam Hart-Davis/Science Photo Library. Mangrove photograph copyright Dave Raffaelli.

very complicated, like the workings of a clock. But however complicated the clock mechanism is, the clock's behaviour is reliably predictable. The clock will always tell the correct time if set up right. This is because once the clock is assembled it cannot change.

Bug-world is very different. Here, the 'cogs' in the machinery are living, they can evolve and they can change their connections with other 'cogs' over time. This means that while for much of the time the behaviour of the system is fairly predictable, there are bound to be surprises now and again. Patterns and processes that we would never have predicted will emerge by themselves from within the system. This is the essence of complexity.

Natural ecosystems are indeed complicated, but they are, of course, bug-worlds and are best described as complex systems. We should expect them to show complex, emergent behaviour, all the more so given that they are a blend not only of the behaviour of natural and physical systems, but also of human social systems. There is an increasing recognition that, for most of the landscapes that we manage, we are dealing with a tightly coupled socio-ecological system. In many parts of the world, the human dimension of the environment is all too apparent, humans having modified the landscape for hundreds if not thousands of years (Chapter 25). In these landscapes, humans are a part of, not apart from, the environment and the appropriate framework for the management of such landscapes is the ecosystem approach described in Chapter 25.

Working at large scales which include both bio-physical and social dimensions immediately presents the challenge of where to draw the study boundaries. The large-scale natural units at which biophysical processes operate are rarely the same as those at which social processes operate. For instance, a natural biophysical unit could be the catchment or river basin, a largely self-contained system which is separate and distinct from other catchments. In contrast, social units are rarely catchment-based. An individual farm or forestry block may straddle the watersheds of adjacent catchments. Local government administrative areas responsible for policy and management often cut across several catchments. For some major rivers, such as the Rhine and the Danube in Europe and the Mekong in South-East Asia, the catchment will cross many different national administrations. This mismatch of social and biophysical boundaries makes environmental management at landscape scales something of a challenge.

One attempt to match social and biophysical scales and extents in the United Kingdom has been the recognition of Natural Character Areas. These are areas defined by their social and geomorphological (and hence biological habitat) features that people may recognise as their 'place' and which are different from other such areas (Figure 24.6). These areas may provide the most obvious natural and social units for landscape management, although there still remains a mismatch between both administrative and other biophysical boundaries, like catchments (Figure 24.6).

Not only are social and biophysical units often spatially mismatched, but the relevant social and biophysical data may be available at very different spatial scales, so that coupling the two is not easy. This is clearly seen for the North York Moors area in the UK, a natural landscape unit (National Character Area) quite distinct in biophysical and

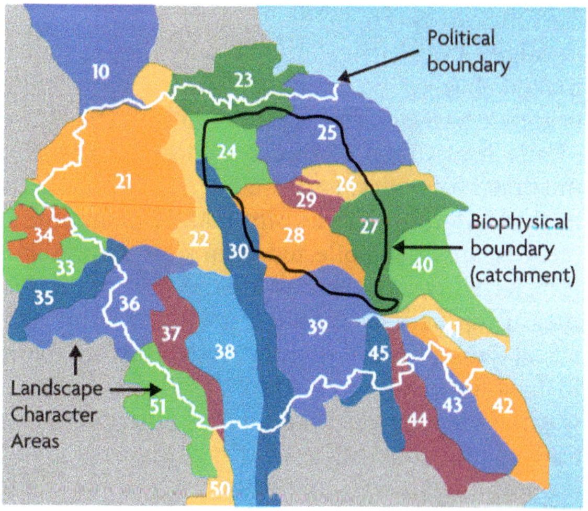

Figure 24.6 The mismatch between an ecological or biophysical unit, the catchment of the Swale–Ouse river, the political administrative boundary of the Yorkshire and Humber Region and natural landscape units, National Character Areas.

Source: The NCA base map is © Natural England. Contains Ordnance Survey data © Crown copyright and database right 2012.

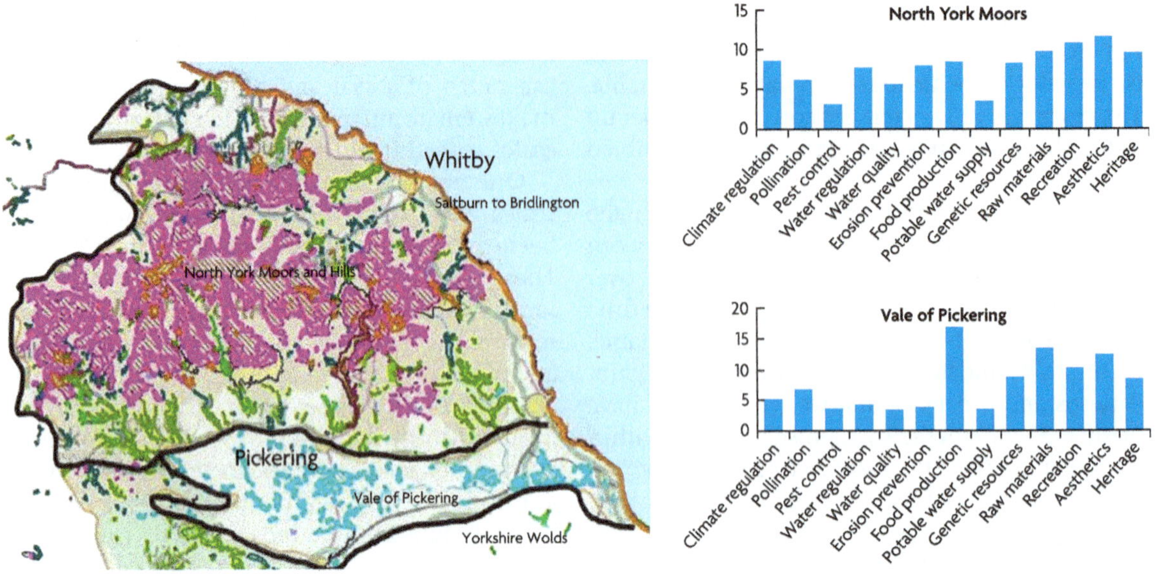

Figure 24.7 Different National Character Areas in Yorkshire, UK, are characterised by different major habitats: upland moorland (bright pink) and wooded valleys (green) for the North York Moors NCA, and river flood plain (blue) and agriculture (white) for the Vale of Pickering. As a consequence, the two areas deliver quite different portfolios of ecosystem services, complicating further ecosystem management at the regional level.

Source: Map © Natural England. Contains Ordnance Survey data © Crown copyright and database right 2012. Figure modified from Raffaelli *et al.* (2010).

social components from adjacent areas, and generating a different bundle of ecosystem services for other areas (Figure 24.7).

Advice from environmental scientists to policy makers has to be based on the best possible evidence base, which means we need to be aware of the limitations of our evidence if there is a mismatch of scales. For instance, in the UK, data on biodiversity, such as bird counts or mammal records, are often collected from spot surveys and transferred onto a regular spatial grid, each cell measuring perhaps 10 km × 10 km. In contrast, social data, such as a population census, are likely to have been collected from a political electoral ward of irregular shape and that is different in size and shape from other wards. Similarly, socio-economic data on farming activity may be available at the level of individual farms that are highly variable in shape and size. Relating these kinds of social data to the biodiversity rectangular grid data will not be straightforward. One solution is to reduce all variables to the same scale and shape, often by generating GIS-type surfaces and then imposing a grid. Such maps are visually appealing, especially to managers and policy makers, but one must never lose sight of the limitations for interpretation from this new, modelled information due to the mismatch of scales and specific locations of the original data.

Lead-in question

What are the scale implications when considering the provision of ecosystem services?

24.6 Scale and the provision of ecosystem services

Issues of scale and spatial arrangements are also important when we have to choose between several policy options or alternative management strategies for a particular landscape. One of the first steps in such a process is to audit that landscape for the range of ecosystem services it currently delivers and that it might deliver in the future (ecosystem services are discussed in more detail in Chapter 25). For instance, when deciding how much of a forest habitat is able to sequester carbon, the spatial scale and arrangement of blocks of forest may not be that important – many small blocks will fix roughly similar amounts of carbon as one large block (Figure 24.8). However, the extent and spatial arrangement of that same forest habitat would affect its value as a recreational service, because larger blocks are preferred for recreation than many small stands of trees. Similarly, the biodiversity value of the forest will be much higher for birds and mammals if it is in large blocks compared to a more fragmented landscape.

An additional spatial factor for managing ecosystem services is where the beneficiaries are located. For the service of shelter and food provided by forests, the beneficiaries will be mostly local, that is, at the same scale as the forest. For the service of flood regulation and water quality provided by forests, the beneficiaries may not be local, but some distance away in a downstream location, increasing the spatial scale (extent) of the area that needs to be studied. At the most extreme, the benefits of climate regulation brought about by forests (apart from the local climate improvement brought about by trees in polluted cities) are felt globally. Matching environmental, ecological and social data for the management of ecosystem services is challenging, but as long as the kinds of issues described above are borne in mind, then sensible decisions can be arrived at.

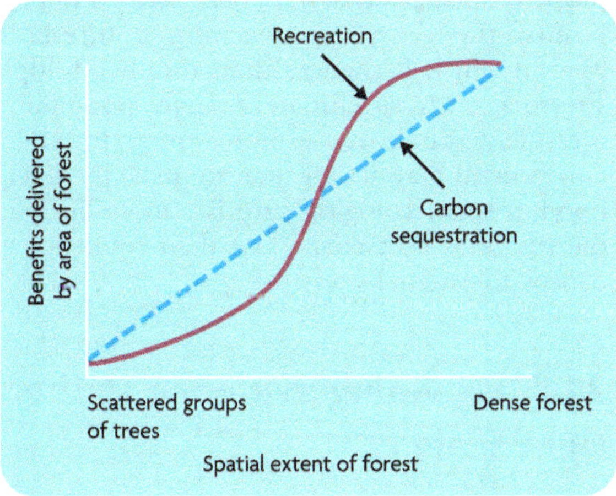

Figure 24.8 The production of ecosystem services can vary quite differently with changes in spatial scale.

24.7 Resilience theory and surprising behaviours of large-scale complex ecosystems

For those responsible for managing large-scale systems, the last thing they need is surprises of the kind that are expected to emerge from complex systems. Yet increasingly it is accepted that such surprises have occurred in the past and are likely to do so in the future. Associated with this is a reduction in ecosystem resilience, the ability to cope with external pressures and shocks. It is not unusual for those managing biodiversity to hold the view that ecosystems inevitably develop over time through orderly successional processes to a predictable end point or mature condition in which it continues in an asymptotic state. Resilience theory challenges this world view by arguing that ecosystems are characterised by environmental thresholds and adaptive cycles (see Walker and Salt (2006) for an excellent introduction to resilience theory). Thresholds are where there is a sudden and abrupt change in the characteristics of an ecosystem, often triggered by very gradual changes in key variables that are so slow they go unnoticed (see below).

Associated with the idea of thresholds is the concept of the adaptive cycle (Figure 24.9). The cycle has four phases: growth, conservation, release and reorganisation. Managing the system in the 'conservation' end-state involves increasing optimisation and specialisation and this makes the system more vulnerable to the effects of stress and shocks. The more we try to manage the system to make it more efficient and produce more, the less stable it becomes. Many believe that collapse of such systems is inevitable. When collapse does occur, the environmental capital locked up in the mature state becomes available to be reorganised, possibly into an ecosystem similar to the original one, but perhaps into a different kind of system.

Adaptive cycles operate at all spatial and temporal scales, from individual leaves growing

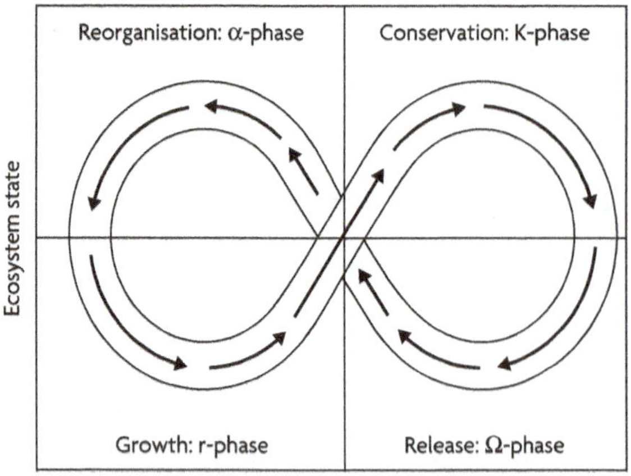

Figure 24.9 The different phases of the adaptive cycle model of ecosystem development and behaviour.

and dying on trees, to the life and death of entire stands of trees and forests. A key consideration for management is to ensure that cycles at different scales do not become synchronised over large areas, which could lead to the wholesale collapse of a resource like a forest and of the social systems that depend on the forest.

A diversity of ecological, social and economic systems seem to behave as predicted by resilience theory which may therefore have important implications for the management of coupled socio-ecological systems. Perhaps it is not sensible to attempt to maintain highly optimised ecosystems such as production landscapes so that they continue in their present, economically profitable state, since doing so may only increase their vulnerability.

One of the best illustrations of this concerns the joint catchment of the Goulburn River and the Broken River in Victoria, Australia (Walker and Salt, 2006). When this catchment was first claimed by Europeans in the 1880s, it was covered by open grassy woodland and hosted a high aboriginal population, indicating a productive landscape. The water table then was at considerable depth, 20–50 m below the soil surface and well below the reach of the deepest plant roots.

Over the next 100 years or so settlers progressively cleared the bush to convert the land for farming. They planted grass-like crops and shortish fruit trees, all of which have shallow roots. Such plants need a great deal of water from irrigation because natural rainfall in this area is too unpredictable for commercial agriculture. Now only about 3 per cent of native vegetation remains and the amount of water needed for irrigation equals that provided by rain, so that the ground now receives about twice as much water compared to the pre-settler period.

The net result is that the water table has gradually crept towards the soil surface. The bad news is that the groundwater is naturally salty, so that when it comes into contact with the roots of plants they die; the ecosystem suddenly starts to collapse and with it the dependent social system. A nasty ecological surprise!

The Goulburn–Broken River story is typical of large-scale landscape nightmares that managers and policy makers have to live with. We now understand the processes that lead up to that particular catastrophe, but only with hindsight: no one was able to forecast what would happen. This is partly because the movement of the water table towards the soil surface occurred at a very slow rate and, up till the point it made contact with the roots of plants, no problem was evident. Identifying such 'slow variables' in ecosystems before they cause problems is extremely challenging and most have been highlighted retrospectively.

A range of ecosystems are now thought to be characterised by the same kinds of dynamical behaviour that leads to sudden changes in socio-ecological systems (Leadley *et al.*, 2010). More properly such changes have been termed tipping points, the ecosystem reaching a threshold beyond which it changes dramatically. Multiple stressors, such as climate change, population pressure, resource over-exploitation and nutrient enrichment, may act together to make the likelihood of such ecological surprises more likely in the future as ecosystems lose their resilience to cope with external stress.

24.8 Providing evidence at the right scale

Putting together an evidence base so that stakeholder groups, scientists and policy makers can arrive at sensible decisions, will often involve very different kinds of evidence. Within the science community, these evidence types can be ranked in terms of their ability to persuade, as follows: anecdotes (least persuasive), common sense arguments, statistical correlations and associations, mathematical modelling, experimental demonstrations and tests (most persuasive). Of all these types of evidence, experimental tests, if done with care and rigour, and at the right scale, can provide the most compelling case for a management strategy or intervention.

When deciding how best to deliver ecosystem services without creating unintended disbenefits for that area or elsewhere, it makes sense to try out a range of management options to see which ones best suit the needs of the area. However, a moment's reflection will reveal that landscape-scale experiments are likely to be very challenging practically, although not impossible. For instance, the effects of replacing traditional agricultural production with biofuels or re-wilding with species lost in the historical past, could be tested by converting large areas of the landscape to each scenario and then comparing the performance of these areas (experimental treatments) with other matched areas that have been left unaltered (control areas). But what size should the areas be, how many replicates of each should there be, and for how long should one run the experiment to ensure that all positive and negative effects are comprehensively recorded?

There are some common-sense rules that one could apply; the areas should be large enough to represent what farmers might manage in the case of biofuels, or to hold a sustainable population of wolves or beavers in the case of re-wilding. The number of replicates required can be estimated statistically using power analysis, based on the size of the effect that the treatment is likely to generate (the difference between the treatment and the control areas) and the known variation between replicates.

With respect to the duration of experiments, there are some rules-of-thumb that can be applied, based on the known dynamics of the processes occurring within the experimental areas, such as the lifecycles of wolves and their prey or the harvest rates of biofuels and the rates of associated nutrient cycles. It is important to get the duration right: trends at large spatial scales are often driven by dynamics which operate over long timescales (see Chapter 25) and short-term outcomes may be quite different from those in the long term (see above). But how long is long? The Canadian ecologist, Peter Yodzis, has come up with a rule-of-thumb for how long experiments to evaluate species removals in control programmes should run for: find the longest pathway between the species to be removed and other species of interest, sum the generation times of all those species in that pathway and double that number. When this rule-of-thumb was applied to a potential large-scale experiment to evaluate the effects on endangered skinks of removing rabbits from New Zealand pasture using a pathogen (Figure 24.10), it was decided to abandon the project – the experiment would have had to run for at least 50 years! Landscape managers and stakeholder groups cannot be expected to put off decisions for quite that long.

However, only rarely do large-scale field experiments take these spatial and temporal considerations into account (Raffaelli and Moller, 2000). The number of replicate areas needed is often guessed at or set by the area of land available or by researcher resource limitations. Durations are also likely to reflect research funding cycles (3–5 years). The size of areas manipulated is often a question of real estate cost rather than based on any consideration of the system dynamics. In many cases, there are no control areas (areas left unchanged) included within the experiment, so that it is not possible to truly assess the effectiveness of the treatment. Controls are basic to any experimental design, whether in the field or the laboratory, and their omission from many landscape-scale experiments is somewhat puzzling. In some instances, there may be a strong political need to apply a treatment everywhere, if not doing so will penalise stakeholders stuck with the control areas, e.g. in the case of pest and predator control.

Finally, one of the most compelling reasons for carrying out experimental scenario-testing at the landscape scale is that such experiments may reveal processes and effects that were never imagined at the design stage and which could not have been

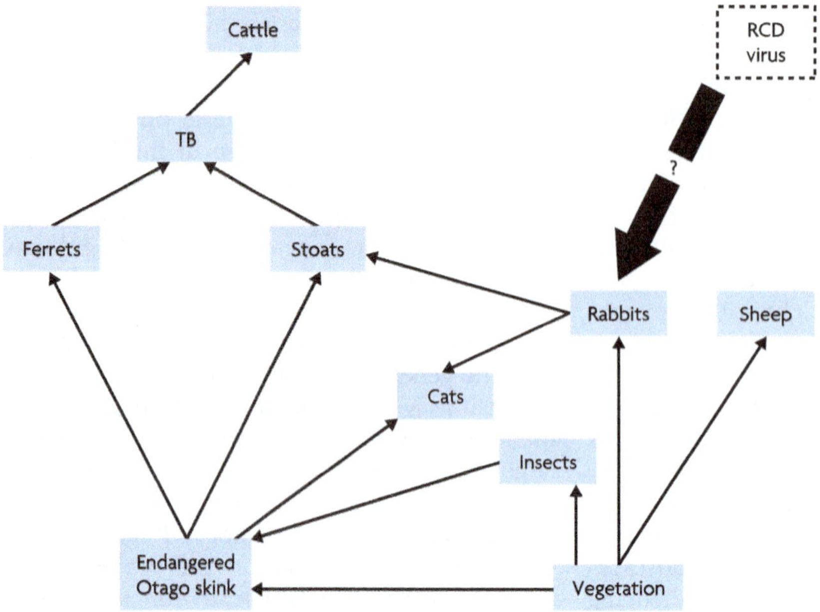

Figure 24.10 Managers may need to be prepared to invest in long-term research if the effects of a planned management intervention are to be adequately evaluated, as shown here for a New Zealand pasture food web. Eradicating rabbits using a viral agent, RCD, might cause cats, stoats and ferrets to switch to preying on the already endangered Otago skink (*Oligosoma otagense*), a species endemic to New Zealand. An experiment to evaluate this would need to run for more than 50 years according to Yodzis' Rule (see text).
Source: Based on Raffaelli and Moller (2000).

predicted from rationale argument, common sense or mathematical modelling. Such 'surprises' are often the catalyst of change, forcing scientists to re-appraise their views about how ecosystems function. As the famous physical scientist Richard Buckminster Fuller once said, there is no such thing as a failed experiment, only one with unintended outcomes!

24.9 Concluding remarks

In concluding this chapter, we hope that the reader will understand the issues of carrying out his or her science at the appropriate scale. In order to address many of the issues in which society has a stake, the environmental scientist must sometimes be prepared to work at the landscape scale. Work at smaller scales is vital for understanding and unravelling mechanisms, but up-scaling such findings to address landscape questions should be done with caution. Ultimately, large-scale tests and demonstrations will provide those involved in management, whether they are government policy makers or individual stakeholders, with the most persuasive evidence of whether a management strategy or intervention will deliver what was intended.

POLICY IMPLICATIONS

Environmental science at the right scale

- Policy makers need to understand that evidence presented to them, or requested by them, needs to be based upon work conducted at appropriate spatial scales.
- Policy makers need to understand that evidence presented to them, or requested by them, needs to be based upon work conducted over appropriate temporal scales.
- Policy makers need to be aware of how inappropriate policy decisions may be made if incorrect scales have been used and data have then been over-interpreted by their advisors.
- Policy makers need to be aware of the intrinsic dangers of using matched pairs of large-scale areas as a surrogate for adequately long experiments.
- Policy makers need to be aware of the importance of adequate replication when making large-scale assessments.
- Policy makers should ensure that adequate controls have been used by experimenters to make sure that their interpretations of trends are unequivocal.
- Policy makers need to be aware of the risk of the unexpected happening when management decisions have been based upon field trials conducted over too short a timescale.

CHAPTER REVIEW EXERCISES

Exercise 24.1

A scientist wants to know the effect of soil pH upon the growth of willow trees as a potential biomass source by coppicing in a specific area near to a power station. He manages to 'borrow', for two years, part of a field 10 m wide and 60 m long and with a shelter belt of trees at one end, from a friendly local farmer. He measures the soil pH and finds it to be 4.5. He also measures how much lime he would need to add to raise the soil pH to 5, 5.5, 6.0, 6.5 and 7 to a depth of 15 cm. He applies the appropriate amounts of lime to 6 plots, each 10 m by 10 m, so the pH goes from 4.5 to 7.0 in steady increments along the 60-m strip of land. A week after applying the lime, which was raked in to the soil surface to get an even spread, he plants 81 equally spaced rooted willow cuttings in each plot.

He allows the plants to establish for 4 months, and then destructively samples 9 plants at random from each plot every 2 months for 18 months.

Critically discuss the strengths and weaknesses (there are several!) of this experiment, with particular reference to spatial and temporal scales and its limitations as a basis for providing advice on growing coppiced willow for biomass provision in the area.

Exercise 24.2

The two plots overleaf illustrate how the mean concentration of nitrate-nitrogen in river water from the River Derwent in North Yorkshire, sampled approximately monthly by the Environment Agency at Malton, varied over each year between 1989 and 2001 and between 1989 and 2006, respectively.

How do you think the response of policy makers might differ if their decision was based only upon the first of these two plots rather than the second? What do the plots tell you about the importance of temporal scale when assessing catchment management effects upon water quality?

Do you think it is adequate to base management policy decisions just upon mean nitrate-N concentration values for each year in this instance?

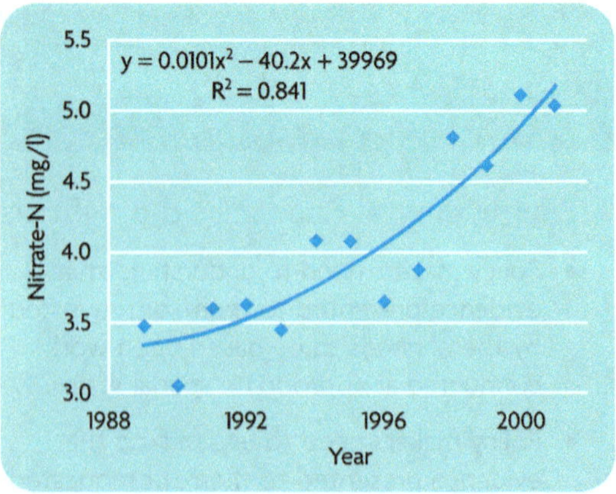

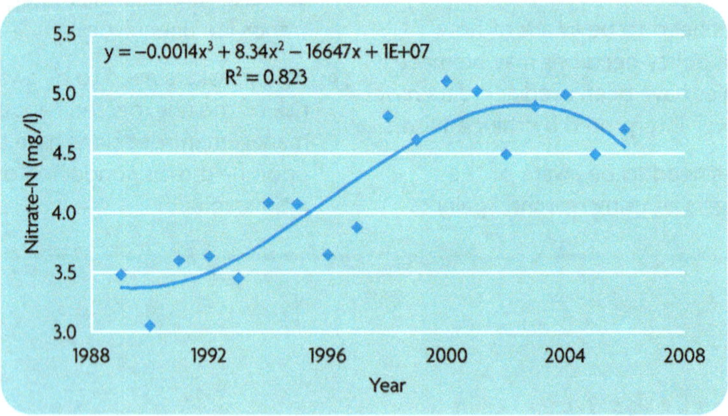

REFERENCES

Leadley, P., Pereira, H.M. *et al.* (2010) Biodiversity Scenarios: Projections of 21st century change in biodiversity and associated ecosystem services. Secretariat of the Convention on Biological Diversity, Montreal. Technical Series no. 50.

Paine, R.T. (1974) Intertidal community structure, experimental studies on the relationship between a dominant competitor and its principal predator. *Oecologia*, **15**, 93–120.

Raffaelli, D. and Moller, H. (2000) Manipulative experiments in animal ecology – do they promise more than they can deliver? *Advances in Ecological Research*, **30**, 299–330.

Raffaelli, D., White, P.C.L. and MacGilvary, A. (2010) *Applying an Ecosystem Approach in Yorkshire and Humber*. Yorkshire Futures, Leeds, 55 pp.

Sutherland, W. (2007) *Ecological Census Techniques*. Blackwell, Oxford, 432 pp.

Walker, B. and Salt, D. (2006) *Resilience Thinking: Sustaining Ecosystems and People in a Changing World*. Island Press, Washington, 174 pp.

CHAPTER 25

Biodiversity: trends, significance, conservation and management

Dave Raffaelli

Learning outcomes

By the end of this chapter you should:

- Understand what the term 'biodiversity' means.
- Be more aware of why 'biodiversity' means different things to different groups of people.
- Know, and be able to explain, why biodiversity is changing.
- Have some insight into the rates at which biodiversity has changed in the distant past and over more recent times.
- Be more aware of why we should be concerned about biodiversity loss.
- Realise that people are components of ecosystems too and why that is important.
- Appreciate the importance and value of the environment for the provision of ecosystem services.
- Understand the importance of inter-linking social and ecological systems in the evaluation of ecosystem health.
- Start to understand how ecosystems may be managed most effectively.

25.1 What is biodiversity?

The basic concept of biodiversity and its importance in environmental studies were introduced in Chapter 6. In the present chapter the understanding and appreciation of biodiversity by wider society is explored. In particular, we will consider why biodiversity is changing and by how much, and the role of biodiversity in delivering the benefits that landscapes provide to people, such as climate regulation, food production, flood mitigation and recreation, to name but a few (Figure 25.1). Many of the activities that environmental researchers are engaged in have societal relevance, such as managing and conserving biodiversity, so it is sensible to ask whether environmental researchers, managers, policy makers and wider society have the same understanding of the term 'biodiversity'.

Lead-in question

How would you define 'biodiversity'?

Figure 25.1 Biodiversity does not simply mean the different kinds of species found in natural systems. It comprises the full range of variation from genes to landscapes, like those shown here, the majority of which have been affected by humans to some degree; managing these landscapes needs to acknowledge that humans are part of the systems. Thus, the area of the temperate rain forest of New Zealand (a) is largely intact in terms of forest but many of the indigenous species have been destroyed by introduced predators. Rocky shores (b) are a source of food and other products for many people throughout the world. A river catchment in Oaxaca, Mexico (c), comprises high altitude commercial forestry, tropical rain forest used for producing shade-grown coffee, lower coastal areas of agriculture and a coastline productive for fisheries, recreation and tourism. A Scottish river, the Ythan (d), typifies the production landscapes of many European ecosystems, mainly cereal and livestock production.

Figure 25.2 Elements of 'biodiversity' that people sketched to express their mental construct of the term. Note the biological elements of trees, deer and eagle, but also the less expected elements of tranquillity, as well as manmade structures like dry-stone walls and a tractor.

Source: Based on Fischer and Young (2007).

There is, in fact, a formal definition of biodiversity provided by the Convention on Biological Diversity (Article 2), which is widely accepted:

biological diversity means the variability among living organisms from all sources, including terrestrial, marine and other aquatic ecosystems and the ecological complexes of which they are part; this includes diversity within species, between species and of ecosystems.

This has been extended to include:

. . . the variety of ecosystems such as those that occur in deserts, forests, wetlands, mountains, lakes, rivers, and agricultural landscapes. In each ecosystem, living creatures, including humans, form a community, interacting with one another and with the air, water, and soil around them.

If you think about it carefully, this is a broad and comprehensive definition ranging from genes to ecosystems, and includes the interactions between all biodiversity elements, including humans. For biodiversity researchers, working in the field, laboratory or the museum, this definition is not really practical and they tend to measure more tangible aspects, such as the number of species and their abundances (Chapter 6). However, the general public may have an even broader idea of biodiversity than that defined by the Convention on Biological Diversity, so that when researchers and the public come together, they may not always be talking about the same thing.

One way to find out what ideas people have in their heads about difficult concepts such as biodiversity is to ask them to draw pictures of those concepts. When this was done in a national park in the Highlands of Scotland, the pictures drawn by the public included, as one might expect, many of the emblematic natural elements of the Scottish uplands, such as red deer, golden eagles, trees and wildflowers (Figure 25.2). But the pictures also included iconic features of the landscape, such as mountains, castles, dry-stone field walls, tractors and representations of solitude and tranquillity (Fischer and Young, 2007), non-natural features that scientists would not include within the term biodiversity. This Scottish study highlights the dangers of each group in the decision-making process thinking that they have shared concepts and then managing the landscape accordingly. Unless such issues are identified early on, the final management plan may not suit any of the groups.

25.2 How and why is biodiversity changing?

The term 'biodiversity' was coined by conservation scientists at about the time of the Earth Summit in Rio de Janeiro in 1992. There, nearly 200 countries signed up to the Convention on Biological Diversity and its articles. 'Biodiversity' resonated well with policy makers and wider society (Takacs, 1996) and it is now a household word. It was acknowledged at Rio that species extinction rates showed little sign of slowing down; by the following Earth Summit in Johannesburg in 2001, the signatories of the Convention agreed:

> *to achieve by 2010 a significant reduction of the current rate of biodiversity loss at the global, regional and national level as a contribution to poverty alleviation and to the benefit of all life on Earth.*

That 2010 deadline has now passed and, while there have been some successes, the overall situation at the end of 2010 was that the target had not been met and that other approaches were required (see below). There are many reasons for that failure; the main drivers involved have been described by the MEA (Millennium Ecosystem Assessment, 2005) in detail and key points are summarised here.

- More land had been converted to cropland in the 30 years after 1950 than in the 150 years between 1700 and 1850. At the ecosystem level, over recent decades, 20 per cent of the world's coral reefs were lost and 20 per cent degraded, and 35 per cent of mangrove area was lost.
- Associated with this, the amount of water in reservoirs had quadrupled since 1960 so that there was 3–6 times more water in reservoirs than in natural rivers, while withdrawals from rivers and lakes had doubled since 1960.
- Chemical fertiliser use had also increased: since 1960, the flows of biologically available nitrogen within terrestrial ecosystems had doubled and the flows of phosphorus had tripled. More than 50 per cent of all the synthetic nitrogen fertiliser ever used had been used since 1985.
- Although some ecosystems in some regions were returning to conditions similar to their pre-conversion states, the rates of ecosystem conversion remained high or were increasing for other ecosystems and regions.
- As a result of these land-use changes and associated processes, the distribution of species on Earth was becoming more homogenous as species were being moved around the world with increases in trade and movement of people.

Humans are clearly responsible for major biodiversity change through over-exploitation, introductions and invasions of alien (introduced) species, and habitat loss and degradation in order to feed an ever-growing human population; now anthropogenic climate change too is becoming an important additional driver (Millennium Ecosystem Assessment, 2005). Most recorded species extinctions since 1500 AD have been on oceanic islands, where populations are often quite small and vulnerable to introduced exotic predators such as rats and cats, as well as to generalist grazers such as pigs and goats. However, since the 1980s about half of recorded extinctions have been from continental land masses.

Putting aside the current science fiction stories of re-creation of extinct species from ancient DNA, we cannot reverse species loss; once a species is lost, it is gone forever. All that can be done is to reduce the impact of the direct drivers of change listed above. To do this we need to engage with the indirect drivers that lay behind those, such as human population growth and demographic changes, economic and socio-political factors, science and technology, and cultural and religious systems. The problem of convincing wider society, especially politicians, of the significance of biodiversity and of the scale of change has many parallels with climate change: hasn't this kind of change always occurred and what does it matter anyway? Here, we will answer those two questions.

25.3 Extinction rates today compared to those from the fossil record

Biodiversity has evolved over long periods of time so that the species, habitats and ecosystems we see around us today are very different from those in the past. As the environment changed and new genetic combinations came into being, some types of biological organisation, such as the dinosaurs, struggled to survive, while others, such as the mammals, seized the opportunities so created to diversify and dominate. Species losses and gains are thus a fundamental part of evolution on Earth, and the rate at which such changes occur can indeed be relatively fast, such as the great extinctions of marine organisms at the Permian–Triassic boundary, about 250 million years ago. This has been termed 'the Great Dying', because 96 per cent of all marine species and 70 per cent of terrestrial vertebrate species became extinct and it is the only known mass extinction of insects (see Benton (2005) for an excellent review). But even when such events are taken into account, present extinction rates are much higher than at other times in the fossil record – about 1000 times higher in fact. Estimates for hundreds of years into the future put the rate up to 10,000 times faster. Humans are responsible for what has become known as the 'Sixth Extinction Event'.

25.4 Biodiversity change in more recent times

There are plenty of examples and case studies of the loss of individual species such as the Dodo or the Great Auk in relatively recent times, but putting together data on all of these species so that trends over time may be quantified has proved more difficult. Perhaps the best known data sets are the Red Data Lists produced by the International Union for Conservation of Nature and Natural Resources (IUCN), which list species of various groups (birds, amphibians, mammals, plants, and so on) that are judged as Extinct, Extinct in the Wild, Critically Endangered, Endangered, Vulnerable, Near Threatened, or Of Least Concern (Table 25.1). In addition, however, there are many taxa classed as Data Deficient or Not Evaluated (www.iucnredlist.org/). By looking at how individual species move from one Red List

Table 25.1 Simplified Red List criteria and thresholds (from Butchart *et al.*, 2005). Any one criterion being met at the level specified in one of the three classification columns (Critically Endangered, Endangered or Vulnerable) is sufficient to place the species in that risk category.

Criterion	Critically endangered	Endangered	Vulnerable
Reduction in population size (over 10 years or 3 generations but where declines are reversible and have ceased)	>90%	>70%	>50%
Small range (plus severe habitat fragmentation, continuing decline or extreme fluctuation in numbers)	<100 km^2	<5000 km^2	<20,000 km^2
Small population of mature individuals and either: • An estimated future decline of >25% within 3 years or 1 generation (if within 100 years) OR • A predicted continuing decline in numbers of mature individuals AND at least one of a/b below: (a) No sub-population estimated to contain >50 mature individuals OR >90% of mature individuals in one sub-population (b) Extreme fluctuations in number of mature individuals	<250	<2500	<10,000
Very small population of mature individuals	<50	<250	<1000
Quantitative analysis (from extinction risk models) suggests population decline of:	>50% in 10 years or 3 generations	>20% in 20 years or 5 generations	>10% in 100 years

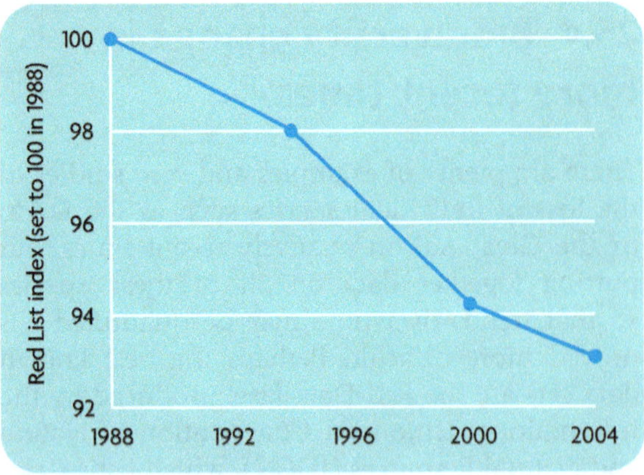

Figure 25.3 Trends in the Red List data for birds (all species).
Source: Based on Butchart *et al.* (2005).

category to the next, conservationists are able to gauge how the statuses of species and entire taxa are changing (Figure 25.3). There are other frameworks that can be used such as the World Wildlife Fund's Living Planet Index (www.panda.org/) and others have been developed for specific taxa, such as birds (see Balmford *et al.*, 2005). But the majority of them have few survey points to work on, usually only a handful, so that there is uncertainty in the trends of some groups over recent time. This creates uncertainty when considering how well governments are doing in meeting specific conservation targets, such as the 2010 targets of the CBD (see above). Nevertheless, there is overwhelming evidence that species are continuing to be lost and that many remain at high risk. The population size or range (or both) of the majority of species across a range of taxonomic groups is declining, so that 10–30 per cent of mammal, bird and amphibian species are currently threatened with extinction with a medium to high certainty (Millennium Ecosystem Assessment, 2005).

25.5 Why we should be concerned about biodiversity loss

Clearly, biodiversity has been, and continues to be, lost at a very high rate. Many people are concerned by this for a variety of reasons, that can be broadly categorised as theocentric, biocentric or anthropocentric.

A theocentric position takes the view that biodiversity is part of creation, in other words that humans have been given stewardship of, or responsibility for, the natural world created by a higher being, or that biodiversity has a spiritual significance without which people cannot fulfil their lives. Many of the world's religions recognise a need to be in harmony with nature, and some religions have explicit statements about environmental stewardship, e.g. Christianity, Islam and Judaism (Posey, 1999; Palmer and Finlay, 2003). Other belief systems have less well defined stewardship demands, but caring for biodiversity will be important for individual well-being, so abusing the environment will have strong spiritual implications. Biodiversity clearly provides a spiritual ecosystem service (see below) for millions of people and charismatic species may be all important for some religions, e.g. crocodiles, scarab beetles, monkeys.

The biocentric basis for biodiversity conservation is the belief that all organisms have inherent value so that human beings should not be seen as different or superior to other species in a moral or ethical sense. In other words, they have as much right to exist as humans do (Taylor, 1986). While this is primarily a philosophical, rather than a religious or spiritual position, it does have clear resonances with theocentrism.

Anthropocentric views of biodiversity are concerned with the value of biodiversity for maintaining the ecological processes and benefits that underpin human well-being. For instance, forests regulate climate and we rely on pollinating insects for the production of many crops. In other words, biodiversity underpins ecosystem services, discussed in more detail below.

Lead-in question

How much biodiversity is needed to make an ecosystem work?

25.6 Biodiversity and the functioning of ecological systems

This area has been touched on in Chapter 6 where the conceptual relationships between

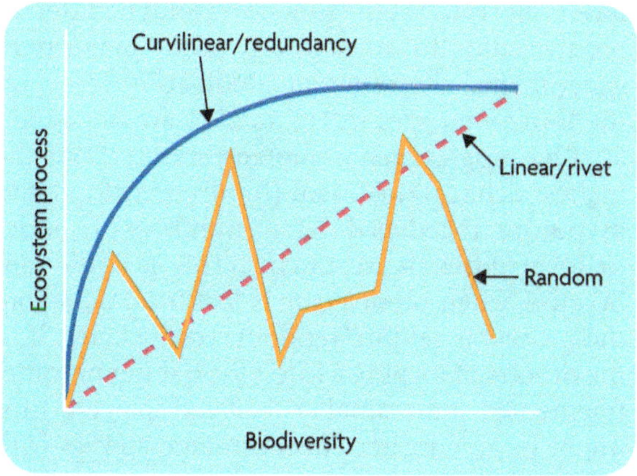

Figure 25.4 The main conceptual models of the relationship between biodiversity and ecosystem functioning.

numbers of species and ecological function were briefly described and the rivet, random and redundancy models were considered (Figure 25.4). A large number of experiments have now been done where areas with different numbers of species (e.g. 1, 2, 4, 8 ... species) representing terrestrial, fresh water and marine communities have been established, and a variety of ecosystem process rates have been measured for each of these species levels in an attempt to find out which of the relationships holds. Given the great range of species used and processes measured, not surprisingly, there are large differences in the findings of all these experiments (Figure 25.5).

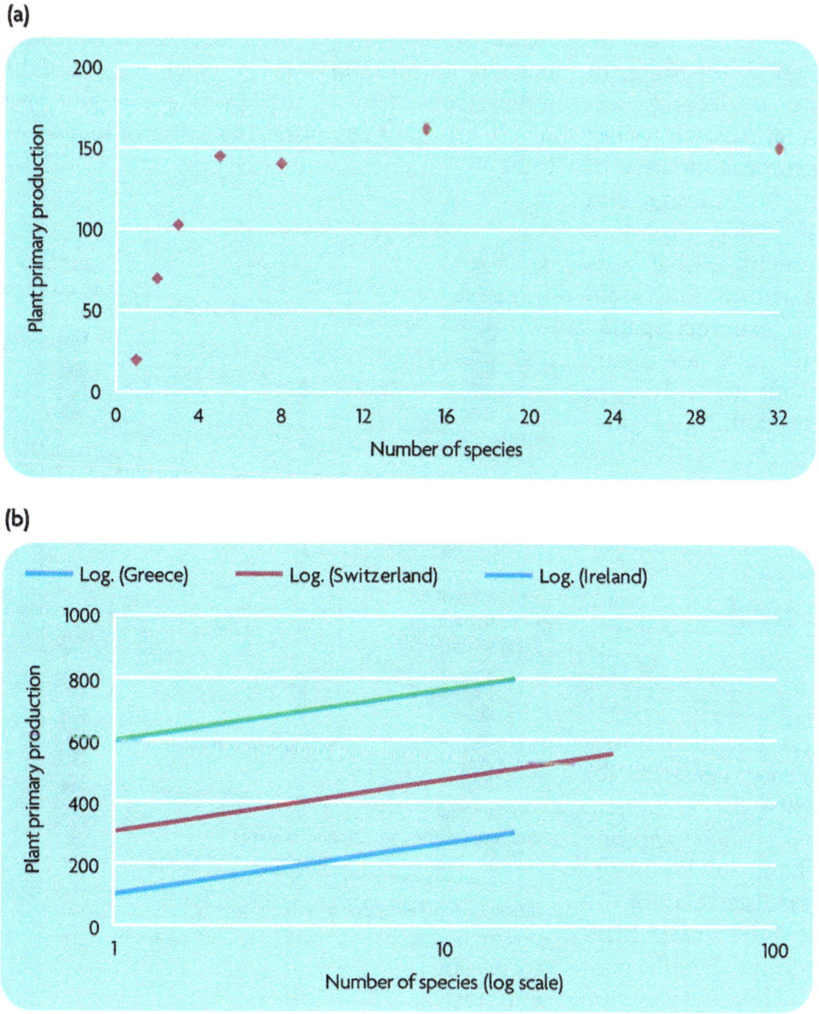

Figure 25.5 Examples of the relationship between an ecosystem process (primary production) and the number of plant species assembled in experimental treatments (plot A) modified from Tilman (1999) and (plot B) modified from Hector *et al*. (1999) Plant diversity and productivity experiments in European grasslands, *Science*, 286, pp. 1123–1127. Reprinted with permission from AAAS. Note the logarithmic scale for number of species in the lower plot which allows linear plots to be drawn.

However, some general trends and patterns can be detected using the statistical technique of meta-analysis which allows all the experiments to be examined together. This approach has shown that there is good evidence of a positive relationship overall between biodiversity and a variety of ecosystem processes (Balvanera *et al.*, 2006; Cardinale *et al.*, 2006), providing support for the argument that biodiversity should be conserved.

An important finding of these analyses is that individual species may not be as important as the functional group they represent. In other words, it may not matter which species of nitrogen-fixing plant or how many of such species are present in a grassland, as long as that function is being performed in the ecosystem. Similar arguments have been made for other kinds of ecosystem engineers, such as how many species of burrowing worm are needed for nutrient cycling in intertidal mudflats or of decomposers in fresh water streams. While this argument implies that there may be some redundancy of species in terms of ecosystem functioning, and therefore species losses may not be as important as claimed, the counter-argument is that it is this very diversity of species within functional groups that provides functional insurance against species loss. Also, of course, there are many other ethical and cultural reasons for wishing to protect species, not only these anthropocentric arguments.

25.7 Biodiversity loss scenarios and ecological functioning

The experimental approaches described above provide powerful and persuasive evidence for the significance of biodiversity for ecosystem functioning, but they are rather artificial and one must be cautious in extrapolating to landscape scales (see also Chapter 24 for discussion of the need to work at the right scale). Their main limitations are that they do not address the effects of biodiversity loss on system function, but describe levels of ecosystem function under different biodiversities. In other words, in the majority of experiments, the different biodiversity levels on the horizontal axes in Figure 25.4 are not generated by progressively removing species from the highest richness treatment (biodiversity loss), but by putting together different numbers of species to form different treatments. This is important because, when biodiversity is lost from a habitat, the process is seldom random. For instance, if in the process of logging a forest the habitat becomes fragmented into smaller isolated parcels, this might have a greater effect on large species with large home ranges such as mammals and birds compared to smaller species like insects. On the other hand, if a habitat becomes polluted it may be the smaller species which are less able to cope and which will therefore be lost first. Few experiments on biodiversity and ecosystem functioning pay attention to the order in which species of different types may be lost from ecosystems, but where this has been done, the order of loss seems to be important (Figure 25.6).

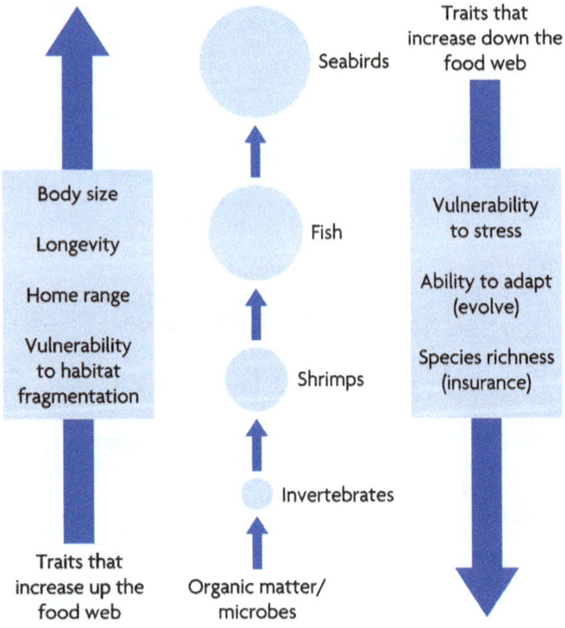

Figure 25.6 The order in which species are lost from ecosystems can have a marked effect on ecosystem functioning.
Source: Based on Raffaelli (2004).

25.8 Problems of low population sizes and the extinction vortex

Some species exist at low population sizes, either because of previous losses or because that's just the way it is; different species have different population sizes. The size at which a population is 'viable' in terms of its ability to persist over time can be calculated for many populations and conservationists try to ensure that the population does not fall below that size. However, this may in itself create additional issues for the species requiring conservation, as discussed in Chapter 13 for disease maintenance thresholds. Thus, the disease Brucellosis, which is devastating to domestic cattle and hence the livelihoods of ranchers, cannot maintain itself in herds of North American bison with fewer than about 200 individuals. Increasing the herd size of bison is desirable from a conservation perspective, but will bring ranchers into conflict with conservationists because of the increased risk of cattle infection (Figure 25.7).

As populations are reduced, they become more vulnerable to random events that may wipe out the population, and hence the species, altogether. A good illustration of this effect is the case of the heath hen (*Tympanuchus cupido*) in North America (Figure 25.8). In the nineteenth century, this species ranged over the entire east coast, but through a combination of hunting and habitat destruction and degradation, numbers were reduced to around 50 individuals in 1908.

Although there was some recovery due to conservation efforts, the low population size made the species vulnerable to random environmental effects and demographic and genetic inbreeding effects, ultimately leading to the bird's extinction in 1932. This process of reduced population size leading to extinction though chance events and demographic and inbreeding issues has become known as the extinction vortex, where species become sucked into an inevitable process of decline from which there is little chance of escape.

Conservationists, however, rarely use the heath hen example as an excuse for inactivity

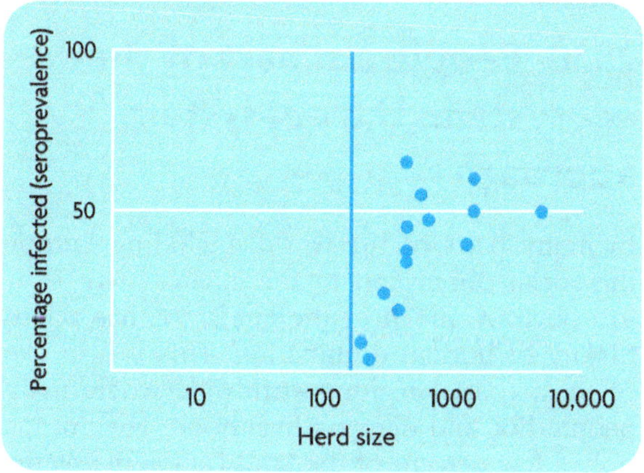

Figure 25.7 Conflicts between conservationists and ranchers arise because, to keep disease (Brucellosis) rates low in cattle, bison herds need to be reduced to below 200 individuals.

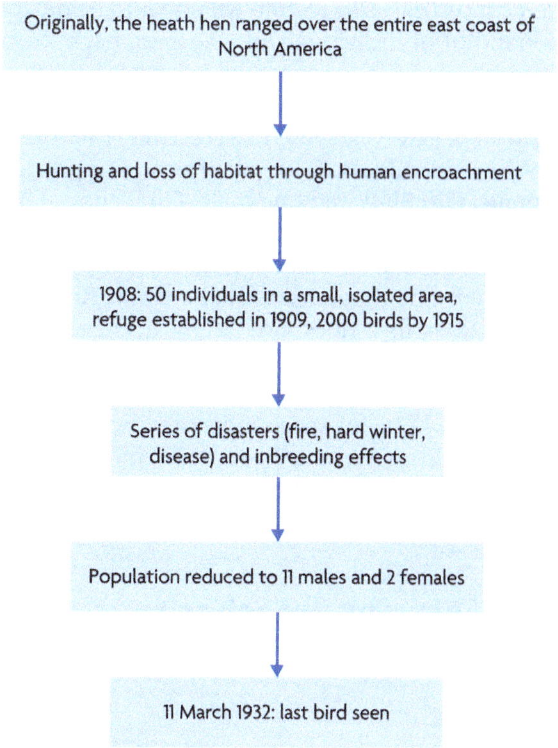

Figure 25.8 The decline of the heath hen through the extinction vortex process.
Source: Based on Krebs (2008).

when faced with low population numbers of species. An excellent example of the dogged determination in the face of a low probability of success comes from New Zealand's Department of Conservation (DOC) heroic struggles to rescue a number of bird species from what seemed like certain extinction, illustrated here by the black robin.

The black robin or Chatham Island robin (*Petroica traversi*) is an endangered bird from the Chatham Islands off the east coast of New Zealand. In 1980, only five individuals of this species were left on Little Mangare Island, the population having been severely reduced by predation by introduced cats and rats and by bush clearance. DOC took the extreme and potentially highly risky decision to transfer what was left of the species to an island free of cats and rats which they also planted up with 100,000 trees (Morris and Smith, 1988). That sounds challenging enough until one reads of the need to leap ashore through surf from a small boat, rope-climb almost vertical cliffs to grab what would be one of a handful of individuals of a species, pack them into a rucksack and repeat the hazardous journey back to the boat. But the action proved successful. Within a few years there were over 150 black robin individuals, but all from one fertile female, Old Blue, and one male, Old Yellow. One might imagine with such an extreme genetic bottleneck that inbreeding issues might have occurred, like those seen for the heath hen, but fortunately this does not seem to have been an issue.

While the black robin success story and others like it give hope for species recovery programmes, there are likely to have been many more species which entered the extinction vortex and never escaped, but which were never documented.

25.9 New approaches to biodiversity conservation are needed

At a meeting of the CBD in 2010 in Nagoya, Japan, there was debate about why the 2010 targets set in Johannesburg 10 years earlier had not been met, and why they were unlikely to be met. One thing is clear: the present institutions and frameworks that have been put in place in order to try to protect the environment generally, and biodiversity in particular, are not working. An excellent review of these frameworks at the international level can be found in Barrett (2003), who also explains why nearly all of the 300 that he examined have not worked! Much of the problem lies in not being able to tackle the root causes of biodiversity loss associated with the unarguable need for development to raise people out of poverty and the pressures on natural systems as they become converted to production systems to feed a rapidly growing world population. In this sense, the Millennium Development Goals and future biodiversity targets and policies need to be more closely aligned. And there is good reason to adopt such an approach; healthy ecological systems rich in biodiversity provide the benefits demanded by development. Development and the sustainable use and conservation of biodiversity go hand-in-hand; one cannot have one without the other.

This was recognised at Nagoya and the CBD emphasis has now changed to reflect the need to maintain biodiversity to ensure that ecosystems continue to deliver the goods and services that society needs. This approach puts people back in the ecosystem and has become known as the ecosystem approach.

25.10 People are part of the ecosystem: the ecosystem approach

In many parts of the world, including Europe, the social dimension to the environment is all too obvious; much of the landscape has a long history of human domination. This is also true of most coastal regions worldwide where many people live and derive a livelihood. For Europe, only a few percent of the landscape remains free from domination by agriculture, forestry or urban development and here at least humans are an integral part of the system. By contrast, in the New World, e.g. North America, there still remain large areas of landscape where humans have left much less of a footprint.

Different historical experiences in different parts of the world inevitably lead to different perspectives on how people (society) fit(s) into the environment. At one end of the spectrum, there is the view that humans are external drivers of change to the environment through their activities as consumers of natural resources. In this world-view, humans are seen as 'apart from' the natural world and their role in environmental studies is one of agents of change. Another world view, long held by many and one which is gaining new currency rapidly amongst the environmental science community (Raffaelli and Frid, 2009), is that humans should be seen not as 'apart from', but as 'part of', the environment. This view of ecosystems is called the 'ecosystem approach' and is encapsulated by the Convention on Biological Diversity's so-called Malawi Principles (Table 25.2).

The ecosystem approach, together with related initiatives such as the Millennium Ecosystem Assessment (2005), has been instrumental in making governments, the science community and society in general much more aware of the dependence of human well-being on the health and integrity of natural systems, in particular, through nature's ability to deliver benefits to people, called ecosystem services (see below). The continued

Table 25.2 The 12 principles of an Ecosystem Approach to environmental management (as adopted by the Convention on Biological Diversity at the 5th meeting of the Conference of the Parties to the CBD, May 2000, Decision V/6. For fuller explanation see http://www.cbd.int/ecosystem/principles.shtml and Annex 4 of UNEP/GPA, 2006).

1	The objective of management of land, water and living resources are matters of societal choice
2	Management should be decentralised to the lowest appropriate level
3	Ecosystem managers should consider the effects (actual or potential) of their activities on adjacent and other ecosystems
4	Recognising potential gains from management, there is usually a need to understand and manage the ecosystem in an economic context. Any such ecosystem-management programme should: • Reduce those market distortions that adversely affect biological diversity • Align incentives to promote biodiversity conservation and sustainable use • Internalise costs and benefits in the given ecosystem to the extent feasible
5	Conservation of ecosystem structure and functioning, to maintain ecosystem services, should be a priority target of the ecosystem approach
6	Ecosystems must be managed within the limits of their functioning
7	The ecosystem approach should be undertaken at the appropriate spatial and temporal scales
8	Recognising the varying temporal scales and lag-effects that characterise ecosystem processes, objectives for ecosystem management should be set for the long term
9	Management must recognise that change is inevitable
10	The ecosystem approach should seek the appropriate balance between, and integration of, conservation and use of biological diversity
11	The ecosystem approach should consider all forms of relevant information, including scientific and indigenous and local knowledge, innovation and practices
12	The ecosystem approach should involve all relevant sectors of society and scientific disciplines

Source: Courtesy of the Secretariat of the Convention on Biological Diversity (CBD).

development of society is increasingly seen as dependent on the wise and sustainable use of natural resources, including biodiversity, implying win–win solutions in environmental management decisions. This is extremely attractive to the political, development and conservation communities who have previously had to deal with much conflict between humans and nature.

The ecosystem approach has, at its heart, two interlinked dimensions, the services that ecosystems provide and the way people value different aspects of the environment. Managing landscapes and the biodiversity those landscapes support means making decisions about what options are best for delivering the benefits that people wish to have (e.g. food, clean air, security from flooding, an attractive place to live, the knowledge that species are protected). It is important that managers and policy makers have a good understanding of the values that people place on these various dimensions of the environment if they are to minimise conflicts which might arise when decisions are made about how landscapes should be used.

25.11 Valuing the environment

A good science evidence base is vital for allowing people to make informed choices between different management options, but we also need to take into account how people value their landscape and environment. Values come in a variety of forms (Figure 25.9) and sometimes these values are at odds with scientific or economic arguments. Often those involved with, or affected by, a management decision hold quite different values; not all these values can be captured using the same

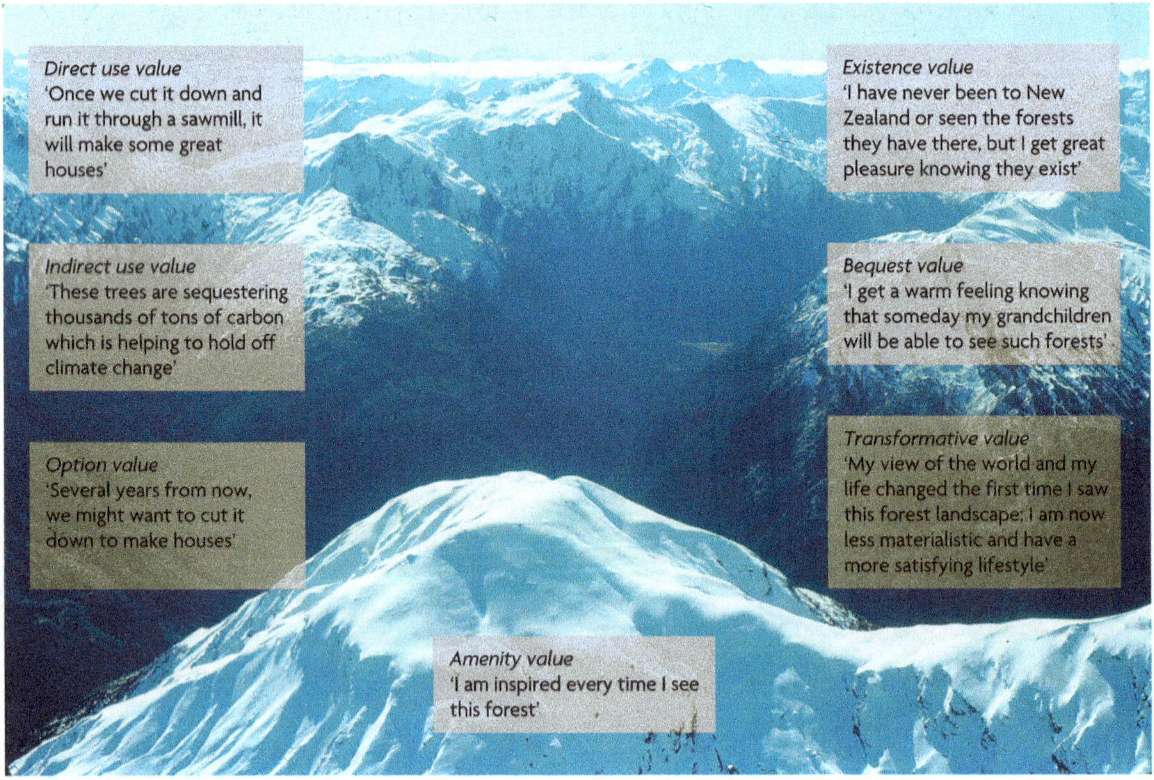
Direct use value
'Once we cut it down and run it through a sawmill, it will make some great houses'
Indirect use value
'These trees are sequestering thousands of tons of carbon which is helping to hold off climate change'
Option value
'Several years from now, we might want to cut it down to make houses'
Existence value
'I have never been to New Zealand or seen the forests they have there, but I get great pleasure knowing they exist'
Bequest value
'I get a warm feeling knowing that someday my grandchildren will be able to see such forests'
Transformative value
'My view of the world and my life changed the first time I saw this forest landscape; I am now less materialistic and have a more satisfying lifestyle'
Amenity value
'I am inspired every time I see this forest'

Figure 25.9 Different types of worth, or value, that people place on biodiversity, in this case a native forest in New Zealand's Southern Alps.

Source: Photograph © Dave Raffaelli. Text modified from Perlman and Adelson (1997), p. 45.

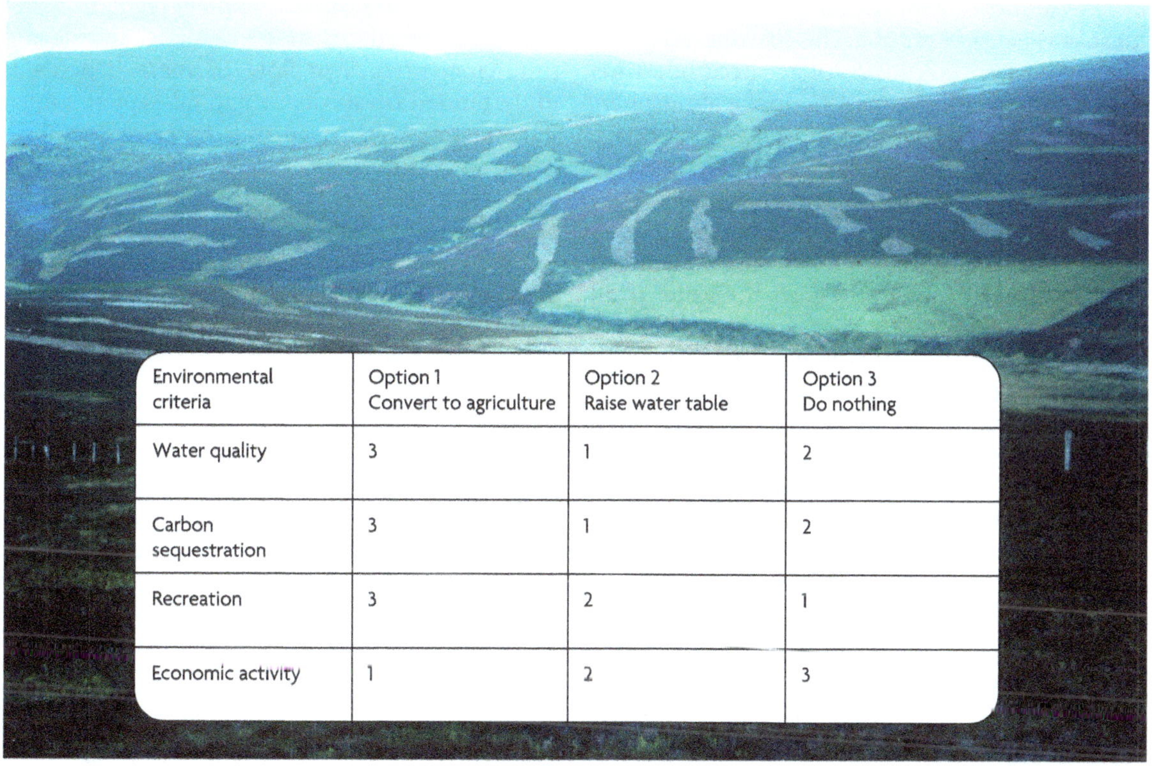

Environmental criteria	Option 1 Convert to agriculture	Option 2 Raise water table	Option 3 Do nothing
Water quality	3	1	2
Carbon sequestration	3	1	2
Recreation	3	2	1
Economic activity	1	2	3

Figure 25.10 Capturing non-commensurate values in decision making. In this hypothetical example, stakeholders ranked different scenarios for upland management of the landscape typified by this one in North Wales, UK.

'currency', which can make logical decisions about trade-offs quite difficult. Thus, monetary value is only one dimension by which people rate the worth of biodiversity (Figure 25.9). Values termed 'non-use', such as amenity, existence, inspirational, bequest and transformative, cannot be easily expressed in monetary terms and there are increasing doubts within the conservation and ecological communities, as well as the economics community, about the limits and applicability of monetary approaches (Raffaelli and Frid, 2009). This is especially true when valuing aspects of the environment for which we can find no substitutes, such as tigers, pandas and scenery.

Making decisions as to how best to manage landscapes often means trying to accommodate the different preferences of different groups of stakeholders. The analysis of these different criteria that people use for making decisions is termed multi-criteria analysis, illustrated here using a simple example (Figure 25.10). Stakeholders were asked to express these preferences by ranking three scenarios for the future management of an upland peatland area in the UK. Such upland areas have fixed carbon for thousands of years and are important carbon stores. Also they provide substantial stocks of potable water, have high recreational potential for hikers and for sport shooting of game birds like grouse, and provide suitable habitat for much plant and bird biodiversity. Correspondingly, the stakeholders could comprise the land owners, conservationists, water companies, farmers, grouse moor managers, hiking associations, or those in local and national government with responsibility for meeting the UK's targets on greenhouse gas emissions. Quite a heady mix!

These peatlands have historically been drained by digging channels through the peat in order to increase the agricultural potential of the land, mainly for sheep and cattle grazing. This management dries out the peat so that it no longer becomes a net sink of carbon and releases CO_2 to the atmosphere through decomposition of the peat. Drainage channels also increase the flashiness of flood events downstream by delivering runoff much faster when it rains heavily. Also, the water is much browner in colour through the increased organic matter entering in the water which costs the water companies money to remove chemically.

While the economic value of the land for marketable goods from agriculture can be readily established, and it should be possible to work out the marginal costs and benefits for the water industry for treating water colour, it is harder even today for carbon storage (there is no market in the UK at present). Recreational value too is very difficult to put into monetary terms, especially for the more aesthetic elements associated, for instance, with landscape painting or poetry. The value of iconic species such as mosses and birds of prey like hen harriers may not be feasible or desirable to capture in monetary terms. Many of the values illustrated in Figure 25.9 will be held by the different stakeholders and they will bring those values to the decision-making table.

In this example, the stakeholders were asked to rate the different scenarios from 1 (most preferred) to 3 (least preferred) for different environmental criteria in order to overcome the issue of different 'value currencies'. While it is unlikely that a clear solution acceptable to all stakeholders will always emerge from such analyses, they do provide a more objective basis for those ultimately required to make that decision. In the present example, an option to raise the water table seemed to have had the most support overall and it is likely that politicians would take that as the least indefensible course of action.

25.12 Ecosystem services and biodiversity

Ecosystem services are the benefits that the natural environment provides for people, such as clean water, clean air, the food we eat, protection from flooding, climate regulation and the enhanced quality of life we derive from beautiful landscapes. The flow of these benefits is underpinned by stocks of natural capital, which include biodiversity, as well as ecosystem processes such as nutrient cycling, soil formation and primary production (Figure 25.11). These in turn generate the services that are of value to people. Landscapes and seascapes can potentially provide a broad range of ecosystem services such as food, wood for fuel and building, climate regulation, water purification and flow regulation, pollination and pest control, inspirational and pleasing scenery, and aspects which have educational and spiritual value. The delivery of these services depends on healthy underlying ecosystem processes, such as soil formation and fertility, and nutrient cycling, and these processes are known as supporting services. There have been many attempts to comprehensively list and classify such services, different schemes being more or less applicable in different contexts. The original scheme produced by the Millennium Ecosystem Assessment serves the present discussion well (Table 25.3).

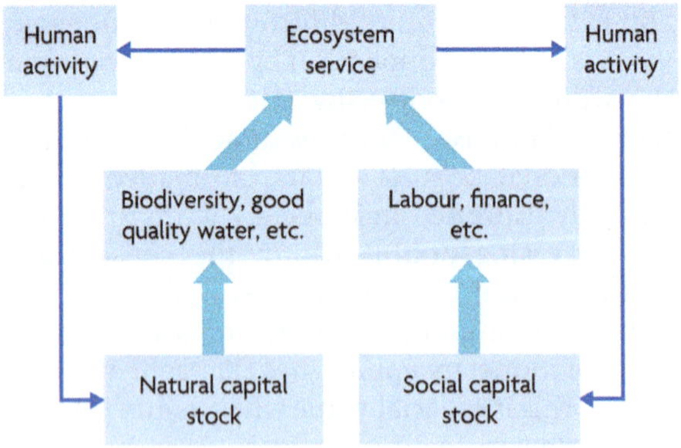

Figure 25.11 Ecosystem services are the flows that arise from stocks of natural capital. Many need similar flows from stocks of social capital (including finance and technology) to generate societal benefits, such as agricultural produce, fish from the sea, and timber from forests.

Table 25.3 Regulatory, provisioning and cultural ecosystem services, modified from various sources.

Regulating services: the benefits obtained from the regulation of ecosystem processes, such as flood control and mitigation, air and water quality.	
Climate regulation	Natural systems affect climate locally (e.g. land cover type can affect temperature and rainfall) and globally (e.g. sequestering carbon).
Pollination	Pollinators of many fruit and other crops are affected by habitat change.
Pest control	Landscape changes affect agricultural pests and diseases.
Water flow regulation	Runoff, flooding and aquifer recharge are affected by land cover type, e.g. the change of wetland and forest into agriculture or urban areas.
Water quality regulation	Soils are significant for the purification of fresh water, including nutrient stripping and the decomposition of organic wastes.
Erosion prevention	Vegetation cover and type affects the stability of soils and the prevention of natural hazards like landslides.
Provisioning services: the products obtained from ecosystems	
Food production	Ecosystems are essential for producing food through agriculture.
Potable fresh water supply	Fresh water comes from natural systems, such as rivers, lakes and aquifers.
Genetic resources	Biodiversity provides the basis of biotechnology and plant breeding.
Raw materials	Natural systems provide timber and wool, and, increasingly, biofuels.
Cultural services: the non-material benefits which people obtain from natural systems through spiritual enrichment, reflection, recreation and aesthetic and inspirational experiences.	
Aesthetics	People find beauty and tranquillity in aspects of nature, reflected in housing locations, support for parks and scenic drives.
Heritage	People place high value on the maintenance of historically important landscapes, their manmade elements and culturally significant species.
Recreation	People choose to spend their leisure time based, at least in part, on the characteristics and attributes of natural areas.

25.13 Using the ecosystem approach for environmental management

The ecosystem approach and the concept of ecosystem services provide an effective and exciting way to manage and conserve biodiversity and the wider environment. The approach provides a holistic framework within which all the costs and benefits of an intervention can be properly accounted for. For instance, introduction of a policy like the European Common Agricultural Policy (CAP), which encouraged the higher production of certain types of crop could, under the ecosystem approach, be expected to take into account the full environmental costs of increased fertiliser application associated with environmental damage and the treatment of drinking water, as well as losses for tourism and related amenity activities. Health, tourism and amenity are more in the social dimension than environmental science in its strictest sense, and the ecosystem approach requires that all these different areas are brought together when formulating policy and making decisions on the ground.

25.14 Bringing social and ecological systems together – ecosystem health

Ecosystem health is a holistic framework that explicitly recognises the inter-relatedness of the social, economic and ecological components of the landscapes and seascapes where people make their livelihoods (Weigand *et al.*, 2010). Each of these components is taken into account when arriving at an assessment of the ecosystem's overall health, interpreted here as its ability to continue to deliver the goods and services on which society depends. The concept has clear resonance with concepts such as sustainability. The general concept is illustrated by one of the earliest quantitative frameworks, the Holistic Ecosystem Health Indicator (HEHI) (Figure 25.12).

HEHI measures how well a socio-ecological system is doing by using ecological, social and interactive (interactions between human and ecological components) indicators, each of which is subdivided into categories tailored to the landscape in question. The ecological component focuses on biophysical aspects of the ecosystem. The social component covers factors fundamental to the exploitation of ecosystem resources and which reflect the economic and social priorities of the communities. The interactive component expresses the connections between people and the other ecosystem components, such as the effectiveness of regulatory agencies, community perceptions, awareness and involvement in management decisions.

By stakeholders giving scores to each indicator according to its health, an overall numerical value can be derived for each of the three components and then these can be combined together to provide a single index. Alternatively, the performance of the system with respect to the separate indicators can be displayed graphically so that stakeholders can appreciate the consequences of trade-offs between ecological and socio-economic imperatives to help them make management decisions (Figure 25.13).

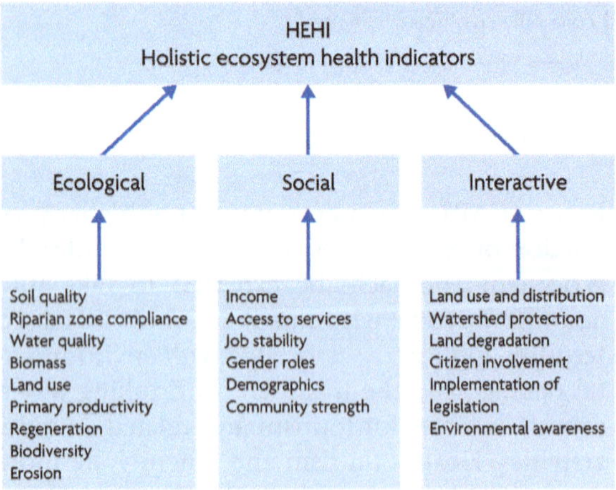

Figure 25.12 The Holistic Ecosystem Health Indicator (HEHI) framework devised by Anguilar (1999) for capturing the health of socio-ecological systems.
Source: Aguilar (1999).

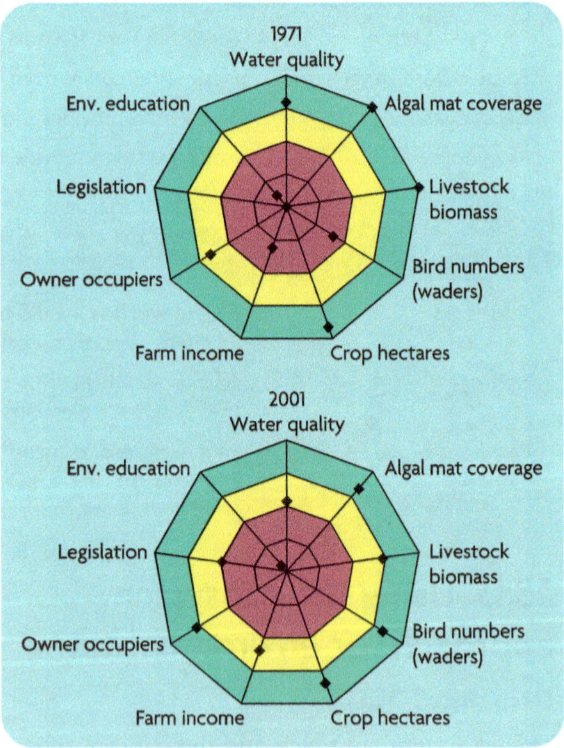

Figure 25.13 Alternative presentation of HEHI-type results for the health of the River Ythan Catchment, Scotland. The spider diagrams show how different ecological, social and interactive components have changed over time using a traffic light system and offer a useful device for working with stakeholders in ecosystem management.
Source: Modified from Wiegand *et al.* (2010).

25.15 Who makes decisions about biodiversity management?

Implementing the ecosystem approach in practice can be challenging. Given that any landscape, including an urban area, has the potential to produce a number of different ecosystem services, decisions to manage that location for particular services have to consider the following:

- Who decides what services are the best ones to encourage or develop (including completely new services)? For example, should this be a government or regulatory agency (a top-down decision) or the local population (a bottom-up decision, in line with Agenda 21, Earth Summit, Rio de Janeiro 1991)?
- What are the consequences of managing for a particular service for the delivery of other services, whose beneficiaries may live far away from the managed location? For example, a decision to harvest all the timber in an upland area could have consequences for future flooding downstream.
- What are the consequences of the management plan for the delivery of services that are not required at present but may be needed by future generations? This is a question of generational equity and sustainability. Managing intensively now for a few services could close down options for the future.

The question of 'who decides?' is vexed and contentious, since different groups of stakeholders are likely to have different preferences for the way they would like their landscape to be managed. Having the best evidence base available is important, so that people can make more rational decisions, but there will always be some uncertainty present. This may be because of uncertainty in measurements and in the data collected, uncertainty in interpretation, uncertainty about likely future processes and, as importantly, uncertainty in the way that people may respond even when they have complete information available to them.

For the ecosystem approach to work effectively there may also need to be cultural changes in some sectors. For instance, a top-down management style characterises many regulatory agencies that are often highly resistant to citizens becoming involved in the decision-making process. There is a diversity of reasons behind this attitude, such as fear that the public may make the 'wrong' decision, or simply the very real challenge of having to make a rapid decision. Bottom-up or participatory processes take more time to work through and require a degree of trust from the regulators. The problem is captured well by Arnstein's ladder of citizen involvement in the social processes (Table 25.4). For a long time, the environmental management decision-making process has remained fairly well towards the bottom of this ladder, with consultation often being tokenist. If the ecosystem approach is to be adopted, a major principle of which is that 'The objectives of management of land, water and

Table 25.4 Arnstein's ladder of citizen involvement in decision making. Readers will need to judge for themselves where environmental decision making comes on the red-green scale, but the author suspects it is still in the amber zone for most decisions. Adapted from various sources.

How much are people involved?	Step on Arnstein's ladder
Citizens are involved	People have control
	Power is delegated to people
	Partnership between people and authorities
Tokenism	Placation
	Consultation
	Informing
Non-participation	Therapy
	Manipulation

living resources are a matter of societal choice' (Table 25.1), the process will need to move much higher up Arnstein's ladder.

25.16 Concluding remarks

Biodiversity is a term that resonates well at many levels of society but, despite this better understanding, biodiversity continues to be lost. Articulating the importance of biodiversity for human well-being is likely to be a more successful strategy to halt biodiversity loss, since development and the sustainable use of natural resources must proceed together. A framework for linking biodiversity and development is the ecosystem approach within which management questions can be posed and addressed, especially through the concept of ecosystem services provided by the landscape. Ecosystem services benefit people and the activities of people can, in turn, reduce service delivery through their negative impacts on ecosystems. Understanding natural environment processes and the motivations of people for using and abusing the environment therefore demands that natural, physical and social scientists work together in an inter-disciplinary way for the sustainable management of ecosystems.

POLICY IMPLICATIONS

Biodiversity

- Policy makers need to be aware in their consultations that diverse groups of society may interpret 'biodiversity' to mean different things.
- Policy makers need to be aware of how and why biodiversity is changing and how much perceived change or changes matter.
- Policy makers need to be aware of how best to involve stakeholders and the general public in the decision making process.
- Policy makers need to understand, and be able to communicate to the public, concepts of ecosystem services.
- Policy makers need to be aware of the values that stakeholders place upon various facets of the environment.
- Policy makers need to base policy decisions related to biodiversity conservation and management upon careful integration of scientific and social aspects.

CHAPTER REVIEW EXERCISES

Exercise 25.1

Ask six friends, relations or colleagues from diverse backgrounds each to draw independently a simple sketch portraying biodiversity in a chosen region with which they are all familiar.

Compare and discuss critically the items each has included in their sketch in the context of each individual's background and life experiences.

How would their sketches compare with a sketch that you might produce for the same area?

Exercise 25.2

Select either three of your own photographs or three pictures from the internet of diverse landscapes. List and discuss the ecosystem services that should be taken into account for each landscape in reaching decisions related to biodiversity management of the landscapes.

Exercise 25.3

Using information from this and earlier chapters, how do you think biodiversity would have changed over the past 10 years, 100 years, and 1000 years within a circle of 10 km radius centred at your home? Explain your conclusions in your answer.

How would the provision of ecosystem services provided by the same area have changed over the same time periods?

REFERENCES

Aguilar, B.J. (1999) Applications of ecosystem health for the sustainability of managed systems in Costa Rica. *Ecosystem Health*, **5**, 1–13.

Balvanera, P., Pfisterer, A., Buchmann, N., He, J-S., Nakashizuka, T., Raffaelli, D. and Schmid, B. (2006) Quantifying the evidence for biodiversity effects on ecosystem functioning and services. *Ecology Letters*, **9**, 1146–1156.

Balmford, A., Bennun, L., Brink, B.T., Cooper, D., Côté, I.M. *et al.* (2005) The convention on Biological Diversity's 2010 target. *Science*, **307**, 212–213.

Barrett, S. (2003) *Environment and Statecraft. The Strategy of Environmental Treaty Making*. OUP, Oxford, 427 pp.

Benton, M.J. (2005) *When Life Nearly Died: The Greatest Mass Extinction of All Time*. Thames & Hudson, London, UK.

Butchart, S.H.M., Stattersfield, A.J., Baillie, J., Bennun, L.A., Stuart, S.N. *et al.* (2005) Using Red List Indices to measure progress towards the 2010 target and beyond. *Philosophical Transactions of the Royal Society, B*, **360**(1454), 255–268.

Cardinale, B.J., Srivastava, D.S., Duffy, J.E., Wright, J.P., Downing, A.L., Sankaran, M. and Jouseau, C. (2006) Effects of biodiversity on the functioning of trophic groups and ecosystems. *Nature*, **443**, 989–992.

Fischer, A. and Young, J.C. (2007) Understanding mental constructs of biodiversity: Implications for biodiversity management and conservation. *Biological Conservation*, **136**, 271–282.

Hector, A., Schmid, B., Beierkuhnlein, C., Caldeira, M.C., Diemer, M., Dimitrakopoulos, P.G. *et al.* (1999) Plant diversity and productivity experiments in European grasslands. *Science*, **286**, 1123–1127.

Krebs, C.J. (2008) *Ecology: The Experimental Analysis of Distribution and Abundance*, 6th edn. Benjamin Cummings, 816 pp.

Millennium Ecosystem Assessment. (2005) Ecosystems and Human Well-being: Synthesis. Island Press, Washington, DC.

Morris, R. and Smith, H. (1988) *Wild South. Saving New Zealand's Endangered Birds*. Random House Press, Auckland, 253 pp.

Palmer, M. and Finlay, V. (2003) *Faith in Conservation: New Approaches to Religions and the Environment*. The World Bank, Washington D.C., 166 pp.

Perlman, D.L. and Adelson, G. (1997) *Biodiversity: Exploring Values and Priorities in Conservation*. Blackwell Science, Oxford, 182 pp.

Posey, D. (1999) *Cultural and Spiritual Values of Biodiversity*. IT Publications and the UN Environment Programme, 752 pp.

Raffaelli, D. (2004) How extinction patterns affect ecosystems. *Science*, **306**, 1141–1142.

Raffaelli, D. and Frid, C.J. (2009) *Ecosystem Ecology: A New Synthesis*. Cambridge University Press, Cambridge, 162 pp.

Takacs, D. (1996) *The Idea of Biodiversity*. Johns Hopkins University Press, Maryland, 393 pp.

Taylor, P. (1986) *Respect for Nature: A Theory of Environmental Ethics*. Princeton University Press, 99 pp.

Tilman, D. (1999) The ecological consequences of changes in biodiversity. *Ecology*, **80**, 1455–1474.

Wiegand, J., Raffaelli, D., Smart, J.C.R. and White, P.C.L.W. (2010) Assessment of temporal trends in ecosystem health using an holistic indicator. *Journal of Environmental Management*, **91**, 1446–1455.

Index

Q

R

S

T

U

V

W

Y

Z

Support your learning with the best-selling textbook in its field

THIRD EDITION

Edited by Joseph Holden

An Introduction to Physical Geography and the Environment

9780273740698

ALWAYS LEARNING   PEARSON

This book offers a comprehensive introduction to the major topics within physical geography and unrivalled support for your learning, with an extensive range of electronic resources.

For further information or to order a copy of this book, please visit: www.pearsoned.co.uk/geography

ALWAYS LEARNING   PEARSON